❖

To Ginny and Stephanie;

And to Austin, Nancy, and R. A.

❖

*B*RIEF TABLE OF CONTENTS

INTRODUCTION

Chapter 1:
An Introduction to Genetics:
A Brief History *9*

PART 1
MENDELIAN GENETICS

Chapter 2:
Fundamentals of
Mendelian Genetics *21*

Chapter 3:
Extensions and Applications of
Mendelian Genetics *45*

Chapter 4:
Chromosomes and Heredity *75*

Chapter 5:
Genetic Linkage *105*

PART 2
MOLECULAR GENETICS

Chapter 6:
Chemistry of the Gene *131*

Chapter 7:
Replication and Recombination
of Genes *153*

Chapter 8:
Transcription and Its Control
in Prokaryotes *181*

Chapter 9:
Eukaryotic Gene Structure
and Expression *203*

Chapter 10:
Translation *237*

Chapter 11:
Gene Mutation I: DNA Damage
and Repair *269*

Chapter 12:
Gene Mutation II: Transposable
Elements and Cancer *295*

Chapter 13:
Genetics of Bacteria and Phages *311*

Chapter 14:
Developmental Genetics *337*

Chapter 15:
Gene Cloning and Manipulation *363*

PART 3
POPULATION GENETICS

Chapter 16:
An Introduction to
Population Genetics *405*

Chapter 17:
Extensions and Applications of
Population Genetics *431*

TABLE OF CONTENTS

Preface *xi*
Acknowledgments *xiv*

INTRODUCTION
Chapter 1:
An Introduction to Genetics: A Brief History *9*

The Three Branches of Genetics   *10*
Transmission Genetics   *10*
  Mendel's Laws of Inheritance   *11*
  The Chromosome Theory
    of Inheritance   *11*
  Genetic Recombination
    and Mapping   *12*
  Physical Evidence for
    Recombination   *13*
Molecular Genetics   *13*
  The Discovery of DNA   *14*
  What Genes Are Made Of   *14*
  The Relationship Between Genes
    and Proteins   *14*
  What Genes Do   *15*
Why Study Genetics?   *16*
Summary   *17*

PART
1
MENDELIAN GENETICS
Chapter 2:
Fundamentals of Mendelian Genetics *21*

Principle of Segregation   *22*
  Mendel's Experimental Results   *22*
  Mendel's Explanation   *27*
Principle of Independent
    Assortment   *29*
Probability   *31*
  Sum Rule   *31*
  Product Rule   *31*
  Conditional Probability   *32*
  Binominal Probability   *33*
  Forked-Line Approach   *35*

Goodness of Fit   *36*
Mendelian Inheritance in Humans   *38*
  Recessive Traits   *39*
  Dominant Traits   *40*
Worked-Out Problems   *41*
Summary   *42*
Problems and Questions   *42*

Chapter 3:
Extensions and Applications of Mendelian Genetics *45*

Dominance   *46*
  Complete Dominance   *46*
  Incomplete Dominance   *46*
  Codominance   *47*
Lethals   *47*
Pleiotropy   *48*
Penetrance and Expressivity   *49*
Multiple Alleles   *51*
Sex-Linked Genes   *53*
Multiple Genes and Epistasis   *56*
Genotype-Phenotype Relationships   *59*
Quantitative Traits   *60*
  Mean and Variance   *62*
  Model of Quantitative Traits   *64*
Estimating Genetic Variance and
    Heritability   *65*
  Race and Intelligence   *67*
Genetic Counseling   *68*
Paternity Exclusion   *69*
Worked-Out Problems   *71*
Summary   *71*
Problems and Questions   *72*

Chapter 4:
Chromosomes and Heredity *75*

Sexual Reproduction   *76*
Chromosomal Morphology   *77*
Mitosis   *79*
  Prophase   *81*
  Metaphase   *81*
  Anaphase   *82*
  Telophase   *82*
Meiosis   *82*
  Meiosis I   *83*
  Meiosis II   *84*
  Spermatogenesis and
    Oogenesis   *84*
  Genes and Chromosomes   *86*
Chromosomal Changes   *89*
  Changes in Chromosomal
    Structure   *89*

Changes in Chromosomal
    Number   *94*
Chromosomes in Different
    Species   *100*
Chromosomes and Sex
    Determination   *100*
Worked-Out Problems   *102*
Summary   *103*
Problems and Questions   *103*

Chapter 5:
Genetic Linkage *105*

Linkage   *106*
  A Physical Explanation   *106*
  Rate of Recombination   *109*
  The Linkage Map   *109*
  Three-Point Crosses   *110*
  Interference   *113*
Linkage in Humans   *113*
  Somatic Cell Hybridization   *115*
Four-Stranded Crossing Over   *116*
  Tetrad Analysis   *119*
  Inversions and Recombination   *121*
Unequal Crossing Over   *121*
Map Distance and
    Physical Distance   *123*
Worked-Out Problems   *126*
Summary   *127*
Problems and Questions   *127*

PART 2

MOLECULAR GENETICS

Chapter 6:
Chemistry of the Gene *131*

The Nature of Genetic Material *132*
  Transformation in Bacteria *132*
  DNA: The Transforming
    Substance *134*
  Further Confirmation *134*
  The Chemical Nature of
    Polynucleotides *136*
DNA Structure *139*
  Experimental Background *139*
  The Double Helix *141*
Physical Chemistry of Nucleic
    Acids *143*
  A Variety of DNA Structures *143*
  Separating the Two Strands of a DNA
    Double Helix *144*
  Rejoining the Separated DNA
    Strands *145*
  Hybridization of Two Different
    Polynucleotide Chains *145*
  DNAs of Various Sizes
    and Shapes *146*
Worked-Out Problems *148*
Summary *149*
Problems and Questions *150*

Chapter 7:
Replication and Recombination of Genes *153*

Mechanism of DNA Replication *154*
  Forming DNA from Nucleotides *154*
  Semiconservative Replication *154*
  Semidiscontinuous Replication *154*
  Priming of DNA Synthesis *155*
  Bidirectional Replication *156*
  "Stitching Together" Nascent
    DNA Fragments *156*
  Putting It All Together *157*

Evidence for the Mechanism of DNA
    Replication *158*
  Semiconservative Replication *158*
  Semidiscontinuous Replication *159*
  RNA Priming *160*
  Bidirectional Replication *160*
  Unidirectional Replication *164*
Enzymology of DNA Replication *164*
  Strand Separation *164*
  Initiation of Replication *166*
  Elongation *167*
  Termination *168*
  Fidelity of Replication *170*
The Mechanism of Recombination *171*
  Forms of Recombination by
    Crossing Over *171*
  A Model for Homologous
    Recombination *171*
  Induction of DNA Strand Breaks *172*
Worked-Out Problems *174*
Summary *176*
Problems and Questions *176*

Chapter 8:
Transcription and Its Control in Prokaryotes *181*

The Mechanism of Transcription *182*
  Initiation *182*
  Elongation *185*
  Termination *185*
Operons *185*
  The *lac* Operon *185*
Temporal Control of
    Transcription *191*
  Modification of the Host RNA
    Polymerase *191*
  The RNA Polymerase Encoded
    in Phage T7 *192*
  Control of Transcription during
    Sporulation *193*
  Infection of *E. coli* by Phage λ *194*
Specific DNA-Protein Interactions *197*
Worked-Out Problems *198*
Summary *199*
Problems and Questions *200*

Chapter 9:
Eukaryotic Gene Structure and Expression *203*

Chromatin Structure *204*
  Nucleosomes: The First Order of
    Chromatin Folding *204*
  Further Folding of the Nucleosome
    String *206*

  Nonhistone Proteins *207*
  Chromatin Structure and Gene
    Expression *208*
RNA Polymerases and
    Their Roles *209*
Promoters *209*
  Promoters Recognized by
    Polymerase II *210*
  Promoters Recognized by
    Polymerase I *212*
  Promoters Recognized by
    Polymerase III *213*
Regulation of Transcription *213*
  Enhancers and Silencers *214*
  Transcription Factors *214*
Additional Complexities of Eukaryotic
    Gene Expression *217*
  Interrupted Genes *219*
  Other RNA Processing Events *222*
Types of Gene Expression Control *228*
  Transcriptional Control *228*
  Posttranscriptional Control *229*
  Translational Control *229*
  Posttranslational Control *230*
Worked-Out Problems *231*
Summary *232*
Problems and Questions *233*

Chapter 10:
Translation *237*

Protein Structure and Function *239*
Genes and Proteins *241*
Discovery of Messenger RNA *242*
Ribosomes: Protein Synthesizing
    Machines *243*
Polysomes *246*
Transfer RNA: The Adapter
    Molecule *247*

Secondary Structure of tRNA   248
Three-Dimensional Structure
    of tRNA   249
Amino Acid Binding to tRNA   250
The Genetic Code   251
Nonoverlapping Codons   251
No Gaps in the Code   251
The Triplet Code   252
Breaking the Code   253
Unusual Base Pairs
    between Codon
    and Anticodon   254
The (Almost) Universal
    Code   255
Mechanism of
    Translation   256
Initiation   256
Elongation: Adding
    Amino Acids to the
    Growing Chain   258
Termination: Release of
    the Finished
    Polypeptide   261
Coding and Anticoding
    Strands of
    DNA   262
The Relationship between
    DNA Size and
    Genetic Capacity   263
DNA Content and the C-Value
    Paradox   263
Worked-Out Problems   264
Summary   265
Problems and Questions   266

Chapter 11:
Gene Mutation I: DNA Damage and Repair   269

Types of Gene Mutations   270
Somatic versus Germ-Line
    Mutations   270
Morphological Mutations   271
Nutritional Mutations   271
Lethal Mutations   272
Conditional Mutations   272
Effects of Mutations on the Genetic
    Material   273
Missense Mutations: A Sickle-Cell
    Disease   273
Other Point Mutations   275
Spontaneous Mutations   275
Chemical Mutagenesis   280
Radiation-Induced Mutations   281
Silent Mutations   283
Reversion   283

Detecting Mutagens   283
DNA Repair   285
Directly Undoing DNA Damage   285
Excision Repair   286
Mismatch Repair   288
Coping with DNA Damage Without
    Repairing It   289
Severe Consequences of Defects in
    DNA Repair Mechanisms   290
Worked-Out Problems   291
Summary   292
Problems and Questions   293

Chapter 12:
Gene Mutation II: Transposable Elements and Cancer   295

Bacterial Transposons   296
Insertion Sequences: The Simplest
    Transposons   296
More Complex Transposons   296
Mechanisms of Transposition   297
Transposons as Mutagenic Agents   298
Eukaryotic Transposable Elements   300
The First Examples of Transposable
    Elements   300
Yeast and *Drosophila* Transposons   302
Retroviruses   303
Transposons and the Mutability
    of Chromosomes   305
Gene Mutations and Cancer   305
Carcinogens as Mutagens   306
Oncogenes   307
The c-Ha-*ras* Gene Product:
    A G Protein   308
Worked-Out Problems   309
Summary   309
Problems and Questions   310

Chapter 13:
Genetics of Bacteria and Phages   311

Bacterial Conjugation   312
Bacterial Genetic Notation   312
Bacterial Exchange of DNA   312
Hfr Strains   314
Mapping by Interrupted
    Conjugation   317
Completion of Conjugation   319
Genetic Mapping with *E. coli*   320
F′ Plasmids   320
The *cis-trans* Complementation
    Test   321
Transduction   323
Genetics of T-Even Phages   326

Structure and Growth of T-Even
    Phages   327
Crossing Mutant T-Even Phages   329
Fine Structure Mapping   331
Applying the *cis-trans* Test to the *rII*
    Region   331
Worked-Out Problems   333
Summary   334
Problems and Questions   335

Chapter 14:
Developmental Genetics   337

The Genetic Basis of Cell
    Differentiation   338
Mosaic Development versus
    Regulative Development   339
Determinants   340
Inducers   342
Control of Gene Expression   342
Mechanisms of Transcription
    Control   343
The Nature of Determination   345
Developmental Activation of Genes
    and DNase Hypersensitivity   345
Control of the 5S rRNA Genes in
    Xenopus laevis   346
Determination of Polymerase II
    Genes   348
The Role of DNA Methylation   349
Determination in the Fruit Fly
    Drosophila   350
Homeoboxes in Other Species   357
Gene Rearrangement during
    Development   358
Rearrangement of Immunoglobulin
    Genes   358
Rearrangement of an Oncogene   360

Rearrangement of Trypanosome
Coat Genes *360*
Sequential Activation of
Members of a Gene
Family *361*
Worked-Out Problems *361*
Summary *362*
Problems and Questions *362*

Chapter 15:
Gene Cloning and
Manipulation *363*

Gene Cloning *364*
The Role of Restriction
Endonucleases *364*
Vectors *366*
Identifying a Specific Clone with a
Specific Probe *372*
cDNA Cloning *373*
Methods of Expressing Cloned
Genes *374*
Expression Vectors *374*
Manipulating Cloned Genes *378*
Protein Engineering with Cloned
Genes *378*
Using Cloned Genes as Probes *380*
Determining the Base Sequence
of a Gene *385*
DNA Sequencing: The Ultimate in
Genetic Mapping *387*
Practical Applications of Gene
Cloning *387*
Proteins: The New Pharmaceutical
Frontier *387*
Improved Vaccines *388*
Intervening in Human Genetic
Disease *389*
Using Cloned Genes in
Agriculture *390*
Protein Products from Transgenic
Organisms *390*
Genetically Engineered Crops *390*
Mapping Genetic Defects in
Humans *393*

Huntington Disease *393*
Restriction Fragment Length
Polymorphisms *394*
Worked-Out Problems *400*
Summary *400*
Problems and Questions *401*

PART
3
POPULATION
GENETICS

Chapter 16:
An Introduction to Population
Genetics *405*

Types of Genetic Variation *406*
Measuring Genetic Variation *409*
The Hardy-Weinberg Principle *410*
The Chi-Square Test *413*
More than Two Alleles *414*
X-Linked Genes *414*
Inbreeding *415*
Self-Fertilization *415*
The Inbreeding Coefficient *416*
Calculating the Inbreeding
Coefficient *417*
Mutation *418*
Genetic Drift *419*
Gene Flow *421*
Natural Selection *422*
Selection against a Homozygote *423*
Heterozygous Advantage *426*
Worked-Out Problems *427*
Summary *428*
Problems and Questions *428*

Chapter 17:
Extensions and Applications
of Population Genetics *431*

Frequency of Human Genetic
Disease *432*
Mutation-Selection Balance *432*
Other Factors Affecting Disease
Frequency *434*
Molecular Evolution *435*
Neutrality *435*
Molecular Clock *437*
Phylogenetic Trees *438*
Use of mtDNA in Studying Genetic
Relationships *439*
Multigene Families *440*
Conservation Genetics *441*
Crop Species *441*
Inbreeding Depression *442*
Loss of Genetic Variation *443*
Animal and Plant Breeding *444*
Artificial Selection *444*
Inbred Lines *447*
Heterosis *448*
Pesticide Resistance *448*
Factors Important in Pesticide
Resistance *449*
Examples of Pesticide Resistance *450*
Control of Pesticide Resistance *451*
Worked-Out Problems *452*
Summary *452*
Problems and Questions *452*

Appendix A: Answers to
End-of-Chapter Problems and
Questions *454*
Appendix B: Selected Readings *470*
Glossary *474*
Index *486*

BOXED READINGS

Chapter 2
2.1 The Father of Genetics   *23*

Chapter 3
3.1 Grandpaternity Exclusion   *70*

Chapter 5
5.1 Multigene Families   *124*

Chapter 6
6.1 Identifying DNA as the Genetic
Material   *135*

Chapter 7
7.1 Biochemists Like to Grind
Things Up   *167*

Chapter 9
9.1 HeLa Cells   *219*
9.2 Splicing, Ribozymes, and the
Origin of Life   *223*

Chapter 12
12.1 Transposition Impacts on
Public Health   *299*

Chapter 13
13.1 Inadvertent Sharing
of a Phage   *330*

Chapter 14
14.1 Genomic Imprinting, Genetic
Disease, and the Battle
of the Sexes   *351*

Chapter 15
15.1 Uses of DNA Fingerprinting   *383*
15.2 Ethical Issues   *395*
15.3 Problems in Genetic
Screening   *398*

Chapter 16
16.1 Industrial Melanism   *425*

Chapter 17
17.1 Lysenko—Soviet
Anti-Geneticist   *449*

"A BASIC LOOK INTO THE CELL" PANELS

Panel 1
Cell Structure   *2*

Panel 2
Cell Cycle   *3*

Panel 2A
A Review of Cell Structure   *4*

Panel 3
Mitosis   *5*

Panel 4
Meiosis   *6–7*

Panel 4A
A Review of Mitosis and Meiosis   *8*

Panel 5
DNA Replication   *151*

Panel 5A
A Review of DNA Replication   *152*

Panel 6
DNA Transcription   *178–179*

Panel 6A
A Review of DNA Transcription   *180*

Panel 7
Translation: Protein Synthesis   *235*

Panel 7A
A Review of the Components of
Translation: Protein Synthesis
Steps 1–3   *236*

PREFACE

The second edition of *Basic Genetics*—like our other book, *Genetics,* second edition—is intended for use in a first college course in genetics. *Basic Genetics,* however, is specifically designed for courses requiring a shorter textbook (less than 550 pages). *Basic Genetics* follows the same organizational approach as *Genetics.* Mendelian, or transmission, genetics is covered first, followed by molecular genetics. However, *Basic Genetics* provides a more balanced, concise approach to a genetics course. In *Genetics,* transmission genetics was covered in a somewhat synopsized way, allowing us to devote more space to the increasingly explosive field of molecular genetics. In *Basic Genetics,* transmission genetics is given a more detailed treatment, while some of the more difficult and detailed molecular topics have been eliminated without sacrificing the timeliness or quality of the topic coverage.

This new edition of *Basic Genetics* features a greater emphasis on problem solving—in both the narrative and in the **worked-out problems** that appear at the end of each chapter. A broader range of problems, many of which are new to this edition, are designed to test a variety of skills.

Numerous illustrations from the first edition of *Basic Genetics* have been redrawn. In addition, because many important genetic concepts and processes are understood more readily if they are illustrated, we have added seven carefully designed **"A Basic Look Into the Cell"** panels to help students understand and review such concepts and processes as cell structure, cell cycle, mitosis, meiosis, replication, transcription, and translation.

As with the first edition of *Basic Genetics,* in fact, even more so this time, we have tried to keep our writing concise and easy to understand. We have attempted to capture the most important basic concepts using language that is economical, and we understand fully that each instructor using this text will shed additional light on those topics he or she feels are in need of further explanation.

ORGANIZATION OF THE CHAPTERS

Like all instructors, we have our prejudices. But in spite of them (or perhaps in recognition of them), we have tried to make this book easy to use no matter what order of topics an instructor prefers. Accordingly, we have organized our book in the following way:

Chapter 1 is a historical introduction to genetics. It provides a few essentials of transmission and molecular genetics. After reading this chapter, students should have some appreciation for the molecular character of genes as they study the subsequent chapters on transmission genetics. If, on the other hand, an instructor chooses to start with molecular genetics, the first four **"A Basic Look Into the Cell"** panels and chapter 1 introduce students to the basics of heredity, providing a context for the molecules.

Chapters 2–5 constitute a unit on transmission genetics. In chapter 2, we use Mendel's experiments to introduce the principles of segregation and independent assortment. At the same time, since it is impossible to discuss these topics without dealing with probabilities, we provide a brief introduction to statistics. In chapter 3, we show that simple Mendelian analysis does not always work, and introduce concepts such as penetrance, expressivity, and epistasis. Chapter 4 deals with chromosomes. The chromosome concept leads naturally to the ideas of linkage and recombination, and how to construct linkage maps with eukaryotes. Each chapter in this first part of the book includes a discussion of the application of genetics to human problems. In chapter 2, for example, students learn to apply the concepts of segregation and independent assortment to pedigree analysis of human genetic diseases. In chapter 4, students see how human chromosomal abnormalities can have severe consequences.

"Brother Mendel! We grow tired of peas!"
Cartoon by J. Chase

The central part of this textbook is a ten-chapter sequence covering molecular genetics. Despite our efforts to balance the coverage of classical and molecular topics, we still believe that the direction of current genetics research demands that the heart of a modern genetics textbook be molecular in nature and up-to-date.

Chapters 6–11 present the fundamentals of molecular genetics. Chapter 6 introduces the structure and properties of DNA and a minimal outline of gene function. In the first edition, chapter 7 consisted of a much more complete introduction to gene function, including gene replication, transcription, translation, and mutation. The following five chapters then expanded on these themes. As a result, some closely related concepts were found in separate chapters—some in chapter 7; others in chapter 11, for example. This "jumping" around to cover some topics in full caused a problem for students. To solve this problem, we have taken the material in chapter 7, **An Introduction to Gene Function,** and distributed it among the appropriate subsequent chapters in this new edition. Chapter 6, **Chemistry of the Gene,** is now followed by five chapters (7–11) that present the concepts of gene function. Chapter 7 describes the replication and recombination of genes. Chapters 8, 9, and 10 describe the process of gene expression. In particular, chapters 8 and 9 cover transcription in prokaroytes and eukaryotes, respectively, and chapter 10 outlines the process of translation. Finally, chapter 11 deals with another characteristic of genes, their capacity for mutation.

Chapters 12–15 further discuss specific aspects of the functions of genes. Chapter 12 expands the discussion of mutagenesis begun in chapter 11, with a focus on two topics, transposable elements and cancer. Chapter 13 covers the transmission genetics of bacteria and phages. It might reasonably be placed in the first part of the textbook, but because so much is known about the molecular basis of prokaryotic genetics, this discussion seems to fit better in the molecular genetics section. Chapter 14 explores the ways that genes govern the development of organisms. The intricate and fascinating development of the fruit fly is the major paradigm for this chapter. Chapter 15 presents the most celebrated practical applications of molecular genetics. It covers gene cloning and manipulation, and shows how molecular genetics has given birth to a kind of engineering. Our experience has been that students have fun with this topic, but because they need to master the basics first, this chapter is the last in this section. On the other hand, manipulation of cloned genes has become such a common and powerful tool of genetic analysis that some instructors may want to introduce at least parts of chapter 15 right after chapter 6, making the references to experiments using cloned genes in subsequent chapters less mysterious.

This textbook concludes with a two-chapter sequence on population genetics. Chapter 16 covers introductory population genetics, including selection, genetic drift, inbreeding, mutation, and gene flow. Chapter 17 integrates these factors to discuss human diseases, molecular evolution, conservation genetics, and pesticide resistance.

The most important underlying theme of this book is its experimental approach. Genetics, like any science, is not merely a collection of facts, but a way of posing and answering questions about nature. Wherever possible, we have included the basic experimental rationale and results that have led to our present understanding of genetics. Furthermore, many of the end-of-chapter worked-out problems and questions require that students examine and interpret hypothetical experimental results, or design experiments to test hypotheses. We know that genetics, especially molecular genetics, can be bewildering to college students. However, we have found that if given a clear understanding of the way geneticists practice their science, students can master the basics of the subject more easily. That is the goal of this textbook.

LEARNING AIDS FOR THE STUDENT

To achieve success in the study of genetics, students must understand the material presented, utilize the text as a tool for learning, and enjoy reading the text. Therefore, we have included many aids to make your study of genetics efficient and enjoyable. Each chapter contains the following:

Chapter Opening Quote
The chapter opening quote is designed to spark student interest and provide a perspective on the chapter's contents.

Chapter Opening Learning Objectives
The chapter opening learning objectives were written to help the student preview the basic concepts in the chapter.

Boldfaced Key Terms
Important terms are emphasized and clearly defined when they are first presented.

Boxed Readings
Most chapters contain a boxed reading. These readings cover a variety of interesting topics, including some of the ethical issues involved with recent developments in molecular biology and gene therapy. A portion of box 14.1, page 351, is shown here.

Box 14.1 Genomic Imprinting, Genetic Disease, and the Battle of the Sexes

n our discussion of transmission genetics, we have assumed that it made no difference which allele came from the female parent and which came from the male. Thus, in garden peas, crossing a homozygous green-seeded and a homozygous yellow-seeded parent will produce all yellow-seeded hybrid offspring, regardless of which parent contributed the gamete that specified yellow seeds. This is indeed generally

true, but there are some important exceptions. These fall into three categories, the first two of which can be explained easily. First, as we saw in chapters 3 and 4, some genes are sex-linked. For example, the outcome of a cross between white-eyed and red-eyed fruit flies is different if the female parent is white-eyed than if the male parent is white-eyed. However, since the eye color gene resides on the X chromosome in fruit flies, and since females have two X chromosomes and males have one X and one Y chromosome, these results still

conform to Mendelian principles. The second category of exceptions includes traits that are carried on genes that lie outside the nucleus—in the mitochondria or chloroplasts of a cell. For instance, the gene for a human neuromuscular disease called encephalomyopathy resides on a mitochondrial gene. Since mitochondria come exclusively from the mother, in the egg cytoplasm, this disease can be inherited only from the mother. As in the sex-linked case, this exception turns out to be readily explainable.

Section Summaries

These summaries help the student review and understand the section's basic concepts before continuing on to the next section of the chapter. For example:

 Meiosis is composed of one replication of DNA and two cell divisions. Meiosis I, the reductional division, results in the separation of homologous chromosomes and the halving of chromosome number. Meiosis II results in the separation of the sister chromatids, with no reduction in chromosome number. In animals, mature male and female gametes are produced by spermatogenesis and oogenesis, respectively.

End-of-Chapter Worked-Out Problems

To help students develop their analytical, quantitative, and problem-solving skills, example problems and detailed solutions are provided. Look for this heading at the end of the chapter:

WORKED-OUT PROBLEMS

End-of-Chapter Summary

This narrative section summarizes the important basic concepts of the chapter.

End-of-Chapter Problems and Questions

The problems and questions are written to aid students in their understanding of basic genetic concepts and processes. They also help students test and develop their analytical problem-solving skills. This section is identified by the following heading:

PROBLEMS AND QUESTIONS

In addition to these chapter aids, this textbook also features the following:

"A Basic Look Into The Cell" Illustrated Overview Panels

Often the best way to understand and review genetic processes such as mitosis, meiosis, DNA replication, transcription, and translation is to study an illustration that summarizes the process. Ahead of relevant chapters within this textbook, students will find **"A Basic Look Into The Cell"** panels, which provide an illustrated overview of basic genetic concepts and processes. These panels were designed to be easily located at any time. They are distinguished from the other text pages by a colored tab that runs down the edge of the panel pages. For an example, please see pages 5 and 6.

Answers to End-of-Chapter Problems and Questions (Appendix A)

The answers to all of the end-of-chapter problems and questions are provided to help students immediately evaluate their problem-solving skills.

Selected Readings for Each Chapter (Appendix B)

References for each chapter are provided to help students investigate and study important topics.

Glossary

Definitions of boldfaced terms are assembled in a glossary at the end of this textbook.

Index

An easy to use, comprehensive index is provided to help students locate information on topics they need to review.

Robert F. Weaver
Philip W. Hedrick

SUPPLEMENTARY MATERIALS

FOR THE INSTRUCTOR

1. An *Instructor's Manual with Test Item File (IM/TIF)* is available free to all adopters of Weaver/Hedrick *Basic Genetics,* second edition. This IM/TIF consists of two sections. The first section is a list of transparencies that accompany the text, a list of software and video suppliers, and thought questions and problems for each text chapter. The second section (TIF) contains objective questions pertinent to the material found in each text chapter. These include multiple choice, true/false, fill-in-the-blank, and short answer questions. (ISBN 13733)

2. *Computerized Testing and Classroom Management Software* is offered free upon request to adopters of this textbook. The TIF from the *Instructor's Manual* is included in the database of this computerized testing software. The software also provides a grade-recording program. The software requires no programming experience and is available in IBM and Macintosh formats. Diskettes are available through your local WCB sales representative. IBM DOS 3.5 (ISBN 13735), IBM DOS 5.25 (ISBN 13736), Windows 3.5 (ISBN 13737, & MAC 3.5 (ISBN 13738)

3. A set of *100 full-color acetate transparencies,* based on illustrations taken from the textbook, may be used to supplement classroom lectures and is available free to adopters (ISBN 13734) Also, please note that a *Student Study Art Notebook* containing these same full-color illustrations is packaged with each new textbook.

4. *Gene Game Software,* by Bill Sofer of Rutgers University, is a Macintosh software game that gives students the opportunity to test their logic and critical thinking skills while attempting to clone a fictitious "Fountain of Youth" gene. Contact your local WCB sales representative for more information on this fun, interactive software game. (ISBN 24893)

5. *Computer-Assisted Academic Programs in Genetics,* by Tully Turney of Hampden-Sydney College, is a Macintosh Hypercard program that features numerous interactive, problem-solving exercises in Mendelian,

molecular, and population genetics. Contact your local WCB sales representative for more information on these programs. (ISBN 13760)

FOR THE STUDENT

1. A *Student Study Art Notebook* containing illustrations from the textbook that correspond to the overhead transparencies, is designed to help you in your study of genetics. A copy of this student study art notebook is packaged with each new textbook. (ISBN 24302)

2. *Genetics Problem Solving Guide,* second edition, by William R. Wellnitz of Augusta College, is written as a companion to accompany any introductory genetics textbook. This book provides problem-solving guidance and practice. It should be available for student purchase at your bookstore. If you don't see a copy of it on your bookstore's shelves, ask the bookstore manager to order a copy for you. (ISBN 13739)

3. *Compendium of Problems in Genetics,* first edition, by John Kuspira and Ramesh Bhambhani of the University of Alberta, is written to complement any introductory genetics textbook. It contains approximately 1,000 thought-provoking genetics problems developed to help you succeed in understanding the basic concepts of genetics. It should also be available for student purchase at your bookstore. If you don't see a copy of it on your bookstore's shelves, ask the bookstore manager to order a copy for you. (ISBN 16734)

ACKNOWLEDGMENTS

We would like to thank the following reviewers of our manuscript for their many helpful suggestions:

First Edition
William S. Barnes
Clarion University of Pennsylvania

J. W. Bennett
Tulane University

Edward Berger
Dartmouth College

Anna W. Berkovitz
Purdue University

Elliott S. Goldstein
Arizona State University

Jeffrey C. Hall
Brandeis University

Duane L. Johnson
Colorado State University

Paul F. Lurquin
Washington State University

Lester J. Newman
Portland State University

Robert R. Schalles
Kansas State University

Thomas P. Snyder
Michigan Technological University

Archie Waterbury
California Polytechnic State University

Oscar H. Will, III
Augustana College

Second Edition
DuWayne C. Englert
Southern Illinois University at Carbondale

Thomas Haresign
Southampton College

Linda C. Martin
Kansas State University

Henri Roger Maurice
Barat College

Thomas P. Snyder
Michigan Technological University

William R. Wellnitz
Augusta College

To the Student

A personal library is a lifelong source of enrichment and distinction. Consider this book an investment in your future, and add it to your personal library.

INTRODUCTION

*Although genetics is a science rich in historical landmark discoveries, it is **not** just a collection of dates and facts to commit to memory. In the study of genetics, one poses questions about heredity and other life processes and then attempts to answer them through scientific study. Knowledge of basic genetic processes, like those illustrated in the following **A Basic Look into the Cell** panels, is fundamental to understanding life itself.*

Chapter 1:
*An Introduction
to Genetics:
A Brief History*

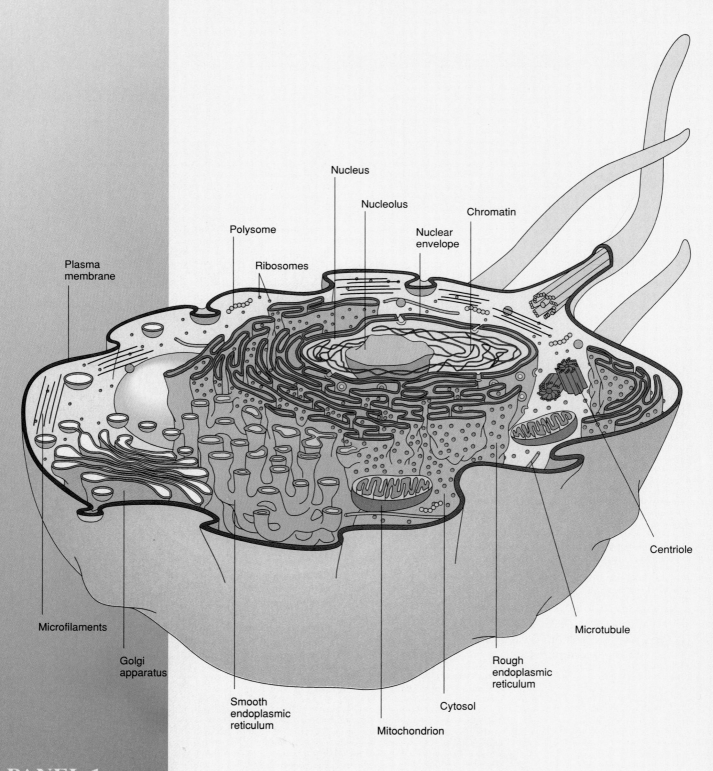

Nucleus

Nucleolus

Chromatin

Polysome

Nuclear
envelope

Ribosomes

Plasma
membrane

Centriole

Microfilaments

Microtubule

Golgi
apparatus

Rough
endoplasmic
reticulum

Smooth
endoplasmic
reticulum

Cytosol

Mitochondrion

CELL STRUCTURE

Generalized representation of an animal (eukaryotic) cell
showing internal structures.

DNA synthesis/replication

Prereplication
stage

Postreplication
stage

S

G_1

G_2

Interphase

Start

G_0

M

Nondividing
cells

Telophase

Anaphase

Metaphase

Prophase

Mitosis

CELL CYCLE

Phases in the cell cycle of a typical eukaryotic cell. The events that occur from the start of one cell division (mitosis), where two daughter cells are produced, to the next cell division are collectively termed the **cell cycle.**

PANEL 2A:
A BASIC
LOOK
INTO THE
CELL

A REVIEW OF CELL STRUCTURE

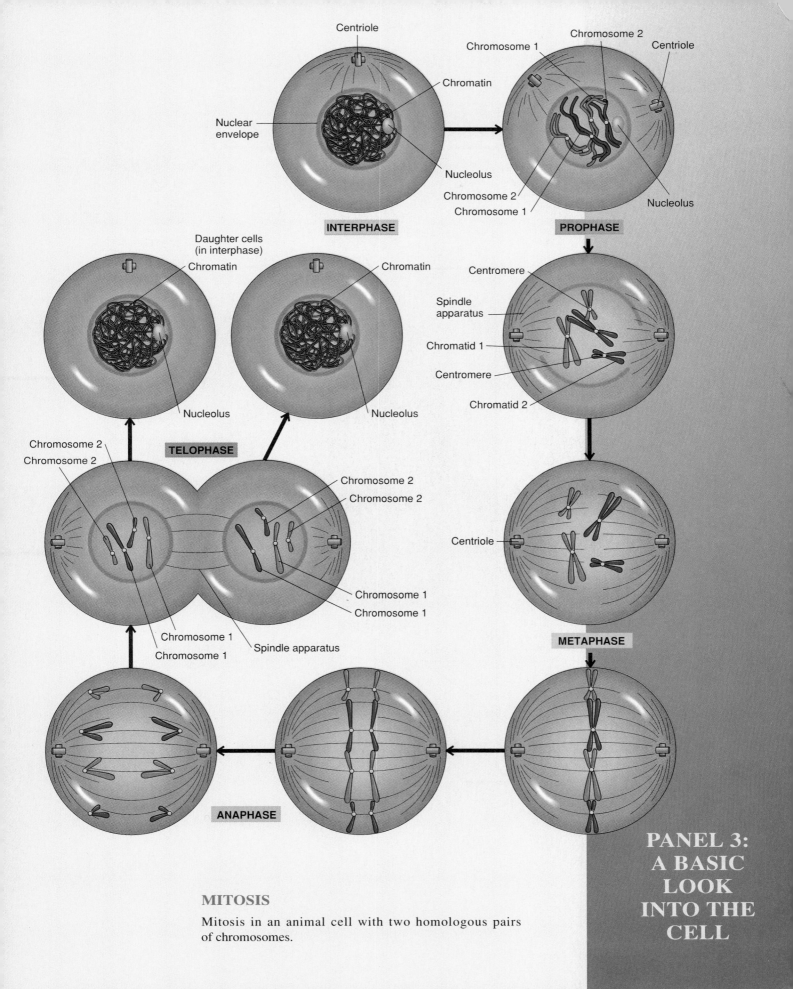

Centriole

Chromosome 1 Chromosome 2 Centriole

Chromatin

Nuclear
envelope

Nucleolus

Chromosome 2
Chromosome 1 Nucleolus

INTERPHASE **PROPHASE**

Daughter cells
(in interphase)
Chromatin Chromatin

Centromere

Spindle
apparatus

Chromatid 1

Centromere

Nucleolus Nucleolus Chromatid 2

Chromosome 2
Chromosome 2

Chromosome 2
Chromosome 2

Centriole

TELOPHASE

Chromosome 1
Chromosome 1

Chromosome 1 Spindle apparatus
Chromosome 1

METAPHASE

ANAPHASE

**PANEL 3:
A BASIC
LOOK
INTO THE
CELL**

MITOSIS

Mitosis in an animal cell with two homologous pairs
of chromosomes.

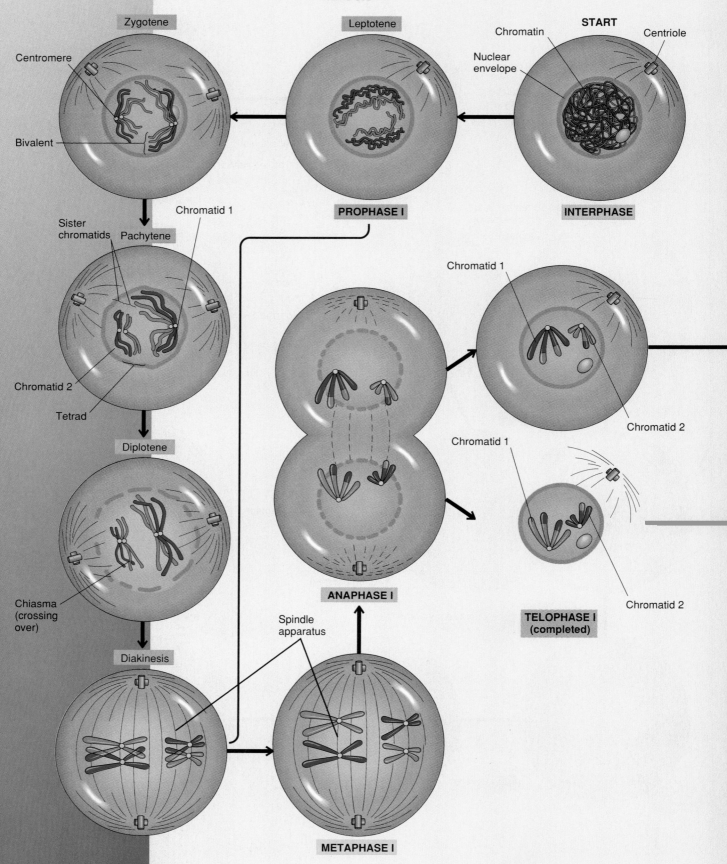

MEIOSIS I

Zygotene

Leptotene

START

Chromatin

Centriole

Centromere

Nuclear envelope

Bivalent

PROPHASE I

INTERPHASE

Sister chromatids

Pachytene

Chromatid 1

Chromatid 1

Chromatid 2

Tetrad

Chromatid 2

Diplotene

Chromatid 1

Chiasma (crossing over)

Chromatid 2

Diakinesis

ANAPHASE I

TELOPHASE I (completed)

Spindle apparatus

METAPHASE I

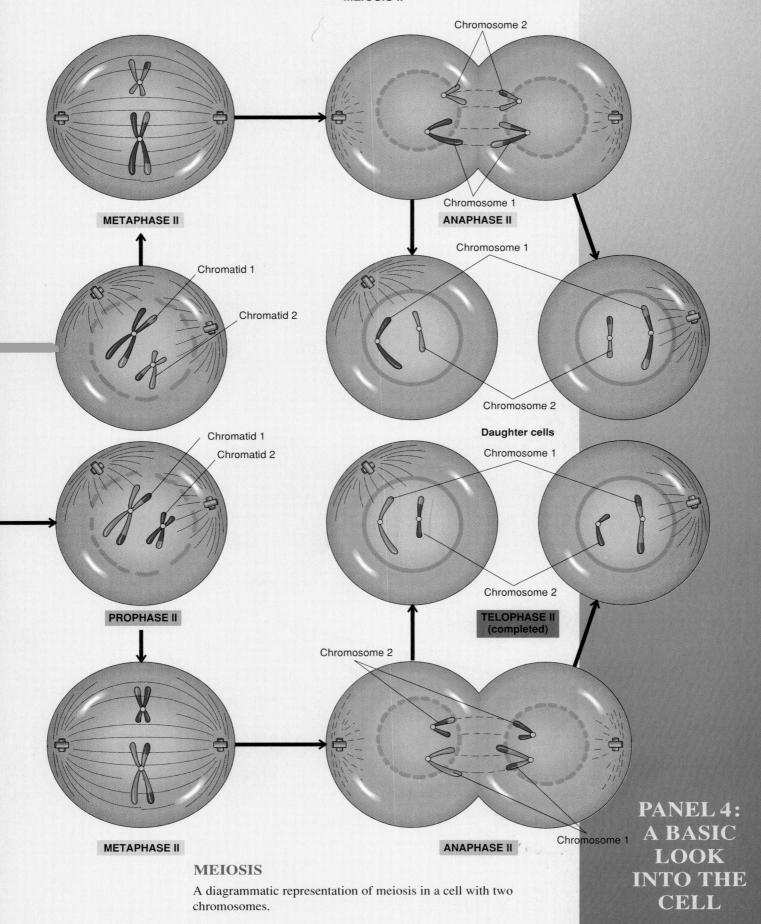

MEIOSIS II

Chromosome 2

Chromosome 1

ANAPHASE II

Chromosome 1

Chromosome 2

Daughter cells

Chromosome 1

Chromosome 2

TELOPHASE II (completed)

Chromosome 2

Chromosome 1

ANAPHASE II

METAPHASE II

Chromatid 1

Chromatid 2

Chromatid 1
Chromatid 2

PROPHASE II

METAPHASE II

MEIOSIS

A diagrammatic representation of meiosis in a cell with two chromosomes.

PANEL 4: A BASIC LOOK INTO THE CELL

MITOSIS

Chromosome 1   Chromosome 2

Chromosome 2
Chromosome 1

_____ phase
_____ phase

_____
_____

_____ phase

_____

_____ phase

_____

_____ cells

_____

_____

MEIOSIS

Chromosome 1

Chromosome 2

_____ phase
_____ phase   **Meiosis I**

_____

_____ phase

_____

_____ phase   **Meiosis II**

_____ phase

_____
_____

_____ phase

_____ phase   _____ phase

_____ phase
_____ phase

_____ phase

_____ cells

**PANEL 4A:
A BASIC
LOOK
INTO THE
CELL**

A REVIEW OF MITOSIS AND MEIOSIS

CHAPTER 1: AN INTRODUCTION TO GENETICS: A BRIEF HISTORY

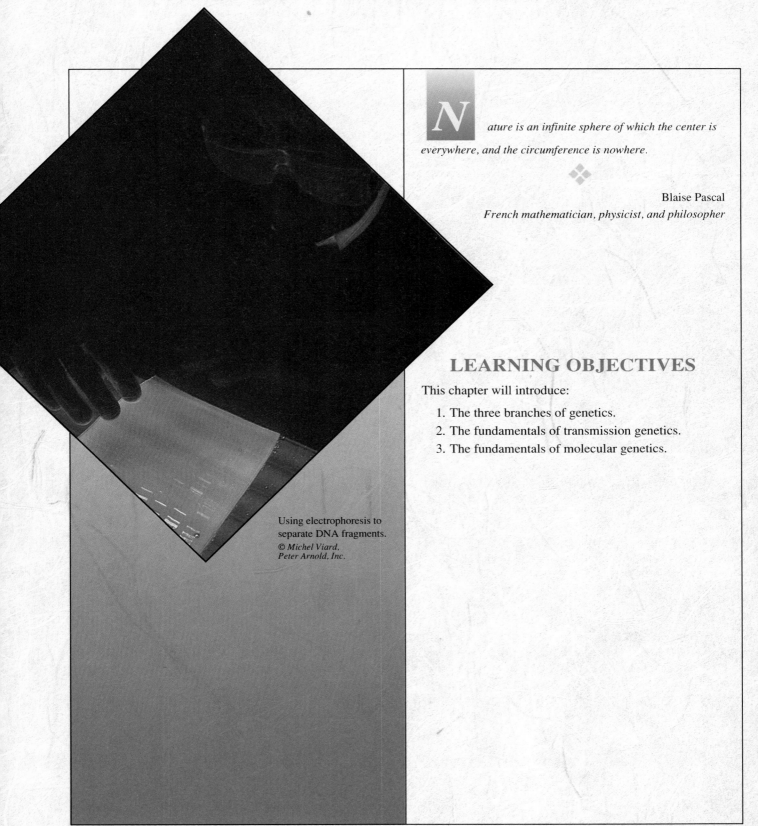

ature is an infinite sphere of which the center is everywhere, and the circumference is nowhere.

Blaise Pascal
French mathematician, physicist, and philosopher

Using electrophoresis to separate DNA fragments.
© *Michel Viard,*
Peter Arnold, Inc.

LEARNING OBJECTIVES

This chapter will introduce:

1. The three branches of genetics.
2. The fundamentals of transmission genetics.
3. The fundamentals of molecular genetics.

T he word "genetics" comes from the Latin *genesis,* which means birth. So genetics is the study of birth, or, more broadly, the study of heredity. Modern genetics has three main branches: (1) transmission genetics, the study of the passing of traits from one generation to the next; (2) molecular genetics, the study of the chemical structure of genes and how they operate at the molecular level; and (3) population genetics, the study of the variation of genes between and within populations.

THE THREE BRANCHES OF GENETICS

The first branch of genetics relies mainly on the same kind of experimental approach used by Gregor Mendel in the middle of the nineteenth century. We mate organisms with differing traits, or **phenotypes,** and then observe the transmission of these traits to the next generation. We can then mate these progeny organisms with others of the same or different phenotypes (or even with themselves, as with Mendel's peas), and again observe the transmission of traits. Because Mendel pioneered this approach to genetics, we frequently call it **Mendelian** or **classical genetics.** Since it deals with the transmission of genes between generations, we also call it **transmission genetics.**

The second branch, **molecular genetics,** approaches the subject from its fundamental base: molecules. Instead of phenotypic characters, molecular geneticists examine the genes themselves. They are concerned with the molecules that compose genes, the molecules that control genes, and the molecules that are the products of genes. The molecular geneticist's job has become much easier (and more fun) since we learned how to clone genes in the 1970s. With these new techniques, we can produce a gene in large quantities in pure form and study its structure and function.

The third branch, **population genetics,** examines the extent of genetic variation within and among populations. This variation was traditionally studied on the phenotypic level, but population geneticists now focus increasingly on molecular variation in a population. Much of population genetics study is aimed at understanding how the observed genetic variation evolved.

Population genetics techniques are also useful in describing the genetic differences between species and in learning about the process of species formation. Remember that a **species** is a reproductively isolated group, such as *Zea mays* (corn) or *Drosophila melanogaster* (fruit fly). A species name has two parts. The first part, which is capitalized (e.g., *Drosophila*), is the **genus,** and may encompass several species. The second part, which starts with a lowercase letter, designates a particular species within the genus.

A great deal of genetics research is now done at the molecular level. The other branches of genetics are in no danger of disappearing, however. We should bear in mind that although molecular techniques can uncover new genes, they frequently cannot reveal a gene's function. A more fruitful approach is this: A researcher first discovers an abnormal organism with interesting characteristics. The abnormality in this organism is caused by a mutated, or altered gene, so the *mutant* is the "beacon" that leads us to a new gene. Once the gene has been identified this way, we can begin to use molecular techniques to examine its structure and function in detail and perhaps to understand the evolutionary significance of variation in the gene. Actually, it is misleading to put too much emphasis on the divisions within the field of genetics. Most geneticists now approach their subject in several ways, using methods from more than one of the three branches.

Teachers of genetics (and writers of genetics textbooks) face an organization problem. Do we start by discussing the molecular aspects of the gene, so that later, when we talk about how genes are transmitted from generation to generation, students can appreciate fully what this means? Do we deal with transmission genetics first and only later explain molecular genetics? Or do we try to examine both at the same time?

In this book, we have decided to use the second approach: transmission genetics first, then molecular genetics, and finally, population genetics. This feels right to us for two reasons. First, transmission genetics is in some ways easiest to understand. It deals with the inheritance of characteristics that we can frequently see, and seeing is a great aid to understanding. Furthermore, inheritance is a concept that college students already appreciate to some degree.

A second reason for treating transmission genetics first is that it developed first historically. We like a good story, and the history of genetics is a very good one. By adopting a historical approach, we get a chance to retell that story.

No matter which part of genetics we cover first, we always wish our students had prior exposure to the other, because it would make their understanding so much richer. This problem is impossible to resolve completely, but we propose to compromise by providing in this introductory chapter a brief overview of transmission and molecular genetics, so that at least the most fundamental concepts of these branches of genetics will be familiar as we look at each in detail in later chapters. Since you do not need an appreciation of population genetics to understand the other two branches, we will not discuss it further in this chapter. In keeping with the basic plan of this book, we will present this overview in historical fashion.

TRANSMISSION GENETICS

In 1866, Gregor Mendel (fig. 1.1) published his findings on the inheritance of seven different traits in the garden pea. Prior to this time, inheritance was considered to occur through a blending of each trait of the parents in the offspring, but Mendel concluded that inheritance was particulate. That is, each parent contributes particles, or genetic units, to the offspring. We now call these particles **genes.** Furthermore, by carefully counting the number of progeny plants having a given **phenotype,** Mendel was able to make some important generalizations. (The word phenotype, by the way, comes from the same Greek root as "phenomenon," meaning "appearance"; thus, a tall pea plant exhibits the tall phenotype, or appearance.)

FIGURE 1.1
Gregor Mendel.
Courtesy of the Department of Library Services, American Museum of Natural History, negative 219467.

MENDEL'S LAWS OF INHERITANCE

Mendel saw that a gene can exist in several different forms called **alleles.** For example, the pea can have either yellow or green seeds. One allele of the gene for seed color gives rise to yellow seeds, the other to green. Moreover, one allele can be **dominant** over the other, **recessive** allele. In this case, yellow is dominant. Mendel showed this when he mated a green-seeded pea with a yellow-seeded pea. All of the progeny in the first *filial,* or offspring, generation (F_1) had yellow seeds (Latin: *filius* = son; *filia* = daughter). However, when these F_1 yellow peas were allowed to self-fertilize, some green-seeded peas reappeared. The ratio of yellow to green seeds in the second filial generation (F_2) was very close to 3:1.

Mendel concluded that the allele for green seeds must have been preserved in the F_1 generation, even though it did not affect the seed color of those peas. His explanation was that each parent plant carried two copies of the gene; that is, the parents were **diploid,** at least for the characteristics he was studying. According to this concept, **homozygotes** have two copies of the same allele, either two alleles for yellow seeds or two alleles for green seeds. **Heterozygotes** have one copy of each allele. The two parents in the first mating above were homozygotes; the resulting F_1 peas were all heterozygotes. Further, Mendel reasoned that sex cells contain only one copy of the gene; that is, they are **haploid.** Homozygotes can therefore produce sex cells, or gametes, that have only one allele, but heterozygotes can produce gametes having either allele.

This explains what happened in the matings of yellow with green peas. The yellow parent contributed a gamete with a gene for yellow seeds; the green parent, a gamete with a gene for green seeds. Therefore, all the F_1 peas got one allele for yellow and one allele for green seeds. They had not lost the allele for green seeds at all, but since yellow is dominant, all the peas were yellow. However, when these heterozygous peas were self-fertilized, they produced gametes with the two alleles in equal numbers, and this allowed the green phenotype to reappear.

Here is how that happened. Assume that we have two sacks, each containing equal numbers of green and yellow marbles. If we take one marble at a time out of each sack and pair it with a marble from the other sack, we will wind up with the following results: ¼ of the pairs will be yellow/yellow; ¼ will be green/green; and the remaining ½ will be yellow/green. The alleles for yellow and green peas work the same way. Recalling that yellow is dominant, you can see that only ¼ of the progeny (the green/green ones) will be green. The other ¾ will be yellow because they have at least one allele for yellow seeds. Hence, there is a 3:1 ratio of yellow to green peas in the second (F_2) generation.

 Genes can exist in several different forms, or alleles. One allele can be dominant over another; thus heterozygotes having two different alleles of one gene will generally exhibit the characteristic dictated by the dominant allele. The recessive allele is not lost; it can still exert its influence when paired with another recessive allele in a homozygote.

THE CHROMOSOME THEORY OF INHERITANCE

Other scientists either did not know about or uniformly ignored the implications of Mendel's work until 1900 when three botanists, who had arrived at similar conclusions independently, rediscovered it. After that time, most geneticists accepted the particulate nature of genes, and the field of genetics began to blossom. One factor that made it easier for geneticists to accept Mendel's ideas was a beginning understanding of the nature of chromosomes, which had occurred in the latter half of the nineteenth century. Mendel had predicted that gametes would contain only one allele of each gene instead of two. If chromosomes carry the genes, their numbers should also be reduced by half in the gametes—and they are. Chromosomes therefore appeared to be the discrete physical entities that carried the genes.

This notion that chromosomes carry genes is the **chromosome theory of inheritance.** It was a crucial new step in genetic thinking. No longer were genes disembodied factors; now they were observable objects in the cell nucleus. Some geneticists, particularly Thomas Hunt Morgan (fig. 1.2), remained skeptical of this idea. Ironically, however, it was Morgan himself who in 1910 provided the first definitive evidence for the chromosome theory.

Morgan worked with the fruit fly (*Drosophila melanogaster*), which was in many respects a much more

FIGURE 1.2
Thomas Hunt Morgan.
Courtesy of the National Library of Medicine.

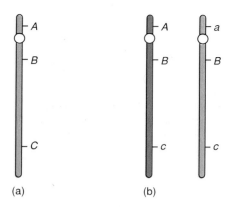

FIGURE 1.3
(*a*) A schematic diagram of a chromosome, indicating the positions of three genes—*A, B,* and *C.* (*b*) A schematic diagram of a diploid pair of chromosomes, indicating the positions of the three genes—*A, B,* and *C*—on each, and the genotype (*A,* or *a, B,* or *b,* and *C* or *c*) at each locus.

convenient organism than the garden pea for genetic studies because of its small size, short generation time, and large number of offspring. When he mated red-eyed flies (dominant) with white-eyed flies (recessive), most but not all of the F₁ progeny were red-eyed. Furthermore, when Morgan mated the red-eyed males of the F₁ generation with their red-eyed sisters, they produced about ¼ white-eyed males, but no white-eyed females. In other words, the eye color phenotype was sex linked. It was transmitted along with sex in these experiments. How could this be?

In hindsight, the answer leaps out at us. Sex and eye color are transmitted together because the genes governing these characteristics are located on the same chromosome—the single X chromosome. (Most chromosomes, called **autosomes,** occur in pairs in a given individual, but the X chromosome is an example of a **sex chromosome,** of which the female fly has two copies and the male has one.) Morgan was reluctant to draw this conclusion, however, until, also in 1910, he observed the same sex linkage with two more phenotypes: miniature wing and yellow body. That was enough to convince him of the validity of the chromosome theory of inheritance.

Before we leave this topic, let us reiterate two crucial points. First, every gene has its place on a chromosome. Figure 1.3*a* depicts a hypothetical chromosome and the positions of three of its genes, called *A, B,* and *C.* Second, diploid organisms such as human beings contain two copies of all chromosomes (except, in some cases, sex chromosomes). That means that they have two copies of most genes, and that these copies can be the same alleles (*homozygous*) or different alleles (*heterozygous*). For example,

figure 1.3*b* shows a diploid pair of chromosomes with different alleles at one locus (*Aa*) and the same alleles at the other two loci (*BB* and *cc*). The **genotype,** or allelic constitution, of this organism with respect to these three genes is *AaBBcc.* Since this organism has two different alleles (*A* and *a*) in its two chromosomes at the *A* locus, it is heterozygous at that locus (Greek: *hetero* = different). Since it has the same, dominant *B* allele in both chromosomes at the *B* locus, it is homozygous dominant at that locus (Greek: *homo* = same). And since it has the same recessive *c* allele in both chromosomes at the *C* locus, it is homozygous recessive there. Finally, because the *A* allele is dominant over the *a* allele, the phenotype of this organism would be the dominant phenotype at the *A* and *B* loci, and the recessive phenotype at the *C* locus.

This discussion of varying phenotypes in *Drosophila* gives us an opportunity to introduce another important genetic concept: **wild-type** versus **mutant.** The wild-type phenotype is the most common, or at least the generally accepted standard phenotype of an organism. To avoid the mistaken impression that a wild organism is automatically a wild-type, some geneticists prefer the term **standard type.** In *Drosophila,* red eyes and full-size wings are wild-type. Mutations in the *white* and *miniature* genes result in mutant flies with white eyes and miniature wings, respectively. Mutant alleles are usually recessive, as in these two examples, but not always.

GENETIC RECOMBINATION AND MAPPING

When two organisms mate, some of the offspring resemble their parents, but some have new combinations of traits not seen in either parent. This can happen in two ways. The first is **independent assortment** of genes on separate chromosomes during formation of gametes (see chapter 2 for details). The second is recombination by **crossing over.**

It is easy to understand that genes on separate chromosomes behave independently in genetic experiments, and that genes on the same chromosome—like the genes for

FIGURE 1.4

Recombination in *Drosophila*. The two X chromosomes of the female are shown schematically. One of them (red) carries two normal genes: (m^+), which results in normal wings, and (w^+), which gives red eyes. The other (blue) carries two mutant genes: *miniature* (*m*) and *white* (*w*). During egg formation, a recombination, or crossover, indicated by the X, occurs between these two genes on the two chromosomes. The result is two recombinant chromosomes with mixtures of the two parental genes. One is m^+w, the other is mw^+.

miniature wing (*miniature*) and white eye (*white*)—behave as if they are linked. However, genes on the same chromosome do not always show perfect genetic linkage. In fact, Morgan discovered this when he examined the behavior of the sex-linked genes he had found. For example, although *white* and *miniature* are both on the X chromosome, they remain linked in offspring only 65.5% of the time. The other offspring have a new combination of alleles not seen in the parents and are therefore called **recombinants.**

How are these recombinants produced? The answer was already apparent by 1910, because microscopic examination of chromosomes during meiosis (gamete formation) had shown crossing over between **homologous** chromosomes (chromosomes carrying the same genes or alleles of the same genes). This resulted in the exchange of genes between the two homologous chromosomes. In the example above, during formation of eggs in the female, an X chromosome bearing the mutant alleles (*w* and *m*) at the *white* and *miniature* loci, respectively, experienced a crossing over with a chromosome bearing the wild-type alleles (*W* and *M*) at these two loci (fig. 1.4). Because the crossing over events occurred between these two genes, it brought together the mutant (*w*) allele and wild-type (*M*) allele on one chromosome and the wild-type (*W*) allele and mutant (*m*) allele on the other. Because it produced a new combination of alleles, we call this process recombination.

Morgan assumed that genes are arranged in a linear fashion on chromosomes, like beads on a string. This, together with his awareness of recombination, led him to propose that the farther apart two genes are on a chromosome, the more likely they are to recombine. This makes sense because there is simply more room between widely spaced genes for a crossing over to occur. A. H. Sturtevant extended this hypothesis to predict that there should be a mathematical relationship between the distance separating two genes on a chromosome and the frequency of recombination between these two genes. Sturtevant collected data on recombination in the fruit fly that supported this hypothesis. This established the rationale for **genetic mapping** techniques still in use today. By the 1930s, other investigators found that the same techniques work with other **eukaryotes** (nucleus-containing organisms; see panel 1), including the mold *Neurospora,* the garden pea, maize (corn), and even human beings. These procedures also apply to **prokaryotes,** organisms that have no nuclei.

FIGURE 1.5
Barbara McClintock.
The Bettmann Archive.

PHYSICAL EVIDENCE FOR RECOMBINATION

Barbara McClintock (fig. 1.5) finally provided a direct physical demonstration of recombination by crossing over in 1931. By examining maize chromosomes microscopically, she could detect recombinations between two easily identifiable features of a particular chromosome (a knob at one end and a long extension at the other). Furthermore, whenever this physical recombination occurred, she could also detect recombination genetically. Thus, there is a direct relationship between a region of a chromosome and a gene.

The chromosome theory of inheritance holds that genes are arranged in linear fashion on chromosomes. The reason that certain traits tend to be inherited together is that the genes governing these traits are on the same chromosome. However, independent assortment or recombination by crossing over during meiosis can scramble the parental alleles to give nonparental combinations. The farther apart two genes are on a chromosome, the more likely recombination by crossing over between them will be.

MOLECULAR GENETICS

Studies such as those just discussed tell us important things about the transmission of genes and even about how to map genes on chromosomes, but they do not tell us what genes are made of or how they work. This has been the province of molecular genetics, which also happens to have its roots in Mendel's era.

THE DISCOVERY OF DNA

In 1869, Friedrich Miescher (fig. 1.6) discovered in the cell nucleus a mixture of compounds that he called nuclein. The major component of nuclein is **deoxyribonucleic acid (DNA).** By the end of the nineteenth century, chemists had learned the general structure of DNA and of a related compound, **ribonucleic acid (RNA).** Both are long polymers—chains of small compounds called nucleotides. Each nucleotide is composed of a sugar, a phosphate group, and a base. The chain is formed by linking the sugars to one another through their phosphate groups.

WHAT GENES ARE MADE OF

By the time the chromosome theory of inheritance was generally accepted, geneticists agreed that the chromosome must be composed of a polymer of some kind. This would accord with its role as a string of genes. But which polymer comprised the genes? There were essentially three choices: DNA, RNA, and **protein.** Protein was the other major component of Miescher's nuclein; its chain is composed of links called amino acids. A single protein chain is often called a **polypeptide.**

Oswald Avery (fig. 1.7) and his colleagues demonstrated in 1944 that DNA is the right choice. These investigators built on an experiment performed earlier by Frederick Griffith in which he transferred a genetic trait from one strain of bacteria to another. The trait was virulence, the ability to cause a lethal infection, and it could be transferred simply by mixing dead virulent cells with live avirulent cells. It was very likely that the substance that caused the transformation from avirulence to virulence in the recipient cells was the gene for virulence, because the recipient cells passed on this trait to their progeny.

What remained was to learn the chemical nature of the transforming agent in the dead virulent cells. Avery and his coworkers did this by applying a number of chemical and biochemical tests to the transforming agent and showing that it behaved just as DNA should, not as RNA or protein should.

THE RELATIONSHIP BETWEEN GENES AND PROTEINS

The other major question in molecular genetics is: How do genes work? To lay the groundwork for the answer to this question, we have to backtrack again, this time to 1902. That was the year Archibald Garrod noticed that the human disease alcaptonuria seemed to behave as a Mendelian recessive trait. It was likely, therefore, that the disease was caused by a defective, or mutated gene. Moreover, the main symptom of the disease was the accumulation of a black pigment in the patient's urine, which Garrod believed derived from the abnormal buildup of an intermediate in a biochemical pathway.

By this time, biochemists had shown that all living things carry out countless chemical reactions, and that these

FIGURE 1.6
Friedrich Miescher.
Courtesy of the National Library of Medicine.

FIGURE 1.7
Oswald Avery.
The Bettmann Archive.

reactions are accelerated or catalyzed by proteins called **enzymes.** Many of these reactions take place in sequence, so that one product becomes the starting material, or substrate, for the next reaction. Such sequences of reactions are called **pathways,** and the products/substrates within a pathway are called **intermediates.** Garrod postulated that an intermediate accumulated to abnormally high levels in alcaptonuria because the enzyme that would normally convert this intermediate to the next was defective. Putting this together with the finding that alcaptonuria behaved genetically as a Mendelian recessive trait, Garrod suggested that a defective gene gives rise to a defective enzyme. To put it another way: A gene is responsible for the production of an enzyme.

Garrod's conclusion was based in part on conjecture; he did not really know that a defective enzyme was involved in alcaptonuria. It was left for George Beadle and E. L. Tatum (fig. 1.8) to prove the relationship between genes and enzymes. They did this using the mold *Neurospora* as their experimental system. *Neurospora* has two enormous advantages over the human being as an experimental system. First, this mold lives most of its life in the haploid state, so there is no wild-type allele to mask the effects of a mutation. Thus, all mutations can be detected after only one generation. A second advantage is that we are not limited to the mutations that nature provides, but can use **mutagens** to introduce mutations into genes and then observe the effects of these mutations on biochemical pathways. Beadle and Tatum found many instances where they could pin the defect down to a single step in a biochemical pathway, and therefore to a single enzyme. They did this by adding the intermediate that would normally be made by the defective enzyme and showing that it restored normal growth. By circumventing the blockade, they discovered where it was. In these same cases, their genetic experiments showed that a single recessive Mendelian gene was involved. Therefore, a defective gene gives a defective (or absent) enzyme. In other words, a gene seemed to be responsible for making one enzyme. This was the one gene-one enzyme hypothesis, which was actually not quite right because an

(a)   (b)

FIGURE 1.8
(*a*) George Beadle; (*b*) E. L. Tatum.
Photos © AP/Wide World Photos.

enzyme can be composed of more than one polypeptide chain, but a gene actually has the information for making only one polypeptide (or RNA) chain.

WHAT GENES DO

We see that genes are made of DNA, and that each gene carries the information for making one polypeptide chain. Let us now return to the question at hand: How do genes work? This is really more than one question because genes are involved in more than one activity. First, they are replicated faithfully; second, they direct the production of RNAs and proteins; third, they accumulate mutations and so allow evolution. Let us look briefly at each of these activities.

How Genes Are Replicated

First of all, how does DNA replicate faithfully? To answer that question, we need to know the overall structure of the DNA molecule as it is found in the chromosome. James Watson and Francis Crick (fig. 1.9) provided the answer in 1953 by building models based on chemical and physical data that had been gathered in other laboratories, primarily X-ray diffraction data collected by Rosalind Franklin and Maurice Wilkins (fig. 1.10).

Watson and Crick proposed that DNA is a **double helix**—two DNA strands wound around one another. More important, the bases of each strand are on the inside of the helix, and a base on one strand pairs with one on the other in a very specific way. There are only four different bases in DNA—adenine, guanine, cytosine, and thymine—which we abbreviate A, G, C, and T. Wherever we find an A in one strand, we always find a T in the other; wherever we find a G in one strand, we always find a C in the other. For example, if the two DNA strands are represented by vertical strings of letters, a DNA duplex might look like this:

A T
G C
T A
C G

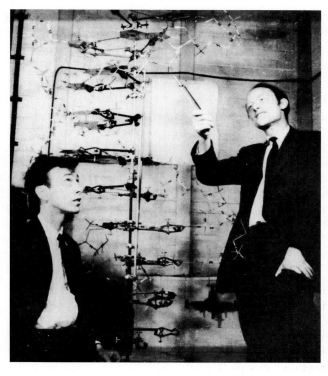

FIGURE 1.9
James Watson (left) and Francis Crick.
The Bettmann Archive.

(a)   (b)

FIGURE 1.10
(*a*) Rosalind Franklin; (*b*) Maurice Wilkins.
Photo (a) from Cold Spring Harbor. Photo (b) courtesy of Professor Wilkins, Biophysics Dept., King's College, London.

In a sense, then, the two strands are complementary. If we know the base sequence of one, we automatically know the sequence of the other. This complementarity is what allows DNA to replicate faithfully. The two strands come apart, and enzymes build new partners for them using the old strands as templates and following the Watson-Crick base-pairing rules (A with T, G with C). Again using the vertical string of letters convention, we can represent the replication of a DNA duplex as follows:

```
AT    A     T    AT     AT
GC →  G  +  C  → GC  +  GC
TA    T     A    TA     TA
CG    C     G    CG     CG
```

How Genes Direct the Production of Proteins

The instructions for making protein are "written" in a **genetic code** that gives the sequence of bases in a DNA strand. Two steps are required to make a protein. First, an enzyme makes a copy of one of the DNA strands; this copy is not DNA but its close cousin RNA. In the second step, this RNA (**messenger RNA**) carries the instructions to the cell's protein factories, called **ribosomes.** The ribosomes "read" the genetic code and put together a protein according to its instructions.

What is the nature of this genetic code? Marshall Nirenberg and Har Gobind Khorana (fig. 1.11), working independently, cracked the code in the early 1960s. They found that three bases constitute a code word, called a **codon,** that stands for one amino acid. Out of the sixty-four possible three-base codons, sixty-one stand for amino acids; the other three are stop signals. The ribosomes scan a messenger RNA three bases at a time and bring in the corresponding amino acids to link to the growing protein chain. When they reach a stop signal, they release the completed protein.

How Genes Accumulate Mutations

Genes change in a number of ways. The simplest is a change of one base to another. For example, if a certain codon in a gene is GAG (for the amino acid called glutamic acid), a change to GTG converts it to a valine codon. The protein that results from this mutated gene will have a valine where it ought to have a glutamic acid. This may be one change out of hundreds of amino acids, but it can have profound effects. In fact, this specific change has occurred in the gene for one of the human blood proteins and is responsible for the genetic disorder we call sickle-cell disease. Genes can suffer more profound changes, such as deletions or insertions of large pieces of DNA. The more drastic the change, the more likely that the gene or genes involved will be totally inactivated.

In recent years, geneticists have learned to isolate genes, place them in new organisms, and reproduce them by a set of techniques collectively known as **gene cloning.** Cloned genes not only give molecular geneticists plenty of raw materials for their studies, but they can also be transplanted to plants and animals, including humans. Thus, they may provide powerful tools for agriculture and for intervening in human genetic diseases.

WHY STUDY GENETICS?

This, then, is a brief introduction to the field of genetics. Some of the major milestones in this story are listed in

FIGURE 1.11
Har Gobind Khorana (left) and Marshall Nirenberg.
The Bettmann Archive.

Most genes are made of DNA arranged in a double helix. Genes engage in three main activities: replication, carrying information, and collecting mutations. The complementary nature of the two DNA strands in a gene allows them to be replicated faithfully by separating and serving as templates for the assembly of two new complementary strands. The sequence of nucleotides in a typical gene is a genetic code that carries the information for making an RNA. Most of these are messenger RNAs that carry the information to protein-synthesizing ribosomes. The end result is a new protein chain made according to the gene's instructions. A change in the sequence of bases constitutes a mutation, which can change the sequence of amino acids in the gene's protein product.

table 1.1. The following chapters will expand on these themes and fill in the large gaps that we have necessarily left.

The study upon which you are embarking is a fascinating one. It is as ancient as our prehistoric ancestors' selection of more fruitful grains to grow and more desirable animals to breed. It is as modern as tomorrow's experimental gene therapy. And it could hardly be more important for promoting an understanding of our future world, since some of the most profound changes in medicine and agriculture over the next few decades are likely to flow from the manipulation of genes. Already, some of the most exciting new pharmaceutical agents, such as human insulin for diabetes, human growth hormone for dwarfism, and interleukin 2 for cancer therapy, are the products of cloned genes. Geneticists also have identified the defective genes responsible for such diseases as cystic fibrosis and Duchenne muscular dystrophy and are beginning to test strategies for curing these diseases by implanting the

Table 1.1 Genetics Time Line

1859	Charles Darwin	Published *On the Origin of Species.*
1865	Gregor Mendel	Advanced the principles of segregation and independent assortment.
1869	Friedrich Miescher	Discovered DNA.
1900	Hugo de Vries, Carl Correns, Erich von Tschermak	Rediscovered Mendel's principles.
1902	Archibald Garrod	Noted the first genetic disease.
1902	Walter Sutton, Theodor Boveri	Proposed the chromosome theory.
1908	G. H. Hardy, Wilhelm Weinberg	Formulated the Hardy-Weinberg principle.
1910	Thomas Morgan	Demonstrated that genes are on chromosomes.
1916	Calvin Bridges	
1913	A. H. Sturtevant	Constructed a genetic map.
1927	H. J. Muller	Induced mutation by X rays.
1931	Harriet Creighton, Barbara McClintock	Obtained physical evidence for recombination.
1941	George Beadle, E. L. Tatum	Proposed the one gene-one enzyme hypothesis.
1944	Oswald Avery, Colin McLeod, Maclyn McCarty	Identified DNA as the material genes are made of.
1953	James Watson, Francis Crick, Rosalind Franklin, Maurice Wilkins	Determined the structure of DNA.
1958	Matthew Meselson, Franklin Stahl	Demonstrated the semiconservative replication of DNA.
1961	Sidney Brenner, François Jacob, Matthew Meselson	Discovered messenger RNA.
1966	Marshall Nirenberg, Har Gobind Khorana	Finished unraveling the genetic code.
1972	Paul Berg	Made the first recombinant DNA in vitro.
1973	Herb Boyer, Stanley Cohen	First used a plasmid to clone DNA.
1977	Walter Gilbert and Frederick Sanger	Worked out methods to determine the sequence of bases in DNA.
1977	Frederick Sanger	Determined the base sequence of an entire viral genome (ϕX174).
1977	Phillip Sharp, Richard Roberts, and others	Discovered interruptions (introns) in genes.
1988	Nancy Wexler, Michael Conneally, and James Gusella	Located the Huntington disease gene on human chromosome 4.
1990	Lap-Chee Tsui, Francis Collins, and John Riordan	Found the gene that is responsible for cystic fibrosis.
1990	James Watson and many others	Launched the Human Genome Project to map the entire human genome and, ultimately, to determine its base sequence.
1991	W. French Anderson and others	First successful gene therapy performed on two girls with severe combined immunodeficiency ("bubble-boy" syndrome).
1993	The Huntington Disease Group	Identification of the Huntington disease gene.

corresponding normal genes. They are also learning how to engineer better crops by adding genes for desirable traits, including resistance to diseases and pests.

Indeed, the very power of such gene manipulation provokes debate about benefits versus risks and even about whether it is appropriate for us to use such awesome tools at all. It is imperative that the public be educated about genetics in order to make intelligent decisions about these issues. In fact, F. H. Westheimer, professor emeritus of chemistry at Harvard University, commented about molecular biology (in this context, molecular genetics) as follows: "The greatest intellectual revolution of the last forty years may have taken place in biology. Can anyone be considered educated today who does not understand a little about molecular biology?"

Summary

The field of genetics consists of three main branches: transmission genetics, the study of the transmission of traits from one generation to the next; molecular genetics, the study of the structure and expression of genes at the molecular level; and population genetics, the study of the variation of genes between and within populations. These branches are not exclusive of one another; rather, they reinforce one another.

Genes can exist in several different forms called alleles. A recessive allele can be masked by a dominant one in a heterozygote, but it does not disappear. It can be expressed again in a homozygote bearing two recessive alleles.

Genes exist in a linear array on chromosomes. Therefore, traits governed by genes that lie on the same chromosome tend to be inherited together. However, crossing over between homologous chromosomes occurs during meiosis, so that gametes bearing nonparental combinations of alleles can be produced. The farther apart two genes lie on a chromosome, the more likely it is that such recombination will take place between them.

Most genes are made of double-stranded DNA arranged in a double helix. One strand is the complement of the other, which means that faithful gene replication requires that the two strands separate and acquire complementary partners. The linear sequence of bases in a gene carries the information for making an RNA, usually an mRNA that holds the information for making a polypeptide, or protein chain. A change (mutation) in this sequence can cause a corresponding change in the protein product.

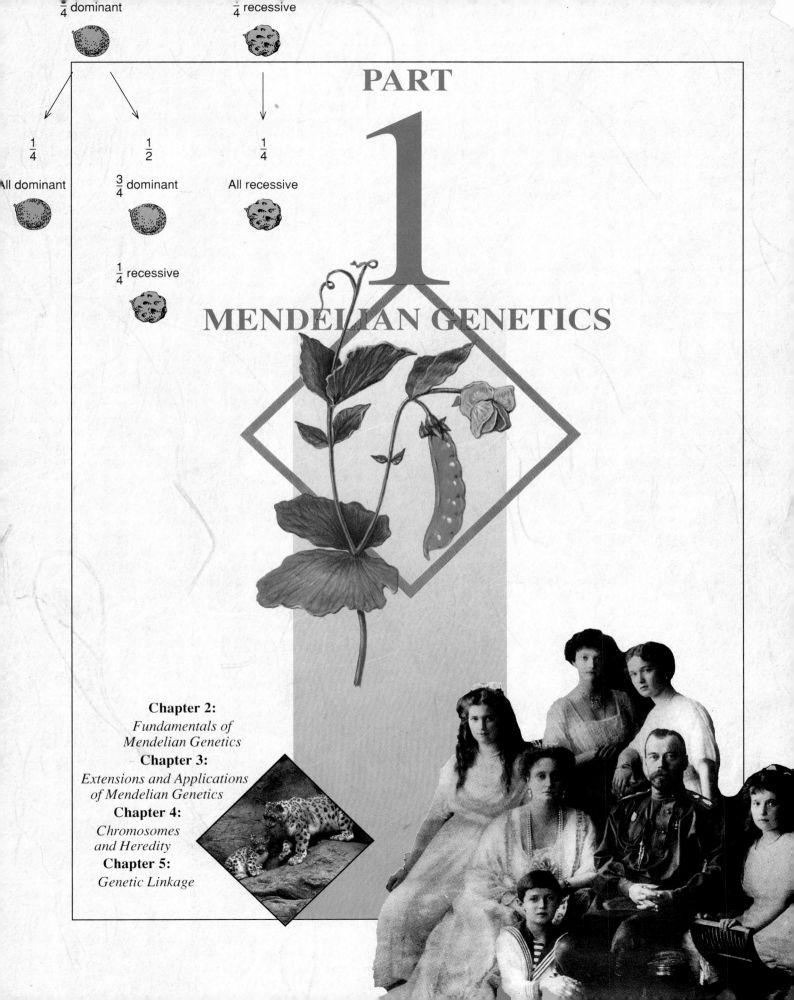

$\frac{3}{4}$ dominant $\frac{1}{4}$ recessive

$\frac{1}{4}$ $\frac{1}{2}$ $\frac{1}{4}$

All dominant $\frac{3}{4}$ dominant   All recessive

$\frac{1}{4}$ recessive

PART

1

MENDELIAN GENETICS

Chapter 2:
*Fundamentals of
Mendelian Genetics*

Chapter 3:
*Extensions and Applications
of Mendelian Genetics*

Chapter 4:
*Chromosomes
and Heredity*

Chapter 5:
Genetic Linkage

CHAPTER 2: FUNDAMENTALS OF MENDELIAN GENETICS

*I*n the ten years G. Mendel worked on his plants in the garden of the monastery, he made the greatest discovery in biology that has been made in the last five hundred years.

❖

Thomas Hunt Morgan
American geneticist

Snow leopard and baby.
© *Dianne Blell,*
Peter Arnold, Inc.

LEARNING OBJECTIVES

In this chapter you will learn:

1. The principle of segregation that governs the transmission of different alleles at a gene from generation to generation.

2. The principle of independent assortment that determines the transmission of genetic variants at different genes from generation to generation.

3. How to describe Mendel's principles using probability models.

4. How to apply the chi-square test to genetic data.

5. The usefulness of genetic models in identifying inheritance patterns in human pedigrees.

or centuries, people realized that individual characteristics were passed on from parent to offspring. However, the mechanism of inheritance was not generally understood until early in this century. The prevailing idea in the mid-nineteenth century was that the characteristics from the parents somehow blended at conception to form the characteristics of the offspring. This theory of **blending inheritance** did account for the fact that offspring possessed the characteristics of both parents. However, the idea of **particulate inheritance,** as proposed by Gregor Mendel in 1866, was able to explain this and many other observations. Mendel's theory suggested that a specific genetic unit controlling a single trait was passed on from generation to generation; this unit we now call a **gene.**

PRINCIPLE OF SEGREGATION

It was not an accident that Mendel was able to explain the phenomenon of inheritance, a subject that had puzzled many previous researchers. His success was based on at least three factors. First, he realized that to understand the basis of inheritance, he had to thoroughly document and quantify the results of his experiments. This inclination apparently arose from his earlier interest in physics, which was even then a quantitative science. Second, Mendel's experimental organism, the garden pea, was well suited for genetic investigation for several reasons: Peas have a number of contrasting characteristics in different individuals (for example, round seeds versus wrinkled seeds); garden pea flowers are large (making them easy to manipulate); pea plants produce large progeny numbers; and many different varieties of peas were available commercially in Mendel's time. Third, Mendel realized that he could best understand the basis of inheritance by concentrating initially on a single trait at a time, rather than on a number of traits as many of his contemporaries had tried to do.

Mendel's choice of the garden pea, *Pisum sativum,* was also fortunate in that varieties of the pea plant are capable of being developed into **true-breeding lines** in which parental plants reproduce offspring all with the same characteristics as themselves. Mendel obtained true-breeding lines for use in his experiments and ensured that his lines were completely true-breeding by growing them for two years in the monastery garden before beginning his experiments. Garden pea varieties are true-breeding because they reproduce primarily by **self-fertilization,** meaning that both the male and female gametes come from the same individual. Self-fertilization occurs when pollen is transferred from the anthers (the "male" part of the flower containing the pollen) to the stigma (the "female" part) of the same flower (fig. 2.1).

In order to study inheritance, Mendel crossed, or hybridized, different true-breeding lines (fig. 2.2). He did this by transferring pollen from the anthers of one plant to the stigma of another plant. The second plant had been emasculated; that is, its anthers had been removed prior to the development of pollen so that it could not self-fertilize.

When Mendel crossed lines that differed in one characteristic, such as round seeds versus wrinkled seeds, he found that the progeny resembled one of the parents. The parental type found in the progeny is called **dominant,** while the alternative characteristic that is not observed is called **recessive.** (The terms *dominant* and *recessive* generally refer to characteristics, but in some cases they also refer to different forms of a gene.)

The characteristics (or traits) that Mendel used to differentiate his pea plants are portrayed in figure 2.3. For each trait, the dominant form is on the left. The physical appearance of an individual plant or animal is called its **phenotype.** For example, an individual plant may have the round seed phenotype and the violet flower phenotype. When Mendel crossed two plants that were alike except that one had round seeds and the other had wrinkled seeds, all the progeny had the dominant round seed phenotype. Crosses between such individuals are called **monohybrid crosses,** because the offspring are hybrid—they are the product of two different varieties, and they have different parental contributions for a single trait. Mendel obtained the same results whether the female parent had round seeds and the male parent wrinkled seeds, or vice versa. In other words, **reciprocal matings** (matings between a female parent of the first type crossed with a male of the second type, or between a female of the second type crossed with a male of the first type) yielded equivalent progeny types. The progeny generation from such a cross is usually referred to as the F_1, or **first filial generation.**

MENDEL'S EXPERIMENTAL RESULTS

What were the experimental results that first allowed Mendel to understand the principles of heredity, concepts that continued to elude other researchers until 1900? After obtaining F_1 progeny, he allowed those individuals to self-fertilize and then observed the phenotypic numbers of

Box 2.1 The Father of Genetics

Gregor Mendel, born in 1822, grew up on his father's farm in a province of Austria. Because his family was poor, he entered a monastery in order to continue his education. During this period, he studied physics and mathematics, subjects that were sound training for his later experiments with inheritance. He was sent by his order to the University of Vienna to take examinations so that he could become an accredited teacher. However, he failed the exams and returned to the monastery where he was an "unaccredited" teacher for many years.

Beginning on his father's farm, Mendel had a lifelong interest in plants and animals, and he generally kept flowers, bees, and mice. He was fascinated by animal breeding, but because his superiors did not think it appropriate for a cleric to breed animals, he concentrated instead on raising garden peas and other plants in the monastery garden (fig. B2.1). One can still visit this tranquil garden in Brno, Czechoslovakia. The garden has a planting of flowers, instead of garden peas, to illustrate Mendel's principles.

Mendel was able to determine the basic principles of inheritance from his plant breeding experiments. The results of his work were reported in 1865 to the Brno Society for the Study of Natural Science and published in German in an 1866 paper entitled "Experiments on Plant Hybrids." The paper was distributed fairly widely through libraries, but Mendel's contemporaries did not appreciate his findings, probably in part because of the mathematical explanations he gave for his results. In addition, most other researchers of the time studied many characteristics simultaneously, which led to such complicated results that the basic principles of inheritance were not discernible. In fact, later on, Mendel himself apparently had some questions about the generality of his principles.

FIGURE B2.1
Gregor Mendel, early discoverer of the principles of genetics, in his garden.
The Bettmann Archive.

When he tried to duplicate his experiments in another plant called hawkweed, the results were completely different from those in garden peas. We now know that these differences occurred because hawkweed, unlike most other plants, requires a pollen grain for the egg to develop but does not include the genetic content of the pollen in the offspring.

Mendel became abbot of the monastery in 1868 and published nothing further on heredity after his 1866 masterpiece. Until his death in 1884, he spent much of his time engaged in administrative conflicts with the local government and also developed a fondness for secular pleasures, including rich food and cigars.

Unfortunately, Mendel died before his work was appreciated by the rest of the scientific community. In 1900, three botanists, Carl Correns of Germany, Hugo de Vries of the Netherlands, and Erich von Tschermak of Austria, rediscovered his work after each had independently reached similar conclusions. This marked the beginning of modern genetic research. Mendel's experimentation is now recognized as a classic example of carefully planned and executed scientific research and his paper as an excellent illustration of scientific genius.

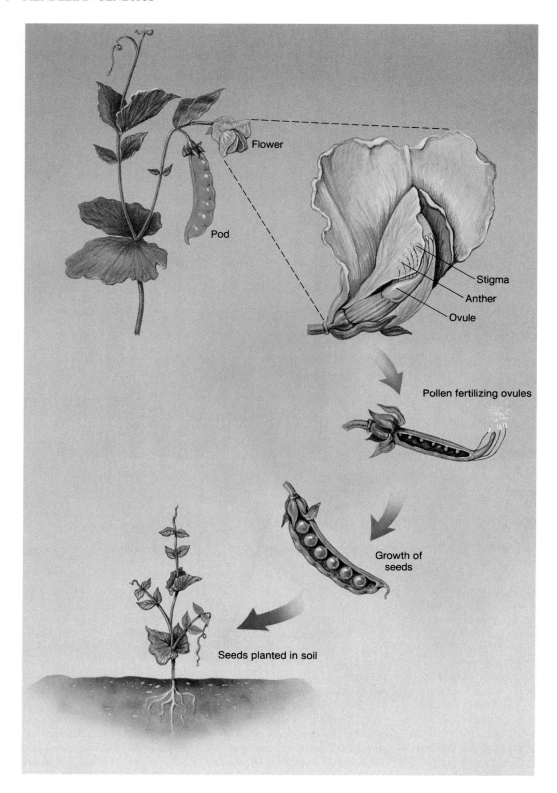

Flower

Pod

Stigma
Anther
Ovule

Pollen fertilizing ovules

Growth of
seeds

Seeds planted in soil

FIGURE 2.1

The anatomy and life cycle of the pea plant. Notice that the pollen
and ovule (egg) are from the same flower; that is, self-fertilization
is occurring.

*From Linda R. Maxson and Charles H. Daugherty, Genetics: A Human Perspective,
3d ed. Copyright © 1992 Wm. C. Brown Communications, Inc., Dubuque, Iowa.
All Rights Reserved. Reprinted by permission.*

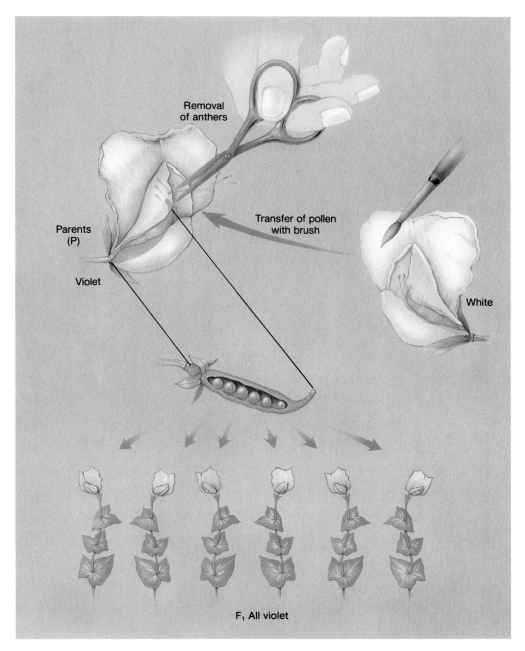

FIGURE 2.2
Artificial cross of a white-flowered (male parent) with a violet-flowered
(female parent) pea plant. In this case, all F_1 progeny are like the dominant
parental type, having violet flowers.

the **second filial generation, or F_2.** In this generation, he
observed both of the original parental phenotypes but found
approximately three times as many of the dominant
parental type as the recessive parental type, or 75% dominant and 25% recessive. These findings allowed Mendel to
reject the theory of blending inheritance because (1) the F_1
was not intermediate in phenotype and (2) the recessive
phenotype had not been lost, since it reappeared in the F_2
progeny.

A summary of Mendel's results from the seven different monohybrid crosses is given in table 2.1. For all of
these crosses, the ratio of dominant to recessive individuals
in the F_2 was very close to 3:1. For example, in the F_2 of
the cross between plants with round versus wrinkled seeds,
there were 5,474 round seeds and 1,850 wrinkled seeds,
giving an F_2 ratio of 2.96 round to 1 wrinkled—extremely
close to 3:1.

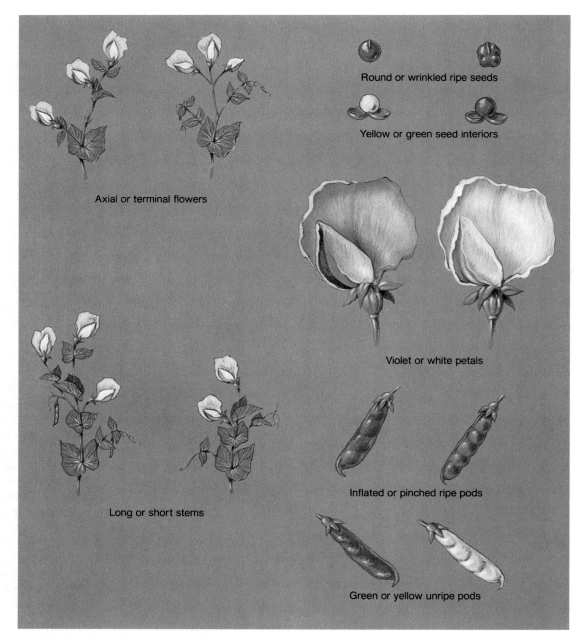

FIGURE 2.3

The seven characteristics in garden peas that Mendel studied. The dominant form is on the left in each case.

Mendel allowed individual F$_2$ plants to self-fertilize and observed the types of progeny in the next (F$_3$) generation. The one-quarter of the F$_2$ plants that bore the recessive phenotype were all true-breeding recessives; that is, all their progeny were recessive. However, of the three-quarters that expressed the dominant phenotype, some were true-breeding dominants, while others were like the F$_1$ individuals in that they had both dominant and recessive progeny. (To help you understand this, we will discuss the principles of probability later in this chapter.) Upon self-fertilization, three-quarters of the progeny of the plants in the latter category were dominant and one-quarter were recessive. Overall then, there were three types of F$_2$ individuals: ¼ true-breeding dominants, ½ dominants that could have both dominant and recessive progeny, and ¼ true-breeding recessives (fig. 2.4).

Table 2.1 Results of Mendel's Monohybrid Crosses

	F₁	F₂		F₂ Ratio
Seeds		*Dominant*	*Recessive*	*Dominant:Recessive*
Round vs. wrinkled	Round	5,474	1,850	2.96:1
Yellow vs. green	Yellow	6,022	2,001	3.01:1
Pods				
Inflated vs. constricted	Inflated	882	299	2.95:1
Green vs. yellow	Green	428	152	2.82:1
Flowers				
Purple vs. white	Purple	705	224	3.15:1
Axial vs. terminal	Axial	651	207	3.14:1
Stem				
Tall vs. short	Tall	787	277	2.84:1

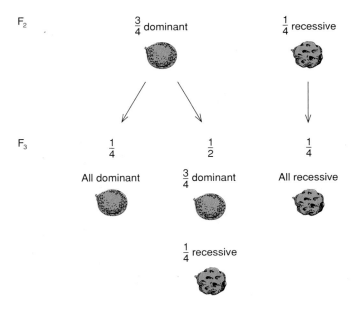

FIGURE 2.4
Proportions of F₃ progeny resulting from selfing F₂ progeny. Note that all F₂ recessive progeny are true-breeding, while only part of the F₂ dominant progeny are true-breeding.

From these experiments, Mendel concluded that particles or factors, which we now call genes, were transmitted intact through the gametes from parents to progeny. Another term we will use to indicate a gene is **locus** (pl. **loci**). More precisely, a locus is the place where the gene resides on the chromosome. Mendel suggested that these factors existed in several alternative forms (which we call **alleles**) that determined the different phenotypes he had observed. He assumed that each individual had two copies of each gene, receiving one copy of the gene (one allele) from the

gamete of the male parent and the other copy of the gene (one allele) from the gamete of the female parent. Although this elegant interpretation seems simple to us today, no one in Mendel's time appreciated it.

MENDEL'S EXPLANATION

Let us examine the hypothesis Mendel developed to explain inheritance for the specific characteristic of round versus wrinkled seeds. We will call the gene determining seed phenotype R and say that it has two allelic forms: R, the allele associated with the dominant, round phenotype, and r, the allele associated with the recessive, wrinkled phenotype. Individuals of three different **genotypes,** or genetic types, are possible. (The genotype of an individual is expressed by a pair of letters or symbols because individuals have two copies of each gene.) Two of the genotypes are **homozygotes,** having both their alleles identical. One would have the RR genotype and the round seed phenotype, while the other has the rr genotype and the wrinkled seed phenotype. The third possible type of individual would be a **heterozygote,** Rr, having one of each type of allele.

The heterozygous genotype has a round seed phenotype because the round allele is dominant over the wrinkled allele. Sometimes the notation $R-$ is used to indicate that the second allele of the genotype may be either R or r; that is, the individual is either genotype RR or Rr, both of which have the dominant round phenotype. If we start out with true-breeding round and wrinkled lines as the parents, then they must be RR and rr, respectively (fig. 2.5). When these parents produce gametes, the RR individuals produce gametes with the R allele, and the rr individuals produce gametes with the r allele. Fertilization results in all progeny (the F₁) being heterozygotes, or Rr, and having the dominant, round phenotype.

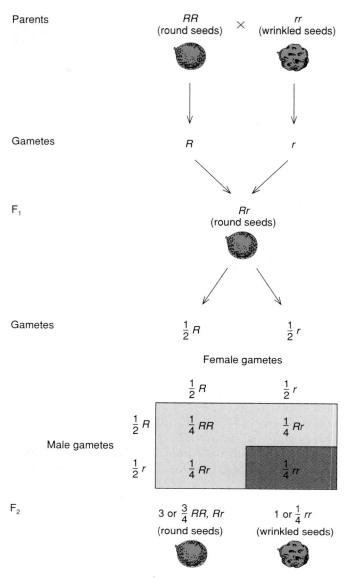

FIGURE 2.5
The genetic basis of the principle of segregation in F$_2$ progeny in a cross between parents with the round and wrinkled seed phenotypes. Notice that half the gametes from the F$_1$ heterozygotes have the R allele and half have the r allele.

To produce the next generation, Mendel allowed these heterozygotes to self-fertilize. In explaining his results, he assumed that *heterozygotes produced equal numbers of gametes having the two different alleles,* an assumption now known as the **principle of segregation,** or sometimes as **Mendel's first law.** (We will not refer to Mendel's principles as laws because there are important exceptions, some of which will be discussed in later chapters.) As indicated in figure 2.5, this occurs both in males and females, so that half of the pollen from a male heterozygote Rr is R and half is r, and half the eggs from an Rr female are R and half are r. In chapter 4 we will see that the mechanistic basis of the principle of segregation lies in the process of gamete formation and cell division called meiosis.

These gametes can combine in four different, equally frequent ways. First, both the male and female gametes may contain the R allele, so that the offspring are homozygous RR with a round phenotype. Second, one gamete may be R and the other r, either with the R allele in the male gamete and r in the female gamete, or with R in the female and r in the male. In either case, the offspring will have a heterozygous Rr genotype with a round seed phenotype. The other possible combination results when both gametes contain r, making all offspring homozygous rr with a wrinkled phenotype. Overall then, ¼ of the progeny are RR, ½ are Rr, and ¼ are rr, but because of the dominance of allele R, ¾ are the round phenotype (RR and Rr) and ¼ are the wrinkled phenotype (rr). In other words, there are three of the dominant type to every one of the recessive type, giving a 3:1 ratio. This explanation is consistent with the results of all the crosses listed in table 2.1. In addition, the dominant F$_2$ plants that were true-breeding in figure 2.4 were genotype RR, while those that produced both types of progeny were heterozygous Rr.

Mendel confirmed the principle of segregation by conducting several other experiments. In one instance, instead of allowing the F$_1$ heterozygous individuals to self-fertilize, he crossed them back to the recessive parent. Such a cross to a recessive is called a **testcross** because it is used to test whether a dominant individual is heterozygous or homozygous. In this case, the cross is also a **backcross** because the F$_1$ is being crossed with an individual having the same genotype as one of its parents. Figure 2.6 gives the genetic basis of this type of cross. It leads to equal proportions of the two progeny types because one parent, the homozygous male, contributes only gametes with the recessive r allele, while the heterozygous female parent gives equal proportions of gametes with the two types of alleles. In other words, segregation occurs only in the production of gametes from one parent rather than in the production of both types of gametes. The progeny are then equally divided between the dominant phenotype (Rr) and the recessive phenotype (rr).

After Mendel's genetic principles were rediscovered in 1900, the principle of segregation was reconfirmed by repeating his crosses and carrying out similar crosses in other plants and animals. For example, a number of researchers repeated the cross between yellow and green peas and determined the types of F$_2$ progeny (table 2.2). As in Mendel's experiments, the ratio of dominant to recessive in all these experiments was very close to 3:1, and the overall ratio for 175,691 individuals was 3.02:1.

Because self-fertilization generally does not occur in animals, two different F$_1$ individuals, or full siblings (brother and sister), had to be crossed in order to examine the segregation ratio in their progeny. The generality of Mendel's principles was confirmed by a number of researchers who crossed full siblings in animals such as mice, pigeons, and *Drosophila*. The first mammalian trait to be analyzed was the albino gene in mice, studied by William Castle and Lucien Cuénot. However, it has been suggested

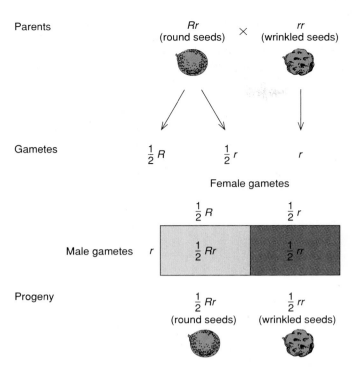

Parents   *Rr* (round seeds) × *rr* (wrinkled seeds)

Gametes $\frac{1}{2}$ *R* $\frac{1}{2}$ *r*   *r*

Female gametes

$\frac{1}{2}$ *R* $\frac{1}{2}$ *r*

Male gametes   *r* $\frac{1}{2}$ *Rr* $\frac{1}{2}$ *rr*

Progeny $\frac{1}{2}$ *Rr* (round seeds) $\frac{1}{2}$ *rr* (wrinkled seeds)

FIGURE 2.6
The genetic basis of the principle of segregation in a testcross. Here segregation occurs only in the female parent, leading to half the progeny having round seeds and half wrinkled seeds.

Table 2.2 Results of Repeat Crosses in Peas

Researcher	Yellow	Green	F_2 Ratio
			(Dominant:Recessive)
Mendel (1866)	6,022	2,001	3.01:1
Correns (1900)	1,394	453	3.08:1
Tschermak (1900)	3,580	1,190	3.01:1
Bateson (1905)	11,902	3,903	3.05:1
Darbishire (1909)	109,060	36,186	3.01:1
Total	131,958	43,733	3.02:1

Table 2.3 Results of Mendel's Dihybrid Cross

Generation	Seed Type	Number	F_2 Ratio
Parents	Round, green vs. wrinkled, yellow	—	—
F_1	Round, yellow	—	—
F_2	Round, yellow	315	9.84
	Wrinkled, yellow	101	3.16
	Round, green	108	3.38
	Wrinkled, green	32	1.0
		556	

"Brother Mendel! We grow tired of peas!"
Cartoon by J. Chase

❖ Mendel was the first scientist to demonstrate that inheritance is based on the segregation of alleles of a single gene. From self-fertilization of F_1 garden pea plants obtained from crosses between plants with different characteristics, he observed a 3:1 ratio of dominant to recessive phenotypes in the F_2 progeny. The principle of segregation he developed states that heterozygous individuals produce equal proportions of gametes containing the two alleles.

that Mendel actually discovered his principles by secretly crossing mice with different coat-color phenotypes, then later validating his findings with peas!

PRINCIPLE OF INDEPENDENT ASSORTMENT

The experiments of Mendel discussed so far have dealt with crosses between lines differing in only one character or in alleles at one gene. What happens if the parents differ in two characters (at two genes)? Mendel examined this situation once he had grasped the pattern of inheritance for a single trait. By performing **dihybrid crosses,** between parents having different parental contributions for two traits, he developed the **principle of independent assortment,** sometimes known as **Mendel's second law.**

In one of Mendel's experiments, he crossed plants that differed in two seed characteristics—round, green seeds in one parent and wrinkled, yellow seeds in the other (table 2.3). In this case, all of the F_1 progeny had the doubly dominant phenotype of round, yellow seeds. When these F_1 individuals were allowed to self-fertilize, four types of progeny were produced in the F_2, with the highest proportion being the dominant type for both traits (round, yellow). The two categories with one dominant type (wrinkled, yellow and round, green) had lower and similar numbers. The fewest had the double recessive phenotype (wrinkled, green). Mendel repeated these experiments using other traits and found that the ratios of the four types were always near 9:3:3:1. For example, if each number in table 2.3

is divided by 32, the number in the double recessive class, the ratio of the types is 9.84 : 3.16 : 3.38 : 1, as shown in the far right column.

Before trying to explain this whole array, let us determine whether the principle of segregation holds in this cross. Examining round versus wrinkled phenotypes in the F_2 reveals 423 round (315 + 108, combining the yellow and green classes) and 133 wrinkled (101 + 32). The ratio of round to wrinkled is then 2.96:1, consistent with expectations using the principle of segregation. Likewise, there are 416 yellow (315 + 101) and 140 green (108 + 32), making the yellow to green ratio 2.97:1, again close to segregation expectations.

Let us go through the model that Mendel developed to show how he could predict a 9:3:3:1 ratio in the F_2 (fig. 2.7). First, note that the F_1 plants are all heterozygous for both genes, $Rr\ Yy$, having received a dominant allele from one parent and a recessive allele from the other. As before, based on the principle of segregation, the frequency of both the R and r alleles in the F_2 gametes is ½. For the other gene as well, the frequency of both the Y and y alleles in the gametes is ½. However, the frequency of a gamete that contains both R and Y is ¼. In a subsequent section of this chapter, we will discuss the use of probability to explain this, but for the time being, we will state that having both R and Y in a gamete is equal to the product of obtaining an R and a Y, or ½ × ½ = ¼.

If we assume that the same process occurs for all four gametes ($R\ Y$, $R\ y$, $r\ Y$, and $r\ y$) and that it occurs in both females and males, we obtain a four-by-four matrix, or grid, as shown in figure 2.7. A matrix like this was first called a **Punnett square** after the British geneticist R. C. Punnett, who employed it for determining the genetic outcome of crosses. In order to determine the proportion of different types of individuals in the F_2 progeny, Mendel assumed that these gametes unite randomly to form genotypes, the same assumption he had made in formulating the principle of segregation.

All of the elements (or cells) in the matrix are equally likely and are, therefore, equal in frequency, ¼ × ¼ = 1/16. In addition, note that the progeny of each genotype is a composite of the two gametes that form it. In other words, the genotype formed from gametes $R\ Y$ and $R\ y$ is $RR\ Yy$. In six cases, two different combinations of gametes yield the same F_2 genotypes. For example, the $RR\ Yy$ genotype results from crossing an $R\ y$ female gamete and an $R\ Y$ male gamete as well as from crossing an $R\ Y$ female gamete and an $R\ y$ male gamete.

If we combine the cells having the same combination of traits, the basis of the 9/16 :3/16 :3/16 :1/16 , or 9:3:3:1, ratio is apparent. First, only one cell of the sixteen is doubly recessive—the lower right one. Second, there are three cells for each of the two types of single dominants; for example, the two $Rr\ yy$ cells and the one $RR\ yy$ cell are all round, green. The remaining nine cells are double dominants, completing the 9:3:3:1 ratio.

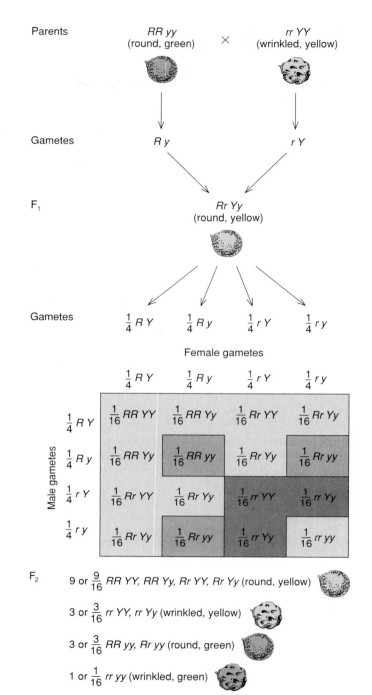

FIGURE 2.7

The genetic basis of the principle of independent assortment in F_2 progeny. Each cell in the 4 × 4 matrix is equally likely. Combine them to find the 9:3:3:1 ratio of F_2 phenotypes.

From An Introduction to Genetic Analysis 3/e by Suzuki, et al. Copyright © 1986, by W. H. Freeman and Company. Reprinted with permission.

Consider the example of corn kernels. A number of genes influence their color and shape. For example, the allele that results in purple kernels is dominant to the normal allele that causes yellow kernels, and the allele that results in shrunken kernels is recessive to that for normal kernels.

FIGURE 2.8
Each corn cob illustrates independent assortment by the color and shape.
Supplied by Carolina Biological Supply Company.

A particularly dramatic example of independent assortment is illustrated in the color and shape of the corn kernels on the cob in figure 2.8. Each kernel is the result of an independent fertilization event, so that the array of kernels represents an F_2 generation. Is the occurrence in this photo of the four phenotypic classes—purple-normal, purple-shrunken, yellow-normal, and yellow-shrunken—close to a 9:3:3:1 ratio?

The mechanistic basis of independent assortment—that genes on different chromosomes randomly go to different gametes during meiosis—was not known to Mendel. In chapter 4 we will discuss the genetic implications of meiosis in some detail. In addition, not all pairs of genes show independent assortment, a fact Mendel also did not discover. In chapter 5 we will discuss an important exception to Mendel's second principle, called linkage, in which pairs of genes on the same chromosome show association.

To better understand both segregation and independent assortment, it is useful to know some facts about probability and statistics. After we discuss these concepts, we will illustrate how they can be applied to Mendel's garden pea experiments.

 Mendel's principle of independent assortment states that alleles at different genes assort independently of each other, resulting in a 9:3:3:1 ratio in F_2 progeny from a dihybrid cross.

PROBABILITY

How can we understand Mendel's findings in a general manner? One way to better grasp the principles of segregation and independent assortment is to relate them to some simple theories of probability. These ideas were known to Mendel and were obviously important in his development of a general model of inheritance.

First, the term **probability** is defined as the number of times a particular event occurs, divided by the total number of opportunities for the event to occur. For example, if a coin is flipped ten times and turns up heads four times and tails six times, the observed probability of heads in these ten tosses is $\frac{4}{10} = 0.4$ and the observed probability of tails is $\frac{6}{10} = 0.6$. The sum of these probabilities is unity; that is, $\frac{4}{10} + \frac{6}{10} = 1$. These are the observed probabilities of heads and tails in a given sample of ten coin tosses. If the coin is "fair"—that is, equally likely to land on heads or tails—and if it is flipped many times, the expected probabilities are 0.5 for heads and 0.5 for tails.

SUM RULE

Heads and tails are examples of **mutually exclusive events.** In other words, if a tail is flipped, that excludes the possibility of a head being flipped on the same coin at the same time. *The combined probability of two (or more) mutually exclusive events occurring is the sum of their individual probabilities.* This is known as the **sum rule.** For example, symbolizing the probability of a head as P(head) and the probability of a tail as P(tail), where capital P is used as the symbol for probability, and assuming a fair coin:

$$P(\text{head or tail}) = P(\text{head}) + P(\text{tail})$$
$$= \frac{1}{2} + \frac{1}{2}$$
$$= 1$$

If there are more than two possible events, the sum of the probabilities of only two of the events should be less than unity, as we will discuss later.

A genetic example of probability occurs when we examine the types of gametes produced by a heterozygote. In this case, because of the genetic consequences of meiosis (see chapter 4), equal numbers of the two gametes having different alleles should be produced. For example, when the parent is heterozygous Rr, the probability of an R gamete is $\frac{1}{2}$ and the probability of an r gamete is also $\frac{1}{2}$. As in the coin example, using the sum rule of probabilities of the two different events, notice that:

$$P(R \text{ or } r) = P(R) + P(r)$$
$$= \frac{1}{2} + \frac{1}{2}$$
$$= 1$$

PRODUCT RULE

Now let us examine the probability of two independent events occurring simultaneously. This probability is expressed by the **product rule,** which states that *the joint probability of both of two independent events occurring is the product of their individual probabilities.* Thus, if we flip two different coins, say a penny and a nickel, what is the probability of obtaining a head from the first and a head from the second? The probability of a head from the penny is $\frac{1}{2}$ and the probability of a head from the nickel is $\frac{1}{2}$. Thus, the joint probability of two heads is:

$$P(\text{two heads}) = P(\text{head on penny}) \times P(\text{head on nickel})$$
$$= \frac{1}{2} \times \frac{1}{2}$$
$$= \frac{1}{4}$$

A genetic example of the product rule occurs when examining the progeny of a cross between two heterozygotes. For example, we would like to know the probability of an *rr* homozygote from the cross *Rr* × *Rr* where the first individual is the female. To obtain an *rr* offspring, both gametes must contain an *r* allele. Based on the principle of segregation, the probability of an *r* from the female is ½ and the probability of an *r* from the male is ½. Therefore, the probability of obtaining an *r* simultaneously from the female and male gametes to yield an *rr* offspring is ½ × ½ = ¼. Mendel used this concept to develop the principle of segregation and the proportions shown in figure 2.5.

We also use the product rule to obtain the probability of different genotypes in a dihybrid cross. First, we assume that the genes assort independently into F_2 gametes so that we can use the product rule to determine the joint probability of alleles from different genes in the same gamete. For example, the probability of an *R* allele and a *Y* allele in the same gamete is:

$$P(R \text{ and } Y) = P(R) \times P(Y) = \tfrac{1}{2} \times \tfrac{1}{2} = \tfrac{1}{4}$$

Likewise:

$$P(R \text{ and } y) = P(R) \times P(y) = \tfrac{1}{2} \times \tfrac{1}{2} = \tfrac{1}{4}$$
$$P(r \text{ and } Y) = P(r) \times P(Y) = \tfrac{1}{2} \times \tfrac{1}{2} = \tfrac{1}{4}$$
$$P(r \text{ and } y) = P(r) \times P(y) = \tfrac{1}{2} \times \tfrac{1}{2} = \tfrac{1}{4}$$

Then, in order to determine the probability of F_2 progeny from these gametes, we again use the product rule. For example, the probability of an *RR YY* genotype is:

$$
\begin{aligned}
P(R\ Y \text{ female gamete and} \\
R\ Y \text{ male gamete}) &= P(R\ Y) \times P(R\ Y) \\
&= \tfrac{1}{4} \times \tfrac{1}{4} \\
&= \tfrac{1}{16}
\end{aligned}
$$

Next, we need to calculate the total probability of two different combined or joint events occurring. To calculate this probability we can again use the sum rule, which states that the probability of either of two mutually exclusive events occurring is the sum of their individual probabilities. Thus, using the coin example again, if we want to know the total probability of obtaining two heads *or* two tails when two coins are flipped:

$$
\begin{aligned}
P(\text{two heads or two tails}) &= P(\text{two heads}) + \\
&\quad P(\text{two tails}) \\
&= P(\text{head}) \times P(\text{head}) + \\
&\quad P(\text{tail}) \times P(\text{tail}) \\
&= \tfrac{1}{2} \times \tfrac{1}{2} + \tfrac{1}{2} \times \tfrac{1}{2} \\
&= \tfrac{1}{4} + \tfrac{1}{4} \\
&= \tfrac{1}{2}
\end{aligned}
$$

In this case, notice that the events—for example, two heads—are actually themselves composed of two coin flips.

Thinking again of the segregation probabilities of progeny from an F_1 cross, the probability of an *RR* offspring is ¼, that of an *Rr* offspring is ½, and that of an *rr* offspring is ¼. Because both genotypes *RR* and *Rr* are dominant, we may want to know the probability of obtaining either *RR* or *Rr*. Using the sum rule:

$$
\begin{aligned}
P(RR \text{ or } Rr) &= P(RR) + P(Rr) \\
&= \tfrac{1}{4} + \tfrac{1}{2} \\
&= \tfrac{3}{4}
\end{aligned}
$$

Both the product rule and the sum rule can be extended to more than two events. For example, the probability of simultaneously obtaining heads on each of four coins, using the product rule, is $\tfrac{1}{2} \times \tfrac{1}{2} \times \tfrac{1}{2} \times \tfrac{1}{2} = (\tfrac{1}{2})^4 = \tfrac{1}{16}$. Likewise, in a standard card deck with thirteen types of cards (aces, twos, threes . . . kings), the probability of drawing a particular type of card (ignoring different suits) is $\tfrac{1}{13}$. The probability of drawing a face card (jack, queen, or king) is $\tfrac{1}{13} + \tfrac{1}{13} + \tfrac{1}{13} = \tfrac{3}{13}$.

CONDITIONAL PROBABILITY

As we have just discussed, more than two outcomes or events may be possible. For example, a fair, six-sided die should have an equal probability of landing with a 1, 2, 3, 4, 5, or 6 showing. In other words, the probability of each of these outcomes is $\tfrac{1}{6}$. Obviously, half the time the die should land with an even number (2, 4, or 6) showing. We can ask, given that an even number is rolled, what is the probability of its being a 2? *A probability that is contingent on given circumstances* is called a **conditional probability.** In other words, the probability that we will obtain a 2, given that an even number is rolled on the die, is:

$$
P(2 \mid \text{even number}) = \frac{P(2)}{P(\text{even})}
$$
$$
P(2 \mid \text{even number}) = \frac{\tfrac{1}{6}}{\tfrac{1}{2}}
$$
$$
= \tfrac{1}{3}
$$

(The vertical line means "given.") In other words, the total probability of a general event, an even number in this case, is subdivided into more specific events.

When Mendel examined his F_2 progeny, he used conditional probability to describe the proportion of heterozygotes among the dominant progeny. In this case, the probability of heterozygotes among dominant progeny is:

$$
P(\text{heterozygous} \mid \text{dominant}) = \frac{P(\text{heterozygous})}{P(\text{dominant})}
$$
$$
P(\text{heterozygous} \mid \text{dominant}) = \frac{\tfrac{1}{2}}{\tfrac{3}{4}}
$$
$$
= \tfrac{2}{3}
$$

In other words, two-thirds of the dominant progeny are expected to be heterozygotes (and one-third are expected to be homozygotes). We will use this same logic later to calculate the probability that normal individuals who have a sibling with a recessive disease are themselves **carriers** of the disease allele; that is, that they are heterozygotes.

BINOMIAL PROBABILITY

Finally, consider the probability that in a group of a given size, say a sibship (a group of brothers and sisters), a certain number of individuals will be of one type and the remainder of another type. To calculate these probabilities, we can use an expansion of the product rule. First, we calculate the probability of the occurrence of a particular sequence of events when there are three different events and only two alternatives for each event. As an example, imagine flipping a coin three consecutive times. The following eight sequences are possible:

	HHT	TTH	
HHH	HTH	THT	TTT
	THH	HTT	

There could be three consecutive heads and no tails, as in the first column; two heads and one tail (second column); one head and two tails (third column); or three tails and no heads (fourth column). In other words, if the order is ignored, four outcomes are possible.

Now, assuming again that the probability of heads is $\frac{1}{2}$ (and that the probability of tails is $\frac{1}{2}$), the probability of their coming up in any one of the orders shown in the second column is, by the product rule:

$$P(HTH) = P(H) \times P(T) \times P(H)$$
$$= \frac{1}{2} \times \frac{1}{2} \times \frac{1}{2}$$
$$= \frac{1}{8}$$

However, to determine the probability of two heads and one tail, independent of order, we have to use the sum rule to get the total probability of the three sequences that give this constitution, or:

$$P(\text{two heads, one tail}) = P(HHT) + P(HTH) + P(THH)$$
$$= \frac{1}{8} + \frac{1}{8} + \frac{1}{8}$$
$$= \frac{3}{8}$$

Thus, the probability of three heads is $\frac{1}{8}$; the probability of two heads (and one tail) is $\frac{3}{8}$; the probability of one head (and two tails) is $\frac{3}{8}$; and the probability of no heads (and three tails) is $\frac{1}{8}$. The sum of all these probabilities is $\frac{1}{8} + \frac{3}{8} + \frac{3}{8} + \frac{1}{8} = 1$, as expected.

Now let us generalize the approach we have just developed. First, assume there are two possible alternatives, say heads or tails, having probabilities p and q, respectively. We know that p + q = 1 because there are only two alternatives. If N different events occur (N being equal to three above because three coins were flipped), what is the probability of x events of the first type and N − x of the second type? In the previous example, x = 2 (two heads) and N − x = 1 (one tail). In general, the **binomial formula** (binomial because there are only two different types of events) gives these probabilities:

P(x type one events
  out of N total events) $= Cp^{x}q^{N-x}$

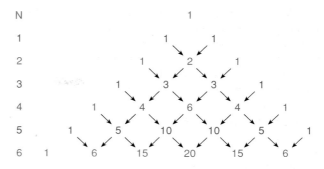

FIGURE 2.9
Pascal's triangle showing the size of the binomial coefficient (C) for different N values, the number of different events. The numbers in a horizontal row indicate C for x = N, N − 1, N − 2, . . . , etc.

In this formula, C is the binomial coefficient and indicates the number of sequences that give the same overall constitution. In the current example, C = 3.

The binomial coefficient for most of the examples we will consider can be obtained using the following formula:

$$C = \frac{N!}{x!(N-x)!}$$

where the symbol ! indicates a factorial. By definition, 0! = 1, 1! = 1, 2! = 2 × 1 = 2, 3! = 3 × 2 × 1 = 6, and so on. For example, if x = 2 and N = 3, then

$$C = \frac{3!}{2!(3-2)!} = \frac{3 \times 2 \times 1}{2 \times 1 \times 1} = 3$$

This is the same answer we found previously by writing out all the possibilities.

Another way to obtain the binomial coefficient is to use Pascal's triangle (fig. 2.9). In this scheme, a given number is the sum of the two numbers above it as indicated by the two arrows. This relationship was named after a seventeenth-century French mathematician and philosopher. It is useful if you forget the formula for the binomial coefficient. Any given number in this array is obtained by summing the two numbers directly above. As an example, let us look at N = 3, row 4 in figure 2.9. The first number in this row, C = 1, is the coefficient for all events of one type (for example, three heads), x = 3; the second number is for x = 2 (two heads), giving C = 3 as before; the third number is for x = 1 (one head), giving C = 3; and the fourth number is for x = 0 (no heads), giving C = 1.

We will use the binomial formula to calculate the probability of a particular number of boys and girls in a sibship. Assuming there are four children, what is the probability that four will be girls? In other words, what is the probability that x = 4 and N − x = 0? Assuming that the probability of a girl is $\frac{1}{2}$ and the probability of a boy is also $\frac{1}{2}$, then:

$$P(\text{four girls of four children}) = C(\tfrac{1}{2})^{4}(\tfrac{1}{2})^{0}$$

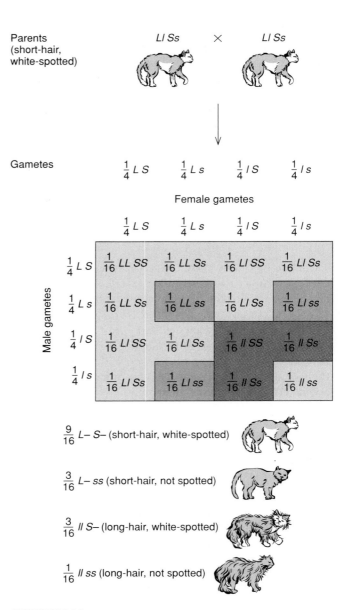

FIGURE 2.10
An orange cat with white spotting and short hair.
© H. Reinhard/OKAPIA, Photo Researchers, Inc.

From the formula for the binomial coefficient:

$$C = \frac{4!}{4!\ 0!} = 1$$

Or, looking at figure 2.9, where N = 4 and x = 4, then C = 1. Given that any number to the zero power is 1:

P(four girls of four children) = $(1)(\frac{1}{2})^4(1)$
= $\frac{1}{16}$

Now let us look at a more complicated situation and calculate the probability of two girls in a sibship of size four. Again, using the formula where x = 2 and N − x = 2:

P(two girls of four children) = $C(\frac{1}{2})^2(\frac{1}{2})^2$

In this case:

$$C = \frac{4!}{2!\ 2!} = 6$$

so that:

P(two girls of four children) = $6(\frac{1}{2})^2(\frac{1}{2})^2$
= $\frac{3}{8}$

C = 6 here because there are six orders, all of which have two girls and two boys: BBGG, BGBG, BGGB, GBBG, GBGB, and GGBB.

We can apply these probability concepts to other situations. For example, a number of genes affect the coats of domestic cats. One gene causes white spotting, primarily on the chest of the cat, where S, the white-spotting allele is dominant over s for no white spotting. Another gene affects hair length, where L, the short-hair allele, is dominant over l, the long-hair allele. Figure 2.10 shows a cat with white

FIGURE 2.11
A cross between two cats doubly heterozygous at the long-hair and white-spotted genes, giving the expected proportions of progeny.

spotting (either genotype SS or Ss) and short hair (genotype LL or Ll). Assuming that two cats, both heterozygous at these genes, mate, what proportion of their progeny will exhibit the different phenotypes? Figure 2.11 gives the expected frequency of the progeny, after segregation and independent assortment. Most of the time ($\frac{9}{16}$), the offspring will have the same phenotype as their parents—white spotting and short hair—but $\frac{1}{16}$ of the time, the offspring will be different from their parents for both characters; that is, they will have no white spotting and long hair.

We can apply probability concepts to the following example: Given these same two cats as parents and a sibship of four kittens, what is the probability that all the kittens will have long hair (*ll*) and white spotting (*S–*)? Because there is only one order in which this can occur and the coat patterns of the different kittens are independent (given these same parental genotypes), then:

$$P(\text{4 of 4 long-hair, white-spotted kittens}) = [P(ll\ S–)]^4$$
$$= (^3\!/_{16})^4$$
$$= 0.0012$$

In this case, the probability is quite low, with such a litter occurring about one in a thousand times. Perhaps the male parent was misidentified and actually had long hair!

Several basic theories of probability are necessary to understand Mendelian genetics: 1) the product rule, which states that the joint probability of two events occurring is the product of their individual probabilities; 2) the sum rule, which states that the probability of either one of two mutually exclusive events occurring is the sum of their individual probabilities; 3) the principle of conditional probability, which is the probability of a specific event divided by the probability of a more general event; and 4) the binomial probability formula, which states the probability of x events of a particular type out of a total of N events.

FORKED-LINE APPROACH

The approach just explained is quite cumbersome when used to calculate the probability of different types of progeny for two or more genes. As a result, the **forked-line approach,** in which the probabilities for each category are determined and combined one gene at a time, is quite useful.

The forked-line approach is based on the assumption of independent assortment of all the genes involved. An example is given in figure 2.12 for the garden pea dihybrid cross discussed previously. First, the progeny are separated into proportions as to round or wrinkled (¾ to ¼) at the first gene. Within each of these categories, they are then separated into proportions at the second gene (¾ yellow to ¼ green). To determine the number in any combined phenotypic class, the probabilities are multiplied over the genes. For example, the expected proportion of wrinkled, yellow offspring is ¼ × ¾ = ³⁄₁₆.

This same approach can be used to calculate proportions for more than two genes. Let us consider cat coat color again and introduce a third gene, *D,* where *dd* homozygotes have dilute (or light) color and *D–* is nondilute or full color. (The difference between gray and black cats is caused by the difference between dilute and nondilute.) Given a cross between two cats that are *Ll ss Dd* and *Ll SS dd,* we can write out all the progeny types using the forked-line approach (fig. 2.13). We can then ask, what is

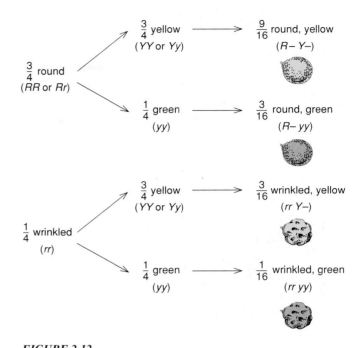

FIGURE 2.12
The forked-line approach for progeny of a dihybrid cross in the garden pea.

the probability of a long-hair, white-spotted, nondilute offspring from this mating? This probability, given in the third row of figure 2.13, is ⅛. Or, assuming that these genes assort independently, we can use the product rule to write:

$$P(\text{long-hair, white-spotted, nondilute offspring}) = P(ll\ S–D–)$$
$$= P(ll) \times P(S–) \times P(D–)$$
$$= ¼ \times 1 \times ½$$
$$= ⅛$$

We could also ask from the same parental cross, what is the probability of an offspring of genotype *Ll Ss Dd?* In this case:

$$P(Ll\ Ss\ Dd) = P(Ll) \times P(Ss) \times P(Dd)$$
$$= ½ \times 1 \times ½$$
$$= ¼$$

The forked-line approach can be extended to calculate the probability of genotypes or phenotypes in offspring for any number of genes as long as they assort independently.

Often the alleles at a gene are symbolized by a capital letter for the dominant allele and a lowercase letter for the recessive allele, as in alleles *R* and *r*. However, to describe all the mutants that have been discovered, the twenty-six letters of the alphabet are not enough. For example, the recessive mutant in *Drosophila melanogaster* that has small, underdeveloped wings is called *vestigial* and is symbolized by *vg*. The normal allele at this gene, instead of being indicated by a capital letter, is symbolized either by *vg*⁺ or just by +, both of which mean the **wild-type,** or nonmutant

FIGURE 2.13
The four progeny categories and their probabilities
for a cross between cats of genotypes *Ll ss Dd* and
Ll SS dd.

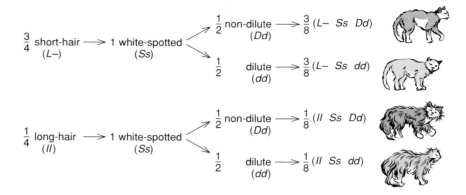

allele. Heterozygotes at this locus having normal wings are
indicated as either vg^+vg or $+vg$. Mutants themselves may
be dominant, such as the curly-wing mutant Cy in
Drosophila, as shown in figure 2.14*a,* where the first letter
of the symbol is capitalized to indicate its dominance. Heterozygotes $Cy+$ have curly wings because Cy is dominant
over the wild-type allele +.

Let us use these symbols to illustrate the forked-line
approach. (Notice that to avoid confusion when using these
symbols we will leave a space between the letters indicating the genotypes for different genes.) If we have the cross
$+vg\ Cy+\ sese \times +vg\ +\ +se$ (*se* indicates the eye color mutant *sepia*), what proportion of progeny would we expect to
be curly-winged—that is, mutant at the second locus and
wild-type at the other two genes? Figure 2.14*b* shows the
eight possible genotypes and their probabilities in the progeny from this cross. The particular category we are interested in is shown in row three of the far right column. To
calculate this:

$$P(+-Cy++-) = P(+-)\times P(Cy+)\times P(+-)$$
$$= \tfrac{3}{4}\times\tfrac{1}{2}\times\tfrac{1}{2}$$
$$= \tfrac{3}{16}$$

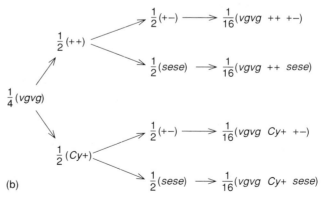

> The forked-line approach, which is based on the product
> rule, can be used to determine the probability of various
> offspring types for any number of genes, assuming they
> assort independently.

GOODNESS OF FIT

How can we use predictions based on probabilities to understand the transmission of genes from one generation to
the next? Remember that Mendel's F_2 data given in table
2.1 very closely fit 3:1 expectations. But what if the numbers had not been so close to the expected ratio? Would this
have been due to chance or because the underlying hypothesis, the principle of segregation, was incorrect? It is useful
to measure the "goodness of fit" to find out whether the observed numbers are adequately close to the numbers expected according to a particular hypothesis. Mendel himself

FIGURE 2.14
(*a*) Curly-wing mutant *Drosophila melanogaster,* $Cy.$ (*b*) The eight progeny
categories and their probabilities for the cross between *Drosophila* with
genotypes $+vg\ Cy+\ sese$ and $+vg\ ++\ +se.$

(a) From Robert H. Tamarin, Principles of Genetics, *4th ed. Copyright © 1993
Wm. C. Brown Communications, Inc., Dubuque, Iowa. All Rights Reserved.
Reprinted by permission.*

Table 2.4 Chi-square Values of Mendel's Backcross Data

Flower Color	Observed Numbers	Expected Numbers	$\frac{(O - E)^2}{E}$
Violet	85	166(0.5) = 83	$\frac{2^2}{83} = 0.05$
White	81	166(0.5) = 83	$\frac{(-2)^2}{83} = 0.05$
	166	166	$\chi^2 = 0.10$

Table 2.5 Probabilities of Different Theoretical Chi-square Values for Given Degrees of Freedom[1]

Degrees of Freedom	Probability				
	0.9	0.5	0.1	0.05*	0.01**
1	0.02	0.46	2.71	3.84	6.64
2	0.21	1.39	4.60	5.99	9.21
3	0.58	2.37	6.25	7.82	11.34
4	1.06	3.86	7.78	9.49	13.28
5	1.61	4.35	9.24	11.07	15.09
6	2.20	5.35	10.64	12.59	16.81
7	2.83	6.35	12.02	14.07	18.48
8	3.49	7.34	13.36	15.51	20.09
9	4.17	8.34	14.68	16.92	21.07
10	4.86	9.34	15.98	18.31	23.21

[1]Probabilities of 0.05 and 0.01 are known as the statistically significant and highly significant levels and are indicated by * and **, respectively.

did not use any statistical tests (they weren't developed until decades later), but such tests have helped identify exceptions to Mendel's principles in other studies.

A test commonly used to measure goodness of fit is the **chi-square test.** In this statistical test, the *observed numbers* in each category, say dominant and recessive progeny, are compared to the *expected numbers* under a certain hypothesis, say Mendel's principle of segregation. Notice that in this test we are not using the observed and expected frequencies but the observed and expected numbers. If the observed number in one category is larger than the expected number, the opposite will be true in another category because the total number over all classes for both observed and expected numbers is the same. To obtain a test that measures the difference between observed and expected numbers, irrespective of the sign of this difference, we use the square of the difference between observed and expected numbers. This value is then standardized by dividing it by the expected number in a category, making the formula for chi-square:

$$\chi^2 = \Sigma \frac{(O - E)^2}{E}$$

where Σ is the Greek letter sigma indicating the sum over all categories, χ is the Greek letter chi, and O and E are the observed and expected numbers.

This statistical technique can be illustrated using the data from one of Mendel's backcross experiments. In this experiment, he crossed F_1 violet-flowered plants (*Vv*) with white-flowered plants (*vv*). Of 166 progeny, 85 had violet flowers and 81 had white flowers. According to the principle of segregation, 0.5 should be violet and 0.5 should be white. Or, using numbers, 166(0.5) = 83 would be expected in each class. Notice that the sum of the expected numbers must always equal the sum of the observed numbers; the sum here is 166. With these observed and expected numbers, we can calculate the chi-square value as in table 2.4. For the violet and white categories, the contributions to chi-square are $(2)^2/83$ and $(-2)^2/83$, respectively, both of which equal 0.05. Notice that if we had not squared O − E, then these two classes would exactly cancel out each other. Summing over both the values, the chi-square is 0.10. Obviously, this is a small value, indicating the similarity of the observed and expected numbers, but how do we know how large or small it really is?

The significance of a chi-square value can be determined by comparing it to a theoretical value from a chi-square table that has the same degrees of freedom. The term **degrees of freedom** refers to an integer that is generally equal to the number of categories in the data, minus one. In other words, the degrees of freedom indicate the number of categories (phenotypic classes) that are independent of each other. For example, if there are two categories, there is only one degree of freedom. This constraint occurs because, given the total number (say 166 as previously) and the number in one class (say 85), the number in the other class (81) is fixed.

Assuming that the Mendelian segregation hypothesis we are examining is correct, if we drew many, many samples and calculated a chi-square value for each, we would obtain a theoretical distribution of chi-square values. Due to chance, the chi-square would be large in a small proportion of the cases because the observed numbers were quite different from the expected numbers. However, most of the time the observed numbers would be close to those expected, making the chi-square small.

Instead of giving the complete theoretical distribution of chi-square values, we will provide a few important values (table 2.5). These critical chi-square values are exceeded by chance only rarely, given that the underlying hypothesis is correct. There are two different critical theoretical values, 0.05 and 0.01, to which the observed (or calculated) chi-square is generally compared. If the calculated chi-square value exceeds the theoretical values given under 0.05 and 0.01 (for given degrees of freedom), the observations are said to be *statistically significant* or *highly statistically significant,* respectively. Such high chi-square values suggest that the hypothesis used to determine the expected numbers is probably incorrect. The values labeled 0.05 are equaled or exceeded by chance only one in twenty times (0.05) even though the underlying hypothesis is correct. In other words, by chance, the observed numbers could be quite different from those expected one in twenty times, making the chi-square value this large or larger. If the observed and expected numbers are very different, the

Table 2.6 Chi-square of a Dihybrid Cross

Seed Type	Observed Numbers	Expected Numbers	$\frac{(O - E)^2}{E}$
Round, yellow	315	556 (9/16) = 312.8	$\frac{(2.2)^2}{312.8} = 0.02$
Wrinkled, yellow	101	556 (3/16) = 104.2	$\frac{(-3.2)^2}{104.2} = 0.10$
Round, green	108	556 (3/16) = 104.2	$\frac{(3.8)^2}{104.2} = 0.14$
Wrinkled, green	32	556 (1/16) = 34.8	$\frac{(-2.8)^2}{34.8} = 0.22$
	556	556	$\chi^2 = 0.48$

chi-square may equal or exceed the numbers in the last column, a value that occurs by chance only one in one hundred times, given a correct hypothesis.

In the example of violet and white flowers, there are two categories, so there is only one degree of freedom. The calculated or observed chi-square value, 0.10, is substantially less than the critical value needed for statistical significance. (With one degree of freedom, it is 3.84.) Therefore, we can say that the observed chi-square value is not significantly different from the expected chi-square value. In other words, the observed numbers are *consistent* with the numbers expected under the principle of segregation. Notice that we did not say this proves the principle of segregation. One cannot prove a hypothesis by such a statistical test. However, because many such crosses have been evaluated and are consistent with the principle of segregation, we can feel confident that this process is responsible for the inheritance of these genes.

Let us go back to the data from Mendel's dihybrid cross and ask whether the observed phenotypic numbers are consistent with the hypothesis of independent assortment. We can use the chi-square test, but first we need to calculate the expected numbers in each category. There were 556 progeny, so if we multiply 556 by 9/16 , 3/16 , 3/16 , and 1/16 , we obtain the expected numbers in the four classes (table 2.6). Using the chi-square formula, and then summing over the four categories, the chi-square value is 0.48. Again, like the single-gene example, this is a very small chi-square value. However, in this case, there are four categories, making the degrees of freedom (number of categories minus one) equal to 3. Looking at table 2.5, we find that the calculated chi-square value would have to be equal to, or greater than 7.82 for the observed values in order to be significantly different from expectations with 3 degrees of freedom. Obviously, 0.48 is much less than 7.82, so we can conclude that the observed numbers are consistent with the expected numbers under the hypothesis of independent assortment.

Remember that when calculating chi-square values, observed and expected numbers, not frequencies, are used.

Also, as a rule of thumb for using a chi-square test, the expected numbers in a category should be five or greater. When the expected number is less than five, a modification of the formula should be used, or else the observed chi-square value may be somewhat inflated.

An interesting historical note is that all of Mendel's results fit expectations very closely and consequently give very low chi-square values. For instance, in the dihybrid example, given a correct hypothesis, the chi-square value should be less than 0.58 only 10% of the time. However, the observed value was 0.48, less than 0.58. Several statisticians have suggested that Mendel (or some of his student helpers) biased the results to conform to expectations of the principles of segregation and independent assortment. We will probably never know whether this is true, but it should not detract from the genius of Mendel's insights into inheritance. We should note too that the workers who repeated Mendel's experiments also obtained results extremely close to expectations (see table 2.2).

A useful statistical test of the goodness of fit between observed and expected numbers is the chi-square. Chi-square values larger than a critical value indicate that the underlying hypothesis used to generate the expected numbers may be incorrect. All of Mendel's results fit expectations very closely and thus yield very low chi-square values.

MENDELIAN INHERITANCE IN HUMANS

The same patterns of inheritance occur in humans as in peas, *Drosophila,* or cats. At this point 2,656 single traits and diseases for humans are categorized in the 1990 edition of Victor McKusick's *Mendelian Inheritance in Man;* another 2,281 loci are not yet fully identified or validated. Determining the mode of inheritance is somewhat more difficult in humans because the sibship sizes are relatively small; that is, there are generally fewer than ten children in a family, whereas hundreds of seeds are produced from a given cross in peas. However, using Mendel's ratios, the same inheritance patterns can be seen even in the pedigrees of small families.

Before looking at Mendelian inheritance in humans, we must introduce the basic symbols used in human pedigree analysis. As shown in figure 2.15, a female or a male in a pedigree chart is indicated by a circle or a square, respectively. Diamonds are used for individuals whose sex has not been determined or in cases where sex is unimportant for genetic analysis. When an individual has a given trait or genetic disease, the symbol is filled in or is a different color. Mating individuals are joined by a single horizontal line. A mating between relatives, called **consanguineous** (Latin: *consanguinitas* = relationship by blood), is indicated by a double horizontal line. Siblings are joined by a line below their parents and are listed in order of birth,

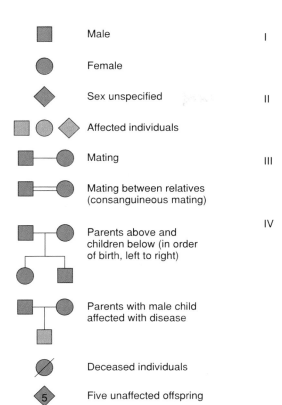

■	Male	I
●	Female	
◆	Sex unspecified	II
■ ● ◆	Affected individuals	
■—●	Mating	III
■═●	Mating between relatives (consanguineous mating)	
■—● (parents above, children below)	Parents above and children below (in order of birth, left to right)	IV
■—● (with affected male child)	Parents with male child affected with disease	
⊘	Deceased individuals	
◆5	Five unaffected offspring	

FIGURE 2.15
Symbols used in human pedigrees.

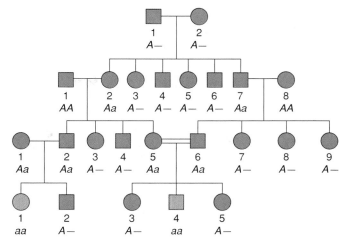

FIGURE 2.16
A human pedigree for a recessive trait, showing the inferred genotypes, where *A* and *a* are the dominant and recessive alleles, respectively. *A*– indicates that state of the second allele is not known.

FIGURE 2.17
A Hopi Indian girl with albinism.
Field Museum of Natural History, Chicago, Neg. No. 118.

from left to right. A pedigree chart begins with the oldest generation at the top; successive generations are identified by Roman numerals (fig. 2.16).

RECESSIVE TRAITS

Let us examine a pedigree exemplifying the inheritance pattern for a recessive trait (fig. 2.16). Two recessive diseases in humans are **albinism,** in which individuals lack the pigment melanin (fig. 2.17), and **cystic fibrosis,** a severe disorder that affects the production of mucus in the pancreas and lungs. (McKusick lists a total of 631 recessive loci.) Cystic fibrosis is the most common recessive disease among Caucasians, occurring in about one in 2,500 births. In the pedigree in figure 2.16, two individuals, IV-1 and IV-4, have the recessive trait or disease. Notice that the parents of the affected individuals, III-1, III-2, III-5, and III-6, are themselves unaffected, a typical observation in pedigrees for recessive traits. Obviously, the trait appeared because the parents are heterozygotes whose recessive alleles were masked by the dominant alleles. Notice also that either males or females can have the recessive phenotype.

In a large proportion of recessive disease cases, the parents of the affected individual are relatives. For example, the parents of individual IV-4 are first cousins; that is, III-5 and III-6 have the same grandparents, I-1 and I-2. (We will discuss the basis for the increase in recessive diseases from consanguineous matings in chapters 16 and 17.) The other grandparents of III-5 and III-6, as well as some other relatives, are omitted in order to simplify the pedigree.

Several other aspects of recessive pedigrees are important. First, if both parents are affected or have the recessive genotype (not shown in figure 2.16), all progeny should be affected. Second, among the sibs of an affected individual, given that both parents are unaffected (heterozygotes), approximately ¼ should be affected, according to the principle of Mendelian segregation. Finally, the probability that an additional child born to this family would have the trait

Calvin and Hobbes

by Bill Watterson

Table 2.7 Patterns for Human Traits

Recessive Traits

(a) Parents are generally unaffected.
(b) Approximately one-quarter of the siblings of an affected individual are affected.
(c) The probability that an additional child will be affected is ¼.
(d) Recessive traits often result from consanguineous mating.
(e) Two affected parents cannot have unaffected offspring.

Dominant Traits

(a) Trait occurs every generation (at least one parent is affected).
(b) When one parent is affected, approximately one-half of progeny are affected.
(c) The probability that an additional child will be affected is ½.
(d) Two unaffected parents do not produce affected offspring.
(e) Two affected parents can produce unaffected offspring.

is also ¼. The attributes of pedigrees for recessive traits are summarized in table 2.7. However, because human families are usually small, these proportions are generally seen only when there are a large number of sibships.

In any given sibship, the numbers of different types of offspring may not be close to Mendelian ratios. To understand this, we can use the expression for binomial probabilities given previously. For example, let us assume we have a sibship of size six in which two individuals are affected. We will designate one affected individual as the **proband** (or index case), the individual who was first identified as having the recessive trait or disease. What is the probability that one of the five remaining sibs will also be affected? Given the Mendelian probabilities of segregation, let us call ¾ dominant (p) and ¼ recessive (q). Thus:

P(4 unaffected,
  1 affected offspring) = $C(\frac{3}{4})^4(\frac{1}{4})^1$

In this case, C = 5, thus:

P(4A−, 1aa) = $5(\frac{3}{4})^4(\frac{1}{4}) = 0.396$

This makes the probability of such a sibship quite large. In fact, it is the most likely type of sibship containing five sibs. The next most probable sibships are P(5A−, 0aa) = 0.237 and P(3A−, 2aa) = 0.264.

Two other points should be made about recessive trait pedigrees before we continue. First, note that unrelated individuals who do not have affected offspring, such as individuals II-1 and II-8 in figure 2.16, are assumed to be homozygous for the dominant allele. Such individuals could possibly be heterozygotes, but because most recessive alleles are rare, it is assumed that unrelated individuals with the dominant phenotype are homozygotes.

Second, given an affected offspring in a family, we know that the unaffected parents are heterozygous (barring mutation, which is extremely rare). Using segregation proportions, we can then calculate the conditional probability that an unaffected sib in such a family is heterozygous. The probability that a sib is heterozygous is ½ and the probability that a sib is unaffected is ¾, so:

$$P(\text{Heterozygous} \mid \text{unaffected}) = \frac{P(\text{heterozygous})}{P(\text{unaffected})}$$
$$= \frac{\frac{1}{2}}{\frac{3}{4}}$$
$$= \frac{2}{3}$$

In other words, two-thirds of the time, unaffected sibs of individuals with a recessive disease are carriers (heterozygotes) for the disease allele.

DOMINANT TRAITS

Now let us examine a pedigree that illustrates the inheritance pattern for a dominant trait. **Achondroplasia,** a common type of dwarfism (fig. 2.18), and **Huntington disease,** a degenerative neurological disorder, are examples of dominant disorders. (McKusick lists a total of 1,864 dominant loci.) For dominant traits, at least one of the parents of an affected individual is also affected, and the trait is present in every generation, as shown in the pedigree in figure 2.19. This pattern results because an offspring must obtain the dominant allele from one of its parents. As a result, unaffected individuals do not produce affected offspring (unless there is a rare mutation). Notice also that either males or females can have the dominant phenotype. Unlike a recessive trait, two affected parents can produce an unaffected offspring, because both parents may be heterozygotes and consequently have recessive alleles. As expected according to Mendelian segregation, when one parent is affected, half the sibs of the proband are expected to be affected. Given such a family, the probability that an additional child will be affected is ½. (See table 2.7 for a summary about dominant traits in pedigrees.) Again, because of the small sibship size in humans, sibships in which the proportion of affected individuals is not ½ occur with probabilities that can be calculated using the binomial formula.

The pedigree patterns listed in table 2.7 are not completely true for some diseases, and there are other types of inheritance patterns as well. We will discuss several such situations in chapter 3 and illustrate how they may affect pedigrees. Furthermore, the pedigrees given in figures 2.16 and 2.19 are large ones, allowing the mode of inheritance to be determined with reasonable ease. In many pedigrees having fewer individuals, the mode of inheritance may not be clear.

 Mendelian principles apply to traits in humans, but because human families are relatively small, determining modes of inheritance can be difficult. However, some general patterns have been discerned regarding the passing on of recessive and dominant traits.

FIGURE 2.18
A boy with achondroplasia, a common type of dwarfism.
Dr. Judith Hall.

FIGURE 2.19
A human pedigree for a dominant trait, showing the inferred genotypes, where *A* and *a* are the dominant and recessive alleles, respectively. *A*– indicates that the second allele is not known.

 WORKED-OUT PROBLEMS

Problem 1
A number of researchers repeated Mendel's crosses in order to confirm his genetic principles. For example, both Correns and Tschermak crossed yellow and green varieties of peas and self-fertilized the subsequent F_1 progeny to get F_2 progeny arrays. Correns observed 1,394 yellow and 453 green, while Tschermak observed 3,580 yellow and 1,190 green. Are these values consistent with the principle of segregation? (Calculate χ^2 values to determine consistency.)

Solution
Set up a table with the observed and expected numbers for the two crosses, then calculate the χ^2 values:

		Observed Numbers	Expected Numbers	$\dfrac{(O-E)^2}{E}$
Correns	Yellow	1,394	$1,847 \times \frac{3}{4} = 1,385.25$	0.055
	Green	453	$1,847 \times \frac{1}{4} = 461.75$	0.166
		1,847	1,847	$\chi^2 = 0.221$
Tschermak	Yellow	3,580	$4,770 \times \frac{3}{4} = 3,577.5$	0.002
	Green	1,190	$4,770 \times \frac{1}{4} = 1,192.5$	0.032
		4,770	4,770	$\chi^2 = 0.074$

In this case, there is 1 degree of freedom (2 categories minus 1 = 1 degree of freedom). As a result, the χ^2 values are not significant because both are less than 3.84 (see table 2.5), making the observations consistent with the principle of segregation.

Problem 2

Using the same traits Mendel used in his experiments to demonstrate independent assortment, assume that a true-breeding pea plant with round, yellow seeds was crossed to one with wrinkled, green seeds. One of the progeny from this cross was then crossed back to plants from the wrinkled, green line. From this mating, the progeny types were 42 round, yellow; 45 round, green; 46 wrinkled, yellow; and 45 wrinkled, green. What are the genotypes of all the phenotypes, parents, and progeny? What are the expected numbers of the four progeny types assuming independent assortment? What is the χ^2 value resulting from this progeny array, and is it statistically significant?

Solution

Parents	Round, yellow $RR\ YY$ × Wrinkled, green $rr\ yy$			
Backcross progeny	*Phenotype*	*Genotype*	*Observed*	*Expected*
	Round, yellow	$Rr\ Yy$	42	44.5
	Round, green	$Rr\ yy$	45	44.5
	Wrinkled, yellow	$rr\ Yy$	46	44.5
	Wrinkled, green	$rr\ yy$	45	44.5

$$\chi^2 = \frac{(42-44.5)^2}{44.5} + \frac{(45-44.5)^2}{44.5} + \frac{(46-44.5)^2}{44.5} + \frac{(45-44.5)^2}{44.5}$$

$$= 0.14 + 0.01 + 0.05 + 0.01$$

$$= 0.21$$

With three degrees of freedom, the number of classes minus one, this value is not statistically significant. That is, 0.21 is less than 7.82, the value in table 2.5 that it would have to exceed to be statistically significant.

Problem 3

A couple had a child with cystic fibrosis, a recessive disease. If we assume that C is the normal allele at this locus and c is the allele that causes cystic fibrosis as a homozygote, what are the genotypes of the unaffected parents and the child? What is the probability that the couple's second child would not have cystic fibrosis?

Treatments are now available for individuals with cystic fibrosis, so that this couple's child could grow up and have children. If we assume that the affected child has children with an individual who does not have the cystic fibrosis allele (as is true for 95% of all people), what is the chance that these children would be affected?

Solution

The parents are both heterozygotes, Cc, and the child is cc. The probability that the second child would be unaffected, that is, either CC or Cc, is 0.75. If the grown-up affected child with the genotype cc mates with an unaffected person with the genotype CC, then all of their children would be Cc, unaffected heterozygotes.

Summary

Mendel first demonstrated that inheritance is based on the segregation of alleles at a gene. Through self-fertilization of F_1 garden pea plants obtained from crosses between plants with different characteristics, he observed a 3:1 ratio of dominant to recessive phenotypes in the F_2 progeny. The principle of segregation he developed from these experiments states that heterozygous individuals produce equal proportions of gametes containing the two alleles. Mendel's second principle, that of independent assortment, states that alleles at different genes assort independently of each other, resulting in a 9:3:3:1 ratio in F_2 progeny from a dihybrid cross.

Several basic theories of probability are useful in understanding Mendelian genetics. These are: (1) the sum rule, which states that the probability of either of two mutually exclusive events occurring is the sum of their individual probabilities; (2) the product rule, which states that the joint probability of two events occurring is the product of their individual probabilities; (3) the theory of conditional probability, which is the probability of a specific event divided by the probability of a more general event; and (4) the binomial probability formula, which figures the probability of x events of a particular type out of a total N.

A useful statistical test of the goodness of fit between the numbers observed and the numbers expected according to Mendel's principles is the chi-square. Chi-square values larger than a critical value indicate that the underlying hypothesis used to generate the expected numbers may be incorrect.

Mendelian principles apply to traits in humans. Human pedigrees show certain patterns that allow us to differentiate recessive and dominant traits.

PROBLEMS AND QUESTIONS

1. What are Mendel's two principles?
2. What characteristics of garden peas made them particularly good organisms for Mendel's experiments?
3. If you had a garden pea plant that had the dominant phenotype R, what cross would you make to determine if it has genotype RR or Rr? Explain your answer.
4. Assume that you have a large container filled with 500 red marbles and 500 blue marbles and that the marbles are thoroughly mixed. (*a*) If you reached in without looking and picked out one marble, what is the probability that it would be red? (*b*) If you put that marble back and picked out two more in the same manner, what is the probability that one would be red and one would be blue?
5. Maple syrup urine disease is a rare genetic disease that runs in families. It is so called because a metabolic error causes the urine of affected babies to have an odor like maple syrup. The babies die at an early age if

not treated. The parents are almost never affected with these symptoms. If we assume that the disease is caused by a single gene, is it a recessive or a dominant gene? Explain your answer.

6. The F_2 numbers of the dominant and recessive phenotypes for Mendel's seven different traits are given in table 2.1. Calculate the chi-square values for axial versus terminal flowers and for tall versus short stems. Are these values consistent with Mendelian expectations using the chi-square test?

7. If the F_1 yellow-seeded plants from Mendel's experiments (see table 2.1) were crossed to green-seeded plants, what proportion of the progeny would be expected to be yellow? If F_2 yellow-seeded plants from self-fertilized F_1 plants (table 2.1) were crossed to green plants, what proportion of the plants would produce only yellow progeny?

8. A student has a penny, a nickel, a dime, and a quarter. She flips them all simultaneously and checks for heads or tails. What is the probability that all four coins will come up heads? She again flips all four coins. What is the probability that she will get four heads both times? What probability rule did you use to determine your answer?

9. The first child of two normally pigmented parents has albinism, a recessive trait that results from a lack of the pigment melanin. (*a*) Given that the normal allele is A and the albino allele is a, draw this pedigree and label both the phenotypes and the genotypes. (*b*) What is the probability that a second child from these parents will be an albino? What is the probability that the second child will be a carrier of the albino allele? Given that the second child is unaffected, what is the probability that he or she is a carrier?

10. Assuming that the probability of producing a female child is 0.5, what is the probability of having three girls and one boy in a family of four? Calculate this value both by writing out all the possible birth orders that can lead to such a family and by using the binomial formula. The actual proportion of girls at birth is about 0.49 in several human populations. Given this value, what is the probability of a family having three girls and one boy?

11. Mendel crossed purebred wrinkled, green-seeded plants with purebred round, yellow-seeded plants. Then he crossed the F_1 progeny to purebred wrinkled, green-seeded plants and observed 31 plants with round, yellow seeds, 26 with round, green seeds, 27 with wrinkled, yellow seeds, and 26 with wrinkled, green seeds. (*a*) Assuming independent assortment, what proportion of these types would be expected? (*b*) Calculate the appropriate chi-square and determine whether it is consistent with Mendelian expectations.

12. Using the forked-line approach and given the cross between two cats with the genotypes *ll Ss dd* and *Ll Ss Dd,* what is the probability of a cat having the genotype *ll ss dd?* What is the probability of a cat having the short-hair, white-spotted, nondilute phenotype?

13. Using the forked-line approach in a trihybrid cross involving three traits, where the parents are both *Aa Bb Cc,* what is the probability of their producing an offspring recessive for all three traits? What is the probability of producing an offspring with the phenotype *A– bb C–?*

14. Given the pedigree shown here for a rare genetic disease, would you conclude that the trait is recessive or dominant? Why? What do you expect the genotypes of individuals II-3, II-4, and III-2 to be, assuming A and a are the dominant and recessive alleles, respectively? Given only this information, what is the probability that I-2 is heterozygous? What is the probability that III-1 will have the trait?

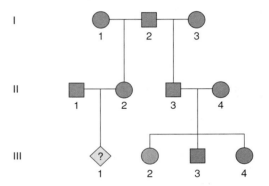

15. Given the following pedigree for a rare genetic disease, would you conclude that the trait is recessive or dominant? Why? What do you expect the genotypes of II-1, II-2, and III-3 to be, assuming that B and b are the dominant and recessive alleles, respectively? What is the probability that a child of II-3 will be affected? Why?

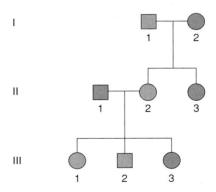

16. A woman with the rare recessive disease phenylketonuria (PKU), who had been treated with a diet having low levels of the amino acid phenylalanine, was told that it was unlikely her children would inherit PKU because her husband did not have it. However, her first child had PKU. What is the most likely explanation? Assuming this explanation is true, what would be the probability of her second child having PKU?

17. In an examination of Mendel's principles, a strain of light brown mice was crossed to a strain of dark brown mice. All F_1 were dark brown. In the F_2, 42 were dark brown and 15 were light brown. Is this consistent with the principle of segregation? (Use a chi-square to check.)

18. The 42 dark brown F_2 mice from question 17 were crossed to light brown mice. Thirty of these mice produced litters composed of both dark and light brown mice, while twelve produced only dark brown mice. Is this consistent with the principle of segregation? (Use a chi-square to check.)

19. What is the probability that, if three identical coins were flipped, all would end up heads? What is the probability that the three coins would *not* be either all heads or all tails?

20. Assume that a cross was made between fruit flies of genotype *AAbb* and those of genotype *aaBB*. (*a*) Give the Punnett square for the expected F_2 progeny types. (*b*) What proportions of *A–B–, A–bb, aaB–,* and *aabb* progeny do you expect in the F_2?

21. In the recessive trait pedigree in figure 2.16, what is the probability that individual IV-2 is a carrier (heterozygote)? What is the probability that a third child of III-1 and III-2 would have the recessive disease?

22. In the dominant trait pedigree in figure 2.19, what is the probability that individual III-2 would be *AA?* If III-1 and III-6 produce an offspring, what is the probability that he or she will have the dominant trait?

23. In a common kind of congenital deaf-mutism, one or both parents are usually affected. Do you think it is a recessive or a dominant trait? What other criteria would you use to distinguish between these two types of deaf-mutism?

24. In the F_2 progeny from two purebred lines of rats, 50 were doubly dominant, *A–B–*. If these were tested by crossing to *aabb* rats, how many would you expect to be double heterozygotes?

25. Assume that one-twelfth of all individuals are born each month. What is the probability that two particular individuals were both born in April? What is the probability that two particular individuals were both born in the same month?

26. A geneticist crossed true-breeding garden peas that were tall (having long stems) with ones that were short (having short stems). All the F_1 offspring were tall. The F_1 were then allowed to self-fertilize. In the F_2 progeny, 153 were tall and 48 were short. Assume that the genotype for the short plants was *aa* and that *A* allele is associated with the tall phenotype. (*a*) What are the genotypes of the parents, the F_1 and the F_2 for the various phenotypes? (*b*) What are the observed proportions of tall to short plants in the F_2 progeny?

27. Some of the F_1 seeds in the experiment discussed in question 26 were crossed to individuals of the short true-breeding line. What are the genotypes in these parents and the expected proportions of genotypes and phenotypes in the progeny?

28. The cob of corn in figure 2.8 represents the F_2 progeny array for two genes. (*a*) First, at the color gene, an *Rr* (purple) plant was self-fertilized to produce an F_2 array expected to be ¾ purple (*RR* or *Rr*) and ¼ yellow (*rr*). Count all the kernels you can see in the photo and calculate the purple to yellow ratio. (*b*) Second, at the second gene, which influences kernel shape, an *Ss* (normal) plant was self-fertilized to produce an F_2 array expected to be ¾ normal (*SS* or *Ss*) to ¼ shrunken (*ss*). Again, count all of the kernels you can see in the photo to calculate normal to shrunken ratio. (*c*) Finally, count all the kernels, categorizing them by both color and shape. Determine the ratio of purple-normal, purple-shrunken, yellow-normal, and yellow-shrunken phenotypes. Do the ratios of (*a*) and (*b*) come close to that expected for segregation? Does the ratio for (*c*) approximate that expected for independent assortment? Calculate the chi-square values for all three.

29. Given that the parent is genotype *RR*, what is the probability of an *R* gamete? Given that the parent is *Rr*, what is the probability of an *R* gamete? Given that there is a mating between two individuals both with genotype *Rr*, what is the probability that an offspring will be *RR?*

30. (*a*) Given that one parent is *Rr Yy*, what is the probability of the gametes *RY, Ry, rY,* and *ry?* (*b*) If the other parent is *rryy*, what is the probability that an offspring will be *RrYy? Rryy? RRYy?*

Answers appear at end of book.

CHAPTER 3: EXTENSIONS AND APPLICATIONS OF MENDELIAN GENETICS

*T*he phenotype of an individual is what is perceived by observation: the organism's structures and functions—in short, what a living being appears to be to our sense organs. . . . The genotype is the sum total of the hereditary materials received by an individual from its parents and other ancestors.

Theodosius Dobzhansky
American geneticist

Tulips
© *Kevin Schafer,*
Peter Arnold, Inc.

LEARNING OBJECTIVES

In this chapter you will learn:

1. The differences between complete dominance, incomplete dominance, and codominance.
2. The effects of pleiotropy and penetrance.
3. How there can be multiple alleles at a gene.
4. The pattern of inheritance of sex-linked genes.
5. The patterns of inheritance when multiple genes affect a trait.
6. The genetic basis of quantitative traits.
7. How to estimate heritability.
8. The basis of genetic counseling.
9. The basis of paternity exclusion.

*T*he elegance of Mendel's experiments was partly due to the complete consistency between his observations and the hypotheses he developed. However, after Mendel's work was rediscovered, it became clear that the simple Mendelian model did not adequately predict experimental observations in all situations. In this chapter, we will examine some of the more complex phenomena that make it necessary to adopt an expanded view of simple transmission, or Mendelian, genetics. In addition, we will introduce the topics of genetic counseling and paternity analysis, which utilize the principles of Mendelian genetics. In chapter 5 we will discuss genetic linkage, which can also be viewed as an extension of Mendelian principles.

DOMINANCE

All seven traits that Mendel selected had one allele dominant over the other, so that when he examined the F_2 progeny, he found a 3:1 dominant to recessive ratio of the phenotypes. (As in chapter 2, we will use the terms *dominant* and *recessive* to refer to alleles as well as to traits.) In fact, Mendel suggested that dominance was a general characteristic of traits, similar to the principles of segregation and independent assortment. However, shortly after 1900, several different patterns of dominance were documented.

COMPLETE DOMINANCE

In most situations, the normal, or wild-type, allele is completely dominant over mutant alleles. The basis for this will be more clear when we later examine the biochemical products of genes and find that one functioning copy of a gene is generally enough to produce sufficient gene product for a wild-type phenotype. For example, the albino allele in humans does not produce functioning gene product, and when an individual is homozygous for this allele, the biochemical pathway for the pigment melanin is blocked. However, one copy of the wild-type allele, as is present in heterozygotes, allows sufficient production of the pigment melanin so that normal pigmentation results.

In some instances, the mutant allele is dominant over the wild-type; that is, the wild-type is recessive. There are 1,864 confirmed loci resulting in dominant traits listed in McKusick's 1990 catalog of human traits, many of them rare diseases. For example, the most common type of dwarfism, achondroplasia, is dominant, so that heterozygotes exhibit the mutant phenotype. In a subsequent section of this chapter, we will discuss some aspects of Huntington disease, a fatal neurological disorder that is also dominant.

INCOMPLETE DOMINANCE

When Carl Correns, one of the rediscoverers of Mendel's principles, experimentally crossed red-flowered four-o'clocks with white-flowered ones, the F_1 plants were unlike either parent, being an intermediate pink instead (fig. 3.1). When he allowed pink F_1 plants to self-fertilize, the F_2 ratio

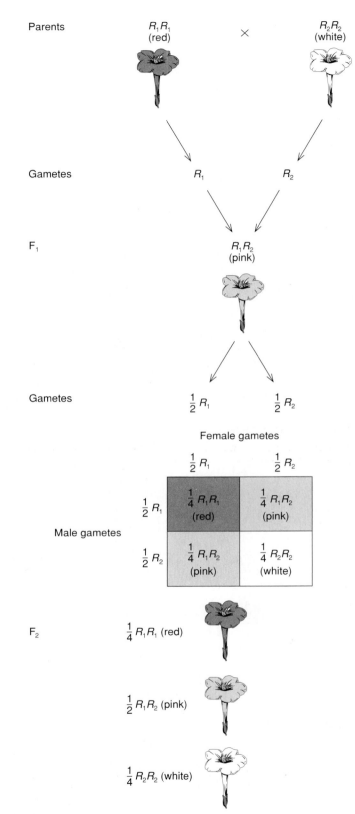

FIGURE 3.1
A cross between red and white four-o'clocks, with the F_2 progeny showing incomplete dominance.

of phenotypes was 1 red : 2 pink : 1 white, or ¼, ½, and ¼. Although at first glance these results appear to deviate from Mendel's, what Correns actually observed in the F_2 was a 1:2:1 phenotypic ratio corresponding to the 1:2:1 genotypic ratio. Because neither parental phenotype was fully expressed in the F_1, the trait is said to exhibit **incomplete dominance** at this gene; that is, all genotypes have different phenotypes, with the heterozygote's phenotype intermediate between the two homozygotes. The notation in figure 3.1 indicates this lack of dominance by using capital letters with different subscripts, R_1 for the red allele and R_2 for the white.

The type of dominance for a given genetic variant is not always easy to designate, however. For example, in what appears to be complete dominance, the observable phenotype of the heterozygote is the same as that of the homozygote, but it may be incomplete dominance on a biochemical level. For example, RR homozygotes and Rr heterozygotes at the gene affecting seed shape in garden peas both have the same round seed phenotype. However, these genotypes differ biochemically: the heterozygote has starch grains that are intermediate between the two homozygotes in type and amount. For this reason, we must be specific about the level of the trait—for example, gross phenotypic or biochemical—we are describing.

CODOMINANCE

In some cases, the traits associated with both alleles are observable in the heterozygote. Such alleles are said to exhibit **codominance.** For example, individuals heterozygous for **red blood cell antigens,** genetically determined chemicals on the membranes of red blood cells, often exhibit properties of both alleles. A well-known case is type AB in the ABO blood group system, which has the antigens from two different alleles (see the discussion of multiple alleles later in this chapter for more about ABO). Another blood group system, called the MN, has two alleles, generally denoted as L^M and L^N, and three genotypes, L^ML^M, L^ML^N, and L^NL^N. (Sometimes these genotypes are called MM, MN, and NN, respectively.) Table 3.1 gives the six possible mating combinations between these genotypes, assuming reciprocal matings are equivalent. The expected proportions of genotypes in the offspring are also shown, and it can be seen that they follow the Mendelian segregation pattern. In chapters 16 and 17, we will discuss naturally occurring allelic forms that determine some biochemical properties of proteins, most of which are also codominant.

Although the example of flower color in four-o'clocks illustrates incomplete dominance on a gross phenotypic scale, close examination of the pink flowers of heterozygotes shows that both red plastids and white plastids are present. (Plastids are a type of cell organelle.) In other words, for this trait we find incomplete dominance on the gross phenotypic level and codominance on a subcellular level.

Table 3.1 Expected Proportions of Offspring for the MN Blood Group System

Parents	Offspring		
	L^ML^M	L^ML^N	L^NL^N
$L^ML^M \times L^ML^M$	1	—	—
$L^ML^M \times L^ML^N$	½	½	—
$L^ML^M \times L^NL^N$	—	1	—
$L^ML^N \times L^ML^N$	¼	½	¼
$L^ML^N \times L^NL^N$	—	½	½
$L^NL^N \times L^NL^N$	—	—	1

 Dominance occurs in different forms. It may be complete, as found by Mendel, where either the mutant or the wild-type allele is dominant. It may be incomplete, where the heterozygote has a phenotype in between the two homozygotes. Or two traits may be codominant, where both alleles are expressed in the heterozygote.

LETHALS

The expectations of segregation given in chapter 2 are true for zygotes (newly fertilized eggs). However, for many traits, the zygotic ratio is modified by differential mortality (that is, survival) as the progeny develop. For example, the initial 3:1 dominant to recessive ratio expected in progeny from two heterozygotes for the sickle-cell allele, an allele that causes sickle-cell disease when homozygous, changes to much less than 25% anemics in adults because of the higher mortality for the sickle-cell homozygotes.

Humans who suffer from genetic diseases have higher mortality than normal individuals in nearly all cases. In other organisms as well, genetic conditions may result in high mortality; for example, genes can cause a lack of chlorophyll in plants resulting in no photosynthesis and mortality of all such seedlings. When a genetic defect causes 100% mortality, it is termed a **lethal allele.** Lethals are generally recessive, resulting in the death of the recessive homozygote. However, some lethals are dominant, as in Huntington disease, where the heterozygote is affected, although generally not until later in life.

In 1904, shortly after the rediscovery of Mendel's principles, the French geneticist Lucien Cuénot, carrying out experimental crosses on coat color in mice, found a gene that was not consistent with Mendelian predictions. He observed from his experiments that the yellow body color allele was dominant, but that crosses between two yellow mice yielded approximately a 2:1 ratio of yellow to wild-type, rather than the expected 3:1. Furthermore, when he crossed yellow individuals to the recessive wild-type (agouti-colored mice), Cuénot found that all yellow mice produced wild-type progeny. He concluded that all yellow mice were heterozygotes and that there were no yellow homozygotes as would have been expected. Later it was

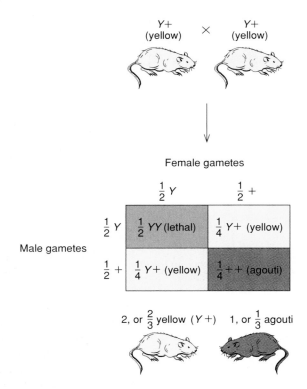

FIGURE 3.2
A cross between two yellow mice, yielding a 2:1 ratio of yellow to agouti-colored mice in the offspring because mice with genotype *YY* die.

suggested that homozygosity for yellow is lethal, and that these individuals died in utero, a fact eventually validated by the histological observation that approximately ¼ of the embryos from yellow × yellow crosses failed to develop.

The explanation for this situation is given in figure 3.2, where *Y* indicates the dominant, yellow mutant, and + indicates the recessive, wild-type allele. If one were to calculate a chi-square using the principle of segregation expectations for such data, it would give a statistically significant value, suggesting that the observations were not consistent with the Mendelian hypothesis. However, the principle of segregation obviously works properly in this case. The observed ratio of phenotypes differs from expectations because yellow homozygotes die before they can be counted. Many dominant disease alleles in humans, such as achondroplastic dwarfism, also appear to be lethals when they are homozygous. So are many dominant mutants in *Drosophila,* such as *Curly* wing, which we discussed in the last chapter.

The lack of a tail (taillessness) in the Manx cat (fig. 3.3) is another trait caused by an allele that has a dominant effect in heterozygotes and is a lethal in homozygotes. This trait is thought to have first occurred as a mutation among domestic cats on the Isle of Man in 1935. The Manx and normal alleles are denoted by *M* and *m,* respectively, so that Manx individuals are the heterozygous genotype *Mm.* True-breeding Manx breeds are not possible because matings between Manx cats, *Mm × Mm,* produce about ⅓ normal-tailed progeny.

Many genes have alleles that affect the rate of mortality but are not lethals. In general, these are termed **deleterious,**

FIGURE 3.3
A tailless Manx cat.
© *Robert Pearcy/Animals Animals.*

or **detrimental, alleles.** They result in various levels of mortality, ranging from a few percent more than the wild-type genotype to nearly lethal (almost 100% mortality). Natural selection (selection that does not result from human intervention), which we will discuss in some detail in chapter 16, results in a lower representation in the progeny of genotypes (and their alleles) having higher mortality than of genotypes with lower mortality.

PLEIOTROPY

The genes we have been discussing are generally well known because they substantially affect an obvious trait such as flower color or seed shape. However, many of these genes may also have secondary, or related, effects. For example, the yellow coat color in mice discussed previously is an allele that affects more than one character; that is, it not only produces yellow body color in heterozygotes but affects survival, causing lethality in homozygotes. Another example of multiple effects we mentioned is the gene affecting seed shape in garden peas; this gene also affects starch grain morphology. In fact, many genes affect more than one trait. Mendel himself noticed that the gene causing the flower colors violet and white also influenced seed color and caused the presence or absence of colored areas on the leaves. The phenomenon in which a single gene affects two or more characteristics is called **pleiotropy.**

Many genetic diseases in humans are caused by genes that have pleiotropic effects or a syndrome of diagnostic characteristics. One such disease is phenylketonuria (PKU), which occurs in individuals homozygous for a defective recessive allele. These people lack the enzyme necessary for the normal metabolism of the amino acid phenylalanine to the next biochemical product. As a result, when normal and PKU individuals are compared, the level of phenylalanine is much higher in the diseased group (fig. 3.4). In addition, this basic biochemical difference results in a suite of other

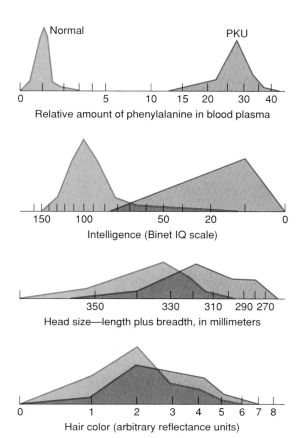

FIGURE 3.4

The values of four different characteristics in PKU and normal individuals, illustrating the pleiotropic effects of this disease. The purple regions show the overlap in the two distributions.

changes in untreated PKU patients, including lower IQ, smaller head size, and lighter hair. All of these pleiotropic effects in PKU can be understood as a consequence of the basic biochemical defect. For example, in PKU patients, a toxic compound accumulates in the head, causing brain damage and leading to a lower IQ and a smaller head size.

Many mutations of *Drosophila* have multiple phenotypic effects. New mutants are generally recognized by a gross change in the phenotype, such as white eyes or vestigial wings. Invariably, after detailed inspection, less extreme variations in body morphology or other characteristics are found. For example, the *w* (or *white*) allele that results in white-eyed flies also causes many changes in the fly's internal organs, and the *vg* (or *vestigial*) allele that results in vestigial, undeveloped wings also alters its halteres (balancers) and modifies the fly's bristles.

PENETRANCE AND EXPRESSIVITY

All of the genes we have considered to this point have a definite genotype-phenotype relationship; for example, a pea plant with genotype *rr* is always wrinkled. In fact, all of the genes Mendel examined, all blood group types, and many human diseases show such an absolute pattern. However, for some genes, a given genotype may or may not show a given phenotype. This phenomenon is described as

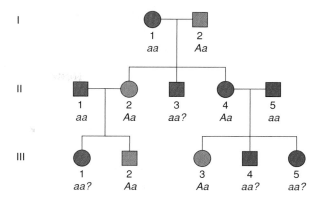

FIGURE 3.5

A pedigree illustrating the pattern of phenotypes that can occur with an incompletely penetrant dominant allele. As before, individuals indicated in purple are affected. Individual II-4 has genotype *Aa* but is not affected by the disease.

the **penetrance** of a gene. The level of penetrance can be calculated as the proportion of individuals with a given genotype who exhibit a particular phenotype. When all individuals of a particular genotype have the same phenotype, the gene shows complete penetrance and the level of penetrance is defined as 1.0. Otherwise, the gene is termed incompletely penetrant and the particular level of penetrance can be calculated. For example, if there are eight individuals of a particular genotype and five of them express a diseased phenotype, the level of penetrance is $\frac{5}{8} = 0.625$.

The presence of incomplete penetrance at a gene may cause a phenotype to skip a generation in a pedigree; that is, a dominant trait present in a given generation may skip the first progeny generation but appear again in the second generation. Such a case is illustrated in figure 3.5, a human pedigree where unaffected individual II-4 is both the daughter and mother of affected individuals. This indicates that she has the genotype *Aa,* but because of incomplete penetrance, she does not show the dominant phenotype. For example, the dominant form of retinoblastoma, a disease that causes malignant eye tumors in children, is only about 90% penetrant.

In addition, given that a particular genotype exhibits the expected phenotype, the level of expression, or **expressivity,** may vary. For example, although a gene causes a detectable disease in most individuals with a given genotype, some may be much more severely affected than others. In other words, individuals of a given genotype may be unaffected, may have intermediate symptoms, or may be severely affected. Figure 3.6 gives the variation seen in spotting in the coats of mice homozygous for the *s* allele. Notice that the phenotype may range from virtually no spotting (bottom right) to completely white.

Huntington disease in humans is a trait that shows variable expressivity in its age of onset. The disease generally occurs in early middle age, with the average age of onset in the thirties, but can occasionally occur very early (at younger than ten years) or in an older individual (fig. 3.7). Some people with the disease genotype die of other causes before they express the disease, resulting in incomplete penetrance as well as variable expressivity for this gene.

FIGURE 3.6
Variations in spotting in the coat of mice homozygous for the *s* allele. All of these mice are genotype *ss,* but their phenotype varies greatly because of variable expressivity.

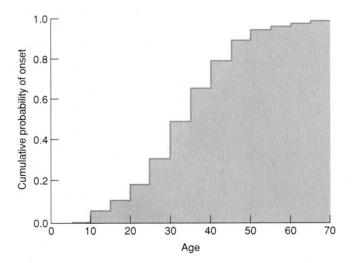

FIGURE 3.7
The cumulative age of onset of Huntington disease in a human population. All individuals are the diseased genotype *Hh,* but some show symptoms of the disease much earlier than others.

What are the causes of incomplete penetrance and variable expressivity? In particular instances, it is often impossible to determine causation, but both environmental factors and other genes are known to influence the penetrance or expressivity of a gene. For example, alleles of a number of genes in the fruit fly *Drosophila melanogaster* affect viability and may be lethal at elevated temperatures (say, greater than 28° C) but have little or no effect at lower temperatures. Such genes, known as **conditional lethals,** generally show their lethal effects in extreme environmental situations.

Temperature can also have a dramatic effect on the phenotypes of plants. For example, primroses have red flowers when grown at 24° C but white flowers when grown above 32° C. The coats of Siamese cats and Himalayan rabbits are darker in color at the paws, ears, and nose because of the lower body temperature at these extremities (fig. 3.8). If the animal is placed in a warmer temperature, the new growth of fur at these points is light in color.

In addition, alleles at other genes may influence the penetrance and expressivity of a gene. For example, the level of expression of a trait is generally more similar among relatives than among unrelated individuals, given that the relatives and unrelated individuals are raised in fairly similar environments. Such genes that have a secondary

FIGURE 3.8
The color of Siamese cats is darkened at the paws, ears, and nose because of a lower temperature at these extremities.
© *Mary Eleanor Browning/National Audubon Society/Photo Researchers, Inc.*

Female gametes

		I^A	I^B	I^O
Male gametes	I^A	$I^A I^A$ (A)	$I^A I^B$ (AB)	$I^A I^O$ (A)
	I^B	$I^A I^B$ (AB)	$I^B I^B$ (B)	$I^B I^O$ (B)
	I^O	$I^A I^O$ (A)	$I^B I^O$ (B)	$I^O I^O$ (O)

FIGURE 3.9
The six genotypes produced by the three alleles in the ABO blood group system. The four corresponding phenotypes are in parentheses.

MULTIPLE ALLELES

All the genes we have discussed so far (and all the traits Mendel studied) have had only two segregating alleles. However, when detailed analysis of a gene is undertaken in a population, generally more than two alleles are discovered. Although some of these alleles may be rare, the existence of more than two alleles per gene (**multiple alleles**) is relatively frequent in a population.

One commonly known genetic variation resulting from multiple alleles is the ABO blood group system in humans. (There are closely related systems in other mammals such as chimpanzees and cattle.) The different ABO blood group types were discovered in 1900, although their genetic basis was not uncovered through pedigree examination until 1925. In the ABO system, there are three major types of alleles (these classes themselves being heterogeneous), which are generally identified as I^A, I^B, and I^O (or sometimes as *A, B,* and *O*). If we assume that female and male gametes can contain one of the three allelic types, then the different genotypes and phenotypes shown in figure 3.9 are possible. Notice that there are six different genotypes—that is, six different ways the three alleles can be combined two at a time.

In general, the number of genotypes possible in a diploid organism with "n" different alleles is found by adding $1 + 2 + 3 + \ldots + n$, where . . . indicates all the integers between 3 and n, or by using the formula $n(n + 1)/2$ (table 3.2). Both of these expressions give the number of different ways n objects (or alleles) can be combined two at a time when order is not important. For the ABO system with three alleles (n = 3), the number of genotypes is either $1 + 2 + 3 = 6$ or $(3)(4)/2 = 6$, as given in figure 3.9. If there were four alleles, there would be $(4)(5)/2 = 10$ possible genotypes. In addition, the number of homozygotes is equal to the number of alleles (n), and the number of heterozygotes is equal to $n(n - 1)/2$ (table 3.2). For example, the ten genotypes resulting from four alleles are composed of four homozygotes and $(4)(3)/2 = 6$ heterozygotes.

Because allele I^O of the ABO system is recessive to both alleles I^A and I^B, there are only four different phenotypes: type A (either genotype $I^A I^A$ or $I^A I^O$), type B ($I^B I^B$

effect on a trait are called **modifier genes** and can substantially influence a phenotype. In many animals, the dilute allele that reduces the intensity of pigmentation when homozygous (such as from black to gray in house cats) is an example of a modifier gene. The extent of taillessness in Manx cats varies from having no tail vertebrae to having a small number of fused tail vertebrae, apparently due to the influence of modifying genes on the expression of the dominant Manx mutant. Mutants of *D. melanogaster* that have been kept in laboratory culture for many years sometimes do not have as extreme a phenotype as when first observed, presumably because of the increased frequency of alleles that modify the mutant phenotype.

A related phenomenon, called a **phenocopy,** occurs when environmental factors induce a particular abnormal phenotype that would usually be genetically determined. In this case, an extreme environment may disrupt normal development in a pattern similar to that of a mutant gene. One tragic example occurred in western Europe in the late 1950s and early 1960s when a number of babies were born with severely shortened limbs, an abnormality that resembled a rare recessive genetic disorder. Only after an intensive investigation was it shown that the defect had been environmentally induced by the sleeping pill medication thalidomide, which the mothers of the affected individuals had taken during early pregnancy.

 Some alleles of genes may be lethal. Pleiotropy occurs when one gene affects several traits. Some genotypes may not always have the same phenotype, a phenomenon resulting from incomplete penetrance of variable expressivity of the genotype.

Table 3.2 Numbers of Homozygotes, Heterozygotes, and Genotypes, Given Multiple Alleles

Alleles	Homozygotes	Heterozygotes	Genotypes
1	1	0	1
2	2	1	3
3	3	3	6
4	4	6	10
5	5	10	15
6	6	15	21
•	•	•	•
•	•	•	•
•	•	•	•
n	n	$\dfrac{n(n-1)}{2}$	$\dfrac{n(n+1)}{2}$

		Recipient			
Donor		A (anti-B)	B (anti-A)	AB	O (anti-A and anti-B)
Genotype	Phenotype				
$I^A I^A, I^A I^O$	A	+	A, anti-A	+	A, anti-A
$I^B I^B, I^B I^O$	B	B, anti-B	+	+	B, anti-B
$I^A I^B$	AB	B, anti-B	A, anti-A	+	A, anti-A B, anti-B
$I^O I^O$	O	+	+	+	+

FIGURE 3.10

The potential for successful blood transfusions, indicated by a + for different ABO donor-recipient combinations. A, anti-A, and B, anti-B, indicate a reaction between A antigens and anti-A antibodies and between B antigens and anti-B antibodies that would result in unsuccessful transfusions.

or $I^B I^O$), type AB ($I^A I^B$), and type O ($I^O I^O$). Alleles I^A and I^B are codominant, illustrating that different dominance relationships are possible among the alleles at a given gene.

Differences among ABO blood group types occur because of the presence of different antigens and antibodies. **Antigens** of the ABO blood group system are molecules composed of a protein-sugar combination that occur on the surface of red blood cells. Individuals with alleles I^A, I^B, and I^O have A, B, and neither A nor B antigens, respectively, on these cells. The A and B antigens differ in the type of terminal sugar on the molecules. The antigen from the I^O allele actually lacks this terminal sugar because when this allele is present, no enzyme to add the terminal sugar is produced.

Antibodies are substances that can be produced by the immune system in large amounts as a response to specific external antigens. (Antibodies are not produced against self antigens.) For example, if cells from a type A individual are injected into rabbits, the immune system of the rabbits will produce anti-A antibodies. In humans, individuals of type A have anti-B antibodies; type B have anti-A antibodies; type AB have neither of these antibodies; and type O have both kinds of antibodies. The presence of antibodies that recognize A and B antigens in individuals lacking these antigens appears to be the result of immune responses to substances produced by bacteria that are similar to these antigens.

In order to transfuse blood successfully from one individual to another, differences between individuals in the ABO blood group system must be taken into account. As an example of transfusion problems that can result from the ABO system, if red blood cells with B antigens (from type B individuals) are transfused to type A individuals, the type B cells will be attacked and eliminated by the anti-B antibodies present in the type A individuals. All combinations of donor and recipient phenotypes are given in figure 3.10. Notice that type O individuals are universal donors; that is, type O blood, because it does not have antigens comparable to the other types, is accepted by all individuals. On the other hand, type AB individuals are called universal recipients, because they do not have

anti-A or anti-B antibodies and can therefore accept A, B, AB, or O blood. Type O recipients given any blood but O will suffer dire consequences because O individuals have both anti-A and anti-B antibodies. Type AB individuals can donate blood only to type AB individuals because other blood types have anti-A, anti-B, or both antibodies.

Some genes may have a series of alleles showing sequential dominance; for instance, a number of alleles can affect color in mammals, insects, or mollusks. In some parts of England, the peppered moth, *Biston betularia,* has at least three such alleles. In chapter 16 we will discuss the evolution of these color types, a phenomenon that has been associated with industrial pollution in England. At this locus, the allele that was characteristic of the species before industrial pollution, the pale typical allele (*m*), is recessive. A second allele, the darkly mottled insularia (*M′*), is dominant to the typical. The third allele, the nearly black melanic (*M*), is dominant to both of the other alleles. In other words, there are six genotypes but only three phenotypes: (1) the melanic phenotype, having three genotypes—homozygous melanic (*MM*) and heterozygous between melanic and the other two alleles (*MM′* and *Mm*); (2) the insularia phenotype, composed of the insularia homozygote (*M′M′*) and the insularia-typical heterozygote (*M′m*); and (3) the typical phenotype, all homozygous recessive typical (*mm*) (fig. 3.11). As an example of the types of mating possible, assume that two heterozygotes, a melanic *Mm* and an insularia *M′m,* were crossed. The expected progeny genotypes are one-quarter each *MM′, Mm, M′m,* and *mm.* However, because of dominance, half the progeny should be the melanic phenotype (*MM′* and *Mm* genotypes), one-quarter insularia (*M′m*), and one-quarter typical (*mm*).

Some genes, such as the human genes for HLA (human leucocyte antigen, which determines antigens that are on the surfaces of nearly all cells), may have many alleles. For example, gene *HLA-B* has more than thirty different antigenically defined alleles in some populations. These alleles are denoted by A1, A2, A3, A11, and so on for *HLA-A,* and by B5, B7, B8, B12, and so on for *HLA-B.* As a result of this diversity, an extraordinary number of genotypes generally

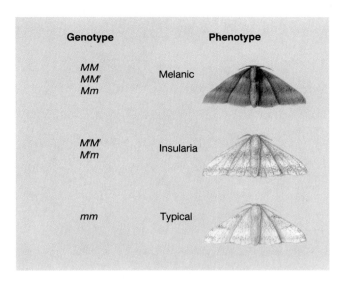

Genotype	Phenotype	
MM MM' Mm	Melanic	
M'M' M'm	Insularia	
mm	Typical	

FIGURE 3.11
The six possible genotypes of the peppered moth from all diploid combinations of three alleles are listed in the left column. The resulting phenotypes are shown in the far right column.

		Recipient		
Donor		A1 A28	A2 A10	A1 A9
A2 A2		A2	+	A2
A1 A28		+	A1, A28	A28
A9 A11		A9, A11	A9, A11	A11
A9 A9		A9	A9	+

FIGURE 3.12
The potential for an organ transplant from four donor individuals to three recipients based on *HLA-A*. The + indicates a match. *A2, A9*, and so on indicate the antigens in the donor that would cause rejection.

occur at this gene in a population. Again, we can use the formulas in table 3.2 to calculate the number of possible genotypes for genes with different numbers of alleles. *HLA-B,* with thirty alleles, potentially has 30 different homozygotes, 435 different heterozygotes [(30)(29)/2], and a total of 465 different genotypes [(30)(31)/2], truly an extensive array of genotypic variation.

HLA genes are important in determining the potential success of organ transplants and are related to susceptibility to a number of autoimmune diseases, such as juvenile diabetes and rheumatoid arthritis. For these genes, all the alleles are codominant, so the number of genotypes equals the number of phenotypes. Matching the genotypes of *HLA-B* (and the *HLA-A* gene, which has nearly twenty different alleles) in the donor and recipient of a kidney increases the likelihood of a successful transplant. Figure 3.12 gives the *HLA-A* genotype for three potential recipients and four potential donors. Only where there is a + is the donor compatible with the recipient. For all other combinations, one or more antigens present in the donor are not present in the recipient, and the likelihood of a successful transplant is much lower. Because there are so many different genotypes, donors and recipients generally do not match. Because relatives are more likely to share HLA types, often close relatives are examined for HLA to find a suitable organ donor. Even with a suitable donor for the major HLA types, it is still necessary to use immunosuppressant drugs to reduce the organ recipient's immune response to foreign antigens from other unmatched genes and thus achieve a successful transplant.

◆ Genes may have more than two alleles. In diploid organisms, many different genotypes can exist when there are multiple alleles.

SEX-LINKED GENES

In chapter 1 we mentioned that most genes are on autosomes—the chromosomes for which each diploid individual has a homologous pair. However, shortly after the rediscovery of Mendel's principles, some experiments with *Drosophila* yielded progeny that differed from Mendelian predictions in the phenotypic proportions of males and females. These initially perplexing results were explained by assuming that the genes in question were not on autosomes but on the X chromosome, which is present in two copies in the female *Drosophila* and in one copy in the male. (These experiments will be discussed in chapter 4.)

Mammals, some insects, and a few plants have this kind of sex chromosome system in which one sex has two homologous chromosomes and the other has two different kinds of chromosomes. (In chapter 4 we will also discuss sex determination.) In mammals, females have two X chromosomes and are designated XX. Males have one X chromosome and one Y chromosome and are designated XY. If we examine the inheritance of genes on these chromosomes, we find patterns different from the autosomal genes examined by Mendel, but the basis of this inheritance is consistent with Mendelian principles.

Let us first examine traits determined by genes on the X chromosome, or **X-linked traits.** McKusick lists 161 identified loci on the human X chromosome. One example is the inheritance of the common type of color blindness, which is caused by a recessive allele located on the X chromosome. We will indicate X chromosomes having the two alternative alleles as X^C and X^c, the normal and color-blind alleles, respectively. In females there are three different genotypes ($X^C X^C$, $X^C X^c$, and $X^c X^c$) in which only the last is color-blind because the normal allele is dominant in heterozygotes. Males have only one X and a Y. Because Y chromosomes do not carry the color blindness gene (the Y has only a few genes allelic to those on the X), the two types of males are $X^C Y$ and $X^c Y$—normal and color-blind, respectively. (Another way to indicate X-linked inheritance is to leave out the X symbol and write the X-linked alleles as we symbolized alleles previously. In other words, a female heterozygous for color blindness would be *Cc*, and a color-blind male would be *cY*.)

(a)

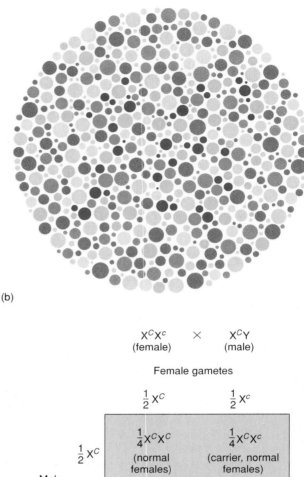

(b)

FIGURE 3.13

Plates to test for color blindness. All individuals with normal sight will read (*a*) as "12" and (*b*) as "16." Color-blind people will not be able to read (*b*) at all.

The above has been reproduced from Ishihara's Tests for Colour Blindness *published by Kanehara & Co., Ltd., Tokyo, Japan, but tests for color blindness cannot be conducted with this material. For accurate testing, the original plates should be used.*

Figure 3.13 shows a test for color blindness. People who have good color vision will see large numbers in both diagrams, while color-blind individuals will not be able to detect the numbers in part (*b*) of the figure. (We should caution that this color vision test is only part of a complete test and that people with some color vision may not be able to distinguish the numbers in part (*b*).)

In females, segregation for X-linked genes is the same as that for autosomal genes, while in males, half of the gametes carry the X and half carry the Y. To illustrate this, let us examine the progeny from a cross between a woman heterozygous for color blindness and a man with normal sight (fig. 3.14). Notice that all male progeny receive a Y chromosome from their father. We would expect half the progeny to be female, all normal-sighted (although half of them are homozygotes and half heterozygotes), and half the progeny to be male (half color-blind and half normal-sighted). The ratio of normal to affected progeny is still 3:1, but unlike the F_2 progeny of a monohybrid cross for an autosomal gene, all the recessive types are males. Obviously, the number of progeny from any one mating in humans is small, but if a number of such matings are lumped together, these expectations are appropriate.

Pedigrees for X-linked traits have particular patterns that generally allow them to be distinguished from pedigrees for autosomal traits. Figure 3.15 shows a pedigree for an X-linked recessive. First, the X-linked recessive affects many more males than females because males need only one copy of the defective allele to express the trait, while females

$$X^C X^c \quad \times \quad X^C Y$$
(female)   (male)

Female gametes

$\frac{1}{2} X^C$ $\frac{1}{2} X^c$

Male gametes

$\frac{1}{2} X^C$ | $\frac{1}{4} X^C X^C$ (normal females) | $\frac{1}{4} X^C X^c$ (carrier, normal females)

$\frac{1}{2} Y$ | $\frac{1}{4} X^C Y$ (normal males) | $\frac{1}{4} X^c Y$ (color-blind males)

3 or $\frac{3}{4} X^C -$ or $X^C Y$ (normal sight)   1 or $\frac{1}{4} X^c Y$ (color-blind)

FIGURE 3.14

Probable outcome of a mating between a female heterozygous for color blindness and a male with normal sight.

need two. When the allele is rare, it is unlikely that homozygous females will be encountered. Second, the X-linked recessive trait skips generations because males can receive an X-linked allele only from their mother and, as just suggested, very few females exhibit X-linked recessive traits.

Several common types of the recessive genetic disease hemophilia, in which blood does not clot normally, are also determined by genes on the X chromosome. As a result, most individuals with X-linked forms of hemophilia are males who have received the hemophilia allele from their unaffected carrier mothers. Hemophilia afflicted a number of the descendants of Queen Victoria of England, suggesting that she was heterozygous for this allele. Figure 3.16

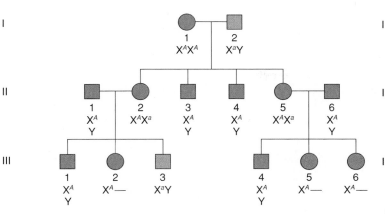

FIGURE 3.15
A pedigree illustrating the pattern of inheritance expected for an X-linked recessive gene.

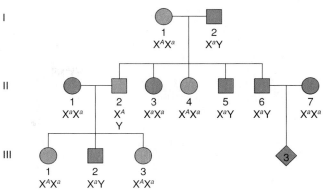

FIGURE 3.17
A pedigree illustrating the pattern of inheritance expected for an X-linked dominant gene.

FIGURE 3.16
A photograph of Czar Nicholas II, the Czarina Alexandra, their four daughters, all with normal blood, and their son, Alexis, who suffered from hemophilia.

The Bettmann Archive.

shows the family of Czar Nicholas II of Russia, whose wife Alexandra was a granddaughter of Victoria. Their son Alexis, pictured in front, had hemophilia and was executed, along with the rest of the family, at age 14.

X-linked dominant traits have a different pattern of inheritance than X-linked recessive traits, as shown in figure 3.17. Notice in this pedigree that both males and females are affected by the trait (sometimes slightly more females are affected), but that the patterns of affected offspring produced by affected individuals differ between the two sexes. Irrespective of sex, approximately half the offspring of affected females, such as I-1 in figure 3.17, are affected, while for affected males, such as II-2, all female offspring and no male offspring are affected.

The last type of sex-linked inheritance we will mention is that expected for genes on the Y chromosome. Although there is little evidence for Y-linked inheritance in humans,

except for a gene or genes that determine maleness, other organisms do carry genes on the Y chromosome. One putative Y-linked gene in humans results in hairy ear rims (fig. 3.18). However, determining the precise mode of inheritance of this trait is complicated by a variable age of expression. The characteristics of Y-linked traits are straightforward: (1) They are expressed only in males, and (2) they are always passed from father to son. An interesting analogy to Y-linked inheritance is the traditional transmission of family names in English-speaking (and many other) societies. A summary of the characteristics of pedigrees for the different types of sex-linked traits is given in table 3.3.

Some traits are expressed differently in the two sexes but are not sex-linked. First, there are **sex-limited** traits, characters expressed only in one sex. Many such traits are related to reproductive characters, such as milk yield in dairy cattle, egg laying in chickens, or oviposition behavior in insects.

Second, other traits are **sex-influenced;** that is, the genotypes determined by autosomal genes are expressed differently in the two sexes. For example, the genes determining the presence of horns in sheep are expressed differently in males and females. In some sheep breeds, such as the Suffolk, neither sex has horns (they are called polled), while in other breeds, such as the Dorset Horn, both sexes have horns, although the horns of the male are much larger than those of the female (table 3.4). When these two breeds are crossed, the hybrids, which are heterozygous for the autosomal *H* gene, are horned if male but polled if female. In the heterozygotes, the presence of male or female hormones determines whether or not horns develop. It is thought that premature pattern baldness (occurring before age thirty-five) follows this same pattern of inheritance; that is, baldness is dominant in male heterozygotes and recessive in female heterozygotes.

Genes on the X and Y chromosomes have different patterns of inheritance than autosomal genes. X-linked recessive traits occur almost entirely in males and generally skip generations in families.

FIGURE 3.18
Men with hairy ear rims, a putative Y-linked trait.
Courtesy of Dr. S. D. Sigamony.

Table 3.3 General Patterns of Sex-linked Inheritance

X-linked Recessive

(a) Usually more males than females are affected.

(b) No offspring of an affected male are affected, making the trait skip generations in the pedigree, always with an unaffected female in the intermediate generation. (An exception to this pattern occurs in the rare instance when the affected male mates with a female carrier, producing affected female offspring.)

X-linked Dominant

(a) Affected males produce all affected female offspring and no affected male offspring.

(b) Approximately half the offspring of affected females are affected, regardless of their sex.

Y-linked Inheritance

(a) Trait is always passed from father to son.

(b) Only males are affected.

Table 3.4 Example of a Sex-influenced Trait in Sheep

	HH (Suffolk)	Hh (Cross)	hh (Dorset Horn)
Males	Polled	Horned	Horned
Females	Polled	Polled	Horned

MULTIPLE GENES AND EPISTASIS

In our discussion of penetrance and expressivity, we mentioned that the phenotype resulting from the genotype at one gene can be altered by alleles at other genes called modifier genes. In fact, many traits are affected by several genes, as was observed by a number of scientists shortly after the rediscovery of Mendel's principles. (Remember that when Mendel examined two genes in his dihybrid crosses, he used genes that affected different traits, not the same trait.)

For a number of traits, different genes affecting the same trait interact so that the phenotype expressed is a function of the particular combination of alleles present at the different genes. The general term for interactions of alleles at different loci to influence a trait is **epistasis** (Greek: *epi* + *stasis* = standing upon). Note that epistasis involves two (or more) genes and differs from dominance, which occurs when the phenotype is determined by different alleles at the same locus.

One of the first observations of epistasis resulted from a cross between two varieties of white-flowered sweet peas, which was done by early geneticists William Bateson and R. C. Punnett. When they crossed the two white-flowered varieties, the F_1 flowers were unexpectedly purple (fig. 3.19). When the F_1 individuals were allowed to self-fertilize, 382 purple and 269 white flowers resulted, which is close to a ratio of 9 purple to 7 white in the F_2 progeny. The explanation for this ratio, confirmed by other crosses, was that homozygosity of recessive alleles of two different genes could result in white flowers. When both genes had one or more dominant alleles, the phenotype was purple. Because nine of the sixteen categories in figure 3.19 are like this, there is a 9:7 ratio in the F_2 progeny.

It is useful to understand the biochemical basis of these phenotypes. Pigmentation in the flowers of sweet peas results from chemicals called *anthocyanins*. Production of anthocyanins occurs through a series of metabolic steps catalyzed by enzymes that are themselves products of genes. Whenever a step in this synthetic process is blocked by the absence of a functional enzyme, pigmentation does not occur. Although the exact details of pigmentation synthesis in sweet peas are not known, the general pattern is illustrated in figure 3.20, where the enzyme products of gene *C* and gene *P* are both necessary for the production of

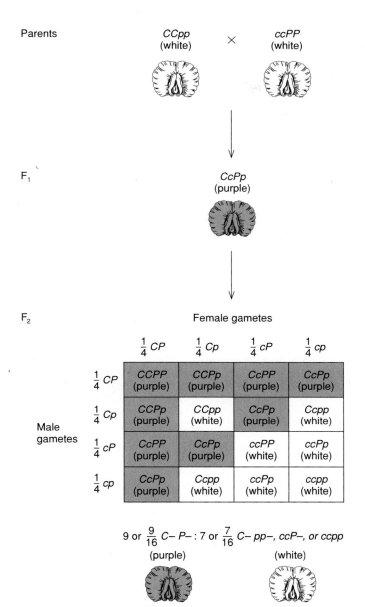

FIGURE 3.19

A cross between two types of white sweet peas, showing the 9:7 ratio that results in the F$_2$ generation.

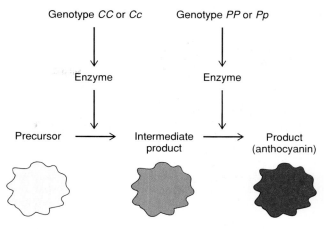

FIGURE 3.20

A diagram illustrating the relationship between genes *C* and *P* and the production of anthocyanins in sweet peas.

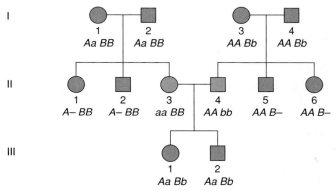

FIGURE 3.21

A pedigree illustrating a recessive disease caused by two genes.

anthocyanins. When the genotype is *cc,* the first enzyme is not produced; consequently, reaction to the intermediate product is stopped. Likewise, if the genotype is *pp,* the lack of the second enzyme halts the second metabolic step. In other words, if the genotype is either *cc* or *pp,* the synthetic pathway is blocked and no pigments are produced, resulting in a white flower. Such an interaction of genes to jointly produce a specific gene product is a type of epistasis called **complementary gene action.**

As we will discuss later, many recessive diseases in humans are caused by metabolic errors, often called inborn errors of metabolism. If two individuals with metabolic errors for the same trait (but in different steps of a biochemical pathway) mate, their offspring should have a normal phenotype. This is what happens when two individuals with

deaf-mutism caused by different recessive genes mate and have offspring. In the pedigree in figure 3.21, II-3 and II-4 are affected with a recessive disease because they are homozygous at different loci. Note that the pedigree is different from that of a single recessive gene because no progeny, instead of all progeny, from a mating of two affected individuals are affected.

The 9:7 ratio is only one type of F$_2$ pattern caused by epistasis. In other situations, the F$_2$ ratios may be 9:3:4, 13:3, or 15:1. Notice that the different categories of a 9:3:3:1 ratio can be combined to obtain these ratios. The biochemical basis for the 15:1 ratio can be understood by examining one step of a pathway in which a dominant allele at either of two genes is enough to produce enzymes for catalysis of a given reaction (fig. 3.22). Only when there is a double recessive, say *aabb,* is the pathway blocked and the aberrant phenotype expressed. When either of two genes can function to produce the dominant phenotype, this type of epistasis is called **duplicate gene action.** Presumably, the two different genes produce similar gene products, and one of them may have arisen by duplication from the other gene (see chapter 4). A classic example of

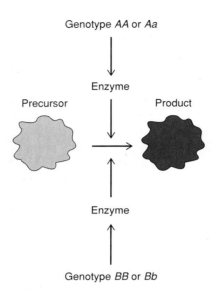

Genotype *AA* or *Aa*

Enzyme

Precursor   Product

Enzyme

Genotype *BB* or *Bb*

FIGURE 3.22
A diagram illustrating duplicate gene action, in which the enzyme from either gene *A* or gene *B* can catalyze the biochemical reaction.

duplicate gene action occurs in shepherd's purse, a widespread weed. Round seed pods result when dominant alleles are present at either of two genes, while narrow seed pods are produced only in the absence of dominant alleles at both loci.

Let us diagrammatically summarize the types of epistasis we have discussed. Where there is dominance at both loci and no epistasis, there are four phenotypes, with a 9:3:3:1 F_2 ratio as shown in row 1 of figure 3.23. The two types of epistasis we have mentioned, complementary and duplicate gene action, are given next, with their 9:7 and 15:1 ratios, respectively. Two other types of epistasis, in which there are three phenotypic classes, are given in the last rows. Can you find any other phenotypic ratios that are possible by combining the different classes—for example, 12:4 or 9:6:1?

Coat color in mammals is a trait that is affected by a number of different genes. In fact, at least 63 different genes with 148 different alleles affect coat color in mice. Some of the variations in mouse coat color are shown in figure 3.24. We have already mentioned two coat-color genes, the yellow gene in mice and the dilute gene in house cats. Other coat-color genes interact in a number of different ways, three of which we will discuss here to illustrate epistatic interaction.

First, the dominant agouti color is controlled by the *A* allele at the *A* gene and results in an overall gray, or "salt and pepper" appearance if you look closely. It is the common color of many mammals in natural populations. In actual fact, the individual hairs of agouti animals have a band of yellow, giving an overall brindled grayish color to the coat. When the genotype is the recessive non-agouti (*aa*), the hair shaft is all one solid color.

Second, the alleles at the *B*, or brown coat color, gene result in different colors, depending upon the genotype at the *A* gene (table 3.5). When the *B* allele is present, the

	A– B–	A– bb	aa B–	aa bb
Dominance with no epistasis	9	3	3	1
Complementary gene action	9	7		
Duplicate gene action	15			1
Dominance with epistasis of *aa* over *B*–	9	3	4	
Dominance with epistasis of *A*– over *bb*	12	3		1

FIGURE 3.23
Examples of epistatic F_2 ratios from the cross *Aa Bb* × *Aa Bb*.

FIGURE 3.24
Mice showing black and brown phenotypes.
© *Runk/Schoenberger/Grant Heilman Photography, Inc.*

Table 3.5 Genotypes and Resulting Coat Color in Mice[1]

A and B Loci		B and C Loci	
Genotype	**Phenotype**	**Genotype**	**Phenotype**
A– B–	Agouti	B– C–	Black
A– bb	Cinnamon	bb C–	Brown
aa B–	Black	B– cc	Albino
aa bb	Brown	bb cc	Albino

[1] A = agouti; B = brown; C = albino

phenotype is agouti if the genotype is *AA* or *Aa,* but black if the genotype is *aa*. On the other hand, when the genotype at the *B* gene is *bb,* the phenotype is cinnamon-colored for genotype *A*– and brown for *aa*.

Finally, a third gene, called *C,* allows the formation of coat pigments. (The symbol *C* is used for this albino gene because *A* already represents the agouti gene.) For the *C* gene, the recessive genotype, *cc,* prevents the formation of color from other genes (table 3.5). For example, the genotype *bb cc* is albino (no pigment) because the *cc* genotype inhibits the formation of the brown color. If two mice of genotype *Bb Cc* are crossed, the 9:3:4 phenotypic ratio in the fourth row of figure 3.23 results. What phenotypic proportions would you expect if *Bb Cc* and *Bb cc* were

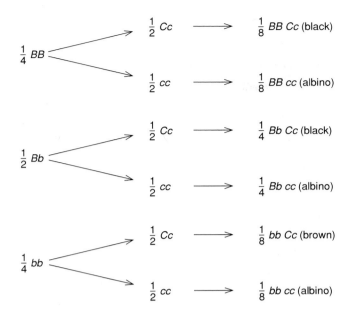

FIGURE 3.25

The expected results of a cross between mouse coat-color genotypes *Bb Cc* and *Bb cc*.

◆◆ More than one gene can affect a single trait. The interaction of two genes to determine a phenotype is called epistasis. Epistatic interaction can result in F_2 ratios different from 9:3:3:1.

crossed? Using the forked-line approach (fig. 3.25), you can see that the expected ratio of phenotypes in the progeny is ⅜ black, ⅛ brown, and ½ albino or a 6:2:8 ratio.

GENOTYPE-PHENOTYPE RELATIONSHIPS

For the traits that Mendel examined, a given genotype always resulted in a certain phenotype. However, in other situations, the phenotype of a given genotype varies, depending upon the penetrance and expressivity of that genotype. Often such variable expression results from environmental factors, and the exact phenotype caused by a given genotype-environment combination is not easily predicted.

In an ideal situation, the extent of the effect of the environment on a given genotype can be determined by looking at the phenotype of genetically identical individuals raised in different environments. In humans, identical, or monozygotic, twins separated at birth provide such a test situation. However, the best indications of environmental effects come from studies of species that can be raised in large numbers in controlled environments.

Working with several plant species, J. Clausen and his coworkers performed a classic example of such a study in the 1940s. To determine the effect of the environment on phenotypes, these researchers grew **clones** (genetically identical individuals) from cuttings at different altitudes. For example, they collected *Potentilla glandulosa,* a member of

the rose family, from three locations: the Pacific coast range at an altitude of 100 meters; midway up the Sierras at 1,400 meters (Mather); and high in the Sierras at 3,040 meters (Timberline). From these samples, they generated clones by vegetative propagation and grew them at the three different locations. In other words, genotypes collected from the three locations were grown in each of the three environmental conditions (100 m, 1,400 m, and 3,040 m).

Figure 3.26 shows the results of these experiments. Each row represents a genotype from the different areas, and each column indicates the environment in which these plants were grown. From this, it appears that both genetic and environmental factors are important in determining plant growth. For example, the coast range genotypes grow well at both low and intermediate altitudes but cannot survive at the high altitude (bottom row), indicating the importance of environmental factors. On the other hand, the timberline genotype (top row) grows much better than the other genotypes at the high altitude (far right column), indicating the importance of genetic factors.

The way the phenotype for a given genotype changes according to its environment is called the **norm of reaction.** Figure 3.27 gives norms of reaction similar to those of the low- and middle-altitude genotypes in figure 3.26. Genotype *A,* like the low-altitude clone, does best at the low end of the environmental spectrum and worst at the high end. But genotype *B,* like the middle-altitude clone, does best at intermediate altitude levels, falling off in phenotypic value in both the low- and high-altitude environments.

Norms of reaction may differ greatly among genotypes for traits involving behavior. Although we have not discussed behavior as a phenotype until now, a number of mutant genes do affect behavior. For example, individuals with untreated PKU have lower than normal IQs, a behavioral trait that can be measured using IQ tests.

The norm of reaction for a behavioral trait can be shown by comparing two strains of rats and their ability to complete a maze in three different environments: restricted, normal, and enriched. In this experiment, the environments are ranked according to the amount of visual stimulus present during the maturation of the rats. The two strains of rats are selected through a number of generations for their increased or decreased maze-learning ability in a normal environment. The lines are therefore called maze-bright and maze-dull, respectively. In the normal environment, animals from the maze-bright strain make an average of fifty fewer errors than animals from the maze-dull strain, reflecting the effectiveness of selection and the genetic differences between the strains (fig. 3.28). However, if the two strains of rats are raised in either restricted or enriched environments, the two strains make a similar number of errors, primarily because the maze-bright strain cannot run the maze well under restricted conditions and the maze-dull strain can perform well under enriched conditions. The implications of these observations are intriguing; they suggest

FIGURE 3.26
The phenotypes of *Potentilla* plants coming from three altitudes, and each cloned and grown at three different altitudes.

FIGURE 3.27
The norms of reaction for two different genotypes: *A*, the low-altitude (coast range) clone, and *B*, the middle-altitude (Mather) clone.

FIGURE 3.28
The numbers of errors in running a maze for two strains of rats, maze-bright and maze-dull, when the two strains are raised in three different environments.

Source: Data from Cooper and Zubek, in Canadian Journal of Psychology, *Vol. 12, 159–164, 1958, Societe Canadienne d'Orientation, Quebec, Canada.*

❖ The relationship of genotype to phenotype may be complicated. The norm of reaction over different environments is one way to quantify this relationship.

that some differences in behavior may be present only in a narrow range of environments and may not be generalizable to a diversity of environments.

QUANTITATIVE TRAITS

For many traits, the genetic basis is not known precisely, and it may not be feasible, or even possible, to determine the number of genes and their individual effects on a particular trait. The phenotypic value of traits such as size, weight, or shape can generally be measured on a quantitative scale; hence, these are called **quantitative traits.** Although this phenotypic value may result from some general underlying genetic determination, for most quantitative traits there is no precise relationship between the phenotypic value and a particular genotype. In general, quantitative traits have these characteristics: (1) They have a continuous distribution;

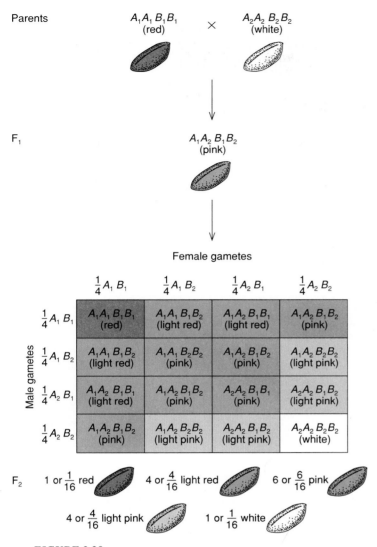

FIGURE 3.29

A cross of red- and white-kerneled wheat differing at two genes, showing the five types of F₂ progeny.

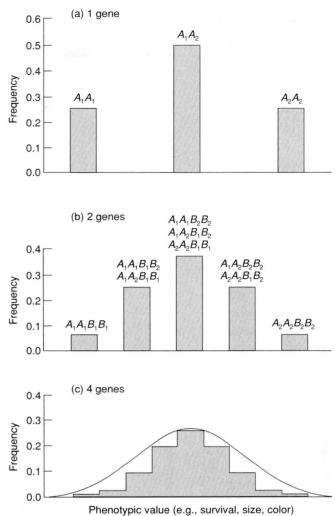

FIGURE 3.30

The influence of (*a*) one, (*b*) two, and (*c*) four genes on the occurrence of a quantitative trait and the resulting phenotypic distribution.

(2) they appear to be affected by many genes; that is, they are **polygenic;** and (3) they are affected by environmental factors.

One of the first good documentations of a polygenic trait was by Swedish geneticist Herman Nilsson-Ehle, who showed in 1909 that the color of wheat kernels was affected by several genes. To understand his findings, we will examine the genetic basis of his crosses, where color differences are determined by two genes. Figure 3.29 gives the F₂ results of a cross between wheat plants with red kernels and those with white kernels. The F₁ plants are all intermediate, having pink kernels, as expected with incomplete dominance. However, instead of just three categories in the F₂, there are five different phenotypes because there are two extra classes, light red and light pink, intermediate between the parents and the F₁. The explanation for the ratios of color phenotypes seen in the F₂ (1:4:6:4:1) is that two genes equally affect this trait, so that red occurs with four

red alleles (indicated by the subscript 1) present (A_1A_1 B_1B_1); light red occurs with three red alleles present at either the A or the B gene (A_1A_1 B_1B_2 and A_1A_2 B_1B_1); and so on. In addition, a third gene affects wheat kernel color. All three genes influence color independently of each other; that is, the phenotype is the sum of the individual effects of the different genes.

To illustrate in general the polygenic nature of quantitative traits, let us assume that we can visualize the influence of multiple genes affecting a trait by comparing the phenotypic distribution that results when one, two, or four genes, each with two alleles, determine the phenotypic value of a trait (fig. 3.30). For simplicity, we assume that heterozygotes are exactly intermediate, that the two alleles at each locus are equal in frequency, and that the environment has no effect on the trait. The frequency of the polygenic genotypes is calculated as the product of the frequencies of the single-locus genotypes. In fact, we have already

Number of individuals	1	0	0	1	5	7	7	22	25	26	27	17	11	17	4	4	1
Height in inches	58	59	60	61	62	63	64	65	66	67	68	69	70	71	72	73	74

FIGURE 3.31

Frequency distribution of heights of male students at Connecticut Agricultural College.

Library of Congress.

shown examples of one-gene and two-gene cases in figures 3.1 and 3.29, respectively. From figure 3.30, it is obvious that as the number of genes affecting the trait increases from one to four, the distribution changes from one of discrete classes to one in which there is no discontinuity. In fact, the exact genotype-phenotype correspondence for all classes disappears when there are only two genes. That is, when two genes affect a trait, the second lowest phenotypic value can result from either genotype $A_1A_1\ B_1B_2$ or genotype $A_1A_2\ B_1B_1$.

Most quantitative traits are influenced by the environment. For example, it is obvious that adult size in most organisms is affected by the amount of available resources. Thus, food availability will determine size in an inbred line of mice, just as light and soil differences will result in phenotypic variation in height among the clones of a fruit tree.

Typically, quantitative traits such as size and weight have a continuous distribution in a population, given that the method of measurement is sufficiently accurate. The distribution of such traits often closely approximates a bell-shaped, or normal, distribution, with a high proportion of the individuals having an intermediate phenotypic value. Both the polygenic character of such traits and the influence of the environment help determine the shape of such a distribution. A classic example of a normal distribution curve is shown in figure 3.31, depicting a group of 175 male students arranged by height. In this case, the students were put into discrete groups for every inch increment in height for ease of measurement even though a trait such as height is really continuous.

MEAN AND VARIANCE

How can we evaluate the phenotypic variation for height in the group of students (or the phenotypic variation in any organism)? To assess the attributes of populations, it is necessary to use some descriptive statistics. Here we will briefly discuss how to calculate the **mean** ($\overline{x}$), a measure of the average of a group of values, and the **variance,** a measure of the variability around that central value.

Let us consider the heights of a sample of n different plants. If we symbolize the height of plant i as x_i, we can calculate the mean plant height ($\overline{x}$) as:

$$\overline{x} = \frac{1}{n} \sum_{i=1}^{n} x_i$$

where Σ (Greek letter sigma) indicates the summation of all n plant heights in the sample. For example, if there are five plants in the sample and their heights in centimeters are $x_1 = 28$, $x_2 = 30$, $x_3 = 27$, $x_4 = 32$, and $x_5 = 33$, then:

$$\overline{x} = \frac{1}{5}(x_1 + x_2 + x_3 + x_4 + x_5)$$
$$= \frac{1}{5}(28 + 30 + 27 + 32 + 33)$$
$$= 30$$

In other words, the mean plant height in this sample of individuals is 30 centimeters.

Where a sample of individuals has a given mean, different amounts of dispersion, or variation, may exist around that mean. Thus, there may be no dispersion if all plants are exactly the same height, or there may be an extreme amount of dispersion if there are equal numbers of small and large individuals. The variance (V_x) is a measure of the dispersion around the mean and is calculated as:

$$V_x = \frac{1}{n-1} \sum_{i=1}^{n} (x_i - \overline{x})^2$$

In other words, the variance of x is the sum of the squared deviations of individual values from the mean, divided by n − 1. The variance for the data given here is:

$$V_x = \frac{1}{4}[(x_1 - \overline{x})^2 + (x_2 - \overline{x})^2 + (x_3 - \overline{x})^2 + (x_4 - \overline{x})^2 + (x_5 - \overline{x})^2]$$
$$= \frac{1}{4}[(28 - 30)^2 + (30 - 30)^2 + (27 - 30)^2 + (32 - 30)^2 + (33 - 30)^2]$$
$$= 6.5$$

For the group of students pictured in figure 3.31, the mean height is 67.3 inches and the variance is 7.3.

The variance is not on the same scale as the mean because the variance is a squared value. Thus, the square root of the variance, or the **standard deviation,** is often used to measure the dispersion around the mean. For example, the standard deviation for the sample of students is:

$$(V)^{\frac{1}{2}} = (7.3)^{\frac{1}{2}}$$
$$= 2.7$$

(The exponent $\frac{1}{2}$ indicates the square root.) If the distribution of observations is bell-shaped, or normal, approximately 95% of the sample should be within two standard deviations of the mean. In this case, two standard deviations is 5.4; therefore, $67.3 - 5.4 = 61.9$ is two standard deviations less than the mean, and $67.3 + 5.4 = 72.7$ is two standard deviations greater. Looking at figure 3.31, we can see that a very high proportion of the students are within these limits. In fact, 168 out of 175 students, or 96%, are within these limits. Only two are below 61.9 inches in height and only five are above 72.7 inches.

However, some polygenic traits have only discrete classes, such as the number of bird eggs in a clutch, the number of scales on a reptile, the number of petals or leaves on a plant, or the number of vibrissae (whiskers) on the face of a mouse. Such discontinuous traits, sometimes called **countable,** or **meristic, traits,** are also assumed to be controlled by many genes and affected by environmental factors. Even more extreme, there may be only two phenotypic categories for a discontinuous trait, such as the presence/absence situation that occurs for certain disease states (diseased/not diseased) or for reproductive states (fertile/sterile). In such cases, the presence of one state or the other is the result of an underlying genetic determination of liability or susceptibility that has a continuous distribution.

The presence of a genetic disease in a particular individual is determined by a combination of the individual's underlying genetic liability and his or her particular environment. Figure 3.32 shows the hypothetical threshold in a normal environment for a disease such as diabetes. Here, a relatively small percentage of the population exceeds the threshold, and thus exhibits diabetes (red region). Given a different environment, in which obesity, alcoholism, or other factors increase the likelihood of diabetes, the threshold is moved to the left (indicated as the stress threshold). This shift results in a higher incidence of the disease, because some individuals, represented by the pink region, would not contract the disease in a normal environment but become afflicted under stress conditions.

Hip dysplasia in dogs, a condition that eventually leads to degenerative joint disease, appears to be determined by a combination of underlying genetic liability (that is, susceptibility) and environmental factors. This trait occurs frequently in many large dog breeds, such as German shepherds and Labrador retrievers, despite substantial efforts to eliminate it. The lack of simple

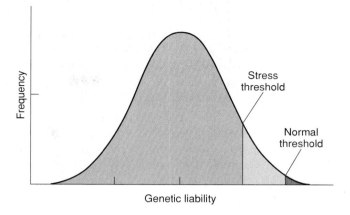

FIGURE 3.32

The genetic liability for a genetic disease trait, with hypothetical thresholds that indicate the value above which individuals exhibit the disease.

inheritance as well as ambiguity in diagnosis (some dogs considered dysplastic by radiographs are clinically unaffected) has made it difficult to reduce the incidence of this condition.

It is important to realize that the genetic determination of quantitative traits may vary considerably among populations. Such a varied response has occurred in populations of insects exposed to various pesticides and may be primarily due to a single gene (or to a number of genes) that confer resistance to a particular insecticide. In addition, the single genes and the method by which they confer resistance may differ among populations. For example, resistance to the insecticide DDT may occur because particular genes influence either detoxification of the insecticide, storage of DDT in fat bodies, or resistance to its diffusion into the circulatory system. (A more detailed discussion of pesticide resistance is given in chapter 17.)

It has been generally believed, and recent molecular genetics evidence has verified, that the genes affecting quantitative traits are essentially like the genes (or are the same genes) that determine single-gene qualitative traits such as color, enzyme, or shape differences. In other words, there is no reason to assume that a different type of gene accounts for quantitative genetic variation; in fact, some genes known for their qualitative effects also affect quantitative traits. Genes affecting quantitative traits follow Mendelian patterns of inheritance, may have multiple alleles, and show various patterns of dominance and epistasis.

 Quantitative traits are affected by many genes as well as by the environment. Polygenic traits generally have a continuous phenotypic distribution in a population, but they may occur in discrete classes, such as diseased or not diseased.

MODEL OF QUANTITATIVE TRAITS

In order to understand and examine the importance of quantitative traits, we need to construct a model that will allow us to break down phenotypic values into genetic and environmental components. We can accomplish this in a simple manner by symbolizing the phenotypic value (P) for a genotype (i) in a particular environment (j) as:

$$P_{ij} = G_i + E_j$$

G_i is the genetic contribution of genotype i, and E_j is the deviation resulting from environment j. E_j may be either positive or negative, depending upon the effect of environment j. This simple model illustrates the basic genetic-environment or nature-nurture dichotomy often discussed in relation to human intelligence and other traits.

In many cases, different populations have different mean phenotypic values. It is generally difficult to determine whether such a difference is the result of genetic factors, environmental factors, or a combination of both. However, if the populations can be grown in the same environment, called a **common garden experiment** for plants, we can make some inferences. For example, if individuals from different populations retain phenotypic differences in a common environment, this is consistent with the hypothesis that the populations are genetically different. But if the phenotypic differences disappear in the common environment, this indicates that environmental factors are important in determining phenotypic differences. For the *Potentilla* plants shown in figure 3.26, both genetic and environmental factors were important.

A particular genotype may do well in a specific environment. If such specific interactions occur between genotypes and environments, the basic model just presented can be expanded to include a term for **genotype-environment interaction,** with the phenotypic value becoming:

$$P_{ij} = G_i + E_j + GE_{ij}$$

where GE_{ij} measures the interaction between genotype i and environment j. As with the environmental deviation, the genotype-environment interaction may be either positive or negative.

Maze-learning ability in the rat experiment (fig. 3.28) is an example of genotype-environment interaction. To further illustrate the complexity introduced by genotype-environment interaction, consider the simple situation of two genotypes in two environments. It is possible that a reversal of the order of the phenotypic values of two genotypes may occur in two different environments, as with a domesticated breed and wild progenitor of a particular species, say pigs or sheep, raised either in association with humans or in the natural environment. The wild progenitor will survive and reproduce better in its natural situation, while the domesticated breed will survive and reproduce better in association with humans (fig. 3.33). As a result, the phenotypic value of a trait like survival is primarily the result of the interaction between the specific genotype and the specific environment and cannot be predicted without information about both the genotype and the environment. Only when the

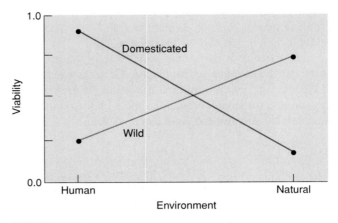

FIGURE 3.33
The viability of domesticated and wild animals in human and natural environments, illustrating the effects of genotype–environment interaction.

norms of reaction are parallel or close to parallel for different genotypes or breeds is there no genotype-environment interaction. If the range of environments examined is limited, then it is probably less likely that genotype-environment interaction will be important.

The genetic model can be more precisely defined by subdividing the genotypic value into two components, with the value of genotype i being:

$$G_i = A_i + D_i$$

where A_i and D_i are the additive and dominant components, respectively. If the heterozygote is exactly intermediate between the homozygotes, the contribution of the dominant component is zero; that is, all of the genetic component is additive. For example, the genetic variation in kernel color in wheat that we discussed earlier is due to the additive component. In other words, the phenotypic value can be symbolized as:

$$P_{ij} = A_i + D_i + E_j$$

assuming that the genotype-environment interaction is negligible.

As pointed out earlier, there is no exact genotype-to-phenotype correspondence for quantitative traits. Therefore, a more practical way to examine quantitative traits is to partition the variation in a population for a particular phenotype into that caused by genetic factors and that caused by environmental factors. If the basic model given above is used to examine the phenotypic variance, then:

$$V_P = V_G + V_E$$

where V_P, V_G, and V_E are the phenotypic, genetic, and environmental variances, respectively.

Adopting the same approach as before, if we use the expression with separate additive and dominant components to examine the phenotypic variance, it becomes:

$$V_P = V_A + V_D + V_E$$

where V_A and V_D are the genetic variances due to additive and dominant effects, respectively. If this expression is divided by V_P, we can obtain the proportion of phenotypic

Separated at birth, the Mallifert twins meet accidentally.
Drawing by Chas. Addams; © 1981 The New Yorker Magazine, Inc.

Table 3.6 Correlations of Four Traits Between Human Twins

	Correlations		Heritability
	Monozygotic Twins (r_M)	*Dizygotic Twins (r_D)*	
Fingerprint-ridge count	0.96	0.47	0.98
Height	0.90	0.57	0.66
IQ score	0.83	0.66	0.34
Social maturity score	0.97	0.89	0.16

variance due to the different components. In particular, the ratio of the additive genetic variance to the phenotypic variance is known as the **heritability** (h^2):

$$h^2 = \frac{V_A}{V_P}$$

The magnitude of the heritability is important in determining the rate and amount of response to directional selection (that is, selecting for an extreme phenotype) and is often estimated by plant and animal breeders prior to initiating a selection program to improve yield or other traits.

 Quantitative traits can be understood by partitioning out the genetic, environmental, and genotype-environment effects. The heritability is the ratio of the additive genetic variance to the phenotypic variance.

ESTIMATING GENETIC VARIANCE AND HERITABILITY

Several different approaches can be used to estimate the amount of genetic variance and heritability. In some organisms, experimental manipulation, or breeding, can either reduce or eliminate the environmental or genetic variance components and make it possible to estimate the complement variance component. In other words, if $V_G = 0$, then $V_P = V_E$, or if $V_E = 0$, then $V_P = V_G$. For example, the genetic variation is zero or near zero in inbred lines, clones, identical twins, and repeated measurements on the same individual. Environmental variance may approach zero if all important environmental factors, such as food, moisture, and temperature, are known and carefully controlled.

Identical (monozygotic) twins have the same genotype; thus any differences between the twins should be the result of environmental factors. In fact, identical twins raised

together share similar environments as well as identical genotypes. One way to determine the extent of similarity is to measure the **correlation** between individuals for different traits. The highest value a correlation can be is 1.0, where all the paired individuals have the same phenotypic value. If the traits are uncorrelated among paired individuals, the correlation is 0.0. The first column of table 3.6 gives the correlation for four different traits in monozygotic twins raised together. The second column gives the correlation for the same traits in dizygotic (nonidentical) twins of the same sex. The difference between these values should be the result of the lower genetic similarity of dizygotic twins, who share (on the average) only half their genes in common. One way to estimate the heritability is from the expression:

$$h^2 = 2(r_M - r_D)$$

where r_M and r_D are the correlations between monozygotic and dizygotic twins, respectively. Using this formula, the heritability for the four traits in table 3.6 varies from 0.98 to 0.16. In other words, most of the phenotypic variation in IQ and social maturity scores appears to be environmental, while most of the variation for fingerprint-ridge count (see fig. 3.36) and weight appears to be genetic.

To change a phenotype, the rate and amount of response to selection depend upon the heritability of the trait under selection. This is quite obvious when h^2 is either 0 or 1. When $h^2 = 0$, none of the phenotypic variation is from the additive genetic variance component; as a result, selecting even extreme individuals as parents would not change the phenotypic mean (fig. 3.34). On the other hand, if $h^2 = 1$, all the phenotypic variation is from the additive genetic variance component, and phenotypically extreme individuals are that way because of their genotypes. The difference in the parental mean and the offspring mean is called the **response (R)** and can be expressed as:

$$R = h^2 S$$

where S is the **selectional differential,** the difference in the mean of the selected parents and the mean of all the individuals in the parental population. This equation may be rearranged so that an estimate of heritability is:

$$h^2 = \frac{R}{S}$$

This expression is also called the **realized heritability** because it is the ratio of observed response over total response possible.

FIGURE 3.34
The change in phenotypic mean for populations with different heritabilities. The shaded area in the parent generation indicates those selected parents. The tic marks indicate mean values of all the parents, the selected parents, and the offspring.

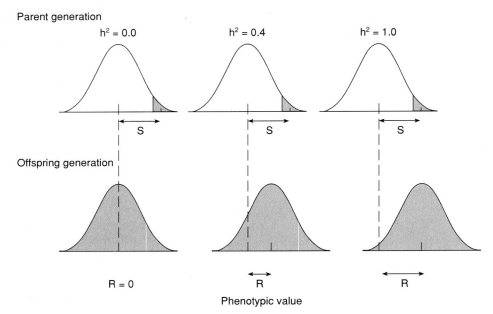

Parent generation

$h^2 = 0.0$ $h^2 = 0.4$ $h^2 = 1.0$

Offspring generation

R = 0   R   R

Phenotypic value

Perhaps the best way to appreciate this effect is to examine the expected change in phenotypic mean from one generation of selection for traits with different heritabilities (fig. 3.34). Notice that even when the selected individuals are quite extreme (blue-shaded region at top of figure), if the heritability is low, the mean of the offspring is similar to that of all the parents (response is low). On the other hand, if the heritability is high, the offspring are much more like the selected parents than the distribution of all parents.

Many directional (or artificial) selection experiments give evidence of a change in phenotypic values over ten to forty generations, often reaching a maximum, or plateau, before the termination of the experiment. Evidence indicates that selection response may continue for many generations, given an adequate initial sample of genetic variation, a reasonable population size during the selection process, and an effective selection scheme. (See chapter 17 for further discussion.)

One of the longest selection experiments ever conducted, for oil and protein percentage in corn, is still running at the University of Illinois. High and low directional selection has been carried out for seventy-six plant generations, and progress is continuing, although large changes in oil and protein percentage have already been made. The data for the high oil and low oil lines are given in figure 3.35. The percentage of oil has increased nearly four-fold in the high line and has been reduced to below 1% in the low line.

Perhaps the most important approach used to measure components of genetic variation and heritability is based on the phenotypic resemblance between relatives. If genetic factors are important, one would expect the offspring of parents with high phenotypic values to have high phenotypic values, and the offspring of parents with low phenotypic values to have low phenotypic values. Obviously, close relatives share more genes with one another than with nonrelatives or distant relatives, and thus should have greater phenotypic resemblance. For example, monozygotic twins share all of their genes; siblings share one-half of

FIGURE 3.35
The percent of high and low oil content in long-term selection experiments in corn.

Reprinted from Proceedings of the International Conference on Quantitative Genetics, *August 16–21, 1976, edited by W. Pollak, O. Kempthorne, and T. B. Bailey, Jr. Copyright © 1977 by Iowa State University Press, Ames, IA.*

their genes (on the average); and first cousins share one-eighth of their genes. As a result, phenotypic values between close relatives should be more highly correlated than those between distant relatives. Using this approach, estimates of heritability for a number of traits in different organisms are given in table 3.7.

Table 3.7 Approximate Values of Heritability of Different Traits in Five Organisms

Organism	Trait	Heritability
Swine	Back fat thickness	0.55
	Body weight	0.3
	Litter size	0.15
Poultry	Egg weight	0.6
	Body weight	0.2
	Viability	0.1
Mice	Tail length	0.6
	Body weight	0.35
	Litter size	0.15
Drosophila melanogaster	Bristle number	0.5
	Body size	0.4
	Egg production	0.2
Corn	Plant height	0.70
	Yield	0.25
	Ear length	0.17

Source: Data from D. S. Falconer, *Introduction to Quantitative Genetics*, Ronald Press, New York.

Fingerprints have been widely used for identification purposes because the fingerprint patterns of an individual are unique. A general way of classifying fingerprints is to count the number of ridges, the raised areas of skin on the fingertips. The three basic configurations of ridges are known as the arch, the loop, and the whorl (fig. 3.36). Fingerprint-ridge count seems to be almost entirely genetic, with heritability estimates near unity (see table 3.6) and with little noticeable influence from the environment.

RACE AND INTELLIGENCE

In the late nineteenth century, some British scientists suggested that the high reproductive rate among the "lower" classes was leading to an overall deterioration of the average intelligence of the British population. The discussion that followed became known as the nature-nurture controversy; that is, given real differences in intellectual ability among social classes, were these differences due to genetic or environmental factors? The controversy was revived again in the late 1960s when it was claimed that the difference in IQ scores among racial groups was genetically determined, and in particular, that lower IQ scores among African-Americans were the result of genetic factors.

Many researchers have suggested that environmental factors have a substantial effect on IQ scores, citing a range of factors that includes nutrition (prenatal, during childhood, and at the time of testing), socioeconomic status of parents, and educational and testing experience. One concern in comparing the IQ scores of blacks and whites is that these groups differ not only ethnically but also in education, upbringing, and other socioeconomic factors. An obvious way to evaluate the effect of socioeconomic factors is to compare blacks and whites within the same socioeconomic category.

Table 3.8 gives the IQ scores of African-American and Caucasian seven-year-olds, some from an upper socioeco-

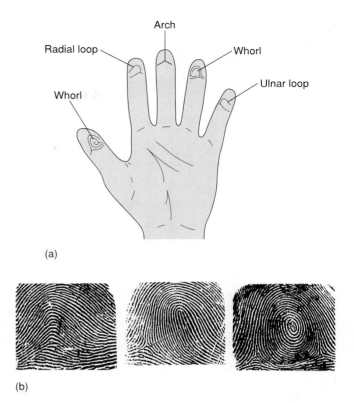

(a)

(b)

FIGURE 3.36

(*a*) The position of fingerprints on a hand and (*b*) examples of the three basic fingerprint types. Fingerprint-ridge count for the loop and whorl is the number of ridges crossed by the drawn lines. Arches have a count of zero.

(*b*) *American Institute of Applied Science, Syracuse, New York.*

Table 3.8 The Effect of Socioeconomic Background on IQ Scores

Socioeconomic Background	African-American	Caucasian
High	100.0 (12.6)[*]	104.2 (13.4)
Low	91.2 (12.1)	95.3 (12.9)

[*]The numbers in parentheses are standard deviations.

nomic background and some from a lower socioeconomic background. Notice first that both groups from the high socioeconomic background score higher than the groups from the low socioeconomic background, suggesting that socioeconomic background has a substantial influence on IQ test scores. Second, the difference between racial groups from the same socioeconomic background is quite small, only 4.2 points in the high group and only 4.1 points in the low group. These differences are in fact much smaller than any of the standard deviations, indicating that the means of the two racial categories are not statistically different within each socioeconomic group. It appears that the differences seen between whites and blacks are probably due in large part to socioeconomic differences between the two groups.

Genetic variance can be estimated by examining identical twins. Realized heritability can be estimated by determining the rate of response from selection. Intelligence is affected by a number of environmental factors; thus any racial differences in IQ are generally confounded with environmental differences.

GENETIC COUNSELING

The importance of genetic disease has increased substantially in the last several decades as the incidence of infectious disease has declined. For example, infectious diseases such as smallpox and diphtheria used to be primary causes of infant mortality, but genetic disease is now a leading factor. Some of these diseases follow a single-gene pattern of inheritance; others are polygenic, with significant environmental components; and still others are the result of aberrant chromosomal numbers or types (see chapter 4).

Table 3.9 Probability of Affected Offspring for an Autosomal Recessive[1]

Parents	Second Child Affected
Both unaffected	¼
One affected	½
Both unaffected; one has affected sib	¼
Both affected	1

[1]Given that one child is already affected.

Table 3.10 Empirical Risks for Congenital Pyloric Stenosis in an English Population

Number of Affected Sibs	Neither Parent Affected	Father Affected	Mother Affected	Both Parents Affected
0	0.003	0.037	0.051	0.298
1	0.030	0.115	0.135	0.365
2	0.119	0.216	0.231	0.431

Source: Data from C. Bonaiti-Pellie and C. Smith, "Risk Tables for Genetic Counseling in Some Common Congenital Malformations" in *J. Med. Genet.* 11:374, 377, 1974.

After a child is born with a genetic disease, the parents may seek advice concerning the probability of a second child having the same disease. Such **genetic counseling** is offered at a number of medical centers. Once the disease is properly diagnosed and its mode of inheritance known, the **recurrence risk,** or probability that the condition will recur in a given family, can be calculated.

A large proportion of known genetic diseases are inherited as autosomal recessives. Many of these are **inborn errors of metabolism**—that is, enzymatic defects that result in the buildup or absence of an important biochemical product. Most of these diseases are quite rare, generally occurring in fewer than 1 in 10,000 individuals. However, because there are a large number of such diseases, their cumulative incidence is substantial.

The probability of a genetic disease occurring in a child depends on whether the parents or other relatives are affected or unaffected, and on whether other offspring have already been affected. Table 3.9 shows the risk of producing affected children for parents with different disease states when a previous child has been affected. These probabilities follow the pattern of inheritance for recessive traits discussed in chapter 2.

Genetic counseling can also be useful when parents strongly wish to have children but there is a possibility that their children will inherit a genetic disease. For example, if both parents are known carriers of a recessive disease allele (heterozygotes can be detected for sickle-cell disease, Tay-Sachs, and a number of other recessive diseases), the probability of a sibship of size N with no affected progeny is $(\frac{3}{4})^N$. With a small family of one or two children, the chances of having no affected offspring are 0.75 and 0.5625, respectively, a risk that some people find acceptable.

A number of diseases, such as diabetes, cleft palate, and spina bifida, have a strong genetic component but do not have a single-gene mode of inheritance. For such diseases, it is assumed that the genetic liability to the disease is polygenically controlled and that, consequently, close relatives of affected individuals are more likely to be affected. Obviously, the closer the affected relative and the more relatives are affected, the higher the genetic liability in an individual.

For such diseases, **empirical recurrence risks** are often calculated from a large number of case studies. These values may be figured where both parents are affected, where one parent is affected, where one sib is affected, or in other situations. Empirical risk values can be used only as a guideline, since specific genetic or environmental factors may either increase or reduce the risk in a particular family.

Pyloric stenosis is a disease in which the pyloric valve, the sphincter valve between the stomach and the small intestine, is not completely opened. Before corrective surgery was introduced in the 1920s, most affected individuals died in infancy because little or no food could pass on to the small intestine. With corrective surgery, individuals born with pyloric stenosis have normal digestion and generally no other related problems. Risk values for this disease are given in table 3.10 with respect to the disease state of the parents and the number of affected sibs. The risk, as intuitively expected, increases as either the number of affected sibs or the number of affected parents increases. For example, with no affected relatives, the risk is only 0.3%, while with two affected sibs and two affected parents, the risk is greater than 40%.

Genetic counseling is used to determine the risks of producing genetically defective infants. Mendelian inheritance probabilities are used to predict single-gene traits, while empirical risk values are used for polygenic diseases.

Table 3.11 Excluded Paternity Types, Given Genotypes of Mother and Child

Mother	Child	Excluded Paternity Types
$L^M L^M$	$L^M L^M$	$L^N L^N$
	$L^M L^N$	$L^M L^M$
$L^M L^N$	$L^M L^M$	$L^N L^N$
	$L^M L^N$	—
	$L^N L^N$	$L^M L^M$
$L^N L^N$	$L^M L^N$	$L^N L^N$
	$L^N L^N$	$L^M L^M$

PATERNITY EXCLUSION

In cases where the father of a child is unknown or disputed, blood groups, enzymes, or HLA types have been used to determine the probability of parentage. In the past, such data were admitted in U.S. courts only when they ruled out an individual as a parent. But in recent years, because of better genetic analysis, courts will often accept information that indicates the probability of parenthood.

When a locus has only a few allelic forms in a population, it is difficult to exclude an individual as a parent, because many individuals have the same alleles, making them all possible parents. As an example, mother-child genotype combinations and the excluded paternal genotypes for the MN blood group locus are listed in table 3.11. Given the genotype of the mother and child, one may exclude certain genotypes as the father. For example, if the genotypes of both mother and child are $L^M L^M$, the father could be either $L^M L^M$ or $L^M L^N$, and only the paternal genotype $L^N L^N$ can be excluded. Notice that only one genotype at most can be excluded and that when both the mother and child are heterozygotes, no paternal genotypes can be excluded.

Even with a large number of loci, the probability of exclusion is not as high as one might wish. The use of HLA **haplotypes** (all the HLA alleles at different genes on a single chromosome) and DNA markers, both of which can have many alleles at a locus, may resolve this difficulty. To illustrate, figure 3.37 gives the HLA haplotypes of a mother, a child, and two possible fathers. For example, the haplotype A1 B17 is on one of the chromosomes in the putative father on the left. (These haplotypes stay intact from generation to generation because the HLA genes are close to each other on the chromosome, that is, they are tightly linked. See chapter 5.) From this information, it is apparent that the mother produced an A1 B8 gamete, and therefore the father produced an A11 B12 gamete. Putative father 1 does not have this haplotype, so he can be excluded. Putative father 2 does, and if there is no other possible male parent, this would indicate that he is the father. If there were other possible fathers, one of those other men could have been the father, so it cannot be stated with complete assurance that putative father 2 is the father.

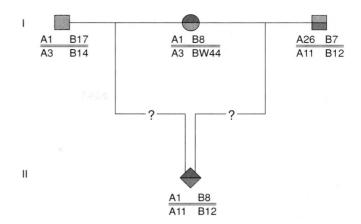

FIGURE 3.37
The HLA haplotypes and genotypes of a mother, child, and two putative fathers.

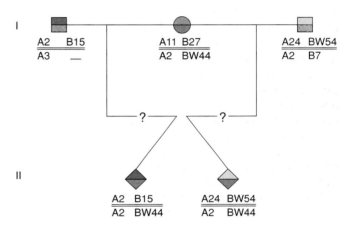

FIGURE 3.38
The HLA haplotypes and genotypes of a mother, dizygotic twins, and the putative fathers of the twins.

In an actual case that proved the usefulness of the HLA system, two different men were identified as the fathers of a set of dizygotic twins. The genotypes of the individuals are given in figure 3.38. The twins both received an A2 BW44 gamete from their mother, but twin 1 received an A2 B15 gamete from putative father 1 and twin 2 received an A24 BW54 gamete from putative father 2. This was not detectable by looking at the ABO blood group system because all of the individuals were type A.

Recently, the identification of a number of loci, each with many alleles, has allowed the development of **DNA fingerprints.** The principles involved are the same as discussed before, except that nearly every individual has a different DNA pattern for these genes. We will discuss some details of the molecular basis of this procedure in chapter 15.

 Paternity exclusion based on genetic variants is used when the father of a given child is unknown. Generally, blood group or HLA alleles are used, but recently, DNA fingerprints have improved the power of this technique.

Box 3.1 Grandpaternity Exclusion

Genetic information has long been used to attempt to identify parents. Recently, the same approach has been extended to grandparents and other relatives. A major impetus for this work was the disappearance of at least nine thousand Argentinians between 1975 and 1983. Among the missing were a number of children who had been abducted by the military and police or who had been born to kidnapped women in captivity. For example, Liliana Pereyra (figure B3.1.1) was abducted when she was five months pregnant and kept alive in a torture center until she gave birth to a boy in February 1978. Her son has not been found and may have been kidnapped by the military.

It appears that many of the Argentinian children are now living with military couples who claim to be their biological parents but may actually be the kidnappers. If all four

grandparents are alive, and if the child has an allele that none of them have, grandpaternity can be excluded. But if the alleles in the child are also present in the grandparents, the probability of grandpaternity can be calculated. For example, if the shared alleles are rare in the population, the probability of grandpaternity is quite high.

Figure B3.1.2 is a pedigree in which a child's HLA types were found in the putative grandparents. In this case, Paula Eva Logares, an Argentinian kidnap victim, was raised by a policeman and his wife. One of her HLA haplotypes, A2 B5, was present in her paternal grandfather. The reconstructed genotype of her maternal grandparent (from his other children) contains the other haplotype, A1 B5. It is very likely that these haplotypes were passed on to her from her deceased parents. These data and probability principles make it 99.9% certain that Paula is the grandchild of the Logares and Grinspon families.

FIGURE B3.1.1
Liliana Pereyra, an Argentinian kidnap victim.

From Dr. Jared M. Diamond. Reprinted by permission from Nature *327:552. Copyright © 1987 Macmillan Magazines Limited.*

FIGURE B3.1.2
A pedigree illustrating how probable grandparenthood can be determined if the HLA haplotypes of some of the individuals involved are known. Slashes indicate deceased individuals.

WORKED-OUT PROBLEMS

Problem 1

A woman with blood type A and a man with blood type B had children that are blood types A, B, AB, and O. What are the genotypes of the parents?

Solution

Without any knowledge about the offspring, the mother could be either $I^A I^A$ or $I^A I^O$, and the father could be either $I^B I^B$ or $I^B I^O$. There are then four possible mating types, with the following possible progeny phenotypes:

Mother	Father	Offspring
$I^A I^A$	$I^B I^B$	All AB
$I^A I^A$	$I^B I^O$	½AB, ½A
$I^A I^O$	$I^B I^B$	½AB, ½B
$I^A I^O$	$I^B I^O$	¼AB, ¼A, ¼B, ¼O

In other words, only if the mother is $I^A I^O$ and the father $I^B I^O$ can they have progeny of all four types.

Problem 2

Figure 3.21 gives the pattern of inheritance that might be expected for a disease that is caused by two different genes, as for example several types of deaf-mutism in humans. Assume that individuals II-3 (who is deaf) and I-3 in this pedigree have a child. What are the possible genotypes from this mating and the probability that the child will be deaf?

Although individuals III-1 and III-2 in this pedigree have normal hearing, their parents were both deaf. What proportion of the progeny of a mating between two individuals with the genotypes of these individuals would be expected to have normal hearing?

Solution

In the mating between II-3 and I-3, one would expect 0.5 of the progeny to be *Aa BB* and 0.5 to be *Aa Bb*. None of these children would be deaf.

There are two ways to calculate this proportion. First, in this mating between individuals of genotypes *Aa Bb* and *Aa Bb*, it is expected that ³⁄₁₆ would be *A– bb,* ³⁄₁₆ would be *aa B–,* and ¹⁄₁₆ would be *aa bb,* all of which would be deaf. Overall then, 1 – ⁷⁄₁₆ or ⁹⁄₁₆ of the progeny would be expected to have normal hearing. Another way to calculate this is to determine that ⁹⁄₁₆ would be expected to be *A– B–,* all of which have normal hearing.

Problem 3

For many traits, the genetic contribution is determined by many genes, and the only way to characterize them is to estimate the proportion of phenotypic variation that is genetic. If such a quantitative trait, say yield in corn, is selected so that the response is 1.4 grams per plant from a selection differential of 3.1 grams per plant, what is the estimated heritability? Do you think that subsequent selection in future generations would be successful in further changing the yield in corn?

Solution

The heritability is $h^2 = \dfrac{R}{S} = \dfrac{1.4}{3.1} = 0.45$. Because nearly half the phenotypic variation for this trait is genetic, one would expect that future generations of selection would significantly increase the yield further.

Summary

Dominance can occur in different forms: (1) complete, as found by Mendel, where either the mutant or the wild-type allele is dominant; (2) incomplete, with the heterozygote having a phenotype in between the two homozygotes; or (3) codominant, in which both alleles are expressed in the heterozygote.

Many variations on Mendel's original findings have been discovered. Some genes may be lethal. Pleiotropy occurs when one gene affects several traits. Some genotypes may not always have the same phenotype, a phenomenon resulting from incomplete penetrance or variable expressivity of the genotype. Genes may have more than two alleles. In diploid organisms, many different genotypes can exist when there are multiple alleles. Genes on the X and Y chromosomes have different patterns of inheritance than do autosomal genes. X-linked recessive traits occur almost entirely in males and generally skip generations in pedigrees.

More than one gene can affect a single trait. When two genes interact to determine a phenotype, this phenomenon is called *epistasis*. Epistatic interaction can result in F_2 ratios different from 9:3:3:1. The relationship of the genotype to the phenotype may be complicated. The norm of reaction over different environments is one way to quantify this relationship. Quantitative traits are affected by many genes and are influenced by the environment. Polygenic traits may occur in a continuous phenotypic distribution in a population or in discrete classes, such as diseased or not diseased.

Quantitative traits can be understood by partitioning out the genetic, environmental, and genotype-environment effects. Heritability is the ratio of the additive genetic variance to the phenotypic variance. Genetic variance can be estimated by examining identical twins. Realized heritability can be estimated by determining the rate of response from selection. Intelligence is affected by a number of environmental factors, and any racial differences in IQ are generally confounded with environmental differences.

Genetic counseling is employed to determine the risks of producing genetically defective infants. The probabilities from Mendelian inheritance are used for single-gene traits, while empirical risk values are used for quantitative traits. Paternity exclusion, based on genetic variants, is applied when the father of a given child is unknown. Generally, blood group or HLA alleles are used, but recently, DNA fingerprints have improved the power of this technique.

PROBLEMS AND QUESTIONS **?**

1. The seven traits that Mendel used in his experiments were all completely dominant. Although this was crucial in his ability to learn about particulate inheritance, we now know that most traits are not completely dominant. What other kinds of dominance are there? Give examples of each.

2. What are penetrance and expressivity? Give examples and discuss them to show that you understand the difference between these concepts.

3. (*a*) If a gene has three alleles, how many genotypes are possible? (*b*) If another gene has five alleles, how many homozygotes and how many heterozygotes are possible?

4. (*a*) What are the differences between sex-linked, sex-limited, and sex-influenced traits? (*b*) Describe how to determine whether a trait is sex-linked or sex-limited.

5. If in his experiments Mendel had used traits that were polygenically determined, what outcome do you think he would have observed in his F_2 progeny?

6. One experiment to elucidate the inheritance of flower color in four-o'clocks crossed two plants with pink flowers. In the progeny from this cross, there were 42 plants with red flowers, 86 with pink flowers, and 39 with white flowers. Using a chi-square test, determine whether those numbers are consistent with Mendelian expectations.

7. A series of matings between individuals where one parent was blood group phenotype N and the other parent was phenotype MN yielded 84 type MN progeny and 91 type N progeny. Is this consistent with expectations? (Use chi-square to test.)

8. Write out all the possible mating types for the ABO blood group system. (Given four phenotypes, there are ten different mating types if it is assumed that reciprocal matings are equivalent.) Specifying genotypes, write out the four mating types designated phenotypically as A × B and the expected proportion of progeny types from each.

9. Trying to understand the inheritance of the dominant yellow gene in mice, researchers mated two yellow heterozygous mice. A typical result was 56 yellow progeny to 31 wild-type. Use a chi-square test to determine if this is consistent with 3:1 Mendelian predictions for a dominant gene. Are these data consistent statistically with the lethal model of Cuénot?

10. (*a*) If two Manx cats are mated (all Manx are genotype *Mm*), what proportion of the progeny should be tailless? (*b*) If a Manx male cat mates with a female cat with a normal tail, what proportion of the progeny should have a tail?

11. Some researchers feel that most genes are pleiotropic for many traits. If you discovered a new genetic disease, how would you go about documenting its pleiotropic effects?

12. Draw a pedigree for a trait such as Huntington disease, illustrating variable expressivity. Assuming *H* and *h* are the dominant and recessive alleles, respectively, give the genotypes and phenotypes for three generations (one set of grandparents in the first generation, at least three progeny, and at least five grandchildren).

13. (*a*) Describe how you would document the importance of environmental factors, such as temperature and diet, on the expression of a recessive mutation in *Drosophila*. (*b*) How would you document the effect of modifier genes on the expression of the same mutation?

14. Which blood group type of the ABO system is known as the universal donor? The universal recipient? Explain your answers.

15. A female *Biston betularia* moth having the typical phenotype is mated to a male melanic. If half the progeny are melanic and half are insularia, what are the genotypes of the two parents?

16. (*a*) Given that two plants with genotypes B_2B_4 and B_1B_5 are mated, what types of progeny (and in what proportions) would you expect? (*b*) For the same gene, if a progeny array from a single mating has equal numbers of B_1B_2 and B_2B_4 individuals (and no other progeny), what are the parents' genotypes?

17. A woman who is blood type A and a man who is blood type B have a child who is blood type O. What are the genotypes of the three individuals? What is the probability that a second child would again be blood type O? Blood type AB?

18. If a color-blind man and a woman with normal color vision have a female child who is color-blind, what are the genotypes of the parents and the child? What proportion of male children from such a mating would be expected to have normal vision?

19. The pedigree in figure 3.17 shows the pattern of inheritance for an X-linked dominant gene. If individuals III-3 and II-6 in that pedigree mate and have children, what would be the proportion of individuals in the four classes: male, unaffected; male, affected; female, unaffected; and female, affected?

20. (*a*) If female *Drosophila* homozygous for the white allele (*ww*) affecting eye color are mated to red-eyed males (w^+Y), what types of F_1 progeny could you expect? (Give both phenotype and sex.) (*b*) F_1 males and females were mated, and in the F_2 progeny there were 40 white-eyed females, 42 red-eyed females,

39 white-eyed males, and 44 red-eyed males. Using a chi-square test, determine whether these results are consistent with the expectations for X-linked genes.

21. Using the forked-line approach, calculate the proportions of the five phenotypes in a cross between two wheat plants with pink kernels and genotype $A_1A_2B_1B_2$. (See figure 3.29 for information on this cross and another way of calculating the proportions.)

22. Assume there is a third gene affecting color in wheat so that the F_1 plants are $A_1A_2B_1B_2C_1C_2$. (*a*) What are expected proportions of the F_2 genotypes using the forked-line approach? (*b*) Assume that gene *C* contributes equally to the other genes for color and that the color of the F_2 can be divided into seven categories: 0 (no "2" alleles, $A_1A_1B_1B_1C_1C_1$); 1 (one "2" allele, $A_1A_2B_1B_1C_1C_1$ or $A_1A_1B_1B_1C_1C_2$); on through to 6 (no "1" alleles, $A_2A_2B_2B_2C_2C_2$). Draw a graph showing the seven phenotypic categories equally spaced on the horizontal axis and their frequencies as obtained from the F_2 progeny on the vertical axis. (*c*) How does this compare to the distribution expected if only one gene or two genes affect color?

23. What is the difference between dominance and epistasis? Why didn't any pairs of genes used by Mendel show epistasis?

24. Bateson and Punnett observed 382 purple-flowered and 269 white-flowered sweet peas in an F_2. Are these numbers consistent with the 9:7 ratio they predicted, using a chi-square test?

25. If the purple F_1 in question 24 were crossed to one of the parental varieties, what proportions of white-flowered progeny would you expect?

26. Assume a mouse from a true-breeding line of albino mice with genotype *BBcc* was crossed to a mouse from a true-breeding brown line with genotype *bbCC*. (*a*) What are the genotype and phenotype of the F_1 mice? (*b*) Assume two F_1 mice were crossed, what are the proportions of the different phenotypes in the F_2 progeny and the genotypes in each color class?

27. Two types of epistasis discussed were duplicate and complementary gene action. How would you distinguish between these types of epistasis in determining purple and white flower color if you could set up any crosses necessary?

28. Graph the effect of the environmental temperature on the phenotype for plant height when the environment affects two genotypes, a short and a tall, equally. Assume that as temperature increases, plant height increases. Next, assume that at high temperature, the tall genotype plant is inhibited in growth so that it has a lower height than the short genotype. Draw the norms of reaction of the two genotypes in this case.

29. Compare the terms *polygenic trait* and *quantitative trait*. Are traits that have only two phenotypes, such as diseased or healthy, due to single genes or are they polygenic? How would you experimentally determine the genetic basis of such a trait?

30. The heritability for plant height in corn is 0.7. Do you expect the rate of response to artificial selection to be fast or slow? Explain your answer.

31. In figure 3.21, what is the probability that a child from an incestuous mating between III-1 and III-2 would be affected with the recessive disease? If individual II-1 has children with II-5, an unrelated individual, what is the probability that their first child would be affected?

32. A woman with blood type M had a child with blood type MN. She said that a certain man who had blood type MN was the father and he claimed he was not. (*a*) With this information, can you exclude this man as the father? (*b*) Further investigation showed that the woman was blood type B, the child blood type O, and the man blood type AB. Can the man be excluded as the father with this information? Explain your answer.

Answers appear at end of book.

CHAPTER 4: CHROMOSOMES AND HEREDITY

I may finally call attention to the probability that the association of paternal and maternal chromosomes in pairs and their subsequent separation during the reducing division . . . constitute the physical basis of the Mendelian law of heredity.

Walter Sutton
American geneticist

Whole chromosome paints. A new technology allows chromosome abnormalities associated with many diseases to be detected more rapidly and accurately.
Dept. of Energy/Photo Researchers, Inc.

LEARNING OBJECTIVES

In this chapter you will learn:

1. How different chromosomes are characterized.
2. How chromosome number is maintained during the process of mitosis.
3. How chromosome number is halved during the process of meiosis.
4. That genes and chromosomes exhibit parallel behavior.
5. The four types of changes in chromosomal structure: duplication, deletion, inversion, and translocation.
6. The two ways in which chromosome number may vary: polyploidy and aneuploidy.
7. That sex determination is generally caused by the chromosomes, although the mechanism varies with the species.

lthough chromosomes had been noticed as cell structures in the nineteenth century, it was only after the rediscovery of Mendel's principles that their fundamental role in heredity was established. In 1902, Walter Sutton in the United States and Theodor Boveri in Germany both suggested that genes might be on chromosomes. They based their hypotheses on the analogous behavior of genes and chromosomes. In fact, the behavior of chromosomes in meiosis, as we will see later, provides a mechanism for the Mendelian principles of segregation and independent assortment.

SEXUAL REPRODUCTION

Before we describe the behavior of chromosomes during cell division, let us briefly discuss sexual reproduction and the overall structure of chromosomes. Reproduction in eukaryotes in general is either asexual or sexual. **Asexual reproduction** occurs when a single individual produces new individuals identical to itself. This is a common mode of reproduction in plants and simple animals. Asexual reproduction using cuttings is the common method of propagating many horticultural plants. **Parthenogenetic** female organisms (such as some aphids) can also produce offspring without fertilization. All progeny produced asexually are genetically identical to their parent (barring a rare mutation).

Sexual reproduction occurs when individuals produce male and female sex cells, or gametes, that in turn unite to form a **zygote,** a single cell from which a new individual develops. Sexual reproduction occurs in nearly all types of organisms, including most simple animals (fig. 4.1), plants, and even bacteria (see chapter 13). Usually the male and female gametes come from different individuals, ensuring that the progeny are different from either parent. An important exception occurs in self-fertilization, as in Mendel's garden peas, where the same individual produces both male and female gametes. In the following discussion, we will concentrate on sexual reproduction and its genetic consequences.

The general scheme of reproduction and growth in a sexual organism is shown in figure 4.2 (see also panels 3 and 4). Let us examine two zygotes, one of which will mature into a female and the other into a male. Each zygote has received a complement of chromosomes of number n in the gamete from each parent, making the total number of chromosomes in each zygote 2n. The chromosomal number in the gamete is called the **haploid** (n) number, and the chromosomal number in the zygote is called the **diploid** (2n) number. The n different chromosomes within each haploid set each have different genes. Each haploid set has one copy of each type of chromosome. The two chromosomes of a particular type present in a diploid set are called **homologous chromosomes.**

The zygote increases in cell number by the process of **mitosis** and cell division. In this process, the chromosomes are replicated so that they remain at the 2n number in the daughter cells. This type of cell division continues until

FIGURE 4.1
Sexual reproduction in barnacles, as shown by two adjacent individuals mating.
© *Heather Angel/Biofotos.*

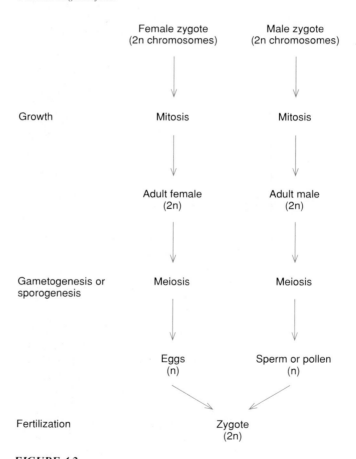

FIGURE 4.2
Outline of growth and reproduction in a sexual organism, indicating the number of chromosomes, diploid (2n) or haploid (n), at various stages.

Table 4.1 Some Organisms and Their Diploid Chromosome Numbers

Common Name (Scientific Name)	Diploid Chromosome Number
Mammals	
Human (*Homo sapiens*)	46
Chimpanzee (*Pan troglodytes*)	48
Dog (*Canis familiaris*)	78
Cat (*Felis catus*)	38
Horse (*Equus caballus*)	64
Donkey (*Equus asinus*)	62
House mouse (*Mus domesticus*)	40
Other Animals	
Chicken (*Gallus domesticus*)	~78
Leopard frog (*Rana pipiens*)	26
Carp (*Cyprinus carpio*)	104
Fruit fly (*Drosophila melanogaster*)	8
Plants	
Corn (*Zea mays*)	20
Tobacco (*Nicotiana tabacum*)	48
Pine (*Pinus* species)	24
Garden pea (*Pisum sativum*)	14
Fungi	
Baker's yeast (*Saccharomyces cerevisiae*)	17*
Black bread mold (*Aspergillus nidulans*)	8*
Pink bread mold (*Neurospora crassa*)	7*

*Haploid number

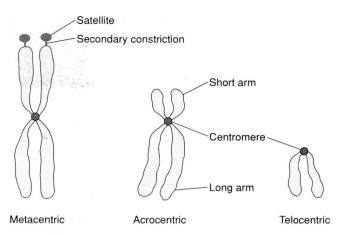

FIGURE 4.3

The three major types of chromosomes as they appear in human karyotypes: metacentric, acrocentric, and telocentric. Satellites may be present on any of the chromosome types.

These chromosomes are generally thought to be devoid of important genes, although in some instances supernumerary chromosomes affect fertility. Other species, including many birds, have a number of very small chromosomes called **microchromosomes.** Although microchromosomes are hard to count exactly because of their number and small size, they are thought to carry genes and to be constant in number, just as the larger chromosomes are.

CHROMOSOMAL MORPHOLOGY

In order to further understand chromosomes and their function, we need to be able to discriminate among different chromosomes. First, chromosomes differ greatly in size. Between organisms the size difference can be over 100-fold, while within a species some chromosomes are often ten times as large as others. In a species **karyotype,** a pictorial or photographic representation of all the different chromosomes in an individual, chromosomes are usually ordered by size and numbered from largest to smallest.

Second, chromosomes may differ in the position of the **centromere,** the place on the chromosome where spindle fibers are attached during cell division. The centromere is identified as a large constriction where the chromosome appears to be pinched. (The term **kinetochore** is used to designate a structure within the centromeric region.) In general, if the centromere is near the middle, the chromosome is **metacentric;** if the centromere is toward one end, the chromosome is **acrocentric** (or **submetacentric**); and if the centromere is very near the end, the chromosome is **telocentric** (fig. 4.3). The centromere divides the chromosome into two **arms,** so that, for example, an acrocentric chromosome has one short and one long arm, while a metacentric chromosome has arms of equal length. All house mouse chromosomes are telocentric, while human chromosomes include both metacentrics and acrocentrics, but no telocentrics. Table 4.2 describes the karyotypes of eight common domestic animals to illustrate the variation in gross chromosomal morphology that occurs among these

reproductive maturity, at which time some cells undergo **meiosis.** Meiosis is a second type of cell division that produces female and male gametes, each having a chromosomal number of n. This process is known as **gametogenesis** in animals and as **sporogenesis** in plants. The female gametes (eggs) and the male gametes (sperm in animals or pollen in plants) then unite to form progeny zygotes by the process of **fertilization.** These new zygotes mature into either males or females, and the whole cycle starts over.

Each species has a characteristic number of chromosomes; that is, virtually all individuals of a given species have the same chromosome number. Generally, this number is expressed as the diploid, or 2n, number. Table 4.1 lists a number of organisms used for genetic research, including corn, with a diploid number of 20 chromosomes, and pink bread mold, with a haploid number of 7 (for this organism, the haploid phase is the major one). The fruit fly has a 2n number of only 8, while carp (and some ferns) have over 100 chromosomes. Notice that closely related species, such as humans and chimpanzees or horses and donkeys, have similar chromosome numbers. In Hymenoptera (bees and wasps), the females are diploid (2n) and the males are haploid (n). The term **haplodiploid** is used to describe such organisms. Certain species have some very small chromosomes, called **supernumerary,** or **B, chromosomes,** that can vary in number among individuals.

Table 4.2 Karyotypes of Eight Common Domestic Mammals

		Autosomal Pairs		Sex Chromosomes	
	Diploid number (2n)	*Number of metacentrics*	*Number of acrocentrics or telocentrics*	*X*	*Y*
Cat (*Felis catus*)	38	16	2	M	M
Dog (*Canis familiaris*)	78	0	38	M	A
Pig (*Sus scrofa*)	38	12	6	M	M
Goat (*Capra bircus*)	60	0	29	A	M
Sheep (*Ovis aries*)	54	3	23	A	M
Cattle (*Bos taurus*)	60	0	29	M	M
Horse (*Equus caballus*)	64	13	18	M	A
Donkey (*Equus asinus*)	62	24	6	M	A

M = metacentric
A = acrocentric
Source: Data from Ohno, 1968, and Eldridge & Blazak, 1976, as appeared in F. W. Nicholas, *Veterinary Genetics.*

species. In addition to a centromere, there may also be a **secondary constriction** on particular chromosomes. The part of the chromosome located distal to a secondary constriction is called a chromosomal **satellite.** The secondary constrictions usually contain genes that help organize or form the **nucleolus,** a structure that is apparent in various stages of cell division. The ends of a chromosome are called **telomeres** and generally appear to have DNA sequences different from the rest of the chromosome.

Third, chromosomes may be identified by regions that stain in a particular manner when treated with various chemicals. In fact, the term *chromosome* literally means "colored body." Several different chemical techniques are used to identify certain chromosomal regions by staining them so that they form chromosomal **bands.** For example, darker bands are generally found near the centromeres or on the ends (telomeres) of the chromosomes, while other regions do not stain as strongly. The positions of the dark-staining, or **heterochromatic,** regions (also called *heterochromatin*) and the light-staining, or **euchromatic,** regions (called *euchromatin*) generally remain constant in different cells or individuals of a given species and are therefore useful in identifying particular chromosomes.

Euchromatic regions often undergo a regular cycle of contraction and extension. The two types of heterochromatin are **constitutive** and **facultative.** Constitutive heterochromatin is a permanent part of the genome and is not convertible to euchromatin. On the other hand, facultative heterochromatin consists of euchromatin that takes on the staining and compactness characteristics of heterochromatin during some phase of development. Evidence suggests that the constitutive heterochromatin is genetically inactive and that the euchromatin contains most of the genes. We should note that the G, Q, and R bands (Giemsa, quinacrine, and reverse bands, respectively), which are chromosomal bands that are observed

using different chemical techniques, are not associated with centromeric heterochromatin.

In humans, a system is used for identifying chromosomes based on chromosomal size, position of the centromere, and banding patterns (fig. 4.4). First, the nonsex chromosomes (autosomes) are numbered 1 to 22 on the basis of length, with the X and Y chromosomes identified separately. Second, the short arm of a chromosome is indicated by p and the long arm by q. Finally, each arm is first divided into regions and then identified by bands. For example, 9q34 indicates the long arm of chromosome 9, region 3, and band 4. This happens to be the location of the gene for ABO blood group types.

In 1956, Joe-Hin Tjio and Albert Levan determined the correct diploid number of human chromosomes (46). (The number had been erroneously thought to be 48 for over thirty years.) Since then, knowledge of human cytogenetics has increased remarkably. These advances resulted mainly from the development of new techniques—ways to culture dividing cells, to swell cells using low osmotic solutions, and to break open cells and spread out the chromosomes. Various new staining techniques have allowed each chromosome, and even small parts of each chromosome, to be identified. Using such biochemical techniques, it is now possible to locate particular genes to certain positions on chromosomes.

Figure 4.5 gives a schematic representation of the X chromosome in *Drosophila melanogaster,* an organism in which a detailed banding pattern is known. The chromosomes in the salivary glands of *Drosophila* and other dipteran insects are extremely large, thus easier to study than other chromosomes. These giant chromosomes, called **polytene chromosomes,** are extended over one hundred times the length of regular chromosomes and include many parallel replicates of the same chromosome held tightly together. As a result, more than five thousand bands have

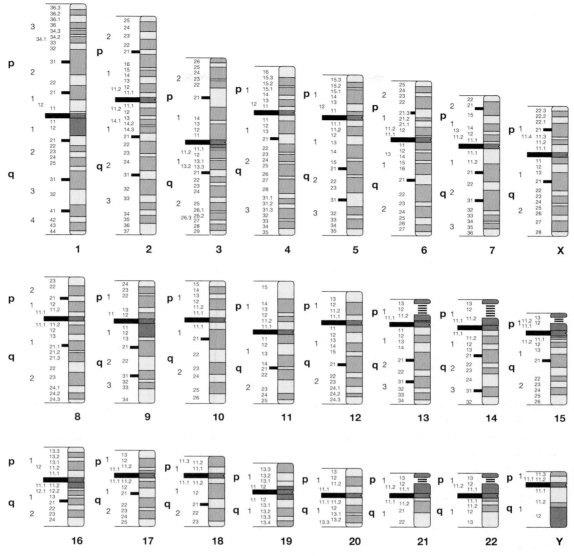

FIGURE 4.4

A schematic representation of the 24 different human chromosomes, in which the heavy black bar indicates the centromeric region, and p and q indicate the short and long arms, respectively. The numbers alongside the chromosomes indicate the regions in each arm. Heterochromatic regions are shown in blue (mainly around centromeres and in satellites), while yellow and green indicate other banded or nonbanded regions.

been identified in *D. melanogaster,* and more than one thousand of them are on the X chromosome alone. Notice in figure 4.5 the variety in the size of the bands and in their position in relation to each other, allowing identification of particular chromosomal regions. This chromosome is telocentric, having the centromere in region 20, on the far right of the drawing. The gene that causes the white-eyed phenotype, which we will discuss later, is at the very left tip of region 1, at the other end from the centromere. This was the first chromosome on which the location of genes was mapped, as we will discuss in the next chapter.

❖ The diploid number of chromosomes varies for different species. Specific chromosomes can be identified according to size, position of the centromere, and banding patterns.

MITOSIS

In order to understand genetics, we must understand the process by which hereditary information is passed from cell to cell and from parents to offspring. Although nineteenth century researchers documented the behavior of chromosomes in cell division, they did not realize the genetic implications of chromosomal duplication and movement. Only after the rediscovery of Mendel's principles was the connection made between the earlier chromosomal observations and the principles of segregation and independent assortment.

We will first trace the steps of mitosis, the process of division in which daughter cells are produced that have chromosomal numbers identical to the parental cell. Mitosis is part of the total **cell cycle** for cells undergoing division

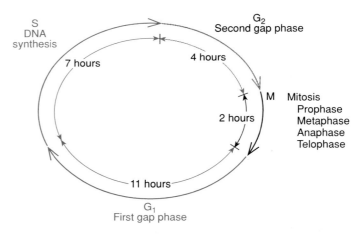

FIGURE 4.5

The chromosome map (top) and a photographic map (bottom) of the X chromosome of *D. melanogaster*. It has been arbitrarily divided into twenty numbered regions, each beginning with a dark band.

From Modern Genetics *by Ayala and Kiger, p. 141, Benjamin/Cummings Publishers, 1980. (From* Handbook of Genetics, *Vol. 3, by R. C. King, Plenum Press, New York, 1975.)*

Figure taken from Lefevre [1976. The Genetics and Biology of Drosophila *(Ashburner and Novitski, eds.). Academic Press, London, New York, San Francisco, Vol. 1a, pp. 31–66].*

(see panel 2). In figure 4.6 notice that the period of mitosis (designated M) is a relatively small part (2 hours out of 24) of the total cell cycle. The times given in figure 4.6 for each stage are for dividing human white blood cells. The duration of the different stages in the cell cycle varies considerably, depending upon the organism, cell type, temperature, and other factors.

The remainder of the cell cycle, called **interphase,** is divided into three parts, termed G_1, S, and G_2. Interphase is generally a time of high metabolic activity, during which DNA, the hereditary material, is synthesized and replicated. Some mature cell types, such as some nerve cells, remain in interphase and never divide further. After a given mitosis is completed, there is an initial time gap, called G_1, in which cells grow, metabolize, and perform their designated functions for the organism. Following the G_1 is a period of DNA synthesis, designated S, in which the DNA is replicated (we will discuss this process in detail in chapter 7). After synthesis, another gap phase, G_2, occurs, followed by the next mitotic division. When an organism is growing, this cycle repeats itself many times, eventually resulting in an individual with billions of cells.

Mitosis itself can be divided into four different stages, listed in chronological sequence as: **prophase, metaphase, anaphase,** and **telophase.** These stages actually merge one

FIGURE 4.6

A schematic representation of the cell cycle for human white blood cells, showing the time involved for each phase.

into another, but each has definite characteristics, particularly in relation to chromosomal behavior, that we can use to identify them. Let us go through these phases to understand the elements of mitosis and cell division and how cells produce identical daughter cells (fig. 4.7 and panel 3).

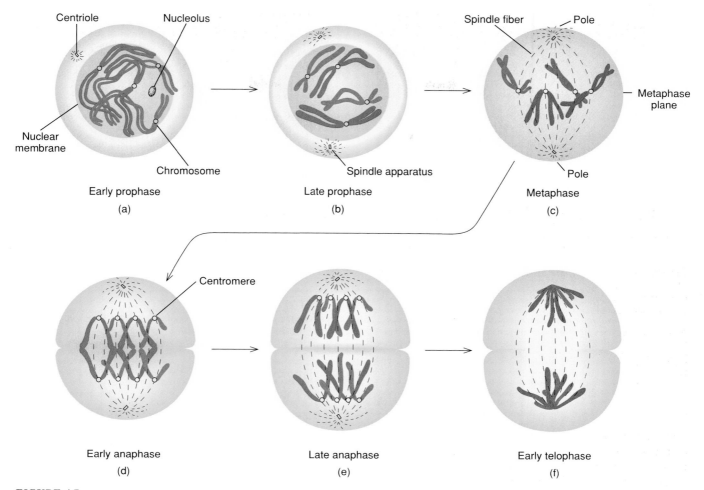

FIGURE 4.7
The behavior of two pairs of chromosomes in mitosis, where pair one is acrocentric (red) and pair two is metacentric (blue).

PROPHASE

At the beginning of mitosis (the end of G_2 interphase), the chromosomes coil and condense, thus becoming visible under the light microscope. The chromosomes have replicated and each has two sister **chromatids** at this stage (fig. 4.7a). The two sister chromatids are attached to each other at the centromere. The chromatids are identical and are the result of the DNA replication in the earlier S phase of the cell cycle. Because the two sister chromatids are attached at the centromeric region, they are considered one chromosome at this stage.

Also during prophase, the **nucleolus,** the largest cell organelle besides the nucleus, usually disappears and the nuclear membrane begins to break down (fig. 4.7b). The **centriole,** an organelle important in organizing cell division, divides. Around it a new structure, the spindle apparatus, which is composed of a series of **spindle fibers** that

stretch to the poles of the cell, also appears (fig. 4.8). The spindle structure is fundamental to the proper movement of the chromosomes to the two poles in the latter part of mitosis.

METAPHASE

The nuclear membrane completely disappears during metaphase, leaving the chromosomes free in the cytoplasm (fig. 4.7c). The spindle fibers attach to the centromeres of the chromosomes, and the chromosomes gradually change their orientation, moving to the **metaphase** (or **equatorial**) **plane,** an imaginary region midway between the two poles. Generally, the chromosomes are most condensed at this stage of mitosis, making it a good stage for preparation of karyotypes. Because sister chromatids are attached at the centromeres, a chromosome at metaphase when it is spread out in a karyotype appears either as an X or a V, depending

(a)

(b)

FIGURE 4.8

Photo (*a*) shows two cells, one in prophase and the other in metaphase. Photo (*b*) shows one cell in anaphase and another in very late telophase. The images in both slides are stained for immunofluorescence microscopy. The green color is from an antibody directed against microtubules, which compose the spindle fibers, and the red is a chemical called propidium iodide, which specifically stains chromosomes.

Courtesy of Mark S. Ladinsky and J. Richard McIntosh, University of Colorado, Boulder.

upon whether it is metacentric or telocentric, respectively. The acrocentric is intermediate between these shapes.

ANAPHASE

Anaphase, nearly always the shortest period of mitosis, begins with the separation of the centromeres (fig. 4.7*d* and panel 3). As a result, each sister chromatid is now called a daughter chromosome. The spindle fibers contract and cause the movement of the centromeres to the poles, with the two identical (homologous) daughter chromosomes moving to opposite poles (fig. 4.7*e* and panel 3). A chromosome appears as a V, J, or I at this stage, depending upon whether it is metacentric, acrocentric, or telocentric, respectively.

TELOPHASE

In telophase, the last stage of mitosis, the two sets of daughter chromosomes have reached opposite poles (fig. 4.7*f* and panel 3) where they begin to uncoil. The nuclear membrane reappears and encloses the chromosomes, and the nucleoli are re-formed.

Next, in animals, the cell membrane develops a furrow; in plants, a cell plate develops that divides the cell in two. This splits the two sets of daughter chromosomes and the cytoplasm between the two daughter cells. The daughter cells have chromosome numbers identical to the original cell.

 Mitosis, cell division that results in the production of two identical daughter nuclei, is divided into four stages: prophase, metaphase, anaphase, and telophase.

MEIOSIS

Before describing meiosis, we will summarize why it is fundamental for maintaining the genetic state of an organism and for generating new genetic types. First, meiosis allows the conservation of chromosome number in sexually reproducing species. If not reduced via meiosis prior to fertilization, the chromosome number would be doubled. Second, meiosis permits the different maternal and paternal chromosomes to randomly assort into each gamete. And finally, the presence of crossing over in meiosis I allows the exchange of genetic material on homologous chromosomes and can result in new combinations of alleles at different genes on a chromosome. (See chapter 5 for a discussion of crossing over.)

Meiosis produces cells with half the parental number of chromosomes through a process involving two cell divisions, meiosis I and meiosis II. **Meiosis I** is also called **reductional division,** because the number of chromosomes is reduced to the haploid number. In meiosis I, the sister chromatids remain attached to each other. **Meiosis II** is called

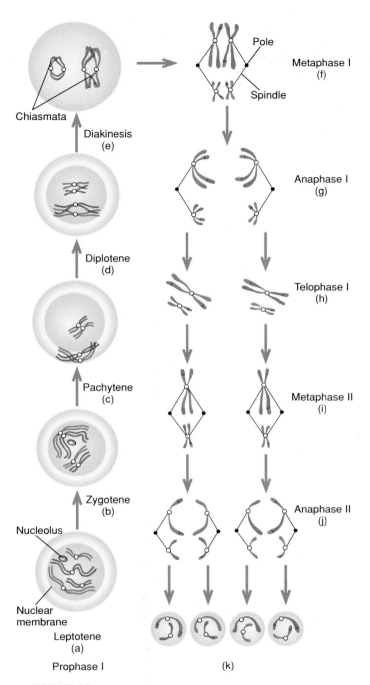

Pole

Metaphase I
(f)

Spindle

Chiasmata

Diakinesis
(e)

Anaphase I
(g)

Diplotene
(d)

Telophase I
(h)

Pachytene
(c)

Metaphase II
(i)

Zygotene
(b)

Nucleolus

Anaphase II
(j)

Nuclear
membrane

Leptotene
(a)

Prophase I

(k)

FIGURE 4.9

A schematic representation of meiosis for two chromosomes. Meiosis I results in the separation of homologous chromosomes, and meiosis II results in the separation of duplicated chromosomes. Here the two larger red and blue chromosomes are homologous, as are the two smaller red and blue chromosomes.

From Modern Genetics, *Second Edition, by Francisco Ayala and John Kiger, Jr. (Menlo Park, CA: Benjamin/Cummings Publishing Company, 1984), p. 15.*

equational division and is much like mitosis in that the centromeres on sister chromatids separate and the chromosome number remains unchanged (fig. 4.9).

MEIOSIS I

As with mitosis, both meiotic divisions can be divided into four stages, based on the position of the chromosomes and other characteristics. The interphase period before meiosis I is similar to that in mitosis, having an S period for DNA synthesis.

Prophase I

The most complex phase of the whole process of meiosis, **prophase I,** is itself broken down into five different parts (fig. 4.9*a–e* and panel 4). The first stage, called **leptotene** (thin-thread), is marked by the appearance of the chromosomes as long threads when seen through a light microscope. This stage is much like the start of prophase in mitosis.

However, in the next stage, **zygotene** (yolk-thread), unlike in mitosis, homologous chromosomes pair side by side and gene by gene with each other. The process of the lateral association of homologues, called **synapsis,** occurs during zygotene and is a key difference between meiosis and mitosis. When the two homologous chromosomes (the large red and blue pair or the small red and blue pair), consisting of four chromatids, are paired, the structure is called a **bivalent.** The next, or third, stage of prophase I, **pachytene** (thick-thread), consists of a shortening and thickening of the bivalents and is the stage during which synapsis is complete. During this stage, portions of homologous chromosomes may be exchanged, a phenomenon called **crossing over** or **recombination.** The structure that allows crossing over is the **synaptonemal complex,** a structure made of protein and DNA that is found between the synapsed homologues. We will discuss the consequences of recombination in chapter 5.

In the fourth stage, **diplotene** (double-thread), the homologues begin to separate, particularly in the region surrounding the centromere. The sister chromatids remain attached at the centromeric region. In addition, the homologous chromosomes generally have several areas, called **chiasmata,** in which they are still close or in contact. Chiasmata are the physical evidence that recombination occurred earlier when the homologues were synapsed. Generally, there is at least one chiasma per chromosome arm, but in long chromosomes, there may be several.

The last stage of prophase I, **diakinesis,** is characterized by shortened chromosomes and the terminalization of the chiasmata; that is, the chiasmata appear to move to the ends of the chromosomes. As in mitosis, at the end of prophase, the nucleolus and the nuclear membrane disappear.

Metaphase I

In **metaphase I** (fig. 4.9*f* and panel 4), the chromosomes have aligned at the equatorial plane as homologous pairs and become attached to the spindle fibers. The bivalents orient so that one centromere is on each side of the metaphase plane,

FIGURE 4.10
(*a*) The numbers of chromosomes per cell in different stages of cell division.
(*b*) The amount of DNA per cell at the various stages of cell division. S indicates DNA synthesis and A indicates anaphase.

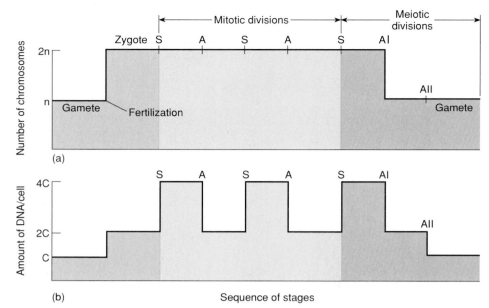

with the chiasmata lined up along it. As a result, metaphase I is different from mitotic metaphase, in which the homologous chromosomes were not associated as bivalents.

Anaphase I

At the stage known as **anaphase I** (fig. 4.9*g* and panel 4), the two sister centromeres in homologous chromosomes of each bivalent move apart, so that one of each homologous chromosome pair goes toward each pole. As the homologues move apart, the chiasmata completely terminalize. In anaphase I, unlike mitotic anaphase, the sister chromatids stay together, joined at the centromere.

Telophase I

In most organisms, after the chromosomes have migrated to the poles, the nuclear membrane forms around them and the cell divides into two daughter cells, just as in mitosis. However, the exact cytological details of **telophase I** (fig. 4.9*h* and panel 4) vary, particularly in plants.

MEIOSIS II

Between meiosis I and meiosis II there is usually only a short interphase called **interkinesis.** No DNA synthesis takes place during this period; at this point, each cell contains only a haploid complement (n) of chromosomes. (Note that each chromosome contains two sister chromatids.) The next stage is **prophase II,** which unlike prophase I is quick, uncomplicated, and generally similar to mitotic prophase. In **metaphase II** (fig. 4.9*i* and panel 4), the centromeres attach to spindle fibers and migrate to the metaphase plate. At the start of **anaphase II** (fig. 4.9*j*), as in mitotic anaphase, the centromeres divide and then separate. One centromere begins to move with one daughter chromosome to one pole, while the other centromere and chromosome go to the opposite pole. There are n chromosomes at

each pole (each chromosome is composed of a single chromatid) when the nuclear membrane forms around them in **telophase II.**

Before we continue, it is useful to compare the changes in chromosome number and amount of DNA that occur during the mitotic and meiotic cycles (see panel 4A). First, as we have already discussed, homologous chromosomes synapse in meiosis I, allowing recombination and precise separation of homologues, but this does not occur in mitosis. Second, at the beginning of both mitosis and meiosis, there are 2n chromosomes in each cell (fig. 4.10*a*). Each chromosome is composed of two sister chromatids, so that the amount of DNA is four times the haploid amount (indicated as 4C in figure 4.10*b*). At the end of mitosis, there are still 2n chromosomes, but the amount of DNA has been reduced to 2C per cell. At the end of meiosis I (indicated by AI, or anaphase I, in figure 4.10), the number of chromosomes has been reduced to n and the amount of DNA to 2C. As in mitosis, meiosis II does not change the number of chromosomes, but the amount of DNA is halved to C as the sister chromatids separate.

SPERMATOGENESIS AND OOGENESIS

The sequence of nuclear events in meiosis that we have described is similar in the production of both male and female gametes; however, there are cytoplasmic differences (fig. 4.11). The production of male gametes in animals, termed **spermatogenesis,** takes place in the **testes,** the male reproductive organs. The process begins with the growth of an undifferentiated diploid cell called a spermatogonium, which then differentiates into a primary spermatocyte. This cell, in turn, undergoes the first meiotic division (the reductional division), becoming two haploid secondary spermatocytes. After meiosis II, these cells each divide into two

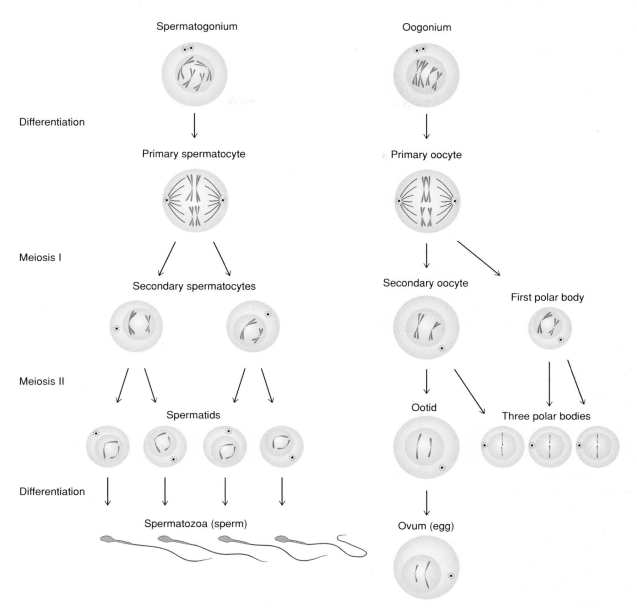

FIGURE 4.11

The stages in the production of mature male and female gametes.

haploid spermatids. The last step is the differentiation of the spermatids into new sperm cells with long tails (fig. 4.12).

Oogenesis, the production of female gametes in animals, occurs in the **ovaries,** the female reproductive organs. The process is parallel to that of spermatogenesis in that it also begins with a diploid cell, or oogonium. However, the cytoplasm is concentrated into only one of the cell products, one of the secondary oocytes, while the other cell products (first and second polar bodies) receive only a small amount of cytoplasm. Meiosis takes place near the cell membrane, allowing the formation of polar bodies with little cytoplasm. The polar bodies remain on the surface of

the egg where they eventually disintegrate. Therefore, only one mature haploid egg is produced in oogenesis, while four mature haploid sperm are produced in spermatogenesis. The concentration of cytoplasm in one egg cell provides nourishment for the developing embryo after fertilization.

Spermatogenesis is continuous in adult animals that reproduce all year round, such as humans, and seasonal in animals that breed seasonally. The capacity for sperm production is extremely high in most animals, with adult males producing many millions of sperm in their lifetimes. On the other hand, egg production in females is generally much lower; for example, a human female can

FIGURE 4.12
Sea urchin sperm on a sea urchin egg. Only one sperm will fertilize this egg.
© *Francis Leroy, Biocosmos/Science Photo Library/Photo Researchers, Inc.*

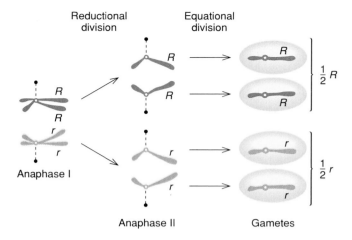

FIGURE 4.13
The chromosomal basis of Mendel's principle of segregation operating in an *Rr* heterozygote. As the diagram shows, ½ *R* gametes and ½ *r* gametes are produced.

produce only around five hundred mature eggs in her lifetime. All these eggs develop into primary oocytes before the female is born and are retained as such in a resting stage until puberty, when they begin maturing into eggs one by one and are ovulated each month. Many other eggs never mature and are reabsorbed.

In higher plants, the formation of male gametes, or pollen, is known as **microsporogenesis.** It is similar to spermatogenesis in animals in that meiosis gives rise to four haploid cells, all of an equal size and all potentially functional. **Megasporogenesis,** the formation of eggs in plants, is also similar to the meiotic process described for animals. However, the products vary considerably among species and are complicated by double fertilization. (It is best to consult a botany text for more details.)

Meiosis is composed of one replication of DNA and two cell divisions. Meiosis I, the reductional division, results in the separation of homologous chromosomes and the halving of chromosome number. Meiosis II results in the separation of the sister chromatids, with no reduction in chromosome number. In animals, mature male and female gametes are produced by spermatogenesis and oogenesis, respectively.

GENES AND CHROMOSOMES

As we mentioned before, several early researchers pointed out the parallel between the behavior of genes and chromosomes. Now that we have discussed meiosis, we can more

precisely show how the behavior of genes and chromosomes is related. For example, figure 4.13 illustrates the principle of segregation in a heterozygote, *Rr* (see also panel 4). In reductional division (meiosis I), the two homologues containing different alleles separate into two different cells. Each of these cells then divides into two cells, which contain the same alleles as the cell produced from meiosis I. As a result, half the gametes are *R* and half are *r*. In other words, equal segregation of alleles into gametes is really the result of reductional division in meiosis.

The chromosomal explanation for independent assortment is given in figure 4.14 and panel 4. Let us examine gamete production in the F_1 double heterozygote *Rr Yy*, where the two genes are on different chromosomes. Two outcomes are equally likely: Either the chromosomes containing *R* and *Y* go to the same pole and the ones containing *r* and *y* go to the other pole, or the *R* and *y* go to one pole and the *r* and *Y* go to the other pole. These two alternatives are equally possible because they result from the chance placement of the centromeres on one side or the other of the metaphase I plane. Each of the four types of cells divides in meiosis II, so that overall, the four gametic types occur in equal proportions. Therefore, the principle of independent assortment also generally results from the behavior of chromosomes in meiosis I. (We should note that this assumes that no recombination takes place between the loci and their centromeres. See chapter 5 for a discussion of recombination.)

The parallel behavior of genes and chromosomes was strong support that genes actually resided in chromosomes, but it was not direct or definitive evidence that specific genes and specific chromosomes were connected. The first evidence of this sort was provided by the experiments of Thomas H. Morgan, who in 1910 described X-linked inheritance in *Drosophila*. Morgan had begun to maintain

FIGURE 4.14
The chromosomal basis of Mendel's principle of independent assortment in an *Rr Yy* double heterozygote. Equal proportions of the four types of gametes are produced. The top and bottom halves of the figure are two equally likely outcomes of independent assortment.

Drosophila cultures shortly after the turn of the century. A momentous event occurred when a white-eyed male appeared in a culture of all wild-type (red-eyed) flies. Using this mutant, Morgan experimentally confirmed that genes are on chromosomes.

Morgan carried out several experiments designed to understand the observed association between the white-eyed mutant and its sex. For example, he crossed red-eyed females and white-eyed males to produce F_1 progeny that were all red-eyed. However, when he crossed the F_1 individuals, the F_2 had the expected 3:1 ratio, but all the white-eyed flies were males (fig. 4.15). Furthermore, the reciprocal cross, white-eyed females and red-eyed males, gave different results (fig. 4.16), with half the F_1 being white-eyed (all of which were males) and half red-eyed (all females). The F_2 progeny from a cross of these F_1 individuals were half white-eyed and half wild-type.

It was known that male and female *Drosophila* differ with respect to their chromosomes: females have two X chromosomes and males have one X and one Y. Morgan suggested that the results of his crosses could be explained if the eye color gene was on the X chromosome and not on the Y chromosome. Obviously, the explanation that the white-eyed mutant was X-linked was consistent with the observations from his experimental crosses.

Although the association of the *white* allele with the X chromosome appeared to be convincing proof that genes

were on chromosomes, another experiment was necessary to exclude all other explanations. Calvin Bridges, a student of Morgan, provided the crucial evidence in 1916, also using the *white* gene. When white-eyed females and red-eyed males were crossed, as in figure 4.17, approximately one in two thousand F_1 offspring had unexpected eye color; that is, they were either white-eyed females or red-eyed males. Bridges suggested that these exceptional flies were the product of an abnormal separation of chromosomes in meiosis called **nondisjunction.** In this case, nondisjunction occurred in females such that the two X chromosomes did not separate properly; instead, both went to the same pole, resulting in eggs with either two X chromosomes or no X chromosomes (see fig. 4.17).

If nondisjunction did occur, and assuming normal meiosis in males, some offspring would be $X^W X^W Y$, with three sex chromosomes. These would be the unexpected white-eyed females, because there is no wild-type allele. Other flies (the red-eyed males) would be X^+, with only one sex chromosome. (We will discuss the chromosomal basis of sex determination in *Drosophila* later in this chapter.) In this explanation, the white-eyed females inherit both of their X chromosomes from their mother, and the red-eyed males inherit their X chromosome from their father. Bridges corroborated his hypothesis by examining the chromosomes of the unexpected flies. He found that the exceptional females were XXY and the exceptional males were

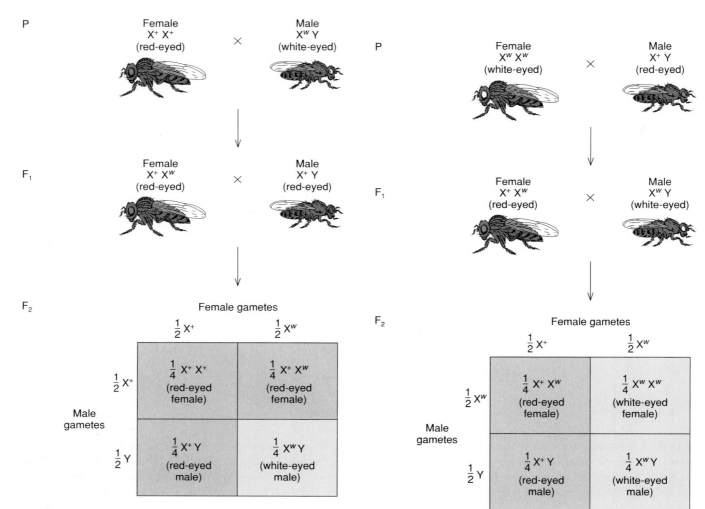

FIGURE 4.15
Results of a cross between a red-eyed female and a white-eyed male to produce an F_2 phenotypic ratio that is sex dependent; that is, all white-eyed flies are male.

FIGURE 4.16
Results of a cross between a white-eyed female and a red-eyed male to produce an F_2 that is different from the reciprocal cross in figure 4.15.

FIGURE 4.17
The results of a cross between white-eyed females and red-eyed males. The * indicates female gametes produced by rare nondisjunction events.

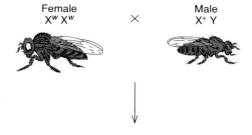

Table 4.3 Types of Chromosomal Changes
Changes in Structure
Duplications
Deletions (or deficiencies)
Inversions
Translocations
Changes in Number
Euploidy level
Autopolyploids
Allopolyploids
Aneuploidy
Monosomies (2n – 1)
Trisomies (2n + 1)

only X. In other words, Bridges's discovery that an abnormal genetic condition and an abnormal chromosomal condition directly corresponded gave convincing proof that genes were indeed on chromosomes.

 The parallel behavior of genes and chromosomes provides a mechanistic explanation for the principles of segregation and independent assortment. The observation of this parallel behavior was also the first indication that genes were on chromosomes. Definitive proof resulted from experiments with X-linked genes in *Drosophila* and from the confirmation that individuals with exceptional eye color were the result of meiotic mistakes that caused abnormal chromosomal numbers.

CHROMOSOMAL CHANGES

Until now, we have been discussing individuals with a normal set of chromosomes. However, an organism's chromosomal constitution may change, or mutate, resulting in differences among individuals of a given species. Chromosomes can change in two basic ways—by altering their structure or by altering their number (table 4.3). Both types of alterations may have consequences besides their immediate effect on the chromosomes. For example, individuals heterozygous for chromosomes with different structures often have lowered fertility, and individuals with altered numbers of chromosomes may be inviable or sterile.

CHANGES IN CHROMOSOMAL STRUCTURE

The four possible types of changes in chromosomal structure are duplications, deletions (or deficiencies), inversions, and translocations. When breaks occur in chromosomes, any two broken chromosomal ends may reunite. Chromosomal changes are often the result of chromosomal breaks

in which the same broken ends do not reunite. Generally, such chromosomal mutations occur infrequently, but some researchers have estimated that more than one in a thousand new gametes may have some type of chromosomal mutation. Note that these chromosomal changes result from chromosomal breaks and abnormal rejoining. The ensuing consequences, which we will discuss in the following section, are the result of the gene-for-gene synapsis of normal and abnormal chromosomes and the subsequent problems that occur in anaphases I and II.

Duplications

When a chromosomal segment is represented twice, it is called a **duplication.** We can categorize duplications by the position and order of the duplicated region (fig. 4.18). First, the duplication may be adjacent to the original chromosomal region. When this occurs, the order may either be the same as the original order, called a **tandem duplication,** or the opposite order, called a **reverse duplication.** Second, the duplicated region may not be adjacent to the original segment, resulting in a **displaced duplication.** In this case, the displaced duplication may still be on the same chromosome, as illustrated in figure 4.18, or it may be on another chromosome.

One possible way a tandem duplication can be generated is illustrated in figure 4.19. It is assumed that the homologues overlap and that there are simultaneous breaks in the two homologous chromosomes at *different* points. If the different homologues reunite, one chromosome will have a tandem duplication and the other a deletion of the duplicated region (fig. 4.19*c*). In this case, the duplication and the deletion are reciprocals of each other. In chapter 5, we will discuss another mechanism (unequal crossing over) by which small duplications (and deletions) can be generated.

When an individual is heterozygous for a duplication and a normal chromosome, the duplicated region does not have a homologous segment to pair with in meiosis. As a result, a loop of the duplicated region may develop (fig. 4.20*a*). In some cases, part of the chromosome may bend back and join with the duplicated sequence on the same chromosome. Individuals having two chromosomal types such as this are called **heterokaryotypic.**

Individuals that are either heterozygous or homozygous for small duplicated segments may be viable, although they often exhibit some phenotypic effects (see the discussion of the *Bar* eye mutations in *Drosophila* in chapter 5). If individuals are viable, there is a potential for further evolutionary change in these extra genes. In fact, it is thought that this happened with the different globin genes, the genes that code for the components of the protein hemoglobin. These genes may have descended from an ancestral gene that was duplicated, and then the duplicate copies diverged in their function.

FIGURE 4.18

A normal chromosome (top), followed by three types of duplication that can result, depending on the position and order of the duplicated segment (shown in green).

Deletions

A missing chromosomal segment is termed a **deletion** or a **deficiency.** We show one way a deletion can occur in figure 4.19. This type of deletion, where an internal part of the chromosome is missing, is called an **interstitial deletion.** If there is only one break and the homologue fails to rejoin, a **terminal deletion** can occur. In this case, the tip of the chromosome is usually lost in cell division because it does not have a centromere. Interstitial and terminal deletions are illustrated in figure 4.21.

When deletions are homozygous, they are often lethal, because essential genes are missing. Even when heterozygous, lethals can cause abnormal development. A well-known example in humans is the deletion of a substantial part of the short arm of chromosome 5 (5p), which when heterozygous causes the *cri du chat* (cry-of-the-cat) syndrome. Infants with this syndrome generally have a characteristic high-pitched, catlike cry as well as microcephaly (small heads) and severe mental retardation. Generally, they die in infancy or early childhood.

In addition, deletion heterozygotes usually show abnormal chromosomal pairing in meiosis. Because the normal chromosome does not have a homologous region to pair with, a deletion loop is formed (fig. 4.20b). This phenomenon may be seen in meiotic chromosomes or in the polytene chromosomes of *Drosophila* and a few other organisms.

Several other characteristics are useful in identifying deletions. First, deletions, unlike other mutations, generally do not revert, or mutate back, to the wild-type chromosome. Second, in deletion heterozygotes, recessive alleles on the normal chromosome are expressed because the deletion chromosome is missing the homologous region. Expression of recessive alleles in such cases, called **pseudodominance,** is useful in defining the length of the deleted segment. For example, let us assume that genes *B* and *C* were deleted on one chromosome as in figure 4.20b. If we have wild-type chromosomes with recessive mutants at different genes, these should be expressed if they are in the deleted region. Figure 4.22 illustrates that the deletion covers genes *B* and *C* but not gene *D*. As will be discussed in chapter 5, deletions can be used to map the sequence of genes on the chromosome.

 Duplications occur when a chromosomal segment is present more than once; deletions occur when a segment is missing. Both can result in phenotypic changes or lethality, although deletions often have greater effects. In meiotic chromosomes, duplication and deletion heterozygotes often form loops.

FIGURE 4.19

(a) Two overlapping homologues (b) break simultaneously, (c) producing a deletion and a duplication.

FIGURE 4.20
Pairing of chromosomes in individuals heterokaryotypic for (*a*) duplication or (*b*) deletion.

FIGURE 4.21
A normal chromosome (top), followed by two types of deletions that may result, depending on the positions of the deleted segment.

FIGURE 4.22
Deletions in heterozygotes that have recessive mutants at the *B, C,* or *D* gene on the normal chromosome.

Inversions

Most of the homologous chromosomes in a population have genes in the same sequence. However, in some instances the sequence may differ on different chromosomes. Alterations in the sequence of genes, called **inversions,** may be of two different kinds relative to the position of the centromere. If the inverted segment does not contain the centromere, it is called a **paracentric inversion** (Greek: *para* = next to), while if the inversion spans the centromere, it is called a **pericentric inversion** (Greek: *peri* = around).

Inversions can be generated by a simultaneous break at two points in a chromosome, followed by an incorrect reunion. For example, figure 4.23 illustrates how a looped chromosome may break and rejoin to generate either a paracentric or a pericentric inversion. Note in this example that the paracentric inversion results in an acrocentric chromosome just like the noninverted chromosome, but that the pericentric inversion results in a metacentric chromosome, because the position of the centromere has been changed.

Individuals heterozygous for an inversion can be recognized by the presence of inversion loops in meiotic pachytene chromosomes. The results of meiosis in inversion heterozygotes will be discussed in chapter 5. These structures occur because of the affinity of the two homologues. The only way the two homologues can pair is if one

twists on itself and makes a loop while the other makes a loop without a twist. Examples of such loops for the two types of inversions are given in figure 4.24. These loops can best be seen in the polytene chromosomes of organisms such as *Drosophila pseudoobscura*. Both a single inversion example, Arrowhead-Standard, and a multiple inversion example, Chiricahua-Standard, are shown in figure 4.25. We will mention these sequences again when we discuss population genetics in chapter 16.

 Inversions occur when the sequence of genes on a chromosome is reversed.

Translocations

A **translocation** is the movement of a chromosomal segment from one chromosome to another, nonhomologous chromosome. Figure 4.26 illustrates two types of translocations, an **interstitial translocation,** involving the one-way movement of a segment, and the more common **reciprocal translocation,** involving a two-way exchange of chromosomal segments. If two of the segments that join in a reciprocal translocation are large and the other two are small, the smaller translocated chromosomes are often lost. In this case, the number of chromosomes is reduced by the chromosomal exchange. Obviously, translocations can change both the size of chromosomes and the position of the centromere.

Even though chromosomal segments have been exchanged between chromosomes in a reciprocal translocation, the affinity of the homologous regions results in pairing during meiosis I. If nearly equal parts of chromosomes are exchanged or not exchanged, the paired chromosomes in a translocation heterozygote have a cross appearance in

FIGURE 4.23
A mechanism for the generation of paracentric and pericentric inversions by two chromosomal breaks.

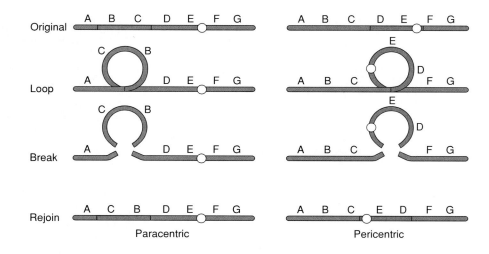

FIGURE 4.24
Loop formed in chromosomal heterozygotes for (a) a paracentric inversion and (b) a pericentric inversion.

(a) Paracentric   (b) Pericentric

FIGURE 4.25
Inversion heterokaryotypes in *Drosophila pseudoobscura*. The Arrowhead-Standard heterokaryotype (left) illustrates a single inversion, while the Chiricahua-Standard heterokaryotype (right) shows a multiple inversion.

From Genes in Populations *by Eliot B. Spiess, p. 71, John Wiley and Sons, Inc., 1977. (From Genetics, 23:28, 1938.)*

FIGURE 4.26
The most common types of events leading to two types of translocations. An interstitial translocation involves the one-way movement of a segment, while a reciprocal translocation involves a two-way exchange of chromosomal segments.

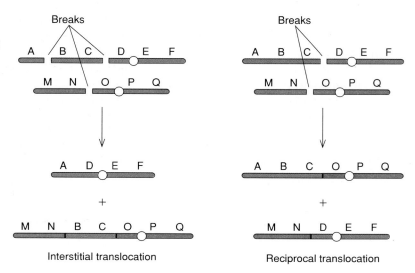

Interstitial translocation   Reciprocal translocation

Metaphase I

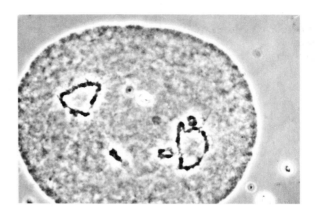

FIGURE 4.28
Cell of *Clarkia*, an annual plant, with two rings of six and three bivalents.
Courtesy of William L. Bloom, University of Kansas.

Anaphase I

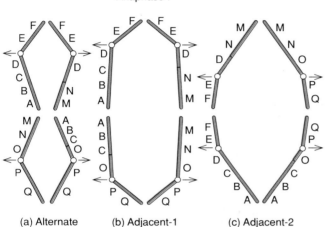

(a) Alternate   (b) Adjacent-1   (c) Adjacent-2

FIGURE 4.27
A heterozygote for a reciprocal translocation and the outcomes of meiotic
division. (*a*) Alternate segregation produces balanced chromosomes. (*b*)
and (*c*) Adjacent-1 and adjacent-2 segregation results in duplications and
deletions.

metaphase I (fig. 4.27). During anaphase I, two major types
of segregation occur: one in which alternate centromeres go
to the same pole and one in which adjacent centromeres go
to the same pole (adjacent-1). The third type of segregation
pictured in figure 4.27 (adjacent-2) is rare, because it requires that homologous centromeres go to the same pole.

When alternate centromeres go to the same pole, the
chromosomes often form a figure eight shape in early
anaphase I. (Remember that the chiasmata terminalize so
that the homologous chromosome ends remain together.)
The products of this event, which is known as **alternate
segregation,** are balanced so that each gamete has a full
complement of chromosomes, either two untranslocated or
two balanced translocated. On the other hand, when adjacent

centromeres segregate together, **adjacent segregation,** the
chromosomes appear as a ring at metaphase I. When this occurs, the products are unbalanced, resulting in duplications
and deletions in the gametes.

Some plants, and also a few animals, have a series of
reciprocal translocations, so that chromosomal heterozygotes also have nearly all the chromosomes associated in a
large ring (or rings) in meiosis (fig. 4.28). However, at
anaphase these chromosomes may undergo an orderly alternate segregation, producing only zygotes with a balanced
chromosomal complement.

Although translocations can result in normal chromosomes, they can also cause several human diseases. For example, about 5% of individuals with **Down syndrome** have
one parent who is heterozygous for a translocation (fig.
4.29). In this instance, chromosome 14 is translocated onto
chromosome 21. Half of the time, the heterozygote produces either the normal set or a balanced translocated set of
chromosomes, making the progeny either normal or
translocation heterokaryotypes, respectively. The other half
of the time, unbalanced chromosomes are produced, either
a 14 without the translocated 21 segment or a translocated
14 with the attached 21 plus a normal 21. In the first case,
offspring get only one 21 chromosome, a lethal chromosomal complement. In the second instance, three 21 chromosomes are received, resulting in Down syndrome.
Overall then, approximately one-third of the live births
from such a translocation heterokaryotype can be expected
to have Down syndrome. In actual fact, the proportion is
less than this, primarily because some Down individuals do
not survive gestation.

FIGURE 4.29
Production of individuals with Down syndrome (trisomy 21) from a translocation heterozygote. Chromosome 14 is indicated in blue and chromosome 21 in red.

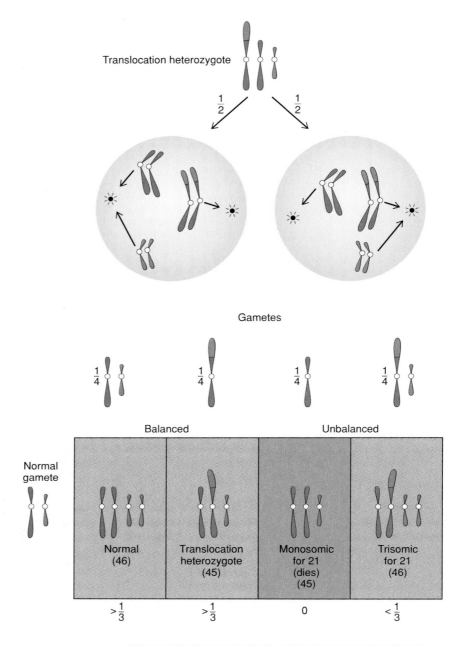

Note that this cause of Down syndrome has implications for genetic counseling. First, Down syndrome could recur in children of a translocation heterokaryotype, whereas normally it does not recur in sibs (see discussion of trisomy later in this chapter). Second, half of the phenotypically normal sibs of Down individuals are themselves translocation heterokaryotypes, and therefore could produce Down progeny.

❖ Translocations are the result of the movement of a chromosomal segment to a nonhomologous chromosome. Normal meiotic products result when there is alternate segregation in translocation heterokaryotypes, but otherwise, duplications and deficiencies occur.

CHANGES IN CHROMOSOMAL NUMBER

The numbers of chromosomes may vary in two basic ways: **euploid** variants, in which the number of chromosomal sets differ, and **aneuploid** variants, in which the number of a particular chromosome is not diploid. As one might expect, changes in chromosome number, either euploid or aneuploid, generally have an even greater effect on survival than do changes in chromosome structure. In fact, in humans, more than half of the spontaneous abortions that occur in the first three months of pregnancy involve fetuses with aneuploidy, polyploidy, or other large chromosomal aberrations.

Before we discuss variants in chromosome number caused by increased or decreased amounts of genetic

FIGURE 4.30
Centric fusion occurs when two nonhomologous acrocentric chromosomes join to form a metacentric chromosome. Centric fission occurs when the metacentric breaks to form two telocentric chromosomes.

material, let us consider the other mechanisms by which chromosome number can change. As already mentioned, if reciprocal translocations occur between two acrocentric chromosomes such that the large segments reattach, the result is a large metacentric chromosome and a small chromosome that may be lost during cell division. In other words, two acrocentric chromosomes may be combined into one metacentric chromosome. This is a mechanism for **chromosomal (or centric) fusion,** in which two nonhomologous chromosomes are joined at their centromeres to form a single metacentric (fig. 4.30). On the other hand, a chromosome may split at its centromere perpendicular to the length of the chromosome, resulting in two smaller chromosomes, a change termed **chromosomal (or centric) fission.** It is generally thought that chromosomal fusions are much more common than chromosomal fissions. Later in this chapter, we will give an example of a fusion that results in the 46 human chromosomes.

Polyploidy (Euploidy Variation)

Organisms with three or more sets of chromosomes are called **polyploids.** If we let the haploid number of chromosomes be x, then organisms with three chromosome sets have 3x chromosomes and are called triploids; those with 4x chromosomes are tetraploids; those with 6x chromosomes are hexaploids; and so on. Earlier in this chapter, we used the symbol n to refer to the haploid number of chromosomes; here, x will stand for the haploid number. However, for organisms that are regularly polyploid, such as many plants, x usually refers to the number of chromosomes in a set and n to the number in a gamete. Thus, in a hexaploid organism with 60 chromosomes, $6x = 2n = 60$, so that $x = 10$ and $n = 30$.

Polyploidy is relatively common in plants but rare in most animals, occurring only in certain beetles, earthworms, salamanders, fishes, and a few other organisms. On the other hand, nearly half of all flowering plants are polyploids, as are many important crops. For example, potatoes are tetraploid ($4x = 48$), bread wheat is hexaploid ($6x = 42$), and strawberries are octoploid ($8x = 56$).

Polyploidy is less frequent in animals than in plants for several reasons. First, sex determination is often more sensitive to polyploidy in animals than in plants. Second, plants can often self-fertilize, so a single new polyploid plant with an even number of chromosomal sets (tetraploid, hexaploid, etc.) can still reproduce. Finally, plants generally hybridize more easily with other related species, an important attribute, because the different sets of chromosomes in a polyploid often have different origins.

We can distinguish two types of polyploids: those that receive all their chromosomal sets from the same species, **autopolyploids,** and those that obtain their chromosomal sets from different species, **allopolyploids.** For example, if an unreduced, or diploid pollen grain from a diploid organism fertilizes a diploid egg of the same species, the offspring are **autotetraploids,** or *AAAA,* where *A* indicates a complete chromosomal set, or **genome,** of type *A.* On the other hand, if diploid pollen of one species fertilizes a diploid egg of another, related species, the offspring are **allotetraploids,** or *AABB,* where *B* indicates a genome from the second species. All the chromosomal sets in an autopolyploid are homologous, just as they are in a diploid. But in allopolyploids, the different chromosomal sets generally vary somewhat and are called **homeologous,** or partially homologous.

Polyploids occur naturally, but in very low frequency, when a cell undergoes abnormal mitosis or meiosis. For example, if during mitosis all chromosomes go to one pole, that cell will have an autotetraploid chromosome number. If abnormal meiosis takes place, an unreduced gamete may result, having 2n chromosomes. However, in most situations this diploid gamete would combine with a normal haploid gamete and produce a triploid. Polyploids can also be produced artificially using colchicine, a chemical that interferes with the formation of spindle fibers. As a result of colchicine application, chromosomes do not move to the poles, and autotetraploids are often formed.

Autopolyploids Triploid organisms are usually autopolyploids (*AAA*) that result from fertilization involving a haploid and a diploid gamete. They are normally sterile because the probability of producing balanced gametes is quite low. In meiosis, the three homologues may pair and form a trivalent, or two may pair as a bivalent, leaving the third chromosome unpaired. Gametes are equally likely to have one or two homologues of a given chromosome

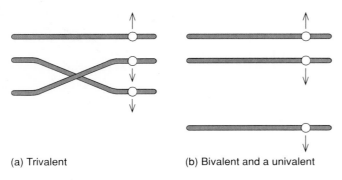

(a) Trivalent   (b) Bivalent and a univalent

FIGURE 4.31
Meiotic pairing and the direction of homologue movement in triploids when (*a*) a trivalent or (*b*) a bivalent and a univalent are formed.

(fig. 4.31). However, because the behavior of nonhomologous chromosomes is independent, the probability of a gamete having exactly n chromosomes is $(1/2)^n$ (using the product rule), and the probability of a gamete having exactly 2n chromosomes is also $(1/2)^n$. All other gametes will be unbalanced and generally nonfunctional, as will the zygotes containing them. For example, most bananas are triploids; they produce unbalanced gametes, and as a result are seedless (they are propagated by cuttings). Other polyploids with odd numbers of chromosome sets (for example, pentaploids) are also usually sterile.

Polyploids are often larger (that is, they produce larger fruit) than related diploids; thus, many food crops are autotetraploids or other types of polyploids. Autotetraploids may have normal meiosis if they form either bivalents or quadrivalents only. However, if the four homologues form a trivalent and a univalent, gametes will generally have either too many or too few chromosomes (fig. 4.32).

Allopolyploids   Most naturally occurring polyploids are allopolyploids, and they may result in a new species. For example, the bread wheat *Triticum aestivum* is an allohexaploid with 42 chromosomes. By examining wild related species, it appears that bread wheat is descended from three different diploid ancestors, each of which contributed two sets of chromosomes (in this case designated as *AABBDD*). Pairing occurs only between the homologous sets, so that meiosis is normal and results in balanced gametes of n = 21.

A particularly interesting allotetraploid was developed by a Russian, G. Karpechenko, in 1928, when he crossed the cabbage and the radish, both of which have a diploid chromosome number of 18. He wanted to produce a hybrid having the leaves of a cabbage and the root of a radish. When he finally recovered seeds from an artificial hybrid, he planted them and found that they had 36 chromosomes. However, instead of the traits he had hoped for, this hybrid had the leaves of a radish and the roots of a cabbage!

A haploid pollen grain with genome *A* may pollinate a flower of a species with genome *B,* resulting in a sterile hybrid of genome constitution *AB*. If mitotic failure subsequently takes place in one branch, *AABB* cells may be produced. If these are self-fertile, an allopolyploid has been formed. Plant breeders use colchicine on sterile hybrids to produce allopolyploids in much the same manner.

> Polyploids, species with more than two chromosomal sets, are most common in plants. They may be either autopolyploids, whose chromosomal sets come from the same species, or allopolyploids, whose chromosomal sets come from different species.

Aneuploidy

We have already discussed the observation by Bridges that *Drosophila* occasionally produce individuals having either one sex chromosome (for example, an X) or three

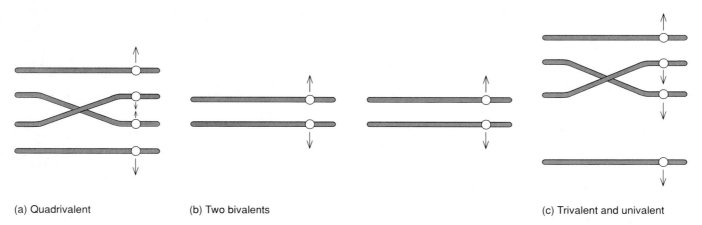

(a) Quadrivalent   (b) Two bivalents   (c) Trivalent and univalent

FIGURE 4.32
Meiotic pairing and the direction of homologue movement in tetraploids for (*a*) a quadrivalent and (*b*) two bivalents. One possible outcome for a trivalent and a univalent is shown in (*c*).

FIGURE 4.33
Nondisjunction (indicated by * for a given chromosome) occurring in meiosis I and in meiosis II, with the resulting gametes.

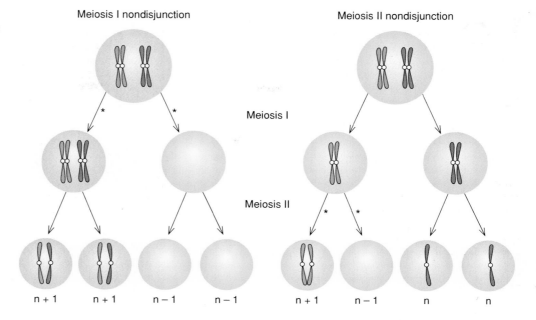

Meiosis I nondisjunction

Meiosis II nondisjunction

Meiosis I

Meiosis II

n + 1   n + 1   n − 1   n − 1   n + 1   n − 1   n   n

sex chromosomes (for example, XXY). The cause of such aneuploidy is nondisjunction; that is, two homologous chromosomes fail to separate properly during meiosis. Nondisjunction in meiosis itself is thought to result from improper pairing of homologues in early meiosis so that the centromeres are not on opposite sides of the metaphase plane, or from failure of chiasma formation. As a result, both chromosomes may go to the same pole, leaving one daughter cell with an extra chromosome and the other daughter cell with no chromosome. When these gametes are fertilized by a normal gamete, they either have an extra chromosome, 2n + 1, termed **trisomy,** or are missing a chromosome, 2n − 1, termed **monosomy.** Nondisjunction is most common in meiosis I, but it can occur in meiosis II as well (fig. 4.33). Nondisjunction can also take place in mitosis, resulting in mosaics for normal and aneuploid cells. Other combinations of extra chromosomes are possible, the most important being a tetrasomic with 2n + 2 chromosomes and a nullisomic with 2n − 2 chromosomes, in which no copies of a particular homologue exist.

Trisomics are known in many different species. They are viable in many plants, but are less frequently viable in animals. For example, among the aneuploids that have been most thoroughly studied are those in the Jimson weed, or thorn apple, *Datura stramonium.* A series of *Datura* mutants with strange properties, studied by Alfred Blakeslee around 1920, turned out to be trisomics for different chromosomes. In fact, a trisomic for each of the twelve different chromosomes was found, and each had a particular phenotype. The effects on the appearance of the seed capsule were quite different for trisomies of the different chromosomes, suggesting that different chromosomes have different hereditary effects on this trait (fig. 4.34).

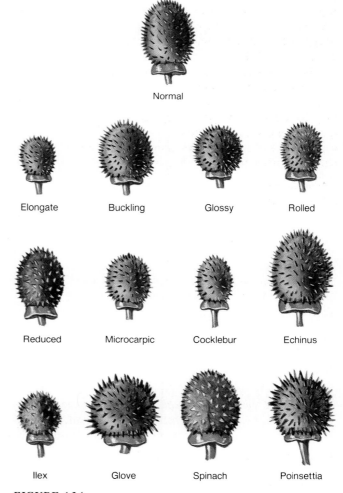

Normal

Elongate   Buckling   Glossy   Rolled

Reduced   Microcarpic   Cocklebur   Echinus

Ilex   Glove   Spinach   Poinsettia

FIGURE 4.34
The normal fruit from *Datura* (top) and the twelve different types of trisomies, each having a different appearance and name.

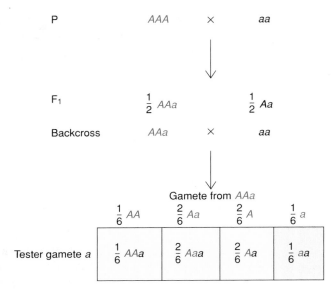

FIGURE 4.35
A cross between an F_1 trisomy and a recessive, giving $^1/_6$ recessive progeny.

FIGURE 4.36
An individual with Down syndrome.
© *M. Coleman/Visuals Unlimited.*

Table 4.4 Frequency and Effects of the Most Common Aneuploidies in Humans

Syndrome	Sex	Chromosomes	Frequency		Expected Viability and Fertility
			Abortuses	*Births*	
Down	M or F	Trisomy 21	1/40	1/700	15 years, rarely reproductive
Patau	M or F	Trisomy 13	1/33	1/15,000	<6 months
Edward	M or F	Trisomy 18	1/200	1/5,000	<1 year
Turner	F	XO	1/18	1/5,000	Sterile
Metafemale	F	XXX (or XXXX)	0	1/700	Nearly all sterile
Klinefelter	M	XXY (or XXXY)	0	1/2,000	Sterile
XYY	M	XYY	?	1/2,000	Normal

Trisomics have been investigated in crop plants such as corn, rice, and wheat in an effort to identify the chromosomes carrying different genes. Crosses involving plants with trisomic chromosomes give unusual segregation ratios. For example, if a homozygous dominant trisomy, *AAA* (the *A* symbol again indicates a dominant allele), is crossed to a recessive diploid, *aa,* half the progeny are trisomic *AAa* and half are diploid *Aa.* When the trisomic progeny are backcrossed to *aa* individuals, approximately one-sixth of the progeny are recessive *aa* (fig. 4.35). If the gene had been on a chromosome that was not trisomic, the F_1 would be *Aa,* and one-half, not one-sixth, of the backcross progeny would be homozygous recessive (*aa*).

In animals, trisomies and other aneuploid chromosomal complements are more unusual. From analysis of the chromosomal constitution of spontaneous abortions in humans, it appears that nearly all monosomics and many trisomics are fetal lethals. However, several trisomies that sometimes come to full term compose a substantial part of congenitally abnormal births. One of the most common is Down syndrome, trisomy of chromosome 21, with a frequency of one in seven hundred live births (table 4.4). Down syndrome, first described nearly 150 years ago, is generally characterized by mental retardation, distinctive palm prints, and a common facial appearance. In general, mortality is higher than normal: the average life span is the middle teens to the forties, depending upon the country, but some individuals live much longer. People with Down syndrome generally have a positive disposition, and some are able to be partially independent (fig. 4.36).

The chromosomal basis of Down syndrome was first discovered in 1959, shortly after the correct human diploid number was determined. Detailed banding of human chromosomes has shown that Down syndrome actually results from a trisomy of the smallest chromosome, which is actually chromosome 22. However, because Down syndrome is known so prevalently as trisomy 21, this association was not changed, and the smallest chromosome is still called chromosome 21. We discussed previously how this condition can

Table 4.5 Gametes and Zygotes Resulting from Normal Meiosis and Nondisjunction

	Sperm	Egg		
		Normal Meiosis	Nondisjunction (Meiosis I or Meiosis II)	
		X	XX	O
Normal Meiosis	X	XX	XXX	XO
	Y	XY	XXY	YO*
Nondisjunction in Meiosis I	XY	XXY		
	O	XO		
Nondisjunction in Meiosis II	XX	XXX	Extremely rare	
	YY	XYY		
	O	XO		

*Inviable zygotes.

FIGURE 4.37
The procedure for amniocentesis.
© Eric Kroll

(a)   (b)   (c)

FIGURE 4.38
Nuclei from cells of (a) a normal female (XX) with one Barr body, (b) a normal male (XY) with no Barr bodies, and (c) an XXX female with two Barr bodies.
Dr. Dorothy Warburton/Peter Arnold, Inc.

be generated from reciprocal translocation heterokaryotypes, but most commonly (in about 95% of cases), it results from meiotic nondisjunction. The other autosomal trisomies are much rarer, mostly because they are not viable as fetuses.

Nondisjunction of the sex chromosomes in humans is the source of several maladies. Table 4.5 shows the gametes and zygotes resulting from normal meiosis and from nondisjunction. Notice that four common viable, but abnormal chromosomal types, XO, XXX, XXY, and XYY, are produced through nondisjunction. The symbol O here indicates the lack of a sex chromosome in a gamete or zygote.

Klinefelter syndrome, XXY, occurs fairly frequently and generally results in a relatively mild abnormality. These individuals are sterile males with some female characteristics. Individuals with **Turner syndrome** (XO) are sterile females, short in stature, with some neck webbing. The frequency of XYY is about one per one thousand males, but such males do not appear to have any congenital problems. At one point, it was suggested that XYY individuals had criminal tendencies, but further study indicates minimal correlation with behavior, if any. The frequency of XYY individuals in prisons is significantly higher than that of the general population; however, less than 5% of all XYY individuals are actually institutionalized.

Abnormal chromosome numbers in a fetus can be diagnosed using **amniocentesis** (fig. 4.37). In this procedure, a sample of fluid is withdrawn from the amniotic sac with a needle. The fetal cells contained in this fluid are cultured for two to three weeks. Dividing cells are then stained, and the chromosomes are examined and counted to check for chromosomal abnormalities.

The X chromosome is different from the other chromosomes in that only one is active in a given cell. Normal males have only one X, which is active in all cells. In normal females, only one X is active in a given cell and the other X is heterochromatinized, or mostly inactive. The mostly inactivated X forms a structure called a **Barr body** (named after its discoverer, Murray Barr) that can be identified in a cell (fig. 4.38). Therefore, normal males and XO individuals have no Barr bodies; normal females and XXY individuals have one; XXX individuals have two; and so on. In other words, by counting the number of Barr bodies in a cell, chromosomal abnormalities involving the X chromosome can be determined.

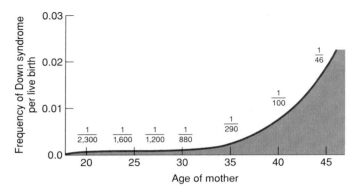

FIGURE 4.39
The frequency of Down syndrome in children as it relates to mothers of different ages.

The incidence of Down syndrome, and to some extent other aneuploidies, increases with the age of the mother (fig. 4.39). The incidence of Down for mothers of age forty-five is nearly 50-fold that for teenage mothers. Although the exact mechanism for this increase is unknown, it appears to be related to the difference in gametogenesis between females and males. In females, oocytes are formed before birth and held in a resting stage (actually prophase of meiosis I) until just before ovulation. In older mothers, an oocyte may remain at this stage for over forty years, during which time it may be affected by environmental factors that may cause a nondisjunction.

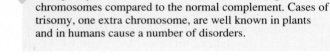

Aneuploidy occurs when there are extra or missing chromosomes compared to the normal complement. Cases of trisomy, one extra chromosome, are well known in plants and in humans cause a number of disorders.

CHROMOSOMES IN DIFFERENT SPECIES

As we discussed earlier in this chapter, variation in chromosomal structure or number may occur within a species. When the structures of chromosomes in related species are compared, they are often quite similar. Early studies of this sort compared the banding patterns of different related *Drosophila* species and found a high degree of homology. For example, species such as *D. melanogaster* and *D. simulans*, which are virtually the same except in the morphology of the male genitalia, have nearly identical banding sequences, differing only in a few inversions.

Recently developed high-resolution techniques for chromosomal banding permit the comparison of chromosomes in many species. Much research has focused on human chromosomes and on those of the related great apes. Humans have a diploid chromosome complement of 46, while chimpanzees, gorillas, and orangutans have 48. The difference in number is explained by a centric fusion of two nonhomologous chromosomes in the great apes to form

human chromosome 2 (fig. 4.40). Notice that the two acrocentric chromosomes in the three great apes that are homologous to the metacentric human chromosome 2 have identical banding patterns, except in the fusion region and in the terminal heterochromatin.

A number of other structural differences exist in the chromosomes among these four species. For example, the banding patterns in chromosome 3 in humans, chimpanzees, and gorillas are virtually identical, while the pattern in orangutans differs by two inversions. The break points of the larger of these inversions are indicated by the arrows in figure 4.40, showing that the inversion is pericentric and changes the chromosome from a metacentric to an acrocentric. Evidence indicates only one translocation difference among these species. It occurs between chromosomes 5 and 17 in the gorilla and is not present in the other species. Overall, the banding patterns for humans are nearly identical to chimpanzees for 13 chromosomes, to gorillas for 9 chromosomes, and to orangutans for 8 chromosomes. Almost all the same bands are present in these species, suggesting that the primary chromosomal differences among them are in the order of the genes, not in the presence or absence of genes.

CHROMOSOMES AND SEX DETERMINATION

In the past, literally hundreds of unscientific theories have attempted to explain sex determination by relating it to phases of the moon, the time of day of fertilization, and other factors. Actually, there is no universal mode of sex determination, but rather many different modes, some of which are environmental. For example, certain reptile eggs that are exposed to high temperatures produce mostly males in particular species and mostly females in others. The sex of some fish is affected by social dominance, and certain plants produce different sexes depending on day length or other factors that affect growth rate.

However, in most organisms, sex is determined by chromosomal differences (table 4.6). The first example of chromosomal determination of sex was documented in 1905 for the bug *Protenor*, in which it was discovered that females have two X chromosomes and males only one. Male gametes were found to be equally likely to have an X or no X chromosome, and female gametes were found to always carry one X, so that equal numbers of progeny are XX (female) and XO (male).

Shortly thereafter, several investigators established the presence of a second sex chromosome, the Y chromosome, in other organisms. In these species, the presence of a Y along with an X results in a male, while two copies of the X produce a female. Again, male gametes are of two types, so that there are equal numbers of X-carrying and Y-carrying sperm or pollen. The sex having both types of sex chromosomes, the XY, is termed the *heterogametic sex*, while the XX is called the *homogametic sex*. In some organisms, such as birds and some reptiles, males are homogametic and females are heterogametic. To avoid confusion, these

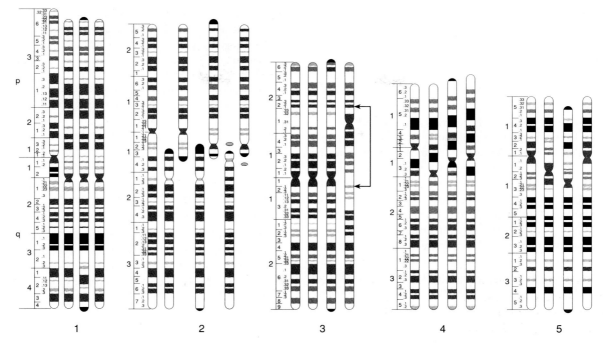

FIGURE 4.40

Comparison of late-prophase banding sequences of the five largest chromosomes (1–5) in humans, chimpanzees, gorillas, and orangutans (left to right).

From J. J. Yunis and O. Prakash, Science, *Vol. 215, March 19, 1982. Copyright © 1982 by the American Association for the Advancement of Science, Washington, DC. Reprinted by permission of the publisher and authors.*

Table 4.6 Different Types of Chromosomal Sex Determination		
Organisms	**Female**	**Male**
Most mammals, some insects, some plants	XX	XY
Protenor, some other insects	XX	XO
Birds, most reptiles, moths	ZW	ZZ
Hymenoptera	Diploid	Haploid

chromosomes are often termed Z and W; thus, homogametic males are ZZ and heterogametic females are ZW. As a result, the females produce two types of gametes, while the male gametes all contain one Z chromosome.

In chickens, it is important to identify the sex of young birds so that, for example, males will not be raised in an egg laying operation. It is difficult and expensive to determine the sex of young birds, and several systems utilizing the sex chromosome difference between males and females are often used. For example, one gene on the Z chromosome has a recessive allele that determines gold feather color (either Z^sZ^s or Z^sW), and a dominant allele that determines silver feather color (Z^SZ^S, Z^SZ^s or Z^SW). As a result, the cross given in figure 4.41 produces all silver males and all gold females, allowing easy sex determination of day-old chicks. Could the opposite cross, Z^SZ^S × Z^sW, yield chicks of different colors that could be used to determine their sex?

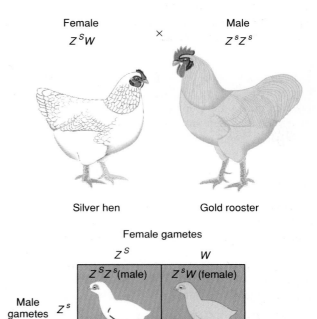

FIGURE 4.41

The results of a cross between a silver hen and a gold rooster to produce all male silver chicks and all female gold chicks.

In nearly all mammals, the presence of a Y chromosome is necessary for the development of the male phenotype. For example, humans that are XO (Turner syndrome) are phenotypically female. Furthermore, individuals that are XXY (Klinefelter syndrome) are phenotypically male, even though they have two X chromosomes. Thus, it appears that the Y chromosome produces a gene product that acts as a necessary switch to begin development toward maleness. Although this switch gene is on the Y chromosome, the rest of the genes involved in the production of sexual phenotypes apparently are on other chromosomes.

Recently, the location of this switch gene on the Y has been identified by examining rare, sex-reversed individuals—that is, XX males and XY females. The XX males actually have a piece of the Y that contains the male-determining gene, and the XY females lack the same region of the Y. In *Drosophila,* there is an autosomal mutant, *tra* (transformer), which when homozygous in XX individuals causes them to be males, not females as expected. The opposite situation is known in humans where individuals with a mutation on the X chromosome have a condition called testicular feminization. In this case, the XY individuals are phenotypically female, having breasts and a vagina, but they are sterile. This mutation is apparently related to a hormone receptor that prevents these individuals from responding to the presence of testosterone; consequently, they do not develop secondary male characteristics.

Aside from this switch function to maleness, there appear to be few active genes on the Y chromosome in humans, according to the following evidence. First, much of the Y chromosome is heterochromatic. Second, the length of the long arm of the Y is quite variable. (Interestingly, the long arm is substantially longer in Japanese males than in males of most other populations, but this variation appears to have no effect on fertility or other phenotypic characteristics.) And finally, there is also extensive variation in the size and intensity of banding on the long arm of the Y in humans without apparent phenotypic effect. These facts support the hypothesis that much of the Y chromosome is inert and does not produce functional gene products.

Although XX individuals in *Drosophila* are female and XY individuals are male, the presence of the Y is not necessary for the male phenotype. For example, the exceptional phenotypes observed by Bridges (see fig. 4.17) were red-eyed XO males and white-eyed XXY females. Later, Bridges developed triploid strains of *Drosophila.* When he

Table 4.7 Sexual Phenotypes of *D. melanogaster*[1]

Number X Chromosomes	Number Autosomal Sets (A)	X/A Ratio	Sexual Phenotype
3	2	1.5	Metafemale
2	2	1.0	Normal female
2	3	0.67	Intersex
1	2	0.5	Normal male
1	3	0.33	Metamale

[1]Expressed as a function of the ratio of the number of X chromosomes to sets of autosomes.

examined individuals with different numbers of X chromosomes, he found that the sexual phenotype in *Drosophila* was a function of the ratio of the number of X chromosomes (X) to the number of sets of autosomes (A). For example, normal females have two X chromosomes and two autosomal sets, making the ratio $X/A = {}^2\!/_2 = 1$ (table 4.7). Normal males have one X and two autosomal sets, so that $X/A = \tfrac{1}{2} = 0.5$. Likewise, his exceptional males and females still had ratios of 0.5 and 1.0, respectively.

When Bridges examined flies with three X chromosomes and two autosomal sets, $X/A = {}^3\!/_2 = 1.5$, he found that they were sterile females (sometimes called **metafemales**), while flies with one X and three autosomal sets, $X/A = {}^1\!/_3 = 0.33$, were sterile males (**metamales**). Finally, flies with two X chromosomes and three autosomal sets, $X/A = {}^2\!/_3 = 0.67$, had phenotypic characteristics of both sexes and were termed **intersexes.** Based on these findings, Bridges proposed that, overall, the X chromosomes were female-determining and the autosomes were male-determining. In other words, when the X/A ratio was 1 or greater, flies would have the female phenotype, and when the X/A ratio was 0.5 or less, flies would have the male phenotype. However, we should note that even though the Y is not necessary for the male phenotype in *Drosophila,* it is necessary for fertility; that is, XO males and metamales are sterile.

Chromosomes in related species are generally quite similar. In most organisms, chromosomal differences are involved in sex determination. The Y chromosome determines maleness in nearly all mammals. In *Drosophila,* the relative proportion of X chromosome to autosomes determines sex.

WORKED-OUT PROBLEMS

Problem 1

How many copies of a given gene does an autohexaploid have? Assuming there are two types of alleles at a gene, *B* and *b,* what are the different possible genotypes?

Solution

Each gene is represented in each chromosomal set, so there would be six copies of each gene. There are seven different possible genotypes: two homozygotes *BBBBBB* and *bbbbbb,* and five types of heterozygotes, *BBBBBb, BBBBbb, BBBbbb, BBbbbb,* and *Bbbbbb.*

Problem 2

Genes on the X chromosome in *Drosophila melanogaster* were used to show that genes are on chromosomes. In one cross, as outlined in figure 4.16, white-eyed females were crossed to red-eyed males and the F_1 progeny of this cross were mated to produce an F_2. In one such array, 40 red-eyed females, 37 white-eyed females, 42 red-eyed males, and 35 white-eyed males were produced. Seeing this array of genotypes and knowing the behavior of the X chromosome, why would you conclude that the white gene is on the X chromosome?

Given this array of phenotypes and the mode of inheritance given in figure 4.16, are the results consistent with the expectations? Use a chi-square test.

Solution

Several observations from this cross point to the fact that genes are on chromosomes. For example, the F_1 males have a Y that they must have received from their father, because their mother does not have a Y chromosome. The males have a white-eye allele (because their phenotype is white) that they must have gotten from their mother. This suggests that the X chromosome that they do have is from their mother and contains the white allele. Likewise, all of the phenotype and sex combinations of the progeny in the F_2 can be explained by assuming that the white allele is on the X chromosome.

The chi-square test can be calculated as follows where the expected number for all classes is 155/4 or ¼ of the total number of progeny.

Phenotype	Sex	Observed	Expected	$(O - E)^2/E$
red-eyed	female	40	38.75	0.04
white-eyed	female	37	38.75	0.08
red-eyed	male	43	38.75	0.47
white-eyed	male	35	38.75	0.36
		155	155	$\chi^2 = 0.95$

With three degrees of freedom, the number of classes minus one, this value is not significant. In other words, these results are consistent with the hypothesis that the white allele is on the X chromosome.

Summary

The diploid number of chromosomes is different in different species. Specific chromosomes can be identified by size, the position of the centromere, and banding patterns. Mitosis, nuclear division that results in the production of two identical daughter nuclei, is divided into four stages: prophase, metaphase, anaphase, and telophase. Meiosis is composed of one replication of DNA and two cell divisions. Meiosis I, the reductional division, results in the separation of homologous chromosomes or a halving of chromosome number. Meiosis II is similar to mitosis and results in the separation of the sister chromatids with no reduction in chromosome number. In animals, mature male and female gametes are produced by spermatogenesis and oogenesis, respectively.

The parallel behavior of genes and chromosomes provides a physical explanation of the principles of segregation and independent assortment. The first indication that genes are on chromosomes came from the parallel behavior of genes and chromosomes. Definitive proof resulted from experiments using X-linked genes in *Drosophila* and the confirmation that individuals with exceptional eye color were the result of meiotic mistakes that produced abnormal chromosomal numbers.

Duplications and deletions occur when a chromosomal segment is present more than once or is missing. Both can result in phenotype changes or lethality, although deletions often have larger effects. In meiotic chromosomes, duplication and deletion heterozygotes often form loops. Inversions occur when the sequence of genes on a chromosome is reversed. When recombination takes place in the inverted region of inversion heterokaryotypes, duplications and deficiencies result in the meiotic products. Translocations are the result of the movement of a chromosomal segment to a homologous chromosome. Normal meiotic products result when there is alternate segregation in translocation heterokaryotypes, but otherwise duplications and deficiencies occur.

Polyploids, species with more than two chromosomal sets, are most common in plants. They may be either autopolyploids, in which all chromosome sets are from the same species, or allopolyploids, in which the chromosomal sets come from different species. Aneuploidy occurs when there are extra or missing chromosomes compared to the normal complement. Cases of trisomy, one extra chromosome, are well-known in plants and in humans cause a number of disorders.

Chromosomes in related species are generally quite similar. In most organisms, chromosomal differences are involved in sex determination. In nearly all mammals, the Y chromosome determines maleness. In *Drosophila*, the relative dose of X chromosome to autosomes determines sex.

PROBLEMS AND QUESTIONS

1. What are the three major morphological classes of chromosomes and how are they distinguished?

2. What are the four phases of the cell cycle, starting with the S phase? What takes place during the S phase?

3. A cross between red-eyed F_1 (heterozygous) female fruit flies and red-eyed male fruit flies yielded 94 red-eyed and 30 white-eyed progeny. Does this fit with expectations? (Use a chi-square test.) Of the red-eyed flies, 56 were females. Does this fit with expectations? (Again, use a chi-square test.)

4. In the cross shown in figure 4.17 involving a group of 4,240 flies, 3 red-eyed males and 2 white-eyed females were observed. From these data, what would be the estimated rate of nondisjunction?

5. Explain why the number of chromosomes per cell (n level) and the amount of DNA per cell (C level) are not synchronous in every stage of cell division.

6. What are two major differences between spermatogenesis and oogenesis in humans and most other mammals?

7. Differentiate between the behavior of chromosomes in metaphase of mitosis, meiosis I, and meiosis II.

8. (a) Given a heterozygote for albinism, *Aa,* what genotypes of sperm would you expect after meiosis? (b) Diagram the chromosomal behavior as in figure 4.13. (c) Extend this analysis to two genes, *Aa,* and another recessive disease heterozygote, *Bb.* Illustrate how chromosomal behavior explains independent assortment by drawing a diagram as in figure 4.14.

9. What are the four types of changes that can take place in chromosome structure? Describe briefly what happens in each.

10. Figure 4.19 shows how a tandem duplication may be generated. (a) Using the same logic, illustrate how a reverse duplication may be generated. (b) Show how a displaced duplication may be generated.

11. Let us assume that we are trying to identify the genes involved in a deletion. Crosses to lines homozygous for recessive mutants at genes *e* and *f* gave offspring that were mutant, while crosses to lines homozygous for recessive mutants at genes *g* and *h* yielded wild-type progeny. Which genes are missing in the deletion?

12. What is the difference between pericentric and paracentric inversions? What are the differences in meiosis and in the meiotic products from heterokaryotypes for paracentric and pericentric inversions?

13. Assume that the breaks for the reciprocal translocation in figure 4.27 were between E and F and between P and Q. (a) What do the chromosomes with the reciprocal translocation look like? (b) Diagram metaphase I and alternate segregation for a heterokaryotype for this reciprocal translocation.

14. Name four common aneuploidy conditions in humans and describe the chromosomal constitutions that are responsible.

15. Some Down syndrome cases are the result of a translocation of chromosome 21 onto chromosome 14. Approximately one-third of the offspring of a translocation heterozygote are expected to have Down syndrome. Explain why. In fact, only about one-sixth of the offspring of translocation heterozygotes actually do have Down. Explain why.

16. (a) Differentiate between the meiotic products expected from nondisjunction in meiosis I and meiosis II. (b) Assuming nondisjunction in meiosis I for the sex chromosomes in a male, what progeny genotypes would you expect? (c) What progeny genotypes would you expect from nondisjunction in meiosis II in males or in females?

17. In an effort to determine on which chromosome a gene is located, a homozygous recessive mutant was crossed to trisomies of two different chromosomes. A backcross of the trisomy F_1 to the recessive homozygote gave 42 wild-type and 38 mutant progeny for chromosome 5 and 56 wild-type and 12 mutant progeny for chromosome 6. On which chromosome do you think the gene was located?

18. How many copies of a gene does an autotetraploid have? Assuming that there are two types of alleles at a gene, *A* and *a,* give the five possible genotypes. If *A* is dominant, how many possible phenotypes would you expect?

19. Given an autotriploid with 3x = 12, what is the probability of producing a gamete having exactly one copy of each of the four homologues? Having exactly two copies of each homologue?

20. In figure 4.40, identify the smaller inversion on chromosome 3 that differs between orangutans and the other great apes. Identify the pericentric inversion that differentiates chromosome 4 for humans and chimpanzees.

21. In chickens, males are the homogametic sex, *ZZ,* and females are the heterogametic sex, *ZW.* In a cross between $Z^A W$ and $Z^A Z^a$, where the superscripts indicate the dominant and recessive alleles, what proportions of progeny would you expect? Give both genotype and sex.

22. Given the rules developed by Bridges to determine sex in *Drosophila,* what would be the sex of an individual with three X chromosomes and three autosomal sets? Of an individual with two X chromosomes and four autosomal sets? Explain why.

23. Using diagrams, show the differences between prophase in mitosis and in meiosis I.

24. Why might allotetraploids have higher rates of growth than either of the parental diploids? If the two sets of chromosomes from different species in an allotetraploid do not pair, would you expect balanced gametes?

25. Two related species of mice have different numbers of chromosomes. Species 1 has twenty pairs of metacentrics, and species 2 has twenty-three pairs of chromosomes, seventeen of which are metacentric and six acrocentric. How many chromosomal fissions would be necessary to produce a karyotype of species 2 from species 1?

Answers appear at end of book.

CHAPTER 5: GENETIC LINKAGE

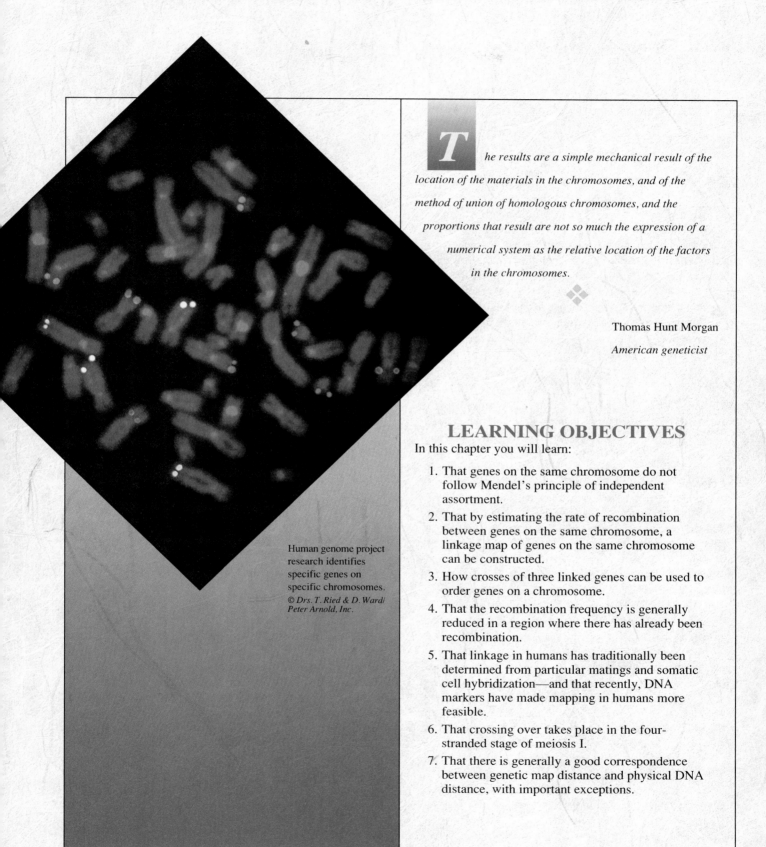

*T*he results are a simple mechanical result of the location of the materials in the chromosomes, and of the method of union of homologous chromosomes, and the proportions that result are not so much the expression of a numerical system as the relative location of the factors in the chromosomes.

Thomas Hunt Morgan

American geneticist

Human genome project research identifies specific genes on specific chromosomes.
© Drs. T. Ried & D. Ward/ Peter Arnold, Inc.

LEARNING OBJECTIVES

In this chapter you will learn:

1. That genes on the same chromosome do not follow Mendel's principle of independent assortment.

2. That by estimating the rate of recombination between genes on the same chromosome, a linkage map of genes on the same chromosome can be constructed.

3. How crosses of three linked genes can be used to order genes on a chromosome.

4. That the recombination frequency is generally reduced in a region where there has already been recombination.

5. That linkage in humans has traditionally been determined from particular matings and somatic cell hybridization—and that recently, DNA markers have made mapping in humans more feasible.

6. That crossing over takes place in the four-stranded stage of meiosis I.

7. That there is generally a good correspondence between genetic map distance and physical DNA distance, with important exceptions.

*T*he principles of segregation and independent assortment are consistent with the fact that genes are on chromosomes and with the behavior of chromosomes in meiosis. We have assumed that the different genes we followed were on different chromosomes. However, shortly after the rediscovery of Mendel's principles, Thomas Hunt Morgan (fig. 5.1) realized that there were more genes than chromosomes in *Drosophila*. In fact, we know today that thousands of genes exist on each of the three major *Drosophila* chromosomes. As a result, the pattern of inheritance predicted by the principle of independent assortment does not apply to genes on the same chromosome. In this chapter, we will examine the pattern of inheritance for two or more genes on the same chromosome. We will also describe how genes can be organized into genetic maps, showing their positions on the various chromosomes.

Because of the human genome project, a recent international effort to map all the genes in the human genome, a great resurgence of interest and advancement in the techniques used to map genes has occurred (see chapter 15). However, many of the fundamental ideas are based on the classic experiments that we will discuss in this chapter.

LINKAGE

In the first decade of this century, William Bateson and R. C. Punnett investigated the inheritance patterns for various traits in the sweet pea. In one study, they examined the simultaneous inheritance of flower color, dominant purple (*P*) versus recessive red (*p*), and pollen shape, dominant long (*L*) versus recessive round (*l*). They carried out a dihybrid cross as Mendel had done for traits in the garden pea; that is, they crossed *PP LL* with *pp ll*, self-fertilized the F$_1$ (*Pp Ll*), and counted the numbers of different types of F$_2$ plants. But instead of the expected 9:3:3:1 ratio, they obtained the results given in table 5.1. Comparing the observed numbers to those expected under a 9:3:3:1 ratio, they found many more phenotypes like the original parents and many fewer single dominant or nonparental types. In this case, there are 3 degrees of freedom, making the chi-square value of 222.0 for the data in table 5.1 highly significant and inconsistent with the hypothesis of independent assortment (see table 2.5).

Bateson and Punnett knew that sometimes F$_2$ ratios were a modification of the 9:3:3:1 expectations because of epistasis, but the F$_2$ numbers they observed in this experiment did not fit a modified Mendelian ratio either (although they did try to fit it to a 7:1:1:7 ratio). Besides, the two genes affected quite different traits, unlike the epistatic examples with modified F$_2$ ratios. What could be the cause of the strange ratios they had found in F$_2$ progeny?

Bateson and Punnett suggested that because the two parental phenotypes were in excess in the F$_2$ progeny, there might be a physical connection between the parental alleles—the dominant alleles from one parent and the recessive alleles from the other parent. This phenomenon, which they

FIGURE 5.1
Thomas Hunt Morgan, a leading geneticist of the early twentieth century, in his fly lab.
Lowe Memorial Library, Columbia University.

Table 5.1 F$_2$ Progeny of a Dihybrid Cross in Sweet Peas			
	Numbers		
Phenotype	*Observed*	*Expected*	$(O - E)^2/E$
Purple, long	296	240.2	13.1
Purple, round	19	80.1	46.5
Red, long	27	80.1	35.1
Red, round	85	26.7	127.3
	427	427	$\chi^2 = 222.0$

termed **coupling,** interfered with independent assortment. On the other hand, they attributed the low numbers of the F$_2$ types, those having a dominant phenotype at one gene and a recessive at the other, to some innate negative affinity of dominant and recessive alleles, which they termed **repulsion.**

A PHYSICAL EXPLANATION

A physical explanation for Bateson and Punnett's observations came later from the research of Morgan, using two genes in *Drosophila*. One gene affected eye color (recessive purple, *pr*, and dominant wild-type red, *pr*$^+$, alleles), and the other gene affected wing development (recessive vestigial, *vg*, and dominant normal, *vg*$^+$, alleles). Instead of examining the F$_2$ progeny of a dihybrid cross, Morgan used a testcross for both genes simultaneously, as outlined in figure 5.2. Note that both wild-type alleles were originally in one parent of the F$_1$ and that the recessives were in the other parent. The value of such a cross is that only one parent, in this case the female, is doubly heterozygous, and therefore, segregation and independent assortment would be expected only in the production of the gametes from that sex.

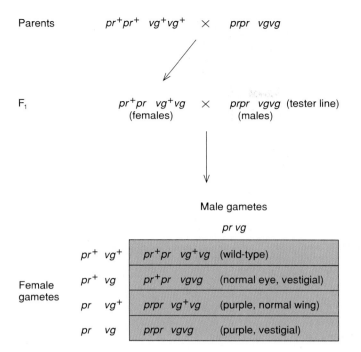

Parents: pr^+pr^+ vg^+vg^+ × $prpr$ $vgvg$

F₁: pr^+pr vg^+vg × $prpr$ $vgvg$ (tester line)
(females) (males)

Male gametes

pr vg

Female gametes	pr^+ vg^+	pr^+pr vg^+vg (wild-type)
	pr^+ vg	pr^+pr $vgvg$ (normal eye, vestigial)
	pr vg^+	$prpr$ vg^+vg (purple, normal wing)
	pr vg	$prpr$ $vgvg$ (purple, vestigial)

FIGURE 5.2
The genotypes and phenotypes from a two-gene testcross using the *vestigial* and *purple* mutants.

Table 5.2 Results of Morgan's First Testcross

	Numbers			
Phenotype	*Gamete*	*Observed*	*Expected*	$(O - E)^2/E$
Wild-type	$pr^+ vg^+$	1,339	709.75	558
Normal, vestigial	$pr^+ vg$	151	709.75	440
Purple, normal	$pr vg^+$	154	709.75	435
Purple, vestigial	$pr vg$	1,195	709.75	332
		2,839	2,839.00	$\chi^2 = 1,765$

If the two genes assorted independently in Morgan's testcross, then the four types of female gametes, and consequently the four different phenotypes, should be equally frequent. However, as shown in table 5.2, the numbers of the parental phenotypes were many times those of the nonparental phenotypes. The chi-square value for these data is very large (with 3 degrees of freedom, a value greater than 11.34 is highly significant), indicating that the independent assortment hypothesis does not hold up. Although the four gametic numbers were not those expected from independent assortment, the two parental gametes, $pr^+ vg^+$ and pr vg, were nearly equal in number, and the two nonparental gametes, $pr^+ vg$ and $pr vg^+$, were also nearly equal in number, but much less frequent than the parental gametes. Today we generally consider gametes with either two wild-type or two mutant alleles *coupling gametes,* and those with one wild-type and one mutant allele *repulsion gametes.* In other words, the two parental gametes in table 5.2 were in coupling, and the two nonparental gametes were in repulsion.

In a second experiment, Morgan crossed two different genotypes, each of which was homozygous for a wild-type allele at one gene and homozygous for a recessive allele at another gene (pr^+pr^+ $vgvg$ × $prpr$ vg^+vg^+). The F₁ flies, double heterozygotes, were then crossed to the tester line *prpr vgvg.* The results of this cross were quite different from the first cross (table 5.3). In this case, the progeny numbers of the single dominants were much larger than expected, while those of the double dominants and double recessives were much lower than expected. In other words,

there were more repulsion types than coupling types. Note that in this cross, the parental gametes were in repulsion and the nonparental gametes were in coupling.

To explain his observations, Morgan suggested that the genes having the alleles for these traits are on the same chromosome. As a result, alleles at these genes tend to remain associated between generations because they are physically linked to each other. A schematic representation of this explanation is given in figure 5.3, where the wild-type alleles are on the blue chromosomes and the recessive alleles are on the red chromosomes. The F₁ flies have one of each type of chromosome. However, instead of these chromosomes always staying intact during the production of gametes, in some chromosomes recombination occurs between the chromosomes. As a consequence, these gametes contain nonparental chromosomes composed of the pr^+ and vg alleles or the pr and vg^+ alleles. A similar explanation is true for the cross discussed in table 5.3, except that the parental chromosomes have one dominant and one recessive allele, rather than two dominants or two recessives.

The intact original chromosomes are often called **parental chromosomal types,** while the new combinations are called **nonparental chromosomal types.** As we mentioned in chapter 4, the physical exchange through which new chromosomal types are formed is termed **crossing over,** or **recombination,** and is often indicated in diagrams by an × joining homologous chromosomes. The other common names for parental and nonparental chromosomal types are **nonrecombinant** and **recombinant,** respectively, reflecting the process involved. We will discuss the biochemical mechanisms of recombination in a subsequent chapter. At this time, we can envision that the homologous chromosomes sometimes break but then reattach to the other member of the homologous pair. The farther apart two genes are on a chromosome, the more likely it is that such breakage–reunion between them will occur.

If a breakage and reunion were to take place between homologous chromosomes, one might expect some cytological manifestation of the event. In fact, it was observed in meiosis that sometimes nonsister chromatids appear to be attached to each other, forming a cross configuration. As we mentioned in chapter 4, this structure (termed a **chiasma**) is a cytological observation resulting from recombination.

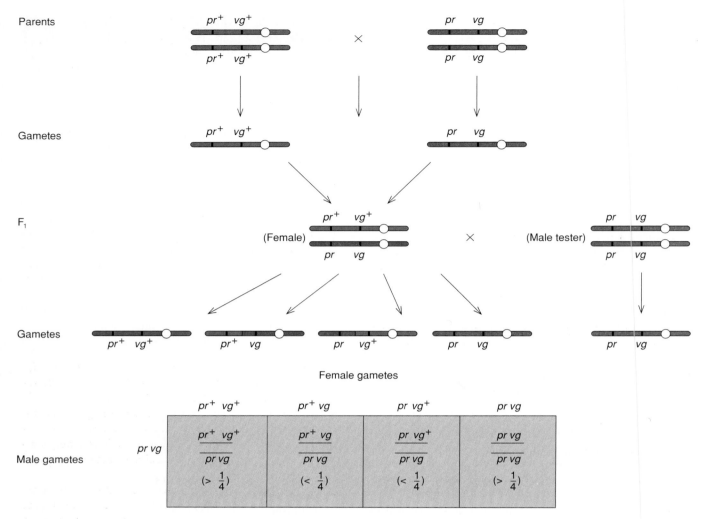

FIGURE 5.3

A schematic representation of the gametes and progeny produced in a testcross, with coupling in the parents.

Table 5.3 Results of Morgan's Second Testcross				
		Numbers		
Phenotype	*Gamete*	*Observed*	*Expected*	$(O - E)^2/E$
Wild-type	$pr^+ vg^+$	157	583.75	312
Normal, vestigial	$pr^+ vg$	965	583.75	249
Purple, normal	$pr vg^+$	1,067	583.75	400
Purple, vestigial	$pr vg$	146	583.75	328
		2,335	2,335.00	$\chi^2 = 1,289$

Genes that are on the same chromosome are said to be **linked,** and the general phenomenon in which genes occur on the same chromosome is called **linkage.** We can refer to alleles that are linked on the same chromosome as being in either the coupling phase—that is, both dominants of pr^+ vg^+ or both recessives as $pr vg$—or in the repulsion phase,

$pr^+ vg$ or $pr vg^+$. Genotypes containing the two different coupling chromosomes are called coupling double heterozygotes and can be indicated by $pr^+ vg^+/pr vg$. Genotypes with two different repulsion chromosomes are called repulsion double heterozygotes and can be expressed as $pr^+ vg/pr vg^+$. Note that the terms *coupling* and *repulsion* came from traits with complete dominance in which the dominant alleles were on the same chromosome. In other words, the definition is based on the phase in these initial crosses, an artifact that becomes apparent when we examine traits having intermediate dominance or codominance.

 Shortly after the rediscovery of Mendel's principles, exceptions to the principle of independent assortment were discovered. The tendency for the alleles at different genes on parental chromosomes to remain together is explained by linkage of genes on a chromosome.

Table 5.4 Proportions of Gametes[1]

Gamete	Parental Genotype	
	Coupling A B/a b	Repulsion A b/a B
A B	½(1 − r)	½r
A b	½r	½(1 − r)
a B	½r	½(1 − r)
a b	½(1 − r)	½r
	1	1

[1] r indicates the rate of recombination.

Table 5.5 Recombination Frequency for Five X-linked Gene Pairs

Genes	Recombination Frequency
yellow (y) — white (w)	$\frac{214}{21,736} = 0.010$
yellow (y) — vermilion (v)	$\frac{1,464}{4,551} = 0.322$
white (w) — vermilion (v)	$\frac{471}{1,584} = 0.297$
vermilion (v) — miniature (m)	$\frac{17}{573} = 0.030$
white (w) — miniature (m)	$\frac{2,062}{6,116} = 0.337$
white (w) — rudimentary (r)	$\frac{406}{898} = 0.452$
vermilion (v) — rudimentary (r)	$\frac{109}{405} = 0.269$

RATE OF RECOMBINATION

When two genes are on different chromosomes, all four gametic types (or phenotypes) are represented equally in the progeny of a testcross. In other words, independent assortment results in equal numbers of parental and nonparental gametes or offspring. However, when two genes are linked on a chromosome, the proportion of recombinant gametes is less that 50%, the specific amount depending upon the physical proximity of the two genes on the chromosome. As an extreme, if the two genes are very, very close to each other, the number of recombinant gametes or offspring should be close to zero. We can measure the **rate of recombination** (r) as the proportion of recombinant offspring observed in a testcross or in other crosses. For example, in table 5.2, 305 (151 + 154) of 2,839 progeny were recombinant, making the rate of recombination 305/2,839 = 0.107, or 10.7%. This value suggests that the genes are relatively close to each other but not exactly at the same location on the chromosome.

A general way to express the proportions of recombinant and nonrecombinant gametes is given in table 5.4. It is assumed that the parent is doubly heterozygous, either in coupling (second column) or in repulsion (third column). For a coupling parent, a proportion (r) of the gametes A b and a B are recombinant. Because we expect equal proportions of these two recombinant classes, the probability of each class is ½r. Likewise, the proportion of nonrecombinant gametes, 1 − r, is equally divided between A B and a b. The same logic applies in determining the expected proportion of gametes for the repulsion genotype, although the recombinant and nonrecombinant gamete categories are reversed.

THE LINKAGE MAP

Alfred Sturtevant, a student of Morgan's, was assigned the project of consolidating the early information on crossing over in D. melanogaster. He conceived an approach in which the recombination data would be used to describe the physical relationship of genes on a chromosome in a linear arrangement called a **linkage map,** or **genetic map.** A genetic map uses the frequency of recombinant gametes (or offspring) from a testcross, or from other crosses, as a measure of the distance between two genes. For example,

10.7% of the offspring from the cross in table 5.2 were recombinants between genes pr and vg, making the **map distance** between them 10.7 **map units.** Morgan and Sturtevant suggested that distance along the chromosome is measured in units determined by the percentage of recombination. In fact, later studies have shown that genetic distance measured by this statistical approach is generally similar to cytologically or biochemically measured distances on a chromosome.

Morgan and his students discovered a number of genes in D. melanogaster that showed X-linked inheritance. When they crossed strains differing in two of these genes and examined their progeny, they found substantial variation in the frequency of recombination for different gene pairs (table 5.5). For example, out of 21,736 potentially recombinant gametes between the genes for yellow and white, only 214, or 1%, were recombinants. On the other hand, 32.2% of the gametes for the genes for yellow and vermilion were recombinants.

Using the data in table 5.5, Sturtevant determined the physical relationship of these five genes and suggested that they formed a linear array. First, y and w had very few recombinant gametes, indicating that they were close to each other. Second, the next lowest level of recombination between w and another gene was with v, which had 0.297, or 29.7%, recombinants. But because the frequency of recombination between y and v is greater than that between w and v, the sequence of these three genes must be y-w-v, with w in between y and v. Next, the gene closest to v is m, only 3 map units away. It must be to the right of v, because the distance w-m is approximately the sum of v-m and w-v. Finally, r is loosely linked to v but shows no evidence of linkage to w in genetic crosses. As a result, r must be to the right of v, making the array of five genes y-w-v-m-r.

	y w		v m		r	Centromere
Sturtevant	0.0 1.0		30.7 33.7		57.6	
Today	0.0 1.5		33.0 36.1		54.5	67.7

FIGURE 5.4
The five X-linked genes studied by Sturtevant, showing the positions he gave and those recognized today.

"Hey! What's this Drosophila melanogaster doing in my soup?"

Sturtevant constructed a map beginning with *y* at 0.0 on the left side and using the frequencies of recombination between adjacent genes (fig. 5.4). That is, he put *w* at 1.0, *v* at 30.7 (1.0 + 29.7), *m* at 33.7 (30.7 + 3.0), and *r* at 57.6 (30.7 + 26.9). Notice in the figure that the positions for *w, v, m,* and *r* today are slightly different from those given by Sturtevant because many intermediate genes have since been used to map them more accurately.

As indicated here, a genetic map can be constructed by estimating the recombination frequency of all pairs of genes. However, a more efficient approach, as we will discuss later, is to use three genes simultaneously. Using more marker genes also allows identification of cases with more than one crossover.

Notice that although both the *white* and *rudimentary* loci are X-linked, the data in table 5.5 indicate a very high recombination frequency between them, nearly that which would be found if the genes were on different chromosomes. This suggests that genes far apart on the same chromosome would not show evidence of linkage in genetic crosses. Genes on the same chromosome, whether or not they show linkage in genetic crosses, are **syntenic.**

People have often wondered why Mendel did not observe linkage in his dihybrid crosses. It turns out that, although he picked seven traits and there are seven chromosomes in garden peas, two of the traits he studied are determined by genes on chromosome 1 and three by genes on chromosome 4 (table 5.6). Thus, out of the

Table 5.6 Chromosomal Locations for the Seven Traits in Mendel's Peas

Character	Chromosome
Seed color (yellow—green)	1
Flowers (violet—white)	1
Pods (smooth—wrinkled)	4
Flowers (axial—terminal)	4
Stem (tall—short)	4
Pods (green—yellow)	5
Seed form (round—wrinkled)	7

twenty-one different possible pairs of genes (the number of combinations of seven genes taken two at a time), four pairs of genes are syntenic. However, three of these pairs are far apart on the chromosome and show no indication of linkage in hundreds of crosses. The remaining pair of genes, containing the alleles that determine pod morphology and stem length, should have shown linkage in genetic crosses but were not investigated by Mendel. In other words, because only one of twenty-one pairs of traits would have shown linkage and Mendel examined only a few pairs, it is not surprising (and perhaps fortunate) that he did not observe patterns differing from the expectations of independent assortment.

 The rate of recombination estimates the linkage of genes and ranges from 0 for very tightly linked loci to ½ for very loosely linked loci (or those on different chromosomes). The relationship of genes on a chromosome can be shown on a genetic map.

THREE-POINT CROSSES

Although the linearity of the chromosome was apparent from the data for pairs of X-linked genes, complete substantiation of this model came by simultaneously crossing individuals differing by three different genes. One test again used the three X-linked genes in *Drosophila*—*y, w,* and *m*. In this experiment, a backcross of a female in coupling for all three genes was done (fig. 5.5), and the gametes in the progeny were scored as receiving either an X chromosome with *y, w,* and *m* or a Y chromosome from the male tester parent. In the progeny gametes, those produced from the F_1 triple heterozygote female, eight types are possible, two of which are parental and six of which are recombinant.

Parents   + + + / + + +   ×   y w m / Y

+ + + / y w m   ×   y w m / Y
(F₁ females)   (tester line males)

FIGURE 5.5

A testcross of females heterozygous for three X-linked genes, giving the number and frequency of different progeny types.

	Gametes	Numbers	Frequency
Parental	+ + +	3,501	
	y w m	3,471	0.664
Recombination of *m* with *y* and *w*	+ + *m*	1,754	
	y w +	1,700	0.329
Recombination of *y* with *w* and *m*	y + +	28	
	+ *w m*	32	0.0057
Recombination of *w* with *y* and *m*	+ *w* +	6	
	y + *m*	3	0.00086
		10,495	1.0

Assuming that the three genes involved are arranged in a linear fashion, there are three possible orders: *y-w-m*, *w-y-m*, and *y-m-w*, each having a different gene in the center. Two of the recombined classes can be generated by a single recombination event, but the third must be the product of two crossovers involving the same chromatids. Because the likelihood of such a double crossover should be small (on the order of the product of the probability of two different crossover events), the smallest recombined class should represent double recombinants. The smallest observed recombinant class in figure 5.5 is that composed of the complementary gametes + *w* + and *y* + *m*. Using the three different gene orders, double recombination can produce these rare gametes only if the order of the genes is *y-w-m* (fig. 5.6). Double recombination for the other two gene orders does not result in + *w* + and *y* + *m* as the rarest class but produces the other, more common recombinant classes.

With three linked genes, there are three different recombination frequencies between pairs of loci: between *w* and *m*, between *y* and *w*, and between *y* and *m*. From figure 5.5, the total frequency of recombination between *w* and *m* is the sum of the two recombinant classes that have recombined in this region, or 0.329 + 0.00086 = 0.330. (Double crossover gametes must have had recombination between *w* and *m* as well.) Likewise, for *y* and *w* the frequency of recombination is the sum 0.0057 + 0.00086 = 0.0066. Between *y* and *m*, the frequency is 0.329 + 0.0057 = 0.335. (Another way to ascertain the order of the genes is to note that the largest of these recombination values must be between the outside genes, *y* and *m* in this case.)

Given that the order of the genes is *y-w-m*, a double recombinant must involve both an exchange between loci *y* and *w* and one between *w* and *m*. Using the recombination frequencies just calculated, the expected frequency of a double recombination event (using the product rule) should

(a)

(b)

(c)

FIGURE 5.6

The three possible orders of genes *y*, *w*, and *m* with gametes produced from double recombination. Order (*a*) with gene *w* in the center is consistent with the data in figure 5.5. The X symbol indicates the position of crossing over.

be (0.0066)(0.330) = 0.00218. The observed frequency of double recombination is 0.00086, which is somewhat lower than that expected (see further discussion later in this chapter). The map distance between *y* and *m* is then calculated as the sum of the recombination between *y* and *w* plus that for *w* and *m*, or 0.0066 + 0.330 = 0.337. Notice that the same information is obtained from one three-point cross as from three crosses involving different pairs of genes. In addition, three-point crosses allow us to identify gametes produced by double recombination.

FIGURE 5.7

The genetic map of *D. melanogaster,* showing some of the important genes on the three large chromosomes.

From W. E. Sinnot, et al., Principles of Genetics, *5th ed. Copyright © 1986 McGraw-Hill, Inc., New York, NY. Reprinted by permission of McGraw-Hill, Inc.*

Using such crosses and related techniques, researchers have constructed a genetic map of *Drosophila melanogaster* that is one of the most complete for any eukaryote. Some of this information is given in figure 5.7, listing the positions of some of the well-known genes on the four different chromosomes. Besides the X chromosome, which is 70 map units long, *Drosophila* has three autosomes. The two large autosomes have 108 and 106 map units, while the small fourth chromosome has only 3 map units. The total length of these chromosomes has been determined by the summation of empirically mapped intervals, as discussed previously. Overall then, the *Drosophila* genome has 287 map units. The linkage maps of other organisms used in genetic research, such as *Neurospora*, corn, and mice, are also quite well known.

 Three-point crosses are used to determine the order of genes on a chromosome.

INTERFERENCE

Recombination frequency is defined as the proportion (or probability) of exchange between two linked genes. As we have suggested, more than one recombinational exchange may occur on a chromosome. Therefore, it is important to know whether these recombinations are independent of each other or whether they somehow influence each other. If different recombinant events are independent, the probability of two such events occurring in the same gamete should be equal to the product of the probability of the separate events. For the *y, w,* and *m* chromosomes discussed previously, the observed frequency of double crossovers was 0.00086, somewhat lower than the 0.00218 proportion expected. Fewer double crossovers than expected are generally found in other organisms as well.

A standard way of describing the difference between the observed and expected numbers of double crossovers was first used by H. J. Muller. The method divides the observed numbers by the expected numbers (or proportions) of double crossovers and subtracts it from unity (one). This value gives a measure of the **interference** (I) of one crossover on another. This expression is:

$$I = 1 - \frac{\text{Observed frequency double recombinants}}{\text{Expected frequency double recombinants}}$$

Thus, for the *y, w,* and *m* data, I = 1 − 0.00086/0.00218 = 1 − 0.372 = 0.628. Positive values such as this one indicate interference between recombination events, while values near zero suggest no interference—that is, that different recombinant events are independent of each other. For closely linked genes, the measure of interference may often be near one, while for widely separated genes on the same chromosome, I is often near zero.

One theory to explain the infrequency of double crossovers is that the high values of interference for closely linked genes appear to be associated with a physical inflexibility of the chromatids, which keeps them from bending enough for two crossovers to occur in a very short region. However, interference may occur in genes up to 30 map units apart, indicating that other factors must be important as well. Interestingly, the centromere for metacentric chromosomes in *Drosophila* (and perhaps other organisms) acts as a barrier for interference. In other words, crossing over in one arm of a metacentric chromosome has no inhibitory effect on recombination in the other arm.

We should also note that when genes are far apart on a chromosome, the observed frequency of recombination may be less than that expected from knowledge of map distances. For example, we know that genes *w* and *r* are 53 map units apart on the X chromosome. First, the maximum proportion of recombinant gametes when the genes are on different chromosomes is 0.5, based on the principle of independent assortment, so we would not expect 0.53 recombinant gametes. Second, multiple even numbers of recombination will reduce the observed recombination frequency. Figure 5.8 illustrates that both double and quadruple crossovers appear as nonrecombinant gametes when there are no intermediate marker genes. As a result, the observed proportion of recombinant gametes for genes 50 map units apart is generally less than 0.5. Remember that the proportion of recombinants between *w* and *r* observed by Sturtevant was 0.452. Only when genes are 70 to 80 map units apart does the observed frequency of recombinants approach 0.5.

 The frequency of double recombination between closely linked genes is often lower than that expected, a phenomenon called interference.

LINKAGE IN HUMANS

Obviously, planned genetic crosses are not possible in humans, the number of progeny from any one mating is small. As a result, until the recent introduction of biochemical techniques, linkage had been determined for only a few human genes, mainly genes on the X chromosome. By studying informative matings (particular pedigrees segregating for the two genes of interest), tight linkage was established only between the ABO blood group system and the nail-patella syndrome (a condition causing abnormal and underdeveloped fingernails and small or absent kneecaps), between the β- and δ-globin genes, and between a few other gene pairs.

Mapping the X chromosome is made easier because recombination between X chromosomes takes place only in females (remember that males have only one X

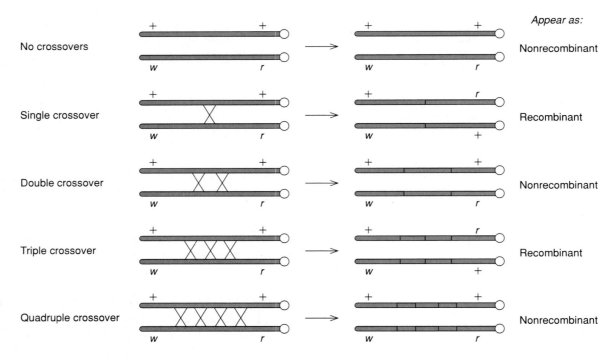

FIGURE 5.8
The types of gametes produced by different numbers of crossovers between two genes.

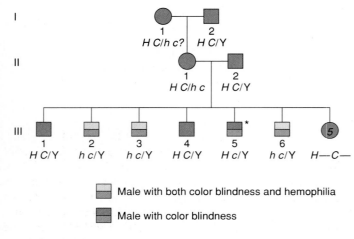

Male with both color blindness and hemophilia

Male with color blindness

FIGURE 5.9
In this pedigree, a female (II-1) heterozygous for two X-linked genes has five nonrecombinant sons (1, 2, 3, 4, and 6) and one recombinant son (5, denoted by an asterisk). In addition, there are five unaffected daughters.

chromosome), and potential recombinant gametes can be directly identified in the male progeny. In other words, if the female parent is a double heterozygote with known gametic types, and if a sizable number of male offspring are available for study, the recombination frequency can be estimated.

An example of linkage is given in figure 5.9 for the X-linked traits color blindness and hemophilia. In this case, a woman (II-1) was unaffected by either recessive trait but produced three sons who had both hemophilia and color blindness, one son who was just color-blind, and two who

were normal. Because her father (I-2) was unaffected by both traits, the woman must have inherited an X chromosome with normal alleles from him and an X with both recessive alleles from her mother. In other words, she was a double heterozygote in coupling phase (*H C/h c*). Therefore, the five sons who were either normal (*H C/Y*) or affected with both traits (*h c/Y*) received a nonrecombinant gamete from her, while the sixth son (III-5) received a recombinant gamete, *H c*. The frequency of recombination in this pedigree is then $^1/_6 = 0.167$. Of course, to obtain a good estimate of the map distance between these genes, many such families must be examined, and the results from different families must then be combined using various statistical techniques. Generally, pedigrees segregating for these two traits do not have recombinants because the genes are closely linked at the end of the long arm of the X chromosome. These genes and a number of other X-linked genes, along with their relative map positions, are shown in figure 5.10.

Interestingly, the recombination rate for autosomal genes in males and females is often different. In fact, the British evolutionary biologist J. B. S. Haldane suggested that in organisms with a chromosomal mechanism of sex determination, recombination is generally higher in the homogametic sex. In mammals, the homogametic sex is usually the female. And in both humans and mice, the recombination rate is higher in females than in males—about 50% to 100% higher for the human gene pairs examined. In *Drosophila*, which also has homogametic females, no recombination occurs in males; but in the silk moth *Bombyx mori*, which has heterogametic females, the females show no recombination.

Xg blood group
Ichthyosis (steroid sulfatase deficiency)
Gonadal dysgenesis (XY female type)
Chronic granulomatous disease (Xk)
Ocular albinism
Retinoschisis

Duchenne muscular dystrophy

Testicular feminization syndrome (TFM)

Phosphoglycerate kinase deficiency (PGK)

Fabry's disease (α galactosidase A deficiency – GALA)
Gout (phosphoribosylpyrophosphate synthetase deficiency – PRPS)

Lesch-Nyhan syndrome (HPRT deficiency)

G-6-PD deficiency
Deutan color blindness
Protan color blindness
Hemophilia A
Adrenoleukodystrophy

FIGURE 5.10

The relative positions of a number of the genes on the human X chromosome, where p and q indicate short and long arms, respectively.

Reprinted with the permission of Macmillan College Publishing Company from Genetics, *3/e by Monroe W. Strickberger. Copyright © 1985 by Monroe W. Strickberger.*

The human genome is not yet mapped in the same detail as the *Drosophila* genome. However, a rough estimate of the total map length for humans can be obtained by observing the average number of chiasmata in meiosis I. The estimated mean number of chiasmata in male meiosis is about fifty per cell, or slightly more than two per bivalent. Assuming that a chiasma occurs approximately every 50 map units, the total map distance is around 2,000 to 2,500 map units, spread over the twenty-three human chromosomes—nearly ten times the map length of the *Drosophila* genome.

For many human genes, it is not known on which chromosome a gene is located, much less what linkage may exist within the chromosome. In some cases, chromosomal abnormalities can give a clue to the location of genes. For example, trisomies are used to locate genes in a number of crops (see fig. 4.35), and in theory, trisomy 21 (Down syndrome) individuals should have a different level of enzymes produced by genes on the affected chromosome. However, Down individuals exhibit many abnormalities, so it is not clear whether any biochemical differences are due to three doses of a gene or to the abnormal physiological condition of these individuals. Likewise, individuals with deletions might also have lower levels of proteins produced by genes in the deleted region.

In both *Drosophila* and prokaryotes (see chapter 13), **deletion mapping** is a technique through which genes can be located to a particular chromosome segment (see also fig. 4.22). For example, assume an individual is heterozygous for a deletion on a given chromosome that has a number of codominant genes (blood group, enzyme, or DNA variants). If different alleles are present at each gene, each genotype should be recognizable as a heterozygote (fig. 5.11a). However, when a gene is in the deleted region,

Chromosome   Genotype

(a) A_1 B_1 C_1 D_1 / A_2 B_2 C_2 D_2 A_1A_2 B_1B_2 C_1C_2 D_1D_2

(b) A_1 C_1 D_1 / A_2 C_2 D_2 B_2 A_1A_2 B_2–C_1C_2 D_1D_2

(c) A_1 B_2 C_1 D_1 / A_2 B_2 C_2 D_2 A_1A_2 B_2B_2 C_1C_2 D_1D_2

FIGURE 5.11

(*a*) A normal heterozygote for codominant alleles; (*b*) a heterozygote missing the gene B region; and (*c*) an individual homozygous for B_2.

a heterozygote should always appear as a homozygote (fig. 5.11b). To distinguish deletion heterozygotes from true homozygotes, some pedigree information is necessary. For example, if a deletion heterozygote B_2–was mated with a B_1B_2, there could be B_1–offspring appearing as B_1B_1 homozygotes. But if the parent really was a B_2B_2 homozygote and was mated with a B_1B_2, the offspring should be half B_1B_2 and half B_2B_2, with no individuals that are B_1–(and appear as B_1B_1 homozygotes).

SOMATIC CELL HYBRIDIZATION

Until recently, it was extremely difficult to locate human genes to specific chromosomes. One ingenious approach researchers used is **somatic cell hybridization.** In this technique, cells from two different species are fused in the laboratory, using a fusion agent such as Sendai virus. The Sendai virus has several points of attachment to its host cell, a characteristic that allows it to serve as a vehicle to fuse two quite different host cells. The chromosomes of the two species in the fused cells and the genetic trait (or traits) of interest are both monitored over time; human cells are usually hybridized with mouse or hamster cells. Eventually, the human chromosomes are more or less randomly lost, so that different cell lines lose different human chromosomes. The chromosome that contains the gene of interest is determined by noting which cell lines still carry the trait and finding a human chromosome that they have in common.

The first gene to be located in this way codes for thymidine kinase, an enzyme important in DNA replication. Mouse cells deficient for this enzyme (symbolized by Tk^-) were hybridized with human cells having the gene for the enzyme (Tk^+). These hybrid cells were then grown on a medium that required the cells to produce thymidine kinase in order to survive. Figure 5.12 shows the chromosomes found for five cell lines that were able to produce thymidine kinase. Notice that in these cell lines, human chromosome 17 is the only chromosome present in all lines. Therefore, these data indicate that the thymidine kinase gene must be on chromosome 17.

This experimental technique can be extended in several ways. For example, some cell lines also have deletions of a particular chromosomal segment. Hybrid cells with a deleted long arm of human chromosome 17 are Tk^-, indicating that the TK gene must be on the long arm of chromosome 17. Furthermore, if several genetic traits are monitored simultaneously, the linkage relationship of the genes can be determined by the joint occurrence or absence of their protein products.

Studies locating genes to particular chromosomes in different organisms have shown that genes in related species are often linked in the same pattern. For example, the same genes appear on the X chromosome among related mammals. Other genes, such as TK, are on the same chromosome in all the great apes, supporting the evidence from chromosomal banding that indicates a high genetic homology for these species (see fig. 4.40). In other words, if either the linkage relationship between two genes or the location of a gene on a chromosome is established in one species, similar species may have the same relationship. Knowledge of this sort could help shortcut the task of determining linkage patterns in each species. For example, the chromosomal location of the albino gene in humans is not known. However, the homologous albino genes in mice and cats are on regions with banding patterns like those on human chromosome 11p.

In the past, linkage of genes in humans generally was determined by accumulating information from informative matings. However, the technique of somatic cell hybridization with mouse cells did allow many more human genes to be assigned to particular chromosomes.

Cell line	A	B	C	D	E
Human chromosomes	1,4,17	2,9,17	2,17	4,11,22,17	5,9,17,20

FIGURE 5.12
Fusion of human and mouse cells as a technique to identify the chromosomal location of the gene for thymidine kinase. The only human chromosome the five cell lines have in common is 17, demonstrating that the Tk gene must be on this chromosome.

FOUR-STRANDED CROSSING OVER

Morgan's theories about the linkage of genes and crossing over were supported by extensive data. However, it was not clear initially whether recombination took place between chromosomes before they replicated (at the two-stranded stage) or after replication (at the four-stranded stage). If crossing over took place before replication, all four chromatids would be recombinants, while if it took place after replication, only two of the four chromatids would be recombinants (fig. 5.13).

To distinguish between these alternatives, a genetic system was necessary in which all of the genetic products from meiosis could be identified. The first such system was developed in *Drosophila,* utilizing the **attached X,** a complex chromosome that has two X chromosomes attached at its centromere. As a result of this attachment, gametes from females having the attached X chromosome and a separate Y chromosome ($\overline{XX}Y$) contain either two X chromosomes or no X chromosomes (just a Y). An example of a cross in which both X chromosomes carry the wild-type allele at the white locus is given in figure 5.14. $\overline{XX}Y$ females with the wild-type allele on both X chromosomes are crossed to normal white-eyed males. The viable offspring are identical to the parents, because the YY flies always die and the $\overline{XXX}$ flies usually die.

When the alleles are different on the two attached X chromosomes, two-stranded and four-stranded crossing over can be differentiated. It is useful to use a mutant that is identifiable as a heterozygote in this experiment. For example, the *Bar* mutant reduces the number of eye facets from

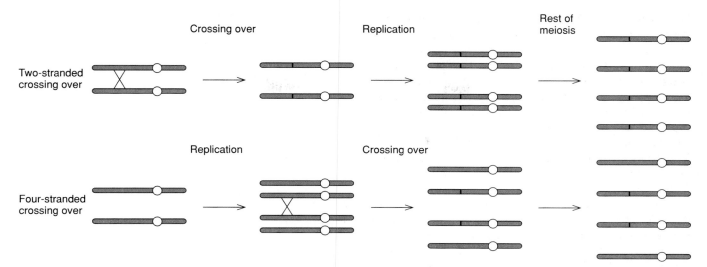

FIGURE 5.13

The different consequences of two-stranded and four-stranded crossing over. Four-stranded crossing over produces both parental and recombinant gametes, while two-stranded crossing over produces only recombinant gametes.

From: An Introduction to Genetic Analysis *3/e by David T. Suzuki, et al. Copyright © 1986, by W. H. Freeman and Company. Reprinted with permission.*

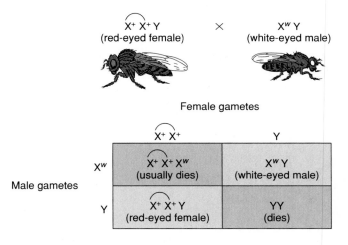

FIGURE 5.14

A cross between an attached X wild-type female and a white-eyed male. Notice that the surviving progeny are exactly like their parents.

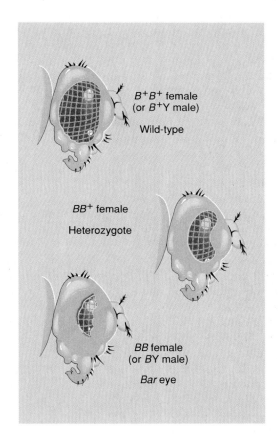

FIGURE 5.15

The phenotypes of wild-type, heterozygous *Bar* eye, and homozygous *Bar* eye.

the normal number of approximately 780 to about 360 in the heterozygotes and about 70 in the homozygote *BB* (fig. 5.15).

The top row of figure 5.16 indicates the types of chromosomes obtained when no crossing over occurs. In this case, all the gametes with X chromosomes have the heterozygous attached X and therefore the heterozygous *Bar* phenotype. Next, if crossing over takes place at the

two-stranded stage (before replication), it will involve both strands. Therefore, when replication occurs later, all gametes are again heterozygous. In other words, the observed outcomes for no crossing over and two-stranded crossing over are the same.

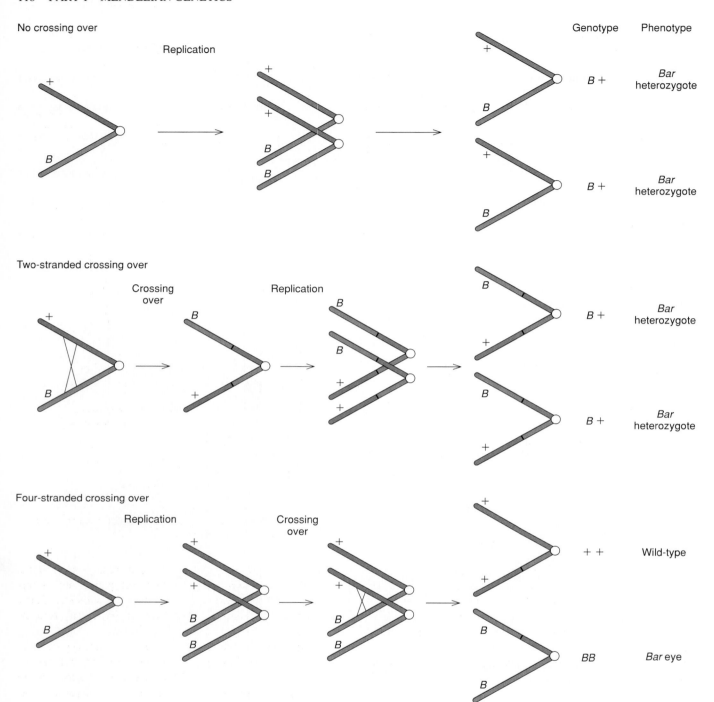

FIGURE 5.16
The gametes produced from an attached X heterozygous for *Bar* with no crossing over (top), followed by the results expected from two- and four-stranded crossing over. In reality, wild-type and *Bar* eye progeny are observed, indicating that crossing over takes place at the four-stranded stage.

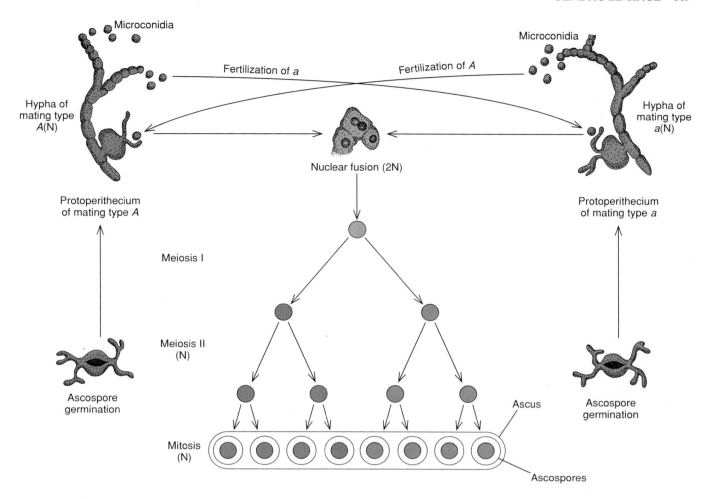

FIGURE 5.17

The production of ascospores following meiosis in *Neurospora crassa.* The ascospores are in a linear array such that each half is descended from the segregants at meiosis I. Each quarter is descended from chromatids associated with a particular centromere.

Recombination at the four-stranded stage yields different results in the progeny. In this instance, only two of the four strands participate in recombination, so that new types are produced. With recombination at the four-stranded stage, both wild-type and *BB* homozygotes should result. In fact, this is the observation from actual experiments, indicating that recombination takes place after replication—that is, at the four-stranded stage.

TETRAD ANALYSIS

Confirmation of four-stranded crossing over was also obtained in *Neurospora crassa,* a fungus with the useful genetic attribute of having its meiotic products arranged in linear fashion. The life cycle of *Neurospora* is given in figure 5.17, illustrating mating and the production of a cell with the diploid chromosome number, which then undergoes meiosis. The meiotic products, called ascospores, are aligned in an ascus (or sac) so that the four on the left half descend from one type of segregants in meiosis I, while the four on the right half descend from the complementary segregants. In addition, within each half of the ascospore, one-half are products from half of meiosis II and the other half

from the complementary half. After meiosis is completed, one mitotic division occurs, so that each pair of ascospores is descended from one of the products of meiosis II. These four products of meiosis, which are linearly arranged in the four pairs of ascospores, are called a **tetrad.**

When there is no crossing over in a heterozygote, one-half of the ascus has one allele and the other half has the second allele (fig. 5.18). All the ascospores on one half of the ascus have chromosomes with one of the centromeres and the other half have the chromosomes with the other centromere. However, when crossing over does occur between the gene and the centromere, the pairs of the ascospores alternate as to which alleles they contain. This observation could be true only if recombination took place at the four-stranded stage. Two-stranded recombination would give results indistinguishable from no crossing over, assuming that there were no other marker genes.

Because the centromeres in the top half of the ascospores in an ascus are descended from one centromere and the bottom half from the other centromere, the proportion of recombinant asci can be used to determine the rate of recombination between the centromere and a given gene.

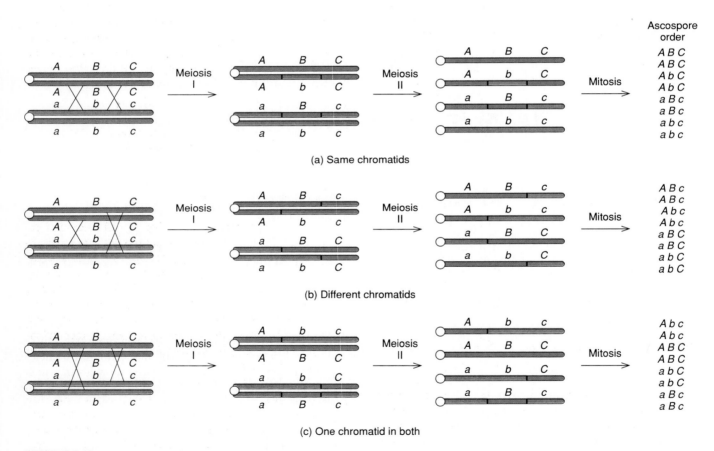

FIGURE 5.18
The ascospore order, illustrating the difference between no crossing over and a single crossover.

FIGURE 5.19
The ascospore order that results from the three different types of double crossovers: (*a*) Involving the same chromatids; (*b*) involving different chromatids; and (*c*) involving one chromatid in both events.

For example, if 84% of a group of scored asci are like the top half of figure 5.18 (all *A* on one side and all *a* on the other side of the ascus) and the remaining 16% are like the bottom half of figure 5.18 (having both *A* and *a* in the top half and both *A* and *a* on the bottom half), the rate of recombination is 0.16.

Other experiments in *Neurospora* have increased our understanding of the fundamental features of meiotic recombination. For example, the three basic ways that nonsister chromatids can have double crossing over are distinguishable in *Neurospora* (fig. 5.19). First, if there are three genes on the chromosome and both crossovers involve the same chromatids, let us say the center ones in figure 5.19*a*, then four types of ascospores are produced, with the parental ones on the outside. This is the type of double recombination event we assumed in our earlier discussion.

Second, different nonsister chromatids may be involved in the two recombination events (fig. 5.19*b*). And finally, one chromatid could be involved in both events, while the second chromatid is different in the two crossovers (fig. 5.19*c*). Notice that in this instance, one parental gametic type (*ab*) is on the outside of the ascus and the other (*AB*) is on the inside. An important observation from these results is that the products of recombination are reciprocal.

The fact that crossing over takes place in the four-stranded stage of meiosis I was first demonstrated using attached X chromosomes in *Drosophila*. Confirmation of this theory was obtained from studies of the fungus *Neurospora*.

INVERSIONS AND RECOMBINATION

When individuals are heterozygous for chromosomal variants, then the products of meiosis may not be viable, a phenomenon that we mentioned in the last chapter. For example, when an individual is heterozygous for an inversion or a translocation, the resulting gametes may have either duplications or deletions. We can understand these problems for inversions by using the information we have just developed on four-stranded recombination. Remember from the last chapter that inversions may either include the centromere (pericentric) or may not include the centromere (paracentric). When an individual is an inversion heterozygote for either type of inversion and recombination occurs only outside the inversion loop, the products of meiosis have the same genes and the same gene sequence as the parental chromosomes. However, if there is crossing over within the inversion loop, the chromosomes may be abnormal. Figure 5.20*a* shows the products of a recombination between genes *B* and *C* within a paracentric inversion loop. At anaphase I, the homologous chromosomes pull apart so that one chromatid involved in recombination is **acentric** (having no centromere) and the other is **dicentric** (having two centromeres). The acentric fragment usually does not go to a pole and becomes lost, while the dicentric fragment

breaks more or less at random between the two centromeres. When the centromere splits in meiosis II, the result is two balanced chromosomes—one with the normal sequence and one with the inverted sequence—and two chromosomes with either a deletion or a deletion and a duplication. In this example, one abnormal chromosome (*EFG*) is missing more than half the total chromosome, while the other is missing only the *A* region but has a duplication of the *D* region. Zygotes formed from gametes containing these chromosomes would in all probability be lethal.

Recombination within a pericentric inversion loop can also result in abnormal chromosomes. As an example, assume that recombination takes place between genes *D* and *E* (fig. 5.20*b*). Again there are two balanced chromosomes, but in this case the other pair has both a duplication and a deficiency. Zygotes with these unbalanced chromosomes would also likely be lethal.

As we have just seen, when recombination takes place within an inversion, the resulting chromosomes have either a deletion or a deletion and a duplication, and they are consequently often lethal. As a result, there appears to be a suppression of recombination in inverted regions, because no recombinant products are observed. Furthermore, in inversion heterokaryotypes, pairing difficulties often reduce recombination near and in the inverted region.

When recombination takes place in the inverted region of inversion heterokaryotypes, duplications and deficiencies show up in the meiotic products.

UNEQUAL CROSSING OVER

Duplication and deletion of a relatively small amount of genetic material can result from a mistake related to recombination. Chromatids are strongly attracted to their homologous counterparts in prophase I, and they pair gene by gene, as discussed earlier. Occasionally, however, pairing does not occur precisely as it should. In particular, when regions already have duplications or repeated sequences, homologous chromosomes sometimes mispair. If recombination occurs within this mispaired region, duplications or deletions can result.

The classic demonstration of this phenomenon is for the *Bar* mutant in *Drosophila melanogaster*. In stocks homozygous for *Bar*, approximately one of 1,600 new chromosomes is either wild-type or a more extreme variant having only twenty-five eye facets, called *Ultrabar* (*B^u*). It was shown by Sturtevant and Morgan that these are not new mutants but the products of recombination.

To understand the origin of *Ultrabar*, Sturtevant and Morgan constructed marked chromosomes that had the *Bar* mutations, as well as other mutations, on either side of the *Bar* region. In particular, these chromosomes had the recessive mutation for forked bristles (*f*) on one side and the recessive mutation causing fusion of two wing veins (*fu*) on

FIGURE 5.20
The meiotic products from a single crossover within (*a*) a paracentric inversion loop and (*b*) a pericentric inversion loop.

the other side (fig. 5.21). The map positions of these genes are 56.7 for *f*, 57.0 for *B*, and 59.5 for *fu*; thus, a relatively small region of about 3 map units on the X chromosome can be examined. Females homozygous at *Bar*, but heterozygous (and in coupling) at these other genes, were crossed with males having the triply mutant chromosome, so that any recombinant chromosomes could be identified. Of the approximately 20,000 progeny examined, most

chromosomes were nonrecombinant; that is, their chromosomes were either + *B* + or *f B fu*, because the marker loci were so tightly linked to *B*. Of the recombinant chromosomes, nearly all had *Bar* eyes, but seven had the wild-type allele and two had *Ultrabar*. In fact, all of the chromosomes that did not have the *Bar* allele were recombinant, three being *f* + +, four + + *fu*, and two *f B^u* +.

Normal pairing and normal crossing over

Mispairing and unequal crossing over

	Number of progeny
	0
	3
	4
	2/9

FIGURE 5.21

Top: The results of normal crossing over. Bottom: The results of mispairing and unequal crossing over that produce both wild-type and *Ultrabar* chromosomes. The number of progeny are those few abnormal progeny observed out of 20,000.

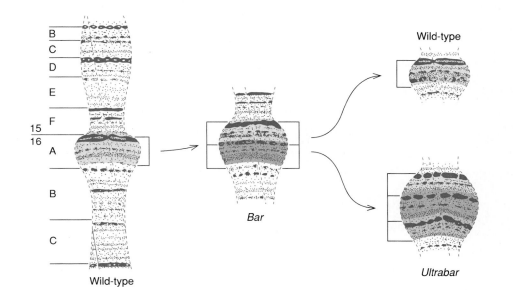

FIGURE 5.22

The banding patterns of homozygous *Bar* eye mutants. Note that wild-type, *Bar,* and *Ultrabar* have one, two, and three copies of band 16A, respectively.

From C. B. Bridges, in Science, *Volume 83, 1936. Copyright © 1936 American Association for the Advancement of Science, Washington, DC. Reprinted by permission of the publisher.*

Confirmation that duplication of the region was responsible for the *Bar* phenotype was obtained by examining in detail the banding pattern of the X chromosome. *Bar* chromosomes have a duplication of region 16A (fig. 5.22), and the subsequent unequal recombinants have either a single copy of the genetic material in this region (wild-type) or three copies (B^u).

In some cases, related genes may be closely linked, such as the different globin genes or the genes of the major histocompatibility complex (MHC) in mammals. (The MHC in humans is called the HLA region.) It is thought that linked genes of similar structure and function are originally generated by duplication resulting from unequal crossing over. A group of related genes is called a **multigene family** (see box 5.1).

❖ In some cases, unequal crossing over occurs, and this can increase the amount of DNA by generating duplications.

MAP DISTANCE AND PHYSICAL DISTANCE

As we have mentioned before, the map distance between two genes estimated by genetic crosses generally corresponds with the physical distance, or length of DNA, that separates the genes. One major exception is the centromeric region, which has no recombination but can contain a sizable length of DNA. Furthermore, it appears that recombination may often be localized into particular regions, and that it is not completely random over the length of the chromosome.

Comparing genetic and physical maps, figure 5.23 gives both the estimated map distance and the number of kilobases in the DNA (physical distance) for the major HLA genes. For example, genes *B* and *C* are close physically and are also only 0.1 map unit apart. On the other hand, genes *DP* and *DR* are physically closer than genes *C* and *A,* but the map distance is greater. This suggests that there may be higher than expected recombination between *DP* and *DR*.

Map distance |← 1.2 →|← 0.3 →|← 0.7 →| 0.1 |← 0.7 →|

Gene   DP   DQ   DR   Com-   B   C   A
                                plement

Physical distance   0   500   1,000   1,500   2,000   2,500   3,000   3,500

FIGURE 5.23

Comparison of map distance (given in map units) and physical distance
(given in kilobases) for the HLA region in humans.

Box 5.1 Multigene Families

Many genes are now known to exist as multigene families. Most of these families have probably arisen as a result of unequal crossing over and subsequent tandem duplication. Generally, the individual genes in a multigene family retain identical or similar functions. However, if the duplication occurred far back in the evolutionary past, the different individual genes in the family may have developed somewhat different functions.

A particularly striking multigene family is the cluster of linked genes comprising the **homeotic** genes in the flour beetle *Tribolium*. Different homeotic genes direct cells in different body segments of insects to differentiate into the appropriate anatomical structures (see figure B5.1.1). Five of these genes have been mapped to a chromosomal segment of 2.6 map units. In addition, the order of the genes is identical to the order of the body segments they affect (see figure B5.1.2).

FIGURE B5.1.1

Normal phenotypes for five closely linked homeotic loci in a flour beetle are shown in (*a*), (*c*), (*e*), and (*h*); mutant phenotypes are shown in (*b*), (*d*), (*f*), (*g*), and (*i*). The five genes that cause the mutant phenotypes are shown in figure B5.1.2.

Source: Richard W. Beeman, USDA.

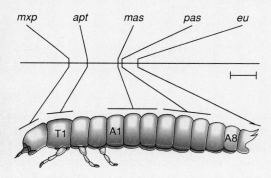

FIGURE B5.1.2

Five genes in *Tribolium*, showing the body segments they affect. (⊢────⊣ = one map unit length.)

Source: Richard W. Beeman, USDA.

Recently, a substantial effort has been focused toward obtaining a genetic map of the human genome (see figure 5.24 and chapter 15). Most of the landmarks on this map are DNA variants called **restriction fragment length polymorphisms (RFLPs);** we will discuss RFLPs more fully in chapter 15. Notice that some chromosomes, such as 1 and 7, are well covered by these genetic variants, while others, such as 14 and 19, have very few markers. Adjacent to the genetic map of each chromosome is the physical map based

on chromosome banding studies. This new map has great potential for helping us determine the location of genes that cause particular diseases and subsequently examine the DNA to see what has caused the defect.

> The genetic map distance and the physical DNA distance generally correspond quite well. Recently, DNA variants have been employed to map the human genome.

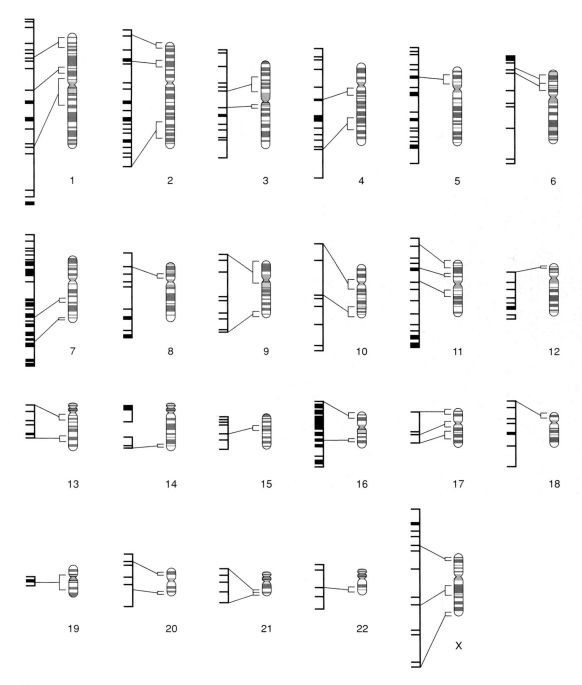

FIGURE 5.24

A preliminary map of the human genome (more detailed maps are now available) in which the left-hand diagram for each chromosome indicates the relative map distance between different RFLPs. The right-hand chromosome gives the major banding patterns along with the general location of some RFLP sites.

WORKED-OUT PROBLEMS

Problem 1

A wild-type strain of flies was crossed to another strain homozygous for the recessive X-linked mutants *m, r,* and *v*. The F_1 females were backcrossed to recessive strain males, and the following numbers of progeny were observed:

+	+	+	331
m	+	+	2
+	*r*	+	73
+	+	*v*	14
m	*r*	+	10
m	+	*v*	81
+	*r*	*v*	0
m	*r*	*v*	309
			820

What is the gene order and the map distance between these mutants?

Solution

There are three possible orders of the genes: *m-r-v, m-v-r,* and *r-m-v*. The two smallest progeny classes must be progeny produced by double recombination; that is, classes *m* + +, and + *rv*. The only way that these progeny can be produced from + *m* + *r* + *v* females by double recombination is if gene *m* is in the center, the gene order *v-m-r*. (See figure 5.6 as an example for genes *y, w,* and *m*.)

To calculate the map distance, rewrite the gene order for these progeny with *m* in the middle, count up all the progeny with recombination in a given region, and divide by the total number of progeny:

Genotype			Numbers	Region of recombination
+	+	+	331	None
+	+	+	2	*v–m, m–r*
+	+	*r*	73	*m–r*
v	+	+	14	*v–m*
+	*m*	*r*	10	*v–m*
v	*m*	*r*	81	*m–r*
v	*m*	+	0	*v–m, m–r*
v	*m*	*r*	309	None
			820	

For the region between *v* and *m*, the rate of recombination is:

$$v - m = (2 + 14 + 10 + 0)/820 = 0.032$$

For the region between *m* and *r*, the rate of recombination is:

$$m - r = (2 + 73 + 81 + 0)/820 = 0.190$$

In other words, the numbers of map units between *v* and *m* is 3.2; between *m* and *r*, 19; and between *v* and *r*, 22.2. The map units between *v* and *m*, as well as between *m* and *r* are calculated by multiplying the rates of recombination by 100.

Problem 2

Let us assume that we have scored a number of *Neurospora* asci for the alleles at two loci that were heterozygous before meiosis. The results of this cross follow, first for the *A* gene only.

Ascus type	Number
AAAA aaaa	56
AAaa AAaa	8
AAaa aaAA	11
aaAA aaAA	12
aaAA AAaa	7
aaaa AAAA	62

What is the rate of recombination between the centromere and the *A* gene?

The types of asci when they are scored for both the *A* and *B* genes are given below. In this case, the two gametes that were fused at the beginning of meiosis were *A B* and *a b*. What is the rate of recombination between these two loci?

Ascus Type			
Top half	*A B*	*a b*	0.5 *A B*, 0.5 *A b*	0.5 *a b*, 0.5 *a B*
Bottom half	*a b*	*A B*	0.5 *a b*, 0.5 *a B*	0.5 *A B*, 0.5 *A b*
Number	54	51	23	28

Solution

The rate of recombination between the centromere and the *A* gene is the number of recombinant asci divided by the total number of asci scored. In this data set, the middle four types were recombinant, i.e., not all *A* or not all *a* on one side of the ascus, so the rate of recombination is (8 + 11 + 12 + 7)/156 = 0.244.

The rate of recombination between loci *A* and *B* is calculated in a similar manner. In this case the first two columns were nonrecombinant asci, so the rate of recombination is the number in the last two columns divided by the total number of asci scored or (23 + 28)/156 = 0.327.

Problem 3

A researcher wanted to find out the map distance between two loci in rats, which we will indicate as *C* and *D*. She crossed inbred lines that were *CC DD* and *cc dd* (Cross 1) and then backcrossed the progeny of this cross to inbred line *cc dd* and scored the progeny as recombinant or nonrecombinant. In another cross between inbred lines *CC dd* and *cc DD* (Cross 2), she also crossed the progeny back to *cc dd* line and scored the progeny. Given the data below from these crosses, what is the estimated rate of recombination between these two loci?

	Numbers	
Progeny type	*Cross 1*	*Cross 2*
Cc Dd	43	12
Cc dd	8	59
cc Dd	11	62
cc dd	51	15

Solution

Because the only recombinant gametes in Cross 1 are *C d* and *c D* (the parental gametes are *C D* and *c d*), the number of recombinant gametes from this cross is 8 + 11 = 19, i.e., the sum of the number of the *Cc dd* or *cc Dd* progeny. Likewise, the number of recombinant gametes from Cross 2 is 12 + 15 = 27, i.e., the number of *Cc dd* or *cc Dd* progeny (the parental gametes in this cross are *C d* and *c D*). Therefore, the total number of recombinant gametes is 19 + 27 = 46. The total number of progeny (or gametes) scored is 113 + 148 = 261. The overall rate of recombination is then 46/261 = 0.176.

Summary

In the early 1900s, exceptions to Mendel's principle of independent assortment were discovered—namely, that alleles at different genes on parental chromosomes tend to remain together because of the distance between genes on a chromosome. The rate of recombination estimates the linkage of genes and ranges from 0 for extremely tightly linked loci to ½ for very loosely linked loci (or for those on different chromosomes). A genetic map illustrates the relationship of genes on a chromosome. Three-point crosses are used to determine the order of genes on a chromosome. In a phenomenon known as interference, the frequency of double recombination between closely linked genes is depressed compared to the frequency expected.

Linkage of genes in humans has been difficult to determine. Originally, it was done by accumulating information from informative matings, but another technique using somatic cell hybridization with mouse cells has located many more human genes on particular chromosomes. The fact that crossing over takes place in the four-stranded stage of meiosis I was first demonstrated in a study of attached X chromosomes in *Drosophila*. The theory was later confirmed using the fungus *Neurospora*.

In some cases, crossing over occurs unequally, thus generating duplications and increasing the amount of DNA. Generally, there is a close correspondence between the genetic map distance and the physical DNA distance. DNA variants are now being used to map the human genome.

PROBLEMS AND QUESTIONS

1. Which principle of Mendel's does linkage violate? Explain, in a sentence, how linkage causes a result different from this principle's expectations.

2. In a cross between two F_1 garden peas having genotype *Pp Ll,* the progeny types were 142 purple, long; 8 purple, round; 11 red, long; and 47 red, round. (This is a progeny array similar to that shown in table 5.1.) Does this fit 9:3:3:1 expectations, using a chi-square test?

3. Using a testcross, a recombination rate of 0.114 was found for the genes for purple and long. In the cross by Bateson and Punnett, what proportions of the four gametes would you expect from F_1 parents?

4. Using the recombination rate in question 3, what proportions of the four phenotypes would you expect

in the F_2 progeny? Use a chi-square test to see if the observed data in table 5.1 are consistent with this explanation.

5. Why did Mendel not discover linkage as perhaps a third principle?

6. Calculate the rate of recombination between gene *pr* and *vg* using the data from Morgan's second testcross (table 5.3). Compare your results to the value calculated from the first testcross in which the parents were in coupling.

7. Show diagrammatically (using chromosomes as in figure 5.3) how gametes for the F_2 are produced when the original parents are $prpr\ vg^+vg^+$ and pr^+pr^+vgvg.

8. In a series of testcrosses, the rate of recombination between genes *a* and *c* is 0.1; between *a* and *b*, 0.27; and between *b* and *c*, 0.2. Determine the order of the three genes. Explain why the sum of rates of recombination for the two closest pairs of genes is not equal to the rate between the two genes farthest apart.

9. Another gene, *d,* had a rate of recombination of 0.15 with gene *b* and a rate of 0.14 with gene *a*. Where would you place this gene on the map constructed for question 8? What cross would you make to confirm your map?

10. Why were three-point crosses a valuable approach to learning about the nature of the linkage of genes?

11. Assume a three-point cross in which the $F_1kk^+ll^+mm^+$ was crossed to *kkllmm* and the offspring had the following genotypes:

$kk^+ll^+mm^+$	621
kk^+ll^+mm	3
kk^+llmm^+	64
$kkll^+mm^+$	103
kk^+llmm	109
$kkll^+mm$	57
$kkllmm^+$	7
$kkllmm$	608

Construct a genetic map of these three loci, including their order and the map distance between them.

12. Using the data in problem 11, calculate the expected proportion of double recombinants. What is the value of interference (I) estimated from these data? Explain what this level of I means.

13. Diagram the types of gametes you would expect from a *y+/+w* female, given the following situations between these loci: no crossovers, one crossover, two crossovers, three crossovers, and four crossovers.

14. Why do you think it was much harder to determine linkage patterns in humans than in other species such as *Drosophila* or corn?

15. In a large family, the mother is heterozygous for both hemophilia and color blindness with repulsion gametes, or genotype *H c/h C*. Two of her sons have hemophilia, two other sons are color blind, and the fifth son is normal. Following the style of figure 5.9, give the genotypes for the sons. What is the estimated rate of recombination between the genes in this family?

16. Several families having a female parent heterozygous for both hemophilia and color blindness were gathered. In these families, the following numbers of recombinants and sons were observed: 1 of 5 sons, 2 of 8, 1 of 4, 1 of 5, and 1 of 6. What is the estimated rate of recombination, using all these data? In fact, the actual rate is much less. Can you give a possible explanation?

17. Assume that a deletion heterozygote (A_1–) was mated with an A_2A_2 individual. What type of progeny, and in what proportions, would you expect? Would this differ from the results expected if the first parent were A_1A_1?

18. Using somatic cell hybridization, a number of cells having a particular human enzyme were generated. Five of these cell lines had human chromosomes: (A) 4, 9, 11; (B) 2, 7, 11, 21; (C) 4, 9, 10, 11, 21; (D) 1, 2, 11, 18; and (E) 4, 11, 15. On which chromosome is the gene for this enzyme?

19. Discuss the difference between two-stranded and four-stranded crossing over. What would proportions (parental and nonparental) of the resulting gametes be in two-stranded crossing over?

20. Give the progeny genotypes and phenotypes of a cross between $\bar{X}^w\bar{X}^+Y$ and X^+Y where the two X chromosomes in the female parent are attached. What types of progeny survive? Why can't this cross be used to differentiate between two- and four-stranded crossing over?

21. Assume that *Neurospora* with allele *D* are crossed to those with allele *d*. If there were no crossing over between the centromere and this gene, what order of ascospores would you expect? If there was one crossover, what order would you expect?

22. Assume that we have two genes, *E* and *F*, which we know are on a particular *Neurospora* chromosome. If we cross *EF* with *ef* and examine the resulting asci, most of them contain the following sequence of ascopores: *EF, EF, EF, EF, ef, ef, ef, ef*. However, some are *EF, EF, eF, eF, Ef, Ef, ef, ef*, or *eF, eF, EF, EF, ef, ef, Ef, Ef*. What is the order of these genes and the centromere?

23. Sturtevant and Morgan used genes *f* (56.7) and *fu* (59.5) as markers in a cross to examine *Ultrabar*. What proportion of recombinant progeny would you expect between these markers? Sturtevant and Morgan found 9 out of approximately 20,000 progeny that were recombinant for the markers and had either wild-type or *Ultrabar* alleles. Assuming that mispairing takes place one out of ten times for *Bar* homozygotes, what is the estimated size of the region that can produce duplications or deficiencies?

24. In this chapter, we have been focusing on genetic or map distance between genes. Generally, map distance is similar to physical distance on the chromosome, but in some cases it is not. Suggest reasons why map and physical distance may not be exactly concordant.

25. Assume that two of the genes Mendel picked for his studies in garden peas were tightly linked. How might this have influenced the principle of independent assortment?

Answers appear at end of book.

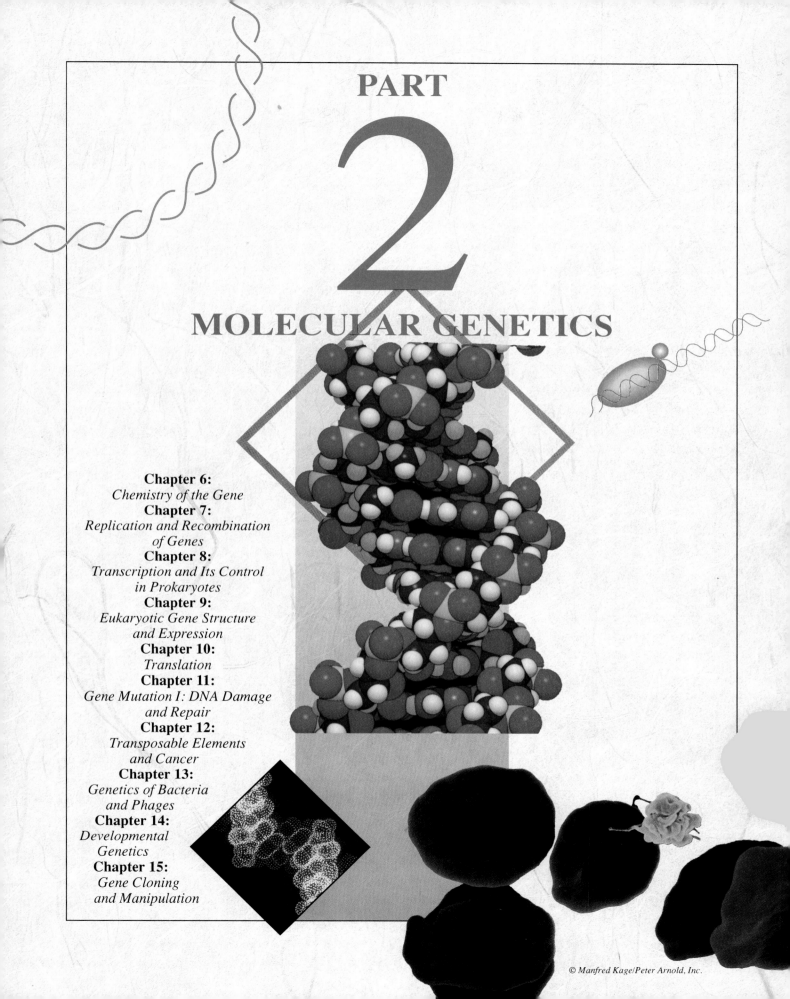

PART

2

MOLECULAR GENETICS

Chapter 6:
Chemistry of the Gene
Chapter 7:
*Replication and Recombination
of Genes*
Chapter 8:
*Transcription and Its Control
in Prokaryotes*
Chapter 9:
*Eukaryotic Gene Structure
and Expression*
Chapter 10:
Translation
Chapter 11:
*Gene Mutation I: DNA Damage
and Repair*
Chapter 12:
*Transposable Elements
and Cancer*
Chapter 13:
*Genetics of Bacteria
and Phages*
Chapter 14:
*Developmental
Genetics*
Chapter 15:
*Gene Cloning
and Manipulation*

CHAPTER 6: CHEMISTRY OF THE GENE

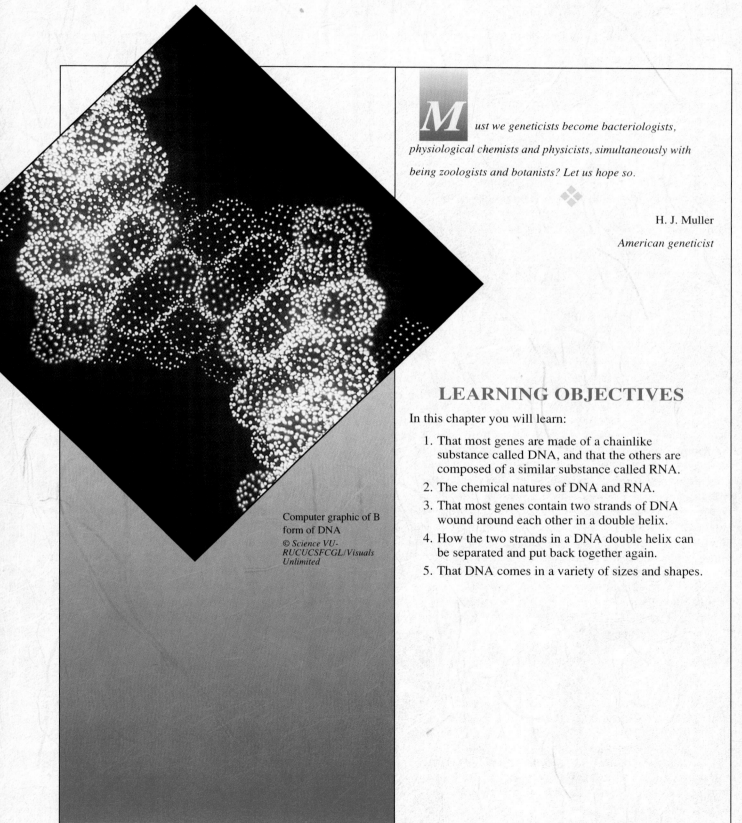

Computer graphic of B form of DNA
© *Science VU-RUCUCSFCGL/Visuals Unlimited*

*M*ust we geneticists become bacteriologists, physiological chemists and physicists, simultaneously with being zoologists and botanists? Let us hope so.

❖

H. J. Muller

American geneticist

LEARNING OBJECTIVES

In this chapter you will learn:

1. That most genes are made of a chainlike substance called DNA, and that the others are composed of a similar substance called RNA.

2. The chemical natures of DNA and RNA.

3. That most genes contain two strands of DNA wound around each other in a double helix.

4. How the two strands in a DNA double helix can be separated and put back together again.

5. That DNA comes in a variety of sizes and shapes.

ife, at its base, is chemistry. All living things, including you, are made of molecules. Moreover, thousands of chemical reactions are occurring in your body all the time; without these, your life would not be possible. In order to understand life at its most fundamental level, we must know something about its chemistry.

This statement applies equally well to genetics. It is certainly possible to do genetic experiments using whole organisms, mating them and determining the phenotypes of their offspring. This can be very fruitful, but it does not tell the whole story. That is because genes themselves are molecules—extremely large, *informational* molecules that contain the instructions for making other biomolecules. To understand genetics fully, we need to understand the structure and function of genes as molecules. This chapter introduces the molecular structure of genes.

THE NATURE OF GENETIC MATERIAL

If we are going to discuss molecular genetics, we must first get to know the molecules. The three molecules at the heart of genetics—DNA, RNA, and protein—have already been mentioned in chapter 1. We will examine DNA and RNA in more detail later in this chapter, but first it will help to introduce a few very basic chemical structures that recur over and over in these molecules.

Substances derived from living things are termed "organic," and chemists had noted even before Mendel's time that organic molecules are rich in carbon. This led to the division of the discipline of chemistry into two parts—organic chemistry, the study of carbon-containing compounds, and inorganic chemistry, the study of all other compounds. This is not to imply that all organic molecules come from living things. Plastics such as polyethylene and nylon are surely organic substances, but they are just as surely synthetic.

The molecules we will be talking about in our study of genetics are organic, or carbon-containing, molecules. The simplest organic compound is the natural biological product natural gas, or methane. Its structure:

$$\begin{array}{c} \text{H} \\ | \\ \text{H}-\text{C}-\text{H} \\ | \\ \text{H} \end{array}$$

contains only hydrogen and carbon, so it is termed a hydrocarbon.

Carbon can bond to a variety of other atoms. In genetics, the most important of these are nitrogen, oxygen, hydrogen, phosphorus, and other carbons. Frequently, these other atoms are present in groups, sometimes called **functional groups,** as illustrated in table 6.1. Notice that a shorthand convention is used to represent most of these

groups. Thus, the methyl group, with a carbon and three hydrogens, is not usually shown with its three hydrogens projecting out as in the structure of methane, but simply as CH_3.

Note that phosphate is written with two negative charges. At very low pH, both of the negatively charged oxygens would be linked to hydrogen atoms. However, at neutral pH, such as we find in living cells, the positively charged hydrogen nuclei (protons) dissociate, leaving negative ions behind. The same principle applies to most carboxyl groups; for example, acetic acid is mostly deprotonated (ionized) at neutral pH, yielding acetate (CH_3—COO^-). These deprotonation reactions are illustrated in figure 6.1.

We will encounter these functional groups repeatedly in this and other chapters. We will also see how some of these groups can link together to form different kinds of compounds.

The studies that have led to our understanding of the gene's chemistry began in Tübingen, Germany, in 1869. There, Friedrich Miescher isolated nuclei from pus cells in waste surgical bandages. He found that these nuclei contained a novel phosphorus-bearing compound that he named **nuclein.** Nuclein is mostly **chromatin,** which is a complex of **deoxyribonucleic acid (DNA)** and chromosomal proteins.

By the end of the nineteenth century, both DNA and **ribonucleic acid (RNA)** had been separated from the protein that clings to them in the cell. This allowed more detailed chemical analysis of these **nucleic acids.** (Notice that the term nucleic acid and its derivatives, DNA and RNA, come directly from Miescher's term "nuclein.") By the beginning of the 1930s, P. Levene, W. Jacobs, and others had demonstrated that RNA is composed of a sugar (ribose) plus four nitrogen-containing bases, and that DNA contains a different sugar (deoxyribose) plus four bases. They discovered that each base is coupled with a sugar-phosphate to form a nucleotide. Furthermore, the four bases in DNA or RNA seemed to be present in roughly equal quantities.

TRANSFORMATION IN BACTERIA

In 1928, Frederick Griffith laid the foundation for the identification of DNA as the genetic material with his experiments on **transformation** in the bacterium *Pneumococcus,* now known as *Streptococcus pneumoniae.* The wild-type organism is a spherical cell surrounded by a mucous coat called a capsule. The cells form large, glistening colonies, characterized as smooth (S) (fig. 6.2). These cells are **virulent**—capable of causing lethal infections upon injection into mice. A mutant strain of *S. pneumoniae* has lost the ability to form a capsule. As a result, it grows as small, rough (R) colonies. More importantly, it is **avirulent;** it has no protective coat, so it is engulfed by the host's white blood cells before it can proliferate enough to do any damage.

Table 6.1 Functional Groups Found in Organic Molecules

			Example
Group Name	Group Structure	Name	Structure
Hydroxyl	— OH	Ethanol (grain alcohol)	CH_3—CH_2—OH, or H—C—C—OH (with H atoms)
Methyl	— CH_3, or —C—H	Methanol (wood alcohol)	CH_3—OH, or H—C—OH
Carbonyl	C=O	Formaldehyde	H_2C=O, or C=O
Carboxyl	—C—OH (with O double bond)	Acetic acid	CH_3—C—OH, or H—C—C—OH
Amino	— NH_2	Methylamine	CH_3—NH_2, or H—C—NH_2
Phosphate	—O—P—O⁻ (with O double bond and O⁻)	Methyl phosphate	CH_3—O—P—O⁻, or H—C—O—P—O⁻

FIGURE 6.1

Deprotonation reactions. (*a*) Methyl phosphoric acid loses two protons (red and blue) in successive deprotonations. (*b*) Acetic acid loses a proton (red) in its conversion to acetate.

(a)

(b)

FIGURE 6.2

Variants of *Streptococcus pneumoniae*. (*a*) The large, glossy colonies contain smooth (S) virulent bacteria; (*b*) the small, mottled colonies are composed of rough (R) avirulent bacteria.

Photographs by Harriett Ephrussi-Taylor.

The key finding of Griffith's work was that heat-killed virulent colonies of *S. pneumoniae* could transform avirulent cells to virulent ones. Neither the heat-killed virulent bacteria nor the live avirulent ones by themselves could cause a lethal infection. Together, however, they were deadly. Somehow the virulent trait passed from the dead cells to the live, avirulent ones. This transformation phenomenon is illustrated in figure 6.3. Transformation was not transient; the ability to make a capsule and therefore to kill host animals, once conferred upon the avirulent bacteria, was passed to their descendants as a heritable trait. In other words, the gene for virulence, missing in the avirulent cells, was somehow restored during transformation. This made it very likely that the transforming agent in the heat-killed bacteria was the gene for virulence itself. The missing piece of the puzzle was the chemical nature of this agent. Whoever discovered this would reveal the nature of genes.

DNA: THE TRANSFORMING SUBSTANCE

Oswald Avery, Colin MacLeod, and Maclyn McCarty supplied the missing piece in 1944. They used a variation of the transformation assay that Griffith had introduced, but they went a step further and took pains to define the transforming substance from virulent cells. First, they destroyed

the protein in the extract with organic solvents and found that it still transformed. Next, they subjected it to digestion with various enzymes. Trypsin and chymotrypsin, which destroy protein, had no effect on transformation. Neither did ribonuclease, which degrades RNA. These experiments ruled out protein or RNA as the transforming agent. On the other hand, Avery and his coworkers found that deoxyribonuclease (DNase), which breaks down DNA, destroyed the transforming ability of the virulent cell extract. Finally, direct physical-chemical analysis showed the purified transforming agent to be DNA. The analytical tools Avery and his colleagues used are described in box 6.1.

FURTHER CONFIRMATION

These findings should have settled the issue, but they had little immediate impact. The mistaken notion that DNA was a monotonous repeat of a four-nucleotide sequence, such as ACTG, persuaded many geneticists that it could not be the genetic material. Furthermore, controversy persisted about possible protein contamination in the transforming agent, whether transformation would work with other characteristics besides R and S, and even whether bacterial genes were like the genes of higher organisms.

Yet, by 1953, when James Watson and Francis Crick published the double-helical model of DNA structure, most geneticists agreed that genes were made of DNA. What had changed? For one thing, Erwin Chargaff had shown in 1950 that the bases were not really found in equal proportions in DNA as previous evidence had suggested, and that the base composition of DNA varied from one species to another. This is in fact exactly what one would expect for genes, which also vary from one species to another. Furthermore, Rollin Hotchkiss had refined and extended Avery's findings. He purified the transforming substance to the point where it contained only 0.02% protein, and showed that it could change other genetic characteristics besides R and S.

Finally, in 1952, A. D. Hershey and Martha Chase performed another experiment that was not definitive by itself but added to the weight of evidence that genes were made of DNA. This experiment involved a **phage** (bacterial virus) called T2 that infects the bacterium *Escherichia coli* (fig. 6.4). During infection, the phage genes enter the host cell and direct the synthesis of new phage particles. The phage is composed of protein and DNA only. The question is: Do the genes reside in the protein or in the DNA? The Hershey-Chase experiment answered this question by showing that most of the DNA entered the bacterium, along with only a little protein; the bulk of the protein stayed on the outside (fig. 6.5). Since the DNA is the major component that gets into the host cells, it is likely that it contains the genes. Of course, this conclusion is not unequivocal; the small amount of protein that entered along with the DNA could conceivably carry the genes. But taken together with the more definitive work that had gone before, this study helped convince geneticists that genes are made of DNA.

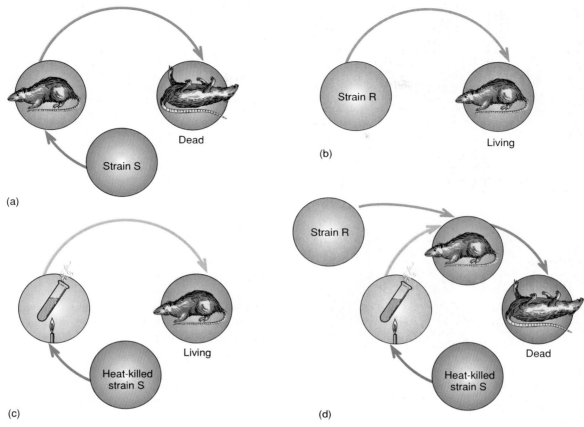

FIGURE 6.3

Griffith's transformation experiments. (*a*) Virulent strain *S. pneumoniae* kill their host; (*b*) avirulent strain R bacteria cannot infect successfully, so the mouse survives; (*c*) strain S bacteria that are heat-killed can no longer infect; (*d*) a mixture of strain R and heat-killed strain S bacteria kills the

mouse. The killed virulent (S) bacteria have transformed the avirulent (R) bacteria to virulent (S).

From R. Sager and J. Ryan, Cell Heredity. Copyright © 1961 John Wiley & Sons, Inc., New York, NY.

Box 6.1 Identifying DNA as the Genetic Material

ere are the experimental methods Avery, MacLeod, and McCarty used to demonstrate that the transforming substance is DNA:

1. *Ultracentrifugation* (spinning the transforming substance in an ultracentrifuge to estimate its size). The material with transforming activity sedimented rapidly (moved rapidly toward the bottom of the centrifuge tube),

suggesting a very high molecular weight, characteristic of DNA.

2. *Electrophoresis* (placing the transforming substance in an electric field to see how rapidly it moved). The transforming activity had a high mobility, also characteristic of DNA.

3. *Ultraviolet absorption spectrophotometry*. A solution of the transforming substance was placed in a spectrophotometer to see what kind of ultraviolet light it absorbed most strongly. Its absorption spectrum matched that of DNA. That is, the

light it absorbed most strongly had a wavelength of about 260 nanometers (nm).

4. *Elementary chemical analysis*. This yielded an average nitrogen/phosphorus ratio of 1.67, about what one would expect for DNA, which is rich in both elements, but vastly lower than the value expected for protein, which is rich in nitrogen but poor in phosphorus. Even a slight protein contamination would have revealed itself by raising the nitrogen/phosphorus ratio.

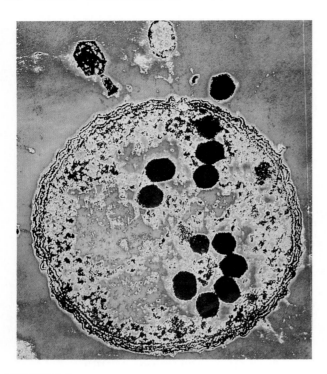

FIGURE 6.4

An electron micrograph of T2 phages infecting an *E. coli* cell. At top, a phage particle is about to inject its DNA into the host cell. However, another T2 phage has already infected the cell, and progeny phage particles are being assembled. The progeny phage heads are readily discernible as dark polygons inside the host cell.

Courtesy of Lee D. Simon, Rutgers University, New Brunswick, NJ. From the cover of Cell, *vol. 41. Reprinted with permission from Cell Press.*

This experiment depended on radioactive labels on the DNA and protein—a different label for each. Such labeling became possible as a by-product of nuclear research before and during World War II. The labels used were phosphorus-32 (^{32}P) for DNA and sulfur-35 (^{35}S) for protein. These choices make sense, considering that DNA is rich in phosphorus while the phage protein has none, and that protein contains sulfur but DNA does not.

The labeled phages were allowed to attach by their tails to bacteria and inject their genes into their hosts. Then the empty phage coats were removed by homogenizing in a Waring blender. (Yes, they had those back then.) Since we know that the genes must go into the cell, our question is: What went in, the ^{32}P-labeled DNA or the ^{35}S-labeled protein? Of course it was the DNA. In general, then, genes are made of DNA.

The genetic system explored by Hershey and Chase was a phage—a bacterial virus. A virus particle by itself is essentially just a bag of genes. It has no life of its own, no metabolic activity; it is inert. But when the virus infects a host cell, it seems to come to life. Suddenly the host cell

begins making viral proteins. Finally, the viral genes are replicated and the newly made genes, together with viral coat proteins, assemble into progeny virus particles. Because of their behavior as inert particles outside, but lifelike activity inside their hosts, viruses resist classification. Some scientists refer to them as "living things" or even "organisms." We prefer a label that, although more cumbersome, is also more descriptive of a virus's less-than-living status; thus, we will call a virus a "genetic system." This term applies to any entity—be it a true organism, a virus, or even bare circles of infectious RNA called **viroids**—that contains genes capable of replicating.

Most genetic systems studied to date contain genes made of DNA. But some viruses, including several bacterial phages, plant and animal viruses (e.g., influenza virus and HIV, the AIDS virus), and all viroids, have RNA genes. Sometimes viral RNA genes are double-stranded, but usually they are single-stranded.

We have already encountered one famous example of the use of viruses in genetics research. We will see many more in subsequent chapters. In fact, without viruses, the field of genetics would be immeasurably poorer.

> Genes are made of nucleic acids, usually DNA, but some simple genetic systems such as viruses have RNA genes.

THE CHEMICAL NATURE OF POLYNUCLEOTIDES

By the mid-1940s, the fundamental chemical structures of DNA and RNA were known. When DNA was broken into its component parts, these constituents were found to be nitrogenous **bases, phosphoric acid,** and the sugar **deoxyribose.** Similarly, RNA yielded bases and phosphoric acid, plus a different sugar, **ribose.** The four main bases found in DNA are **adenine (A), cytosine (C), guanine (G),** and **thymine (T).** RNA contains the same bases, except that **uracil (U)** replaces thymine. The structures of these bases, shown in figure 6.6, reveal that adenine and guanine are derived from the parent molecule, purine. Therefore, we refer to these compounds as **purines.** The other bases resemble pyrimidine, so they are called **pyrimidines.** These structures constitute the alphabet of genetics.

The sugars found in nucleic acids are depicted in figure 6.7. Notice that they differ only in one place. Where ribose contains a hydroxyl (OH) group in the 2-position, deoxyribose lacks the oxygen and simply has a hydrogen (H) represented by the vertical line. Hence the name *deoxyribose.* The bases and sugars in RNA and DNA are joined

(a)

(b)

FIGURE 6.5

The Hershey-Chase experiment. Phage T2 contains genes that allow it to replicate in *E. coli*. Since the phage is composed of DNA and protein only, its genes must be made of one of these substances. To discover which, Hershey and Chase performed a two-part experiment. In the first part (*a*), they labeled the phage protein with ^{35}S (blue), leaving the DNA unlabeled (black). In the second part (*b*), they labeled the phage DNA with ^{32}P (red), leaving the protein unlabeled (black). Since the phage genes must enter the cell, the experimenters reasoned that the type of label found in the infected cells would indicate the nature of the genes. Most of the labeled protein remained on the outside and was stripped off the cells by blending (*a*), while most of the labeled DNA entered the infected cells (*b*). The conclusion was that the genes of this phage are made of DNA.

FIGURE 6.6

The bases of DNA and RNA. The parent bases, purine and pyrimidine, on the left, are not found in DNA and RNA. They are shown for comparison with the other five bases.

FIGURE 6.7

The sugars of nucleic acids. Note the OH in the 2-position of ribose.

together into units called **nucleosides** (fig. 6.8). The names of the nucleosides derive from the corresponding bases:

Base	Nucleoside (RNA)	Deoxynucleoside (DNA)
Adenine	Adenosine	Deoxyadenosine
Guanine	Guanosine	Deoxyguanosine
Cytosine	Cytidine	Deoxycytidine
Uracil	Uridine	Not usually found
Thymine	Not usually found	(Deoxy)thymidine

Because thymine is not usually found in RNA, the "deoxy" designation for its nucleoside is frequently assumed, and the deoxynucleoside is simply called thymidine. There does not seem to be much rhyme or reason to these names, but they are important. It is also worth learning the numbering of the carbon atoms in the sugars of the nucleosides (fig. 6.8). Note that the ordinary numbers are used in the bases, so the carbons in the sugars are called by primed numbers. Thus, for example, the base is linked to the 1′-position of the sugar, and the 2′-position is deoxy in deoxynucleosides.

The structures in figure 6.6 were drawn using an organic chemistry shorthand that leaves out certain atoms

FIGURE 6.8

Two examples of nucleosides.

FIGURE 6.9

The structures of (*a*) adenine and (*b*) deoxyribose. Note that the structures on the left omit most or all of the carbons and some of the hydrogens. These are included in the structures on the right, in red and blue, respectively.

for simplicity's sake. Figures 6.7 and 6.8 use a slightly different convention in which a straight line with a free end denotes a C–H bond. Figure 6.9 shows the structures of adenine and deoxyribose, first in shorthand, then with every atom included.

The subunits of DNA and RNA are **nucleotides,** which are nucleosides with one or more phosphate groups attached through phospho-ester bonds (fig. 6.10). An ester is an organic compound formed from an alcohol (bearing a hydroxyl group) and an acid. In the case of a nucleotide, the alcohol group is the 5′-hydroxyl of the sugar, and the acid is phosphoric acid, which is why we call the ester a phospho-ester. The structure of one of the four precursors of DNA synthesis, deoxyadenosine-5′-triphosphate (dATP), is also shown in figure 6.10. Upon incorporation of dATP into DNA, two phosphate groups are removed, leaving deoxyadenosine-5′-monophosphate (dAMP). The other three nucleotides in DNA (dCMP, dGMP, and dTMP) have analogous structures and names.

FIGURE 6.10

Three nucleotides. The 5′-nucleotides of deoxyadenosine are formed by phosphorylating the 5′-hydroxyl group. The addition of one phosphate results in deoxyadenosine-5′-monophosphate (dAMP). One more phosphate yields deoxyadenosine-5′-diphosphate (dADP). Three phosphates give deoxyadenosine-5′-triphosphate (dATP).

We will discuss the synthesis of DNA in detail in chapter 7. For now, it will be sufficient to point out that the nucleotides are joined together in DNA or RNA by **phosphodiester bonds.** These bonds are called phosphodiesters because they involve phosphoric acid linked to *two* sugars—one through the 5′-group, the other through the 3′-group (fig. 6.11). You will notice that the bases have been rotated in this picture, relative to their positions in previous figures. This is more like their geometry in DNA or RNA. Note also that this **trinucleotide**, or string of three nucleotides, has polarity: The top of the molecule bears a free 5′-phosphate group, so it is called the **5′-end.** The bottom, with a free 3′-hydroxyl group, is called the **3′-end.**

 DNA and RNA are chainlike molecules composed of subunits called nucleotides. The nucleotides, in turn, contain a base linked to the 1′-position of a sugar (ribose in RNA or deoxyribose in DNA) and a phosphate group. The phosphate joins the nucleotides in a DNA or RNA chain through their 5′- and 3′-hydroxyl groups by phosphodiester bonds.

DNA STRUCTURE

All the facts about DNA and RNA just mentioned were known by the end of the 1940s, and by that time, it was also becoming clear that DNA was the genetic material and that it therefore stood at the very center of the study of life. Yet the detailed structure of DNA was unknown. For these reasons, several researchers dedicated themselves to solving the riddle of DNA.

EXPERIMENTAL BACKGROUND

One of the scientists interested in DNA structure was Linus Pauling, a theoretical chemist at the California Institute of Technology. He was already famous for his studies on

FIGURE 6.11

A trinucleotide. This little piece of DNA contains only three nucleotides linked together by phosphodiester bonds (red) between the 5′- and 3′-hydroxyl groups of the sugars. The 5′-end of this DNA is at the top, where there is a free 5′-phosphate group (blue); the 3′-end is at the bottom, where there is a free 3′-hydroxyl group (blue). The sequence of this DNA could be read as 5′pdTpdCpdA3′. This would usually be simplified to TCA.

Table 6.2 Composition of DNA in Moles of Base/Mole of P

| | Human | | | | Yeast | | Avian Tubercle Bacilli | Bovine | | | | |
| | Sperm | | Thymus | Liver carcinoma | | | | Thymus | | | Spleen | |
	#1	#2		#1	#2			#1	#2	#3	#1	#2
A:	0.29	0.27	0.28	0.27	0.24	0.30	0.12	0.26	0.28	0.30	0.25	0.26
T:	0.31	0.30	0.28	0.27	0.25	0.29	0.11	0.25	0.24	0.25	0.24	0.24
G:	0.18	0.17	0.19	0.18	0.14	0.18	0.28	0.21	0.24	0.22	0.20	0.21
C:	0.18	0.18	0.16	0.15	0.13	0.15	0.26	0.16	0.18	0.17	0.15	0.17
Recovery:	0.96	0.92	0.91	0.87	0.76	0.92	0.77	0.88	0.94	0.94	0.84	0.88

Source: Data from E. Chargaff, *Expermentia*, 6:206–207.

chemical bonding and for his discovery of the α-helix, an important feature of protein structure. Indeed, the α-helix, held together by hydrogen bonds, laid the intellectual groundwork for the double helix model of Watson and Crick. Another group toiling on the DNA problem was composed of Maurice Wilkins, Rosalind Franklin, and their colleagues at King's College in London. They were using X-ray diffraction to reveal the three-dimensional structure of DNA. Finally, there were James Watson and Francis Crick. Watson, in his early twenties and already holding a Ph.D. degree from Indiana University, had come to the Cavendish Laboratories in Cambridge, England, to learn about DNA. There he met Crick, a physicist who at age thirty-five was retraining as a molecular biologist. Watson and Crick performed no experiments themselves. Their tactic was to use other groups' data to build a DNA model.

Erwin Chargaff was another very important contributor. We have already seen how his 1950 paper helped identify DNA as the genetic material, but the paper contained another piece of information that was even more significant. Chargaff's studies of the base compositions of DNAs from various sources revealed that the content of purines was always equal to the content of pyrimidines. Furthermore, the amounts of adenine and thymine were always equal, as were the amounts of guanine and cytosine. These findings, known as Chargaff's rules, provided a valuable confirmation of Watson and Crick's model. Table 6.2 presents Chargaff's data. You will notice some deviation from the rules due to incomplete recovery of the bases, but overall the pattern is clear.

Perhaps the most crucial piece of the puzzle came from an X-ray diffraction picture of DNA taken by Franklin in 1952—a picture that Wilkins shared with James Watson over dinner in London on the night of January 30, 1953. The X-ray technique worked as follows: The experimenter made a very concentrated, viscous solution of DNA, then reached in with a needle and pulled out a fiber. This was not a single molecule, but a whole batch of

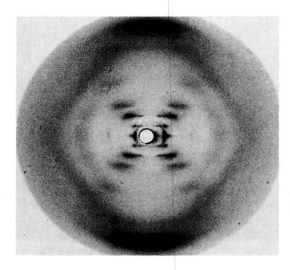

FIGURE 6.12

Franklin's X-ray picture of DNA. The regularity of this pattern indicated that DNA is a helix. The spacing between the strong bands at the top and bottom gave the spacing between elements of the helix (base pairs) as 3.4 angstroms. The spacing between neighboring lines in the pattern gave the overall repeat of the helix (the length of one helical turn) as 34 angstroms. © *M. H. F. Wilkins, Biophysics Department. King's College, London.*

DNA molecules, forced into a side-by-side arrangement by the pulling action. Given the right relative humidity, this fiber was enough like a crystal that it diffracted X rays in an interpretable way. In fact, the X-ray diffraction pattern in Franklin's picture (fig. 6.12) was so simple—a series of spots arranged in an X shape—that it required the DNA structure itself to be very simple. By contrast, a complex, irregular molecule like a protein gives a complex X-ray diffraction pattern with many spots, rather like a surface peppered by shotgun pellets. Since DNA is very large, it can only be simple if it has a regular, repeating structure. And the simplest repeating shape that a long, thin molecule can assume is a corkscrew, or helix.

FIGURE 6.13

The base pairs of DNA. A guanine-cytosine pair (G–C), held together by three hydrogen bonds (dashed lines), has almost exactly the same shape as an adenine-thymine pair (A–T), held together by two hydrogen bonds.

THE DOUBLE HELIX

Franklin's X-ray work strongly suggested that DNA was a helix. Not only that, it gave some important information about the size and shape of the helix. In particular, the spacing between adjacent spots in an arm of the X is related to the overall repeat distance in the helix (34Å), and the spacing from the top of the X to the bottom is related to the spacing (3.4Å) between the repeated elements (**base pairs**) in the helix. (Note the inverse relationship between spacing in the X-ray diffraction pattern and spacing in the molecule.) However, even though the Franklin picture told much about DNA, it presented a paradox: DNA was a helix with a regular, repeating structure, but for DNA to serve its genetic function, it must have an *irregular* sequence of bases.

Watson and Crick saw a way to resolve this contradiction and satisfy Chargaff's rules at the same time: DNA must be a **double helix** composed of two strands with their sugar-phosphate backbones on the outside and their bases on the inside. Moreover, the bases must be paired, with one purine always across from one pyrimidine. This way the helix would be uniform; it would not have bulges where two large purines are paired or constrictions where two small pyrimidines are paired. We sometimes refer to a double-stranded DNA as a DNA **duplex.** Watson has joked about the reason he seized on a double helix: "I had decided to build two-chain models. Francis would have to agree. Even though he was a physicist, he knew that important biological objects come in pairs."

But Chargaff's rules went further than this. They decreed that the amounts of adenine and thymine were equal and so were the amounts of guanine and cytosine. This fit very neatly with Watson and Crick's observation that an adenine-thymine base pair held together by hydrogen bonds has almost exactly the same shape as a guanine-cytosine base pair (fig. 6.13). So Watson and Crick postulated that adenine must always pair with thymine, and guanine with cytosine. This way, the double-stranded DNA will be uniform, composed of very similarly shaped base pairs, regardless of the unpredictable sequence of either DNA strand by itself. This was their crucial insight, the key to the structure of DNA.

The double helix, often likened to a twisted ladder, is presented in several ways in figure 6.14. The curved sides of the ladder represent the sugar-phosphate backbones of the two DNA strands; the rungs are the base pairs. The spacing between base pairs is 3.4 angstroms, and the overall helix repeat distance is about 34 angstroms, meaning that there are about ten base pairs per turn of the helix. (One angstrom is one ten-billionth of a meter or one-tenth of a nanometer.) The arrows indicate that the two strands are **antiparallel.** If one has 5′→3′ polarity from top to bottom, the other must have 3′→5′ polarity. In solution, DNA has a structure very similar to the one just described, but the helix contains about 10.4 base pairs per turn.

Watson and Crick published the outline of their model in *Nature,* back to back with papers by Wilkins and Franklin and their coworkers showing the X-ray data. The Watson-Crick paper is a classic of simplicity—only nine hundred words, barely over a page long. It was published very rapidly, less than a month after it was submitted. Actually, Crick wanted to spell out the biological implications of the model, but Watson was uncomfortable doing that. They compromised on a sentence that is one of the great understatements in scientific literature: "It has not escaped our notice that the specific base pairing we have proposed immediately suggests a possible copying mechanism for the genetic material."

As this cryptic sentence indicates, Watson and Crick's model does indeed suggest a copying mechanism for DNA. Since one strand is the **complement** of the other, the two strands can be separated, and each can then serve as the template for building a new partner. Figure 6.15 shows schematically how this could be accomplished.

The Watson-Crick model of DNA structure has the following features: 1) The DNA molecule is double-helical. 2) The double helix has sugar-phosphate backbones on the outside and base pairs on the inside. 3) The bases pair in a specific way: adenine (A) with thymine (T), and guanine (G) with cytosine (C). 4) The two strands are antiparallel: One has 5′ to 3′ polarity top to bottom; the other has 3′ to 5′ polarity top to bottom. 5) The base pairs lie parallel to each other and perpendicular to the helical axis. 6) The distance from one base pair to the next is 3.4 angstroms. 7) The helix makes one complete turn in 34 angstroms. 8) The helix is about 20 angstroms wide.

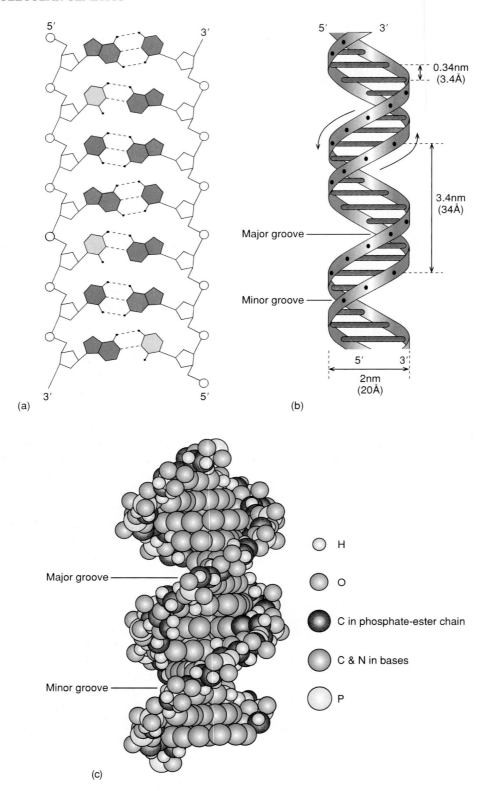

FIGURE 6.14

Three models of DNA structure. (*a*) The helix is straightened out to show the base pairing in the middle. Each type of base is represented by a different color, with the sugar-phosphate backbones in black. Note the three hydrogen bonds (dashed lines) in the G–C pairs and the two in the A–T pairs. (A = green, T = yellow, G = red, C = blue.) (*b*) The DNA double helix is represented as a spiral ladder whose sides represent the sugar-phosphate backbones of the two strands and whose rungs represent base pairs. The two strands are antiparallel; as suggested by the curved arrows, one strand (red) runs 5′→3′ top to bottom, and the other strand,

(blue) runs 5′→3′ bottom to top. (*c*) A space-filling model. The sugar-phosphate backbones appear as strings of mostly black, red, white, and yellow spheres, while the base pairs are rendered as horizontal flat plates composed of blue spheres. Note the major and minor grooves in the helix depicted in (*b*) and (*c*).

(a) Adapted from "The Synthesis of DNA" *by Arthur Kornberg. Copyright © 1968 by Scientific American, Inc. All rights reserved. (b) Reprinted from* Nature, *vol. 171, p. 737. Copyright © 1953 Macmillan Magazines, Ltd. Reprinted by permission of the publisher and authors. (c) Reprinted from* Nature, *vol. 175, p. 834. Copyright © 1955 Macmillan Magazines, Ltd.*

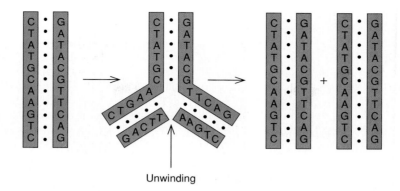

Unwinding

FIGURE 6.15

Replication of DNA. For simplicity, the two parental DNA strands (blue) are represented as parallel lines. During replication these parental strands separate, or unwind, and new strands (red) are built with bases complementary to those of the parental strands. The end result is two double-stranded molecules of DNA identical to the original. Each daughter DNA duplex therefore ends up with one parental strand (blue) and one new strand (red). Since only one parental strand is conserved in each of the daughter duplexes, this mechanism of replication is called semiconservative.

PHYSICAL CHEMISTRY OF NUCLEIC ACIDS

We conclude this chapter by briefly discussing the variety of structures available to DNA and RNA molecules and the behavior of DNA under conditions that encourage the two strands to separate and then come together again.

A VARIETY OF DNA STRUCTURES

The DNA structure proposed by Watson and Crick is right-handed; the helix runs clockwise away from you whether you look at it from the top or the bottom. But Alexander Rich and his colleagues discovered in 1979 that DNA does not always have to be right-handed. They showed that double-stranded DNA containing strands of alternating purines and pyrimidines (e.g., poly [dG-dC] • poly[dG-dC]):

- - - GCGCGCGC - - -
- - - CGCGCGCG - - -

can exist in an extended left-handed helical form. Because of the zigzag look of this DNA's backbone when viewed from the side, it is often called **Z-DNA.** Figure 6.16 presents a picture of Z-DNA along with B-DNA for comparison's sake. Although Rich discovered Z-DNA in studies of model compounds like poly[dG-dC] • poly[dG-dC], this structure seems to be more than just a laboratory curiosity. There is evidence that living cells contain a small component of Z-DNA. Figure 6.17 presents computer graphic models of the A and Z forms of DNA.

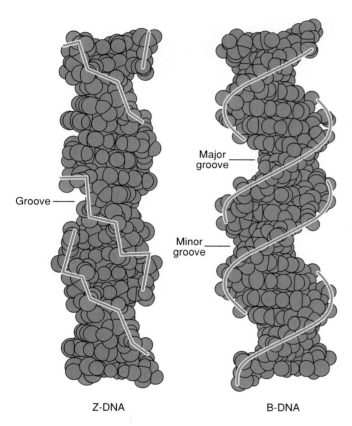

Z-DNA   B-DNA

FIGURE 6.16

Structures of Z- and B-DNA. Note the left-handed nature of the Z-DNA helix, in which the backbone (red line connecting red phosphate groups) goes up from right to left. By contrast, the B-DNA is clearly right-handed. Note also the contrast between the zigzag appearance of the Z-DNA backbone and the smooth B-DNA backbone.

From A. H. J. Wang, et al., Nature 282:684, 1979. Copyright © 1979 Macmillan Magazines Limited, London, England.

❖ In the cell, DNA may exist in the common B form, or a small fraction of the DNA may assume an extended left-handed helical form called Z-DNA.

(a)   (b)

FIGURE 6.17

Computer graphic models of B- and Z-DNA. (*a*) B-DNA. Note the familiar right-handed helix. (*b*) Z-DNA. Note that the base pairs are grouped in twos in an alternating side-to-side pattern. This gives Z-DNA its characteristic zigzag shape. Carbons are dark blue; nitrogens, light blue; oxygens, red; and phosphorous atoms, yellow. Green represents a bromine atom added to aid the X-ray crystallography.

© Dr. Nelson Max, Lawrence Livermore National Laboratory, University of California.

SEPARATING THE TWO STRANDS OF A DNA DOUBLE HELIX

When the temperature of a DNA solution rises high enough, the forces that hold the two strands together weaken and finally break. When this happens, the two strands come apart in a process known as **DNA denaturation** or **DNA melting.** The temperature at which the DNA strands are half denatured is called the melting temperature, or T_m. Figure 6.18 contains a melting curve for DNA from *Streptococcus pneumoniae.* The amount of strand separation, or melting, is measured by the absorbance of the DNA solution at 260 nm. Nucleic acids absorb light at this wavelength because of the electronic structure in their bases, but when two strands of DNA come together, the close proximity of the bases in the two strands quenches some of this absorbance. When the two strands separate, this quenching disappears and the absorbance rises 20% to 30%. The precipitous rise in the curve shows that the strands hold fast until the temperature approaches the T_m and then rapidly let go.

The GC content of a DNA has a significant effect on its T_m. In fact, as figure 6.19 shows, the higher a DNA's

FIGURE 6.18

Melting curve of *Streptococcus pneumoniae* DNA. The DNA was heated, and its melting was measured by the increase in absorbance at 260 nm. The point at which the melting is half complete is the melting temperature, or T_m. The T_m for this DNA under these conditions is about 85° C.

GC content, the higher its T_m. Why should this be? Recall that one of the forces holding the two strands of DNA together is hydrogen bonding. Remember also that G-C pairs form three hydrogen bonds, whereas A-T pairs have only two. It stands to reason, then, that two strands of DNA rich in G and C will hold to one another more tightly than those of AT-rich DNA. Consider two pairs of embracing centipedes. One pair has two hundred legs each, the other three hundred. Naturally, the latter pair will be harder to separate.

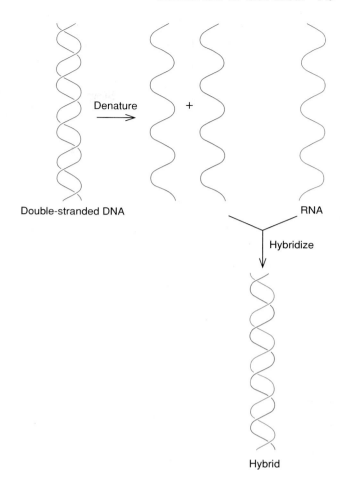

FIGURE 6.19

Relationship between DNA melting temperature and GC content. AT-DNA refers to synthetic DNAs composed exclusively of A and T (GC content = 0).

From Paul Doty, Harvey Lectures *55:121, 1961. Copyright © 1961 Academic Press, Orlando, FL. Reprinted by permission of the author and publisher.*

Heating is not the only way to denature DNA. Organic solvents such as dimethyl sulfoxide and formamide disrupt the hydrogen bonding between DNA strands and thereby promote denaturation. Lowering the salt concentration of the DNA solution also aids denaturation by removing the ions that shield the negative charges on the two strands from one another. At low ionic strength, the mutually repulsive forces of these negative charges are strong enough to denature the DNA at a relatively low temperature.

 The melting temperature (T_m) of a DNA is the temperature at which the two strands are half-dissociated, or denatured. The GC content of a DNA can have a strong effect on the physical properties of the DNA, in particular on its melting temperature, which increases linearly with GC content. Low ionic strength and organic solvents also promote DNA denaturation.

REJOINING THE SEPARATED DNA STRANDS

Once the two strands of DNA separate, they can, under the proper conditions, come back together again. This is called **annealing** or **renaturation.** One of the most important factors in DNA annealing is temperature. The best temperature for renaturation of a DNA is about 25° C below its T_m. This temperature is low enough that it does not promote denaturation, but high enough to allow rapid diffusion and weaken the transient bonding between mismatched sequences and short intrastrand base-paired

FIGURE 6.20

Hybridizing DNA and RNA. First, the DNA at upper left is denatured to separate the two DNA strands (blue). Then the DNA strands are mixed with a strand of RNA (red) that is complementary to one of the DNA strands. If this hybridization reaction is carried out under appropriate conditions (relatively high temperature, for example), an RNA-DNA hybrid will form. This duplex has one DNA strand (blue) and one RNA strand (red).

regions. This suggests that rapid cooling following denaturation would frustrate renaturation. Indeed, a common procedure to ensure that denatured DNA stays denatured is to plunge the hot DNA solution into ice. This is called **quenching.**

HYBRIDIZATION OF TWO DIFFERENT POLYNUCLEOTIDE CHAINS

Consider a piece of single-stranded DNA forming a double-stranded structure with a single-stranded RNA of complementary sequence (fig. 6.20). Because we are putting together a **hybrid** of two different kinds of polynucleotide chain, we call this **hybridization.** The two chains do not have to be as different as DNA and RNA. If we put together two different strands of DNA having complementary or nearly complementary sequences, we can call it hybridization—as long as the strands are of different origin. The difference between the two complementary strands may be very subtle; for example, one may be radioactive,

Table 6.3 Sizes of Various DNAs

Source	Molecular Weight	Base Pairs	Length
Subcellular Genetic Systems:			
SV40 (mammalian tumor virus)	3.5×10^6	5,226	1.7 μm
Bacteriophage φX174 (double-stranded form)	3.2×10^6	5,386	1.8 μm
Bacteriophage λ	33×10^6	5×10^4	13 μm
Bacteriophage T2 or T4	1.3×10^8	2×10^5	50 μm
Mouse mitochondria	9.5×10^6	1.4×10^4	5 μm
Prokaryotes:			
Hemophilus influenzae	8×10^8	1.2×10^6	300 μm
Escherichia coli	2.8×10^9	4.2×10^6	1.4 mm
Salmonella typhimurium	8×10^9	1.1×10^7	3.8 mm
Eukaryotes (content per haploid nucleus):			
Saccharomyces cerevisiae (yeast)	1.2×10^{10}	1.8×10^7	6.0 mm
Neurospora crassa (pink bread mold)	1.9×10^{10}	2.7×10^7	9.2 mm
Drosophila melanogaster (fruit fly)	1.2×10^{11}	1.8×10^8	6.0 cm
Mouse	1.5×10^{12}	2.2×10^9	75 cm
Human	1.9×10^{12}	2.8×10^9	94 cm
Zea mays (corn, or maize)	4.4×10^{12}	6.6×10^9	2.2 m
Frog	1.4×10^{13}	2.3×10^{10}	7.7 m
Lilium longiflorum (lily)	2×10^{14}	3×10^{11}	100 m

while the other is not. As we will see later, especially in chapter 15, molecular hybridization is an extremely valuable technique. In fact, it would be difficult to overestimate the importance of such hybridization in genetics.

DNAs OF VARIOUS SIZES AND SHAPES

Table 6.3 shows the sizes of the DNAs of several organisms and viruses. The sizes are expressed three ways: molecular weight, number of base pairs, and length. These are all related, of course. We already know how to convert number of base pairs to length, since there are ten base pairs per helical turn, which is 34 angstroms long. To convert base pairs to molecular weight, we simply need to multiply by 660, the approximate molecular weight of one average nucleotide pair.

How do we measure these sizes? For small DNAs, this is fairly easy. For example, consider phage PM2 DNA, which contains a double-stranded, circular DNA. How do we know it is circular? The most straightforward way to find out is simply by looking at it. We can do this using an electron microscope, but first we have to treat the DNA so that it stops electrons and will show up in a micrograph just as bones stop X rays and therefore show up in an X-ray picture. The most common way of doing this is by "shadowing" the DNA with a heavy metal such as platinum. We place the DNA on an electron microscope grid and bombard it with minute droplets of metal from a shallow angle. This makes the metal pile up beside the DNA like snow behind a fence. We rotate the DNA on the grid so that it becomes shadowed all around. Now the metal will stop the electrons in the electron microscope and make the DNA appear as light strings against a darker background. Printing reverses this image to give a picture such as figure 6.21, which is an

electron micrograph of PM2 DNA in two forms: an open circle (lower right) and a **supercoil** (upper left), in which the DNA coils around itself rather like a twisted rubber band. You can also use pictures like these to measure the length of the DNA. This is more accurate if you include a DNA of known length in the same picture.

The size of a very small DNA can be estimated by **gel electrophoresis.** In this technique, we make a gel with slots in it, as shown in figure 6.22. We put a little DNA in a slot and run an electric current through the gel. The negatively charged DNA migrates toward the positive pole (the anode) at the end of the gel. The secret of the gel's ability to distinguish DNAs of different sizes lies in its porosity. Small DNA molecules can fit easily through the minute holes in the gel, so they migrate rapidly. Large DNAs, by contrast, move with difficulty through the gel's pores, so their mobility is retarded. The result is that the electric current will distribute the DNA fragments according to their sizes: the largest near the top, the smallest near the bottom. Finally, we stain the DNA with a fluorescent dye and look at the gel under ultraviolet illumination. Figure 6.23 depicts the results of such analysis on fragments of phage DNA of known size. We plot the mobilities of these fragments versus the log of their molecular weights (or number of base pairs). Now we can electrophorese any unknown DNA in parallel with the standard fragments, and determine its size if it falls within the range of the standards. For example, a DNA with a mobility of 20 mm in figure 6.23 (marked with a Δ) would contain about 820 base pairs (bp), or 0.82 **kilobase pairs** (**kbp,** or simply **kb**).

The size of a larger DNA is not as easy to determine. One reason is that double-stranded DNA is a relatively rigid rod—very long and thin. The longer it is, the more fragile it is. In fact, large DNAs break very easily; even

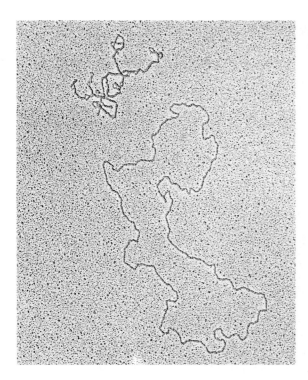

FIGURE 6.21
Autoradiograph of phage PM2 DNA in open circular and supercoiled forms.
© *Dr. Jack Griffith.*

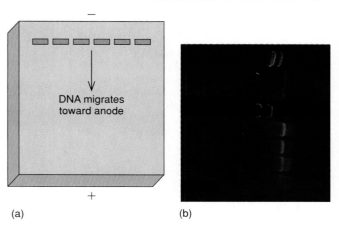

(a)   (b)

FIGURE 6.22
DNA gel electrophoresis. (*a*) Scheme of the method: This is a horizontal gel made of agarose (a substance derived from seaweed, and the main component of agar). The agarose melts at high temperature, then gels as it cools. A "comb" is inserted into the molten agarose; after the gel cools, the comb is removed, leaving slots, or wells (purple). The DNA is then placed in the wells, and an electric current is run through the gel. The DNA electrophoreses, or migrates, toward the positive pole, or anode. (*b*) A photograph of a gel that reveals DNA as pink bands. The DNA binds to a dye that fluoresces under ultraviolet light.
Photo (b) © Hank Morgan/Photo Researchers, Inc.

(a)   (b)

FIGURE 6.23
Gel electrophoresis of φX174 phage DNA fragments. (*a*) Autoradiograph of radioactive fragments. The fragments are numbered from 1 to 10 in order of decreasing size. (*b*) Plot of migration of the DNA fragments versus the log of their sizes in kilobase pairs (kb). The Δ marks the mobility of a hypothetical DNA that migrated 20 mm.
Photo (a) from Lee and Robert Sinsheimer, Proceedings of the National Academy of Science, *vol. 71, p. 2884, 1974.*

FIGURE 6.24
Pulsed-field gel electrophoresis of yeast chromosomes.
Identical samples of yeast chromosomes were
electrophoresed in ten parallel lanes and stained with
ethidium bromide. The bands represent chromosomes
having sizes ranging from 0.2 megabases (at bottom) to 2.2
megabases (at top). Original gel is about 13 cm wide by
12.5 cm long.
Bio-Rad Labs, CHEF-DR ® II pulsed-field electrophoresis system.

seemingly mild manipulations, like swirling in
a beaker or pipetting, create shearing forces
sufficient to fracture them. To visualize this,
think of DNA as a piece of uncooked spaghetti.
If it is short—say an inch or two—you can treat
it roughly without harming it, but if it is long,
breakage becomes almost inevitable.

In spite of these difficulties, geneticists have
recently developed a new kind of gel electrophoresis that can separate DNA molecules up
to several million base pairs (**megabases, Mb**)
long. Instead of a constant current through the
gel, this method uses pulses of current. This
pulsed-field electrophoresis is valuable for measuring
the sizes of DNAs even as large as some of the chromosomes found in yeast. Figure 6.24 presents the results
of pulsed-field gel electrophoresis on yeast chromosomes.
The sixteen visible bands represent chromosomes containing
0.2 to 2.2 Mb.

❖ Natural DNAs come in sizes ranging from a few kilobases to
hundreds of megabases. We can measure these sizes (up to
several Mb) by electrophoresis, or, for small DNAs, by
electron microscopy. Circular DNAs frequently twist around
themselves, forming supercoils.

WORKED-OUT PROBLEMS

Problem 1
A certain phage has a double-stranded DNA with 45,000
base pairs (45 kb). (*a*) Approximately how many double-
helical turns does this DNA contain? (*b*) How long is the
DNA in microns (1 micron = 10^4 angstroms)? (*c*) What is
the molecular weight of this DNA? (*d*) How many phosphorous atoms does the DNA contain?

Solution
(*a*) One double-helical turn encompasses about 10 base
pairs, so all we have to do is divide the total number of
base pairs by 10:

45,000 bp × 1 helical turn/10 bp = 4,500 helical turns

(*b*) One base pair is about 3.4 angstroms, or 3.4×10^{-4}
microns long, so we just need to multiply the total
number of base pairs by 3.4×10^{-4}:

45,000 bp × 3.4×10^{-4} microns/bp = 15.3 microns

(*c*) One base pair has a molecular weight of about 660, so we
need to multiply the total number of base pairs by 660:

45,000 bp × 660/bp = 2.97×10^7

(*d*) Each base pair has two phosphorous atoms, one on
each strand. Therefore, we simply multiply the total
number of base pairs by 2:

45,000 bp × 2 phosphorous atoms/bp = 90,000
phosphorous atoms

Problem 2
One strand of a double-stranded DNA has this sequence:
5′-TCCGATGCGA-3′. (*a*) Write the sequence in double-
stranded form. (*b*) Write the sequence of the opposite
strand in the 3′ to 5′ direction. (*c*) Write the sequence of
the opposite strand in the standard 5′ to 3′ direction.

Solution

(*a*) Remembering that T always pairs with A and C with G, we can build the opposite strand as follows:

5′-TCCGATGCGA-3′
3′-AGGCTACGCT-5′

Notice that the two strands are antiparallel, as usual. The top strand reads 5′ to 3′ left to right, and the bottom strand reads 3′ to 5′ in the same direction.

(*b*) The answer to part (*a*) immediately shows us that the opposite (bottom) strand reads AGGCTACGCT in the 3′ to 5′ direction.

(*c*) To find the sequence of the opposite (bottom) strand in the standard 5′ to 3′ direction, we simply read it backward: 5′-TCGCATCGGA-3′.

Problem 3

Here are the outlines of two DNA base pairs, with the bases identified as *a, b, c,* and *d.* What are the real identities of these bases?

Solution

Base *a* is a purine (composed of a six-membered ring and a five-membered ring), so we know it must be either adenine (A) or guanine (G). Furthermore, it is joined to its partner by three H-bonds (dashed lines). G–C pairs are joined by three H-bonds, while A–T pairs are joined by only two. Therefore, base *a* must be guanine (G). Its partner, *b,* must therefore be cytosine (C). Base *c* is a purine that is joined to its partner by only two H-bonds. Since, A–T pairs are joined by two H-bonds, base *c* must be adenine (A). Its partner, *d,* must therefore be thymine (T).

Problem 4

Following is a graph showing the results of a gel electrophoresis experiment on double-stranded DNA fragments having sizes between 0.2 kb and 1.2 kb. On the basis of this graph, answer the following questions: (*a*) What is the size of a fragment that migrated 16 mm in this experiment? (*b*) How far would a 0.5 kb fragment migrate in this experiment?

Solution

(*a*) We draw a vertical dashed line from the 16 mm point on the X axis up to the experimental line. From the point where that vertical line intersects the experimental line, we draw a horizontal dashed line to the Y axis. This line intersects the Y axis at the 0.9 kb point. This tells us that fragments that migrate 16 mm in this experiment are 0.9 kb (or 900 base pairs) long.

(*b*) We draw a horizontal dashed line from the 0.5 kb point on the Y axis across to the experimental line. From the point where that horizontal line intersects the experimental line, we draw a vertical dashed line down to the X axis. This line intersects the X axis at the 28 mm point. This tells us that 0.5 kb fragments migrate 28 mm in this experiment.

Summary

Genes of all true organisms are made of DNA; certain viruses and all viroids have genes made of RNA. DNA and RNA are chainlike molecules composed of subunits called nucleotides. DNA in most genetic systems is a double-helical molecule with sugar-phosphate backbones on the outside and base pairs on the inside. The bases pair in a specific way: A with T and G with C. When DNA replicates, the parental strands separate; each then serves as the template for making a new, complementary strand.

The G + C content of a DNA can vary from about 25% to about 75%, and this can have a strong effect on the physical properties of DNA, particularly its melting temperature. The melting temperature (T_m) of a DNA is the temperature at which the two strands are half-dissociated, or denatured. Separated DNA strands can be induced to re-nature, or anneal. Complementary strands of polynu-cleotides (either RNA or DNA) from different sources can form a double helix in a process called hybridization. Natural DNAs vary widely in length. The size of a DNA can be estimated by a variety of methods, including electron microscopy and gel electrophoresis.

PROBLEMS AND QUESTIONS

1. What evidence suggested that the transforming agent in Griffith's experiments was really a gene or genes?

2. One possible explanation for Griffith's transformation results was that avirulent (R) bacteria somehow restored the dead virulent (S) cells to life. How did the experiments of Avery et al. eliminate this unlikely possibility?

3. Which experimental approach more clearly points to DNA as the genetic material—that of Avery et al. or that of Hershey and Chase? *Hint:* Which experiment is easier to criticize on the basis of possible impurities in the genetic material that enters the test cell?

4. Would ^{14}C and ^{3}H have been suitable radioactive tracers for the Hershey-Chase experiment? Why, or why not?

5. If one strand of double-stranded DNA has the following sequence: 5′-AGCATTCG-3′, what is the sequence of the opposite strand in *its* 5′ to 3′ direction? In its 3′ to 5′ direction?

6. Look at figure 6.14a. (*a*) Cover up the legend and write the sequence of bases in the left-hand strand, top to bottom. (*b*) Write the sequence of bases in the right-hand strand, top to bottom. See the worked-out problems for guidance.

7. One strand of a double-stranded DNA has the sequence TCCGATTGAC. Write the sequence in double-stranded form.

8. If bovine DNA has an adenine content of 30% and a guanine content of 20%, what is its content of the other two bases?

9. According to the graph in figure 6.19, what is the approximate G + C content of a DNA with a T_m of 90° C? Round your answer off to the nearest 5%.

10. An animal virus contains a circular double-stranded DNA with 120,000 base pairs. (*a*) Approximately how many helical turns does this DNA contain? (*b*) How long is the DNA in microns (1 micron = 10^4 angstroms)? (*c*) What is the approximate molecular weight of this DNA? (*d*) How many phosphorus atoms does the DNA contain?

11. A certain DNA fragment has a mobility of 30 mm in the electrophoresis experiment depicted in figure 6.23. What is its size in kilobase pairs?

12. Using the graph in worked-out problem 4, what is the size of a fragment that has a mobility of 40 mm?

Answers appear at end of book.

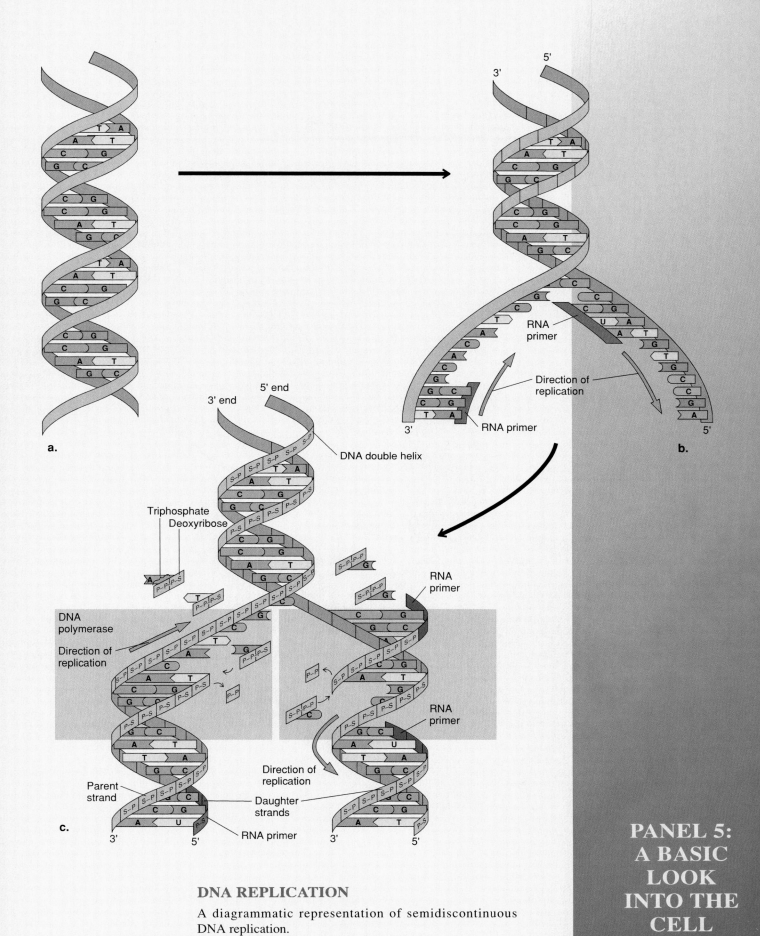

DNA REPLICATION

A diagrammatic representation of semidiscontinuous DNA replication.

**PANEL 5:
A BASIC
LOOK
INTO THE
CELL**

a.

b.

c.

A REVIEW OF DNA REPLICATION

CHAPTER 7: REPLICATION AND RECOMBINATION OF GENES

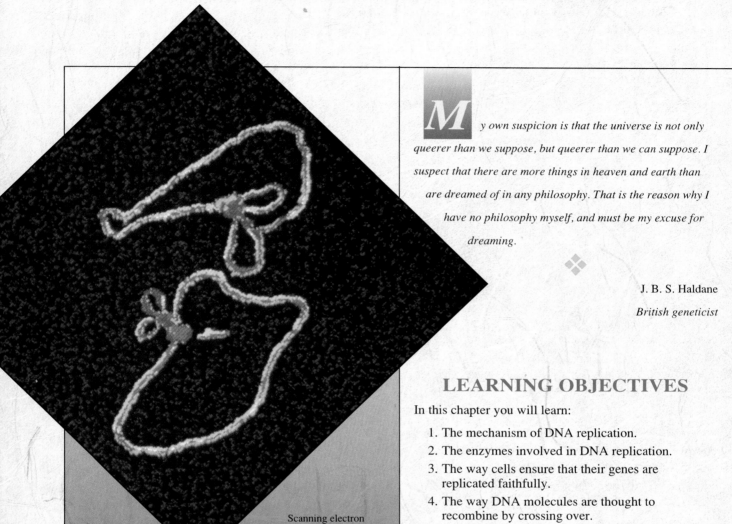

*M*y own suspicion is that the universe is not only queerer than we suppose, but queerer than we can suppose. I suspect that there are more things in heaven and earth than are dreamed of in any philosophy. That is the reason why I have no philosophy myself, and must be my excuse for dreaming.

❖

J. B. S. Haldane

British geneticist

LEARNING OBJECTIVES

In this chapter you will learn:

1. The mechanism of DNA replication.
2. The enzymes involved in DNA replication.
3. The way cells ensure that their genes are replicated faithfully.
4. The way DNA molecules are thought to recombine by crossing over.

Scanning electron micrograph of replicating DNA.
© *Dr. Jack Griffith/ Peter Arnold, Inc.*

ow that we have covered the structure and function of genes as molecules, it is time to examine the basic functions or activities of genes. A gene participates in three major activities:

(1) A gene can be replicated. This duplication is very faithful, so the gene information can be passed essentially unchanged from generation to generation.

(2) A gene is a repository of information. That is, it holds the instructions for making one of the key molecules of life: an RNA. The sequence of bases in the RNA depends directly on the sequence of bases in the gene. Most of these RNAs, in turn, serve as templates for making other critical cellular molecules: proteins. Thus, most genes can be thought of as blueprints for making proteins.

(3) A gene can accept occasional changes, or mutations. This allows organisms to evolve.

In this chapter and the following four chapters, we will examine each of these activities in detail.

MECHANISM OF DNA REPLICATION

In chapter 6 we introduced a simple model of DNA replication: The two parental strands separate and progeny strands are built according to the Watson-Crick base-pairing rules of A with T and G with C. This is an elegant concept, but it really only hints at the complexity of DNA replication. Our job in this chapter will be to fill in some of the gaps we have left.

Since Watson and Crick's original papers appeared, other investigators have performed experiments that revealed the following interesting and sometimes unexpected general features of DNA replication: (*a*) As Watson and Crick predicted, DNA replicates by separation of parental strands and formation of new strands complementary to the separated parental strands. (*b*) Replication is at least half discontinuous; that is, the DNA is made in short pieces that are later stitched together. (*c*) DNA replication requires an RNA primer. (*d*) Replication is usually bidirectional; that is, it proceeds in both directions away from a common starting point.

FORMING DNA FROM NUCLEOTIDES

Before we examine the four aspects of DNA replication listed above, let us consider the basic process by which nucleotides are put together to make a strand of DNA. Figure 7.1 illustrates the core of this process: the formation of a phosphodiester bond between a nascent (growing) DNA strand and an incoming deoxyguanine nucleotide (see also panel 5). The nucleotides that participate in DNA synthesis are the four deoxynucleoside triphosphates: dATP, dCTP, dGTP, and dTTP. In this example, only dGTP is considered. As the phosphodiester bond forms, the two external phosphate groups of dGTP are lost, leaving only the internal one to form the bond. The net reaction is the addition of one nucleotide, dGMP in this case, to the nascent DNA. This process is repeated over and over as the DNA chain lengthens.

Now let us discuss the other features of DNA replication, and then examine the experimental evidence for them.

SEMICONSERVATIVE REPLICATION

The Watson-Crick model for DNA replication introduced in chapter 6 assumes that as new strands of DNA are made, they follow the usual base-pairing rules of A with T and G with C. This is essential because the DNA-replicating machinery must be capable of discerning a good pair from a bad one, and the Watson-Crick base pairs give the best fit. The model also presupposes that the two parental strands separate and that each then serves as a template for a new progeny strand. This is called **semiconservative** replication because each daughter double helix has one parental strand and one new strand (fig. 7.2*a*). In other words, only one of the parental strands is "conserved" in each daughter double helix. This is not the only possibility. Another potential mechanism (fig. 7.2*b*) is **conservative** replication, in which the two parental strands stay together and somehow produce another daughter helix with two completely new strands. Semiconservative replication is a much more attractive alternative because it takes advantage of the complementary nature of the two DNA strands. Each separated parental strand serves as the template for making a new, complementary partner. As we will see shortly, experimental evidence also favors this scheme.

SEMIDISCONTINUOUS REPLICATION

If we were charged with the task of designing a DNA replicating machine, we might come up with a system such as the one pictured in figure 7.3. DNA would replicate continuously on both strands, following in the same direction as the moving fork. However, there is one fatal flaw in this scheme. It demands that the replicating machine be able to make DNA in both the $5' \rightarrow 3'$ and $3' \rightarrow 5'$ directions. That is because of the antiparallel nature of the two strands of DNA; if one goes $5' \rightarrow 3'$, the other must go $3' \rightarrow 5'$ (recall chapter 6). But the DNA-making enzyme (**DNA polymerase**) can go only one way: $5' \rightarrow 3'$. That is, it inserts the $5'$-most nucleotide first and extends the chain toward the $3'$-end.

Following this line of reasoning, Reiji Okazaki concluded that both strands could not replicate continuously. DNA polymerase could theoretically make one strand (the **leading strand**) continuously in the $5' \rightarrow 3'$ direction, but the other strand (the **lagging strand**), in order to be made $5' \rightarrow 3'$, would have to be made **discontinuously** as shown in figure 7.3*b* and *c* and in panel 5. The discontinuity of synthesis of the lagging strand comes about because its direction of synthesis is opposite to that of the replicating fork. Therefore, as the fork opens up and exposes a new region of DNA to replicate, the lagging strand is growing in the wrong direction, away from the fork. The only way to replicate this newly exposed region is to restart DNA

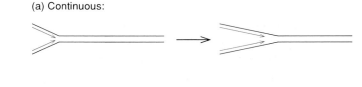

FIGURE 7.1

Phosphodiester bond formation in DNA synthesis. (*a*) Simplified reaction showing only the phosphorus atoms. Note that the phosphorus atom (red) closest to the deoxyguanosine is retained in the phosphodiester bond. (*b*) More complete structures showing all the atoms participating in phosphodiester bond formation as a deoxyguanine nucleotide is added to a growing DNA strand. Note the phosphorus atom (red) on the dGTP and the oxygen atom (blue) in the hydroxyl group at the 3′-end of the nascent DNA strand, both of which are retained in the phosphodiester bond in the product.

FIGURE 7.2

Two hypotheses for DNA replication. (*a*) Semiconservative replication (see also fig. 6.15) gives two daughter duplex DNAs, each of which contains one old strand (blue) and one new strand (red). (*b*) Conservative replication yields two daughter duplexes, one of which has two old strands (blue) and one of which has two new strands (red).

FIGURE 7.3

Continuous, semidiscontinuous, and discontinuous models of DNA replication. (*a*) Continuous model. As the replicating fork moves to the right, both strands are replicated continuously in the same direction, left to right (blue arrows). (*b*) Semidiscontinuous model. Synthesis of one of the new strands (the leading strand, bottom) is continuous (blue arrow), as in the model in (*a*); synthesis of the other (the trailing strand, top) is discontinuous (red arrows), with the DNA being made in short pieces. (*c*) Discontinuous model. Both leading and trailing strands are made in short pieces (i.e., discontinuously; red arrows).

synthesis at the fork, behind the piece of DNA that has already been made. This starting and restarting of DNA synthesis occurs over and over again. The short pieces of DNA thus created would of course have to be joined together somehow to produce the continuous strand that is the final product of DNA replication. These short pieces of DNA are joined together by DNA ligase, as we will see shortly.

PRIMING OF DNA SYNTHESIS

If we supply a DNA polymerase with all the nucleotides and other small molecules it needs to make DNA, then add either single-stranded or double-stranded DNA with no strand breaks, the polymerase will make no DNA. What is missing?

We now know that the missing component here is a **primer,** a piece of nucleic acid that the polymerase can "grab onto" and extend by adding nucleotides to its 3′-end. Furthermore, this primer is not DNA at all, but a little piece of RNA. Imagine that DNA is a sentence. We need to start the sentence with a capital letter, but DNA polymerase "writes" only lowercase letters. Therefore, an RNA-making enzyme must intervene to put in a capital letter and get the

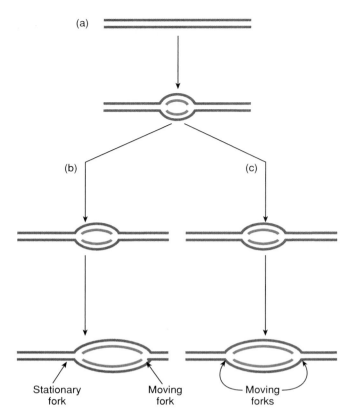

FIGURE 7.4
Priming in DNA synthesis. (*a*) The two parental strands (black) separate. (*b*) Short RNA primers (red) are made. (*c*) DNA polymerase uses the primers as starting points to synthesize progeny DNA strands (blue).

FIGURE 7.5
Unidirectional versus bidirectional replication. (*a*) The two parental strands (blue) open up, forming a "bubble" and allowing synthesis of progeny strands (red). (*b*) In unidirectional replication, one fork (the left one here) remains stationary, while the other supports DNA synthesis. (*c*) In bidirectional replication, both forks are active, with DNA replication proceeding outward in both directions away from a starting point in the middle.

sentence started. Figure 7.4 shows a simplified version of this process. First, a replicating fork opens up; next, short RNA primers are made; next, DNA polymerase adds deoxyribonucleotides to these primers, forming DNA, as indicated by the arrows (see panel 5).

BIDIRECTIONAL REPLICATION

As we learned earlier in this chapter, semiconservative replication requires that the two parental strands separate and serve as templates for the production of new, progeny strands. Since DNA replication never starts at the end of a DNA, the separation of parental strands must create a bulge, or "bubble," somewhere in the middle (fig. 7.5*a*). Now new strands can be made, and these will of course be complementary to the separated parental strands. Figure 7.5*a* shows short progeny strands inside the bubble formed by the separated parental strands. Notice that the two ends of this bubble are both forks at which DNA replication can occur. But does it occur at both forks, or just at one? Figure 7.5*b* and *c* illustrates these two possibilities. If DNA replication is **unidirectional** (7.5*b*), only one fork operates, and DNA replication proceeds away from the other, stationary fork. If replication is **bidirectional** (7.5*c*), both forks operate and replication proceeds in both directions away from a

common starting point. We will see examples of both unidirectional and bidirectional replication later in this chapter. However, bidirectional replication is far more common.

"STITCHING TOGETHER" NASCENT DNA FRAGMENTS

The mechanism we have described so far for DNA replication leaves us with nascent DNA fragments of the lagging strand that need to be linked together. An enzyme called **DNA ligase** serves this linking function, but two chores must be accomplished before ligase can operate. First, the RNA primers have to be removed so that the final DNA product is not interrupted with patches of RNA. Second, the gaps left by removing the primers need to be filled in.

Figure 7.6 demonstrates how a single enzyme performs both these tasks in *E. coli*. This multipurpose enzyme is **DNA polymerase I,** the first DNA-synthesizing enzyme discovered by Arthur Kornberg in his pioneering work on DNA replication. DNA polymerase I, in addition to its $5' \rightarrow 3'$ polymerizing activity, has two nuclease activities. One degrades polynucleotides in the $5' \rightarrow 3'$ direction, the

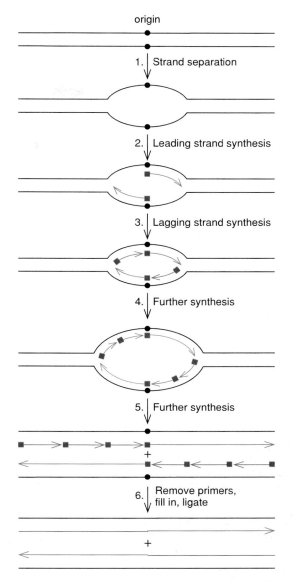

FIGURE 7.6

Removing primers and joining nascent DNA fragments. (*a*) We start with two adjacent progeny DNA fragments, the right-hand one containing a primer (red) at its 5′-end. The two fragments are separated by a single-stranded break called a nick. (*b*) DNA polymerase I binds to the double-stranded DNA at the nick. (*c*) DNA polymerase I simultaneously removes the primer and fills in the resulting gap by extending the left-hand DNA fragment rightward. It leaves degraded primer in its wake. (*d*) DNA ligase seals the remaining nick by forming a phosphodiester bond between the left-hand and right-hand progeny DNA fragments.

FIGURE 7.7

Summary of the mechanism of DNA replication. The parental strands are shown in black, with the origin of replication denoted by black circles. The RNA primers are represented by purple squares; the leading and lagging strands are denoted by red and blue arrows, respectively.

other in the 3′ → 5′ direction. The 5′ → 3′ nuclease activity destroys the RNA primers; then the DNA polymerizing activity fills in the gap. In fact, these two activities even seem to work in concert, with primer-destruction and gap-filling going on simultaneously. In this respect, DNA polymerase resembles a highway paving machine that tears up old pavement at its front end and lays down new at its rear.

After DNA polymerase has done its job, we have the situation depicted in figure 7.6*c*. All the nucleotides have been replaced, but a single-stranded break called a **nick** remains in the DNA. Notice that the nick has effectively moved from its position in figure 7.6*a,* rightward to its position in figure 7.6*c*. This movement or "translation" of the nick to a new location gives rise to the term **nick translation** to describe this process. Nick translation can also occur in nicked double-stranded DNA in the absence of RNA primers. DNA polymerase, for all its versatility, cannot repair nicks. That task is performed by DNA ligase, which forms a new phosphodiester bond to seal the DNA. This process is repeated as many times as necessary to stitch together all the fragments into a single, covalently closed DNA duplex.

PUTTING IT ALL TOGETHER

Figure 7.7 focuses on a short stretch of DNA to summarize the elements of DNA replication we have just discussed (see also panel 5). First, the parental strands separate at the origin of replication. Next, leading strand synthesis takes place; both parental strands serve as templates in this and subsequent steps. Next, lagging strand synthesis occurs. In real DNA synthesis, these events would occur more or less simultaneously and continuously. We have presented them stepwise for clarity. Next, the replicating "bubble" expands in both directions and further leading and lagging strand synthesis occurs. This process repeats itself once more, which leads to complete separation of the parental strands. Finally, the

FIGURE 7.8

Principle of cesium chloride density gradient centrifugation. As a solution of cesium chloride spins in the ultracentrifuge, the dense cesium chloride sediments toward the bottom until equilibrium is established between the centrifugal force and diffusion forces. At equilibrium there is a gradient of cesium chloride, and therefore of density, from high density at the bottom of the tube to relatively low density at the top. If DNA is included in the solution, it will also seek an equilibrium position—at the point in the tube where its density exactly matches that of the cesium chloride solution. Thus, two DNAs of different density will form two separate bands in the tube.

RNA primers are removed, the gaps are filled in, and the fragments are ligated together to complete the progeny strands.

◆ DNA replication is semiconservative. The two parental strands separate and serve as templates to make complementary progeny strands. DNA replication is also semidiscontinuous. That is, one strand (the leading strand) is made continuously in one long piece, but the other strand (the lagging strand) is made in short pieces. DNA replication also requires RNA primers to get started. Most DNAs replicate bidirectionally; both forks move away from a common starting point. In order to make a long DNA strand from RNA-primed short fragments, the cell removes the primers, fills in the resulting gaps, and ligates the fragments together.

EVIDENCE FOR THE MECHANISM OF DNA REPLICATION

All of the features of DNA replication we have just discussed were demonstrated through experiments performed on both prokaryotes and eukaryotes. We will examine some of the most important of these experimental findings in this section.

SEMICONSERVATIVE REPLICATION

In 1958, Matthew Meselson and Franklin Stahl worked out a clever procedure to distinguish between conservative and semiconservative DNA replication, using a nonradioactive heavy isotope of nitrogen. Ordinary nitrogen, the most abundant isotope, has an atomic weight of 14, so it is called ^{14}N. A relatively rare isotope (^{15}N) has an atomic weight of 15. Meselson and Stahl found that if bacteria grow in a

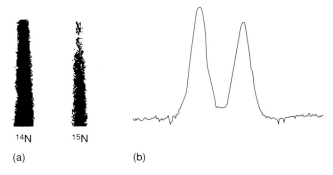

^{14}N ^{15}N

(a)   (b)

FIGURE 7.9

Separation of DNAs by cesium chloride density gradient centrifugation. DNA containing the normal isotope of nitrogen (^{14}N) was mixed with DNA labeled with a heavy isotope of nitrogen (^{15}N) and subjected to cesium chloride density gradient centrifugation. The two bands have different densities, so they separate cleanly. (a) A photograph of the spinning tube under ultraviolet illumination. The two dark bands correspond to the two different DNAs that absorb ultraviolet light. (b) A graph of the darkness of each band, which gives an idea of the relative amounts of the two kinds of DNA.

From Meselson and Stahl, in Proceedings of the National Academy of Science *44:671–82, 1958. Copyright © 1958 Meselson and Stahl. Reprinted by permission of the authors.*

medium enriched in ^{15}N, they incorporate the heavy isotope into their DNA, which becomes denser than normal. This labeled DNA clearly separates from ordinary DNA in a gradient of cesium chloride (CsCl) spun in an ultracentrifuge. CsCl is used because it is a very dense salt and therefore makes a dense enough solution that DNA (a dense material itself) will form a band somewhere in the middle, rather than sinking to the bottom. Figure 7.8 shows a schematic diagram of this experimental procedure. Figure 7.9, from Meselson and Stahl's 1958 paper, demonstrates that the normal ^{14}N- and heavy ^{15}N-tagged DNAs can be separated.

The crux of the experiment was to grow ^{15}N-labeled bacteria in ^{14}N-medium for one or more generations and then to look at the density of the DNA. If replication is conservative, the two heavy parental strands will stay together after one generation and the two newly made light strands in the other progeny DNA will appear as a separate band in the CsCl gradient (fig. 7.10). On the other hand, if replication is semiconservative, each heavy parental strand will take a new, light partner. These heavy/light hybrid duplexes will have a density halfway between the heavy/heavy parental DNA and light/light ordinary DNA. This is exactly what happened; after the first DNA doubling, a single band appeared midway between the labeled heavy/heavy DNA and a normal light/light DNA. This ruled out conservative replication, but was consistent with semiconservative replication. Another mechanism, called dispersive replication, was not ruled out, but it is so far-fetched that we will not consider it here.

 DNA replicates in a semiconservative manner. When the parental strands separate, each serves as the template for making a new, complementary strand.

Parental   First generation

Semiconservative:

HH   HL   LH

Expected density gradient results:

HH   HL

(a)

Conservative:

Parental   First generation

HH   HH   LL

Expected density gradient results:

HH   LL HH

(b)

FIGURE 7.10

Two replication hypotheses. (*a*) The semiconservative model predicts that after one generation there will be one kind of duplex DNA, with one heavy (parental) strand and one light (progeny) strand. During density gradient ultracentrifugation, this DNA will form a single band halfway between the bands expected for light/light and heavy/heavy DNAs. (*b*) The conservative model predicts that after one generation there will be equal amounts of two different DNAs: one heavy/heavy (parental) and one light/light (progeny). Meselson and Stahl's results confirmed the former prediction, ruling out the conservative mechanism.

SEMIDISCONTINUOUS REPLICATION

Okazaki's model of semidiscontinuous replication makes two predictions that his research team tested experimentally: (1) Since at least half of the newly synthesized DNA appears first as short pieces, we ought to be able to label and catch these before they are stitched together by allowing very short periods of labeling with radioactive DNA precursor. These short bursts of labeling with a radioactive substance are called **pulses.** (2) If we eliminate the enzyme (DNA ligase) that is responsible for stitching together the short pieces of DNA, we ought to be able to see these pieces even with long pulses of DNA precursor.

For his model system, Okazaki chose replication of bacteriophage T4 DNA. This had the advantage of simplicity, as well as the availability of mutants of T4 DNA ligase. To test the first prediction, Okazaki gave shorter and shorter pulses of tritium-labeled thymidine to *E. coli* cells that were replicating T4 DNA. To be sure of catching short pieces of DNA before they could be joined together, he even administered pulses as short as two seconds. Finally,

he measured the approximate sizes of the newly synthesized DNAs by ultracentrifugation. Large DNA species sediment faster toward the bottom of the centrifuge tube than smaller species do.

Figure 7.11*a* shows the results. Already at two seconds, some labeled DNA is visible in the gradient; within the limits of detection, it appears that all of the label is in very small DNA pieces, 1,000–2,000 bases long, which remain near the top of the tube. With increasing pulse time, another peak of labeled DNA appears much nearer the bottom of the centrifuge tube. This is the result of attaching the small, newly formed pieces of labeled DNA to much larger, preformed pieces of DNA that were made before labeling began. These large pieces, because they were unlabeled before the experiment began, do not show up until enough time has elapsed for DNA ligase to join the smaller, labeled pieces to them; this takes only a few seconds. Appropriately enough, the small pieces of DNA that are the initial products of replication have come to be known as **Okazaki fragments.** This serves as a fitting memorial to Okazaki, a survivor of Hiroshima, who died of leukemia at the pinnacle of his career in cancer research.

The discovery of Okazaki fragments provided powerful evidence for at least partially discontinuous replication of T4 DNA. The clincher was the demonstration that these small DNA fragments accumulate to very high levels when the stitching enzyme, DNA ligase, does not operate. Okazaki's group performed this experiment with the T4 mutant containing a defective DNA ligase gene. Figure 7.11*b* shows that the peak of Okazaki fragments grows to a tremendous size in this mutant. Even after a full minute of labeling, these fragments are virtually the only species of labeled DNA.

The latter experiment suggests that Okazaki fragments are an authentic initial product of DNA synthesis and not just an artifact of very short labeling times. Indeed, the fact that *only* small pieces of DNA were labeled either with short pulses of radioactive precursor or when DNA ligase did not operate suggested that replication proceeded discontinuously on *both* strands (see fig. 7.3c). Subsequent experiments have shown that half the Okazaki fragments seen in the early studies were probably artifacts caused by breakage of the continuously replicating strand. Thus, replication is really **semidiscontinuous:** One strand replicates continuously, the other discontinuously.

◆ To demonstrate that DNA replication is at least semidiscontinuous, Okazaki showed that short DNA fragments are made, especially when DNA ligase is inactive.

(a)   (b)

FIGURE 7.11

Experimental demonstration of at least semidiscontinuous DNA replication. (*a*) Okazaki and his colleagues labeled replicating phage T4 DNA with very short pulses of radioactive DNA precursor and separated the product DNAs according to size by ultracentrifugation. At the shortest times, the label went primarily into short DNA pieces (found near the top of the tube), as the discontinuous model predicted. (*b*) When a mutant phage with a defective DNA ligase gene was used, short DNA pieces accumulated even after relatively long labeling times (one minute in the results shown here).

RNA PRIMING

The first line of evidence supporting RNA priming was the finding that replication of M13 phage DNA by an *E. coli* extract is inhibited by the antibiotic rifampicin. This was a surprise because rifampicin inhibits RNA polymerase, not DNA polymerase, and RNA polymerase makes RNA, not DNA. The explanation is that M13 uses *E. coli* RNA polymerase to make the RNA that primes its DNA synthesis. However, this is not a general phenomenon. Even *E. coli* does not use its own RNA polymerase for priming; it has a special enzyme system, tailor-made for this purpose.

Perhaps the best evidence for RNA priming was the discovery by Tuneko Okazaki, Reiji Okazaki's wife and scientific colleague, that the DNA-degrading enzyme DNase cannot completely destroy Okazaki fragments. It leaves little pieces of RNA 10–12 bases long. Other work has shown that these RNA **oligonucleotides** are covalently linked, not just hydrogen-bonded, to the 5′-end of Okazaki fragments.

 Tuneko Okazaki isolated the DNA replication primers from *E. coli* and showed that they were RNAs 10–12 bases long.

BIDIRECTIONAL REPLICATION

One direct approach to investigating DNA replication is to catch DNA molecules in the act of replicating and take pictures of them. This has been done, and we will see examples of this kind of experiment later in this chapter. The early studies used a related technique: **autoradiography.** Autoradiography involves incorporating a radioactive label into a substance and then placing the radioactive substance in contact with a photographic emulsion. Since the radioactive emissions expose the emulsion, the radioactive substance "takes a picture of itself." That is how this process got its name.

John Cairns exploited this technique in the early 1960s to investigate the replication of the *E. coli* chromosome.

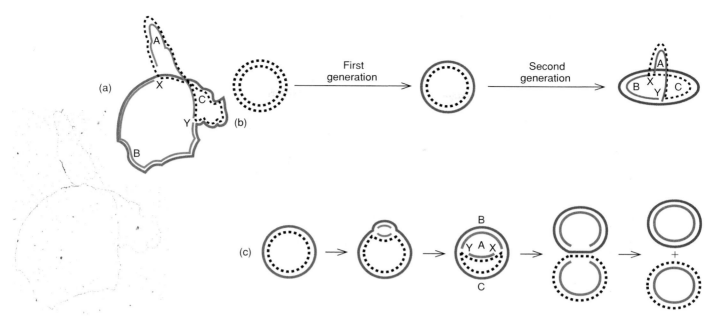

FIGURE 7.12

The theta mode of DNA replication in *Escherichia coli*. (*a*) An autoradiograph of replicating *E. coli* DNA with an interpretive diagram. The DNA was allowed to replicate for one whole generation and part of a second in the presence of radioactive nucleotides. That part of the DNA that has replicated only once will have one labeled strand (blue) and one "cold" strand (dashed, black). The DNA that has replicated twice will have one part that is doubly labeled (loop B blue and red) and one part with only one labeled strand (loop A red and black). Cairns uses a dotted line to show unlabeled DNA and a solid line for labeled DNA. Note that

loop B contains two solid lines, indicating that it is the part of the DNA that has become doubly labeled. The fact that loop B is darker than loops A and C fits with this interpretation. (*b*) Interpretation of Cairns's autoradiograph, using the same dotted and solid line and color conventions as in (*a*). (*c*) More detailed description of the theta mode of DNA replication. The dotted and solid lines and colors have the same meaning as in (*a*) and (*b*).

(*a*) *From John Cairns,* Symposia on Quantitative Biology *(Cold Spring Harbor)* 28:44, 1963.

Figure 7.12*a* shows an autoradiograph of replicating *E. coli* DNA, along with Cairns's interpretation, which is amplified in figure 7.12*b*. Figure 7.12*a* and *b* also attempts to explain the difference in labeling of the three loops (A, B, and C) of the DNA as follows: The DNA underwent one round of replication and part of a second. That means all the DNA should have at least one labeled strand. During the second round of replication, the unlabeled parental strand will serve as the template to make a labeled partner; this duplex will therefore become singly labeled. The labeled parental strand will obtain a labeled partner, making this duplex doubly labeled. This doubly labeled part (loop B) should expose the film more and therefore appear darker than the rest of the DNA (loops A and C).

The structure represented in figure 7.12*a* is a so-called theta structure because of its resemblance to the Greek letter θ. Since it may not be immediately obvious that the DNA in figure 7.12*a* looks like a theta, figure 7.12*c* provides a schematic diagram of the events in the second round of replication that led to the autoradiograph. This drawing shows that DNA replication begins with the creation of a "bubble" or small region where the parental strands have separated and progeny DNA has been synthesized. As the bubble expands, the replicating DNA begins to take on the theta shape. We can now recognize the autoradiograph as representing a structure shown in the middle of figure 7.12*c,* where the crossbar of the theta has grown long enough to extend above the circular part.

The theta structure contains two **forks,** marked X and Y in figure 7.12. This raises an important question: Does one of these forks, or do both, represent sites of active DNA replication? In other words, is DNA replication unidirectional, with one fork moving away from the other, which remains fixed at the **origin of replication?** Or is it bidirectional, with two replicating forks moving in opposite directions away from the origin? Cairns's autoradiographs were not designed to answer this question, and in fact, they led to the erroneous conclusion that replication in *E. coli* is unidirectional. A subsequent study performed by Elizabeth Gyurasits and R. B. Wake showed clearly that DNA replication in *Bacillus subtilis* is bidirectional.

These investigators' strategy was to allow *B. subtilis* cells to grow for a short time in the presence of a weakly radioactive DNA precursor, then for a short time with a more strongly radioactive precursor. The labeled precursor was the same in both cases: ^{3}H-thymidine. Tritium (^{3}H) is the radioactive isotope of hydrogen. It is especially useful for this type of autoradiography because its radioactive emissions are so weak that they do not travel far from their point of origin before they stop in the photographic emulsion and create silver grains. This means that the pattern of silver grains in the autoradiograph will bear a close relationship to the shape of the radioactive DNA. It is important to note that autoradiography reveals the shape of radioactive substances only; unlabeled DNA will not show up. The pulses of label in this experiment were short

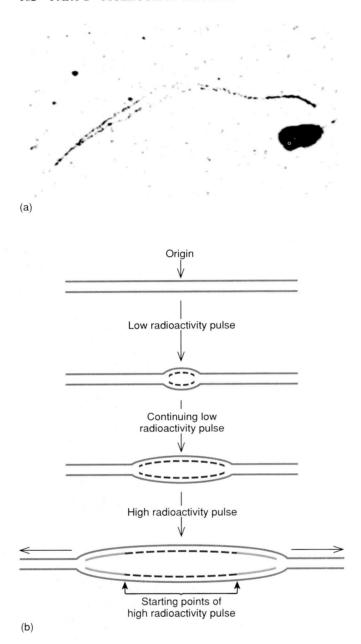

(a)

Origin

Low radioactivity pulse

Continuing low
radioactivity pulse

High radioactivity pulse

Starting points of
high radioactivity pulse

(b)

FIGURE 7.13

Experimental demonstration of bidirectional DNA replication.
(a) Autoradiograph of replicating *Bacillus subtilis* DNA. Dormant
bacterial spores were germinated in the presence of a low radioactivity
DNA precursor, so the newly formed replicating bubbles immediately
became slightly labeled. After the bubbles had grown somewhat, a more
radioactive DNA precursor was added to label the DNA for a short period.
(b) Interpretation of the autoradiograph. The purple color represents the
slightly labeled DNA strands produced during the low radioactivity pulse.
The orange color represents the more highly labeled DNA strands
produced during the later, high radioactivity pulse. Since both forks picked
up the high radioactivity label, both most have been functioning during the
high radioactivity pulse. Therefore, DNA replication in *B. subtilis* is
bidirectional.

*Photo (a) from E. B. Gyurasits and R. B. Lake. Journal of Molecular Biology,
Vol. 73, 1973, facing pg. 58. Reprinted by permission of Academic Press.*

enough that only the replicating bubbles are visible
(fig. 7.13a). You should not mistake these for whole bacterial
chromosomes such as in figure 7.12.

If you look carefully at figure 7.13a, you will notice
that the pattern of silver grains is not uniform. There are
concentrations of silver grains near both forks in the bubble. This extra labeling identifies the regions of DNA that
were replicating during the "hot," or high radioactivity
pulse period (fig. 7.13b). Both forks incorporated extra
label, telling us that they were both operating during the hot
pulse. Therefore, DNA replication in *B. subtilis* is bidirectional; two forks arise at a fixed starting point—the origin of
replication—and move in opposite directions around the circle until they meet on the other side. Later experiments employing this and other techniques have shown that the
E. coli chromosome also replicates bidirectionally from a
single origin of replication.

J. Huberman and A. Tsai have performed the same
kind of autoradiography experiments in a eukaryote, the
fruit fly *Drosophila melanogaster*. Here, the experimenters
gave a pulse of strongly radioactive (high specific activity)
DNA precursor, followed by a pulse of weakly radioactive
(low specific activity) precursor. Alternatively, they reversed the procedure and gave the low specific activity label
first, followed by the high. Then they autoradiographed the labeled insect DNA. The spreading of DNA in
these experiments did not allow the replicating bubbles to
remain open; instead, they collapsed and appear on the autoradiographs as simple streaks of silver grains.

One end of a streak marks where labeling began, the
other shows where it ended. But the point of this experiment is that the streaks always appear in pairs (fig. 7.14a).
The pairs of streaks represent the two replicating forks that
have moved apart from a common starting point. Why
doesn't the labeling start in the middle, at the origin of
replication, the way it did in the experiment with *B. subtilis*
DNA? In the *B. subtilis* experiment, the investigators were
able to synchronize their cells by allowing them to germinate from spores, all starting at the same time. That way
they could get label into the cells before any of them had
started making DNA (i.e., before germination). That was
not tried in the *Drosophila* experiments, where it would
have been much more difficult. As a result, replication usually began before the label was added, so there is a blank
area in the middle where replication was occurring but no
label could be incorporated.

Notice the shape of the pairs of streaks in figure 7.14a.
They taper to a point, moving outward, rather like a waxed
mustache. That means the DNA incorporated highly radioactive label first, then more weakly radioactive label,
leading to a tapering off of radioactivity as we move outward in both directions from the origin of replication. The
opposite experiment—"cooler" label first, followed by
"hotter" label—gives a reverse mustache, with points on
the inside. It is possible, of course, that closely spaced, independent origins of replication gave rise to these pairs of
streaks. But we would not expect that such origins would

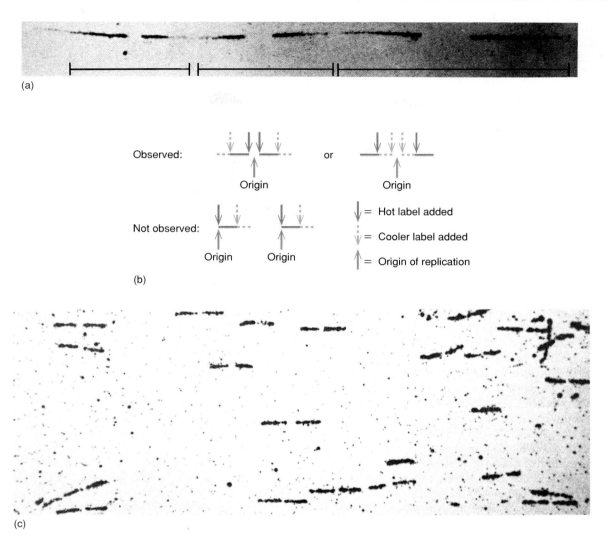

FIGURE 7.14

Bidirectional DNA replication in eukaryotes. (*a*) Autoradiograph of replicating *Drosophila melanogaster* DNA, pulse-labeled first with high radioactivity DNA precursor, then with low. Note the pairs of streaks (denoted by brackets) tapering away from the middle. This reflects the pattern of labeling of a replicon with a central origin and two replicating forks. (*b*) Idealized diagram showing the patterns observed with hot-then-cold or cold-then-hot labeling, respectively, and the pattern expected if the pairs of streaks represent two independent unidirectional replicons whose replicating forks move in the same direction. The latter pattern was not observed. (*c*) Autoradiograph of replicating embryonic *Triturus vulgaris* DNA. Note the constant size and shape of the pairs of streaks, suggesting that all the corresponding replicons are replicating at the same time.

Photo (a) from J. A. Huberman and A. Tsai, Journal of Molecular Biology, *Vol. 75, 1973, facing pg. 8(A). Reprinted by permission of Academic Press. Photo (c) from H. G. Callan,* Symposia on Quantitative Biology *(Cold Spring Harbor), vol. 38, p. 196, 1973.*

always give replication in opposite directions. Surely some would lead to replication in the same direction, producing asymmetric autoradiographs such as the hypothetical one in figure 7.14*b*. But these are not seen. Thus, these autoradiography experiments confirm that each pair of streaks we see really represents *one* origin of replication, rather than two that are close together. It therefore appears that replication of *Drosophila* DNA is bidirectional.

These experiments were done with *Drosophila* cells originally derived from mature fruit flies and then cultured in vitro. H. G. Callan and his colleagues performed the same type of experiment using highly radioactive label and embryonic amphibian cells. These experiments (with embryonic cells of the newt) gave the striking results shown in

figure 7.14*c*. In contrast to the pattern in adult insect cells, the pairs of streaks here are all the same. They are all approximately the same length and they all have the same size space in the middle. This tells us that replication at all these origins began simultaneously. This must be so, because the addition of label caught them all at the same point—the same distance away from their respective origins. This phenomenon probably helps explain how embryonic newt cells complete their DNA replication so rapidly (in as little as one hour, compared to forty hours in adult cells). Replication at all origins begins simultaneously, rather than in a staggered fashion.

This discussion of origins of replication helps us define an important term: **replicon.** The DNA under the control of one origin is called a replicon, even though it may require two replicating forks. The *E. coli* chromosome is a single replicon because it replicates from a single origin. Eukaryotic chromosomes are much bigger and therefore have many replicons; otherwise, it would take far too long to replicate a whole chromosome. This would be true even at the breakneck speed of DNA synthesis in prokaryotes— about 150 nucleotides per second. In fact, eukaryotic DNA replicates more slowly than this, which makes the need for multiple replicons even more obvious.

UNIDIRECTIONAL REPLICATION

Do all genetic systems replicate bidirectionally? This is certainly the most common scheme, but there are exceptions. For example, consider the replication of a plasmid called **colicin E1 (colE1)**. A **plasmid** is a circular piece of DNA that replicates independently of the cell's chromosome; some strains of *E. coli* harbor colE1, which explains how this plasmid got the "coli" part of its name. Lovett used an electron microscope to examine replicating molecules of colE1 and found that only one fork moves.

Of course, if the experimenters had simply looked at circular replicating molecules, they would have seen a replication bubble with two forks, but there would have been no way of knowing whether replication was occurring at both forks. We need some kind of marker in the DNA that we can use as a guide to locating the two forks at different times during replication. Then, if both forks move relative to the marker, we know that they are both active and DNA replication is bidirectional. On the other hand, if only one fork moves, replication must be unidirectional.

The marker Lovett and colleagues chose is the site at which colE1 DNA is cut by the restriction endonuclease *Eco*RI. As explained in chapter 15, *Eco*RI recognizes the sequence: 5′-GAATTC-3′. There is only one such recognition site for *Eco*RI in the whole colE1 molecule, so cutting the DNA with this enzyme will create a linear molecule whose ends represent the *Eco*RI cutting site. Figure 7.15 shows the results of the experiment. Clearly, only one fork moves relative to the ends of the linearized DNA molecule. Therefore, replication of colicin E1 DNA is unidirectional.

 Experiments employing autoradiography have shown that most eukaryotic and prokaryotic DNAs replicate bidirectionally. ColE1 is an example of a DNA that replicates unidirectionally.

ENZYMOLOGY OF DNA REPLICATION

Over thirty distinct polypeptides cooperate in replicating *E. coli* DNA. In the following section, we will examine the activities of some of these proteins.

STRAND SEPARATION

In the foregoing discussion of the general features of DNA replication, we have been assuming that the two DNA

FIGURE 7.15

Unidirectional colicin E1 DNA replication. Molecule 1 has not begun to replicate, so no replicating bubble is seen. Molecules 2–8 show replicating bubbles of progressively increasing size. It is important to notice that the unreplicated stretch of DNA above the bubble in each picture remains the same size, while that below the bubble grows shorter as the bubble grows larger. Clearly, only the lower fork is replicating; the other is stationary.
Michael Lovett, Leonard Katz, and Donald Helinski. Reprinted by permission from Nature, *Vol. 251 p. 338, 1974. 1991 Macmillan Magazines Limited.*

strands at the fork must somehow come unwound. This does not just happen automatically as DNA polymerase does its job; the two parental strands hold tightly to one another, and it takes energy and enzyme action to separate them.

Helicases

The enzyme that harnesses the chemical energy of ATP to separate the two parental DNA strands at the replicating fork is called a **helicase.** ATP is the primary source of chemical energy in a cell. When it breaks down to ADP and phosphate, considerable energy is released. Many enzymes are able to trap this energy and use it to perform work. Helicases contain an ATPase activity that breaks down two molecules of ATP for every base pair separated, which means that duplication of DNA sequences requires tremendous amounts of energy.

Single-Strand-Binding Protein

Once the parental DNA strands are locally separated by one or more helicases, another kind of protein binds to the newly exposed single-stranded regions and prevents their reassociation. This protein is called a **single-strand-binding protein** or **SSB.** Actually the SSB of *E. coli* also binds to double-stranded DNA, but the mode of binding is different. When SSB binds to double-stranded DNA, only a single molecule binds. When it encounters single-stranded DNA, it binds cooperatively. In other words, once one molecule binds, it helps more molecules attach. The result is

FIGURE 7.16

Cairns's swivel concept. As the closed circular DNA replicates, the two strands must separate at the fork (F). The strain of this unwinding would be released by a swivel mechanism. Cairns actually envisioned the swivel as a machine that rotated actively and thus drove the unwinding of DNA at the fork.

FIGURE 7.17

Rubber band model of supercoiling in DNA. If you cut the rubber band and twist one free end through one complete turn while preventing the other end from rotating, the rubber band will relieve the strain by forming one supercoil.

much greater binding of SSB to single- than to double-stranded DNA. Eukaryotic SSBs appear similar to the *E. coli* protein with one important exception: They do not bind cooperatively to single-stranded DNA. Thus, they must bind preferentially to single strands by another mechanism.

Topoisomerases

Sometimes we refer to the separation of DNA strands as "unzipping." We should not forget, when using this term, that DNA is not like a zipper with straight, parallel sides. It is a helix. Therefore, when the two strands of DNA separate, they must rotate around one another. Helicase could handle this task alone if the DNA were linear and unnaturally short, but closed circular DNAs, such as the *E. coli* chromosome, present a special problem. As the DNA unwinds at the replicating fork, a compensating winding up of DNA will occur elsewhere in the circle. This tightening of the helix will create intolerable strain unless it is relieved somehow. Cairns recognized this problem when he first observed circular DNA molecules in *E. coli* in 1963, and he proposed a "swivel" in the DNA duplex that would allow the DNA strands on either side to rotate to relieve the strain (fig. 7.16). We now know that an enzyme known as **DNA gyrase** serves the swivel function. DNA gyrase belongs to a class of enzymes called **topoisomerases** that introduce transient single- or double-stranded breaks into DNA and thereby allow it to change its form or topology.

To understand how the topoisomerases work, we need to look more closely at the phenomenon of supercoiled, or superhelical DNA mentioned in chapter 6. All naturally occurring, closed circular, double-stranded DNAs studied so far exist as supercoils. Closed circular DNAs are those with no single-strand breaks, or nicks. When a cell makes such a DNA, it causes some unwinding of the double helix; the DNA is then said to be "underwound." As long as both strands are intact, there can be no free rotation around the bonds in either strand's backbone, so there is no way the DNA can relieve the strain of underwinding except by supercoiling. The supercoils introduced by underwinding are called negative, by convention. This is the only kind of supercoiling found so far in nature; positive supercoils exist only in the laboratory.

You can visualize the supercoiling process with a Mr. Wizard-type experiment (fig. 7.17). Take a medium-sized rubber band, hold it at the top with one hand, and cut through one of the sides about halfway down. With your other hand, twist one free end of the rubber band one full turn and hold it next to the other free end. You should notice that the rubber band resists the turning as strain is introduced, then relieves the strain by forming a supercoil.

FIGURE 7.18
Nicking one strand relaxes supercoiled DNA. A nick in one strand of a supercoiled DNA (left) allows free rotation around a phosphodiester bond in the opposite strand (middle). This releases the strain that caused the supercoiling in the first place and so allows the DNA to relax to the open circular form (right).

The more you twist, the more supercoiling you will observe: one superhelical turn for every full twist you introduce. Reverse the twist and you will see supercoiling of the opposite handedness or sign. DNA works the same way. Of course, the rubber band is not quite a perfect model. For one thing, you have to hold the severed ends together; in a circular DNA, chemical bonds join the two ends. For another, the rubber band is not a double helix.

If you release your grip on the free ends of the rubber band in figure 7.17, of course the superhelix will relax. In DNA, it is only necessary to cut one strand to relax a supercoil because the other strand can rotate freely, as demonstrated in figure 7.18.

Underwinding can occur as follows: A nicked DNA is partially unwound, then the nick is sealed to give a closed circular DNA. However, unwinding DNA at the replicating fork would have the opposite effect if there were no way to relax the strain. The reason is that here we are *permanently* unwinding one region of the DNA without nicking it; this means that the rest of the DNA would become *over*wound, and therefore positively supercoiled, to compensate. To appreciate this, look at the circular arrow ahead of the replicating fork in figure 7.16. Notice how twisting the DNA in the direction of the arrow causes unwinding behind the arrow but overwinding ahead of it. Therefore, unwinding the DNA at the replicating fork introduces *positive* superhelical strain that must be constantly relaxed so that replication will not be retarded. You can appreciate this when you think of how the rubber band increasingly resisted your twisting as it became more tightly wound. In principle, any enzyme that creates transient breaks (either single- or double-stranded) in DNA should be able to relax this strain and thereby serve as a swivel. In fact, of all the topoisomerases in an *E. coli* cell, only one, DNA gyrase, appears to perform this function.

There is direct evidence that DNA gyrase is crucial to the DNA replication process. First of all, mutations in the genes for the two polypeptides of DNA gyrase are lethal and they block DNA replication. Second, antibiotics such as novobiocin, coumermycin, and nalidixic acid attack DNA gyrase and thereby inhibit replication.

> One or more enzymes called helicases use ATP energy to separate the two parental DNA strands at the replicating fork. As helicase unwinds the two parental strands of a closed circular DNA, it introduces a compensating positive supercoiling force into the DNA. The stress of this force must be overcome or it will resist progression of the replicating fork. The name given to this stress-release mechanism is the swivel. DNA gyrase is the leading candidate for this role. Once the two strands are separated, single-strand-binding protein (SSB) keeps them that way.

INITIATION OF REPLICATION

We have already seen that a crucial aspect of initiation in DNA replication is synthesis of a short RNA primer. (There are a few exceptions to this rule, notably adenovirus, which uses a special protein to prime its DNA synthesis.) A rich variety of mechanisms exist to form these RNA primers, and each genetic system uses one of them. A large aggregate of polypeptides combines to make the primers in *E. coli* DNA replication. This **primosome** is so complex that the identification of its components would have taken much longer were it not for the existence of single-stranded DNA phages that use only parts of the primosome. These have been used as models for unraveling the mechanism of primer formation in *E. coli* itself.

Making Primers in *E. coli*
More than twenty-five elements combine to make up the DNA replicating machinery of *E. coli*. For the moment, let us concentrate on some of the proteins involved in the synthesis of primers, including the primer-synthesizing enzyme, or **primase.**

During replication of double-stranded DNA, the primosome moves **processively** along the template for the lagging strand in the direction of motion of the replication fork. "Processively" means that it moves continuously along the DNA without dissociating from it. It is advantageous for the primosome to bind to the lagging strand template since that is where DNA is being made discontinuously and new primers are needed for each new Okazaki fragment. However, this presents a problem for the primase, because the direction the replicating fork is moving on the lagging strand template is $5' \rightarrow 3'$, which means a primer synthesized complementary to that strand would be made in the $3' \rightarrow 5'$ direction. And RNAs, even primers, have to be made in the $5' \rightarrow 3'$ direction. Actually, the primer *is* made in the $5' \rightarrow 3'$ direction, even though the primosome moves in the opposite direction (fig. 7.19). Just how this is accomplished is not known.

Kornberg likens the primosome to a locomotive moving along the railroad track of DNA in the direction of the

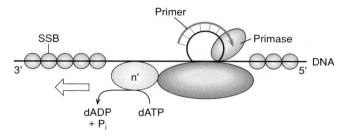

FIGURE 7.19
Primosome forming a primer on φX174 DNA. The complex primosome is moving in the 5′ → 3′ direction on the template strand (open arrow), while making a primer in the opposite direction (red arrow). The actual synthesis of the primer is performed by the primase (orange). The n′ protein (yellow) provides energy for this process by cleaving energy-rich ATP or (as shown here) dATP. The rest of the primosome is represented by a green oval.

replicating fork. The **n′** protein provides the energy for this process by cleaving ATP and thus plays the engine role in the locomotive. This protein also serves as a "cow-catcher" by bumping SSB molecules off the DNA "track" in advance of the moving primosome.

According to Kornberg's metaphor, if the n′ protein is the power plant, then the dnaB protein is the engineer. It has binding sites for many of the other components involved in the primosome, including double- and single-stranded DNA, ATP, the dnaC protein, and the primase. Thus, it can act as a primosome organizer. Furthermore, it has the ability to harness the energy in ATP to change the conformation of DNA so that it is receptive to primase activity.

The primosome is a complex structure containing about twenty polypeptides that makes primers for *E. coli* DNA replication. Some of the proteins involved in the primosome are: the n′ protein, which uses energy from ATP to pull the primosome along behind the replicating fork, removing SSB as it goes; the dnaB protein, which organizes some of the other components of the primosome and prepares the DNA strand for primase activity; and the primase itself, which forms the primer. In *E. coli* DNA replication, the primosome moves along the template for the lagging strand and makes primers in the opposite direction.

ELONGATION

Once the primosome has made a primer, it is time for elongation to begin. The key enzyme involved in the elongation process is **DNA polymerase III holoenzyme,** a complex of seven different polypeptides. The term holoenzyme simply means "the whole thing," including the active center (for DNA synthesis in this case) as well as any attendant factors that participate in the reaction. We do not know for sure, but we strongly suspect that the DNA polymerase III holoenzyme joins with the primosome and helicases to make a **replisome** that then carries out elongation of the DNA chain.

Aid from the *E. coli* Replisome

A major question immediately arises concerning the process of elongation: How does the replisome coordinate the synthesis of the leading and lagging strands? It appears at first that each replicating fork would either need two replisomes, one for each strand, or that the replisome would have to be very flexible to carry out DNA synthesis in opposite directions on the two strands.

However, single DNA strands are considerably more flexible than proteins, so it is not unreasonable to expect that the lagging strand template loops out of the replisome as shown in figure 7.20. If this happens, as Kornberg suggests, then the replisome could read both strands simultaneously because one (the leading strand template) would be read as it fed into the top of the replisome, while the other (the lagging strand template) would be read as it fed into the bottom of the replisome. This would allow the replisome to make both leading and lagging strands in the 5′ → 3′ direction.

This model has the great advantage that it permits one replisome to replicate both strands processively, without having to leave the replicating fork. This is important because once dissociated from the DNA, the replisome would waste much time rebinding. To put this in perspective, consider that the rate of elongation in *E. coli* is about 150 nucleotides per second, and the rate of formation of a complex capable of initiation is one or two minutes. This means that while the replisome paused to rebind to the DNA, it could have been elongating the DNA chain by at least 9,000 nucleotides. Some time will be lost even in the looping and releasing process described here, but the intramolecular transfer involved only requires two to five seconds—quite a saving compared to one to two minutes.

Elongation during DNA synthesis in *E. coli* apparently requires a replisome made up of the primosome plus a DNA polymerase III holoenzyme composed of at least ten different polypeptides. Kornberg's model for simultaneous replication of both strands calls for a double-headed replisome with two DNA polymerase centers. One would make the leading strand continuously; the other would synthesize the lagging strand discontinuously, but without having to leave the replicating fork.

Rolling Circle Replication

Certain circular DNAs replicate, not by the theta mode we have already examined, but by a mechanism called **rolling circle** replication. Lambda (λ) phage actually uses both means of replication. During the early phase of λ DNA replication, the phage follows the theta mode of replication to produce several copies of circular DNA. These circular DNAs are not packaged into phage particles; they serve as templates for rolling circle synthesis of linear λ DNA molecules that *are* packaged.

Figure 7.21 shows how this rolling circle operates. Here, the replicating fork resembles that in *E. coli* DNA replication, with continuous synthesis on the leading strand (the one going around the circle) and discontinuous synthesis on the lagging strand. It is not difficult to see how the name "rolling circle" arose. In fact, you can almost think of the replicating DNA as a roll of toilet paper unrolling as it speeds across the floor. The unrolled part represents the growing double-stranded progeny DNA. In λ, this progeny DNA reaches lengths that are several genomes long before it is packaged. These multiple-length DNAs are called **concatemers.** The packaging mechanism is designed to accept only one genome's worth of linear DNA into each phage head, so the concatemer must be cut during packaging. This rolling circle mechanism of DNA replication is also sometimes called the sigma mode because the intermediate resembles the lowercase Greek letter sigma (σ).

FIGURE 7.20
A model for simultaneous synthesis of both DNA strands. (*a*) The lagging template strand (blue) has formed a loop through the replisome (gold), and a new primer, labeled 2 (red), has been formed by the primase. A previously synthesized Okazaki fragment (green, with red primer labeled 1) is also visible. The leading strand template and its progeny strand are shown at left (gray), but the growth of the leading strand is not considered here. (*b*) The lagging strand template has formed a bigger loop by feeding through the replisome as shown by the arrows. The motion of the lower part of the loop (lower arrow) allows the second Okazaki fragment to be elongated. (*c*) Further elongation of the second Okazaki fragment brings its end to a position adjacent to the primer of the first Okazaki fragment. (*d*) The replisome releases the loop, including Okazaki fragments 1 and 2, and forms a new loop, which permits the primase to form a new primer (number 3). The process can now begin anew.

 Circular DNA can replicate by a rolling circle, or sigma mechanism. One strand of a double-stranded DNA is nicked and the 3′-end is extended, using the intact DNA strand as template. This displaces the 5′-end. In phage λ, the displaced strand serves as the template for discontinuous, lagging strand synthesis.

TERMINATION

Let us return to the problem of filling in the gaps left when RNA primers are removed. With circular DNAs, such as those in bacteria, there is no problem filling all the gaps because there is always another DNA 3′-end upstream to serve as primer (fig. 7.22*a*). But consider the problem faced by eukaryotes, with their linear chromosomes. Once the

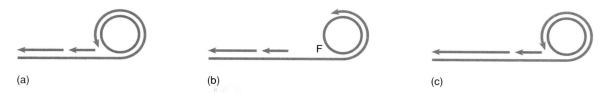

(a)   (b)   (c)

FIGURE 7.21

Rolling circle model for phage λ DNA replication. (*a*) The circle, containing the leading strand (red) and its template (blue), is rolling to the right. As it rolls, the leading strand is continuously elongated in the direction of the arrow. The newly synthesized leading strand in turn serves as the template for the lagging strand (blue arrows), which is made discontinuously. (*b*) The circle has rolled farther to the right, exposing more leading and lagging strand templates at the replicating fork (F).

(*c*) The leading strand (red) has been elongated, the two original Okazaki fragments have been ligated (long blue arrow), and a new Okazaki fragment of lagging strand (short blue arrow) has been made. These drawings suggest a discontinuity to elongation of the leading strand. In reality, this strand would be elongated continuously as the circle rolls. The discontinuity is pictured here to help illuminate the concept of the rolling circle.

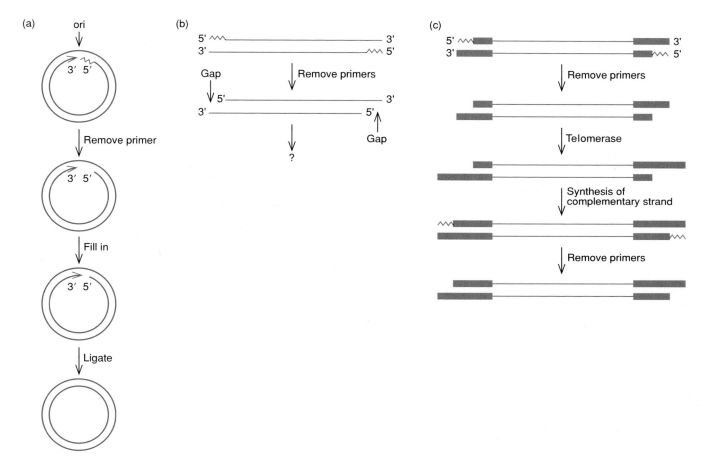

FIGURE 7.22

Coping with the gaps left by primer removal. (*a*) In prokaryotes, the 3′-end of a circular DNA strand can prime the synthesis of DNA to fill in the gap left by the first primer. For simplicity, only one replicating strand is shown. (*b*) Hypothetical model to show what would happen if primers were simply removed from the 5′-ends of linear DNA strands with no telomerase action. The gaps at the ends of chromosomes would grow longer each time the DNA replicated. (*c*) How telomerase can solve the

problem. In the first step, the primers (red) are removed from the 5′-ends of the daughter strands, leaving gaps. In the second step, telomerase adds extra telomeric DNA (green) to the 3′-ends of the other daughter strands. In the third step, DNA synthesis occurs, using the newly made telomeric DNA as a template. In the fourth step, the primers used in step three are removed. This leaves gaps, but the telomerase action has ensured that no net loss of DNA has occurred.

FIGURE 7.23

Proofreading in DNA replication. (*a*) An adenine nucleotide (red) has been mistakenly incorporated across from a guanine. This destroys the perfect base pairing required at the 3′-end of the primer, so the replicating machinery stalls. (*b*) This then allows polymerase III holoenzyme to use its 3′ → 5′ exonuclease function to remove the offending nucleotide. (*c*) With the appropriate base pairing restored, polymerase III holoenzyme is free to continue replication.

first primer on each strand is removed (fig. 7.22*b*), there is no way to fill in the gaps, since DNA cannot be extended in the 3′ → 5′ direction, and there is no 3′-end upstream as there would be in a circle. If this were actually the situation, the DNA strands would get shorter every time they replicated. This is a termination problem in that it deals with formation of the ends of the DNA strands. But how is this problem solved?

Elizabeth Blackburn and her colleagues have provided the answer, which is summarized in figure 7.22*c*. The telomeres, or ends of eukaryotic chromosomes, are composed of repeats of short, GC-rich sequences. These are added to the very 3′-ends of DNA strands, not by semiconservative replication, but by an enzyme called **telomerase**. The exact sequence of the repeat in a telomere is species-specific. In *Tetrahymena,* a ciliated protozoan, it is TTGGGG/AACCCC; in vertebrates, including humans, it is TTAGGG/AATCCC. Blackburn showed that this specificity resides in the telomerase itself, and is due to a small RNA in the enzyme. For example, the *Tetrahymena* telomerase has an RNA 159 nucleotides long that contains the sequence CAACCCCAA. Thus, it can serve as the template for the TTGGGG repeats in *Tetrahymena* telomeres. Now the termination problem is solved; the telomerase adds many repeated copies of its characteristic sequence to the 3′-ends of chromosomes. Priming can then occur within these telomeres. There is no problem when the primer is removed and not replaced, because only telomere sequences are lost, and these can always be replaced by telomerase and another round of replication.

The ends of eukaryotic chromosomes are formed by an enzyme called telomerase, which adds DNA to the 3′-ends of the chromosomes according to the instructions of its own RNA template.

FIDELITY OF REPLICATION

By this time, you may be asking yourself: "Why is DNA replication so complex? Why go to all the trouble of semi-discontinuous replication with RNA priming?" The answer seems to be that both of these complex mechanisms contribute to something very important to life: faithful DNA replication. Consider the alternative to this fidelity. What if every time our chromosomes replicated, the DNA synthesizing machinery made many mistakes? These mistakes are mutations, which would accumulate with every new cell division. We will see in later chapters that mutations can be very deleterious; it is clear that life cannot tolerate too many of them. In other words, if our replicating machinery were much less faithful, our genome would have to be much smaller in order to minimize the number of mutations occurring each generation. Complex life forms, relying on many genes, would be impossible.

How good are DNA polymerases at pairing up the right bases during replication? The *E. coli* DNA polymerase III holoenzyme, operating in vitro, makes about one pairing mistake in one hundred thousand—not a very good record, considering that even the *E. coli* genome contains about three million base pairs. At this rate, replication would introduce errors into a significant percentage of genes. Fortunately, the DNA polymerase III holoenzyme includes its own **proofreading** activity—a 3′ → 5′ exonuclease—that allows it to synthesize DNA with only one error in at least ten million nucleotides polymerized.

Figure 7.23 shows the proofreading process. The DNA polymerase III holoenzyme in the replisome can add a nucleotide to the growing strand only if the previously incorporated nucleotide is properly base-paired. This is a critical requirement. It means that the polymerase cannot add to a mismatched nucleotide. Therefore, if the replisome adds the wrong nucleotide, it will stall until the mismatch is repaired. The 3′ → 5′ exonuclease activity of DNA polymerase III holoenzyme can remove the mismatched nucleotide, perhaps along with several others. With the proper base-paired primer restored, the replisome can continue.

Now consider the implications of this proofreading mechanism. It immediately explains the need for a primer. If DNA polymerase needs a base-paired nucleotide to add to, it clearly cannot start a new DNA chain unless a primer is there. But why primers made of RNA? The reason seems to be the following: Primers are made with more errors, since their synthesis is presumably not subject to proofreading. Making primers out of RNA guarantees that they will be recognized, removed, and replaced with DNA by extending the neighboring Okazaki fragment. The latter process is, of course, relatively error-free. Besides these theoretical arguments, there is the simple fact that RNA synthesizing enzymes can start without primers, whereas DNA polymerase cannot.

The proofreading mechanism also helps rationalize semidiscontinuous replication. If there were a polymerase capable of synthesizing DNA in the $3' \rightarrow 5'$ direction, thus making fully continuous replication possible, we would need a whole new $5' \rightarrow 3'$ proofreading function, with the ability to restore the 5′-triphosphates that would inevitably be destroyed during proofreading. Semidiscontinuous replication, as cumbersome as it seems, is probably simpler.

The DNA replicating machinery has a built-in proofreading system that requires priming. Only a base-paired nucleotide can serve as a primer for the replicating enzyme; therefore, if the wrong nucleotide is incorporated by accident, replication stalls until a nuclease erases the error by removing the most recently added nucleotide. The fact that the primers are made of RNA may mark them for degradation.

THE MECHANISM OF RECOMBINATION

We have already seen many examples of progeny organisms with different combinations of genes than those found in either parent. Two mechanisms can explain how this happens. First, as we learned in chapter 2, genes on different chromosomes can simply assort independently. Second, as shown in chapter 4, **crossing over** can occur between homologous chromosomes. Indeed, DNA replication in meiotic cells is usually followed by such crossing over. This process scrambles the genes of maternal and paternal chromosomes, so that nonparental combinations occur in the progeny. This scrambling is valuable because the new combinations sometimes allow the progeny organisms a better chance of survival than their parents.

Both of these processes, independent assortment and crossing over, are central features of the science of genetics. Independent assortment is relatively easy to understand, but the mechanism of recombination by crossing over is not so simple. Now that we have become well acquainted with DNA by studying its replication, we should pause to answer the question: How does such recombination work?

FORMS OF RECOMBINATION BY CROSSING OVER

The examples of recombination shown in figure 7.24 always involve a **strand exchange** or **crossing over** event that joins DNA segments previously separated. This does not mean the two segments must start out on separate DNA molecules. Recombination can be intramolecular, in which case crossing over between two sites on the same chromosome either removes or inverts the DNA segment

in between. On the other hand, bimolecular recombination involves crossover events between two independent DNA molecules. Ordinarily, recombination is **reciprocal**—a "two-way street" in which the two participants trade DNA segments.

What does recombination look like at the molecular level? First, consider that there are three fundamentally different kinds of recombination: (1) **Site-specific recombination** depends on limited sequence similarity between the recombining DNAs and always involves the same DNA regions, which explains why it is called site-specific. A prominent example of this kind of recombination is the insertion of λ phage DNA into the host DNA (chapter 8). (2) A related form is **illegitimate recombination.** This also requires little if any sequence similarity between the recombining DNAs, but it is not site-specific. Transposable elements (chapter 12) usually use the illegitimate recombination pathway. (3) By far the most common form of recombination depends on extensive sequence similarity between the participating DNA molecules. This is the kind of recombination that occurs by crossing over in meiosis. It is called **generalized** or **homologous recombination.**

A MODEL FOR HOMOLOGOUS RECOMBINATION

Our current models for the mechanism of homologous recombination share the following essential steps, although not necessarily in this order: (1) pairing of two homologous DNA duplexes; (2) breaking two of the homologous DNA strands; (3) re-forming phosphodiester bonds to join the two homologous strands (or regions of the same strand during recombination within one strand); (4) breaking the other two strands and joining them. The basic mechanism, originally proposed by Robin Holliday, is presented in figure 7.25.

The first step in the Holliday model is the pairing of two homologous DNA duplexes. The mechanism of this pairing is still not well understood. The second step is creation of nicks in corresponding positions in two homologous DNA strands. These nicks could theoretically come from a variety of sources. They could be leftover breaks that occurred during discontinuous DNA replication; they could be the result of random breaking of the DNA strands; they could be introduced by the recombination machinery itself; or a random break in one strand could provoke a break in the corresponding position of the homologous strand by a "strand invasion" process that we will explain later in this chapter. Once both homologous strands are broken, the free ends can cross over and join with ends in the other duplex, forming intermolecular bonds, rather than rejoining to form the old intramolecular bonds.

When the two strands have crossed over and DNA ligase has sealed the new intermolecular phosphodiester bonds, we have a **half chiasma,** or **chi structure,** a cross-shaped structure pictured in several ways in figure 7.25.

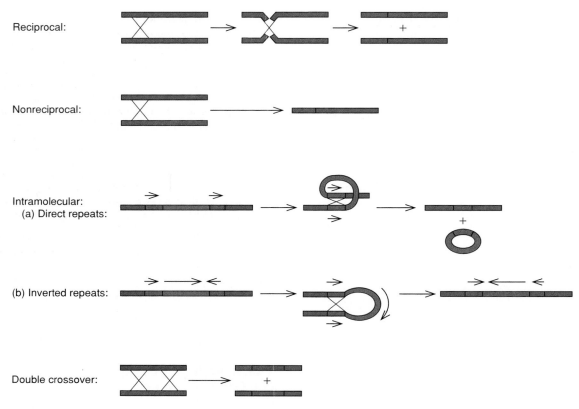

FIGURE 7.24

Examples of recombination. The ×'s represent crossovers between the two chromosomes or parts of the same chromosome. To visualize how these crossovers work, look at the intermediate form of the reciprocal recombination on the top line. Imagine the DNAs breaking and forming

new, interstrand bonds as indicated by the arms of the ×. This same principle applies to all the examples shown; in the nonreciprocal case, one of the recombinant DNAs is lost.

The branch in the half chiasma can migrate in either direction simply by breaking old base pairs and forming new ones, but this **branch migration** is not essential to the recombination process. The chi structure is first shown with two of its strands crossed, but a 180-degree rotation of either the top or bottom half uncrosses these strands to give an intermediate with a diamond shape at the center where all the strands meet. If a few base pairs break at the junction point, the diamond, which is composed of single-stranded DNA, will expand. Figure 7.26 is an electron micrograph of recombining plasmid DNAs, which demonstrates beautifully that such structures really do form.

Of course, the two duplexes that are participating in the chi structure cannot remain linked indefinitely. The other two DNA strands must also break and rejoin to resolve the chi structure into two independent DNA duplexes. This can happen in two different ways. If the same strands that broke the first time break again and cross back, no true recombination results. What we get instead are two DNA duplexes with patches of **heteroduplex,** containing one strand from each of the DNAs involved in the chi structure. On the other hand, if the strands that did *not* break the first time are involved in the second break, the chi structure resolves into two recombinant molecules.

INDUCTION OF DNA STRAND BREAKS

One problem with Holliday's model is that it assumes two breaks in exactly corresponding positions in homologous DNA strands. Single-strand breaks occur frequently, so there is no problem in imagining one break appearing eventually in any given position in a DNA molecule. Nevertheless, it seems unlikely that two breaks, conveniently located right across from one another in two paired molecules, would occur spontaneously. Other investigators have therefore proposed alternate models to explain the origin of the second break. One of these is the Meselson-Radding model (fig. 7.27), which suggests that the free end created by the first break "invades" the other duplex, base-pairing with one of the strands and displacing the other. The displaced strand forms a loop called a **D-loop,** where D stands for "displacement." According to this model, the act of forming the D-loop breaks the looping strand, allowing formation of the chi structure. Hotchkiss proposed a similar model, which includes the notion that the breakage results, not from D-loop formation itself, but from extension of the loop by DNA replication.

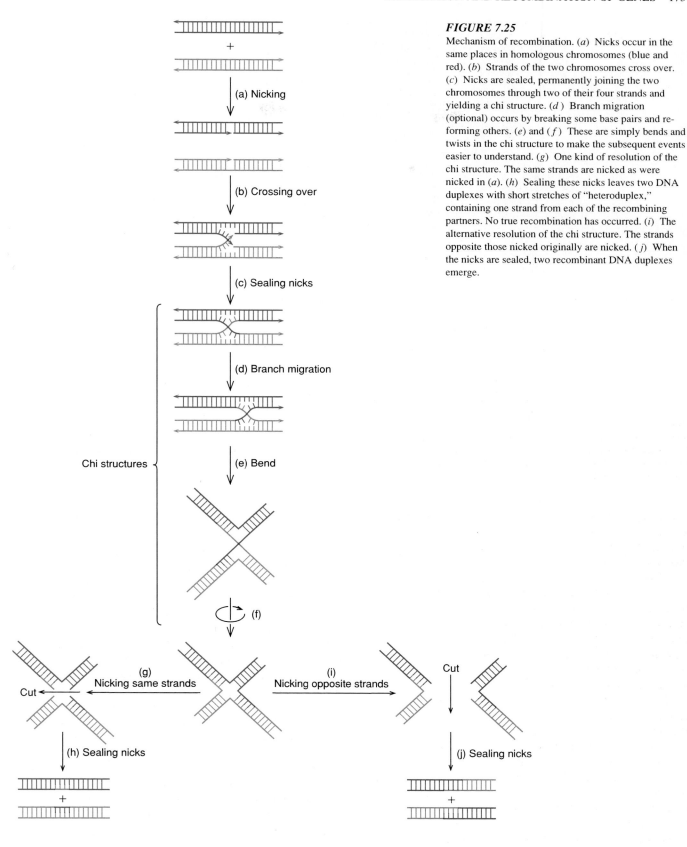

FIGURE 7.25
Mechanism of recombination. (*a*) Nicks occur in the same places in homologous chromosomes (blue and red). (*b*) Strands of the two chromosomes cross over. (*c*) Nicks are sealed, permanently joining the two chromosomes through two of their four strands and yielding a chi structure. (*d*) Branch migration (optional) occurs by breaking some base pairs and re-forming others. (*e*) and (*f*) These are simply bends and twists in the chi structure to make the subsequent events easier to understand. (*g*) One kind of resolution of the chi structure. The same strands are nicked as were nicked in (*a*). (*h*) Sealing these nicks leaves two DNA duplexes with short stretches of "heteroduplex," containing one strand from each of the recombining partners. No true recombination has occurred. (*i*) The alternative resolution of the chi structure. The strands opposite those nicked originally are nicked. (*j*) When the nicks are sealed, two recombinant DNA duplexes emerge.

FIGURE 7.26
Chi structure generated during recombination. This corresponds to the intermediate generated in figure 7.25*f*, after some base pairs have opened up around the central "diamond" to produce short stretches of single-stranded DNA.
Courtesy of Dr. Huntington Potter and Dr. David Dressler.

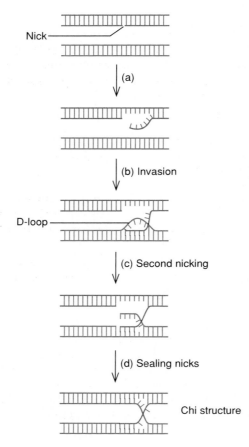

FIGURE 7.27
The Meselson-Radding model of nicking the second strand during recombination. (*a*) The first strand breaks. (*b*) A loose end generated by this first break invades the second DNA duplex, forming a D-loop. (*c*) This invasion creates a nick in the second duplex. (*d*) The two nicks are sealed, forming the chi structure.

In homologous recombination, the two homologous DNA duplexes first align themselves. Then single-stranded breaks occur in two homologous strands, allowing the two free ends to cross over and form new, interstrand bonds. This yields a chi structure, which is resolved when the two other strands break and cross over.

WORKED-OUT PROBLEMS

Problem 1

Draw schematic diagrams of the DNA molecules in a Meselson-Stahl experiment after two generations in ^{14}N-medium assuming: (*a*) conservative; and (*b*) semiconservative replication. Use solid lines for ^{15}N DNA, and dotted lines for ^{14}N DNA.

Solution

(a) Conservative replication:

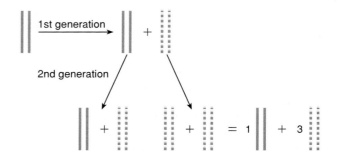

(b) Semiconservative replication:

Problem 2

Draw diagrams of the bands of DNA in CsCl gradients that would arise from the two types of replication in the previous question. Show the positions of the heavy/heavy (H/H), heavy/light (H/L), and light/light (L/L) DNAs.

Solution

(a) Conservative replication:

The L/L band is three times darker than the H/H band because there is three times as much of it.

(b) Semiconservative replication:

Problem 3

(*a*) Draw a schematic diagram of a replicating bubble in *Drosophila* DNA, showing both parental and progeny strands after all Okazaki fragments have been ligated together.

(*b*) Assuming that replication began before labeled thymidine was added, show on the schematic where labeled and unlabeled DNA will be found.

(*c*) What would an autoradiograph of the bubble in (*b*) look like?

(*d*) What would this autoradiograph look like if the two daughter duplexes wound around each other (i.e., the bubble collapsed)?

Solution

(*a*) Representing parental DNA with dark blue lines and progeny DNA with light blue lines, we can draw the replicating bubble as follows:

There are no breaks in the progeny strands because all Okazaki fragments have been joined.

(*b*) Representing labeled DNA with jagged lines, we have the following:

The progeny DNA in the middle is unlabeled because it replicated before the labeled thymidine was added.

(*c*)

Note: We see only labeled regions of the DNA (i.e., jagged lines) on an autoradiograph. Unlabeled DNA has no way of exposing the film and therefore remains invisible.

(*d*)

——— — ———

If the bubble collapses, the upper and lower labeled segments on each side will be superimposed and appear as one streak each on the film. Thus we will see two streaks (regions of developed grains) instead of four on the autoradiograph.

Problem 4

How long would it take *E. coli* to replicate 15,000 base pairs of DNA under the following conditions:

(*a*) Assuming no pauses.

(*b*) Assuming a pause of 2 seconds after completing every Okazaki fragment (as required by the looping model in figure 7.26), and assuming that Okazaki fragments average 1,000 nucleotides long.

(*c*) Assuming that the replisome completely dissociates from DNA after every Okazaki fragment and must spend one minute rebinding.

Solution

(*a*) *E. coli* DNA elongates at the rate of about 150 nucleotides per second per fork. Since DNA replication is bidirectional and two forks are used, the effective rate of elongation is 300 nucleotides per second. If there were no pauses, 15,000 base pairs could replicate in 15,000/300, or 50 seconds.

(b) It is perhaps easiest to work out this problem assuming only one fork is operating, then divide by two to take account of the fact that there are actually two forks, so the job will be completed in half the time. If a two-second pause is required after completion of every Okazaki fragment, the total time required will be the sum of two terms: the time taken to replicate the DNA with no pauses plus the time taken for the pauses. The time taken to replicate the DNA (assuming one fork and no pauses) would be 15,000/150, or 100 seconds. The time required for the pauses would be the time for each pause (2 seconds) × the total number of pauses. Since a pause occurs at the end of each Okazaki fragment, the total number of pauses is the same as the number of Okazaki fragments, or the total length of the DNA divided by the length of an average Okazaki fragment, or 15,000/1,000 = 15. The time needed for the pauses is therefore 2 seconds × 15 = 30 seconds. Therefore, the total time required, assuming one fork, is 100 + 30 = 130 seconds. With two forks operating, the time required would be 65 seconds.

(c) Again, assuming only one fork: If the replisome had to spend one minute rebinding after every Okazaki fragment, it would spend 100 seconds + [1 minute × (15,000/1,000)] = 100 seconds + 15 minutes = 16 minutes, 40 seconds. Two forks will therefore do the job in half the time, or 8 minutes, 20 seconds.

Problem 5

How long a streak on an autoradiograph would be produced by colicin E1 DNA replicating at a single fork for 50 seconds, assuming no pauses?

Solution

A single fork moves at the rate of 150 nucleotides a second, so it will move 7,500 nucleotides in 50 seconds. Each nucleotide (or base pair) is 0.34 nm long; so 7,500 nucleotides would be 7,500 × 0.34 nm, or 2,550 nm, or 2.55 microns long. (1 micron = 1,000 nm.)

Summary

The DNA that replicates under the control of a single origin of replication is called a replicon. The chromosomes of simple genetic systems like plasmids, viruses, and even bacteria contain just one replicon. DNA replication in most genetic systems is bidirectional. The replication machinery starts at a fixed point—the origin of replication—and two replicating forks move in opposite directions away from this point. DNA replication in prokaryotes and eukaryotes is semidiscontinuous. The leading strand grows continuously, $5' \rightarrow 3'$, in the direction of motion of the replicating fork. The other, trailing strand grows discontinuously, in pieces (Okazaki fragments) that are made, $5' \rightarrow 3'$, in a direction opposite to the motion of the replicating fork. DNA polymerase is incapable of initiating DNA synthesis by itself. It needs a primer to supply a free 3′-end upon which it

can build the nascent DNA. Very short pieces of RNA serve this priming function. After Okazaki fragments are made, their primers are removed, the resulting gaps are filled in, and the remaining nicks are sealed.

As helicase unwinds the two parental strands at the replicating fork of a closed circular DNA, it introduces superhelical strain into the DNA. DNA gyrase releases this strain, permitting replication to continue. When replication of a double-stranded circular DNA approaches completion, the two daughter molecules are still entwined. A topoisomerase is presumably needed to separate them.

Circular DNA can replicate by a rolling circle mechanism. One strand of a double-stranded DNA is nicked, and the 3′-end is extended using the intact DNA strand as template. This displaces the 5′-end. In phage λ, the displaced strand serves as the template for discontinuous, lagging strand synthesis.

The ends of eukaryotic chromosomes are formed by an enzyme called telomerase, which elongates the 3′-ends of DNA strands according to the instructions of its own RNA template.

DNA polymerase, like all enzymes, is not perfect. If left to itself, it would make about one mistake per one hundred thousand nucleotides of DNA. To improve on this intolerably poor performance, the DNA replicating machinery has a built-in proofreading system that depends on primers. Only a base-paired nucleotide can serve as a primer for the replicating enzyme; therefore, if the wrong nucleotide is incorporated by accident, replication stalls until a nuclease erases the error by degrading the most recently made stretch of DNA. Thus, the error can be repaired.

DNA replication in meiotic cells is usually followed by recombination, using the generalized recombination mechanism. First, homologous regions of two DNA duplexes line up, then single-strand breaks occur in two homologous strands. The free ends of the broken strands cross over and form new, interstrand bonds, yielding a chi structure. This cross-shaped structure is resolved in two recombinant DNA products when the other two strands break and cross over.

PROBLEMS AND QUESTIONS

1. Draw schematic diagrams of the DNA molecules in a Meselson-Stahl experiment after *three* generations in ^{14}N-medium assuming: (a) conservative and (b) semiconservative replication. Use solid lines for ^{15}N-labeled DNA and dotted lines for ^{14}N-labeled DNA.

2. Draw diagrams of the bands of DNA in CsCl gradients that would arise from the two types of replication in question 1. Show the positions of fully ^{15}N-labeled DNA and fully ^{14}N-labeled DNA.

3. (*a*) Draw a schematic diagram of a replicating bubble in *Drosophila* DNA, showing both parental and progeny strands after all Okazaki fragments have been ligated together. (*b*) Assuming that replication began before labeled thymidine was added, show on the schematic where labeled and unlabeled DNA will be found.

4. Assume the following sequence of events: Replication begins; DNA is pulse-labeled with low specific activity thymidine; DNA is pulsed with high specific activity thymidine. Then show what the resulting autoradiogram will actually look like if the two daughter duplexes twist around one another to form a straight line instead of a bubble.

5. What could we conclude from an autoradiographic pattern in which all pairs of silver streaks fall into two categories: —— —— or —— ——?

6. Colicin E1 is a small, double-stranded circular DNA that replicates unidirectionally. Consider an autoradiography experiment in which you label colicin E1 DNA uniformly with low specific activity thymidine. Then, *after* a round of replication begins, you label with high specific activity thymidine for one-sixth of a replication cycle. Draw a picture of the expected autoradiogram.

7. An extraterrestrial being brings you a vial of bacteria from the planet Mongo and informs you that all living things on Mongo replicate their DNA fully continuously. (*a*) Describe the results of a pulse labeling experiment you would perform that verifies this assertion. (*b*) What unearthly characteristic must the DNA polymerase of these bacteria have?

8. What role do helicases play in DNA replication? Why is an ATPase a necessary part of this role?

9. Contrast the function of a helicase with that of a topoisomerase.

10. Why does the single-strand-binding protein (SSB) of *E. coli* bind so much better to a single-stranded DNA than to double-stranded DNA?

11. Would a swivel be required in replicating a linear chromosome? Why, or why not?

12. If there were no swivel, what kind of supercoils would develop in a replicating closed circular DNA?

13. The *E. coli* primosome binds to the template for which strand? Why does it seem more useful for the primosome to bind to this strand?

14. What advantage would a double-headed replisome with two DNA polymerase centers offer? What structural shift would the DNA have to make in order to accommodate such a double-headed replisome?

15. Why is primer synthesis inherently more error-prone than DNA elongation?

Answers appear at end of book.

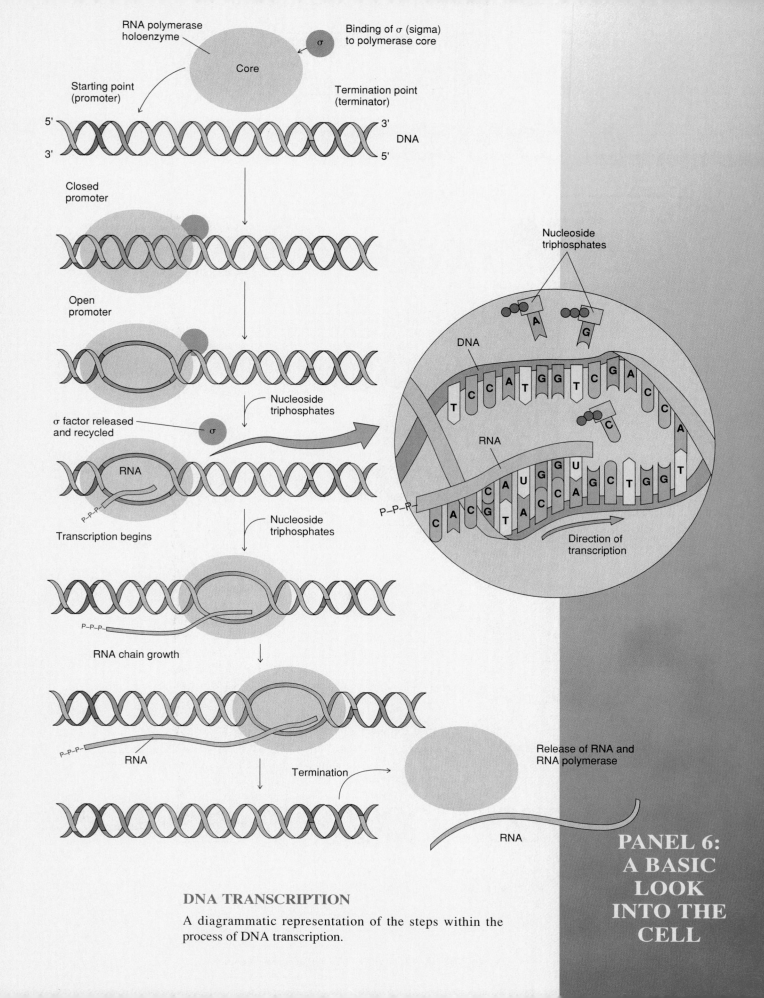

DNA TRANSCRIPTION

A diagrammatic representation of the steps within the process of DNA transcription.

PANEL 6:
A BASIC
LOOK
INTO THE
CELL

A REVIEW OF DNA TRANSCRIPTION

CHAPTER 8: TRANSCRIPTION AND ITS CONTROL IN PROKARYOTES

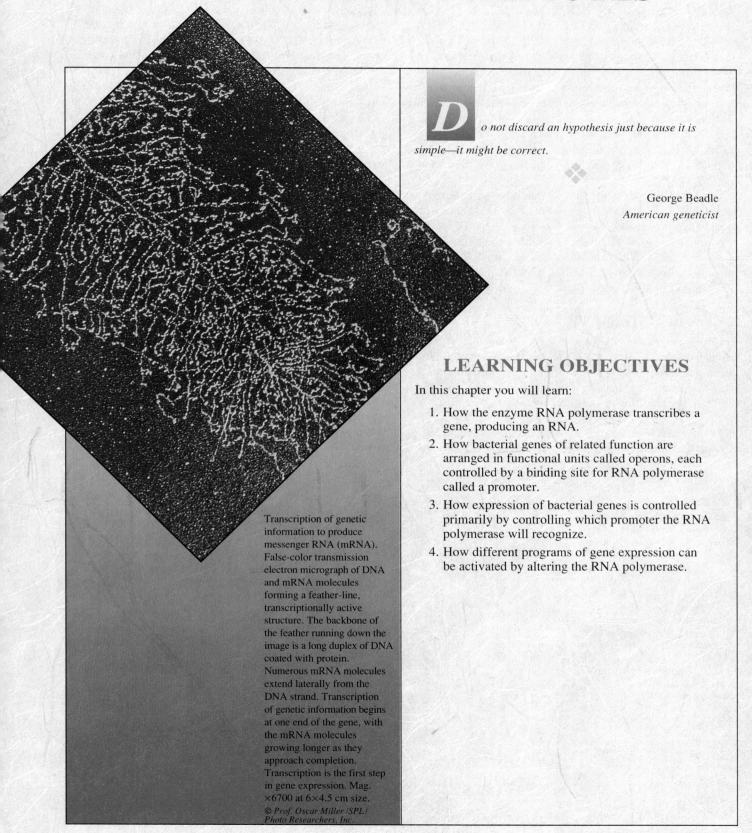

Transcription of genetic information to produce messenger RNA (mRNA). False-color transmission electron micrograph of DNA and mRNA molecules forming a feather-line, transcriptionally active structure. The backbone of the feather running down the image is a long duplex of DNA coated with protein. Numerous mRNA molecules extend laterally from the DNA strand. Transcription of genetic information begins at one end of the gene, with the mRNA molecules growing longer as they approach completion. Transcription is the first step in gene expression. Mag. ×6700 at 6×4.5 cm size.
© Prof. Oscar Miller /SPL/ Photo Researchers, Inc.

D o not discard an hypothesis just because it is simple—it might be correct.

❖

George Beadle
American geneticist

LEARNING OBJECTIVES

In this chapter you will learn:

1. How the enzyme RNA polymerase transcribes a gene, producing an RNA.
2. How bacterial genes of related function are arranged in functional units called operons, each controlled by a binding site for RNA polymerase called a promoter.
3. How expression of bacterial genes is controlled primarily by controlling which promoter the RNA polymerase will recognize.
4. How different programs of gene expression can be activated by altering the RNA polymerase.

ene expression is the process by which gene products are made. The information in the gene is decoded and expressed as a product. Most genes hold the information for making vital cellular molecules called proteins. The expression of these genes requires two steps, transcription and translation. **Transcription** means making an RNA copy of a gene. This RNA, called **messenger RNA (mRNA),** carries the coded information to the cell's protein factories, the ribosomes. The ribosomes carry out the translation step, in which they follow the instructions in the mRNAs to make proteins. Our goal in this chapter is to understand how prokaryotes perform the first step—transcription. We will see that bacteria and their phages have evolved intricate and fascinating methods of controlling their gene expression by regulating transcription.

In chapter 9, we will examine transcription in eukaryotes. Finally, in chapter 10, we will look at the second step in gene expression—translation. Messenger RNA is not the only kind of RNA a cell makes. Two other classes of RNA, transfer RNA (tRNA) and ribosomal RNA (rRNA), also participate in protein synthesis. We will discuss these RNAs more fully in chapter 10.

THE MECHANISM OF TRANSCRIPTION

Transcription, in its simplest terms, is the sequential joining of nucleotides to form an RNA chain. Figure 8.1 demonstrates the essential step in transcription, the formation of a phosphodiester bond between nucleotides. The nucleotides that participate in this process can be any of the four RNA nucleotides: ATP, GTP, CTP, or UTP. In this case, ATP and GTP come together to form the shortest possible RNA, a dinucleotide. In transcription, phosphodiester bond

formation occurs over and over again; each time, a new nucleotide link joins the growing RNA chain. Figure 8.2 shows this process repeating three times, producing a short RNA containing only four nucleotides (a tetranucleotide). The participants in this process are (1) the DNA template, shown in red; (2) RNA nucleotides, shown in blue; and (3) the enzyme, RNA polymerase, which catalyzes the reaction and makes it specific, so that the RNA product is a faithful copy of one of the strands of the DNA template.

As figure 8.2 suggests, transcription follows the same base-pairing rules as DNA replication: T,G,C, and A in the DNA pair with A,C,G, and U, respectively, in the RNA product. (Notice that uracil appears in RNA in place of thymine in DNA.) This base-pairing pattern ensures that an RNA transcript is a faithful copy of one of the strands of the gene.

Of course, highly directed chemical reactions such as transcription do not happen by themselves—they are enzyme-catalyzed; the enzyme that directs transcription is called **RNA polymerase.** Figure 8.3 presents a schematic diagram of *E. coli* RNA polymerase at work (see also panel 6). There are three phases of transcription: initiation, elongation, and termination. Since RNA polymerase transcribes only one of a gene's strands, we say that transcription is **asymmetric.**

INITIATION

First, the enzyme recognizes a region called a **promoter,** which lies just "upstream" from the gene. The polymerase binds tightly to the promoter and causes localized melting, or separation of the two DNA strands within the promoter. Approximately ten base pairs are melted. Next, the polymerase starts building the RNA chain. The substrates, or building blocks, it uses for this job are the four **ribonucleoside triphosphates:** ATP, GTP, CTP, and UTP. The first, or initiating substrate is usually a purine nucleotide. After

FIGURE 8.1
Phosphodiester bond formation in RNA synthesis. (*a*) Simplified reactions showing only the phosphorus atoms. Note that the phosphorus atom (red) closest to the guanosine is retained in the phosphodiester bond. The other two phosphorus atoms (black) are removed as a by-product. (*b*) More complete structures showing all the atoms participating in phosphodiester bond formation. Note the phosphorous atom (red) in the GTP and the oxygen atom (blue) in the ATP hydroxyl group, both of which are retained in the phosphodiester bond.

Table 8.1 Ability of Core and Holoenzyme to Transcribe DNAs

DNA	Transcription Activity	
	Core	Holoenzyme
T4 (intact)	0.5	33.0
Calf thymus (nicked)	14.2	32.8

the first nucleotide is in place, the polymerase binds the second nucleotide and joins it to the first nucleotide, forming the initial phosphodiester bond in the RNA chain. At this point, the initiation process is complete.

This initiation process is specific. The polymerase can select which promoters to use, and which to ignore. Richard Burgess, Andrew Travers, and their colleagues discovered in 1969 that they could physically separate the *E. coli* RNA polymerase holoenzyme into two components: a polypeptide called **sigma** (σ), and the remainder—a complex of four polypeptides collectively known as the **core polymerase,** or simply the **core.** In the process, they caused a profound change in the enzyme's activity (table 8.1). Whereas the holoenzyme could transcribe intact phage T4 DNA in vitro quite actively, the core enzyme had little ability to do this. On the other hand, core polymerase retained its basic RNA polymerizing function, since it could still transcribe highly nicked templates, such as commercial calf thymus DNA, very well. (As will be explained, this transcription of nicked DNA has no biological significance.)

Adding σ back to the core reconstituted the enzyme's ability to transcribe un-nicked T4 DNA. Even more significantly, the holoenzyme transcribed only a certain class of T4 genes (called immediate early genes), but the core enzyme showed no such specificity; it transcribed the whole T4 genome, though weakly. Furthermore, transcription of T4 DNA in vitro by the holoenzyme was asymmetric—just as it is in vivo—but the weak transcription by the core enzyme was symmetric. In other words, not only did the core

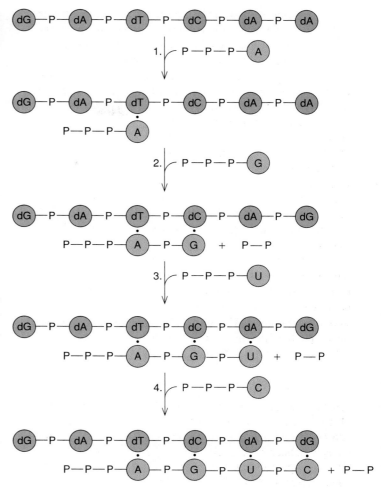

FIGURE 8.2

Synthesis of RNA on a DNA template. Step 1: RNA polymerase initiates RNA synthesis by matching an RNA nucleotide, an ATP (blue) in this case, with a dT (red) in the DNA template. Transcription in vivo never starts right at the beginning of the DNA, which is why we do not initiate with a CTP to match the dG at left. Step 2: RNA polymerase completes the initiation step by forming the first phosphodiester bond—between the ATP and an incoming GTP. Step 3: RNA polymerase elongates the chain to a trinucleotide by adding a UTP. Step 4: The RNA chain (a tetranucleotide) is completed by the addition of a CTP.

(a) RNA polymerase binds to promoter:

(b) Initiation:

(c) Elongation:

(d) Termination:

FIGURE 8.3

Transcription. (*a*) RNA polymerase (red) binds tightly to the promoter and "melts" a short stretch of DNA. (*b*) In the initiation step, the polymerase joins the first two nucleotides of the nascent RNA (blue) through a phosphodiester bond. The first nucleotide retains its triphosphate group. (*c*) During elongation, the melted bubble of DNA moves with the polymerase, allowing the enzyme to "read" the bases of one strand and make a complementary RNA. (*d*) Termination occurs when the polymerase reaches a termination signal at the end of a gene, causing the RNA and the polymerase to fall off the DNA template.

enzyme read the whole genome, it read both strands. Clearly, depriving the holoenzyme of its σ subunit leaves a core enzyme with basic RNA synthesizing capability but lacking specificity. Adding σ back restores specificity. In fact, σ was named only after this characteristic came to light. The σ, or Greek letter *s*, was chosen to stand for "specificity."

 The key player in the transcription process is RNA polymerase. The *E. coli* enzyme is composed of a core, which contains the basic transcription machinery, and a σ factor, which directs the core to transcribe specific genes.

Binding RNA Polymerase to Promoters

Why was core enzyme still capable of transcribing nicked DNA, but not intact DNA? Nicks and gaps in DNA provide initiation sites for RNA polymerase, even core polymerase, but this kind of initiation is necessarily nonspecific. There were few nicks or gaps on the intact T4 DNA, so the core polymerase encountered only a few such artificial initiation sites. On the other hand, when σ was present, the holoenzyme could recognize the authentic RNA polymerase binding sites—the promoters—on the intact T4 DNA and begin transcription there. This initiation is specific and mirrors the initiation that would occur in vivo. Thus, σ operates by directing the polymerase to initiate at specific promoters.

RNA polymerase can exist in one of two states: one appropriate for tight binding to promoters (the initiation state) and one appropriate for loose binding (the elongation state). The tight binding, or initiating, state can occur only in the presence of σ; after initiation is achieved, σ dissociates and leaves the core enzyme to carry out elongation. Furthermore, tight binding of holoenzyme to promoters involves local melting of the DNA to form a so-called **open promoter complex.** This conversion of a loosely bound polymerase in a **closed promoter complex** to the tightly bound polymerase in the open complex requires σ, and this is what allows transcription to begin (fig. 8.4 and panel 6). We can now appreciate how σ fulfills its role in determining specificity of transcription. It selects the promoters to which RNA polymerase will bind tightly. The genes adjacent to these promoters will then be transcribed.

 The σ factor allows initiation of transcription by causing the RNA polymerase holoenzyme to bind tightly to a promoter. This tight binding depends on local melting of the DNA to form an open promoter complex, and it is possible only in the presence of σ. The σ factor can therefore select which genes will be transcribed. After σ has participated in initiation, it dissociates from the core polymerase, leaving the core to carry on with the elongation process.

Common Base Sequences in *E. coli* Promoters

What is the special nature of a prokaryotic promoter that attracts RNA polymerase? David Pribnow compared several *E. coli* and phage promoters and discerned a region they

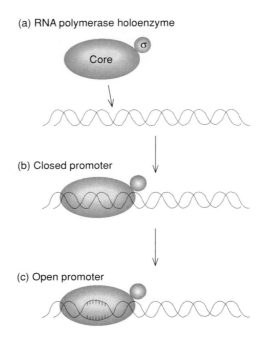

FIGURE 8.4
RNA polymerase-promoter binding. (*a*) The polymerase holoenzyme searches for a binding site on the DNA. (*b*) The holoenzyme has found a promoter and has bound loosely, forming a closed promoter complex. (*c*) The holoenzyme has bound tightly, melting a local region of DNA and forming an open promoter complex.

held in common: a sequence of six or seven bases centered approximately ten bases upstream from the start of transcription. This was originally dubbed the "Pribnow box," but is now commonly referred to as the **−10 box.** Later, another short sequence, centered approximately thirty-five bases upstream from the transcription start, was found in a variety of *E. coli* and phage promoters; it is usually called the **−35 box.** More than one hundred promoters have now been examined, and a "typical" sequence for each of these boxes has emerged (fig. 8.5).

These so-called **consensus sequences** represent probabilities. The capital letters in figure 8.5 denote bases that have a high probability of being found in the given position. The small letters correspond to bases that are usually found in the given position, but at a lower frequency than those denoted by capital letters. The probabilities are such that you rarely find −10 or −35 boxes that match the consensus sequences perfectly. However, when such perfect matches are found, they tend to occur in very strong promoters that initiate transcription unusually actively. In fact, mutations that destroy matches with the consensus sequences tend to be **down mutations.** That is, they make the promoter weaker, resulting in less transcription. For example, mutation of a −10 box sequence from TATAAT to TGTAAT would usually be a down mutation because it would weaken the promoter. By contrast, mutations that make promoter sequences more like the consensus sequences usually make the promoters stronger; these are called **up mutations.** For instance, mutation of a −10 box sequence from TATCTT to TATATT would probably be an up

– 35 box	– 10 box	Transcription

————TTGACa————————————————————TAtAaT————
————AACTGt————————————————————ATaTtA————

Unwound region

FIGURE 8.5

A prokaryotic promoter. The positions of the –10 and –35 boxes and the unwound region are shown relative to the start of transcription for a typical *E. coli* promoter. Capital letters denote bases found in those positions in more than 50% of promoters examined; small letters denote bases found in those positions in 50% or fewer of promoters examined.

mutation. In chapter 10 we will see that eukaryotic promoters have their own consensus sequences, one of which resembles the –10 box quite closely.

 Prokaryotic promoters contain two regions centered at –10 and –35 bases upstream from the transcription start site. In *E. coli*, these bear a greater or lesser resemblance to two consensus sequences: TATAAT and TTGACA, respectively. In general, the more closely regions within a promoter resemble these consensus sequences, the stronger that promoter will be.

ELONGATION

During the elongation phase of transcription, RNA polymerase directs the sequential binding of ribonucleotides to the growing RNA chain in the $5' \rightarrow 3'$ direction (from the 5′-end toward the 3′-end of the RNA as in panel 6). As it does so, it moves along the DNA **template,** and the "bubble" of melted DNA moves with it. This melted region exposes the bases of the template DNA one-by-one so they can pair with the bases of the incoming ribonucleotides. As soon as the transcription machinery passes, the two DNA strands wind around one another again, re-forming the double helix. This points to two fundamental differences between transcription and DNA replication: (a) RNA polymerase makes only one RNA strand during transcription, which means that it copies only one DNA strand. Transcription is therefore said to be asymmetric. This contrasts with replication, in which both DNA strands are copied. (b) In transcription, DNA melting is limited and transient. Only enough strand separation occurs to allow the polymerase to "read" the DNA template strand. However, during replication, the two parental DNA strands separate permanently.

TERMINATION

Just as promoters serve as initiation signals for transcription, other regions at the ends of genes, called **terminators,** signal termination. These work in conjunction with RNA polymerase, sometimes aided by another protein, to loosen the association between RNA product and DNA template. The result is that the RNA dissociates from the RNA polymerase and DNA; transcription is thereby terminated.

A final, important note about conventions: RNA sequences are usually written 5′ to 3′, left to right. This feels natural to a molecular geneticist because RNA is made in a 5′ to 3′ direction, and as we shall see, mRNA is also translated 5′ to 3′. Thus, since ribosomes read the message 5′ to 3′, it is appropriate to write it 5′ to 3′ so that *we* can read it like a sentence.

Genes are also usually written so that their transcription proceeds in a left to right direction. This "flow" of transcription from one end to the other gives rise to the term "upstream," which refers to the DNA close to the start of transcription (near the left end when the gene is written conventionally). Thus, we can describe most promoters as lying just upstream from their respective genes and the genes, therefore, downstream from their promoters.

 Transcription takes place in three stages: initiation, elongation, and termination. Initiation involves binding RNA polymerase to the promoter, local melting, and forming the first phosphodiester bond; during elongation, the RNA polymerase links together ribonucleotides in the $5' \rightarrow 3'$ direction to make the rest of the RNA; finally, in termination, polymerase and RNA product dissociate from the DNA template.

OPERONS

In chapter 1 we introduced biochemical pathways that are catalyzed by groups of enzymes. The product of one step in the pathway becomes the starting material for the next step. Obviously, all of the enzymes in the pathway will be needed in roughly equal amounts, since one enzyme deficiency would stall the whole process. Therefore, if four enzymes are needed for a pathway to operate, it would be wasteful to turn on the genes for three of them and leave the other turned off. In other words, for greatest efficiency, the genes for proteins that function together should be controlled together. One mechanism that allows coordinate control is to group functionally related genes together so they can be regulated together easily. Such a group of contiguous, coordinately controlled genes is called an **operon.**

THE *lac* OPERON

The first operon to be discovered has become the paradigm of the operon concept. It contains three genes that code for

FIGURE 8.6
The β-galactosidase reaction. The enzyme breaks the β-galactosidic bond (gray) between the two sugars, galactose (red) and glucose (blue), that compose lactose.

Lactose → β-galactosidase → Galactose + Glucose

the enzymes that allow *E. coli* cells to use the sugar **lactose,** hence the name *lac* **operon.** Consider a flask of *E. coli* cells growing on the sugar **glucose.** We take away the glucose and replace it with lactose. Can the cells adjust to the new nutrient source? For a short time it appears that they cannot; they stop growing as soon as the glucose disappears and behave as if the lactose is no help at all. Then, after the lag period, growth resumes. During the lag, the cells have been turning on the *lac* operon and beginning to accumulate the enzymes they need to metabolize lactose.

What are these enzymes? First, *E. coli* needs an enzyme to transport the lactose into the cells. The name of this enzyme is **galactoside permease.** Next, the cells need an enzyme to break the lactose down into its two component sugars, galactose and glucose. Figure 8.6 shows this reaction. Since lactose is composed of two simple sugars, we call it a disaccharide. As mentioned previously, these are the six-carbon sugars galactose and glucose; they are joined together by a linkage called a β-galactosidic bond. Lactose is therefore called a β-galactoside, and the enzyme that cuts it in half is called **β-galactosidase.** The genes for these two enzymes, galactoside permease and β-galactosidase, are found side-by-side in the *lac* operon, along with another structural gene—for **galactoside transacetylase**—whose function in lactose metabolism is still not clear. These names give some insight into enzyme nomenclature. The "ase" suffix is an almost universal feature of enzyme names; the rest of the name usually gives a clue to the reaction the enzyme catalyzes.

The three genes coding for enzymes that carry out lactose metabolism are grouped together in the order: β-galactosidase (*lacZ*), galactoside permease (*lacY*), galactoside transacetylase (*lacA*). They are all transcribed together on one messenger RNA, called a **polycistronic message,** starting from a single promoter. Thus, they can all be controlled together simply by controlling that promoter. The term "polycistronic" comes from "cistron," which is a synonym for "gene." A polycistronic message is therefore simply a message with information from more than one gene.

 Lactose metabolism in *E. coli* is carried out by two enzymes, with possible involvement by a third. The genes for all three enzymes are clustered together and transcribed together from one promoter, yielding a polycistronic message. Therefore, these three genes, linked in function, are also linked in expression. They are turned off and on together.

Negative Control of the *lac* Operon

Figure 8.7 illustrates one aspect of *lac* operon regulation: negative control. The term **negative control** implies that the operon is turned on unless something intervenes to stop it. The "something" that can turn off the *lac* operon is the *lac* **repressor.** This repressor, the product of the *lacI* **gene** shown at the extreme left in figure 8.7, is a tetramer of four identical protein chains; it binds to the **operator** just to the right of the promoter. When the repressor is bound to the operator, the operon is **repressed.** That is because the operator and promoter are contiguous, and when the repressor occupies the operator, it blocks RNA polymerase's access to the promoter, or at least its proper function there. If RNA polymerase cannot transcribe the *lacZ, Y,* and *A* genes, the operon is off, or repressed. The *lacZ, Y,* and *A* genes are sometimes called **structural genes** to distinguish them from *lacI,* which is a **regulatory gene.** All of these code for proteins, but the *lacI* product, unlike the others, has a regulatory function. This same pattern of nomenclature also applies to the genes of other operons.

The *lac* operon is repressed as long as glucose is present and no lactose is available. That is economical; it would be wasteful for the cell to produce enzymes needed to use an absent sugar, especially when an alternate sugar, glucose, is available. In fact, if all of an *E. coli* cell's genes were turned on all the time, this would drain the cell of so much energy that it could not survive. Thus, control of gene expression is essential to life.

On the other hand, when all the glucose is gone and lactose is at hand, there should be a mechanism for removing the repressor so the operon can turn on and take advantage of the new nutrient. How does this work? The repressor is a so-called **allosteric protein,** which means that it changes its conformation, and therefore its function, when it binds to a certain small molecule. The small molecule in this case is called the **inducer** of the *lac* operon, because it binds to the repressor, causing the protein to change to a conformation that favors dissociation from the operator, thus inducing the operon (fig. 8.7).

What is the nature of this inducer? It is actually an alternate form of lactose called **allolactose** (Greek: *allo* = other). When β-galactosidase cleaves lactose to galactose plus glucose, it rearranges a small fraction of the lactose to allolactose. Figure 8.8 shows that allolactose is just galactose linked to glucose in a different way than in lactose. (In lactose, the linkage is through a β-1,4 bond; in allolactose, the linkage is β-1,6.)

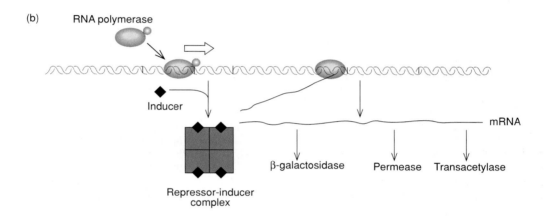

FIGURE 8.7

Negative control of the *lac* operon. (*a*) Repression. The *lacI* gene produces repressor (green), which binds to the operator and prevents RNA polymerase (red and blue) from binding to the promoter.

(*b*) Derepression. Inducer (black) binds to repressor, changing it to a form (green squares) that no longer binds well to the operator. Now polymerase can bind successfully to the promoter and transcribe the operon.

You may be asking yourself: How can lactose be metabolized to allolactose if there is no permease to get it into the cell and no β-galactosidase to perform the metabolizing, since the *lac* operon is repressed? The answer is that nothing in life is perfect; even repression is somewhat leaky, and a low basal level of the *lac* operon products is always present. This is enough to get the ball rolling by producing a little inducer. It does not take much inducer to do the job, since there are only about ten tetramers of repressor per cell. Furthermore, the derepression of the operon will snowball as more and more operon products are available to produce more and more inducer.

The Operon Hypothesis

The development of the operon concept by François Jacob and Jacques Monod and their colleagues was a classic triumph of the combination of genetic and biochemical analysis. The story begins in the 1940s, when Monod began studying the inducibility of lactose metabolism in *E. coli*.

He chose this inducible system over others, such as the maltose system, because of the availability of a variety of galactosides that could serve as inducers without themselves being metabolized. This allowed the two crucial phenomena of induction and metabolism to be analyzed separately.

Monod knew from the outset that a key feature of lactose metabolism was β-galactosidase, and that this enzyme was inducible by lactose and by other galactosides. Furthermore, he and Melvin Cohn had used an anti-β-galactosidase antibody to measure the amount of β-galactosidase protein, and they showed that this increased upon induction. Therefore, since more gene product appeared in response to lactose, it appeared that the β-galactosidase gene itself was being induced.

To complicate matters, certain mutants were found that could make β-galactosidase but still could not grow on lactose. What was missing in these "cryptic" mutants? To

FIGURE 8.8

Conversion of lactose to allolactose. A side reaction carried out by β-galactosidase rearranges lactose to the inducer, allolactose. Note the change in the galactosidic bond from β-1,4 to β-1,6.

Lactose
(β-1,4 linkage)

β-galactosidase

Allolactose
(β-1,6 linkage)

answer this question, Monod and his coworkers added a radioactive galactoside to wild-type and cryptic bacteria. They found that uninduced wild-type cells did not take up the galactoside, and neither did the cryptic mutants, even if they were induced. Of course, previously induced wild-type cells did accumulate the galactoside. This revealed two things: First, a substance (galactoside permease) is induced along with β-galactosidase in wild-type cells and is responsible for transporting galactosides into the cells (the gene encoding galactoside permease is called Y); second, the cryptic mutants seem to have a defective gene (Y^-) for this substance.

Monod named this substance "galactoside permease," and then endured criticism from his colleagues for naming a protein before it had been isolated. He has since remarked, "This attitude reminded me of that of two traditional English gentlemen who, even if they know each other well by name and by reputation, will not speak to each other before having been formally introduced." In their efforts to purify galactoside permease, Monod and his colleagues identified another protein, galactoside transacetylase, which is induced along with β-galactosidase and galactoside permease, and whose function in the *lac* operon remains somewhat obscure to this day.

By the late 1950s, Monod knew that three enzyme activities (and therefore at least three genes) were induced together by galactosides. He had also found some mutants, called **constitutive mutants,** that needed no induction. They produced the three gene products all the time. Monod realized that further progress would be greatly accelerated by genetic analysis, so he teamed up with François Jacob, who was working just down the hall at the Pasteur Institute.

In collaboration with Arthur Pardee, Jacob and Monod created merodiploids (partial diploids) carrying both the wild-type (inducible) and constitutive alleles of the *lac*

operon. The inducible allele proved to be dominant, demonstrating that wild-type cells produce some substance that turns the *lac* genes off. Because this substance turned off the genes from the constitutive as well as the inducible parent, it made the merodiploids inducible. Of course, this substance is the *lac* repressor.

The existence of a repressor required that there be some specific receptor to which the repressor would bind. Jacob and Monod called this the operator. The specificity of this interaction suggested that it should be subject to genetic mutation; that is, some mutations in the operator should abolish its interaction with the repressor. That means these would also be constitutive mutations, so how are we to distinguish them from constitutive mutations in the repressor gene?

Jacob and Monod realized that they could do this by determining whether the mutation was dominant or recessive. Jacob explained this with an analogy. He likened the two *lac* operators in a merodiploid bacterium to two radio receivers, each of which controls a separate door to a house. The repressor genes in a merodiploid are like little radio transmitters, each sending out an identical signal to keep the doors closed. A mutation in one of the repressor genes would be like a breakdown in one of the transmitters; the other transmitter would still function and the doors would both remain closed. In others words, both *lac* operons in the merodiploid would still be repressible. Thus, such a mutation should be recessive (fig. 8.9a), as we have already observed it is.

On the other hand, a mutation in one of the operators would be like a breakdown in one of the receivers. Since it could no longer receive the radio signal, its door would swing open; the other, with a functioning receiver, would remain closed. In other words, one of the *lac* operons in the merodiploid would be constantly derepressed (turned on), while the other would remain repressible. Therefore, this

FIGURE 8.9

Effects of regulatory mutations in the *lac* operon in merodiploids. (*a*) This merodiploid has one wild-type operon (top) and one operon (bottom) with a mutation in the repressor gene (*I*⁻). The wild-type repressor gene (*I*⁺) makes enough normal repressor (green) to repress both operons, so the *I*⁻ mutation is recessive. (*b*) This merodiploid has one wild-type operon (top) and one operon (bottom) with a mutation in the operator (*O*ᶜ) that makes it incapable of binding repressor (green). The wild-type operon remains repressible, but the mutant operon is not; it makes *lac* products even in the absence of lactose. Since only the operon connected to the

mutant operator is affected, this mutation is *cis*-dominant. (*c*) This merodiploid has one wild-type operon (top) and one operon (bottom) with a mutant repressor gene (*I*ˢ) whose product (yellow) cannot bind inducer. The mutant repressor therefore binds irreversibly to both operators and renders both operons uninducible. This mutation is therefore dominant. Notice that these repressor tetramers containing some mutant and some wild-type subunits behave as mutant proteins. That is, they remain bound to the operator even in the presence of inducer.

mutation would be dominant, but only with respect to the *lac* operon under the control of the mutant operator (fig. 8.9*b*). We call such a mutation ***cis*-dominant** because it is dominant only with respect to genes on the same piece of DNA (***in cis;*** Latin: *cis* = here). Jacob and Monod did indeed find such *cis*-dominant mutations, and they defined the operator. These mutations are called *O*ᶜ, for **operator constitutive.**

What about mutations in the repressor gene that render the repressor unable to respond to the inducer? Such mutations should make the *lac* operon uninducible and should be dominant both *in cis* (on the same chromosome) and ***in trans*** (on a different chromosome; Latin: *trans* = across) because the mutant repressor will remain bound to both operators even in the presence of the inducer (fig. 8.9*c*). We might liken this to a radio signal that cannot be turned off.

Monod and his colleagues found two such mutants, and Suzanne Bourgeois later found many others. These are named I^s to distinguish them from constitutive repressor mutants (I^-), which make a repressor that cannot recognize the operator.

Both kinds of constitutive mutants (I^- and O^c) affected all three of the *lac* genes in the same way. The genes had already been mapped and were found to be adjacent on the chromosome. These findings strongly suggested that the operator lay near the three structural genes, although it was still formally possible that the repressor acted on a polycistronic mRNA rather than on the DNA itself.

Notice that Jacob and Monod, by skillful genetic analysis, were able to develop the operon concept. They predicted the existence of two key control elements: the repressor gene and the operator. Deletion mutations revealed a third element (the promoter) that was necessary for expression of all three structural genes. Furthermore, they could conclude that all three genes were clustered into a single control unit: the *lac* operon. Subsequent biochemical studies have amply confirmed Jacob and Monod's beautiful hypothesis.

Negative control of the *lac* operon occurs as follows: The operon is turned off as long as repressor binds to the operator, because the repressor interferes with RNA polymerase's binding to the adjacent promoter. When the supply of glucose is exhausted and lactose is available, the few molecules of *lac* operon enzymes produce a few molecules of allolactose from the lactose. The allolactose acts as an inducer by binding to the repressor and causing a conformational shift that encourages dissociation from the operator. With the repressor removed, RNA polymerase is free to bind to the *lac* promoter and transcribe the three structural genes.

Positive Control

As you might predict, positive control is the opposite of negative control; that is, the operon is turned off unless something intervenes to turn it on. If negative control is like the brake of a car, positive control is the ignition switch. In the case of the *lac* operon, this means that removing the repressor from the operator (releasing the brake) is not enough to turn on the operon. An additional, positive factor (turning on the ignition) is needed.

What is the selective advantage of this positive control system? Isn't negative control enough? Negative control alone can respond only to the presence of lactose. If there were no positive control, the presence of lactose alone would suffice to activate the operon. But that would be inappropriate if glucose were still available, because *E. coli* cells metabolize glucose more easily than lactose; it would therefore be wasteful for them to turn on the *lac* operon in the presence of glucose. In fact, *E. coli* cells keep the *lac* operon turned off as long as glucose is present. Geneticists

Cyclic AMP

FIGURE 8.10
Cyclic AMP. Note the cyclic 5′–3′ phosphodiester bond (blue).

have long suspected that this selection in favor of glucose metabolism and against use of other energy sources is due to the influence of some breakdown product, or catabolite, of glucose. It is therefore known as **catabolite repression.**

The ideal positive controller of the *lac* operon would be a substance that sensed the lack of glucose and responded by activating the *lac* promoter so that RNA polymerase could bind and transcribe the structural genes (assuming, of course, that lactose is present and the repressor is therefore gone). One substance that responds to glucose concentration is a nucleotide called **cyclic AMP (cAMP)** (fig. 8.10). The lower the level of glucose drops, the higher the concentration of cAMP rises.

Is cAMP the positive effector, then? Not exactly. The positive controller of the *lac* operon is a complex composed of two parts: cAMP and a binding protein called **CAP,** for **catabolite activator protein** by its discoverers, Geoffrey Zubay and his colleagues. The protein binds cAMP, and the resulting complex binds to the *lac* promoter and turns it on by helping RNA polymerase bind. Figure 8.11 illustrates this mechanism. Notice that the *lac* promoter is divided into two parts, the CAP binding site on the left and the RNA polymerase binding site on the right.

How does binding CAP plus cAMP to one side of the promoter facilitate RNA polymerase binding to the other? X-ray diffraction studies of the CAP/cAMP/promoter complex have shown that when CAP/cAMP binds, it causes the promoter to bend very noticeably. This bending might make it easier for the RNA polymerase to separate the two DNA strands, forming an open promoter complex.

CAP and cAMP also stimulate transcription of other inducible operons, including the well-studied *ara* and *gal* operons. Just as the *lac* operon makes possible the metabolism of lactose, these other operons code for enzymes that break down the alternative sugars arabinose and galactose. Thus, for greater efficiency, all three operons should remain turned off as long as the cell has glucose available. Since cAMP responds to glucose concentration, it is not surprising that all three operons share a common regulatory mechanism involving cAMP. The CAP-cAMP complex binds at or near each operon's promoter and facilitates the binding of RNA polymerase. However, the exact location and character of the CAP binding site varies from one operon to the next.

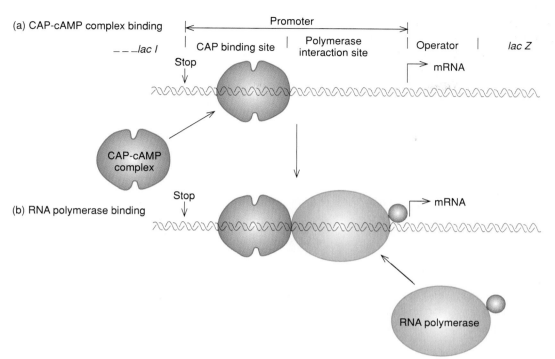

(a) CAP-cAMP complex binding

Promoter

___lac I

CAP binding site | Polymerase interaction site | Operator | lac Z

Stop

mRNA

CAP-cAMP complex

(b) RNA polymerase binding

Stop

mRNA

RNA polymerase

FIGURE 8.11
Positive control of the *lac* operon. (*a*) CAP-cAMP complex (green) binds to the CAP binding site part of the promoter. (*b*) The CAP-cAMP complex facilitates binding of RNA polymerase (red and blue) to the polymerase-interaction-site part of the promoter. Transcription can now begin.

◆ Positive control (catabolite repression) of the *lac* operon works as follows: A complex composed of cAMP plus a protein known as catabolite activator protein (CAP) binds to the upstream part of the promoter and facilitates binding of RNA polymerase to the downstream part. This greatly enhances transcription of the operon. The physiological significance of this positive control mechanism is that it can only operate when cAMP concentration is elevated; this occurs when glucose concentration is low and there is a corresponding need to use an alternate energy source.

TEMPORAL CONTROL OF TRANSCRIPTION

Up to this point we have been discussing the ways in which bacteria control the transcription of a very limited number of genes at a time. For example, when the *lac* operon switches on, only three structural genes are activated. There are other times in a bacterium's life when more radical shifts in gene expression take place. For example, when a phage infects a bacterium, it usually subverts the host's transcription machinery to its own use; further, it establishes a time-dependent, or temporal program of transcription. In other words, the phage early genes are transcribed first, then the later genes. By the time phage T4 infection of *E. coli* reaches its late phase, there is essentially no more host transcription—only phage transcription. This massive shift in specificity would be hard to explain by the operon mechanisms we have looked at so far. Instead, it is engineered by

a fundamental change in the transcription machinery: a change in RNA polymerase itself. Another profound change in gene expression occurs during sporulation in bacteria such as *Bacillus subtilis*. Here, genes that are needed in the vegetative phase of growth switch off, and other, sporulation-specific genes switch on. Again, this switch is accomplished by changes in RNA polymerase. The remainder of this chapter is devoted to this important aspect of gene control.

MODIFICATION OF THE HOST RNA POLYMERASE

What part of RNA polymerase would be the logical candidate to change the specificity of the enzyme? We have already seen that σ determines the specificity of T4 DNA transcription in vitro, so σ is the obvious answer to our question. And experimentation has shown it to be the correct answer. It is interesting, however, that these experiments were not done first with the *E. coli*-T4 system, which turned out to be very grudging about yielding further secrets. A more favorable experimental climate was found with *B. subtilis* and its phages, especially phage SPO1.

SPO1, like T4, is a large phage with many genes. It has a temporal program of transcription as follows: In the first five minutes or so of infection, the early genes are expressed; next, the middle genes turn on (about five to ten minutes post-infection); from about the ten-minute point until the end of infection, the late genes switch on. Since the phage has a large number of genes, it is not surprising that it uses a fairly elaborate mechanism to control this program. J. Pero and her colleagues have been the leaders in developing the model illustrated in figure 8.12.

The host RNA polymerase holoenzyme handles transcription of early SPO1 genes, which is analogous to the T4 model, where the earliest genes are transcribed by the host holoenzyme. This is necessary because the phage does not carry its own RNA polymerase. Therefore, when the phage first infects the cell, the host holoenzyme is the only RNA polymerase available. The *B. subtilis* holoenzyme resembles the *E. coli* enzyme closely. Its core consists of two large β and β′ and two small α polypeptides; its σ factor has a molecular weight of 43,000, somewhat smaller than *E. coli*'s σ. One of the genes transcribed in the early phase of SPO1 infection is called gene 28. Its product, **gp28,** associates with the host core polymerase, displacing the host σ (σ^{43}). With this new, phage-coded polypeptide in place, the RNA polymerase changes specificity. It begins transcribing the phage middle genes instead of the phage early genes and host genes. In other words, gp28 is a novel σ factor that accomplishes two things: It diverts the host's polymerase from transcribing host genes, and it switches from early to middle transcription.

The switch from middle to late transcription occurs in much the same way, except that two polypeptides team up to bind to the polymerase core and change its specificity. These are **gp33** and **gp34,** the products of two phage middle genes (genes 33 and 34, respectively). These proteins replace gp28 and direct the altered polymerase to transcribe the phage late genes in preference to the middle genes. Note that the polypeptides of the host core polymerase remain constant throughout this process; it is the progressive substitution of σ factors that changes the specificity of the enzyme and thereby directs the transcriptional program.

How do we know that the σ-switching model is valid? Two lines of evidence, genetic and biochemical, support it. First, genetic studies have shown that mutations in gene 28 prevent the early-to-middle switch, just as we would predict if the gene 28 product is the σ factor that turns on the middle genes. Similarly, mutations in either gene 33 or 34 prevent the middle-to-late switch, again in accord with the model. The biochemical studies have relied on purified RNA polymerases that contain one or another of the σ factors in question. When polymerase bearing the host σ encounters phage DNA in vitro, it transcribes only the early genes; when the polymerase bears gp28, it transcribes the middle genes preferentially, and when it is bound to gp33 and gp34, it is specific for the late genes.

Transcription of phage SPO1 genes in infected *B. subtilis* cells proceeds according to a temporal program in which early genes are transcribed first, then middle genes, and finally late genes. This switching is directed by a set of phage-encoded σ factors that associate with the host core RNA polymerase and change its specificity from early to middle to late. The host σ is specific for the phage early genes; the phage gp28 switches the specificity to the middle genes; and the phage gp33 and gp34 switch to late specificity.

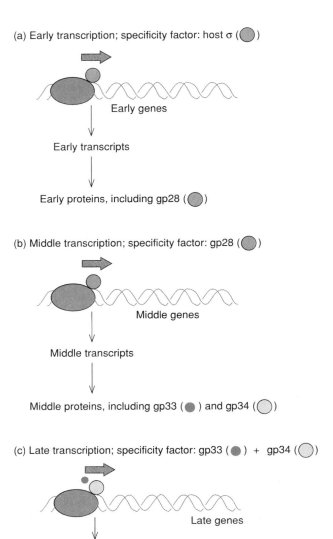

(a) Early transcription; specificity factor: host σ ( )

Early genes

Early transcripts

Early proteins, including gp28 ( )

(b) Middle transcription; specificity factor: gp28 ( )

Middle genes

Middle transcripts

Middle proteins, including gp33 ( ) and gp34 ( )

(c) Late transcription; specificity factor: gp33 ( ) + gp34 ( )

Late genes

Late transcripts ———→ Late proteins

FIGURE 8.12
Temporal control of transcription in phage SPO1-infected *B. subtilis*. (*a*) Early transcription is directed by the host RNA polymerase holoenzyme, including the host σ factor (blue); one of the early phage proteins is gp28 (green), a new σ factor. (*b*) Middle transcription is directed by gp28, in conjunction with the host core polymerase (red); two middle phage proteins are gp33 and gp34 (purple and yellow, respectively); together, these constitute yet another σ factor. (*c*) Late transcription depends on the host core polymerase plus gp33 and 34.

THE RNA POLYMERASE ENCODED IN PHAGE T7

Phage **T7** belongs to a class of relatively simple *E. coli* phages that also includes T3 and φII. These have a considerably smaller genome than SPO1 and, therefore, much fewer genes. In these phages we distinguish three phases of transcription, called classes I, II, and III. (They could just as easily be called early, middle, and late, to conform to SPO1 nomenclature.) One of the five class I genes (gene 1) is necessary for class II and class III gene expression. When it is mutated, only the class I genes are transcribed. Having just learned the SPO1 story, you are probably expecting to hear that gene 1 codes for a σ factor directing the host RNA polymerase to the later phage

(a) Early transcription; specificity factor: host σ ( )

Class I genes

↓

Class I transcripts

↓

Class I proteins, including phage RNA polymerase ( )

(b) Late transcription; phage RNA polymerase ( )

Class II and III genes

↓

Class II and III transcripts

↓

Class II and III proteins

FIGURE 8.13
Temporal control of transcription in phage T7-infected *E. coli.*
(*a*) Early (class I) transcription depends on the host RNA polymerase holoenzyme, including the host σ factor (blue); one of the early phage proteins in the T7 RNA polymerase (green). (*b*) Late (class II and III) transcription depends on the T7 RNA polymerase.

(a)

(b)

FIGURE 8.14
(*a*) *B. subtilis* vegetative cells and (*b*) a sporulating cell, with an endospore developing at the left end.
Dr. Kenneth Bott.

genes. In fact, this was the conclusion reached by the first workers on T7 transcription, but it was erroneous.

The gene 1 product is actually not a σ factor, but a whole phage-specific RNA polymerase contained in one polypeptide. This polymerase, as you might expect, reads the T7 phage class II and III genes specifically, leaving the class I genes completely alone. Indeed, this enzyme is unusually specific; it will transcribe the class II and III genes of phage T7, and virtually no other natural template. The switching mechanism in this phage is thus quite simple (fig. 8.13). When the phage DNA enters the host cell, the *E. coli* holoenzyme transcribes the five class I genes, including gene 1. The gene 1 product—the phage-specific RNA polymerase—then transcribes the phage class II and class III genes.

 Phage T7, instead of coding for a new σ factor to change the host polymerase's specificity from early to late, encodes a whole new RNA polymerase with absolute specificity for the later phage genes. This polymerase, composed of a single polypeptide, is a product of one of the earliest phage genes, gene 1. The temporal program in the infection by this phage is simple. The host polymerase reads the earliest (class I) genes, one of whose products is the phage polymerase, which then reads the later (class II and class III) genes.

CONTROL OF TRANSCRIPTION DURING SPORULATION

We have already seen how phage SPO1 changes the specificity of its host's RNA polymerase by replacing its σ factor. In the following section, we will show that the same kind of mechanism applies to changes in gene expression in the host itself during the process of sporulation. *E. coli,* the bacterium we have examined in greatest detail, lives its whole life in the **vegetative,** or growth state. If you starve a culture of *E. coli,* the cells stop growing and then begin to die; they have little defense against adverse conditions. But there are other bacteria that sense the onset of hard times and protect themselves by **sporulating.** *B. subtilis,* for example, forms **endospores**—tough, dormant bodies that can survive until favorable conditions return (fig. 8.14).

Gene expression must change during sporulation; cells as different in morphology as vegetative and sporulating cells must contain at least some different gene products. In fact, when *B. subtilis* cells sporulate, they activate a whole new set of sporulation-specific genes. The switch from the vegetative to the sporulating state is accomplished by a complex σ-switching scheme that turns off transcription of some vegetative genes and turns on sporulation-specific transcription.

(a) Lytic

Reinfection

Infection

Phage DNA cyclizes

Phage DNA replicates (rolling circle)

Cell lysis

Phage heads,
tails, and DNA
assemble into
progeny phages

(b) Lysogenic

Cell division

UV light induction (rare)

Integration of phage DNA

Cell division

FIGURE 8.15
Lytic versus lysogenic infection by phage λ. (*a*) Lytic infection;
(*b*) lysogenic infection.

Sporulation is a fundamental change, involving large numbers of old genes turning off and new genes turning on. Furthermore, the change in transcription is not absolute; some genes that are active in the vegetative state remain active during sporulation. How is the transcription-switching mechanism able to cope with these complexities?

As you might anticipate, more than one new σ factor is involved in sporulation. In fact, there are at least three: σ^{29}, σ^{30}, and σ^{32} (now called σ^E, σ^H, and σ^C, respectively), in addition to the vegetative σ^{43} (now called σ^A). These polypeptides have molecular weights indicated in their names. Each recognizes a different class of promoter with different −10 and −35 boxes. For example, the vegetative σ^{43} recognizes promoters that are very similar to the promoters recognized by the *E. coli* σ factor, with a −10 box that looks something like TATAAT and a −35 box having the consensus sequence TTGACA. By contrast, the sporulation-specific factors σ^{32} and σ^{29} have their own recognition sequences.

When the original σ was discovered in *E. coli* in 1969, Burgess and Travers speculated that many different σ's would be found, and that these would direct transcription of different classes of genes. This principle was first verified in *B. subtilis,* as we have seen, but it has also proven true in *E. coli.* For example, a distinct *E. coli* σ (σ^{32}, or σ^H) directs the transcription of a class of genes called **heat-shock genes** that switch on when the bacteria experience high temperature or other environmental insults.

When the bacterium *B. subtilis* sporulates, a whole new set of sporulation-specific genes turns on, and many, but not all vegetative genes turn off. This switch takes place largely at the transcription level. It is accomplished by several new σ factors that displace the vegetative σ factor from the core RNA polymerase and direct transcription of sporulation genes instead of vegetative genes. Each σ factor has its own preferred promoter sequence. *E. coli* also has multiple σ factors with their own specificities.

INFECTION OF *E. coli* BY PHAGE λ

Many of the phages we have studied so far (T2, T4, T7, and SPO1, for example) are **virulent** phages. When they replicate, they kill their host by **lysing** it, or breaking it open. On the other hand, **lambda** (λ) is a **temperate** phage; when it infects an *E. coli* cell, it does not necessarily kill it. In this respect, λ is more versatile than many phages; it can follow two paths of replication (fig. 8.15). The first is the **lytic** mode, in which infection progresses just as it would with a virulent phage. It begins with phage DNA entering the host cell and then serving as the template for transcription by host RNA polymerase. Phage mRNAs are translated to yield phage proteins, the phage DNA replicates, and progeny phages assemble from these DNA and protein components. The infection ends when the host cell lyses to release the progeny phages.

In the **lysogenic** mode, something quite different happens. The phage DNA enters the cell, and its early genes are transcribed and translated, just as in a lytic infection. But then a phage protein (the λ **repressor**) appears and binds to the two phage operator regions, shutting down transcription of all genes except for *cI*, the gene for the λ repressor itself. (The *I* is a Roman numeral one, not a capital I.) Under these conditions, with only one phage gene active, it is easy to see why no progeny phages can be produced. Furthermore, when lysogeny is established, the phage DNA integrates into the host genome. A bacterium harboring this integrated phage DNA is called a **lysogen.** The integrated DNA is called a **prophage.** The lysogenic state can exist indefinitely, and it should not be considered a big disadvantage for the phage, because the phage DNA in the lysogen replicates right along with the host DNA. In this way, the phage genome multiplies without the necessity of making phage particles; thus, it gets a "free ride." Under certain conditions, such as when the lysogen encounters mutagenic chemicals or radiation, lysogeny can be broken and the phage enters the lytic phase.

Phage λ can replicate in either of two ways: lytic or lysogenic. In the lytic mode, almost all of the phage genes are transcribed and translated, and the phage DNA is replicated, leading to production of progeny phages and lysis of the host cells. In the lysogenic mode, the λ DNA is incorporated into the host genome; after that occurs, only one gene is expressed. The product of this gene, the λ repressor, binds to the two early phage operators and prevents transcription of all the rest of the phage genes. However, the incorporated phage DNA (the prophage) still replicates, since it has become part of the host DNA.

Lytic Replication of Phage λ

The lytic replication cycle of phage λ resembles that of the virulent phages we have studied in that it contains three phases of transcription. These phases are called **immediate early, delayed early,** and **late.** These genes are sequentially arranged on the phage DNA, which helps explain how they are regulated, as we will see. Figure 8.16 shows the λ genetic map in two forms: linear, as the DNA exists in the phage particles, and circular, the shape the DNA assumes shortly after infection begins. The circularization is made possible by 12-base overhangs, or "sticky" ends, at either end of the linear genome. These cohesive ends go by the name **cos.** Note that cyclization brings together all the late genes, which had been separated at the two ends of the linear genome.

As usual, the program of gene expression in this phage is controlled by transcriptional switches, but λ uses a switch we have not seen before: **antitermination.** Figure 8.17 outlines this scheme. Of course, the host RNA polymerase holoenzyme transcribes the immediate early genes first. There are only two of these genes, *cro* and *N,* which lie immediately downstream from the rightward and leftward promoters, P_R and P_L, respectively. At this stage in the lytic cycle, no repressor is bound to the operators that govern these promoters (O_R and O_L, respectively), so

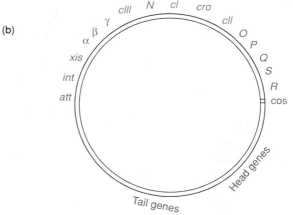

FIGURE 8.16
Genetic map of phage λ. (*a*) The map is shown in linear form, as the DNA exists in the phage particles; the cohesive ends (cos) are at the ends of the map. The genes are grouped primarily according to function. (*b*) The map is shown in circular form, as it exists in the host cell during a lytic infection after annealing of the cohesive ends.

transcription proceeds unimpeded. When the polymerase reaches the ends of the immediate early genes, it encounters ρ-dependent terminators and stops short of the delayed early genes.

The products of both immediate early genes are crucial to further expression of the λ program. The *cro* gene product is an **antirepressor** that turns off transcription of the repressor gene, *cI,* and therefore prevents synthesis of repressor protein. This is, of course, necessary if any of the other phage genes are to be expressed. The *N* gene product, **pN,** is an **antiterminator** that permits RNA polymerase to ignore the terminators at the ends of the immediate early genes and continue transcribing right on into the delayed early genes. When this happens, the delayed early phase begins. Note that the same promoters (P_R and P_L) are used for both immediate early and delayed early transcription. The switch does not involve a new σ factor or RNA polymerase that recognizes new promoters and starts new transcripts, as we have seen with other phages; instead, it involves an extension of transcripts controlled by the same promoters.

The delayed early genes are important not only in continuing the lytic cycle, but also in establishing lysogeny. The delayed early genes that help continue the lytic cycle are all on the rightward side, downstream from *cII.* Genes *O* and *P* code for proteins that are necessary for phage DNA replication, a key part of lytic growth. The *Q* gene product (**pQ**) is another antiterminator, which permits transcription of the late genes.

FIGURE 8.17

Temporal control of transcription during lytic infection by phage λ.
(*a*) Immediate early transcription (red) starts at the rightward and
leftward promoters (P_R and P_L, respectively) that flank the repressor
gene (*cI*); transcription stops at the ρ-dependent terminators (t) after
the *N* and *cro* genes. (*b*) Delayed early transcription (blue) begins at
the same promoters, but bypasses the terminators by virtue of the *N*
gene product, pN, which is an antiterminator. (*c*) Late transcription
(gray) begins at a new promoter ($P_{R'}$); it would stop short at the
terminator (t) without the *Q* gene product, pQ, another
antiterminator.

The late genes are all transcribed in the rightward direction, but not from $P_{R'}$, as you may be thinking. The late
promoter, $P_{R'}$, lies just downstream from *Q*. Transcription
from this promoter terminates after only 194 bases, unless
pQ intervenes to prevent termination. The *N* gene product
will not do; it is specific for antitermination after *cro* and *N*.
The late genes code for the proteins that make up the phage
head and tail, and for proteins that lyse the host cell so the
progeny phages can escape.

The immediate early/delayed early/late transcriptional
switching in the lytic cycle of phage λ is controlled by
antiterminators. One of the two immediate early genes is
cro, which codes for an antirepressor that allows the lytic
cycle to continue. The other, *N,* codes for an antiterminator,
pN, that overrides the terminators after the *N* and *cro* genes.
Transcription then continues on into the delayed early genes.
One of the delayed early genes, *Q,* codes for another
antiterminator (pQ) that permits transcription of the late
genes from the late promoter, $P_{R'}$, to continue without
premature termination.

Establishing Lysogeny

We have mentioned that the delayed early genes are required not only for the lytic cycle, but also for establishing
lysogeny. The delayed early genes help establish lysogeny
in two ways: (1) Most of the delayed early gene products
are needed for integration of the phage DNA into the host
genome, a prerequisite for lysogeny. (2) The products of
the *cII* and *cIII* genes allow transcription of the *cI* gene
and therefore production of the λ repressor, the central
component in lysogeny.

There are two promoters for the *cI* gene (fig. 8.18): P_{RM}
and P_{RE}. The $_{RM}$ in P_{RM} stands for repressor maintenance.
This is the promoter that is used *during* lysogeny to ensure a
continuing supply of repressor to maintain the lysogenic state.
It has the peculiar property of requiring its own product—
repressor—for activity. Therefore, this promoter cannot be
used to establish lysogeny because there is no repressor to
activate it at the beginning of infection. Instead, the other
promoter, P_{RE}, is used. The $_{RE}$ in P_{RE} stands for repressor
establishment. P_{RE} lies to the right of both P_R and *cro*. It directs transcription leftward through *cro* and then through *cI*.
Of course, the natural direction of transcription of *cro* is
rightward from P_{RE}, so the leftward transcription from P_R
gives an anticoding transcript of *cro*. This does not matter,
since the purpose of this transcription is to express *cI*, not *cro*.

Phage λ establishes lysogeny by causing production of
enough repressor to bind to the early operators and prevent
further early RNA synthesis. The promoter used for
establishment of lysogeny is P_{RE}, which lies to the right of
P_R and *cro*. Transcription from this promoter goes leftward
through the *cI* gene. Another promoter, P_{RM}, comes into
play later, to maintain lysogeny.

Lysogen Induction

We mentioned that a lysogen can be induced by treatment
with mutagenic chemicals or radiation. The mechanism of
this induction is as follow: *E. coli* cells respond to environmental insults, such as mutagens or radiation, by inducing a
set of genes whose collective activity is called the **SOS response;** one of these genes, in fact the most important, is
recA. The *recA* product participates in recombination repair

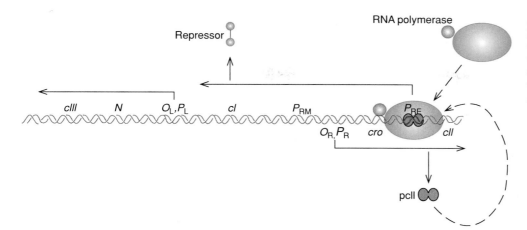

FIGURE 8.18
Establishing lysogeny. Delayed early transcription from P_R gives *cII* mRNA that is translated to pcII (purple). pcII allows RNA polymerase (blue and red) to bind to P_{RE} and transcribe the *cI* gene, yielding repressor (green).

of DNA damage (which explains part of its usefulness to the SOS response), but environmental insults also induce a new activity in the RecA protein. It stimulates a latent protease, or protein-cleaving activity in the λ repressor. This protease then cleaves the repressor in half and releases it from the operators, as shown in figure 8.19. As soon as that happens, transcription begins from P_R and P_L. One of the first genes transcribed is *cro,* whose product shuts down any further transcription of the repressor gene. Lysogeny is broken.

Surely λ would not have evolved with a repressor that responds to RecA by chopping itself in half unless there were an advantage to the phage. That advantage seems to be this: The SOS response signals that the lysogen is under some kind of DNA-damaging attack. It is expedient under those circumstances for the prophage to get out by inducing the lytic cycle, rather like rats deserting a sinking ship.

❖ When a lysogen suffers DNA damage, it induces the SOS response. The initial event in this response is the appearance of co-protease activity in the RecA protein. This causes the λ repressors to cut themselves in half, removing them from the λ operators and inducing the lytic cycle. In this way, λ phages can escape the potentially lethal damage that is occurring in their host.

SPECIFIC DNA-PROTEIN INTERACTIONS

We have seen several examples of proteins that bind to specific regions of DNA. These include the λ repressor and Cro, which bind to their respective operators, and CAP, which binds to the CAP binding site in the *lac* promoter. The binding of these proteins to their targets is very specific. They can locate and bind to one particular short DNA sequence among a vast excess of unrelated DNA.

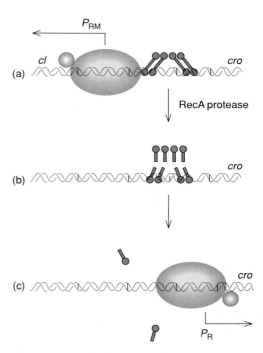

FIGURE 8.19
Inducing the λ prophage. (*a*) Lysogeny. Repressor (green) is bound to O_R (and O_L) and *cI* is being actively transcribed from the P_{RM} promoter. (*b*) The RecA protease (activated by ultraviolet light or other mutagenic influence) cleaves the repressor. (*c*) The repressor falls off the operator, allowing polymerase (red and blue) to bind to P_R and transcribe *cro.* Lysogeny is broken.

How do these proteins accomplish such specific binding? All the proteins just listed have a similar structural motif: two alpha helices connected by a short protein "turn." This **helix-turn-helix** motif allows the second helix (the **recognition helix**) to fit snugly into the major groove of the target DNA site, as illustrated in figure 8.20.

Helix 2

Turn

Helix 3

(a)

(b)

FIGURE 8.20
The helix-turn-helix motif as a DNA-binding element. (*a*) The helix-turn-helix motif of the λ repressor. (*b*) The fit of the helix-turn-helix motif of one repressor monomer with the λ operator. Helix 3 (blue) lies in the major groove of its DNA target; some of the amino acids on the back of this helix (away from the viewer) are available to make contacts with the DNA.

Furthermore, the amino acids of the recognition helix of a DNA-binding protein have chemical groups that fit hand-in-glove with specific atoms in the major DNA groove at that protein's binding site. Other proteins with helix-turn-helix motifs will not be able to bind as well at that same site, because they do not have the correct amino acids in their recognition helices.

Many DNA-binding proteins in prokaryotes employ a helix-turn-helix binding motif. The second (recognition) helix of each motif fits into the major groove of the target site on the DNA, and certain key amino acids in this helix make intimate contact with specific atoms in the DNA.

WORKED-OUT PROBLEMS

Problem 1

The table below gives the genotypes (with respect to the *lac* operon) of several partial diploid *E. coli* strains. Fill in the phenotypes, using a "+" for β-galactosidase synthesis and "−" for no β-galactosidase synthesis. Glucose is absent in all cases. Give a brief explanation of your reasoning.

Genotype	Phenotype for β-galactosidase Production	
	No Inducer	*Inducer*
a. $I^+O^+Z^+/I^+O^+Z^+$		
b. $I^+O^+Z^-/I^+O^+Z^+$		
c. $I^-O^+Z^+/I^+O^+Z^+$		
d. $I^sO^+Z^+/I^+O^+Z^+$		
e. $I^+O^cZ^+/I^+O^+Z^+$		
f. $I^+O^cZ^-/I^+O^+Z^+$		
g. $I^sO^cZ^+/I^+O^+Z^+$		

Solution

Genotype	Phenotype for β-galactosidase Production	
	No Inducer	*Inducer*
a. $I^+O^+Z^+/I^+O^+Z^+$	−	+
b. $I^+O^+Z^-/I^+O^+Z^+$	−	+
c. $I^-O^+Z^+/I^+O^+Z^+$	−	+
d. $I^sO^+Z^+/I^+O^+Z^+$	−	−
e. $I^+O^cZ^+/I^+O^+Z^+$	+	+
f. $I^+O^cZ^-/I^+O^+Z^+$	−	+
g. $I^sO^cZ^+/I^+O^+Z^+$	+	+

Reasoning:

a. The genotypes of both halves of the partial diploid are wild-type. Therefore, these cells will show the wild-type *lac* phenotype, which is repressed (turned off) in the absence of inducer and derepressed (turned on) in the presence of inducer. This means that these cells make no β-galactosidase without inducer, but they do make β-galactosidase in the presence of inducer.

b. This partial diploid has one mutant allele and one wild-type allele at the β-galactosidase (*Z*) locus. In this case the wild-type allele is dominant, since the mutant allele does not inhibit the function of the wild-type allele. Thus, these cells show the wild-type phenotype, as in (*a*), above.

c. This partial diploid has one constitutive (I^-) and one wild-type allele at the *I* locus. Again, the wild-type allele is dominant, since the mutant allele cannot inhibit the function of the wild-type allele. Thus, these cells show the wild-type phenotype, as in (*a*).

d. This partial diploid has one wild-type and one (I^s) allele at the *I* locus. The mutant allele is dominant in this case, since it produces repressor that is incapable of binding to inducer. Thus, this repressor will remain bound to the operator and prevent induction of the *lac* operon, regardless of the fact that the wild-type allele still makes wild-type repressor. Therefore, the operon remains repressed even in the presence of inducer.

e. This partial diploid has one wild-type and one constitutive (O^c) allele at the O locus. The constitutive allele is dominant because this operator cannot bind repressor even though the wild-type operator can. Thus, these cells will produce β-galactosidase in the presence or absence of inducer.

f. One half of this partial diploid is double mutant (constitutive in the operator and defective in the β-galactosidase [Z] locus). The other half is wild-type at both loci. When these two mutations are found *in cis,* as they are here, they cancel each other. The O^c mutation is dominant, but it is only *cis*-dominant. That is, it is only dominant with respect to the operon to which it is attached. In this case, that operon has a defective Z gene, so no β-galactosidase will be produced under any circumstances. On the other hand, there is nothing wrong with the other half of the partial diploid, so it will give these cells a wild-type phenotype.

g. One half of this partial diploid is double mutant (I^s at the I locus, and constitutive at the O locus). Since the O^c operator cannot bind repressor, it will not bind the defective repressor that is the product of the I^s gene. Thus, this operon is turned on in the presence or absence of inducer, even though the wild-type operon is always turned off because of the mutant repressor provided *in trans.*

Problem 2

(*a*) In the genotype listed in the table below, the letters A, B, and C correspond to the *lacI, lacO,* and *lacZ* genes, though not necessarily in that order. From the mutant phenotypes exhibited by the first three genotypes listed in the table below, deduce the identities of A, B, and C as they correspond to the three genes of the *lac* operon. The minus superscripts (e.g., A^-) can refer to any aberrant function (e.g., I^-, I^s, or O^c).

(*b*) Determine the genotypes, in conventional *lac* operon genetic notation, of the partial diploid strains shown in lines 4 and 5 of the table below.

	Phenotype for β-galactosidase Production	
Genotype	No Inducer	Inducer
1. $A^+B^+C^-$	+	+
2. $A^-B^+C^+$	+	+
3. $A^+B^+C^-/A^+B^+C^+$	+	+
4. $A^-B^+C^+/A^+B^+C^+$	–	–
5. $A^-B^+C^-/A^+B^+C^+$	–	+

Solution

(*a*) Since the phenotype of Mutant #1 is constitutive, its genotype must carry a mutation in either the I or the O gene, i.e., I^- or O^c (products are made without inducer). Therefore C is either I or O. Mutant #2 is also constitutive, so its genotype is also either I^- or O^c. Therefore, A is either I or O. So far we know that A and C are I and $O,$ though not necessarily in that order. This means that B must correspond to gene Z, by elimination. Now we need to determine whether $A = I$ and $C = O$, or vice-versa. The phenotype of Mutant #3 gives the answer. Because it is constitutive and the mutant form is dominant, C^- must be O^c. If it were I^-, the other constitutive possibility, the mutation would have been recessive, and no β-galactosidase would have been produced in the absence of inducer. Thus, $A = I, B = Z,$ and $C = O$.

(*b*) The partial diploid in line 4 fails to make β-galactosidase even in the presence of inducer; that is, the mutation (A^-) is dominant, and A^- must, therefore, be I^s. On the other hand, the phenotype of the strain in line 5 is similar to that of a wild-type strain. Therefore, the mutation (A^-) in the genotype of the strain shown in line 5 must be recessive, i.e., $A^- = I^-$. Therefore, the two genotypes are

$$I^sZ^+O^+/I^+Z^+O^+$$
and
$$I^-Z^+O^+/I^+Z^+O^+$$

or in standard order,

$$I^sO^+Z^+/I^+O^+Z^+$$
and
$$I^-O^+Z^+/I^+O^+Z^+$$

Summary

RNA polymerase is the central figure in the transcription process. The *E. coli* enzyme is composed of a core, which contains the basic transcription machinery, and a polypeptide called σ, which directs the core to transcribe specific genes. σ allows initiation of transcription by causing the RNA polymerase holoenzyme to bind tightly to a promoter, melting a local region of DNA. After initiation, σ dissociates from the core polymerase, which continues to elongate the RNA chain. Finally, the core polymerase encounters a DNA site called a terminator, and transcription stops.

Bacterial genes that are linked in function are frequently clustered together in operons and are controlled together from a common promoter. For example, the genes for the three enzymes involved in lactose metabolism in *E. coli* are clustered together in the *lac* operon. This operon is controlled both positively (by cAMP plus catabolite activator protein) and negatively (by the *lac* repressor).

Gene expression is controlled primarily at the transcription level in prokaryotes. The temporal control of transcription in *B. subtilis* cells infected by phage SPO1 is directed by a set of phage-encoded σ factors that associate with the host core RNA polymerase and change its specificity from early to middle to late. Phage T7, instead of coding for a new σ factor to change the host polymerase's specificity from early to late, encodes a whole new RNA polymerase with absolute specificity for the later phage genes. Transcriptional switching in the lytic cycle of phage λ is controlled by antiterminators that override ρ-dependent terminators and allow transcription to continue into the next class of genes.

Phage λ can replicate in either of two ways: lytic or lysogenic. In the lytic mode, almost all of the phage genes are transcribed and translated, and the phage DNA is replicated, leading to production of progeny phages and lysis of the host cells. In the lysogenic mode, the λ DNA is incorporated into the host genome; after that occurs, only the λ repressor gene (*cI*) is expressed. The product of this gene, the λ repressor, binds to the two early phage operators and prevents transcription of all the rest of the phage genes. However, the incorporated phage DNA (the prophage) still replicates, since it has become part of the host DNA.

When a lysogen suffers DNA damage, it induces the SOS response. The initial event in this response is the appearance of a co-protease activity in the RecA protein. This unmasks a latent protease activity in the λ repressors, which cut themselves in half, removing them from the λ operators and inducing the lytic cycle. In this way, λ phages can escape the damage that is occurring in their host.

Several prokaryotic DNA-binding proteins, such as the λ repressor and Cro, bind to their DNA targets via helix-turn-helix motifs that insert specifically into the DNA major groove. Amino acids in the second (recognition) helix in these motifs make specific contact with atoms in the major DNA groove.

PROBLEMS AND QUESTIONS ❓

1. Contrast the activities of RNA polymerase holoenzyme and core polymerase on un-nicked phage DNA in vitro.

2. Consider a –10 box with the following sequence: CATAGT. (*a*) Would a C → T mutation in the first position be likely to be an up or a down mutation? (*b*) Would a T → A mutation in the last position be likely to be an up or a down mutation?

3. What structural characteristic of an operon ensures that the genes contained in the operon will be coordinately controlled?

4. Consider an *E. coli* cell with the following mutations:

   (*a*) a mutant *lac* operator (the O^c gene product) that cannot bind repressor

   (*b*) a mutant *lac* repressor (the I^- gene product) that cannot bind to the *lac* operator

   (*c*) a mutant *lac* repressor (the I^s gene product) that cannot bind to allolactose

   (*d*) a mutant *lac* promoter that cannot bind CAP plus cAMP

   What effect would each mutation have on the function of the *lac* operon (assuming no glucose is present)?

5. The table below gives the genotypes (with respect to the *lac* operon) of several strains of *E. coli*. Fill in the phenotypes, using a "+" for β-galactosidase synthesis and a "−" for no β-galactosidase synthesis. Glucose is absent in all cases. Give a brief explanation of your reasoning.

Genotype	Phenotype for β-galactosidase Production	
	No Inducer	*Inducer*
a. $I^+O^+Z^+$		
b. $I^+O^+Z^-$		
c. $I^+O^cZ^+$		
d. $I^-O^+Z^+$		
e. $I^+O^cZ^-$		
f. $I^-O^cZ^+$		
g. $I^sO^+Z^+$		
h. $I^sO^cZ^+$		

6. The table below presents the genotypes and phenotypes of several *lac* mutants, but with one missing notation (represented by an ×) in each case. From the information given, fill in the missing notation for each genotype. Give your reasoning. (Sometimes × can represent wild-type.)

Genotype	Phenotype for β-galactosidase Production	
	No Inducer	*Inducer*
a. $I^×O^+Z^+$	+	+
b. $I^×O^+Z^+/I^+O^+Z^+$	−	−
c. $I^×O^+Z^+/I^+O^+Z^+$	−	+ (2 possible answers)
d. $I^×O^+Z^+/I^-O^+Z^+$	+	+
e. $I^+O^×Z^+/I^+O^+Z^+$	+	+
f. $I^+O^×Z^+/I^+O^+Z^+$	−	+
g. $I^+O^+Z^×/I^+O^+Z^-$	−	−

7. Consider mutations involving the promoter of the *lac* operon. These mutations prevent RNA polymerase binding and are designated P^-. Predict the phenotypes of the following partial diploids, assuming no glucose in the growth medium. $Z = lacZ$, the gene for β-galactosidase.

	Phenotype for β-galactosidase Production	
Genotype	*No Inducer*	*Inducer*
a. $I^+O^+P^-Z^+/I^+O^+P^+Z^+$		
b. $I^+O^+P^-Z^+/I^+O^+P^-Z^+$		
c. $I^+O^+P^-Z^+/I^+O^+P^+Z^-$		
d. $I^-O^+P^-Z^+/I^-O^+P^-Z^+$		

8. A hypothetical structural gene (G) is controlled by an operator (O) and a repressor gene (R). List the genotypes that would produce the following phenotypes:

(a) G product made in the presence of inducer, but not in its absence.

(b) G product made in the presence or absence of inducer.

(c) G product never made.

9. You form partial diploids by adding the genes $G^-O^+R^+$ to the bacteria with each of the eleven genotypes in the answer to problem 8 (see answer at end of book). Give the genotype of each partial diploid, and the phenotype of each, with respect to G product, in the presence and absence of inducer. Present your answer in the form of a table as indicated below. Use the symbol (+) if the product is made, and the symbol (−) if it is not. The top line is already filled in.

	Phenotype for G product	
Genotype	+ inducer	− inducer
$G^+O^+R^+/G^-O^+R^+$	+	−

10. Which phage genes would be transcribed in *B. subtilis* cells infected with SPO1 phages with mutations in the following genes? Why?

(*a*) Gene 28.

(*b*) Gene 33.

(*c*) Gene 34.

11. What would be the effect of a mutation in gene 1 of phage T7? Why?

12. What would be the effect of a mutation in the *B. subtilis* gene for σ^{29} or σ^{32}? Why?

13. Which phage gene would *not* be expressed after the λ lytic cycle is underway?

14. Which phage genes would *not* be expressed in a λ lysogen?

15. pN is an immediate early product that is required for delayed early transcription, yet it is very unstable. Why does this not pose a problem for continued delayed early expression in the lytic cycle?

16. RNA polymerase transcribing from P_{RE} traverses *cro*. Can this result in *cro* expression? Why, or why not?

17. Would you classify the antiterminator activity of pN as positive or negative control? Why?

Answers appear at end of book.

CHAPTER 9: EUKARYOTIC GENE STRUCTURE AND EXPRESSION

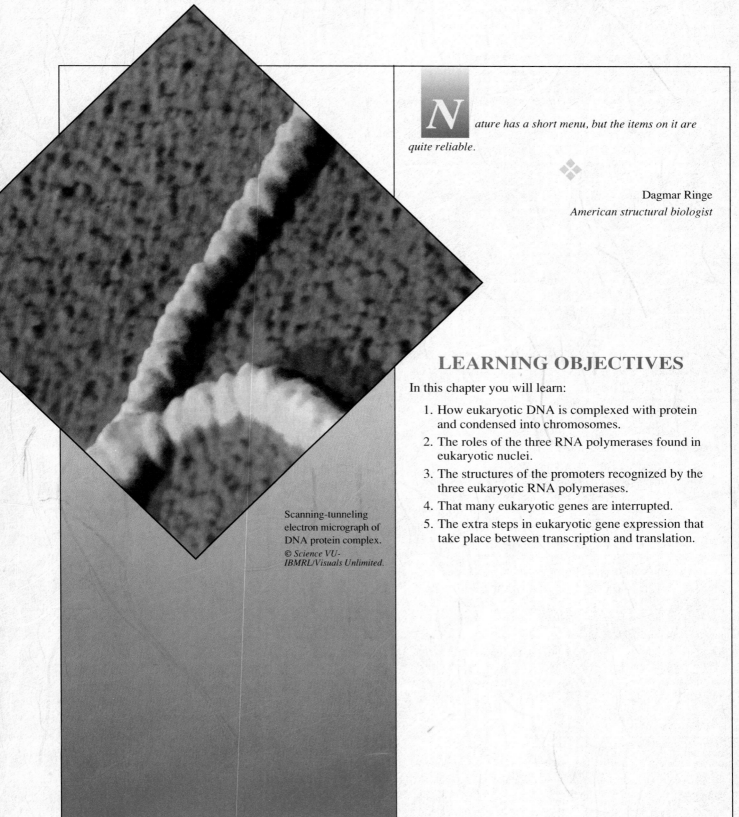

*N**ature has a short menu, but the items on it are quite reliable.*

❖❖

Dagmar Ringe
American structural biologist

Scanning-tunneling
electron micrograph of
DNA protein complex.
© *Science VU-
IBMRL/Visuals Unlimited.*

LEARNING OBJECTIVES

In this chapter you will learn:

1. How eukaryotic DNA is complexed with protein and condensed into chromosomes.
2. The roles of the three RNA polymerases found in eukaryotic nuclei.
3. The structures of the promoters recognized by the three eukaryotic RNA polymerases.
4. That many eukaryotic genes are interrupted.
5. The extra steps in eukaryotic gene expression that take place between transcription and translation.

e turn now to the subject of gene expression control in eukaryotes. In the last chapter we dealt with the same topic—control of gene expression—but in prokaryotes rather than in eukaryotes. Why do we need a whole separate chapter for the eukaryotes? We will see that the process of gene expression is much more complex in eukaryotes. To begin with, the prokaryotic chromosome is a circle of DNA encumbered with relatively little protein, while the eukaryotic chromosome contains a large amount of protein. This extra protein gives eukaryotic chromosomes their structure and also plays a regulatory role. Also, while the genes of prokaryotes are almost always uninterrupted, the genes of eukaryotes, especially higher eukaryotes, tend to be interrupted by regions of nonsense. In this chapter we will examine the structure of eukaryotic chromosomes and the genes they contain. Then we will study the mechanisms eukaryotes have evolved to deal with their more complex genomes.

CHROMATIN STRUCTURE

Have you ever dealt with a length of kite string, or fishline, or thread, or even a garden hose? If so, you know that long, thin things tend to get hopelessly tangled. One look at figure 9.1 should convince you that the DNA in a chromosome is longer in relationship to its width than any kite string. In fact, you would have to fly a kite from sea level to an altitude far above Mt. Everest to make the length/width ratio of the string approach that of the DNA in an average human chromosome (more than ten million to one). How do those immensely long molecules stay untangled? As we will see, they are complexed with protein (forming a mixture called **chromatin**) and coiled up in a highly organized way.

When chromatin is lightly treated with a DNase called *micrococcal nuclease,* DNA-protein particles called **nucleosomes** are released (fig. 9.2a). These nucleosomes, ubiquitous in the eukaryotic kingdom, are remarkably uniform in size, about 11 nanometers (nm) in diameter, and they contain about 200 base pairs of DNA.

The abundance and uniformity of nucleosomes suggests that they are a repeating unit of chromatin structure, and there is direct evidence for this supposition. Consider, for example, the "minichromosome" of the mammalian tumor virus SV40. When this virus infects a host cell, its circular DNA attracts host proteins to form a miniature chromosome. If we lower the salt concentration of a solution of such minichromosomes, we reveal a beaded structure (fig. 9.2b). The beads are nucleosomes.

NUCLEOSOMES: THE FIRST ORDER OF CHROMATIN FOLDING

Nucleosomes contain, in addition to the 150 base pairs of DNA, a set of basic proteins called **histones.** Most eukaryotic cells contain five different kinds of histones called **H1,**

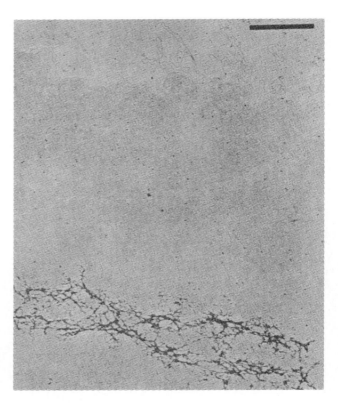

FIGURE 9.1

DNA in a human chromosome. A metaphase chromosome from HeLa (human) cells was extracted to remove most of its protein. This allowed the DNA to uncoil, leaving a scaffolding in the shape of the chromosome behind. The tangled web of DNA looks like a gray cloud throughout the lower part of the figure, but the individual DNA duplexes near the top are visible. Bear in mind that this is a single duplex molecule of DNA. The bar represents one micron.

Courtesy of James R. Paulson. (From James R. Paulson and U. K. Laemmli, Cell *12:820, 1977.) Reprinted with permission from Cell Press.*

H2A, H2B, H3, and **H4.** Figure 9.3 shows how these five species can be separated by gel electrophoresis. The "bead" of a nucleosome is actually a ball of histones with the DNA wound twice around the outside. Each ball contains exactly eight histone molecules, a pair each of H2A, H2B, H3, and H4. A single molecule of histone H1 binds to DNA outside the ball and can be removed by more stringent DNase treatment, which yields a nucleosome *core particle.* Figure 9.4 shows the probable arrangement of histones and DNA in the nucleosome core particle. It appears that the core histones are really not arranged in pairs. H3 and H4 form a tetramer at the center of the particle, and this tetramer is flanked by H2A–H2B dimers at top left and lower right.

What is the function of nucleosomes? Whatever it is, two properties of these particles suggest that their function is fundamentally important. The first is the universality of nucleosomes in eukaryotes; the second is the extreme evolutionary conservation of most of the histones. The classic example of this conservation is that histone H4 from the cow differs from H4 from the pea plant by only two amino acids out of 102, and these are **conservative substitutions** that should not change the protein structure significantly. For example, one basic amino acid substitutes for another.

(a)

(b)

FIGURE 9.3

Separation of histones by electrophoresis. The histones of calf thymus were separated by polyacrylamide gel electrophoresis.

From S. Panyim and R. Chalkley, Archives of Biochemistry and Biophysics *130:343, fig. 6(A), 1969.*

FIGURE 9.2

Nucleosomes. (*a*) Nucleosomes (arrow) released from chicken red blood cells by treatment of chromatin with micrococcal nuclease. The bar represents 250 nm. (*b*) Nucleosomes in the minichromosome of SV40 virus. The minichromosome has been relaxed by suspending it in low-salt buffer, which reveals the beaded structure. The bar represents 100 nm.

Photo (a) from P. Oudet, M. Gross-Bellard, and P. Chambon, Cell *4:287, 1975. Reprinted with permission from Cell Press. Photo (b) © Dr. Jack Griffith.*

In other words, in all the eons that have intervened since the cow and pea lines diverged from a common ancestor, only two amino acids in histone H4 have changed. Obviously, the histones, and the nucleosomes of which they are a part, play a vital role, or their structures would have been allowed more evolutionary latitude.

   The most obvious role for nucleosomes is, of course, a structural one; nucleosomes provide the first order of condensing, or coiling, of the extremely long, thin chromosomal fiber. Figure 9.5 makes this point very clearly. SV40 DNA that has been stripped of its protein is shown next to an SV40 minichromosome at the same magnification. The apparent decrease in length of the DNA in the minichromosome is due to the formation of nucleosomes. In fact, the DNA achieves about a 6-fold condensation by coiling up into nucleosomes.

FIGURE 9.4

One interpretation of nucleosome structure, based on X-ray and neutron diffraction studies. The DNA (red) wraps around the core of histones, composed of the (H3–H4)$_2$ tetramer in the center (yellow) and an H2A–H2B dimer at each end (purple).

Eukaryotic DNA combines with basic protein molecules called histones to form structures known as nucleosomes. These structures contain four pairs of histones (H2A, H2B, H3, and H4) in a ball, around which is wrapped a stretch of about 200 base pairs of DNA. Histone H1, bound to DNA outside the ball, completes the nucleosome structure. The first order of chromatin folding is represented by a string of nucleosomes.

FIGURE 9.5

Condensation of DNA in nucleosomes. Deproteinized SV40 DNA is shown next to an SV40 minichromosome (inset). The condensation of DNA afforded by nucleosome formation is apparent.

Jack Griffith, "Chromatin Structure: Deduced from a Minichromosome." Science, March 28, 1975, Vol. 187, p. 1202, figure 1. Copyright 1975 by AAAS.

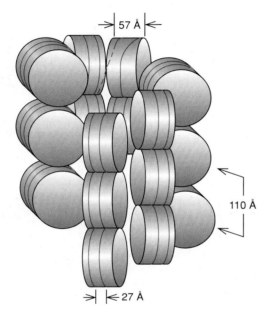

FIGURE 9.6

The solenoid model of chromatin folding. A string of nucleosomes coils into a hollow spiral to form the solenoid. Each nucleosome is represented by a blue cylinder with a wirelike DNA (red) coiled around it.

From Widom and Klug, in Cell 43:210, 1985. Copyright © 1985 Cell Press, Cambridge, MA. Reprinted by permission of the publisher and authors.

FURTHER FOLDING OF THE NUCLEOSOME STRING

Chromatin in interphase nuclei takes on various thicknesses. One of these, 11 nm in thickness, simply represents a string of nucleosomes, or **nucleosome fiber.** The next thickness, about 30 nm, is due to further winding of the nucleosome string to form a hollow coil called a **solenoid** (fig. 9.6). Aaron Klug, who first described the solenoid, showed how it forms by making electron micrographs of chromatin in solutions of increasing salt concentration (fig. 9.7). At very low salt concentration, the chromatin appears as a string of nucleosomes, much like the SV40 minichromosome in figure 9.2. As the salt

FIGURE 9.7

Formation of the solenoid. (*a*) A string of nucleosomes, extended due to very low ionic strength environment. (*b*) As the salt concentration is raised, the string begins to coil. (*c*) Still higher salt concentration, with more coiling. (*d*) At physiological salt concentration, the strings of nucleosomes have coiled all the way into tightly packed solenoids. The bar represents 200 nm.

F. Thoma, T. Koller, A. Klug. Reproduced from Journal of Cell Biology 83:403–27, Rockefeller University Press, 1979. By copyright permission of the Rockefeller University Press.

concentration rises, coiling takes place, until the typical solenoid structures appear. The DNA achieves another 6–7-fold condensation in coiling up into the solenoid. We know that histone H1 participates in this coiling, because coiling cannot occur in salt-treated chromatin that lacks H1.

Further observations of chromatin suggest at least one more possible order of folding in the chromosome. First of all, it has been observed that eukaryotic chromosomes undergo supercoiling. We learned in chapter 7 that supercoiling can occur in bacterial and certain viral DNAs because the circular structure of these molecules prevents the release of strain. But eukaryotic chromosomes are linear, so how can the DNA in them be supercoiled? One way this could happen would be if the chromatin fiber were looped and held fast at the base of each loop (fig. 9.8). Each loop would be the functional equivalent of a circle, at least for supercoiling purposes, and the winding of DNA around the nucleosomes would provide the strain necessary for supercoiling.

The direct visualization of a human chromosome in figure 9.1 also supports the loop idea, although the DNA is so tightly bunched in that picture that the looping is difficult to see. Figure 9.9, also depicting a deproteinized

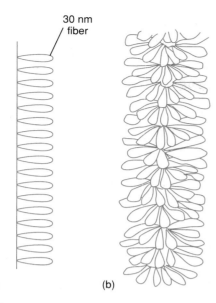

(a)   (b)

FIGURE 9.8

Radial loop models of chromatin folding. (*a*) This is only a partial model, showing some of the loops of chromatin attached to a central scaffold; of course, all the loops are part of the same continuous DNA molecule. (*b*) A more complete model, showing how the loops are arranged in three dimensions around the central scaffold.

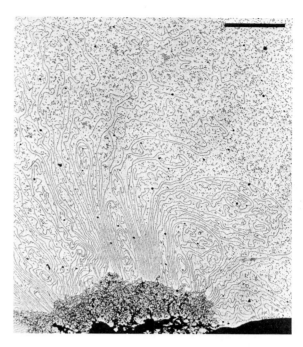

FIGURE 9.9

Deproteinized chromosome showing looped structure. The scaffold is at the bottom, with loops of DNA extending upward. The bar represents 2 microns.

Courtesy of James R. Paulson. (From James R. Paulson and U. K. Laemmli, Cell *12:823, 1977.) Reprinted with permission from Cell Press.*

chromosome, makes it very clear that loops begin and end at adjacent points on the chromosomal scaffold. Figure 9.10, showing cross sections through several swollen chromosomes, further supports the notion of radial loops. Figure 9.11 summarizes our current thinking about chromatin folding, from the 2-nm bare DNA fiber to the postulated radial looped chromosome.

> The second order of chromatin folding involves a coiling of the string of nucleosomes into a 25-nm-thick fiber called a solenoid. Histone H1 is thought to participate in this folding by interacting with other H1 molecules. The third order of folding is probably looping of the 30-nm fiber into a brushlike structure with the loops anchored to a central matrix.

NONHISTONE PROTEINS

Of course, histones are not the only proteins involved in chromatin. Another, more complex class of proteins also associates with eukaryotic chromosomes. Since little is known about many of these proteins, they are usually called by a name that tells us not what they are, but simply what they are not: **nonhistone proteins.** RNA polymerase is an example of a nonhistone protein about which we know quite a bit. There are many other nonhistones, some that undoubtedly play structural roles in the chromosome and others that help control gene expression. We will focus on some possible regulatory molecules later in the chapter.

FIGURE 9.10

Swollen chromosomes showing radial loops. The chromatids in the lower portion of the photograph are shown in cross section. The chromatid at the top is cut lengthwise.

M. P. F. Marsden and U. K. Laemmli, Cell *17:851, 1979. Reprinted with permission from Cell Press.*

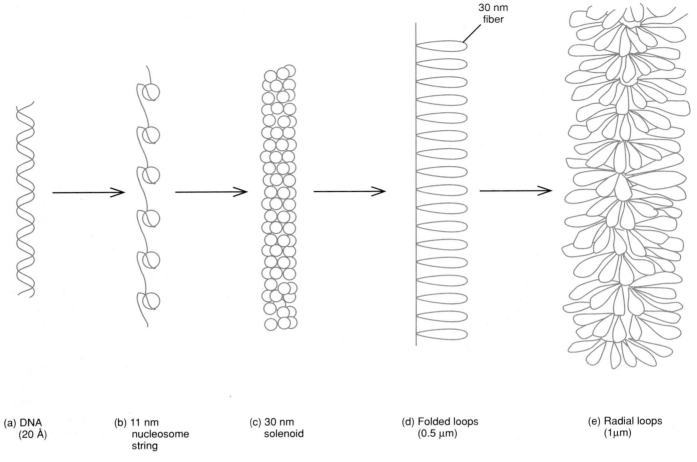

(a) DNA
    (20 Å)

(b) 11 nm
    nucleosome
    string

(c) 30 nm
    solenoid

(d) Folded loops
    (0.5 µm)

(e) Radial loops
    (1µm)

FIGURE 9.11

Orders of chromatin folding. Structures (*a*), (*b*), and (*c*) rest on solid experimental ground; (*d*) and (*e*), also shown in figure 9.8, are more speculative.

Reproduced, with permission, from the Annual Review of Genetics, *Volume 12,* © *1978 by Annual Reviews, Inc.*

CHROMATIN STRUCTURE AND GENE EXPRESSION

A strong relationship exists between the physical state of the chromatin in the vicinity of a particular gene and the activity of that gene. To begin with, there are two fundamental classes of chromatin structure: **euchromatin** and **heterochromatin.** Euchromatin is like the string of nucleosomes in figure 9.11*b*—relatively extended and open. By contrast, heterochromatin is very condensed and its DNA is inaccessible. It even appears as clumps when viewed microscopically (fig. 9.12). One example of heterochromatin is found at or near the centromeres of chromosomes. The DNA associated with the centromeres belongs to a class of DNA that is repeated many times in the genome, in contrast to structural genes, which are usually found in only one or a few copies each. This

pericentromeric DNA is complexed with protein and coiled tightly; obviously, DNA in such a state is not available to RNA polymerase and is not transcribed.

This does not imply that the relatively extended euchromatin is automatically transcribed, just because it does not physically exclude RNA polymerase. On the contrary, the majority of the euchromatin in a given cell is inactive. Why is this? That, of course, is an extremely important question; it is the same as asking what controls eukaryotic genes. Later in this chapter and in chapter 14, we will examine this question in more detail.

◆ Chromatin exists in eukaryotic cells in two forms: heterochromatin, which is condensed and transcriptionally inactive, and euchromatin, which is extended and at least potentially active.

FIGURE 9.12

Interphase nucleus showing heterochromatin. Bat stomach lining cell with nucleus at center. Dark areas around periphery of nucleus are heterochromatin (H).

© Dr. Keith Porter.

Table 9.1 Roles of Eukaryotic RNA Polymerases	
Polymerase	**RNA Product**
I	large rRNA precursor
II	mRNA precursor
III	tRNA precursor; 5S rRNA precursor

RNA POLYMERASES AND THEIR ROLES

In chapter 8 we learned that prokaryotes have just one RNA polymerase (not counting primase), which makes all the different types of RNA the cell needs. True, the σ factor can change to meet the demands of a changing environment, but the core enzyme remains essentially the same. Furthermore, the normal holoenzyme can make all three major types of RNA: messenger RNA, ribosomal RNA, and transfer RNA. (The latter two classes of RNA are used in translation and will be discussed in chapter 10.) Quite a different situation prevails in the eukaryotes. Robert Roeder and William Rutter showed in 1969 that eukaryotic cells have not one but three different RNA polymerases in their nuclei. They named these **polymerase I, polymerase II,** and **polymerase III.** Furthermore, these three enzymes have distinct roles in the cell.

It is intriguing that there are three different kinds of RNA polymerase and three different kinds of RNA to make: mRNA, rRNA, and tRNA. In the fourteenth century, William of Occam developed a maxim now called "Occam's razor," which states in effect that when you are confronted with several possible explanations for a phenomenon, you should choose the simplest. The simplest

explanation for the existence of three kinds of polymerase and three kinds of RNA is that each polymerase makes one of the classes of RNA. In fact, there *is* such a one-to-one correspondence, and in simplified form it goes like this: Polymerase I makes ribosomal RNA (the 18S, 28S, and 5.8S varieties); polymerase II makes messenger RNA; polymerase III makes transfer RNA.

Of course, there are some complicating factors. One of these is that the three kinds of RNA are not originally transcribed in mature form; instead, they are made as larger precursors that must later be cut down to size. So it is more correct to say, for example, that polymerase I makes an rRNA *precursor,* not rRNA itself. We have also glossed over the fact that there is a small 5S rRNA, in addition to the varieties made by polymerase I. This 5S rRNA is made by polymerase III. Thus, polymerase III specializes in synthesizing small RNAs. Table 9.1 summarizes the roles of the three polymerases.

Eukaryotic cells contain three distinct RNA polymerases, named polymerase I, polymerase II, and polymerase III. The roles of these three enzymes are as follows: Polymerase I makes a large rRNA precursor; polymerase II makes mRNA precursors; and polymerase III makes tRNA and 5S rRNA precursors.

PROMOTERS

As we saw in chapter 8, prokaryotic promoters recognized by different σ factors vary quite a bit. For example, the −10 and −35 boxes of σ^{43}-specific promoters in *B. subtilis* are so different from those of σ^{32}-specific promoters that they would not be recognized by an RNA polymerase bearing the latter σ factor. Still, the basic architecture of the prokaryotic polymerases remains the same: core plus σ; and so does the structure of the promoters they recognize: −10 and −35 boxes, lying upstream from the gene(s) they control. No evidence exists for σ factors in eukaryotes, and the subunit structures of the three RNA polymerases in a given cell vary considerably. It should come as no surprise then that these three quite different RNA polymerases recognize quite different promoters.

PROMOTERS RECOGNIZED BY POLYMERASE II

We start with the promoters recognized by polymerase II because they will be most familiar to you; they bear the closest resemblance to prokaryotic promoters. In general, these promoters have more than one element.

The TATA Box

Closest to the transcription start site, centered approximately at position −30, is an A–T-rich sequence that is a very common feature of polymerase II promoters. Its consensus sequence is TATAAAA, although T's frequently replace the A's in the fifth and seventh positions. Its name, **TATA box,** is derived from its first four bases.

You may have noticed the close similarity between the eukaryotic TATA box and the prokaryotic −10 box. But notice that it is found farther upstream than its prokaryotic counterparts (−30 compared to −10). As is the case with all consensus sequences, there are exceptions to the rule. Sometimes G's and C's creep in, as in the "TATA" box of the rabbit β-globin gene, which starts with a C (CATA). Sometimes there is no recognizable TATA box at all.

Another conserved region in polymerase II promoters is not as prevalent as the TATA box. It is usually found about fifty base pairs farther upstream, approximately at position −75, and it has the consensus sequence GGC-CAATCT. This sequence commonly goes by the name CAAT box or, more often, **CCAAT box** (pronounced "cat box").

It is one thing to find conserved sequences upstream from the transcription start site; it is another to demonstrate that they actually function as parts of the promoter. The latter task has been accomplished for both the TATA box and the CCAAT box. The experimental plan has been to clone (reproduce in large quantity) a particular gene transcribed by polymerase II, introduce mutations into its promoter, and then observe the effects of these mutations on transcription of the gene. These mutations differ from those that occur naturally at random in that we can choose just which bases we want to change and how. Thus, we call this process "site-directed mutagenesis." We will discuss this topic in greater detail in chapter 15.

To test for the effect of a mutation in the TATA box, we have to be able to add our cloned genes (mutant or wild-type) to some kind of transcription system so we can observe the difference in the way these genes are transcribed. Figure 9.13 illustrates examples of systems geneticists use to study transcription in vitro and in vivo. To make an in vitro system, we start by grinding up the cells, then slightly purifying their contents to yield a cell-free extract. This extract usually contains RNA polymerases and all the other factors needed to transcribe the DNAs

that we add, as long as these DNAs contain promoters the extract can recognize. In vivo systems come in several types, but all are living cells to which we can add DNAs. We then check by one means or another to see if our added DNA is transcribed. (Both terms are derived from Latin: *vitro* = glass; *vivo* = life.)

When such tests were applied to several genes with mutations in their TATA boxes, something surprising happened. Instead of eliminating transcription, as one might have predicted, TATA box mutations in these genes caused an increase in the number of different transcription start sites. Figure 9.14 shows the results Pierre Chambon and his colleagues got when they tested TATA box mutants of the SV40 early promoter using an in vitro transcription system. Whereas the wild-type DNA gave transcription from three tightly clustered start sites, the TATA-less mutant showed starts at more than twenty different sites.

On the other hand, mutating the TATA box in certain other genes (the rabbit β-globin gene, for example) did not change the transcription initiation site. Instead, it caused a profound decrease in efficiency of transcription. Thus, the TATA box always seems to be important, but the exact effect of its absence varies from one promoter to another.

Upstream Promoter Elements

Similar studies have been performed in which sequences upstream from the TATA boxes in a variety of genes have been deleted or changed. The CCAAT box, for example, seems to influence promoter activity. Thus, removing the CCAAT box from the rabbit β-globin promoter destroys most of that promoter's activity. The CCAAT box plays a similar role in the promoter for the thymidine kinase gene of herpes simplex virus (HSV). The SV40 early promoter also contains an important upstream element; instead of a CCAAT box, it is a 21-base-pair sequence that is repeated three times between positions −40 and −103 just upstream from the TATA box. Each of these 21-bp repeats contains two copies of a **GC box** with the sequence GGGCGG on the coding strand. The HSV thymidine kinase promoter also contains GC boxes: one on each side of the CCAAT box. HSVs, by the way, cause cold sores and genital herpes infections in humans.

As a general rule then, promoters recognized by RNA polymerase II contain a TATA box near the transcription start site and at least one other important element farther upstream. We will see later in this chapter that these upstream elements serve as binding sites for transcription factors that activate the attached genes.

Many eukaryotic genes contain other sequences in the flanking regions (or even within the genes) that are crucial to active transcription. But these are not strictly part of the promoters. We call them enhancers, and we will discuss

(a) In vitro:

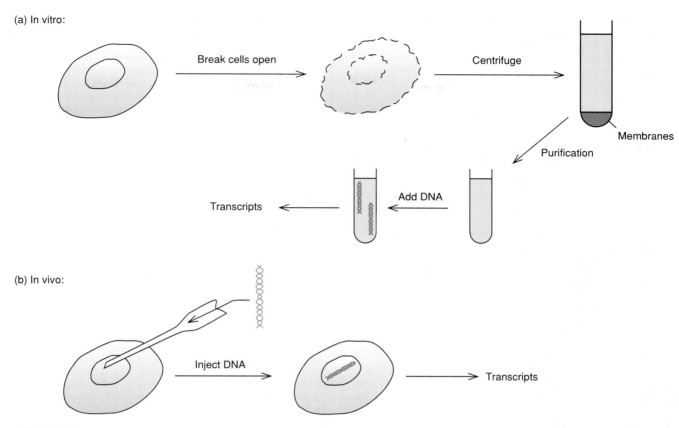

Break cells open

Centrifuge

Membranes

Purification

Transcripts

Add DNA

(b) In vivo:

Inject DNA

Transcripts

FIGURE 9.13

In vitro and in vivo transcription systems.
(*a*) In vitro systems use components from broken cells; (*b*) in vivo
systems use whole cells. In both cases, we introduce foreign DNA (red)
into the system, and measure its transcription.

(a) Wild-type gene:

Transcription

"TATA"

Initiation
sites

FIGURE 9.14

Effect of TATA box on transcription
initiation. (*a*) In the wild-type SV40 early
gene that contains a TATA box,
transcription starts from three tightly
clustered sites. (*b*) In the mutant that lacks
a TATA box, we observe a multitude of
transcription initiation sites.

(b) Gene lacking TATA box:

Transcription

Initiation sites

(a) SV40 early promoter:

21 bp   21 bp   21 bp

(b) Herpes simplex virus thymidine kinase promoter:

(c) Typical polymerase II promoter + enhancer:

Enhancer   Upstream element

FIGURE 9.15

Structures of promoters recognized by RNA polymerase II. (*a*) The SV40 early promoter contains a TATA box (red) and an upstream element composed of three 21-bp repeats. Each of these repeats contains two 6-base GC boxes (green). (*b*) The HSV thymidine kinase promoter contains a TATA box (red) plus three upstream elements: two GC boxes (green) and a CCAAT box (purple). (*c*) A typical polymerase II promoter has a TATA box and at least one upstream element. It is also usually associated with an enhancer (blue), which generally lies upstream from the promoter.

them in detail later in the chapter. Figure 9.15*a* and *b* shows the arrangement of the promoter elements for the SV40 early and HSV thymidine kinase genes. Figure 9.15*c* presents a generalized view of the control region of a typical gene recognized by RNA polymerase II, including both promoter and enhancer. This is the arrangement of these elements in the SV40 early region—a favorite example of a polymerase II promoter-enhancer complex.

> Many eukaryotic promoters that are recognized by RNA polymerase II have at least one feature in common: a consensus sequence found about twenty-five base pairs upstream from the transcription start site and containing the motif TATA. This TATA box acts either to position the start of transcription about 30 base pairs downstream, or to increase transcription efficiency, or both. Polymerase II promoters also contain important sequences upstream from the TATA box. Examples are the CCAAT box and the GC box.

PROMOTERS RECOGNIZED BY POLYMERASE I

Several groups of investigators have searched for conserved sequences and have applied site-directed mutagenesis techniques to elucidate the nature of the promoter recognized by RNA polymerase I. Promoter is written here in the singular because each species should have only one kind of gene recognized by polymerase I: the rRNA precursor gene. It is true that this gene is present in hundreds of copies in each cell, but each copy is virtually the same

UCE   CORE

−156   −107   −45   +1   +20

FIGURE 9.16

The human rRNA promoter. Two DNA elements, UCE (green) and core (red), are important for efficient transcription by RNA polymerase I.

as the others, and they all have the same promoter sequence. Robert Tjian and his colleagues identified two important regions of the human rRNA promoter by mutagenesis (fig. 9.16). By replacing 10 bp at a time with unrelated DNA and observing the effects of these replacements on transcription efficiency, they found a **core element** between positions − 45 and +20, and an **upstream control element (UCE)** between positions −156 and −107, that were required for efficienct transcription. These DNA elements resemble those of polymerase II promoters only in that they appear in the 5′-flanking region of the gene (that is, upstream from the start of transcription); their base sequences do not remind us at all of TATA, GC, or CCAAT boxes. Notice that the end of the gene where transcription begins is called the 5′-end. This refers to the 5′-end of the transcript, or of the coding DNA strand.

> Promoters recognized by polymerase I, like polymerase II promoters, are found in the 5′-flanking region of the gene (the rRNA precursor gene). Two DNA blocks, a core sequence near the transcription start site and an upstream control element (UCE) are required for efficient transcription.

(a) 5S rRNA:

Intermediate element

Box A   Box C

(b) tRNA:

Box A   Box B

FIGURE 9.17
Promoters of two "classical" polymerase III genes. The promoters, or ICRs, are depicted as red boxes within the genes they control.

PROMOTERS RECOGNIZED BY POLYMERASE III

Polymerase III ordinarily transcribes genes that are much shorter than those transcribed by the other two polymerases. Should this make a big difference in the nature of the polymerase III promoters? You might not expect so, but the polymerase III promoters that direct transcription of the tRNA and 5S rRNA genes are radically different from those recognized by polymerases I and II and by bacterial polymerases. Instead of being located in the 5′-flanking region of the gene, as all these other promoters are, these polymerase III promoters have elements located *within* the genes they control.

Donald Brown and his coworkers performed the first experiments leading to this astonishing conclusion using a cloned 5S rRNA gene from *Xenopus laevis,* the South African clawed frog. The experimental plan was to mutate the 5S rRNA gene and measure the effects of these mutations on transcription of the gene in vitro. Correct transcription was scored by measuring the size of the transcripts by gel electrophoresis. An RNA of approximately 120 bases (the size of 5S RNA) was deemed an accurate transcript, even if it did not have the same sequence as real 5S rRNA. We have to allow for incorrect sequence in the transcript for the following reason: By changing the internal sequence of the gene in order to disrupt the promoter, we automatically change the sequence of the transcript.

The surprising result was that the entire 5′-flanking region of the gene could be removed without affecting transcription very much. Furthermore, big chunks of the gene could be removed, or extra DNA could be inserted into the gene, and a transcript of about 120 bases would still be made. However, there was one sensitive spot in the gene where one could not remove or insert DNA and still retain promoter function. This was a region between bases 50 and 83 of the transcribed sequence. This is the location of the promoter, or **internal control region (ICR)** of the *Xenopus* 5S gene.

More detailed analysis has shown that the 5S rRNA promoter is actually split into two major elements, called **box A** and **box C,** as well as a short sequence in the middle, called the **intermediate element.** Figure 9.17 summarizes the results of these experiments and similar ones performed on tRNA genes. The tRNA promoter is more obviously split than the 5S rRNA promoter. Like the 5S rRNA ICR, it has a box A, but it has no intermediate element or box C. Instead, it has a **box B.** The space in between these two elements can be altered without destroying promoter function, but there are limits to such alteration; if one inserts too much DNA between boxes A and B, the efficiency of transcription suffers.

The 5S rRNA and tRNA genes are "classical" polymerase III genes. Polymerase III also transcribes genes encoding other small RNAs, including one called 7SL RNA. Strangely enough, these "nonclassical" genes do not have ICRs. Instead, their promoters lie in the 5′-flanking regions of the genes, and at least one even has a TATA box.

 RNA polymerase III transcribes a set of short genes, including those for 5S rRNA and tRNA. The promoters for these two genes contain internal control regions (ICRs) that lie wholly within the genes. The 5S rRNA gene has a three-part, divided ICR lying between bases 50 and 83 of the transcribed region. The tRNA genes have two-part, divided ICRs.

REGULATION OF TRANSCRIPTION

In chapter 8 we learned that control of gene expression in bacteria is a prerequisite for life; if all genes were switched on all the time, this would consume far more energy than the bacterial cell could provide. If anything, this statement is even more true of eukaryotes. To begin with, there are many more genes in a eukaryotic cell. Moreover, in higher eukaryotes with their complex organs, cells are much more specialized. As we will see in chapter 14, the cells of different tissues have different capabilities *because* different genes are expressed in each. Thus, complex organisms would not be possible if gene expression were uncontrolled.

As we have seen, bacteria control their gene expression primarily at the level of transcription. In general, if a cell does not need the product of a gene, it will not even transcribe that gene. Although there are exceptions, this generalization holds true for eukaryotes. This means that the initiation of transcription is the most important control point of eukaryotic gene expression. It is vital to know, therefore, what causes transcription to initiate. This section will show that many promoters are associated with distinct *cis*-acting DNA regions called enhancers that can greatly stimulate transcription, and some are linked to negative control elements called silencers. Moreover, transcription of eukaryotic genes depends not only on RNA polymerase but on a collection of *trans*-acting proteins called transcription factors that interact with promoters and enhancers.

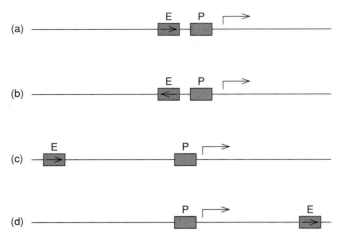

FIGURE 9.18
Enhancers are orientation- and position-independent. (*a*) The wild-type relationship between a typical enhancer (blue) and its associated promoter (red). (*b*) The enhancer has been inverted. (*c*) The enhancer has been moved a few thousand base pairs upstream from the promoter. (*d*) The enhancer has been moved a few thousand base pairs downstream from the promoter. In all four cases, the enhancer functions normally.

ENHANCERS AND SILENCERS

The early site-directed mutagenesis experiments on polymerase II promoters also pointed to very important areas of the 5′-flanking regions, aside from the TATA box and upstream promoter elements we have discussed. One of the best studied of these upstream sequences is a 72-base-pair repeat in the early region of the monkey tumor virus **SV40,** or simian virus number 40. The early region of SV40 codes for a protein called **T antigen** that is needed for viral DNA synthesis and also allows SV40 virus to transform normal cells to malignant ones. In other words, the early region is what makes SV40 a tumor virus. The 72-bp repeat is located just upstream from the three 21-bp repeats in the 5′-flanking region of the SV40 early gene.

The 72-bp repeat is just what its name implies: a side-by-side duplication of a sequence 72 base pairs long. When this element was deleted from the SV40 early region, transcription was greatly depressed. This is exactly the behavior we would expect from part of a promoter, so why did we not include the 72-bp repeat in our discussion of polymerase II promoters? Because this DNA element exhibits two characteristics that are very unpromoterlike: First, as figure 9.18 illustrates, it can be moved to a new location thousands of base pairs away from the SV40 early promoter (either in front of it or behind it) and still enhance early transcription. Second, it can be inverted without altering its effect. For this reason, we classify the 72-bp repeat and other elements with the same characteristics as **enhancers,** which stimulate transcription from promoters but are not really part of the promoters themselves. The term "promoter" is reserved for elements that have a relatively fixed spatial relationship with the genes they control.

Enhancers have been found to be associated with a variety of genes—in eukaryotes as well as their viruses and in genes transcribed by all three RNA polymerases. They are

usually tissue-specific; that is, a gene with a given enhancer can function only in certain kinds of cells. For example, a mouse immunoglobulin enhancer can function in cells that express the immunoglobulin genes (the cells of the immune system), but it does not function in mouse connective tissue cells. Sometimes enhancers have sequences in common, but there are no real consensus sequences shared by all enhancers.

Enhancers are not the only DNA elements that can act at a distance to control transcription of a gene. **Silencers** also do this, but—as their name implies—they prohibit rather than stimulate transcription. The mating system (*MAT*) of yeast provides a good example. The yeast chromosome III contains three loci of very similar sequence: *MAT, HML,* and *HMR*. Though *MAT* is expressed, the other two loci are not, and silencers located at least 1 kb away seem to be responsible for this genetic inactivity. We know that something besides the inactive genes themselves is at fault, because active yeast genes can be substituted for *HML* or *HMR* and the transplanted genes become inactive. Thus, they seem to be responding to some external negative influence. How do silencers work? The best guess at present is that they somehow cause the chromatin to coil up into a condensed, inactive form, thereby preventing transcription of neighboring genes.

> Enhancers are DNA elements that can greatly stimulate transcription from the eukaryotic promoters with which they associate. They differ from promoter elements in that they work in either orientation and can operate at some distance from the genes they control. Silencers also work at a distance and have an effect opposite that of enhancers.

TRANSCRIPTION FACTORS

We learned in chapter 8 that prokaryotic RNA polymerase holoenzyme is sufficient for specific transcription in vitro. For example, recall the experiments performed in the late 1960s showing that purified *E. coli* holoenzyme could transcribe the immediate early genes of purified T4 phage DNA but leave the other genes alone. People working on eukaryotic RNA polymerase in that same era were disappointed to discover that their systems did not work that way. Instead, purified eukaryotic RNA polymerases, when presented with purified viral DNA, responded by transcribing the whole genome promiscuously. Indeed, transcription did not even begin at specific sites but more or less at random. Obviously, something was missing.

In other words, eukaryotic RNA polymerases, unlike their prokaryotic counterparts, are incapable of binding by themselves to their respective promoters. Instead, they rely on proteins called **transcription factors** to show them the way. There are two classes of such factors: **general transcription factors,** and **gene-specific transcription factors.** The general transcription factors, by themselves, can attract the RNA polymerases to their respective promoters. This seems to be sufficient for active transcription of polymerase I and III promoters, but not for promoters

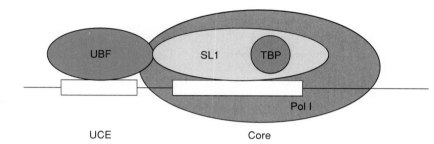

FIGURE 9.19
The class II preinitiation complex. The sizes of the factors are not necessarily to scale, and their arrangement is hypothetical.

FIGURE 9.20
The class I preinitiation complex. The two parts of the rRNA promoter (UCE and Core) are represented by white boxes and are labeled at the bottom.

recognized by polymerase II. Polymerase II promoters are indeed turned on by general transcription factors, but only to a weak extent. They need gene-specific transcription factors to boost their transcription to high levels and thus exert the exquisitely fine control that is typical of these genes.

The general transcription factors combine with their three respective RNA polymerases to form three different **preinitiation complexes,** so-called because they are poised to initiate transcription. Let us examine these preinitiation complexes in the same order we introduced their promoters.

The Class II Preinitiation Complex

The great variety of promoters recognized by RNA polymerase II suggests that there should be a similar variety of polymerase II (class II) general transcription factors, and indeed there is. These factors are involved in transcription of most, and probably all, polymerase II (class II) genes. Because they seem to have such broad specificity, we would predict that they interact with something found in most polymerase II promoters (the TATA box), or with other factors, or with polymerase II itself. Indeed, we will soon see that this prediction has proven true. The class II general transcription factors that cooperate with polymerase II are **TFIIA, TFIIB, TFIID, TFIIE, TFIIF, TFIIH,** and **TFIIJ.**

The best studied of these factors is TFIID. In higher eukaryotes, including mammals, this factor contains many polypeptide subunits, but the most important is the **TATA box-binding protein,** or **TBP.** The other subunits in TFIID are known as **TBP-associated factors,** or **TAFs.** TBP does what its name suggests; it binds to the TATA box, bringing the rest of TFIID (the TAFs) along with it. Once bound to the promoter, TBP (or TFIID) serves as a nucleus around which all the other factors can organize. The factors bind in this order: D, B, F + polymerase II, E, H, and J. TFIIA can apparently bind anytime after TFIID. Polymerase II and TFIIF bind simultaneously. Figure 9.19 shows one possible arrangement of factors in the class II preinitiation complex.

Some class II promoters lack TATA boxes. However, TFIID (with TBP) is still involved in the preinitiation complexes involving these promoters. Presumably, it still plays its organizing role.

> RNA polymerase II is assisted by at least seven general transcription factors, TFIIA, B, D, E, F, H, and J, as well as by gene-specific transcription factors, most of which bind to enhancers associated with certain genes. TFIID contains the TATA box-binding protein, which binds to the promoter's TATA box (if any) and serves as an organizing center for the formation of the preinitiation complex.

The Class I Preinitiation Complex

The preinitiation complex that forms at rRNA promoters is much simpler than the polymerase II preinitiation complex we have just discussed. It involves polymerase I, of course, in addition to just two transcription factors, called **SL1** and an **upstream binding factor, UBF.** Robert Tjian discovered SL1 in 1985 and showed that it does not bind by itself to the rRNA promoter. Later experiments demonstrated that purified polymerase I also cannot bind by itself to the promoter, even with the help of SL1. Finally, in 1988 Tjian and coworkers showed that another factor, UBF, binds to the upstream control element (UCE) of the rRNA promoter and causes polymerase I to bind near the transcription start site. Once UBF binds, SL1 can join the complex, strengthening polymerase I binding, and therefore promoting formation of the preinitiation complex. Interestingly enough, SL1 is similar to TFIID in that it contains TATA box-binding protein (TBP), even though rRNA promoters do not have TATA boxes. In addition to TBP, SL1 has three TAFs, which are different from the TAFs associated with TBP in TFIID. Figure 9.20 shows the class I preinitiation complex.

The class I preinitiation complex contains RNA polymerase I plus two transcription factors. The upstream binding factor (UBF) binds to the UCE and provokes binding of the polymerase I to the core element near the start of transcription. SL1, a TBP-containing factor, strengthens polymerase I binding to the promoter.

The Class III Preinitiation Complex

Donald Brown and Robert Roeder and their colleagues, working independently, have supplied most of our knowledge of the factors that interact with RNA polymerase III to effect the transcription of the 5S rRNA and tRNA genes. Three factors are involved: *TFIIIA, TFIIIB,* and *TFIIIC.* At first, TFIIIA received most of the attention because it bound to the well-known ICR of the 5S rRNA gene. However, it now appears that TFIIIB is the true transcription initiation factor for RNA polymerase III; the other two are assembly factors that allow TFIIIB to bind to its DNA site, which actually lies upstream from the transcription initiation site. TFIIIB then helps RNA polymerase III bind just downstream, in a position to start transcribing the gene. Thus, these three factors participate in a binding cascade, as illustrated in figure 9.21.

First, TFIIIA and C (in the case of the 5S rRNA genes) or TFIIIC alone (in the case of the tRNA genes) bind to the ICR and, presumably by protein–protein interaction, allow TFIIIB to bind near the transcription start site. TFIIIB then positions RNA polymerase III at the transcription start site.

TFIIIB, like SL1 and TFIID, contains TBP plus some TAFs. This is another example of TBP in a preinitiation complex lacking a TATA box (although at least one non-classical class III gene does have a TATA box). Thus, TBP is a universal transcription factor with a general role to play besides TATA box binding. This role is probably in organizing the preinitiation complex.

Classical class III promoters (tRNA and 5S rRNA promoters) rely on two kinds of general transcription factors: assembly factors, TFIIIA and C (5S rRNA gene) or TFIIIC alone (tRNA genes); and an initiation factor TFIIIB, that actually positions polymerase III for transcription initiation.

Gene-Specific Transcription Factors

In addition to these general transcription factors, class II genes interact with a diverse group of *promoter specific factors.* Some of these bind to sequences such as GC boxes and

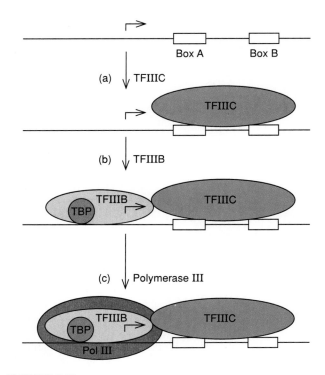

FIGURE 9.21

Hypothetical scheme for assembly of the preinitiation complex on a classical polymerase III promoter (tRNA). (*a*) TFIIIC (green) binds to the internal promoter's A and B boxes. (*b*) TFIIIC promotes binding of TFIIIB (yellow), with its TBP (blue) to the region upstream of the transcription start site. (*c*) TFIIIB promotes polymerase III (red) binding at the start site, ready to begin transcribing.

CCAAT boxes that are present in many mammalian promoters. Therefore, these factors are found in a wide variety of cells—at least in mammals.

For instance, Robert Tjian (pronounced "Tee-jun" by his American colleagues) has purified a protein factor called **Sp1** from human tumor cells (HeLa cells) that greatly stimulates transcription from the SV40 early promoter and from the HSV thymidine kinase promoter. At the same time, this factor actually seems to decrease transcription from the major late promoter of another human virus called **adenovirus.** Thus, this factor has the ability to select certain promoters for transcription while ignoring others.

Structural studies of Sp1 have shown that it contains a repeating domain that is finger-shaped (fig. 9.22). Because

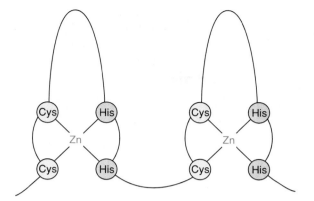

FIGURE 9.22
Postulated structure of two of the nine metal fingers of
TFIIIA. The zinc (magenta) in each finger is bound to
four amino acids: two cysteines (yellow) and two
histidines (blue), holding the finger in the proper shape
for DNA binding. These four key amino acids are
found in the same place in each finger.

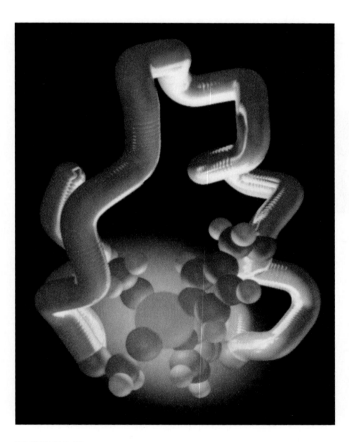

FIGURE 9.23
Three-dimensional structure of one of the zinc fingers of the *Xenopus*
protein *Xfin*. The zinc is represented by the turquoise sphere at bottom
center. The sulfurs of the two cysteines are represented by the yellow
spheres. The two histidines are represented by the blue and green
structures at lower right. The backbone of the finger is represented by the
purple tube.

*Cover of Science from M. S. Lee: Three-Dimensional Solution Structure of a Single
Zinc Finger DNA-Binding Domain published 08/11/89. Vol. 245, beginning on page
635. Copyright 1989 by the AAAS. Photograph provided by Michael Pique and Peter
E. Wright, Dept. of Molecular Biology, Research Institute of Scripps Clinic, La Jolla,
California.*

each of these three finger domains binds a zinc ion, they are
known as **zinc fingers.** Similar zinc fingers have been
found in a variety of DNA binding proteins, including
TFIIIA, the protein in which zinc fingers (nine of them)
were first discovered, and in *Xfin,* an activator of *Xenopus*
polymerase II promoters (fig. 9.23). It appears that zinc fingers assist in DNA binding by inserting into the major
groove of the DNA double helix.

The CCAAT boxes in the HSV thymidine kinase promoter and in other promoters attract their own transcription
factor, called the **CCAAT box transcription factor
(CTF).** Figure 9.24 depicts interactions between two well-
studied polymerase II promoters (SV40 early and herpes
simplex virus [HSV1] thymidine kinase) and their respective transcription factors. Other transcription factors are
more specific. They bind to target sequences found in a relatively small number of genes. They do not necessarily
bind to promoter elements; many bind to enhancers (or
silencers) associated with promoters.

ADDITIONAL COMPLEXITIES OF EUKARYOTIC GENE EXPRESSION

In our discussion of transcription of eukaryotic genes, we
have thus far been ignoring a central fact: Many eukaryotic
genes are fundamentally different from their prokaryotic
counterparts. A prokaryotic structural gene is simple. Once
it begins, it continues uninterrupted until it ends. In fact,
this seems such an obvious way of doing things that nobody imagined eukaryotes would behave any differently,
but they do. Eukaryotic genes are frequently arranged in
pieces; in other words, they are interrupted. In addition,
eukaryotes add extra nucleotides to both ends of many of
their transcripts. In this section we will consider these
extra complexities of eukaryotic gene expression.

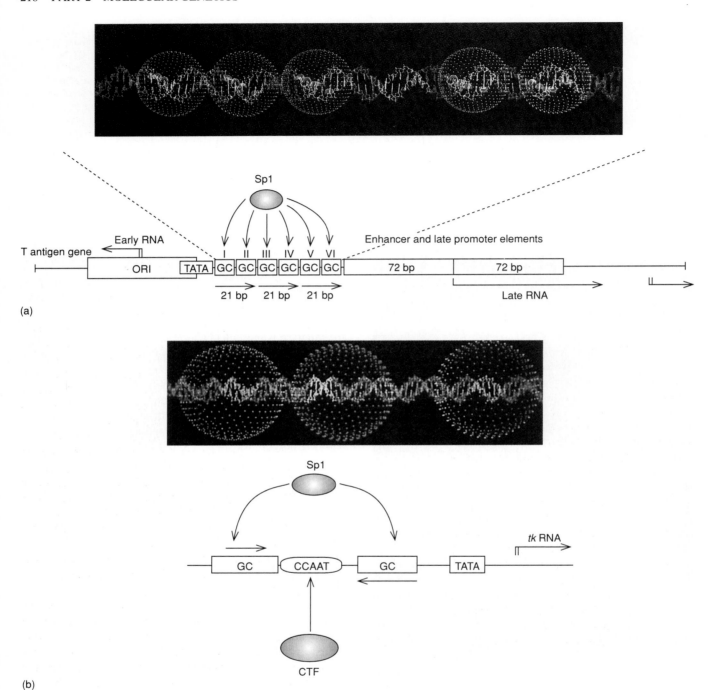

FIGURE 9.24
Interactions of transcription factors with two promoters. (*a*) The drawing
shows the whole SV40 control region, including both late and early
promoters. Note that the early promoter is in the opposite orientation from
that in figure 9.15. The drawing also depicts the interaction between the
transcription factor Sp1 and the six GC boxes. (Box IV shows little if any
interaction under normal conditions.) (*b*) The drawing shows the HSV
thymidine kinase promoter elements and their interaction with the
transcription factors Sp1 and CTF (CCAAT box transcription factor).
From Steven McKnight and Robert Tjian, Cell 46:796, 799, 1986. Copyright © 1986
Cell Press, Cambridge, MA. Reprinted by permission of the publisher and authors.

Box 9.1 *HeLa Cells*

When people see the term *HeLa cells* for the first time, they may think it is a misprint. Shouldn't the *L* be lowercase? No, HeLa is correct; the reason for capitalizing the *H* and the *L* is that the term stands for the name of the woman who donated the cells: Henrietta Lacks. In 1952, George Gey was trying to grow human tumor cells in culture at Johns Hopkins University. No one had

succeeded in this enterprise before, but Gey managed to get human cervical cancer cells to grow in a serum clot. One of these specimens came from Henrietta Lacks, so Gey called the culture HeLa cells. Ms. Lacks died of her cancer not long afterward, but she has achieved a kind of immortality in that her cells are still growing in laboratories all over the world. They are one of the favorite sources of human cells for molecular genetics investigations. In fact, they

grow so easily—almost like weeds—that many researchers have cultured them inadvertently. That is, their cultures of other cells became contaminated with HeLa cells, which then outgrew their neighbors until only HeLa cells remained. This change sometimes went undetected, and several studies, ostensibly on the behavior of other kinds of cells, have actually been investigations of already well-studied HeLa cells.

INTERRUPTED GENES

If we expressed the sequence of the human β-globin gene as a sentence, this is how it might read:

This is bhgty the human β-globin qwtzptlrbn gene.

There are obviously two regions within the β-globin gene that make no sense; they contain sequences totally unrelated to the globin coding sequences surrounding them. These are sometimes called **intervening sequences,** or **IVSs,** but they usually go by the name Walter Gilbert gave them: **introns.** Similarly, the parts of the gene that make sense are sometimes called coding regions, or expressed regions, but Gilbert's name for them is more popular: **exons.** Some genes have no introns at all; others have an abundance. The current record (at least sixty introns) is held by the dystrophin gene, whose failure causes Duchenne muscular dystrophy.

The usefulness of introns remains something of a mystery; fortunately, we can do a lot better at describing them—how we know they are there and what their implications are. Consider the major late locus of adenovirus, a favorite system for studying gene expression and one of the first places introns were found by Phillip Sharp and his colleagues in 1977. Several lines of evidence converged at that time to show that the genes of the adenovirus major late locus are interrupted, but perhaps the easiest to understand comes from studies using a technique called *R-looping.*

In R-looping experiments, we **hybridize** RNA to its DNA template. In other words, we separate the DNA strands and form a double-stranded hybrid between one of these strands and the RNA product. Such a hybrid double-stranded polynucleotide is actually a bit more stable than a double-stranded DNA. After we form the hybrid, we examine it by electron microscopy. There are two basic ways to do these experiments: (1) using DNA whose two strands are separated only enough to let the RNA hybridize, or (2) completely separating the two DNA strands before hybridization. Sharp used the latter method, hybridizing single-stranded adenovirus DNA to mature mRNA for one of the viral coat proteins: the hexon protein. Figure 9.25 shows the results.

What would we expect to see if there were no introns in the hexon gene? In that case, there should be a smooth, linear hybrid where the mRNA lined up with its DNA template. But what if there *are* introns in this gene? Clearly, there can be no introns in the mature mRNA, or nonsense would appear in the protein product. Therefore, introns are sequences that occur in the DNA but are missing from mRNA. That means the hexon DNA and hexon mRNA will not be able to form a smooth hybrid. Instead, the intron regions of the DNA will not find partners in the mRNA and so will form unhybridized loops. Of course, that is exactly what happened in the experiment shown in figure 9.25.

The electron micrograph shows an RNA-DNA hybrid interrupted by three single-stranded DNA loops (labeled A, B, and C). These loops represent the introns in the hexon gene. Each loop is preceded by a short hybrid region, and the last loop is followed by a long hybrid region. This means there are four exons in the gene: three short ones near the beginning, followed by one large one. The three short exons are transcribed into **leader** regions that appear at the 5′-end of the hexon mRNA before the coding region; the long exon contains the coding region of the gene.

So far we have only discussed introns in mRNA genes, but many tRNA genes also have introns, and even rRNA genes sometimes do. The introns in both these latter types of genes are a bit different from those in mRNA genes. For example, tRNA introns are relatively small, ranging in size from four to about fifty bases long. Not all tRNA genes have introns; those that do have only one, and it is always in the same place.

 Most eukaryotic genes coding for mRNA and tRNA, and a few coding for rRNA, are interrupted by unrelated regions called intervening sequences, or introns. The other parts of the gene, surrounding the introns, are called exons; the exons contain the sequences that finally appear in the mature RNA product. Genes for mRNAs have been found with anywhere from zero to more than sixty introns. Transfer RNA genes have either zero or one.

FIGURE 9.25

R-looping experiments reveal introns in adenovirus. (*a*) Electron micrograph of a cloned fragment of adenovirus DNA containing the 5′-part of the late hexon gene, hybridized to mature hexon mRNA. The loops represent introns in the gene that cannot hybridize to mRNA. (*b*) Interpretation of the electron micrograph, showing the three intron loops (labeled A, B, and C), the hybrid (heavy red line), and the unhybridized region of DNA upstream from the gene (upper left). The fork at the lower right is due to the 3′-end of the mRNA, which cannot hybridize because the 3′-end of the gene is not included. Therefore, the mRNA forms intramolecular double-stranded structures that have a forked appearance. (*c*) Linear arrangement of the hexon gene, showing the three short leader exons, the two introns separating them (A and B), and the long intron (C) separating the leaders from the coding exon of the hexon gene. All exons are represented by red boxes.

Photo (a) from Susan M. Berget, Claire Moore, and Phillip Sharp. Proceedings of the National Academy of Sciences, vol. 74, p. 3173, 1977. (b) and (c) from Susan M. Berget, et al., Proceedings of the National Academy of Sciences, vol. 74, p. 3173, 1977. Reprinted by permission of the author.

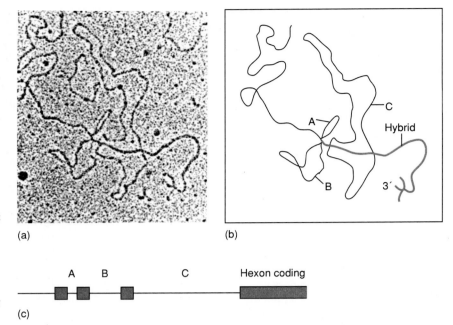

(a)   (b)

(c)

RNA Splicing

Consider the problem introns pose. They are present in genes but not in mature RNA. How is it that the information in introns does not find its way into the final products of the genes? There are two main possibilities: (1) The introns are never transcribed; the polymerase somehow jumps from one exon to the next and ignores the introns in between. (2) The introns are transcribed, yielding a **primary transcript,** an overlarge gene product that is cut down to size by removing the introns. As wasteful as it seems, the latter possibility is the correct one. The process of cutting introns out of immature RNAs and stitching together the exons to form the final product is called **RNA splicing.** The splicing process is outlined in figure 9.26, although, as we will see later in the chapter, this picture is considerably oversimplified. The mRNA precursors belong to a class of large nuclear RNAs called **heterogeneous nuclear RNA (hnRNA).**

 Messenger RNA synthesis in eukaryotes occurs in stages. The first stage is synthesis of the primary transcription product, an mRNA precursor that still contains introns copied from the gene, if any were there. This precursor is part of a pool of large nuclear RNAs called hnRNAs. The second stage is mRNA maturation. Part of the maturation of an mRNA precursor is the removal of its introns in a process called splicing. This yields the mature-sized mRNA.

Splicing Signals

Pause to consider how important it is that splicing be accurate. If too little RNA is removed from an mRNA precursor, the mature RNA will be interrupted by "garbage." If too much is removed, important sequences may be left out.

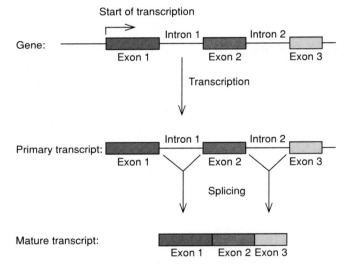

FIGURE 9.26

Outline of splicing. The introns in a gene are transcribed along with the exons (colored boxes) in the primary transcript. Then they are removed as the exons are spliced together.

Given the importance of accurate splicing, signals must occur in the mRNA precursor that tell the splicing machinery exactly where to "cut and paste." What are these signals? One way to find out is to look at the base sequences of a number of different genes, locate the intron boundaries, and see what sequences they have in common. In principle, these common sequences could be part of the signal for splicing. The most striking observation, first made by Chambon, is that almost all introns in nuclear mRNA precursors begin and end the same way:

exon/GU-intron-AG/exon

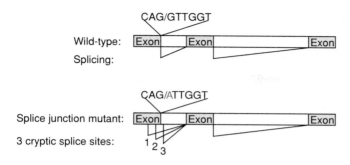

Cryptic site 1: AAG/GTGAAC
Cryptic site 2: GTG/GTGAGG
Cryptic site 3: AAG/GTTACA
Wild-type: CAG/GTTGGT

FIGURE 9.27
Importance of consensus sequences at splice junctions. With the wild-type splice junctions, splicing proceeds normally; when the first G of the intron changes to A (blue), splicing no longer occurs at that location, but at three alternate, cryptic sites. The DNA sequences of the three cryptic splice junctions are shown, with the wild-type sequence below for comparison. Since these are DNA sequences, the introns begin with GT instead of GU.

In other words, the first two bases in the intron of a transcript are GU and the last two are AG. This kind of conservation does not occur by accident; surely the GU—AG motif is part of the signal that says: "Splice here." However, GU– – –AG motifs occur too frequently. A typical intron will contain several GU's and AG's within it. Why are these not used as splice sites? The answer is that GU– – –AG is not the whole story; there is more to it than that. By this time, the sequences of many intron/exon boundaries have been examined and a consensus sequence has emerged:

$$5'\text{-}{}^C_A\,AG/GU\,{}^A_G\,AGU\text{-intron-}Y_N NAG/G\text{-}3'$$

where Y_N denotes a string of about nine pyrimidines (U's and C's) and N is any base.

Again, finding consensus sequences is one thing; showing that they are really important is another. Several research groups have found ample evidence supporting the importance of these splice junction consensus sequences. Their experiments were of two basic types. In one, they mutated the consensus sequences at the splice junctions in cloned genes, then checked whether proper splicing still occurred. In the other, they collected defective genes from human patients with presumed splicing problems and examined the genes for mutations near the splice junctions. Both approaches gave the same answer: Disturbing the consensus sequences usually destroys normal splicing.

For example, consider the β-globin gene in several cases of human β-**thalassemia,** an often fatal disease in which no functional β-globin mRNA appears in the cytoplasm. This can be for a number of different reasons; in the case cited here, it is because the patient carried a β-globin gene with defective splice junctions. Figure 9.27 shows the sequence of the mutated β-globin gene of this thalassemia

patient and the consequences of this mutation. In this case, a single base change destroyed a normal splice site at the beginning of the first intron, so the splicing system used a series of alternate sites ("cryptic sites") where splicing does not normally occur. Two of these are upstream from the normal site, within the first exon; the other is downstream, in the intron.

The sequences of the cryptic sites resemble that of the normally used site. In this way, the splicing machinery is like a child who always buys the same candy bar until one day that particular brand is sold out. For the first time, the child notices other brands that don't look too bad, and buys not one, but three different kinds. These alternate brands had been cryptic, or hidden, to the child before—even though they were in plain sight—because the favorite was there.

Unfortunately for thalassemia patients, these cryptic splice sites are in the wrong places, so the wrong sequences are removed from the mRNA precursors. In the first two cases, parts of exon 1 are spliced out; in the third case, part of intron 1 is left in. In none of these cases is functional mRNA produced. Thus, even though the globin coding sequence may have been perfectly normal in this mutated gene, the fact that its transcript could not be spliced properly rendered it nonfunctional, with life-threatening consequences.

 The splicing signals in nuclear mRNA precursors are remarkably uniform. The first two bases of the intron are almost always GU, and the last two are almost always AG. In addition, the 5′- and 3′-splice sites have consensus sequences that are important to proper splicing; when they are mutated, abnormal splicing occurs.

Small Nuclear RNAs
Joan Steitz and her coworkers first focused attention on a class of **small nuclear RNAs (snRNAs)** as potential agents in splicing. These snRNAs exist in the cell coupled with proteins in complexes called **small nuclear ribonucleoproteins (snRNPs),** usually referred to informally as "snurps." Steitz noticed that one snRNA, called U1, has a region whose sequence is almost perfectly complementary to both 5′- and 3′-splice site consensus sequences. In principle, this would allow base pairing between U1 RNA and an mRNA precursor. In practice, such base pairing does occur, but probably only at the 5′-splice site.

How do we know that snRNPs are involved in splicing? One line of evidence comes from in vitro splicing studies using antibodies directed against snRNPs. Antibodies react with a specific substance and frequently interrupt its normal function. Therefore, if snRNPs are needed for splicing, one would expect that antibodies directed against snRNPs would prevent splicing. This is what the experiments showed.

We also know that splicing of nuclear mRNA precursors occurs on a complex particle called a **spliceosome.** The mammalian spliceosome is about the size of a large ribosomal subunit (50–60S). It includes the mRNA precursor as well as a set of snRNPs: U2, U4, U5, and U6. The fact

FIGURE 9.28
Lariat splicing model. In reaction 1, the bond between the first exon (blue) and the intron breaks, and the terminal G of the intron bonds to an A within the intron, forming a lariat-shaped intermediate. The branched nucleotide (A) in the intron is shown in parentheses at right. In reaction 2, the first exon (blue) is spliced to the second (yellow), liberating the lariat-shaped intron (red).

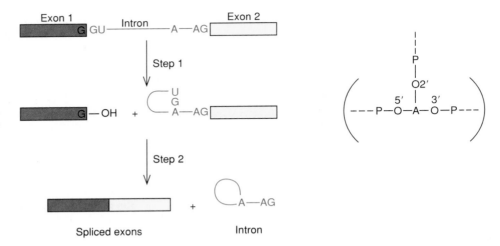

that U1 snRNP is absent from the spliceosome does not mean U1 is not involved in splicing; excellent evidence indicates that it *is* involved and has already served its function by the time the spliceosome forms.

A Lariat-Shaped Intermediate

The detailed mechanism by which cells splice nuclear mRNA precursors seems at first quite illogical. Figure 9.28 illustrates this two-step **lariat model** of mRNA splicing. The first step is the formation of a looped intermediate that looks like a lariat or lasso. This occurs when a nucleotide in the middle of the intron attacks the G at the beginning of the intron, forming the loop, and simultaneously breaking the phosphodiester bond between the first exon and the intron. The splicing is completed when the 3′-end of the first exon attacks the 5′-end of the second exon, forming the exon-exon phosphodiester bond and releasing the intron, in lariat form, at the same time.

 A class of small nuclear RNAs (snRNAs) appears to be important in splicing nuclear mRNA precursors. In particular, U1 snRNA has a region that is complementary to the consensus splice site sequences in mRNA precursors. This RNA has also been shown to base-pair with mRNA precursors at the 5′-splice site in vivo. The mechanism of mRNA precursor splicing involves intramolecular attack by a nucleotide within the intron on the G at the 5′-end of the intron, breaking the 5′-exon/intron bond and creating a lariat-shaped intermediate. Then the two exons are joined, releasing the intron, still in its lariat shape.

OTHER RNA PROCESSING EVENTS

Splicing is one example of something that happens to RNA between the time it is transcribed and the time it is exported to the cytoplasm for use in protein synthesis. But several

other things occur during this **posttranscriptional** period. Some of them are peculiar to mRNA; others pertain to rRNA and tRNA only; and some involve all three types of RNA. All these events can be lumped together under the heading **RNA processing.** Among these processing events are: (1) trimming RNA precursors outside the coding regions; (2) methylating (adding methyl groups to) RNA; (3) adding nucleotides to the 3′-ends of RNAs; and (4) blocking the 5′-ends of mRNAs with special structures called caps.

Trimming Ribosomal RNA Precursors

The ribosomal RNA genes in eukaryotes are repeated several hundred times and clustered together in the nucleolus of the cell. Their arrangement in amphibians has been especially well studied, and as figure 9.29a shows, they are separated by regions called **nontranscribed spacers.** This is to distinguish them from **transcribed spacers,** regions of the gene that are transcribed as part of the rRNA precursor and then removed during processing of the precursor to mature rRNA species.

This clustering of the reiterated rRNA genes in the nucleolus made them easy to find and therefore provided Oscar Miller and his colleagues with an excellent opportunity to observe genes in action. These workers looked at amphibian nuclei with the electron microscope and uncovered a visually appealing phenomenon, shown in figure 9.29b. The DNA containing the rRNA genes can be seen winding through the picture, but the most obvious feature of the micrograph is a series of "Christmas tree" structures. These include the rRNA genes (the trunk of the tree) and growing rRNA transcripts (the branches of the tree). We must remember that these transcripts are actually rRNA precursors, not mature rRNA molecules (more about this later). The spaces between "Christmas trees" are of course the nontranscribed spacers. You can even

Box 9.2 Splicing, Ribozymes, and the Origin of Life

Self-Splicing rRNA Precursors

The protozoan *Tetrahymena* is unusual in that one of its rRNA genes contains an intron. As we will see, the large rRNAs from all eukaryotes are encoded in a common gene. In other words, the 5.8S, 18S, and 28S rRNAs, or their equivalents, are transcribed together on a precursor RNA, which is cut, or **processed,** to liberate the mature RNAs. Part of this processing involves removing a spacer RNA between the 18S and 28S species, but this spacer is not an intron, and its removal is not splicing. The difference is that the two RNA species remain independent after processing; they do not join together the way exons do. However, the 26S part of the *Tetrahymena* rRNA precursor does contain an intron, and it can be spliced in a very shocking way: The RNA splices itself, without any help from outside enzymes!

Figure B9.2.1*a* shows the *Tetrahymena* rRNA precursor. It includes the 17S, 5.8S, and 26S rRNA sequences, as well as regions in between that must be removed. It also includes the intron within the 26S region. To examine the processing of this molecule, Thomas Cech (pronounced "Check") and his collaborators cloned a piece of its gene (indicated with a bracket in figure B9.2.1*a*) and transcribed this gene fragment with *E. coli* RNA polymerase. This gave a piece of RNA containing the intron. And, much to everyone's surprise, this little piece of RNA could splice itself.

Figure B9.2.1*b* presents the splicing scheme. In the first step, a guanosine nucleotide (GMP, represented here as GOH) attacks the 5′-end of the intron, breaking the exon/intron bond. Then the 3′-end of the first exon (UOH) attacks the 5′-end of the second exon, joining the two exons (the splicing step) and removing the intron in linear form. The boundaries of this intron do not look at all like those of a nuclear mRNA intron.

The intron in the *Tetrahymena* nuclear rRNA precursor belongs to a class called **group I introns.** These all self-splice with the help of a guanosine

FIGURE B9.2.1

Self-splicing of *Tetrahymena* rRNA precursor. (*a*) Structure of the rRNA precursor, containing the 17S, 5.8S, and 26S sequences. Note the intron within the 26S region (red). The cloned segment used in subsequent experiments is indicated by a brace. (*b*) Self-splicing scheme. In the first step (top), GMP attacks the adenine nucleotide at the 5′-end of the intron, releasing exon 1 (black) from the rest of the molecule and generating the hypothetical intermediates shown in brackets. In the second step, exon 1 (black) attacks exon 2 (blue), performing the splice reaction that releases a linear intron (red) and joins the two exons together. Finally, in a series of reactions not shown here, the linear intron loses 19 bases from its 5′-end.

nucleotide, which initiates the first step. Another class of introns, called **group II introns,** also self-splice, but they do not need assistance from guanosine to start the reaction. Instead, the initiating entity is an adenosine nucleotide within the intron of the RNA itself. When this attacks the bond between the first exon and the intron, it forms a lariat, just as we have observed for nuclear mRNA splicing. The main difference between nuclear mRNA precursors and RNAs bearing group II introns is that the latter can handle the splicing all by themselves, without spliceosomes.

It is always easier to believe something if we can see it with our own eyes. In the case of group II introns, Arnberg and colleagues have provided us with electron micrographs (fig. B9.2.2) that actually show the lariat-shaped splicing intermediate (intron plus exon 2) and product (intron). These RNAs come from splicing an rRNA precursor in yeast. This is not the rRNA precursor of the yeast nuclei; instead, it comes from the mitochondria, the energy-producing organelles of the cell.

(*continued on next page*)

Box 9.2 (continued)

Self-splicing RNAs, besides tickling the fancy of molecular geneticists, have a number of interesting implications. One of these is the fact that an RNA can act like an enzyme, at least in that it has catalytic activity. It had been widely assumed that, of the biological molecules, only proteins had this ability. This in turn may help explain how life originated. If we assume that only proteins can act as enzymes, the origin of life is plagued by a catch-22: The first organism needed proteins to catalyze the replication of its genes, but it could not make proteins without catalysts. In other words, the origin of life would require the simultaneous and spontaneous appearance of both nucleic acids and proteins. Catalytic nucleic acids help explain away this paradox: RNA could have served as both genome and catalyst in the earliest forms of life.

Cech has extended his earlier studies to show that the intron, once excised from the rRNA precursor, still has enzymatic activity. It can both form and break phosphodiester bonds in RNA molecules, and it can do this in a sequence-specific manner. For instance, the only RNAs that can be cleaved are those that have a sequence complementary to a special "guide sequence" in the catalytic intron. Since this intron can catalyze a chemical reaction and is not itself changed in the process, it is a true enzyme. However, since it is made of RNA instead of protein, we call it a **ribozyme.**

Amber Beaudry and Gerald Joyce have taken this process one step further and have replicated the catalytic intron in vitro under conditions that favored

mutation. This produced a heterogeneous mixture of introns having slightly differing sequences. Then they selected for mutant RNAs that had the ability to cut DNA, in addition to RNA, in a sequence-specific manner. The result was an altered ribozyme that could cut DNA about a hundred times better than the original intron. This looks like evolution in vitro.

Even more exciting achievements no doubt lie ahead. One of the most electrifying would be to produce an RNA molecule with the ability to replicate itself. This would not only be a model for the origin of life on earth, it would satisfy some people's definition of creating life in a test tube.

FIGURE B9.2.2
Lariat-shaped splicing intermediate and product. Electron micrographs of intermediate and product of yeast mitochondrial rRNA splicing provide direct visual evidence for lariats. (Top) Two lariat-shaped introns. (Bottom) Two lariat-shaped intermediates containing an intron plus the second exon.
Annika Arnberg, Gerda Van der Horst, and Henk F. Tabak, Cover of Cell, *Vol. 44, No. 2, January 31, 1986.*

tell the direction of transcription from the lengths of the transcripts within a given gene; the shorter RNAs are at the beginning of the gene and the longer ones are at the end.

Ribosomal and transfer RNAs are sometimes called **stable RNAs** because they are so much longer-lived than mRNA in the cell. It is sometimes useful to lump rRNA and tRNA together this way, because they behave differently than mRNA in several respects. One of these is in processing. We have seen that mRNA precursors frequently require splicing but no other trimming. On the other hand,

rRNAs and tRNAs first appear as precursors that sometimes need splicing, but they also have excess nucleotides at their ends, or even between regions that will become separate mature RNA sequences (e.g., 5.8S, 18S, and 28S rRNA sequences). These excess regions must also be removed.

The rRNA precursors of vertebrates are more typical than the one found in *Tetrahymena* because they contain no introns. Figure 9.30 depicts the structure and scheme of

(a)

(b)

FIGURE 9.29

Transcription of rRNA precursor genes. (*a*) A portion of the newt (amphibian) rRNA precursor gene cluster, showing the alternating rRNA genes (blue) and nontranscribed spacers (NTS, red). (*b*) Electron micrograph of part of a newt nucleolus, showing rRNA precursor transcripts (T) being synthesized in a "Christmas tree" pattern on the tandemly duplicated rRNA precursor genes (G). At the base of each transcript is a polymerase I, not visible in this picture. The genes are separated by nontranscribed spacer DNA (NTS).

(b) From O. L. Miller, Barbara R. Beatty, Barbara Hankalo, and C. A. Thomas. Symposia on Quantitative Biology *(Cold Spring Harbor), vol. 35, p. 506, 1970.*

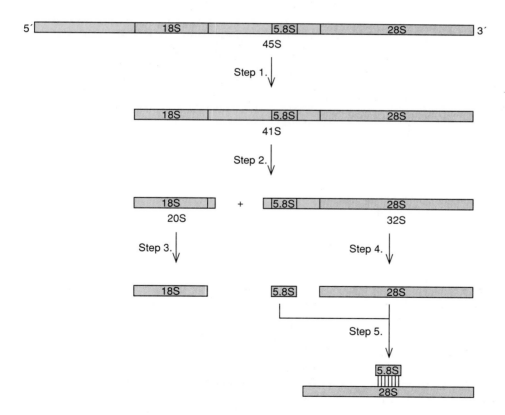

FIGURE 9.30

Processing scheme of 45S human (HeLa) rRNA precursor. Step 1: The 5′-end of the 45S precursor RNA is removed, yielding the 41S precursor. Step 2: The 41S precursor is cut into two parts, the 20S precursor to the 18S rRNA, and the 32S precursor to the 5.8S and 28S rRNAs. Step 3: The 3′-end of the 20S precursor is removed, yielding the mature 18S rRNA. Step 4: The 32S precursor is cut to liberate the 5.8S and 28S rRNAs. Step 5: The 5.8S and 28S rRNAs associate by base pairing.

processing of the 45S human rRNA precursor. It contains the 28S, 18S, and 5.8S sequences, embedded between transcribed spacer RNA regions. The processing of this precursor occurs in the nucleolus. The first step is to cut off the spacer at the 5′-end, leaving a 41S intermediate. The next step involves cleaving the 41S RNA into two pieces, 32S and 20S, that contain the 28S and 18S sequences, respectively. The 32S precursor also retains the 5.8S sequence. Finally, the 32S precursor is split to yield the mature 28S and 5.8S RNAs, which base-pair with each other, and the 20S precursor is trimmed to mature 18S size.

The details of this processing scheme are not universal; even the mouse does things a little differently, and the frog precursor is only 40S, which is quite a bit smaller than 45S. Still, the basic mechanism of rRNA processing, including the order of mature sequences in the precursor, is preserved throughout the eukaryotic kingdom.

Poly(A) Tails
We have already seen that hnRNA is a precursor to mRNA. One finding that suggested such a relationship between these two types of RNA was that they shared a unique structure at their 3′-ends: a long chain of AMP residues called **poly(A)**. Neither rRNA nor tRNA has a poly(A) tail. James Darnell and his coworkers performed much of the early work on poly(A) and the process of adding poly(A) to RNA (called **polyadenylation**). They released the poly(A) intact from the rest of the mRNA and found it to be an average of 150 to 250 nucleotides in length.

It is apparent that the poly(A) goes on the 3′-end of the mRNA or hnRNA because it can be released very quickly with an enzyme that degrades RNAs from the 3′-end inward. We also know that poly(A) is added posttranscriptionally, because there are no runs of T's in the DNA long enough to encode it. Instead, there is an enzyme in nuclei called **poly(A) polymerase** that adds AMP residues one at a time to mRNA precursors once they are complete. We know that poly(A) is added to mRNA precursors because it is found on hnRNA. When we look closely, we can even see that specific unspliced mRNA precursors (the 15S mouse globin mRNA precursor, for example) contain poly(A). Once in the cytoplasm, the poly(A) on an mRNA "turns over"; in other words, it is constantly being broken down by RNases and rebuilt by a cytoplasmic poly(A) polymerase.

Most mRNAs contain poly(A). One famous exception is the histone mRNAs, which somehow manage to perform their function without detectable poly(A) tails. The near-universality of poly(A) suggests that it has an important function. Actually, poly(A) seems to have two functions. It helps protect the vulnerable mRNA from degradation by ribonucleases lurking in the cell, and it stimulates translation of many mRNAs.

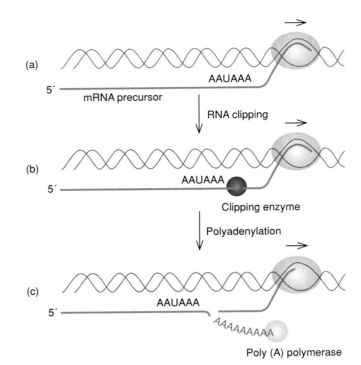

FIGURE 9.31
Polyadenylation. (*a*) The mRNA precursor (blue) is being synthesized by RNA polymerase II (orange). (*b*) A clipping enzyme (purple) recognizes the polyadenylation signal (AAUAAA) and cuts the growing RNA about twenty bases downstream. (*c*) The poly(A) polymerase (yellow) adds AMP residues (red) stepwise to the new 3′-end, as transcription continues. The transcript beyond the polyadenylation site is presumably discarded.

Most eukaryotic mRNAs carry a poly(A) tail about 150–250 bases long on their 3′-ends. Poly(A) polymerase adds this tail to mRNA precursors in the nucleus as soon as transcription is complete, and the poly(A) apparently remains with the message through splicing and transport to the cytoplasm. Poly(A) seems to protect mRNA from degradation.

Polyadenylation
It would be logical to assume that poly(A) polymerase simply waits for a transcript to be finished, then adds poly(A) to its 3′-end. But this is not the way it happens, at least in most cases. Instead, the mechanism of polyadenylation usually involves clipping an mRNA precursor, sometimes even before transcription has terminated, and then adding poly(A) to the newly exposed 3′-end (fig. 9.31). This means that, contrary to expectations, polyadenylation and transcription termination are not necessarily linked. RNA polymerase II can still be elongating an RNA chain as it approaches a termination signal, while the polyadenylating apparatus has already located a polyadenylation signal, cut the growing RNA, and polyadenylated it.

FIGURE 9.32

Cap 1 structure at the 5′-end of an mRNA. The methyl groups of the cap are shown in red. X and Y represent the bases on the second and third nucleotides, respectively. Cap 0 would have no methyl group on the penultimate ribose. Cap 2 (rare) would have both methyl groups of cap 1, plus another on the ribose linked to base Y.

FIGURE 9.33

Capping and methylating mechanism. Reaction 1: GTP caps the mRNA precursor; the blue p indicates that only one of GTP's phosphate groups is preserved in the cap. Reaction 2: A methyl transferase transfers a methyl group (red) from SAM to the 7-position of the capping guanine. Reaction 3: Another methyl transferase methylates (gray) the 2′-hydroxyl group of the penultimate nucleotide.

What constitutes a polyadenylation signal? One consensus sequence appears in the transcript about twenty bases upstream from the polyadenylation site. This is usually AAUAAA in higher eukaryotes, although deviations, especially AUUAAA, are also found. When the base sequences of 269 mRNAs were compared, the "correct" bases were found in the consensus region with the following frequencies:

$$A_{98}A_{86}U_{98}A_{98}A_{95}A_{96}$$

In other words, some "incorrect" sequences are tolerated, the most common of which, AUUAAA, is found in 12% of

polyadenylation signals. Another common feature is a U,G-rich sequence surrounding the polyadenylation site itself.

> Polyadenylation occurs by clipping a transcript upstream from its 3′-end, sometimes while transcription is still in progress, then adding poly(A) to the newly created 3′-end. The signal for such clipping and polyadenylation, at least in higher eukaryotes, includes the consensus sequence AAUAAA about 20 bases upstream from the polyadenylation site, plus a U,G-rich sequence near the polyadenylation site.

Eukaryotic mRNA Cap Structures

The 3′-end of a eukaryotic messenger is not the only place that differs from its prokaryotic counterpart. The 5′-end also has something different—a **cap.** The structure of a cap can vary somewhat, but its typical architecture is shown in figure 9.32. The cap contains a methylated guanine nucleotide linked by a triphosphate to the penultimate nucleotide, or next-to-last nucleotide. The penultimate nucleotide is frequently methylated also, but the methyl group is on the 2′-hydroxyl group of the ribose instead of on the base.

The capping process is illustrated in figure 9.33. The growing RNA has a triphosphate group at its 5′-end, just as a prokaryotic mRNA does. Then, before the RNA has grown to fifty bases long, a capping enzyme adds a GTP to its 5′-end. This reaction is unlike the normal RNA polymerase reaction in several respects: (1) The incoming nucleotide adds to the 5′-end of the RNA chain instead of the 3′-end. (2) Instead of a simple phosphodiester bond forming between the two end nucleotides, a triphosphate linkage forms. (3) The triphosphate does not link the 3′-site of one nucleotide with the 5′-site of the next; instead, the two end nucleotides are linked through their 5′-sites. Thus, the capping nucleotide "backs in" to join the growing RNA chain. Two enzymes, called **methyl transferases,** perform the methylation that finishes capping. They transfer methyl groups from the methyl donor, **S-adenosyl methionine (SAM),** to the appropriate positions on the cap. Note that, strictly speaking, capping cannot be considered a posttranscriptional process, because it occurs before transcription is complete. It might be better to call it a "cotranscriptional" process.

Yasuhiro Furuichi and Aaron Shatkin and their colleagues first discovered caps in viral mRNA in 1974. As frequently happens when such a surprising discovery occurs, one wondered whether it would be a general phenomenon or an isolated curiosity. It turns out that caps are ubiquitous among eukaryotes. What, then, is the function of these widespread structures? Furuichi showed that caps have two important roles. First, they protect mRNA from degradation, just as poly(A) does. Caps and poly(A) can be envisioned as tough blocking structures at each end of the mRNA, protecting the tender midsection from attack by ribonucleases. Second, caps are necessary for binding mRNA to the ribosome. Little if any translation of decapped messages occurs.

FIGURE 9.34
Levels of gene expression in eukaryotes.

The 5'-ends of eukaryotic mRNAs are blocked with structures called caps. A cap is added by the capping enzyme, which joins a GTP through a 5'-5'-triphosphate linkage to the penultimate nucleotide. Methyl transferases then add methyl groups to the 7-nitrogen of the terminal guanine and to the 2'-hydroxyl group of the penultimate ribose. Caps serve a dual purpose: They protect messages from degradation, and they allow them entry to ribosomes for translation.

TYPES OF GENE EXPRESSION CONTROL

By now, you can readily appreciate that gene expression in eukaryotes is much more complex than in prokaryotes. This extra complexity gives a eukaryotic cell more opportunities to control gene expression. In principle, control could occur at the transcriptional, posttranscriptional, or translational level.

TRANSCRIPTIONAL CONTROL

We will see examples of eukaryotic gene control at all three levels (fig. 9.34), but it seems that the most common level in the eukaryotes, as in the prokaryotes, is transcription. Why is this? One reason is that transcription is by far the most economical level at which to control genes. It allows for genes to be transcribed only if their products are needed. Consider the alternatives. If there were no transcriptional control, all genes would be transcribed equally. Then those unneeded transcripts would either fail to be processed properly or, if they were processed, would fail to be translated. In either case, the surplus transcripts would be degraded. This would be grossly wasteful, especially considering the energy it takes to produce RNA. In fact, it is doubtful that an organism could survive by transcribing all its genes all of the time.

However, nature does not always do things in what we would regard as the most sensible way, so we need to ask: What is the evidence that transcription is the primary level of control in eukaryotes? The best evidence comes from direct examination of transcription products in differentiated cells. To begin with, we can look at the levels of mature mRNA in the cytoplasms of various differentiated cells.

For example, we can show that red blood cells contain about 50,000 molecules of globin mRNA, compared to less than one molecule in other types of cells. Does this mean that the globin genes are not turned on in the other types of cells? It would be tempting to say so, but there are other possible explanations. For example, perhaps the globin gene is active in all types of cells, but the globin RNA is degraded before it reaches the cytoplasm of all cells except red blood cells.

Table 9.2 Synthesis of Ovalbumin mRNA in Various Tissues[1]

Nuclei From	Ovalbumin mRNA (% of Total)
Oviduct + DES	0.238
Oviduct − DES	0.002
Spleen	<0.001
Liver	<0.001

[1]The limit of detectability was 0.001%.
Data from Swaneck et al., *Proceedings of the National Academy of Sciences,* 76:1049, 1979.

Table 9.3 Effect of Prolactin on Half-Life of Casein mRNA[1]

Species	RNA Half-Life (H_r)	
	Without Prolactin	*With Prolactin*
rRNA	>90	>90
Poly(A)$^+$ RNA (short-lived)	3.3	12.8
Poly(A)$^+$ RNA (long-lived)	29	39
Casein mRNA	1.1	28.5

[1]Prolactin has relatively little effect on rRNA and total mRNA (poly[A]$^+$RNA), but causes a great increase in the half-life of casein mRNA.
Data from Guyette, et al., *Cell* 17:1013, 1979.

In order to prove our thesis, we need to examine newly synthesized RNA before it has a chance to be degraded. That means looking in the nucleus, not in the cytoplasm. In most cases where this has been done, we find evidence for transcriptional control. For example, red blood cells make vast quantities of globin (about ten million molecules per cell) but no ovalbumin. Accordingly, their nuclei contain many transcripts of the globin genes, but almost no transcripts of the ovalbumin gene. Conversely, oviduct cells make great amounts of ovalbumin, but no globin; as expected, their nuclei contain abundant ovalbumin transcripts, but almost no globin transcripts.

In table 9.2 we see that spleen and liver nuclei support no detectable ovalbumin mRNA synthesis. On the other hand, oviduct nuclei from chickens exposed to the hormone diethylstilbestrol (DES) make a considerable amount of ovalbumin mRNA. Nuclei from chickens that have been withdrawn from DES for three days drop back to a barely detectable level of ovalbumin mRNA synthesis. These results are certainly not compatible with a model in which all genes are transcribed equally in all cells. On the contrary, it appears that DES stimulates transcription of the ovalbumin gene in oviduct cells by at least a factor of 100. In other words, a great deal of transcriptional control occurs.

 Specialized genes of eukaryotes are controlled, at least primarily, at the transcription level. They are significantly active only in those differentiated cells where their protein products are produced.

POSTTRANSCRIPTIONAL CONTROL

Cells could, in principle, control their gene expression at the posttranscriptional level in several different ways. For example, they could add caps or poly(A) tails to certain mRNAs, but not to others; or they could transport certain mRNAs, but not others, to the cytoplasm; or they could selectively degrade certain mRNAs or their precursors. In practice, cells do not seem to use all of these potential control mechanisms: (1) Cells do not withhold capping from certain types of messages. If they did, we would expect to find some uncapped messages, or at least message precursors, but virtually all mRNA precursors are capped at a very early time after transcription initiation. (2) With one exception, cells seem not to withhold polyadenylation from certain types of messages. The lone exceptions are the histone mRNAs, which never seem to be polyadenylated in higher eukaryotes.

One way that posttranscriptional control does appear to occur is by selective degradation of mRNAs in the cytoplasm. The response of breast tissue to the hormone prolactin provides a good example of this mechanism. When cultured breast tissue is stimulated with prolactin, it responds by producing the milk protein casein. One would expect an increase in casein mRNA concentration to accompany this casein buildup, and it does. The number of casein mRNA molecules increases about 20-fold in twenty-four hours following the hormone treatment. But this does not mean the rate of casein mRNA synthesis has increased 20-fold. In fact, it only increases about 2- to 3-fold. The rest of the increase in casein mRNA level depends on an approximately 20-fold increase in stability of the casein mRNA.

An experiment was performed to measure the **half-life** of casein mRNA. The conclusion, shown in table 9.3, was that the half-life of casein mRNA increased dramatically, from 1.1 hours to 28.5 hours, in the presence of prolactin. At the same time, the half-life of total polyadenylated mRNA increased only 1.3- to 4-fold in response to the hormone. It appears prolactin causes a selective stabilization of casein mRNA that is largely responsible for the enhanced expression of the casein gene.

 Of the possible mechanisms of posttranscriptional control, there is good evidence only for altering mRNA stability. Differential capping and polyadenylation appear not to be significant means of controlling gene expression.

TRANSLATIONAL CONTROL

Maternal messages in eggs provide good examples of how genes can be controlled at the translational level. Maternal messages are mRNAs made in great quantity during **oogenesis,** the development of an egg, then stored, complexed with protein, in the egg cytoplasm. These stored mRNAs are abundant enough that they can support the

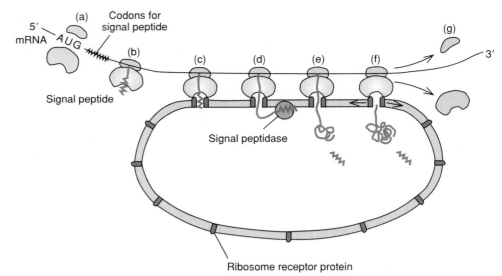

FIGURE 9.35

Signal peptide hypothesis. (*a*) A ribosome (yellow) attaches to the 5′-end of an mRNA. (*b*) The first part of the protein to be made is the signal peptide (blue) at the amino terminus. (*c*) The signal peptide attaches to ribosome receptor proteins (green) on the outside of the endoplasmic reticulum (large oval) and begins to draw the growing protein into the interior. (*d*) The signal peptide is removed by a signal peptidase (orange) within the endoplasmic reticulum. (*e*) The protein (red) continues to grow and fold. (*f*) The completed protein moves completely into the endoplasmic reticulum. (*g*) The ribosome dissociates from the mRNA.

From P. Walter, et al., Cell *38:6, 1984. Copyright © 1984 Cell Press, Cambridge, MA. Reprinted by permission of the publisher and authors.*

early development of the embryo even in the absence of transcription. Maternal messages in sea urchin eggs are translated only poorly until fertilization, when their translation increases at least 50-fold. We know that this represents enhanced translation of maternal messages, because the increase occurs even when new RNA synthesis is blocked, or even in cells that lack nuclei.

Surf clam eggs are also interesting in this regard. The translation rate does not change after fertilization, but the spectrum of messages translated does change. We can assay the total population of maternal messages in these eggs before and after fertilization by isolating them, translating them in vitro, and displaying their protein products by gel electrophoresis. The same bands appear before and after fertilization, showing that the same mRNAs are present. Next, we can look at the proteins produced from these mRNAs in vivo before and after fertilization. The spectrum of proteins made in vivo changes after fertilization, even though the mRNA population does not change. In other words, something is controlling which messages are read.

❖❖ The expression of some genes, at least in early embryogenesis, can be controlled at the level of translation.

POSTTRANSLATIONAL CONTROL

In principle, at least, there is one more level at which gene expression can be controlled. This is the **posttranslational** level, or what happens to a protein after it is released from the ribosome. Cells modify and decorate proteins in a variety of ways. Proteins can be clipped by proteases and thus converted from immature to mature form, or they can be coupled to sugars, phosphate groups, or other compounds.

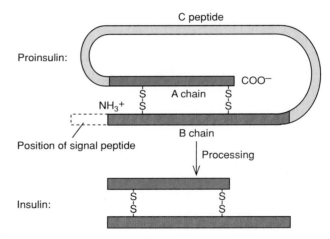

FIGURE 9.36

Converting proinsulin to insulin. Proinsulin consists of the B chain (red), the C peptide (yellow), plus the A chain (blue). The A and B chains are linked through two disulfide bonds (S–S). The signal peptide (dashed lines) has already been removed during proinsulin synthesis. The conversion of proinsulin to insulin is accomplished by excising the C peptide.

One type of clipping actually occurs *during* synthesis of certain proteins. It involves the removal of a stretch of about twenty amino acids at the protein's **amino terminus**—the end with the free amino group, where protein synthesis began. This portion of the protein, called the **signal peptide,** anchors the nascent protein and its polysome to membranes called **endoplasmic reticulum (ER),** which simply means "network in the cell." As figure 9.35 demonstrates, the signal peptide also serves as a "needle" to pull the protein "thread" into the interior of the ER. Once inside, the signal peptide is removed. The shortened protein moves through the ER to the **Golgi apparatus,** where it is packaged for export. Therefore, this type of processing is characteristic of proteins that are destined for export from a cell.

The classic example of **proteolytic** (protein-cleaving) processing is provided by the hormone insulin, which appears first as a precursor called **preproinsulin.** After the signal peptide of preproinsulin is removed, we call it **proinsulin,** because there is still clipping to be done. During this latter process, an enzyme clips out a part of the proinsulin called the **C peptide,** leaving two pieces, the **A peptides** and **B peptides** that comprise mature insulin (fig. 9.36). If these clipping steps did not occur, active insulin would not

be produced, and expression of the insulin gene would be blocked as surely as if it were never transcribed. Insulin gene expression is apparently not controlled this way, but expression of some other genes does seem to be.

Many proteins experience posttranslational alterations such as proteolytic clipping or addition of sugars. Because these alterations are frequently important for the formation of an active protein, they are potential steps at which gene expression can be controlled.

WORKED-OUT PROBLEMS

Problem 1

Here is the sequence of the 5′-end of the coding strand of a hypothetical mammalian gene recognized by RNA polymerase II:

5′CGATGCCGTATATAACGATGCATCCAGTACG-TAGCATCCTCACTGA

Identify the most likely transcription start site (cap site).

Solution

First find the TATA box. There is a sequence (TATATAA) near the 5′-end of the sequence that matches closely the TATA box consensus sequence (TATAAAA). Next, since the cap site of a gene transcribed by mammalian RNA polymerase II usually lies about 30 bases downstream (to the 3′-side) of the TATA box, count 30 bases to the 3′-side of the middle of this sequence (the second A). This brings us to a region (CCTCACTGA) in which the boldfaced **A** is exactly 30 bases from the middle of the TATA box. Since transcription usually begins with a purine, this is the most likely start site, although the G and A just downstream are also candidates, and could be alternate start sites in a real gene.

Problem 2

Here is the structure of a simple gene with one intron.

The boxes represent exons, and the line, the intron. It is possible to hybridize the DNA from such a gene (including DNA sequences surrounding the gene) to the mRNA product of the gene, then visualize the hybrid by electron microscopy. In the following sections (*a*)–(*d*), show schematically what the hybrids would look like. Label the RNA and DNA parts of the hybrid, and show the locations of the exons and introns. Ignore the poly(A) in parts (*a*)–(*c*). (*a*) If this hybridization were done with single-stranded DNA and mature, spliced mRNA, what would the hybrid look

like? (*b*) If the hybridization were done with double-stranded DNA and mature, spliced mRNA, what would the hybrid look like? (*c*) If the hybridization were done with double-stranded DNA and the unspliced mRNA precursor, what would the hybrid look like? (*d*) What would the hybrid in part (*c*) look like if the poly(A) on the mRNA precursor is considered?

Solution

(*a*) Since the intron has been spliced out of the mRNA, the intron in the single-stranded DNA has nothing to hybridize with, and it will loop out as follows:

(*b*) Since the intron has been spliced out of the mRNA, the intron in the double-stranded DNA has nothing to hybridize with, and it will loop out. On either side of this double-stranded intron loop, we can see RNA hybridizing to one strand of the DNA and displacing the other DNA strand in what we call an R-loop.

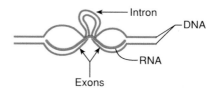

(*c*) Since the mRNA precursor has not been spliced, the intron still remains in the RNA, and it will hybridize to the intron in the DNA. Of course, the exons will also hybridize, so the hybrid will show one long R-loop.

(d) The hybrid will be the same as in part (c), but the mRNA will have a poly(A) tail that will not hybridize to the DNA, because it is not encoded in the DNA. Thus, the hybrid will look like this:

Problem 3

Here is the structure of a gene you are studying:

The boxes represent exons (e_1, e_2, and e_3) and the single lines represent introns (i_1 and i_2). Draw the results of an R-looping experiment performed on this gene, using mature mRNA and single-stranded DNA, encompassing the whole gene as well as its 5′- and 3′- flanking regions. Provide two diagrams: (a) a simple version of how the actual experimental result would appear, with single-stranded DNA represented as a thin line and RNA-DNA hybrid as a thick line; and (b) an interpretive diagram showing the DNA as a solid line and the mRNA as a dashed line. Ignore the poly(A). Label the parts of your diagrams (e_1, i_1, etc.).

Solution

Messenger RNA is already processed to remove the introns, so it will hybridize only to the exons in the single-stranded DNA. The introns in the DNA will therefore form loops whose sizes correspond to the sizes of the respective introns.

(a) The electron micrograph of the RNA-DNA hybrid will show the loops, but it will not show the difference between the DNA and RNA strands. Thus, the R-loops will look like this:

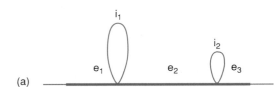

Notice that the exons are represented by thick lines, which denote RNA-DNA hybrids, whereas the introns are thin lines, denoting single-stranded DNA that had no RNA to which they could hybridize. Notice also that the relative sizes of the hybrid regions and loops correspond to the relative sizes of the exons and introns.

(b) The interpretive diagram shows the mRNA (dashed line) hybridizing to the exons, but not the introns:

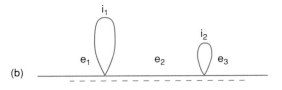

Summary

Eukaryotic chromosomes consist of chromatin: DNA combined with five basic proteins called histones and a large group of nonhistone proteins. This chromatin is organized by several types of folding. The first order of chromatin folding is represented by a string of nucleosomes. These structures contain four pairs of histones (H2A, H2B, H3, and H4) in a ball, around which are wrapped about 200 base pairs of DNA. The fifth histone, H1, binds to DNA outside the ball. The second order of chromatin folding involves a coiling of the string of nucleosomes into a 30-nm-thick fiber called a solenoid. This folding depends on histone H1. The third order of folding is probably looping of the 30-nm fiber into a brushlike structure with the loops anchored to a central matrix. Chromatin exists in eukaryotic cells in two forms: heterochromatin, which is condensed and transcriptionally inactive, and euchromatin, which is extended and at least potentially active.

Eukaryotic cells contain three distinct RNA polymerases, named polymerase I, polymerase II, and polymerase III. The roles of these three enzymes are as follows: Polymerase I makes a large rRNA precursor; polymerase II makes mRNA precursors; and polymerase III makes tRNA and 5S rRNA precursors. Eukaryotic promoters that are recognized by RNA polymerase II have at least one feature in common: a consensus sequence found about 30 base pairs upstream from the transcription start site and containing the motif TATA. This TATA box acts either to position the start of transcription about 30 base pairs downstream, or to increase efficiency of transcription, or both. Polymerase II promoters generally contain other important sequences upstream from the TATA box. These sequences are typically recognition sites for specific activating proteins called transcription factors. Many eukaryotic genes also contain, or reside close to, position- and orientation-independent elements called enhancers. These are not promoter elements, but they strongly stimulate transcription of nearby genes. Polymerase I promoters also lie in the 5′-flanking regions of the genes they control, but the classical polymerase III promoters lie wholly within their genes.

Most eukaryotic genes coding for mRNA and tRNA, and a few coding for rRNA, are interrupted by unrelated regions called intervening sequences, or introns. The other parts of the gene, surrounding the introns, are called exons. Part of the maturation of an mRNA precursor is the removal of its introns in a process called splicing.

The splicing signals in eukaryotic mRNA precursors are remarkably uniform. The first two bases of the intron are almost always GU and the last two are almost always AG. In addition, there are consensus sequences at the 5′- and 3′-splice sites.

The rRNA precursor of the protozoan *Tetrahymena* contains an intron in the 28S part of the molecule. This intron can be spliced out in vitro by the RNA precursor itself. Some mitochondrial mRNA precursors are also self-splicing.

Most eukaryotic mRNAs carry a poly(A) tail about 150–250 bases long on their 3′-ends. Poly(A) polymerase adds this tail to mRNA precursors in the nucleus, and the poly(A) apparently remains with the message through splicing and transport to the cytoplasm. Poly(A) seems to protect mRNA from degradation. The 5′-ends of eukaryotic mRNAs are blocked with structures called caps. Caps serve a dual purpose; they protect messages from degradation and they allow them entry to ribosomes for translation.

Specialized genes of eukaryotes are controlled primarily at the transcription level, by interacting with regulatory proteins. Other levels at which gene expression is controlled are posttranscription and translation. In principle, it can also be controlled at the posttranslation level.

PROBLEMS AND QUESTIONS

1. Why do changes in polymerase III promoters change the sequence of the gene product?

2. Cite two lines of evidence suggesting that nucleosomes are important to life.

3. Without histone H1, would the 11-nm chromatin fiber be possible? The 30-nm fiber? What is the rationale for your predictions?

4. In each somatic cell of a female mammal, one X chromosome is euchromatic and the other is heterochromatic. Which chromosome carries the active genes?

5. What important characteristic do eukaryotic transcription factors have in common with prokaryotic σ factors? In what important respect do they differ?

6. Here is the structure of a gene you want to examine:

The boxes represent exons and the single lines, introns. Draw a picture of the results of an experiment on this gene in which you hybridize mature, spliced mRNA and single-stranded DNA. Label the parts of your diagram (e_1, i_1, etc.).

7. What would the experimental results look like if the gene in problem 6 were left in double-stranded form and hybridized to mRNA? Assume that all the introns are big enough to produce a visible loop.

8. What would the results be if the experiment in problem 6 were run with an unspliced mRNA precursor?

9. If mRNA is hybridized to the gene that encodes it, does the poly(A) hybridize to the gene? Why, or why not?

10. Draw a diagram of the "Christmas tree" pattern of transcription of the rRNA gene cluster in amphibian nucleoli. Point out: (*a*) an rRNA precursor gene; (*b*) a nontranscribed spacer; (*c*) an rRNA precursor being synthesized; (*d*) an RNA polymerase; (*e*) the direction of transcription.

11. What would a cap look like in the absence of methylation?

12. If you are interested in demonstrating whether control of expression of a given gene is exerted at the transcription level, why is it not sufficient to sample mRNA levels in the cytoplasm before and after that gene's expression is activated?

13. Here is the sequence of the 5′-end of the coding strand of a hypothetical eukaryotic gene transcribed by RNA polymerase II:

5′AATTGCGTATAAATTCGAGACTACTGATCCTACG-
GCTACCCGTTAGCT

Identify the most likely start (cap) site.

Answers appear at end of book.

Initiation components

Initiation

1. mRNA binds to small subunit

Initiation

2. Initiator fMet-tRNA binds to mRNA codon

Elongation

3. Large subunit binds to initiation complex

Elongation

4. Second aminoacyl-tRNA enters A site

Peptide bond formation
catalyzed by peptidyl transferase

5. Peptide bond formation

Elongation

6. Translocation of dipeptidyl-tRNA and mRNA. Deacylated tRNA leaves P site.

EFG + GTP → EFG + GDP + P

7. Third aminoacyl-tRNA enters A site

Elongation

8. Repetition of peptide bond formation to add third amino acid

9. Polypeptide chain completed

Termination

10. Termination: Deacylated tRNA and polypeptide chain released

11. Dissociation of components

TRANSLATION: PROTEIN SYNTHESIS

A diagrammatic representation of the process of protein synthesis in *E. coli*.

PANEL 7: A BASIC LOOK INTO THE CELL

_____

1. _____

_____

2. _____

_____

3. _____

**A REVIEW OF THE COMPONENTS OF
TRANSLATION: PROTEIN SYNTHESIS
STEPS 1–3**

CHAPTER 10: TRANSLATION

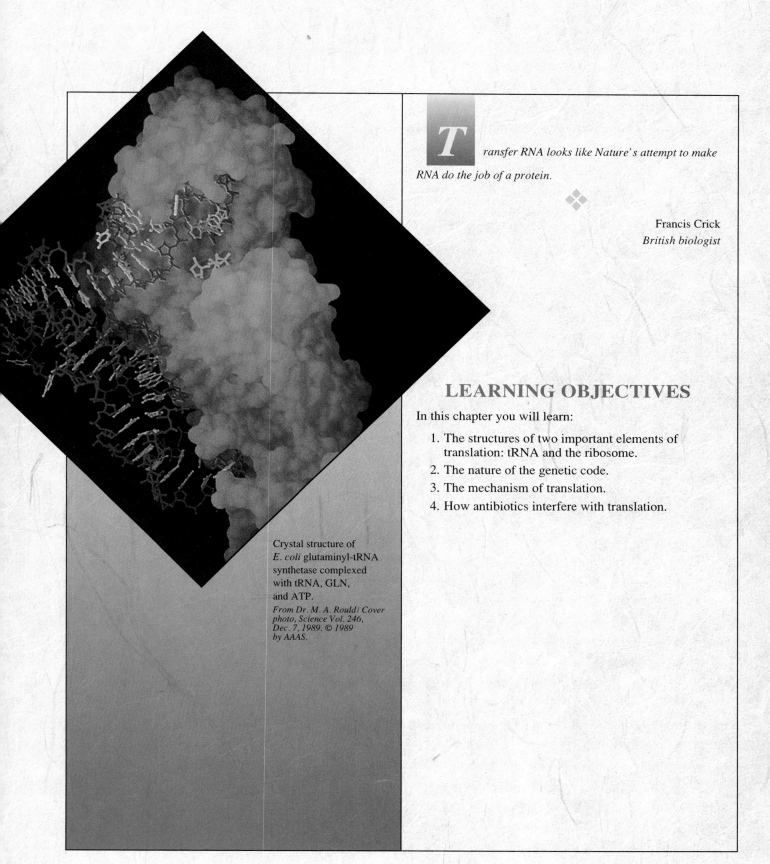

Crystal structure of
E. coli glutaminyl-tRNA
synthetase complexed
with tRNA, GLN,
and ATP.
*From Dr. M. A. Rould/ Cover
photo, Science Vol. 246,
Dec. 7, 1989. © 1989
by AAAS.*

*T ransfer RNA looks like Nature's attempt to make
RNA do the job of a protein.*

❖

Francis Crick
British biologist

LEARNING OBJECTIVES

In this chapter you will learn:

1. The structures of two important elements of
   translation: tRNA and the ribosome.

2. The nature of the genetic code.

3. The mechanism of translation.

4. How antibiotics interfere with translation.

FIGURE 10.1
Amino acid structure. (*a*) The general structure of an amino acid. It has both an amino group (NH_3^+; red) and an acid group (COO^-; blue); hence the name. Its other two positions are occupied by a proton (H) and a side chain (R, gray). (*b*) There are twenty different side chains on the twenty different amino acids, all of which are illustrated here.

FIGURE 10.2

Formation of a peptide bond. Two amino acids with side chains R and R′ combine through the acid group of the first and the amino group of the second to form a dipeptide, two amino acids linked by a peptide bond. One molecule of water also forms as a by-product.

Peptide bond

ost genes are blueprints for proteins; that is, each gene carries the coded information for making one protein chain. The expression of this genetic information as a protein is a two-step process: **transcription,** or RNA synthesis, and then **translation,** or protein synthesis. In this chapter we will examine translation in more detail. To introduce this topic, let us briefly explore the nature of proteins.

PROTEIN STRUCTURE AND FUNCTION

Proteins, like nucleic acids, are chainlike polymers of small subunits. In the case of DNA and RNA, the links in the chain are nucleotides. The chain links of proteins are **amino acids.** Whereas DNA contains only four different nucleotides, proteins contain twenty different amino acids. The structures of these compounds are shown in figure 10.1. The amino acid subunits join together in proteins via **peptide bonds,** as shown in figure 10.2. Another name for a chain of amino acids is **polypeptide.**

The linear order of amino acids constitutes a protein's **primary structure.** The way these amino acids interact with their neighbors gives a protein its **secondary structure.** The α-*helix* is a common form of secondary structure. It results from hydrogen bonding among near-neighbor amino acids, as shown in figure 10.3. Another common secondary structure found in proteins is the β-pleated sheet. This involves extended protein chains, packed side-by-side, that interact by hydrogen bonding. The packing of the chains next to each other creates the sheet appearance. Silk is a protein very rich in β-pleated sheets. A third example of secondary structure is simply a turn. Such turns connect the α-helices and β-pleated sheet elements in a protein.

The total three-dimensional shape of a polypeptide is its **tertiary structure.** Figure 10.4 illustrates how the protein myoglobin folds up into its tertiary structure. Elements of secondary structure are apparent, especially the several α-helices of the molecule and the turns that connect them. Note the overall roughly spherical shape of myoglobin. Most polypeptides take this form, which we call **globular.** Many proteins are large enough that they contain more than one compact structural region. Each of these regions is called a **domain**. Antibodies (the proteins our white blood cells make to repel invaders) provide a good example of domains. Each of the four polypeptides in the IgG-type antibody contains globular domains, as shown in figure 10.5.

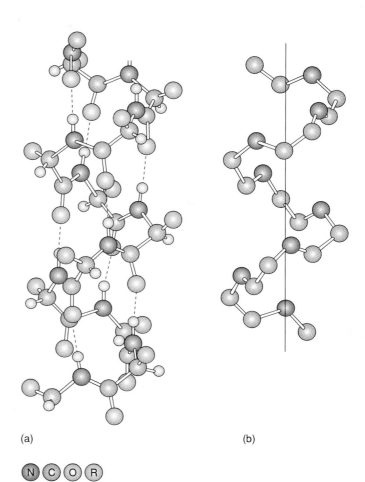

(a) (b)

Ⓝ Ⓒ Ⓞ Ⓡ

FIGURE 10.3

An example of protein secondary structure: the α-helix. (*a*) The positions of the amino acids in the helix are shown, with the helical backbone in black and red. The dotted lines represent hydrogen bonds between hydrogen and oxygen atoms on neighboring amino acids. The small white circles represent hydrogen atoms. (*b*) A simplified rendition of the α-helix, showing only the atoms in the helical backbone.

Reprinted from Linus Pauling, The Nature of the Chemical Bond, *3rd edition. Copyright 1939 and 1940, third edition © 1960 by Cornell University. Used by permission of the publisher, Cornell University Press.*

This figure also illustrates the highest level of protein structure: **quaternary structure.** This is the way two or more individual polypeptides fit together in a complex protein. Note that the word "protein" is somewhat ambiguous. It can refer to a single polypeptide, or a collection of polypeptides that associate to perform their function. The term "polypeptide" is more precise. It refers to a single chain of amino acids.

FIGURE 10.4

Tertiary structure of myoglobin. The several α-helical regions of this protein are represented by blue-green corkscrews. The overall molecule seems to resemble a sausage, twisted into a roughly spherical or globular shape. The heme group is shown in red, bound to two histidines (green polygons) in the protein.

From R. E. Dickerson, The Proteins, 2d ed. Copyright © 1964 Academic Press, Orlando, FL. Reprinted by permission of the publisher and author.

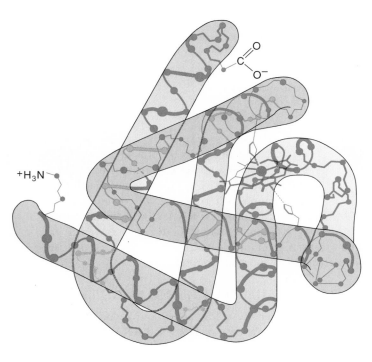

FIGURE 10.5

The globular domains of immunoglobulin. (*a*) Schematic diagram, showing the four polypeptides that comprise the immunoglobulin: two light chains (L) and two heavy chains (H). The light chains each contain two globular regions, while the heavy chains have four globular domains apiece. (*b*) Space-filling model of immunoglobulin. The colors correspond to those in (*a*). Thus, the two H-chains are in red and blue, while the L-chains are in green and yellow. A complex sugar attached to the protein is shown in gray. Note the globular domains in each of the polypeptides. Also note how the four polypeptides fit together to form the quaternary structure of the protein.

(a)

(b)

FIGURE 10.6
Pathway of phenylalanine breakdown. Alcaptonuria patients are defective in the enzyme that converts homogentisate to 4-maleylacetoacetate.

Scientists have long assumed that a protein's amino acid sequence determines all of its higher levels of structure, much as the linear sequence of letters in this book determines word, sentence, and paragraph structure. However, this is an oversimplification. Many proteins cannot fold properly by themselves and require other proteins called chaperones to help them assume their proper shape.

Why are proteins so important that most genes are dedicated to their creation? Some proteins provide the structure that helps give cells integrity and shape. Others serve as

hormones to carry signals from one cell to another. For example, the pancreas secretes the hormone insulin that signals liver and muscle cells to take up sugar from the blood. Proteins can bind and carry substances. The protein hemoglobin carries oxygen from the lungs to remote areas of the body; myoglobin stores oxygen in muscle tissue until it is used. Proteins also bind to genes and control their activity, as we have seen in chapters 8 and 9. Perhaps most important of all, proteins serve as enzymes that catalyze the myriad chemical reactions necessary for life. Left to themselves, these reactions would take place far too slowly at ordinary body temperatures; without enzymes, life as we know it would be impossible.

> Proteins are polymers of amino acids linked through peptide bonds. The sequence of amino acids in a polypeptide (primary structure) gives rise to that molecule's local shape (secondary structure), overall shape (tertiary structure), and interaction with other polypeptides (quaternary structure). Proteins serve many vital functions, as catalysts, carriers, hormones, and structural elements.

GENES AND PROTEINS

Our knowledge of the gene-protein link dates back as far as 1902, when Archibald Garrod, a physician, noticed that a human disease, alcaptonuria, behaved as if it was caused by a single recessive gene. Fortunately, Mendel's work had been rediscovered two years earlier and provided the theoretical background for Garrod's observation. Patients with alcaptonuria excrete copious amounts of a compound called homogentisate, which has the startling effect of coloring their urine black. Garrod reasoned that the abnormal buildup of this substance resulted from a defective metabolic pathway. Somehow, a blockage somewhere in the pathway was causing the intermediate, homogentisate, to accumulate to abnormally high levels, much as a dam causes water to accumulate behind it. Several years later, Garrod postulated that the problem came from a defect in the pathway that degrades the amino acid phenylalanine (fig. 10.6).

By that time, metabolic pathways had been studied for years and were known to be controlled by enzymes—one enzyme catalyzing each step. Thus, it seemed that alcaptonuria patients carried a defective enzyme. And since the disease was inherited in a simple Mendelian fashion, Garrod concluded that a gene must control the enzyme's production. When that gene is defective, it gives rise to a defective enzyme. This established the crucial link between genes and proteins.

George Beadle and E. L. Tatum carried this argument a step further with their studies of a common bread mold, *Neurospora*, in the 1940s. They performed their experiments as follows: First, they bombarded the peritheca (spore-forming parts) of *Neurospora* with X rays or

FIGURE 10.7

Pathway of pantothenic acid synthesis. The last step (step 3), formation of pantothenic acid from the two half-molecules, pantoate acid (blue) and β-alanine (red), was blocked in one of Beadle and Tatum's mutants. The enzyme that carries out this step must have been defective.

ultraviolet radiation to cause mutations. Then, they collected the spores from the irradiated mold and germinated them separately to give pure strains of mold. They screened many thousands of strains to find a few mutants. The mutants revealed themselves by their inability to grow on minimal medium composed only of sugar, salts, inorganic nitrogen, and the vitamin biotin. Wild-type *Neurospora* grows readily on such a medium; the mutants had to be fed something extra—a vitamin, for example—in order to survive.

Next, Beadle and Tatum performed biochemical and genetic analyses on their mutants. By carefully adding substances, one at a time, to the mutant cultures, they pinpointed the biochemical defect. For example, the last step in the synthesis of the vitamin pantothenate involves putting together the two halves of the molecule: pantoate and β-alanine. One "pantothenateless" mutant would grow on pantothenate, but not on the two halves of the vitamin. This demonstrated that the last step in the biochemical pathway leading to pantothenate was blocked, so the enzyme (enzyme 3 in fig. 10.7) that carries out that step must have been defective.

The genetic analysis was just as straightforward. *Neurospora* is an ascomycete, in which nuclei of two different mating types fuse and undergo meiosis to give eight haploid ascospores, borne in a fruiting body called an ascus (see chapter 5). A mutant can therefore be crossed with a wild-type strain of the opposite mating type to give eight spores (fig. 10.8). If the mutant phenotype is due to a mutation in a single gene, then four of the eight spores should be mutant and four should be wild-type. Beadle and Tatum collected the spores, germinated them separately, and checked the phenotypes of the resulting molds. Sure enough, they found that four of the eight spores gave rise to mutant molds, demonstrating that the mutant phenotype was controlled by a single gene. This happened over and over again, leading these investigators to the conclusion that each enzyme in a biochemical pathway is controlled by one gene. Subsequent work has shown that many enzymes contain more than one polypeptide chain and that each polypeptide is usually encoded in one gene. This is the **one gene–one polypeptide hypothesis.**

 Most genes contain the information for making one polypeptide.

FIGURE 10.8

Sporulation in the mold *Neurospora crassa.* (*a*) Two haploid nuclei, one wild-type (yellow) and one mutant (blue), have come together in the immature fruiting body of the mold. (*b*) The two nuclei begin to fuse. (*c*) Fusion is complete, and a diploid nucleus (green) has formed. One haploid set of chromosomes is from the wild-type nucleus and one set is from the mutant nucleus. (*d*) Meiosis occurs, producing four haploid nuclei. If the mutant phenotype is controlled by one gene, two of these nuclei (blue) should have the mutant allele and two (yellow) should have the wild-type allele. (*e*) Finally, mitosis occurs, producing eight haploid nuclei, each of which will go to one ascospore. Four of these nuclei (blue) should have the mutant allele and four (yellow) should have the wild-type allele. If the mutant phenotype is controlled by more than one gene, the results will be more complex.

DISCOVERY OF MESSENGER RNA

Let us consider the expression of genes whose products are proteins. The term "code" and its derivatives, aside from providing a useful metaphor to describe the language of the gene, are legitimate scientific terms. Thus, when a gene contains the information for making a certain protein, we say that it **encodes** or **codes for** that protein. These genes are expressed in two major steps: transcription and translation.

Transcription means making an RNA copy of a gene. Translation means that the information in this RNA, still "written" in genetic code, is used to make protein. Again, the terms are apt. In everyday language, transcription refers to making a close copy of an original, just as RNA is very similar to DNA. By contrast, translation means rendering something in a different language, which is exactly what happens when the genetic code in RNA is translated to the amino acid code of protein.

The RNA copy, or **transcript,** of a structural gene is called a **messenger RNA (mRNA).** It carries the information, or **message,** for making the protein from the gene to the cell's protein factories, the **ribosomes.**

The concept of a messenger RNA carrying information from gene to ribosome developed in stages during the years following the publication of Watson and Crick's DNA model. In 1958, Crick himself proposed that RNA serves as an intermediate carrier of genetic information. He based his hypothesis in part on the fact that the DNA resides in the nucleus of eukaryotes, whereas proteins are made in the cytoplasm. This means that something must carry the information from one place to the other. Crick noted that ribosomes contain RNA and suggested that this ribosomal RNA (rRNA) is the information bearer. But rRNA is an integral part of ribosomes; it cannot escape. Therefore, Crick's hypothesis implied that each ribosome, with its own rRNA, would produce the same kind of protein over and over. This turned out to be wrong.

François Jacob and Jacques Monod proposed an alternate hypothesis calling for nonspecialized ribosomes that translate unstable RNAs called **messengers.** The messengers are independent RNAs that bring genetic information from the genes to the ribosomes. In 1961, Jacob, along with Sydney Brenner and Matthew Meselson, probed the validity of the messenger hypothesis. This study used the same bacteriophage (T2) that Hershey and Chase had employed almost a decade earlier to show that the genes were made of DNA. The premise of the experiments was this: When phage T2 infects *E. coli,* it subverts its host from making bacterial proteins to making phage proteins. If Crick's hypothesis were correct, this switch to phage protein synthesis should be accompanied by the production of new ribosomes equipped with phage-specific RNAs.

To distinguish new ribosomes from old, these investigators labeled the ribosomes in uninfected cells with heavy isotopes of nitrogen (^{15}N) and carbon (^{13}C). This made "old" ribosomes heavy. Then they infected these cells with phage T2 (or the related phage T4) and simultaneously transferred them to medium containing light nitrogen (^{14}N) and carbon (^{12}C). Any "new" ribosomes made after phage infection would therefore be light and would separate from the old, heavy ribosomes upon density gradient centrifugation. Brenner et al. also labeled the infected cells with ^{32}P to tag any phage RNA as it was made. Now we can pose the question: Was the radioactively labeled phage RNA associated with new or old ribosomes? Figure 10.9 shows that the phage RNA was found on old ribosomes whose rRNA was made before infection even began. Clearly, this old rRNA cannot carry phage genetic information; by extension, it was very unlikely that it could carry host genetic information either.

Other workers had already identified a better candidate for the messenger: a class of unstable RNAs that associate transiently with ribosomes. Interestingly enough, in phage T2-infected cells, this RNA had a base composition very similar to that of phage DNA—and quite different from that of bacterial DNA and RNA. This is exactly what we would expect of phage messenger RNA, and that is exactly what it is. On the other hand, host mRNA, unlike host rRNA, has a base composition similar to that of host DNA. This lends further weight to the hypothesis that it is the mRNA, not the rRNA, that is the "instructional" molecule.

 Most genes are expressed in a two-step process: transcription, or synthesis of an mRNA copy of the gene, followed by translation of this message to protein.

RIBOSOMES: PROTEIN SYNTHESIZING MACHINES

Figure 10.10 shows the probable shapes of the *E. coli* ribosome and its two subunits, the **50S** and **30S** particles (see also panel 7). The numbers 50S and 30S refer to the **sedimentation coefficients** of the two subunits. These coefficients are a measure of the speed with which the particles sediment in an ultracentrifuge. The 50S particle, with a larger sedimentation coefficient, migrates more rapidly to the bottom of the centrifuge tube under the influence of a centrifugal force. The coefficients are functions of the mass and shape of the particles. Heavy particles sediment more rapidly than light ones; spherical particles migrate faster than extended or flattened ones—just as a skydiver falls more rapidly in a tuck position than with arms and legs extended. The 50S particle is actually about twice as massive as the 30S. Together, the 50S and 30S subunits comprise a 70S ribosome. Notice that the numbers do not add up. This is because the sedimentation coefficients are not proportional to the particle weight; in fact, they are roughly proportional to the $\frac{2}{3}$-power of the particle weight.

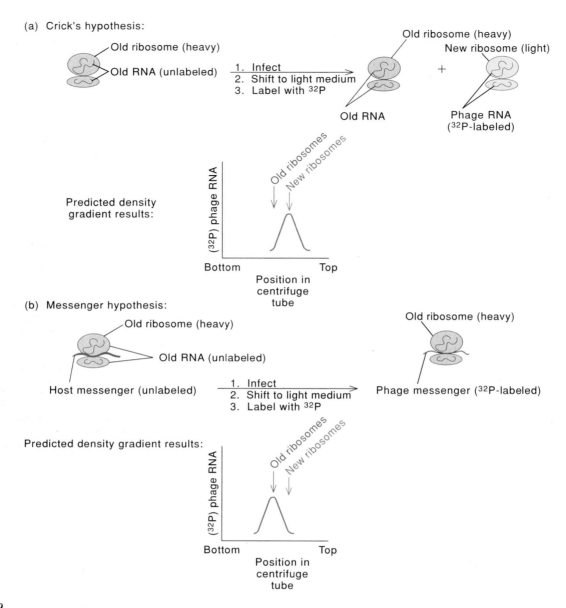

FIGURE 10.9

Experimental test of the messenger hypothesis. *E. coli* ribosomes were made heavy by labeling the bacterial cells with heavy isotopes of carbon and nitrogen. The bacteria were then infected with T2 or T4 phage and simultaneously shifted to "light" medium containing the normal isotopes of carbon and nitrogen, plus some ^{32}P to make the phage RNA radioactive. (*a*) Crick had proposed that ribosomal RNA carried the message for making proteins. If this were so, then whole new ribosomes with phage-specific ribosomal RNA would have been made after phage infection. In that case, the new, ^{32}P-labeled RNA (red) should have moved together with the new, light ribosomes (red). (*b*) Jacob and Monod had proposed that a messenger RNA carried genetic information to the ribosomes. According to this hypothesis, phage infection would cause the synthesis of new, phage-specific messenger RNAs that would be ^{32}P-labeled (red). These would associate with old, heavy ribosomes (blue). Therefore, the radioactive label would move together with the old, heavy ribosomes in the density gradient. This was indeed what happened.

Each ribosomal subunit contains RNA and protein. The 30S particle includes one molecule of **ribosomal RNA (rRNA)** with a sedimentation coefficient of 16S, plus twenty-one ribosomal proteins. The 50S particle is composed of two rRNAs (23S + 5S) and thirty-four proteins (fig. 10.11).

We know exactly how many ribosomal proteins there are, because several research groups, primarily Masayasu Nomura's, have taken ribosomes apart and put them back together again. This work, in the best biochemical tradition of separating and reconstituting, allows us to identify the components that are necessary to yield an active ribosome.

Figure 10.12 demonstrates the resolution of the proteins from the 30S particle by polyacrylamide gel electrophoresis. The proteins are named S1–S21, the S standing for "small," since they derive from the small ribosomal subunit. The proteins of the large subunit are therefore named L1–L34, where the L, of course, stands for "large."

(a)

(b)

FIGURE 10.11

Composition of the *E. coli* ribosome. The arrows at the top denote the dissociation of the 70S ribosome into its two subunits when magnesium ions are withdrawn. The lower arrows show the dissociation of each subunit into RNA and protein components in response to the protein denaturant urea. The masses (M_r in daltons) of the ribosome and its components are given in parentheses.

Reprinted from Nature, *vol. 331, p. 225. Copyright © 1988 Macmillan Magazines, Ltd. Reprinted by permission of the publisher and author.*

FIGURE 10.10

E. coli ribosome structure. (*a*) The 70S ribosome is shown from the "side" with the 30S particle (yellow) and the 50S particle (red) fitting together. (*b*) The 70S ribosome is shown rotated 90° relative to the view in (*a*). The 30S particle (yellow) is in front, with the 50S particle (red) behind.

James A. Lake, Journal of Molecular Biology *105(1976):131–59. Reproduced by permission of Academic Press.*

In some cases, two or more proteins electrophoresed together, but further purification separated these proteins completely. Once the proteins were completely resolved, they could be added back to 16S rRNA to produce fully active 30S particles. The test for activity was to combine these reconstituted 30S particles with 50S particles to give 70S ribosomes that could carry out translation in vitro.

The 50S particle proved much more difficult to reconstitute than the small particle. In fact, the first experiments failed because the temperatures needed for reassociation of the 50S particle were so high that some of the proteins denatured. Later, tricks were discovered that could allow 50S ribosomes from *E. coli* to reassemble at a lower temperature. This means that the whole ribosome is a **self-assembly system:** it can be reconstituted from its component parts, with no intervention from outside enzymes or factors.

FIGURE 10.12

Resolution of *E. coli* 30S ribosomal proteins. The proteins were electrophoresed on a polyacrylamide gel, which gave only partial resolution (fewer than twenty bands appear, so some contain more than one protein). The proteins that were incompletely resolved here were separated completely by chromatography.

From P. Traub and M. Nomura. Laboratory of Genetics, University of Wisconsin, Madison. Reprinted from Proceedings of the National Academy of Science, *Vol. 59, No. 3, fig. 2, pp. 777–84, March 1968.*

FIGURE 10.13

A polysome from the midge *Chironomus*. The 5´-end of the mRNA is at lower right; it runs across and off the picture on the left, loops around and re-enters the picture at top left, and finally ends at top right. The dark blobs attached to the mRNA are ribosomes; the fact that there are many of them is the reason for the name *polyribosome*, or *polysome*. Nascent polypeptides are visible, especially on the top part of the polysome; they grow longer as the ribosomes approach the 3´-end of the mRNA. The faint blobs on the nascent polypeptides are not individual amino acids but domains containing groups of amino acids.

E. V. Kiseleva FEBS letters 257:252(1989). Reprinted by permission from Nature *344:585 (1990) photo by E. V. Kiseleva.*

We assume that a similar assembly process occurs in vivo in prokaryotes. But what about assembly of eukaryotic ribosomes? Do they also self-assemble, and if so, where? We cannot answer the first part of that question because the complex eukaryotic ribosome has not been self-assembled in vitro. However, we do know the site of assembly of the eukaryotic ribosome: the nucleolus. The large ribosomal RNA genes are located in the nucleolus, and therefore the 5.8S, 18S, and 28S rRNAs are made there. The ribosomal proteins are made on cytoplasmic ribosomes, of course, so they must be imported into the nucleolus. There they join with rRNAs and build new ribosomes.

> Ribosomes are the cell's protein factories. Bacteria contain 70S ribosomes with two subunits, called 50S and 30S. Each of these contains ribosomal RNA and many proteins. Ribosomal subunits can be dissociated into their component parts—RNAs and proteins—and then reassociated. Thus, the ribosome is a self-assembly system that needs no help from the outside.

POLYSOMES

We have seen in chapters 8 and 9 that more than one RNA polymerase can transcribe a gene at one time. The same is true of ribosomes and mRNA. In fact, it is common for many ribosomes to be traversing the same mRNA in tandem at any given time. The result is a polyribosome, or **polysome,** such as the one pictured in figure 10.13. Only one mRNA is present in this photograph, although it loops out of the picture to the left. Even in the visible part of the polysome we can count over 70 ribosomes translating the mRNA simultaneously. We can also tell which end of the polysome is which, by looking at the nascent polypeptide chains. These grow longer and longer as we go from the 5´-end (where translation begins) to the 3´-end (where translation ends). Therefore the 5´-end is at lower right, while the 3´-end is at upper right.

The polysome in figure 10.13 is from a eukaryote (the midge). Since transcription and translation occur in different compartments in eukaryotes (transcription in the nucleus, translation in the cytoplasm), polysomes will always occur in the cytoplasm, independent of the genes themselves. Prokaryotes also have polysomes, but the picture in these organisms is complicated by the fact that transcription and translation of a given gene and its mRNA occur simultaneously, and in the same location. Thus, we can see nascent mRNAs being synthesized and being translated by ribosomes at the same time. Figure 10.14 shows just such a situation in *E. coli*. We can see two segments of the bacterial chromosome running parallel from left to right. Only the segment on top is being transcribed. We can tell that transcription is occurring from left to right in this picture because the polysomes are getting longer as we move in that direction; as they get longer, they have room for more and more ribosomes. Do not be misled by the difference in scale between figures 10.13 and 10.14; the ribosomes are smaller, and the nascent protein chains are not visible in the latter picture. Remember also that the strands running

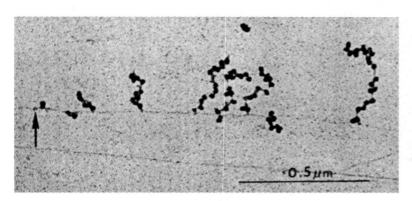

FIGURE 10.14

Simultaneous transcription and translation in *E. coli.* Two DNA segments stretch across the picture from left to right. The top segment is being transcribed from left to right. As the mRNAs grow, more and more ribosomes attach and carry out translation. This gives rise to polysomes, which are arrayed perpendicular to the DNA. The arrow at left points to a faint spot, which may be an RNA polymerase just starting to transcribe the gene. Other such spots denoting RNA polymerase appear at the bases of some of the other polysomes, where the mRNAs join the DNA.

1 Figure, Vol. 169, page 394, 24 July 1970, "Visualization of Bacterial Genes in Action." *Copyright 1970 by the AAAS.*

across figure 10.14 are DNA, whereas those in figure 10.13 are mRNA. The mRNAs are more or less vertical in figure 10.14.

 Most mRNAs are translated by more than one ribosome at a time; the result, a structure in which many ribosomes translate an mRNA in tandem, is called a polysome. In eukaryotes, polysomes are found in the cytoplasm. In prokaryotes, transcription of a gene and translation of the resulting mRNA occur simultaneously. Therefore, many polysomes are found associated with an active gene.

TRANSFER RNA: THE ADAPTER MOLECULE

The transcription mechanism was easy for molecular geneticists to predict. RNA resembles DNA so closely that it follows the same base-pairing rules. By following these rules, RNA polymerase produces replicas of the genes it transcribes. But what rules govern the ribosome's translation of mRNA to protein? This is a true translation problem. A nucleic acid language must be translated to a protein language. Francis Crick suggested the answer to this problem in a 1958 paper before much experimental evidence was available to back it up. What is needed, Crick reasoned, is some kind of adapter molecule that can recognize the nucleotides in the RNA language as well as the amino acids in the protein language. He was right. He even noted that there was a type of small, soluble RNA of unknown function that might play the adapter role. Again, he guessed right. Of course, there were some bad guesses in this paper as well, but even they were important. By their very creativity, Crick's ideas stimulated the research (some from Crick's own laboratory) that led to solutions to the puzzle of translation.

The adapter molecule in translation is indeed a small RNA that recognizes both RNA and amino acids; it is called **transfer RNA (tRNA).** Figure 10.15 shows a schematic diagram of a tRNA that recognizes the amino acid phenylalanine (Phe). Later in this chapter we will discuss the structure and function of tRNA in detail. For the

AOH 3′ (Amino acid binding site)

FIGURE 10.15

Transfer RNA. A schematic representation of the yeast phenylalanine tRNA (tRNA^{Phe}), the tRNA that functions in yeast to insert the amino acid phenylalanine into proteins. The tRNA does not really look like this in three dimensions; this diagram simply points out the secondary structure of the molecule—how base pairing occurs within the RNA. The four ladderlike stems show this base pairing. For example, observe the G–C pair between the 5′-terminal G and the C that is five bases from the 3′-end. The 3′-end of a tRNA always contains the same three bases: CCA. The hydroxyl (OH) group of the terminal A (top, blue) is the site of attachment of the amino acid—phenylalanine in this case. The other business end of the tRNA is at the bottom of the diagram. This is the anticodon (red), which pairs with a three-base phenylalanine codon in a messenger RNA.

present, the "cloverleaf" model, though it bears scant resemblance to the real shape of tRNA, will serve to point out the fact that the molecule has two "business ends." One end (the top of the model) attaches to an amino acid. Since this is a tRNA specific for phenylalanine (tRNA^{Phe}), only phenylalanine will bind. An enzyme called phenylalanine-tRNA synthetase catalyzes this binding. The generic name for such enzymes is **aminoacyl-tRNA synthetase.**

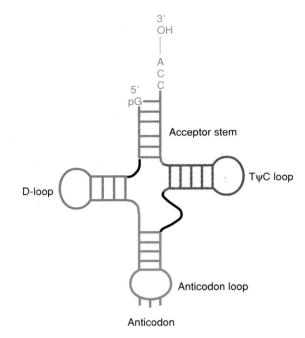

FIGURE 10.16

Codon-anticodon recognition. The recognition between a codon in an mRNA and a corresponding anticodon in a tRNA obeys essentially the same Watson-Crick rules that apply to other polynucleotides. Here, a 5′-mGAA3′ anticodon (red) on a tRNA^{Phe} is recognizing a 5′UUC3′ codon (gray) for phenylalanine in an mRNA. The ^mG denotes a methylated G, which base-pairs like an ordinary G. Notice that the mRNA is pictured backwards (3′ → 5′) relative to normal convention, which is 5′ → 3′, left to right. That was done to put its codon in the proper orientation (3′ → 5′, left to right) to base-pair with the tRNA, shown conventionally with its anticodon reading 5′ → 3′, left to right. Remember that the two strands of DNA are antiparallel; this applies to any double-stranded polynucleotide, including one as small as the three-base codon-anticodon pair.

FIGURE 10.17

Cloverleaf structure of yeast tRNA^{Phe}. At top is the acceptor stem (red), where amino acid binds to 3′-terminal adenosine. At left is the dihydro U loop (D-loop, blue), which contains at least one dihydrouracil base. At bottom is the anticodon loop (green), containing the anticodon. The TψC loop (right, gray) contains the virtually invariant sequence TψC. Each loop is defined by a base-paired stem of the same color.

The other end (the bottom of the model) contains a three-base sequence that pairs with a complementary three-base sequence in an mRNA. Such a triplet in mRNA is called a **codon;** naturally enough, its complement in a tRNA is called an **anticodon.** The codon in question here has attracted the anticodon of a tRNA bearing a phenylalanine. That means that this codon says: "Please insert one phenylalanine into the protein at this point." A schematic of codon-anticodon recognition appears in figure 10.16. The recognition between codon and anticodon, mediated by the ribosome, obeys the same Watson-Crick rules as any other double-stranded polynucleotide, at least in the case of the first two base pairs. The third pair is allowed somewhat more freedom, as we will see later in this chapter.

It is apparent from figure 10.16 that UUC is a codon for phenylalanine. This implies that the **genetic code** contains three-letter words, as indeed it does. We can predict the number of possible three-letter code words as follows: The number of permutations of four different bases taken three at a time is 4^3, which is 64. But there are only twenty amino acids. Are some code words not used? Actually, three of the possible codons (UAG, UAA, and UGA) code for termination; all of the other codons specify amino acids. This means that most amino acids have more than one codon; the genetic code is therefore said to be **degenerate.**

 The two business ends of tRNAs allow them to recognize both amino acids and nucleic acids. They are therefore capable of serving the adapter role postulated by Crick and are the key to the mechanism of translation.

SECONDARY STRUCTURE OF tRNA

Figure 10.17 shows again the familiar cloverleaf structure of tRNA. Remember that the molecule does not really look like this in three dimensions; the cloverleaf model simply points out the **secondary structure** of tRNA—that is, the base-paired regions of the molecule. For example, at the top of the diagram lie the two ends of the RNA: the 5′-end on the left and the 3′-end on the right. These two ends base-pair with one another to form the so-called **acceptor stem.** Four other base-paired stems occur, defining three loops: the **dihydrouracil loop (D-loop),** named for the modified uracil bases this region always contains; the **anticodon loop,** named for the all-important anticodon near its apex; and the **TψC loop,** which takes its name from a nearly invariant sequence of three bases, TψC. The psi (ψ) stands for a modified nucleoside in tRNA, **pseudouridine.** It is the same as normal uridine, except that the base is linked to the ribose through the 5-carbon of the base instead

FIGURE 10.18
Three-dimensional structure of tRNA. (*a*) A planar
projection of the three-dimensional structure of yeast
tRNA^Phe. The various parts of the molecule are color-
coded to correspond to the diagrams in figure 10.17 and
to (*b*) and (*c*) of this figure. (*b*) Familiar cloverleaf
structure of tRNA with same color scheme as (*a*).
Arrows indicate the contortions this cloverleaf would
have to go through to achieve the approximate shape of
a real tRNA, shown in (*c*).

(a) *From Gary Quigley and Alexander Rich,* Science *185:436,
1974. Copyright © 1974 by the American Association for the
Advancement of Science, Washington, D.C. Reprinted by
permission of the publisher and authors.*

of the 1-nitrogen. The region between the anticodon loop
and the TψC loop in figure 10.17 is called the **variable
loop** because it varies in length from four to thirteen
nucleotides; some of the longer variable loops contain
base-paired stems.

Since tRNAs contain many modified bases in addition
to dihydrouracil, a reasonable question would be: Is the
tRNA made with modified bases, or are the bases modified
after synthesis of the tRNA is finished? The answer is that
tRNAs are made with the same four ordinary bases as any
other RNA. Then, once transcription is complete, multiple
enzyme systems modify the bases. What effects, if any, do
these modifications have on tRNA function? We know they
are important, because at least two tRNAs have been made
in vitro with the four normal, unmodified bases, and they
were unable to bind amino acids. On the other hand, D.
Wang and a team of over two hundred Chinese colleagues
chemically synthesized a fully modified yeast alanine
tRNA and found that it was active. This demonstrated that
it was not the in vitro synthesis of the inactive tRNAs that
was at fault, but the fact that they were unmodified.

The cloverleaf model of tRNA shows its secondary (base-
pairing) structure and reveals four base-paired stems that
define three loops of RNA: the D-loop, the anticodon loop,
and the TψC loop. Transfer RNAs contain many modified
bases, at least some of which are vital to tRNA activity.

THREE-DIMENSIONAL STRUCTURE OF tRNA

Alexander Rich and his colleagues used X-ray diffraction
techniques to reveal the three-dimensional structure, or **tertiary structure,** of tRNAs. Since all tRNAs have essentially the same secondary structure, represented by the
cloverleaf model, it is perhaps not too surprising that they
all have essentially the same tertiary structure as well. This
L-shaped structure is shown in figure 10.18.

The key to understanding the three-dimensional structure of tRNA is to realize that the longer the base-paired regions in a tRNA are, the more stable they will be. The centipede metaphor we used in chapter 6 helps illuminate this

FIGURE 10.19

Stereo view of tRNA. To see the molecule in three dimensions, use a stereo viewer, or force the two images to merge either by relaxing your eyes as if focusing on something in the distance or by crossing your eyes slightly. Each colored dot indicates a base pair in the acceptor stem that is especially easy to see. It may take a little time for the three-dimensional effect to develop.

From Gary J. Quigley and Alexander Rich, "Structural Domains of Transfer RNA Molecules" in Science, *194:796–806, Figure 2, November 19, 1976. Copyright © 1976 by the American Association for the Advancement of Science, Washington, D. C. Reprinted by permission of the publisher and authors.*

point. Each base pair is like a pair of legs on two embracing centipedes; the more legs, the tighter the grip. However, this imagery does not convey the idea that ten legs in a row are much better than two separated sets of five, which may not be the case with centipede legs, but is certainly true of base pairs. Transfer RNA maximizes the length of its base-paired stems by lining them up in two extended regions. One of these lies at the top of the molecule and encompasses the acceptor stem and the TψC stem; the other forms the vertical axis of the molecule and includes the D stem and the anticodon stem. Even though the two parts of each stem are not aligned perfectly and the stems therefore bend slightly, the alignment confers stability.

Figure 10.19 illustrates the double-helical nature of these extended double stems. This is a stereo diagram; you can see it in three dimensions with a stereo viewer. Without a viewer, you can force the two images to merge by staring at a point far behind the page or by crossing your eyes slightly (although the eye-crossing technique interchanges the front and back of the diagram). This may take a little practice, and some people never get the hang of it, but the fun of seeing the helical stems of tRNA in three dimensions is worth some effort. If all else fails, you can get an idea of the extent of the helical regions by observing the base pairs in one of the two-dimensional drawings in figure 10.19. The base pairs are the flat structures seen edge-on near the centers of the two helices. They are more or less parallel over the length of each helix—a feature that is especially noticeable in the horizontal helix at the top, which includes the TψC stem plus the acceptor stem.

Even in two dimensions, figure 10.19 demonstrates clearly that the bases of the anticodon do not extend inward to base-pair with other parts of the tRNA molecule. Rather, they protrude outward to the right so they can base-pair with a codon in an mRNA. Notice that the three bases of the anticodon are stacked parallel to one another, suggesting

FIGURE 10.20

Linkage between tRNA and an amino acid. Most amino acids are bound by an ester linkage to the 2´-hydroxyl group of the terminal adenosine of the tRNA as shown, but some bind to the 3´-hydroxyl group.

a helical arrangement. In fact, the structure of the tRNA twists the anticodon into a helical shape that greatly facilitates its base pairing with the codon. The ribosome presumably makes this binding even easier.

> The three-dimensional structure of tRNA consists of an L-shaped framework in which the four stems stack upon one another to produce two long, double-helical regions that help stabilize the molecule. The bases of the anticodon protrude to one side and are arranged in a helical shape so they can more easily form a partial double helix by pairing with the three-base codon.

AMINO ACID BINDING TO tRNA

When an amino acid binds to the 3´-end of a tRNA, we say the tRNA is **charged** with that amino acid. All tRNAs have the same three bases (CCA) at their 3´-ends, and the terminal adenosine is the target for charging. Figure 10.20 gives the chemical structure of the end of this terminal CCA and shows an amino acid attached through its acid group to the 2′-hydroxyl group of the adenosine's ribose. As we learned in chapter 6, this kind of linkage between an acid group and a hydroxyl group is called an ester bond.

Charging takes place in two steps, both catalyzed by the enzyme **aminoacyl-tRNA synthetase.** In the first step, the amino acid is activated, using energy from ATP; the product of the reaction is aminoacyl-AMP. The pyrophosphate by-product is simply the two end phosphate groups from the ATP, which were lost in forming AMP:

amino acid + ATP → aminoacyl-AMP

+ pyrophosphate

The bonds between phosphate groups in ATP (and the other nucleoside triphosphates) are "high energy bonds." When they are broken, this energy is released. In this case, the energy is trapped in the aminoacyl-AMP, which is why we call this an activated amino acid. The principle is the same as burning wood to release energy, which is then used to boil water. The boiling water is activated in much the same way as the aminoacyl-AMP. In the second reaction of charging, the energy in the aminoacyl-AMP is used to transfer the amino acid to a tRNA, forming **aminoacyl-tRNA.**

aminoacyl-AMP + tRNA → aminoacyl-tRNA

Just like other enzymes, an aminoacyl-tRNA synthetase plays a dual role. Not only does it catalyze the reaction leading to an aminoacyl-tRNA, but it also determines the specificity of this reaction. Only twenty synthetases exist, one for each amino acid, and they are very specific. Each will almost always place its amino acid on the right kind of tRNA.

Consider what would happen if these enzymes made many mistakes. For example, if serine-tRNA synthetase, which attaches the amino acid serine to its cognate tRNA (tRNA^{Ser}), suddenly started attaching lysine instead, the result would be catastrophic. The lysine-tRNA^{Ser} would still recognize serine codons, because its anticodon would be unchanged and the amino acid plays no role in codon-anticodon recognition. But this aminoacyl-tRNA contains lysine instead of serine, so it would insert lysines into proteins where serines ought to go. Since a protein's structure and function depend on its primary structure, these changes would in all likelihood destroy the protein's usefulness.

 Each aminoacyl-tRNA synthetase couples its specific amino acid to one of its cognate tRNAs. The reaction, which is very specific for both amino acid and tRNA, requires activation of the amino acid by ATP to form aminoacyl-AMP, then transfer of the activated amino acid to the appropriate tRNA.

THE GENETIC CODE

The term "genetic code" refers to the set of three-base code words (codons) in mRNAs that stand for twenty amino acids in proteins. Like any code, this one had to be broken before we knew what the codons stood for. Indeed, before 1960, other more basic questions about the code included: Do the codons overlap? Are there gaps, or "commas," in the code? How many bases make up a codon? These questions were answered in the 1960s by a series of imaginative experiments, which we will describe here.

NONOVERLAPPING CODONS

A. Tsugita and H. Frankel-Conrat reasoned that if the code were nonoverlapping, a change of one base in an mRNA (a **missense mutation**) would be expected to change no more than one amino acid in the resulting protein. For example, consider the following "micromessage":

AUGCUA

Assuming that the code is triplet (three bases per codon) and this message is read from the beginning, the codons will be AUG and CUA if the code is nonoverlapping. A change in the third base (G) would change only one codon (AUG) and therefore at most only one amino acid. On the other hand, if the code were overlapping, base G could be part of three adjacent codons (AUG, UGC, and GCU). Therefore, if the G were changed, up to three adjacent amino acids could be changed in the resulting protein. But when the investigators introduced one-base alterations into mRNA from tobacco mosaic virus (TMV), they found that these caused changes in only one amino acid. Hence, the code must be nonoverlapping.

NO GAPS IN THE CODE

If the code contained untranslated gaps, or commas, mutations that add or subtract a base from a message might change a few codons, but we would expect the ribosome to get back on track after the next comma. In other words, these mutations might frequently be lethal, but there should be many cases where the mutation occurs just before a comma in the message and therefore has little if any effect. If there were no commas to get the ribosome back on track, these mutations would be lethal except when they occur right at the end of a message.

Such mutations do occur, and they are called **frameshift mutations;** they work as follows. Consider another tiny message:

AUGCAGCCAACG

If translation starts at the beginning, the codons will be AUG, CAG, CCA, and ACG. If we insert an extra base (X) right after base U, we get:

AUXGCAGCCAACG

1. Wild-type: CAT |CAT |CAT |CAT |CAT

2. Add a base: CAG |TCA |TCA |TCA |TCA

3. Delete a base: CAT |CTC |ATC |ATC |ATC

4. Cross #2 and #3: CAG |TCT |CAT |CAT |CAT

5. Add 3 bases: CAG |GGT |CAT |CAT |CAT

FIGURE 10.21

Frameshift mutations. Line 1: An imaginary mRNA has the same codon, CAT, repeated over and over. The vertical dotted lines show the reading frame, starting from the beginning. Line 2: Adding a base, G (red), in the third position changes the first codon to CAG and shifts the reading frame one base to the left so that every subsequent codon reads TCA. Line 3: Deleting the fifth base, A (marked by the triangle), from the wild-type gene changes the second codon to CTC and shifts the reading frame one base to the right so that every subsequent codon reads ATC. Line 4: Crossing the mutants in lines 2 and 3 sometimes gives pseudo-wild-type revertants with an insertion and a deletion close together. The end result is a DNA with its first two codons altered, but every other one put back into the correct reading frame. Line 5: Adding three bases, GGG (red), after the first two bases disrupts the first two codons, but leaves the reading frame unchanged. The same would be true of deleting three bases.

Now this would be translated from the beginning as AUX, GCA, GCC, AAC. Notice that the extra base changes not only the codon (AUX) in which it appears, but every codon from that point on. The **reading frame** has shifted one base to the left; whereas C was originally the first base of the second codon, G is now in that position. This could be compared to a filmstrip viewing machine that slips a cog, thus showing the last one-third of one picture together with the first two-thirds of the next.

On the other hand, a code with commas would be one in which each codon is flanked by one or more untranslated bases, represented by Z's in the following message. The commas would serve to set off each codon so the ribosome could recognize it:

AUGZCAGZCCAZACGZ

Deletion or insertion of a base anywhere in this message would change only a single codon. The comma (Z) at the end of the damaged codon would then put the ribosome back on the right track. Thus, addition of an extra base (X) to the first codon would give the message:

AUXGZCAGZCCAZACGZ

The first codon (AUXG) is now wrong, but all the others, still neatly set off by Z's, would be translated normally.

When Francis Crick and his colleagues treated bacteria with acridine dyes that usually cause single-base insertions or deletions, they found that such mutations were very severe; the mutant genes gave no functional product. This is what we would expect of a "comma-less" code with no gaps; base insertions or deletions cause a shift in the reading frame of the message that persists until the end of the message.

Moreover, Crick found that adding a base could cancel the effect of deleting a base, and vice versa. This phenomenon is illustrated in figure 10.21, where we start with an artificial gene composed of the same codon, CAT, repeated over and over. When we add a base, G, in the third position, we change the reading frame so that all codons thereafter read TCA. When we start with the wild-type gene and delete the fifth base, A, we change the reading frame in the other direction, so that all subsequent codons read ATC. Crossing these two mutants sometimes gives a recombined "pseudo-wild-type" gene like the one on line 4 of the figure. Its first two codons, CAG and TCT, are wrong, but thereafter, the insertion and deletion cancel, and the original reading frame is restored; all codons from that point on read CAT.

Finally, Crick and Leslie Barnett discovered that a set of three adjacent insertions or deletions could produce a wild-type gene (fig. 10.21, line 5). This of course demands that a codon consist of three, or a multiple of three, bases. As Crick remarked to Barnett when he saw the experimental result: "We're the only two [who] know it's a triplet code!"

THE TRIPLET CODE

We have been assuming that the codons contain three letters, not a multiple of three as Crick and Barnett's experiment still allowed. How do we know this is true? Marshall Nirenberg and Johann Heinrich Matthaei performed the experiment that laid the groundwork for answering this question and for breaking the genetic code itself. The experiment was deceptively simple; it showed that synthetic RNA could be translated in vitro. In particular, when Nirenberg and Matthaei translated poly(U), a synthetic RNA composed only of U's, they made polyphenylalanine. Of course, that told them that a codon for phenylalanine contained only U's. This finding by itself was important, but the long-range implication was that one could design synthetic mRNAs of defined sequence and analyze the protein products to shed light on the nature of the code. Har Gobind Khorana and his colleagues were the chief practitioners of this strategy.

Here is how Khorana's synthetic messenger experiments demonstrated that the codons contain three bases: First, if the codons contain an odd number of bases, then a repeating dinucleotide [poly(UC) or UCUCUCUC . . .] should contain two alternating codons (UCU and CUC, in this case), no matter where translation starts. The resulting protein would be a repeating dipeptide—two amino acids alternating with each other. If codons have an even number of bases, only one codon (UCUC, for example) should be repeated over and over. Of course, if translation started at the second base, the single repeated codon would be different (CUCU). In either case, the resulting protein would be a homopolypeptide, containing only one amino acid repeated over and over. Khorana found that poly(UC) translated to a repeating dipeptide, poly(serine-leucine) (fig. 10.22a), proving that the codons contained an odd number of bases.

Repeating triplets translated to homopolypeptides, as had been expected if the number of bases in a codon was three or a multiple of three. For example, poly(UUC)

(a) UCUCUCUCUCUC
     Ser Leu Ser Leu

(b) UUCUUCUUCUUC   or   UUCUUCUUCUUC   or   UUCUUCUUCUUC
     Phe Phe Phe Phe   Ser Ser Ser   Leu Leu Leu

(c) UAUCUAUCUAUC
     Tyr Leu Ser Ile

FIGURE 10.22

Coding properties of several synthetic mRNAs. (*a*) Poly(UC) contains two alternating codons, UCU and CUC, which code for serine (Ser) and leucine (Leu), respectively. Thus, the product is poly(Ser-Leu). (*b*) Poly(UUC) contains three codons, UUC, UCU, and CUU, which code for phenylalanine (Phe), serine (Ser), and leucine (Leu), respectively. Therefore, the product is poly(Phe), or poly (Ser), or poly (Leu), depending on which of the three reading frames the ribosome uses. (*c*) Poly(UAUC) contains four codons in a repeating sequence: UAU, CUA, UCU, and AUC, which code for tyrosine (Tyr), leucine (Leu), serine (Ser), and isoleucine (Ile), respectively. The product is therefore poly(Tyr-Leu-Ser-Ile).

translated to polyphenylalanine plus polyserine plus polyleucine (fig. 10.22*b*). The reason for three different products is that translation can start at any point in the synthetic message. Therefore, poly(UUC) can be read as UUC, UUC, etc., or UCU, UCU, etc., or CUU, CUU, etc., depending on where translation starts. In all cases, once translation begins, only one codon is encountered, as long as the number of bases in a codon is divisible by three.

The clincher was a repeating tetranucleotide, which translated to a repeating tetrapeptide. For example, poly(UAUC) yielded poly(tyrosine-leucine-serine-isoleucine) (fig. 10.22*c*). As an exercise, you can write out the sequence of such a message and satisfy yourself that it is compatible with codons having three bases, or nine, or even more, but not six. Since it is not likely that codons would be as cumbersome as nine bases long, three is the best choice. Look at the problem another way: Three is the lowest number that gives enough different codons to specify all twenty amino acids. (The number of permutations of four different bases taken three at a time is 4^3, or 64.) There are only sixteen two-base codons ($4^2 = 16$), not quite enough. But there would be over a quarter of a million ($4^9 = 262{,}144$) nine-base codons. Nature is usually more economical than that.

The genetic code is a set of three-base code words, or codons, in mRNA that instruct the ribosome to incorporate specific amino acids into a polypeptide. The code is nonoverlapping; that is, each base is part of only one codon. It is also devoid of gaps, or commas; that is, each base in the coding region of an mRNA is part of a codon.

BREAKING THE CODE

Obviously, Khorana's synthetic mRNAs gave strong hints about some of the codons. For example, since poly(UC) yields poly(serine-leucine), we know that one of the codons (UCU or CUC) codes for serine and the other codes for

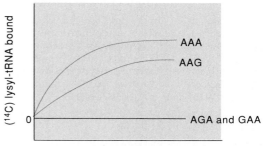

FIGURE 10.23

Binding of lysyl-tRNA to ribosomes in response to various codons. Lysyl-tRNA was labeled with radioactive carbon (^{14}C) and mixed with *E. coli* ribosomes in the presence of the following trinucleotides: AAA, AAG, AGA, and GAA. Lysyl-tRNA-ribosome complex formation was measured by binding to nitrocellulose filters. (Unbound lysyl-tRNA does not stick to these filters, but a lysyl-tRNA-ribosome complex does.) AAA was a known lysine codon, so binding was expected with this trinucleotide.

leucine. The question remains: Which is which? Nirenberg developed a powerful assay to answer this question. He found that a trinucleotide was usually enough like an mRNA to cause a specific aminoacyl-tRNA to bind to ribosomes. For example, the triplet UUU will cause phenylalanyl-tRNA to bind, but not lysyl-tRNA or any other aminoacyl-tRNA. Therefore, UUU is a codon for phenylalanine. This method was not perfect; some codons did not cause any aminoacyl-tRNA to bind, even though they were authentic codons for amino acids. But it provided a nice complement to Khorana's method, which by itself would not have given all the answers either, at least not easily.

Here is an example of how the two methods could be used together: Translation of the polynucleotide poly(AAG) yielded polylysine plus polyglutamic acid plus polyarginine. There are three different codons in that synthetic message: AAG, AGA, and GAA. Which one codes for lysine? All three were tested by Nirenberg's assay, yielding the results shown in figure 10.23. Clearly, AGA and GAA cause no binding of (^{14}C)lysyl-tRNA to ribosomes, but AAG does. Therefore, AAG is the lysine codon in poly(AAG). Something else to notice about this experiment is that the triplet AAA also causes lysyl-tRNA to bind. Therefore, AAA is another lysine codon. This illustrates a general feature of the code: In most cases, more than one triplet codes for a given amino acid. In other words, the code is **degenerate.**

Figure 10.24 shows the entire genetic code. As predicted, there are sixty-four different codons and only twenty different amino acids, yet all of the codons are used. Three are "stop" codons found at the ends of messages, but all the others specify amino acids, which means that the code is highly degenerate. Leucine, serine, and arginine have six different codons; several others, including proline, threonine, and alanine, have four; isoleucine has three; and many others have two. Just two amino acids, methionine and tryptophan, have only one codon.

The genetic code was broken by using either synthetic messengers or synthetic trinucleotides and observing the polypeptides synthesized or aminoacyl-tRNAs bound to ribosomes, respectively. There are sixty-four codons in all. Three are stop signals, and the rest code for amino acids. This means that the code is highly degenerate.

UNUSUAL BASE PAIRS BETWEEN CODON AND ANTICODON

How does an organism cope with multiple codons for the same amino acid? One way would be to have multiple tRNAs (**isoaccepting species**) for the same amino acid, each one specific for a different codon. In fact, this is part of the answer, but there is more to it than that; we can get along with considerably fewer tRNAs than that simple hypothesis would predict. Again, Francis Crick anticipated experimental results with insightful theory. In this case, Crick hypothesized that the first two bases of a codon must pair correctly with the anticodon according to Watson-Crick base-pairing rules, but the last base of the codon can "wobble" from its normal position to form unusual base pairs with the anticodon. This proposal was called the **wobble hypothesis.** In particular, Crick proposed that a G in an anticodon can pair not only with a C in the third position of a codon, but also with a U. This would give the **wobble base pair** shown in figure 10.25b. Notice how the U has wobbled upward from its normal position to form this base pair.

Furthermore, Crick noted that one of the unusual nucleosides found in tRNA is **inosine (I),** which has a structure similar to that of guanosine. This nucleoside can ordinarily pair like guanine, so we would expect it to pair with C (Watson-Crick base pair) or U (wobble base pair) in the third position of a codon. But Crick proposed that inosine could form still another kind of wobble pair, this time with A in the third position of a codon (figure 10.25c). That means an anticodon with I in the first position can potentially pair with three

Second position

		U	C	A	G	
U	UUU } Phe UUC UUA } Leu UUG	UCU UCC } Ser UCA UCG	UAU } Tyr UAC UAA } STOP UAG	UGU } Cys UGC UGA STOP UGG Trp	U C A G
C	CUU CUC } Leu CUA CUG	CCU CCC } Pro CCA CCG	CAU } His CAC CAA } Gln CAG	CGU CGC } Arg CGA CGG	U C A G
A	AUU AUC } Ile AUA AUG Met	ACU ACC } Thr ACA ACG	AAU } Asn AAC AAA } Lys AAG	AGU } Ser AGC AGA } Arg AGG	U C A G
G	GUU GUC } Val GUA GUG	GCU GCC } Ala GCA GCG	GAU } Asp GAC GAA } Glu GAG	GGU GGC } Gly GGA GGG	U C A G

First position (5′-end) / Third position (3′-end)

FIGURE 10.24

The genetic code. All sixty-four codons are listed, along with the amino acid for which each codes. To find a given codon, ACU for example, we start with the wide horizontal row labeled with the name of the first base of the codon (A) on the left border. Then we move across to the vertical column corresponding to the second base (C). This brings us to a box containing all four codons beginning with AC. It is now a simple matter to find the one among these four we are seeking, ACU. We see that this triplet codes for threonine (Thr), as do all the other codons in the box: ACC, ACA, and ACG. This is an example of the degeneracy of the code. Notice that three codons (red) do not code for amino acids; instead, they are stop signals.

different codons ending with C, U, or A. The wobble phenomenon reduces the number of tRNAs needed to translate the genetic code. For example, consider the two tyrosine codons: UAU and UAC. According to the wobble hypothesis, they can both be recognized by the anticodon 3′AUG5′. The G in the 5′-position of the anticodon could form a Watson-Crick G–C base pair with the C in UAC, or a G–U wobble base pair with the U in UAU. Similarly, the two lysine codons, AAA and AAG, can both be recognized by the anticodon 3′UUU5′. The U in the 5′-position can form a Watson-Crick A–U pair with the A in AAA, or a G–U wobble pair with the G in AAG.

Part of the degeneracy of the genetic code is accommodated by isoaccepting species of tRNA that bind the same amino acid but recognize different codons. The rest is handled by wobble, in which the third base of a codon is allowed to move slightly from its normal position to form a non-Watson-Crick base pair with the anticodon. This allows the same aminoacyl-tRNA to pair with more than one codon. The wobble pairs are G–U (or I–U) and I–A.

Anticodon (first base)   Codon (third base)

(a) Standard Watson-Crick base pair (G–C):

G   C

(b) G–U (or I–U) wobble base pair:

G   U

(c) I–A wobble base pair:

I   A

FIGURE 10.25

Wobble base pairs. (*a*) Relative positions of bases in a standard (G–C) base pair. The base on the left in (*a*) and in the wobble base pairs (*b*) and (*c*) is the first base in the anticodon. The base on the right is the third base in the codon. (*b*) Relative positions of bases in a G–U (or I–U) wobble base pair. Notice that the U has to "wobble" upward to pair with the G (or I). (*c*) Relative positions of bases in an I–A wobble base pair. The A has to rotate (wobble) counterclockwise in order to form this pair.

THE (ALMOST) UNIVERSAL CODE

In the years after the genetic code was broken, all organisms examined, from bacteria to humankind, were shown to share the same code. It was therefore generally assumed that the code was universal, with no deviations whatsoever. This apparent universality led in turn to the notion of a single origin of present life on earth. In this section we will see that there are a few deviations from the "universal" genetic code, but these are not serious enough to overturn the hypothesis of a single origin of present life.

The reasoning for the single origin idea goes like this: There is nothing inherently advantageous about each specific codon assignment we see. There is no obvious reason, for example, why UUC should make a good codon for phenylalanine, whereas AAG is a good one for lysine. Rather, in a sense, the genetic code may be an "accident"; it just happened to evolve that way. However, once these codons were established, there is a very good reason why they should not change: A change that fundamental would almost certainly be lethal.

Consider, for instance, a tRNA for the amino acid cysteine and the codon it recognizes, UGU. In order for that relationship to change, the anticodon of the cysteinyl-tRNA would have to change so it can recognize a different codon, say UCU, which is a serine codon. At the same time, all the UCU codons in that organism's genome that code for important serines would have to change to alternate serine codons so they would not be recognized as cysteine codons. The chances of all these things happening together, even over vast evolutionary time, are negligible. That is why the genetic code is sometimes called a "frozen accident"; once it was established, for whatever reasons, it had to stay that way. So a universal code would be powerful evidence for a single origin of life. After all, if life started independently in two places, we would hardly expect the two lines to evolve the same genetic code by accident!

In light of all this, it is remarkable that the genetic code is *not* absolutely universal; there are some exceptions to the rule. The first of these to be discovered were in the genomes of mitochondria. In mitochondria of the fruit fly

Table 10.1 Deviations from the "Universal" Genetic Code

Source	Codon	Usual Meaning	New Meaning
Fruit fly mitochondria	UGA	Stop	Tryptophan
	AGA & AGG	Arginine	Serine
	AUA	Isoleucine	Methionine
Mammalian mitochondria	AGA & AGG	Arginine	Stop
	AUA	Isoleucine	Methionine
	UGA	Stop	Tryptophan
Yeast mitochondria	CUN*	Leucine	Threonine
	AUA	Isoleucine	Methionine
	UGA	Stop	Tryptophan
Protozoa nuclei	UAA & UAG	Stop	Glutamine
Higher plants	UGA	Stop	Tryptophan
	CGG	Arginine	Tryptophan
Mycoplasma	UGA	Stop	Tryptophan

*N = Any base

D. melanogaster, UGA is a codon for tryptophan rather than for "stop." Even more remarkably, AGA in these mitochondria codes for serine, whereas it is an arginine codon in the universal code. Mammalian mitochondria show some deviations, too. Both AGA and AGG, though they are arginine codons in the universal code, have a different meaning in human and bovine mitochondria; there they code for "stop." Furthermore, AUA, ordinarily an isoleucine codon, codes for methionine in these mitochondria.

These aberrations might be dismissed as relatively unimportant, occurring as they do in mitochondria, which have very small genomes coding for only a few proteins and therefore more latitude to change than nuclear genomes. But there are exceptional codons in nuclear genomes and prokaryotic genomes as well. In at least three ciliated protozoa, including *Paramecium,* UAA and UAG, which are normally stop codons, code for glutamine. In the prokaryote *Mycoplasma capricolum,* UGA, normally a stop codon, codes for tryptophan. All the known deviations from the standard genetic code are summarized in table 10.1.

Clearly, the so-called universal code is not really universal. Does this mean that the evidence now favors more than one origin of present life on earth? If the deviant codes were radically different from the standard one, this might be an attractive possibility, but they are not. In many cases, the novel codons are stop codons that have been recruited to code for an amino acid: glutamine or tryptophan. There is a well-established mechanism for this sort of occurrence, as we will see later in the chapter. The only examples of codons that have switched their meaning from one amino acid to another occur in mitochondria. Again, mitochondrial genomes, because they code for far fewer proteins than nuclear genomes or even prokaryotic genomes, might be expected to change a codon safely every now and then.

In summary, even if there is no universal code, there is a standard code from which the deviant ones almost certainly evolved.

The genetic code is not strictly universal. In certain eukaryotic nuclei and mitochondria and in at least one bacterium, codons that cause termination in the standard genetic code can code for amino acids such as tryptophan and glutamine. In several mitochondrial genomes, the sense of a codon is changed from one amino acid to another. These deviant codes are still closely related to the standard one from which they probably evolved.

MECHANISM OF TRANSLATION

The mechanism of translation is considerably more complex than that of transcription, as one might guess from the relative complexities of the ribosome and RNA polymerase. Translation, like transcription, is divided into three phases: initiation, elongation, and termination. We will deal with each in turn.

INITIATION

Initiation of translation in *E. coli* takes place in several steps (fig. 10.26a). First, a ribosome that has finished translating an mRNA must be separated into its component subunits. An **initiation factor, IF-3,** effects this step, joining the 30S subunit and preventing its reassociation with free 50S subunits. **IF-1** is also thought to participate in this dissociation process, although its role is not as clear. Next, the 30S subunit binds a new mRNA in a reaction directed by IF-3. **Initiation factor 2 (IF-2)** then joins the growing collection of molecules and attaches the first aminoacyl-tRNA

(a)

(b)

$$H_3C - S - CH_2 - CH_2 - \overset{\overset{\textstyle H}{|}}{\underset{\underset{\textstyle NH}{|}}{C}} - \overset{\overset{\textstyle O}{\|}}{C} - OH$$
$$\underset{\underset{\textstyle H}{|}}{\overset{|}{C=O}}$$

FIGURE 10.26

Initiation of translation in *E. coli.* (*a*) Initiation sequence. First, the 70S ribosome dissociates into its component subunits (pink and blue). The initiation factor IF-3 (yellow) aids this dissociation by binding to the free 30S particle and preventing its reassociation with the 50S subunit. IF-1, not shown, is also thought to participate at this point. Next, IF-3 promotes binding of mRNA (red) to the 30S particle. In the next step, IF-2 (purple)

binds fMet-tRNA and attaches it to the 30S particle; GTP also binds in this reaction, completing the initiation complex. Finally, the 50S particle joins the initiation complex, and the two initiation factors dissociate; at the same time, GTP is broken down to GDP and inorganic phosphate. (*b*) Structure of N-formyl methionine, with formyl group in red.

to the 30S subunit. GTP also binds to this **initiation complex.** Next, the 50S subunit joins the complex, and IF-3 dissociates. Finally, GTP is broken down to GDP and phosphate, and IF-2 dissociates, leaving the complete ribosome bound to a fresh mRNA and the first aminoacyl-tRNA. Initiation is now complete (see panel 7).

In prokaryotes, the first amino acid, the one that initiates protein synthesis, is always the same: **N-formyl methionine (fMet).** The structure of fMet is given in figure 10.26*b*. N-formyl methionyl-tRNA (fMet-tRNA) forms in two steps. First, ordinary methionine binds to a special tRNA, then the methionine on the tRNA receives the formyl group. This unique amino acid participates only in initiation; it never goes into the interior of a polypeptide.

A Special tRNA for Initiation

The codon for fMet is AUG (or sometimes GUG). But these codons cause initiation only when they occur near the

beginning (the 5′-end) of an mRNA. If they appear in the middle of a message, they simply cause elongation, coding for methionine and valine, respectively. Thus, something special about the context of the codons AUG and GUG at the beginning of a message identifies them as initiation codons. These initiation codons are recognized by a special tRNA called **tRNA$_F^{Met}$**, which bears fMet. AUG and GUG in the interior of a message attract tRNA$_M^{Met}$ and tRNA^{Val}, bearing methionine and valine, respectively.

Eukaryotes perform the initiation step by a mechanism similar to that used by prokaryotes, but ordinary methionine, instead of fMet, plays the initiating role. The methionine may be ordinary, but it is set apart by being bound to a special tRNA, analogous to tRNA$_F^{Met}$, that participates only in initiation. Having served as the initiating amino acid, the methionine is usually, if not always, removed from eukaryotic proteins. In prokaryotes, the formyl group on the initiating fMet is always removed, and the whole fMet is removed from most proteins.

Initiation consists of binding an mRNA and the first aminoacyl-tRNA to the ribosome to form an initiation complex. This complex is built in stages by three initiation factors (IF-1, IF-2, and IF-3). Prokaryotes use a special amino acid (N-formyl methionine, fMet) and tRNA ($tRNA_F^{Met}$) for initiation. Eukaryotes use a unique tRNA, but unmodified methionine. AUG is ordinarily the initiating codon in prokaryotes and eukaryotes.

mRNA Ribosome Binding Sites

We have mentioned that there must be something special about the context of AUG codons at the beginning of a message that identifies them as initiation codons rather than ordinary methionine codons. Presumably, that "something" could be either a special primary structure (sequence of bases) or a characteristic secondary structure of the mRNA, such as a hairpin loop. Whatever the signal, it could be expected to be part of the ribosome binding site, the place on the mRNA where ribosomes first bind to initiate translation.

Using the RNA phage R17 as a model, Joan Steitz and her colleagues showed at the end of the 1960s that there is no secondary structure common to all initiation codons, even in this one phage. Furthermore, relaxing the secondary structure of the phage RNA increases, rather than decreases, its overall efficiency of translation. Therefore, it appears that ribosomes do not recognize a common secondary structure as a binding signal.

The alternative was to look for a common primary structure. J. Shine and L. Dalgarno pointed out that all the initiation regions recognized by *E. coli* ribosomes contained at least part of the sequence 5′-AGGAGGU-3′ a few bases upstream from the initiation codon. This sequence is complementary to a sequence contained at the very 3′-end of *E. coli* 16S rRNA: 5′-CACCUCCUUA-3′. Base pairing could therefore occur as follows:

16S rRNA: 3′-AUUCC UCC AC-5′

Initiation site: 5′- AGGAGGU-3′

This would provide a convenient mechanism for binding an mRNA to a ribosome. The 16S rRNA in the ribosome would simply base-pair with the **Shine-Dalgarno sequence (SD sequence)** in the initiation region of the mRNA. Most mRNAs do not have such a good fit with the 16S rRNA sequence. The R17 coat gene, for instance, has only three bases that can base-pair, yet it is still translated actively. But mRNAs that can form fewer than three base pairs are not translated well.

Eukaryotic mRNAs do not contain a sequence analogous to the Shine-Dalgarno sequence of prokaryotes. Instead, the cap at the 5′-end of a eukaryotic mRNA

(chapter 9) seems to provide a "docking site" for ribosomes. First, a **cap binding protein (CBP)** attaches to the cap; then initiation factors bind to CBP and promote 40S ribosomal subunit binding to form an initiation complex.

Most prokaryotic mRNAs have a sequence of bases (the Shine-Dalgarno sequence) just to the 5′-side of the initiation codon that is more or less complementary to a sequence at the very 3′-end of the 16S rRNA in the ribosome. Base pairing between these two regions of RNA presumably helps the ribosome recognize and bind to the initiation regions of mRNAs. Eukaryotic mRNAs have no Shine-Dalgarno sequences. Instead, the cap ensures binding of mRNA to ribosomes.

ELONGATION: ADDING AMINO ACIDS TO THE GROWING CHAIN

As suggested in the preceding discussion, ribosomes translate an mRNA in the 5′ → 3′ direction, the same direction in which the RNA was made. We will see that the protein grows in the amino (N) to carboxyl (C) direction. In other words, the N-terminal amino acid is added first, and the C-terminal amino acid is added last. Therefore, the 5′-end of the mRNA encodes the amino end of the protein.

The elongation step in protein synthesis is better visualized than described verbally (see panel 7). Figure 10.27 schematically depicts two rounds of elongation (adding two amino acids to a growing polypeptide chain) in *E. coli*. We start with fMet-tRNA bound to a ribosome. There are two binding sites for aminoacyl-tRNAs on the ribosome, called the **P (peptidyl) site** and the **A (aminoacyl) site.** In our schematic diagram, the P site is on the left and the A site is on the right. The fMet-tRNA is in the P site. Here are the elongation events:

1. To begin elongation, we need another amino acid to join with the first. This second amino acid arrives bound to a tRNA, and the nature of this aminoacyl-tRNA is dictated by the second codon in the message. The second codon is in the A site, which is otherwise empty, so our second aminoacyl-tRNA will bind to this site. Such binding requires a protein **elongation factor (EF)** known as **EF-Tu,** and GTP for energy.

2. Next, the first peptide bond forms. An enzyme called **peptidyl transferase**—an integral part of the large ribosomal subunit—transfers the fMet from its tRNA in the P site to the aminoacyl-tRNA in the A site. This forms a two-amino acid unit called a dipeptide linked to the tRNA in the A site. The whole assembly is called a dipeptidyl-tRNA. The tRNA in the P site, minus its amino acid, is said to be deacylated. The peptidyl transferase step in prokaryotes is inhibited by an important antibiotic called **chloramphenicol** (fig. 10.28).

FIGURE 10.27

Elongation in translation. Note first of all that this is a highly schematic view of protein synthesis. For example, tRNAs are represented by forklike structures that show only the two business ends of the molecule. Upper left: A ribosome with an mRNA attached is shown to illustrate the two sites, P and A, roughly delineated with dotted lines. Round 1: (*a*) EF-Tu brings in the second aminoacyl-tRNA (yellow) to the A site on the ribosome. The P site is already occupied by fMet-tRNA (red).

(*b*) Peptidyl transferase forms a peptide bond between fMet-tRNA and the second aminoacyl-tRNA. (*c*) In the translocation step, EF-G shifts the message and tRNAs one codon's width to the left. This moves the dipeptidyl-tRNA into the P site, bumps the deacylated tRNA aside, and opens up the A site for a new aminoacyl-tRNA. In round 2, these steps are repeated to add one more amino acid (green) to the growing polypeptide.

This drug has no effect on most eukaryotic ribosomes, which makes it selective for bacterial invaders in higher organisms. However, the mitochondria of eukaryotes have their own ribosomes, and chloramphenicol does inhibit their peptidyl transferase. Thus, chloramphenicol's selectivity for bacteria is not absolute.

3. In the next step, called **translocation,** the mRNA with its peptidyl-tRNA attached in the A site moves one codon's length to the left. This has the following results: (a) The deacylated tRNA in the P site (the one that lost its amino acid during the peptidyl transferase step when the peptide bond formed) leaves the ribosome. (b) The dipeptidyl-tRNA in the A site, along with its corresponding codon, moves into the P site. (c) The codon that was "waiting in the wings" to the right moves into the A site, ready to interact with an aminoacyl-tRNA. Translocation requires an elongation factor called EF-G plus energy from GTP.

FIGURE 10.28
Chloramphenicol.

The process then repeats itself to add another amino acid: (1) EF-Tu, in conjunction with GTP, brings the appropriate aminoacyl-tRNA to match the new codons in the A site. (2) Peptidyl transferase brings the dipeptide from the P site and joins it to the aminoacyl-tRNA in the A site, forming a tripeptidyl-tRNA. (3) EF-G, using energy from GTP, translocates the tripeptidyl tRNA, together with its mRNA codon, to the P site.

We have now completed two rounds of peptide chain elongation. We started with an aminoacyl-tRNA

(a)

Tyrosyl-tRNA

Puromycin

(b)

Polypeptide

Puromycin (puro-NH₂)

Peptidyl transferase

FIGURE 10.29

Puromycin structure and activity. (*a*) Comparison of structures of tyrosyl-tRNA and puromycin. Note the rest of the tRNA attached to the 5′-carbon in the aminoacyl-tRNA, where there is only a hydroxyl group in puromycin. (*b*) Mode of action of puromycin. First, puromycin (puro-NH₂) binds to the open A site on the ribosome. (The A site must be open in order for puromycin to bind.) Next, peptidyl transferase joins the peptide in the P site to the amino group of puromycin in the A site. Finally, the peptidyl-puromycin dissociates from the ribosome, terminating translation prematurely.

(fMet-tRNA) in the P site and we have lengthened the chain by two amino acids to a tripeptidyl-tRNA. This process continues over and over until the ribosome reaches the last codon on the message. The protein is now complete; it is time for chain termination.

Some controversy continues about the two-site model presented here, but there is general agreement about the existence of at least the A and P sites, based on experiments with the antibiotic **puromycin** (fig. 10.29). This drug has a structure that resembles an aminoacyl-tRNA; as such, it can be accepted by the A site of the ribosome. Then, since puromycin has an amino group in the same position as an aminoacyl-tRNA would, peptidyl transferase forms a peptide bond between the peptide in the P site and puromycin in the A site, yielding a peptidyl-puromycin. At this point, the ruse is over. Puromycin has no association with the mRNA and cannot be translocated to the P site. Therefore, the peptidyl-puromycin dissociates from the ribosome, aborting translation prematurely. This is why puromycin kills bacteria and other cells.

The link between puromycin and the two-site model is this: Before translocation, because the A site is occupied by a peptidyl-tRNA, puromycin cannot bind and release the peptide; after translocation, the peptidyl-tRNA has moved to the P site, and the A site is open. At this point, puromycin can bind and release the peptide. We therefore see two states the ribosome can assume: puromycin-reactive and puromycin-unreactive. Those two states require at least two binding sites on the ribosome for the peptidyl-tRNA.

The model in figure 10.27 is simplified for ease of understanding; for example, it pictures the ribosome as a two-dimensional entity, with the mRNA traveling in only one dimension. This does not accurately reflect the real situation, which must exist in three dimensions.

Elongation takes place in three steps: (1) EF-Tu, with energy from GTP, binds an aminoacyl-tRNA to the ribosomal A site. (2) Peptidyl transferase forms a peptide bond between the peptide in the P site and the newly arrived aminoacyl-tRNA in the A site. This lengthens the peptide by one amino acid and shifts it to the A site. (3) EF-G, with energy from GTP, translocates the growing peptidyl-tRNA, with its mRNA codon, to the P site.

TERMINATION: RELEASE OF THE FINISHED POLYPEPTIDE

In our discussion of the genetic code, we saw that three codons (UAG, UAA, and UGA), instead of coding for amino acids, cause termination of protein synthesis. In the early 1960s, Seymour Benzer and his colleagues discovered the first stop codon (later determined to be UAG) as a mutation in phage T4 that rendered the phage incapable of growing, except in special **suppressor** strains of *E. coli* that counteract the effect of the mutation. These workers called the mutation **amber,** and that name has stuck to the codon **UAG** ever since. The reason for the lethality of the amber mutation is that it causes premature termination in the middle of messages. This was first demonstrated with amber mutations in the alkaline phosphatase gene of *E. coli.* Such mutants produce no full-length alkaline phosphatase; instead, they make fragments that begin at the amino terminus of the protein and stop somewhere in the middle.

Later, a new mutation occurred that could not grow in **amber suppressor** strains, but could grow in certain other suppressor strains. In keeping with the colorful name amber, these new mutants were called **ochre;** they involve the codon **UAA,** which also causes premature termination. The third stop codon, **UGA,** has been named in kind: **opal.**

The sequences of the amber and ochre codons were discovered in 1965, not by sequencing mRNA, but by sequencing protein products. The rationale was as follows: Amber mutants regularly reverted to a phenotype that no longer required the amber suppressor. But these revertants did not usually give a product exactly like that in wild-type cells. Instead, one amino acid, the one corresponding to the position of the amber mutation, was different. Martin Weigert and Alan Garen studied such revertants of an amber mutation at one position in the alkaline phosphatase gene of *E. coli.* The amino acid at this position in wild-type cells was tryptophan, whose sole codon is UGG. Since the amber mutation originated with a one-base change, we already know that the amber codon is related to UGG by a one-base change. To find out what this change was, Weigert and Garen determined the amino acids inserted in this position by several different revertants. The amino acids found there were serine, tyrosine, leucine, glutamic acid, glutamine, and lysine. The only codon that can yield at least one codon for each of those amino acids, including tryptophan, by a one-base change is UAG (fig. 10.30).

By the same logic, including the fact that amber mutants can mutate by single-base changes to ochre mutants, Sydney Brenner and his colleagues reasoned that the ochre codon must be UAA. Severo Ochoa and his group verified that UAA is a stop signal when they showed that the synthetic message AUGUUUUAAA- - - directed the synthesis and release of the dipeptide fMet-Phe. (AUG codes for

FIGURE 10.30

The amber codon is UAG. The amber codon (middle) came via a one-base change from the tryptophan codon (UGG), and the gene reverts to a functional condition in which one of the following amino acids replaces tryptophan: serine, tyrosine, leucine, glutamic acid, glutamine, or lysine. The red represents the single base that is changed in all these revertants, including the wild-type revertant that codes for tryptophan.

fMet; UUU codes for Phe; and UAA codes for stop.) With UAG and UAA assigned to the amber and ochre codons, respectively, UGA must be the opal codon. These three codons, because they code for no amino acid, are frequently referred to as **nonsense codons.** This is really a misnomer. They make perfect sense; they say, "Stop."

How do suppressors overcome the lethal effects of premature termination signals? A clue came from an experiment showing that an amber mutation could be suppressed in vitro by purified tRNA from an amber suppressor strain. Therefore, tRNA is the suppressor molecule. Furthermore, suppressor strains insert amino acids into polypeptides in the position corresponding to the stop codon. Together, these data suggest that a tRNA from the suppressor strain recognizes the stop codon and inserts an amino acid, thus preventing termination. Proof of this mechanism came from experiments on an amber suppressor strain that inserts a tyrosine in the position corresponding to an amber codon. A tRNA^Tyr from this strain has an altered anticodon. Instead of 3′AUG 5′, which would respond to the tyrosine codon UAC, it has the anticodon 3′AUC 5′, which responds to the amber codon UAG. This tRNA can insert a tyrosine in response to the stop codon UAG, blocking termination (fig. 10.31). Thus, the suppressor is itself a mutant. Its mutation is in a gene for a tRNA; in particular, in the part of the gene that specifies the anticodon.

We have seen that three codons, UAG, UAA, and UGA, can cause premature termination when they occur in the middle of a message. Are these actually the termination signals at the ends of messages? Every gene sequenced to date (except in certain mitochondria) has at least one of these codons at the end, just after the codon for the last amino acid in the protein. Sometimes there are two stop codons in a row, such as the UAAUAG sequence at the end of the coat gene in R17 phage.

Because a special tRNA participates in initiation of translation, it would seem logical to propose a similar mechanism for termination, especially since the termination signals are three-base codons just like all other codons. Indeed, it was suggested that special tRNAs recognize the

FIGURE 10.31

Mechanism of suppression. Top: The original codon in the wild-type *E. coli* gene was CAG, which was recognized by a glutamine tRNA. Middle: This codon mutated to UAG, which was translated as a stop codon by a wild-type strain of *E. coli*. Notice the tyrosine tRNA, whose anticodon (AUG) cannot translate the amber codon. Bottom: A suppressor strain contains a mutant tyrosine tRNA with the anticodon AUC instead of AUG. This altered anticodon recognizes the amber codon and causes the insertion of tyrosine (gray) instead of allowing termination.

Gln

GUC

Original codon: CAG ⟶ Gln

Mutation

UAG

Tyr

AUG

Translation in wild-type strain: UAG ⟶ (Stop)

Tyr

AUC

Translation in suppressor strain: UAG ⟶ Tyr

termination signals. However, when Mario Capecchi devised an assay for termination in 1967, he discovered that when he primed the ribosome with a synthetic mRNA containing an early stop codon, termination was triggered, not by RNA, but by protein factors. These factors, termed **release factor-1** and **release factor-2** (**RF-1** and **RF-2**), recognize the stop codons as follows: RF-1 reacts to UAA and UAG; RF-2 reacts to UAA and UGA. Both cause breakage of the bond between the polypeptide and its tRNA, releasing the product from the ribosome. Another factor, **RF-3,** appears to cooperate with both of the others. In eukaryotes, a single release factor, **eRF,** causes termination. This reaction, unlike the termination reactions in prokaryotes, requires GTP.

 Translation release factors, RF-1 and RF-2 recognize the stop signals UAA, UAG, and UGA, and, along with RF-3, cause termination in prokaryotes. Along with GTP, eRF performs the same function in eukaryotes. When a mutation creates one of these stop signals in the middle of a message, premature termination occurs. Such mutations can be overridden by suppressors—tRNAs that respond to termination codons by inserting an amino acid, thus preventing termination.

CODING AND ANTICODING STRANDS OF DNA

Now that you know the relationship between the sequence of bases in a gene and the sequence of amino acids in its polypeptide product, you can appreciate the nomenclature we apply to the two DNA strands. Figure 10.32 summarizes the transcription-translation process and introduces the coding-anticoding nomenclature we apply to the strands of DNA. Notice that the mRNA has the same sequence (except that U's substitute for T's) as the top strand (blue) of the DNA. In turn, the mRNA codes for protein, in this case the tetrapeptide fMet-Ser-Asn-Ala. Since the top DNA strand codes for protein in the same way that mRNA does, we call it the **coding strand.** The opposite DNA strand is therefore the **anticoding strand.** A possible source of confusion is the fact that the anticoding DNA strand is the one that is copied, or transcribed, into mRNA. Just remember that the transcribed strand is complementary, or opposite, to the mRNA, and is therefore the anticoding strand.

Since it serves as the template for making mRNA, the transcribed strand is also called the **template strand.** Its partner is therefore the **nontemplate strand.** Many geneticists use the terms *sense strand* and *antisense strand* synonymously with *coding strand* and *anticoding strand.*

Gene: ATGAGTAACGCG Coding strand
         TACTCATTGCGC Anticoding strand

↓ Transcription

mRNA:AUGAGUAACGCG

↓ Translation

Protein: fMetSerAsnAla

FIGURE 10.32

Outline of gene expression. In the first step, transcription, the anticoding strand is transcribed into mRNA. Note that the coding strand (blue) of the DNA has the same sequence (except for the T–U change) as the mRNA (red). In the second step, the mRNA is translated into protein (gray). This little "gene" is only twelve base pairs long and codes for only four amino acids (a tetrapeptide). Real genes are much larger.

However, some geneticists use these terms in exactly the opposite way. Thus, to avoid confusion we will not use these terms in this book.

THE RELATIONSHIP BETWEEN DNA SIZE AND GENETIC CAPACITY

How many genes are in a given DNA? It is impossible to tell just from the sizes of the DNAs, because we do not know how much of a given DNA is devoted to genes and how much is intergenic space, or even intervening sequences within genes. However, we can estimate an upper limit on the number of genes a DNA can hold. We start with the assumption that an average protein has a molecular weight of about 40,000. How many amino acids does this represent? The molecular weights of amino acids vary, but they average a bit over 100. To simplify our calculation, let us assume that it is 100. That means our average protein contains 40,000/100, or 400 amino acids. Since each amino acid requires three base pairs to code for it, a protein containing 400 amino acids needs a gene of about 1,200 base pairs.

Consider a few of the DNAs listed in table 6.3. The *E. coli* chromosome contains 4.2×10^6 base pairs, so it could encode about 3,500 average proteins. The phages T2 and T4, which infect *E. coli,* have only 2×10^5 base pairs, so they can code for only about 160 proteins. The smallest double-stranded DNA on the list, belonging to the mammalian tumor virus SV40, has a mere 5,226 base pairs. In principle, that is only enough for four proteins, but the virus squeezes in some extra information by overlapping its genes.

DNA CONTENT AND THE C-VALUE PARADOX

You would probably predict that complex organisms like vertebrates would need more genes than simple organisms like yeast. Therefore, they should have higher **C-values,** or DNA content per haploid cell. In general, your prediction

would be right; mouse and human haploid cells contain more than one hundred times more DNA than yeast cells. In addition, yeast cells have about five times more DNA than *E. coli* cells, which are even simpler. However, this correspondence between an organism's physical complexity and the DNA content of its cells is not perfect. Consider, for example, the frog. There is no reason to suspect that an amphibian has a great many more genes than a human, yet the frog has seven times more DNA per cell. Even more dramatic is the fact that the lily has one hundred times more DNA per cell than a human.

This perplexing situation is called the **C-value paradox.** It becomes even more difficult to explain when we look at organisms within a group. For example, some amphibian species have a C-value one hundred times higher than that of others, and the C-values of flowering plants vary even more widely. Does this mean that one kind of higher plant has one hundred times more genes than another? That is simply unbelievable. It would raise the questions of what all those extra genes are good for and why we do not notice tremendous differences in physical complexity among these organisms. The more plausible explanation of the C-value paradox is that organisms with extraordinarily high C-values simply have a great deal of extra, noncoding DNA. The function, if any, of this extra DNA is still mysterious.

In fact, even mammals have a lot more DNA than they need for genes. Applying our simple rule (dividing the number of base pairs by 1,200) to the human genome yields an estimate of about two million for the maximum number of genes. We learned in chapter 9 that most genes in higher eukaryotes actually have within them one or more noncoding introns. However, even assuming five to ten times as much intron DNA as coding DNA leaves us with an estimate of about 200,000 to 400,000 genes, which is still almost certainly too high.

If we cannot use the C-value as a reliable guide to the number of genes in higher organisms, how can we make a reasonable estimate? One approach has been to count the number of different mRNAs in a given mammalian cell type, which gives an estimate of about 10,000. Of course, mammals have over one hundred different cell types, but they all produce many common mRNAs. Therefore, we cannot simply multiply 10,000 times 100. In fact, the overlap in genes expressed in different mammalian cells is so extensive that we probably need to multiply 10,000 by a number less than 10. By this reasoning, the upper limit on the number of functional mammalian genes is about 100,000.

 There is a rough correlation between the DNA content and the number of genes in a cell or virus. However, this correlation breaks down in several cases of closely related organisms where the DNA content per haploid cell (C-value) varies widely. This C-value paradox is probably explained, not by extra genes, but by extra noncoding DNA in some organisms.

WORKED-OUT PROBLEMS

Problem 1

Table 6.3 shows that the DNAs of bacteriophages T2 and T4 contain 2×10^5 base pairs. How many genes of average size (encoding proteins of about 40,000 molecular weight) can these phages contain?

Solution

Since the average molecular weight of an amino acid is about 100, a protein with a molecular weight of 40,000 contains about 400 amino acids. It takes three base pairs to encode one amino acid, so each protein will be encoded in about 1,200 base pairs ($3 \times 400 = 1,200$). Thus, the number of genes of average size in a DNA having 2×10^5 base pairs will be $2 \times 10^5 / 1.2 \times 10^3 = 1.67 \times 10^2$, or 167.

Problem 2

Below is a double-stranded bacterial DNA. The top strand is the coding strand. Find and write the sequence of the open reading frame in the mRNA that would be transcribed from this gene.

5′ATCCGATGAAACCGTGGACACCCAGATAAATCG3′
3′TAGGCTACTTTGGCACCTGTGGGTCTATTTAGC5′

Solution

Remembering that T in DNA replaces U in RNA, we need to look for an ATG initiation codon near the 5′-end of the top (coding) strand. Such a sequence begins with the A in the sixth position from the 5′-end of the coding strand. Now write the sequence of bases in groups of three until you encounter a stop codon (TAG, TAA, or TGA, corresponding to UAG, UAA, or UGA). There is a TAA beginning with the T that is the seventh base from the 3′-end. Writing the sequence in groups of three (as codons) as follows will help you verify that there are no intervening stop codons:

5′ AUG AAA CCG UGG ACA CCC AGA UAA 3′

Problem 3

(a) Use the genetic code to predict the amino acid sequence of the small polypeptide encoded in this mini-message, assuming that only physiologically significant initiation signals are used.

5′AUGAGAUACCAUGGGCUAAUGUGAAAA3′

What amino acid sequences would result if the following changes occurred in the message?

(b) The first C is changed to a G.

(c) The first U is changed to a G.

(d) The first C is changed to a U.

(e) The second G is changed to an A.

(f) The first C is deleted.

(g) An extra G is added after the first G.

Solution

(a) Since physiologically significant initiation signals (AUGs) are the only ones used, translation of this message will begin with the first codon (AUG). In order to simplify the decoding of the message, it helps to write the codons in order, separated by spaces, as follows:

AUG AGA UAC CAU GGG CUA AUG UGA AAA

Next, we consult the genetic code for the meaning of each of these codons, i.e., the amino acids specified by these codons.

fMet Arg Tyr His Gly Gln Met Stop Lys
AUG AGA UAC CAU GGG CUA AUG UGA AAA

The next-to-last codon (UGA) is a stop signal, so translation will stop after the methionine (Met) codon (AUG). Thus, the last codon (AAA, lysine) will remain untranslated, and the amino acid sequence of the peptide encoded in this little message is: fMet-Arg-Tyr-His-Gly-Gln-Met.

(b) The first C is changed to a G. The new message will read AUG AGA U**A**G etc., where the boldface indicates the altered base. Now UAG is a stop codon, so only the first two codons will be translated, and the product will be fMet-Arg.

(c) The first C is changed to a U. The new message will read AUG AGA UA**U** etc., where the boldface indicates the altered base. UAU codes for tyrosine (Tyr), just as UAC does, so there will be no change in the amino acid sequence.

(d) The second G is changed to an A. The new message will read AUG A**A**A etc., where the boldface indicates the altered base. AAA codes for lysine (Lys), so the product will have Lys in the second position, and the rest will remain the same: fMet-Lys-Tyr-His-Gly-Gln-Met.

(e) The first G is changed to a U. This changes the first codon from AUG to A**U**U, which is no longer an initiation codon. Thus, initiation will occur at the next AUG to the right. Note that this does not have to be in the original reading frame, and in fact it is not in this case. The first AUG in this new message begins at the 11th base: AUUAGAUACC AUG GGC UAA UGU GAA AA. The coding sequence and the corresponding amino acid sequence of this new open reading frame will be:

fMet Gly   Stop
AUG GGC UAA UGU GAA AA

Thus, this message codes for fMet-Gly.

(f) The first C is deleted. The new message will read:

AUG AGA U**AC** AUG GGC UAA UGU GAA AA

where the boldfaced bases lie on either side of the deleted base. The coding sequence and corresponding amino acid sequence are as follows:

fMet Arg Tyr Met Gly Stop Cys Glu

AUG AGA UAC AUG GGC UAA UGU GAA AA

Thus, this message codes for fMet-Arg-Tyr-Met-Gly.

(g) An extra G is added after the first G. The new message will read:

AUG **G**AG AUA CCA UGG GCU AAU GUG AAA A

where the boldface represents the inserted base. This message will have the following coding properties:

fMet Glu Ile Pro Trp Ala Asn Val Lys

AUG GAG AUA CCA UGG GCU AAU GUG AAA A

Thus, this message codes for fMet-Glu-Ile-Pro-Trp-Ala-Asn-Val-Lys.

Problem 4

A certain amber suppressor inserts tryptophan (Trp) in response to the amber codon. What change in the anticodon of tRNA^{Trp} probably created this suppressor strain?

Solution

The tryptophan codon is 5´UGG3´, which is recognized by the anticodon 3´ACC5´ of the wild-type tRNA^{Trp}. The amber codon is 5´UAG3´, which is most likely recognized by the anticodon 3´AUC5´of the amber suppressor tRNA. Thus, comparing this anticodon (3´AUC5´) with the wild-type one (3´ACC5´) shows that the change in the tRNA^{Trp} anticodon that created the amber suppressor was 3´ACC5´ → 3´AUC5´.

Summary

Proteins, or polypeptides, are polymers of amino acids linked through peptide bonds. Most genes contain the information for making one polypeptide, and are expressed in a two-step process: transcription or synthesis of an mRNA copy of the gene, followed by translation of this message to protein. Translation takes place in three stages: initiation, elongation, and termination, on complex structures called ribosomes, the cell's protein factories. Translation also requires adapter molecules that can recognize both the genetic code in mRNA and the amino acids the mRNA encodes. Transfer RNAs (tRNAs), with their two business ends, fill this role.

The cloverleaf model of tRNA shows its secondary (base-pairing) structure and reveals four base-paired stems that define three loops of RNA: the D-loop, the anticodon loop, and the TψC loop. Transfer RNAs contain many modified bases, at least some of which are vital to tRNA activity. The three-dimensional structure of tRNA consists of an L-shaped framework, with the bases of the anticodon protruding to one side and arranged in a helical shape for easy pairing with the three-base codon. Each aminoacyl-tRNA synthetase couples a specific amino acid to its cognate tRNA.

The genetic code is a set of three-base code words, or codons, in mRNA that instruct the ribosome to incorporate specific amino acids into a polypeptide. The code is nonoverlapping; that is, each base is part of only one codon. It is devoid of gaps, or commas; each base in the coding region of an mRNA is part of a codon.

The code is also degenerate. More than one codon can code for one amino acid. Part of the degeneracy of the code is accommodated by isoaccepting species of tRNA that bind the same amino acid but recognize different codons. The rest is handled by wobble, in which the third base of a codon is allowed to move slightly from its normal position to form a non-Watson-Crick base pair with the anticodon.

Ribosomal subunits can be dissociated into their component parts, RNAs and proteins, and then reassociated. Thus, the ribosome is a self-assembly system that needs no help from the outside.

Initiation of translation consists of binding the first aminoacyl-tRNA to the ribosome-mRNA complex. Prokaryotes use a special amino acid (N-formyl methionine, fMet) and tRNA (tRNA$_F^{Met}$) for initiation. A sequence of bases (the Shine-Dalgarno sequence) just to the 5´-side of the initiation codon in prokaryotic mRNAs is more or less complementary to a sequence at the very 3´-end of the 16S rRNA in the ribosome. Base pairing between these two regions of RNA presumably helps the ribosome recognize and bind to the initiation regions of the mRNAs.

Elongation takes place in three steps: (1) EF-Tu, with energy from GTP, binds an aminoacyl-tRNA to the ribosomal A site. (2) Peptidyl transferase forms a peptide bond between the peptide in the P site and the newly arrived aminoacyl-tRNA in the A site. This lengthens the peptide by one amino acid and shifts it to the A site. (3) EF-G, using energy from GTP, translocates the growing peptidyl-tRNA, with its mRNA codon, to the P site.

Translation release factors RF-1 and RF-2 recognize the stop signals UAA, UAG, and UGA, and, along with RF-3, cause termination in prokaryotes. Together with GTP, eRF performs the same function in eukaryotes. When a mutation creates one of these stop signals in the middle of a message, premature termination occurs. Such mutations can be overridden by suppressors—tRNAs that respond to termination codons by inserting an amino acid, thus preventing termination.

There is a rough correlation between the DNA content and the number of genes in a cell or virus. However, this correlation does not hold in several cases of closely related organisms where the DNA content per haploid cell (C-value) varies widely. This C-value paradox is probably explained by extra noncoding DNA in some organisms.

PROBLEMS AND QUESTIONS ?

1. An animal virus contains a circular double-stranded DNA with 120,000 base pairs. How many proteins of average size (400 amino acids) can the virus code for?

2. You have isolated a *Neurospora* mutant that cannot make amino acid z. Normal *Neurospora* cells make z from a common cellular substance x by a pathway involving three intermediates, *a, b,* and *c:* $x \rightarrow a \rightarrow b \rightarrow c \rightarrow z$. How would you establish that your mutant contains a defective gene for the enzyme that catalyzes the $b \rightarrow c$ reaction?

3. Here is a schematic diagram of a double-stranded DNA being transcribed.

    1. _____
    2. _____
    3. --------------- →

    Strand 1 is the coding strand. Strand 3 is the nascent RNA. (*a*) Label the 5′- and 3′- ends of the RNA. (*b*) Label the 5′- and 3′- ends of the coding strand. (*c*) Label the 5′- and 3′- ends of the anticoding strand.

4. If a mutant *E. coli* cell had an aminoacyl-tRNA synthetase that should attach the amino acid phenylalanine to its tRNA (tRNA^{Phe}), but at the elevated temperature of 42° C attached the amino acid arginine instead, what consequence would this have for the proteins that cell made at the elevated temperature? Would this affect the function of these proteins? Why, or why not?

5. Here is the sequence of the anticoding strand of a DNA fragment:

    5′GTCAATCGTAGCGGCCATAT3′

    (*a*) Write the sequence of this DNA in double-stranded form. (*b*) Assuming that transcription of this DNA begins with the first nucleotide and ends with the last, write the sequence of the transcript of this DNA in conventional form (5′ to 3′ left to right).

6. Would you predict that two tRNAs with complementary anticodons would bind to one another less tightly, more tightly, or the same as two complementary trinucleotides? Why? *Hint:* Free trinucleotides have no special shape, but tRNAs do.

7. If you charged a methionine tRNA with methionine, then chemically modified the amino acid to alanine and used this alanyl-tRNA^{Met} in an in vitro translation reaction, what do you predict would be the result? In other words, would the alanine go into the protein where an alanine is supposed to go or where a methionine is supposed to go? Why?

8. What does the answer to question 7 say about the importance of accurate aminoacyl-tRNA synthetases in faithful translation?

9. What is the product of the reaction catalyzed by aminoacyl-tRNA synthetase?

10. Consider this short message: 5′-AUGGCAGUGCCA-3′. Answer the following questions, assuming first that the code is overlapping and then that it is nonoverlapping. (*a*) How many codons would be represented in this oligonucleotide? (*b*) If the second G were changed to a C, how many codons would be changed?

11. What would be the effect on reading frame and gene function if: (*a*) two bases were inserted into the middle of a message? (*b*) three bases were inserted into the middle of a message? (*c*) one base was inserted in one codon and one subtracted from the next?

12. If codons were six bases long, what kind of product would you expect from a repeating tetranucleotide such as poly(UUCG)?

13. What would be the effect of a mutation in the initiating codon of a gene?

14. (*a*) Use the genetic code to predict the amino acid sequence of a peptide encoded in this mini-message: 5′-AUGUUCAAGAUGGUGACUUGGUAAAUC-3′. (*b*) What amino acid sequence would result if the first G in this message were changed to C, assuming that only physiologically meaningful start codons are used? (*c*) What amino acid sequence would result if the first C were changed to a U? (*d*) If the first C were changed to a G? (*e*) If the last G were changed to an A?

15. (*a*) Which three amino acids have codons with 6-fold degeneracy? (*b*) Which two amino acids have only one codon each?

16. How many different codons would exist in a genetic code with four-base codons?

17. Assuming wobble can occur, what two codons could be recognized by the anticodon 3′-GAG-5′? Write the codons in the 5′ → 3′ direction, and remember that the pairing between codon and anticodon is antiparallel.

18. What anticodon could recognize all three codons for isoleucine, assuming wobble? Write the anticodon in the 3′ → 5′ direction.

19. A ribonuclease called cloacin DF13 specifically cleaves off 49 bases from the 3′-end of 16S rRNA in the intact ribosome. What effect would you expect this RNase to have on translation? Why?

20. Where would chain elongation be stalled if mutations were to occur in the genes for the following proteins: (*a*) EF-G; (*b*) EF-Tu; (*c*) peptidyl transferase?

21. Of all the aminoacyl-tRNAs, only one enters the P site on the ribosome. Which one is it? Why is it important that this aminoacyl-tRNA not stay in the A site?

22. A bacterium makes a protein that contains 403 amino acids, counting the terminal methionine that is retained. How many peptidyl transferase reactions occur during synthesis of this protein?

23. If every stop signal in phage QT were UAGUGA, what would happen to all the phage QT proteins made in an amber suppressor strain in which the suppressor tRNA is tRNA^{Tyr}? Why?

24. A certain ochre suppressor inserts glutamine (Gln) in response to the ochre codon. What is the likeliest change in the anticodon of a tRNA^{Gln} that created this suppressor strain?

Answers appear at end of book.

CHAPTER 11: GENE MUTATION I: DNA DAMAGE AND REPAIR

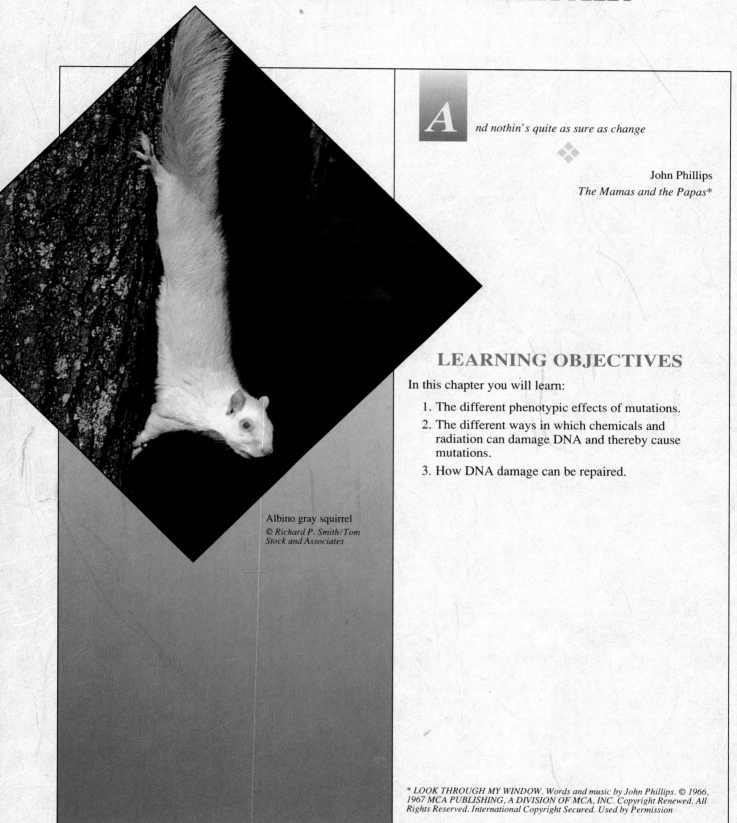

A nd nothin's quite as sure as change

John Phillips
*The Mamas and the Papas**

Albino gray squirrel
© *Richard P. Smith/Tom
Stock and Associates*

LEARNING OBJECTIVES

In this chapter you will learn:

1. The different phenotypic effects of mutations.
2. The different ways in which chemicals and radiation can damage DNA and thereby cause mutations.
3. How DNA damage can be repaired.

*C*hapters 7–10 have treated in detail the first two fundamental characteristics of genes: They replicate faithfully, and they store information for the production of RNAs and proteins. The third characteristic of genes is that they accumulate alterations, or mutations. We have already mentioned mutations several times in this book; indeed, there could hardly be a science of genetics without mutations, since we usually do not notice a gene until a mutation changes it. So far, however, we have not said much about the mechanisms by which genes are changed. That will be the main purpose of this chapter.

Sometimes a mutation goes beyond the alteration of one or a few bases and involves a larger part of a chromosome. Such changes are called chromosome mutations. Genomic mutations, or aneuploidies, are even more noticeable and involve loss or gain of whole chromosomes. We have already discussed chromosome mutations and aneuploidy in chapter 4 and we will not consider them further here. Instead, we will focus on **gene mutations,** alterations that are confined to a single gene.

TYPES OF GENE MUTATIONS

We can look at gene mutations in several different ways. For example, we can classify them according to their effects on the phenotype of the mutant organism or according to their effects on the genetic material itself. Let us first examine the various ways that mutations affect phenotype.

SOMATIC VERSUS GERM-LINE MUTATIONS

Multicellular organisms that reproduce sexually can experience mutations in their sex cells, in which case the gametes can be altered and the mutation passed on to the progeny. Such mutations are called **germ-line,** or **germinal mutations.** A germ-line mutation probably happened to Queen Victoria of Great Britain (chapter 3), because some of her male descendants were afflicted with hemophilia, a sex-linked blood clotting disorder, yet none of her ancestors apparently had this disease.

On the other hand, mutations in the nonsex cells, or **somatic cells,** may change the phenotype of the individual that suffers the mutation, but the mutant trait will stop with that individual. Since the mutation does not affect the gametes, it cannot be passed on.

Examples of such **somatic mutations** abound. Consider, for instance, a kernel of Indian corn such as the one pictured in figure 11.1. A wild-type kernel would be solid purple, but a mutation has inactivated the color gene, so this kernel is mostly white. However, you can see that the kernel is not solid white, but speckled. Each spot of color represents a back-mutation, or reversion, that reactivated the color gene in one cell, which then grew into a patch of colored cells. If the mutation takes place early in kernel development (fig. 11.2*a*), when there are few cells, the

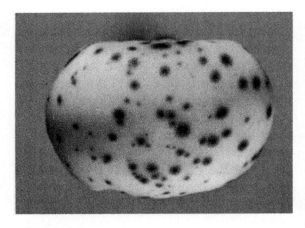

FIGURE 11.1

Somatic mutation. Each colored spot on this speckled kernel of corn represents a group of cells descended from one that experienced a back-mutation in a color gene. Without the back-mutations, the kernel would be all white.

From "Transposable Genetic Elements in Maize" by Nina V. Fedoroff. Copyright © 1984 by SCIENTIFIC AMERICAN, Inc. June, p. 86, fig. H. All rights reserved.

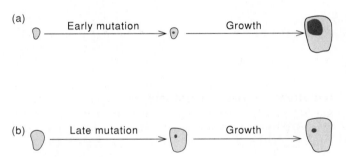

FIGURE 11.2

Influence of timing on the effect of somatic mutation. (*a*) The mutation occurs early in development of the seed, so many colored cells descend from the one with the back-mutation, producing a large colored spot. (*b*) The mutation occurs late in seed development, so not many more cell divisions take place. As a result, the colored spot is small.

mutant cell will give rise to a large group of colored progeny cells, so the colored patch will be large and will sometimes encompass the entire kernel. On the other hand, if the mutation takes place late in seed development (fig. 11.2*b*), when there are many cells and few cell divisions left to go, the colored spot will be tiny. It is obvious that these mutations must have occurred during the lifetime of this corn plant and in the endosperm tissue, which is made up of somatic cells. Therefore, they must be somatic mutations; we will see in chapter 12 that these particular corn mutations are caused by mobile pieces of DNA called transposable genetic elements.

It is also likely that somatic mutations play a major role in cancer. Many cancers are derived from a single cell that has changed its behavior from normal to malignant. It appears that somatic mutations frequently contribute to such behavior changes. We will explore this topic further in chapter 12.

FIGURE 11.3

An albino mouse (top) with a wild-type littermate.

R. L. Brinster, from the cover of Cell, *vol. 27, November 1981. Reprinted with permission from Cell Press.*

 Germ-line mutations affect the gametes, so the mutation can be transmitted to the next generation. Somatic mutations affect the nonsex, or somatic cells. Their effects are felt in the organism that experiences the mutation, but are passed no further.

MORPHOLOGICAL MUTATIONS

Complex multicellular organisms carrying **morphological mutations,** or **visible mutations,** can usually be distinguished readily from wild-type organisms because of their altered appearance. For example, **albino** mammals, including humans, have a mutation in a gene that is responsible for dark coat (or skin) pigment. This is ordinarily the gene for tyrosinase, the key enzyme in the pathway that leads to melanin, the black pigment in hair, eyes, and skin. A mutated tyrosinase gene may produce no active enzyme, so no melanin can be made. As a result, albino humans have very fair skin and hair, and light blue eyes; albino mice have white fur and pink eyes (fig. 11.3).

Many morphological mutants of the fruit fly *Drosophila* have been observed. Among these are *white* flies, with white eyes instead of the usual red; *miniature,* with tiny wings; *Bar eye,* with bar-shaped eyes instead of round; and *Curly,* with curly wings instead of straight (fig. 11.4).

Microorganisms and even viruses also show a few morphological mutations. The mutant bacterium *Streptococcus pneumoniae,* used by Avery and his colleagues to demonstrate that DNA is the genetic material (chapter 6), had a different morphology (rough) than the wild-type (smooth). A mutant yeast is called *petite* because it forms much smaller than normal colonies. The bacterial virus T4 can be mutated to the *rapid lysis,* or *r* form, which clears out a bigger and clearer hole, or plaque, in a lawn of infected bacteria than does the wild-type phage (chapter 13). Thus, the altered morphology we observe in microorganisms is usually not in a single organism, but in colonies of organisms—or, in the case of a virus, in plaques.

(a)

(b)

FIGURE 11.4

Morphological mutation in *Drosophila.* (*a*) A wild-type fly with red eyes is contrasted with (*b*) a white-eyed mutant fly.

Gerald M. Rubin, from the cover of Cell, *vol. 40, April 1985. Reprinted with permission from Cell Press.*

 Morphological mutations cause a visible change in the affected organism.

NUTRITIONAL MUTATIONS

Although some morphological mutations can be distinguished in microorganisms, these are necessarily few in number. That is because these simple organisms have a limited number of observable morphological features. Thus, we must usually rely on more subtle mutant phenotypes. The most popular of these is the inability to grow on simple growth media. Certain wild-type organisms can grow on "minimal medium," which contains just salts and an energy source such as glucose. These organisms, called **prototrophs,** can make all the amino acids, nucleotides, vitamins, and other substances they need to sustain life. Nutritional mutants, also called **auxotrophs,** need an extra substance in order to survive, so they die on minimal medium.

For example, a *bio* mutant of *E. coli* will die unless it is supplied with the vitamin biotin, while a *leu* mutant will die without the amino acid leucine. Similarly, a pantothenateless mutant of *Neurospora* will die without the vitamin pantothenate. The reason for these deficiencies, of course, is that genes for enzymes in the biotin, leucine, and pantothenate synthesis pathways, respectively, are defective. Since these genes do not make active enzymes, no biotin, leucine, or pantothenate can be made and the cells die. Such mutations are also sometimes called biochemical mutations, but that is not a very useful term, since all mutations have biochemical effects.

 Nutritional mutations render an organism dependent on added nutrients that a wild-type organism does not require, since it can make these substances itself.

LETHAL MUTATIONS

Some mutations are so severe that an organism carrying them cannot survive at all; these mutations are called **lethals** (chapter 3). An example would be an inactivating mutation in the gene for one of the subunits of RNA polymerase. If the mutation rendered the RNA polymerase inactive, the mutant organism could make no RNA and therefore could not live. Haploid organisms with lethal mutations die immediately because they have no wild-type gene to compensate. But diploid organisms can carry a lethal mutation, masked by a wild-type allele, for generations with no ill effects. It is only when two heterozygotes mate that we may discover the lethal consequences of the mutation. In chapter 3 we saw one such example in yellow mice. The heterozygotes are viable, but homozygous yellow mice are never seen—they die in the mother's uterus. Thus, most lethal mutations in diploid organisms are recessive. If one parent contributes a defective gene for an essential protein and the other contributes a wild-type gene, the latter will usually allow the cell to make enough protein to compensate. It is only when two defective genes are brought together in a homozygote that lethality ordinarily results.

Huntington disease in humans is an example of a dominant lethal. You may be wondering how a dominant lethal mutation allows the person to live long enough to have a disease, let alone pass the mutant gene along to his or her children. We do not know how the disease operates, but we do know that heterozygotes typically live normal lives well past the childbearing years, then begin to show signs of the degenerative disease that ultimately kills them. This is the fate that befell folk singer Woody Guthrie.

 Lethal mutations occur in indispensable genes. Haploid organisms with lethal mutations die immediately, but lethal mutations are usually recessive, so diploid organisms can tolerate them in heterozygotes.

CONDITIONAL MUTATIONS

As we have seen, lethal mutations in haploid organisms usually cause death immediately. Hence, they are of little use to geneticists who are interested in studying the altered gene product. With no organism, there can be no gene product. One way around this problem is to look for **conditional lethals.** These are mutations that are lethal under certain circumstances, but not under others.

Consider, for example, a **temperature-sensitive (ts) mutation.** This allows growth at low temperature (the **permissive temperature**), but not at normal growth temperature (the **restrictive,** or **nonpermissive temperature**), so lethality in this case is conditional on temperature. It is important to realize that the gene itself is not temperature-sensitive; its protein product is. The mutation creates an altered protein that is more easily unfolded, or **denatured,** than the wild-type protein. Since heat is a powerful protein denaturant, the mutant protein will therefore be heat-sensitive. One has only to consider what happens to a boiled egg to appreciate heat denaturation of protein: The transparent solution of protein (mostly ovalbumin) in the egg white becomes a rubbery, opaque mass upon heating. In effect, this is what happens to the protein product of a temperature-sensitive gene at the nonpermissive temperature. It can happen to the wild-type protein too, but at a considerably higher temperature.

Another example of a conditional mutation is the amber mutation introduced in chapter 10. This mutation creates a premature translation stop signal—a "nonsense" mutation—in the middle of a gene, an event that would be lethal in a vital gene of an ordinary, nonpermissive strain. It is only conditionally lethal because the gene will usually still work in an amber suppressor strain (a permissive strain). Recall from chapter 10 that such a strain contains a tRNA with the ability to recognize an amber codon and insert an amino acid, thus preventing termination. When a phage bearing an amber mutation infects an amber suppressor strain, it will therefore usually survive.

Mutations can be conditional without being conditional lethal. An example is the Siamese cat (fig. 3.8). These animals have a mutation in the gene for dark coat color. But Siamese cats are not pure white like albino mice. Instead, they have dark patches on their feet, noses, and ears—areas of the body where the temperature is somewhat lower than the rest. This indicates that the mutation in the color gene is temperature-sensitive. Most of the cat's coat is warm enough to inactivate the color-producing enzyme, so it is white, but the extremities are cool enough that the enzyme operates, producing the dark color.

Conditional mutations affect the mutant organism only under certain conditions. Temperature-sensitive mutations, for example, cause a protein to be inactive at high temperature but active at lower temperature. Amber mutations do not produce active product under ordinary circumstances but do so in organisms with an amber suppressor tRNA. If a conditional mutation causes death under nonpermissive conditions, it is a conditional lethal.

EFFECTS OF MUTATIONS ON THE GENETIC MATERIAL

A second way of classifying mutations is to consider their effects on DNA. The mutations we are considering in this chapter involve the alteration, insertion, or deletion of one or a few bases at a time, and so are called **point mutations.** Many of these are **missense mutations.** In a missense mutation, a base change alters the sense of a codon from one amino acid to another. This causes an improper amino acid to be inserted into the protein product of the mutated gene. For example, a missense mutation might change the proline codon CCG to the arginine codon CGG.

MISSENSE MUTATIONS: SICKLE-CELL DISEASE

An excellent example of a missense mutation is **sickle-cell disease,** a true genetic disease. People who are homozygous for this condition have normal-looking red blood cells when their blood is rich in oxygen. The shape of normal cells is a biconcave disc; that is, the disc is concave viewed from both the top and bottom. However, when these people exercise, or otherwise deplete the oxygen in their blood, their red blood cells change dramatically to a sickle, or crescent shape (fig. 11.5). This has dire consequences. The sickle cells cannot fit through tiny capillaries, so they clog and rupture them, starving parts of the body for blood and causing internal bleeding and pain. Furthermore, the sickle cells are so fragile that they burst, leaving the patient anemic. Without medical attention, people undergoing a sickling crisis are in mortal danger.

What causes this sickling of red blood cells? The problem is in **hemoglobin,** the red, oxygen-carrying protein in the red blood cells. Normal hemoglobin remains soluble under ordinary physiological conditions, but the hemoglobin in sickle cells precipitates when the blood oxygen level falls, forming long, fibrous aggregates that distort the blood cells into the sickle shape.

What is the difference between normal hemoglobin (HbA) and sickle-cell hemoglobin (HbS)? Vernon Ingram answered this question in 1957 by determining the amino acid sequences of parts of the two proteins using a process that was invented by Frederick Sanger and is known as **protein sequencing.** Ingram focused on the β-globins of the two proteins. β-globin is one of the two different polypeptide chains found in the tetrameric (four-chain)

(a)

(b)

FIGURE 11.5
Normal red blood cells and a sickle cell. These scanning electron micrographs contrast (*a*) the regular, biconcave shape of normal cells with (*b*) the distorted shape of red blood cells from a sickle-cell disease patient. Magnifications: (*a*) ×3,333; (*b*) ×5,555.
(*a*) © Jeroboam/Photo Researchers, Inc.; (*b*) © Omikron/Photo Researchers, Inc.

hemoglobin protein. First, Ingram cut the two polypeptides into pieces with an enzyme that breaks selected peptide bonds. These pieces, called peptides, can be separated by paper chromatography run in two dimensions (fig. 11.6). One solvent is used to cause partial resolution of the peptides in the first dimension; the paper is then turned 90 degrees and rechromatographed with another solvent to separate the peptides still further. The peptides usually appear as spots on the paper; different proteins, because of their different amino acid compositions, give different patterns of spots. These patterns are aptly named **fingerprints.**

When Ingram compared the fingerprints of HbA and HbS, he found that all the spots matched except for one (fig. 11.7). This spot had a different mobility in the HbS fingerprint than in the normal HbA fingerprint, which indicated that it had an altered amino acid composition. Ingram checked the amino acid sequences of the two peptides in these spots. He found that they were the amino-terminal peptides located at the very beginning of both proteins. And

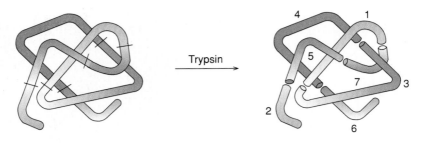

(a) Cutting protein to peptides

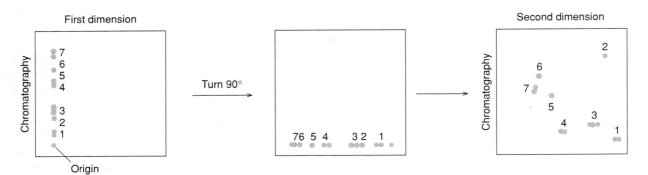

(b) Two-dimensional separation of peptides

FIGURE 11.6

Fingerprinting a protein. (*a*) A hypothetical protein, with its six trypsin-sensitive sites indicated by slashes. After digestion with trypsin, seven peptides are released. (*b*) These tryptic peptides separate partially during chromatography in the first dimension, then fully after the paper is turned 90 degrees and rechromatographed in the second dimension with another solvent.

FIGURE 11.8

Sequences of amino-terminal peptides from normal and sickle-cell β-globin. The numbers indicate the positions of the corresponding amino acids in the mature protein. The only difference is in position 6, where a valine (Val) in HbS replaces a glutamic acid (Glu) in HbA.

Hemoglobin A

Hemoglobin S

FIGURE 11.7

Fingerprints of hemoglobin A and hemoglobin S. The fingerprints are identical except for one peptide (circled), which shifts up and to the left in hemoglobin S.

From Biochemistry, *1/e, p. 92, by Lubert Stryer.* © *W. H. Freeman and Company Publishers.*

he found that they differed in only one amino acid. The glutamic acid in the sixth position of HbA becomes a valine in HbS (fig. 11.8). This is the only difference in the two proteins, yet it is enough to cause a profound distortion of the protein's behavior.

Knowing the genetic code, we can ask: What change in the β-globin gene caused the change Ingram detected in its protein product? The two codons for glutamic acid (Glu) are GAA and GAG; two of the four codons for valine (Val) are GUA and GUG. If the glutamic acid codon in the HbA gene is GAG, a single base change to GTG would alter the mRNA to GUG, and the amino acid inserted into HbS would be valine instead of glutamic acid. A similar argument can be made for a GAA → GTA change. Notice that, by convention, we are presenting the DNA strand that has the same sense as the mRNA (the coding strand). Actually,

FIGURE 11.9
The sickle-cell mutation and its consequences. The GAG in the sixth codon of the coding strand of the normal gene changes to GTG. This leads to a change from GAG to GUG in the sixth codon of the β-globin mRNA of sickle cells. This, in turn, results in insertion of a valine in the sixth amino acid position of sickle-cell β-globin, where a glutamic acid ought to go.

the opposite strand (the anticoding strand), reading CAC, is transcribed to give a GUG sequence in the mRNA. Figure 11.9 presents a summary of the mutation and its consequences. We can see how changing the blueprint does indeed change the product.

Sickle-cell disease is a very common problem among people of central African descent. The question naturally arises: Why has this deleterious mutation spread so successfully through the population? The answer seems to be that although the homozygous condition can be lethal, heterozygotes have little if any difficulty because their normal genes make enough product to keep their blood cells from sickling. Moreover, heterozygotes are at an advantage in central Africa, where malaria is rampant, because HbS helps protect against replication of the malarial parasite in their blood cells. In chapter 15 we will see how genetic engineering techniques offer hope of combating sickle-cell disease and related genetic disorders.

> ◆◆ Sickle-cell disease is a human genetic disorder. It results from a single base change in the gene for β-globin. The altered base causes insertion of the wrong amino acid into one position of the β-globin. This altered protein results in distortion of red blood cells under low-oxygen conditions.

OTHER POINT MUTATIONS

Sometimes, as we saw in chapter 10, a codon can be converted to a nonsense codon, which causes termination of translation. For example, the tryptophan codon UGG can be converted by a one-base change to either UGA or UAG, both of which are stop (or nonsense) codons. Of course, the mutation occurs at the DNA level, so TGG would be converted to TGA or TAG in this example. We present the RNA codons here because they are the familiar way of writing codons. Sometimes stop codons are mutated to sense codons. In such a case, translation could continue, producing a longer protein. In chapter 10 we also learned that single-base insertions or deletions can alter the translation reading frame in the middle of a gene; these are called **frameshift mutations.**

We can also consider the chemistry of a single base change. Viewed in the simplest terms, such changes fall into two classes: (1) **transitions,** in which a pyrimidine replaces a pyrimidine (C → T or T → C) or a purine replaces a purine (A → G or G → A), or (2) **transversions,** a more drastic kind of change, in which a purine replaces a pyrimidine or vice versa.

Why do we say that a transversion is "more drastic" than a transition? This makes sense chemically, because one purine resembles another purine (or one pyrimidine another pyrimidine) far more than a purine resembles a pyrimidine. It also makes sense genetically, because of the degeneracy of the genetic code. Recall from chapter 10 that two related codons ending in a pyrimidine, such as UUU and UUC, are more likely to code for the same amino acid (phenylalanine, in this case) than are two related codons, one of which ends in a pyrimidine and the other of which ends in a purine (e.g., UUU and UUG, which code for phenylalanine and leucine, respectively). Similarly, two related codons ending in a purine are likely to code for the same amino acid (e.g., GAA and GAG both code for glutamic acid). Thus, changing one pyrimidine to another or one purine to another, at least in the third position, is less likely to change the sense of a codon than is changing a purine to a pyrimidine, or vice versa.

> ◆◆ Besides missense mutations, DNA can experience single-base changes that create stop codons within genes, leading to premature termination. The change of one pyrimidine to another, or one purine to another, is called a transition. The conversion of a purine to a pyrimidine, or vice versa, is called a transversion.

SPONTANEOUS MUTATIONS

All mutations have a cause, of course, but sometimes mutations occur without our having to apply a **mutagen,** a mutation-causing agent. Such spontaneous mutations have several causes.

Mutations Caused by the DNA Replication Machinery
We saw in chapter 7 that replication is very faithful, but it is not perfect, so a certain number of mutations will occur simply due to fallible DNA synthesis.

We can see the effects of unfaithful replication especially clearly in mutant strains of *E. coli* called **mutators.** These bacteria make more than the usual number of mistakes during DNA replication, so their mutation rates are higher than normal. Mutator mutations map in several

FIGURE 11.10

Spontaneous mutation induced by tautomerization. (*a*) The normal (keto) tautomer of thymine (left) base-pairs with adenine. After a rare conversion to the enol tautomer (right), the thymine base-pairs with guanine. The dotted lines represent hydrogen bonds between bases. (*b*) The normal (amino) tautomer of adenine (left) base-pairs with thymine. After a rare conversion to the imino tautomer (right), the adenine base-pairs with

cytosine. (*c*) Left: The normal (amino) tautomer of cytosine can sometimes convert to the rare imino tautomer, which pairs with adenine (not shown). Right: The normal (keto) tautomer of guanine sometimes converts to the rare enol tautomer, which pairs with thymine (not shown). The properties of these altered base pairs are the same as those shown in (*b*) and (*a*), respectively.

different genes, some of which have been identified. For example, *mutD* encodes the epsilon (ε) subunit of the DNA polymerase III holoenzyme. This is the polypeptide that, together with the α subunit, gives the holoenzyme its 3′ → 5′ exonuclease activity. Without this activity, proofreading cannot occur, so the newly replicated DNA is left with an excess of mutations (chapter 7). Mutations in *mutH, mutL,* and *mutS* are also mutator mutations. They impair mismatch repair, a mechanism for repairing mismatches that the proofreading system missed. This mechanism will be discussed later in the chapter.

As unlikely as it may seem, there are also mutations that have the opposite effect—they make DNA replication even more faithful than normal. This raises a question: If a more faithful DNA replication system is available, why has *E. coli* evolved with a less effective one? The answer is probably that mutants with these extra-faithful systems have too slow a rate of evolution; they are not flexible enough to compete with more plastic organisms in adapting to a changing environment.

Mistakes in Replication Caused by DNA Bases

The bases in DNA ordinarily exist in one of two possible forms, or **tautomers.** For example, the base thymine usually assumes the **keto** form, as illustrated in figure 11.10a. This is the structure presented in chapter 6 that base-pairs naturally with adenine. Occasionally, the keto form of thymine switches to the **enol** form, also shown in figure 11.10. Note that both tautomers have the same atoms but are arranged slightly differently. The enol form pairs naturally with guanine instead of adenine, so if a thymine happens to be in the enol form at the moment it takes a partner during replication, a guanine will be inserted in place of an adenine. If this error goes uncorrected, DNA replication will perpetuate it and a mutation will result.

Figure 11.10b shows the two tautomeric forms of adenine. The **amino** form is by far the more common one; as we have seen, it base-pairs with thymine. Occasionally, adenine can assume the alternate, **imino** form, which base-pairs with cytosine. If this happens to an adenine in DNA just as DNA polymerase is supplying it with a partner, the polymerase will insert a C instead of a T. Again, if this mistake persists until the next round of replication, it will cause a mutation that will last: a GC pair instead of an AT pair. Figure 11.10c shows the **tautomerization** of cytosine to its rare imino form and of guanine to its rare enol form. By inspecting these rare tautomers you can satisfy yourself that they form abnormal CA and GT base pairs, respectively.

So far, we have only considered what would happen if the bases in the replicating DNA strand form the rare tautomer during base pairing. It is also possible that a newly inserted base would exist in the rare tautomeric form just at the moment of base pairing. If so, it would result in the same kind of faulty base pairing we have just observed.

Met Ala Leu Trp Ile Arg Phe Ile Arg
ATGGCCCTGTGGATCCGCTTCATTAGG---

(a)

Met Ser Pro Val Asp Pro Leu His Stop
ATGAGCCCTGTGGATCCGCTTCATTAGG--- —— New reading frame
                                 —— Old reading frame
Met   Ala Leu Trp Ile Arg Phe Ile Arg

(b)

FIGURE 11.11

Frameshift mutation. (*a*) Normal reading frame of part of a gene. (*b*) The adenine nucleotide inserted in the fourth position of the coding region (gray) shifts the reading frame one base to the left. The new reading frame with the corresponding codon meanings is shown by the brackets above the base sequence (red); the old reading frame, which corresponds to the unmutated base sequence, is denoted by brackets below the sequence (blue). A deletion of a base, instead of an insertion, would have shifted the reading frame one base to the right.

Spontaneous Frameshift Mutations During Replication

Sometimes DNA replication causes the insertion or deletion of one or more bases in the middle of a coding region, which changes the translational reading frame from that point on (chapter 10). These **frameshift mutations** are very severe because they change every codon from the point of the mutation to the end of the mRNA. Figure 11.11 shows how this works: (*a*) is a hypothetical gene fragment, showing the translation of each codon. (*b*) is the same gene fragment, showing how insertion of a single A in the fourth position of the gene shifts the reading frame one base to the left. All the codons after that point are different. In fact, one of these new codons, at the far right, is a stop codon, so protein synthesis will stop prematurely.

If small insertions and deletions cause frameshift mutations, the next question is: How do these insertions and deletions occur? We do not understand exactly how this works, but figure 11.12 presents one hypothesis that calls for the DNA replication machinery to "slip a cog" every now and then. The idea is that a base in one strand sometimes fails to pair with its partner in the complementary strand. This looping out of a base seems especially likely in stretches of DNA where one base is repeated over and over. Note that if a base in the template strand loops out, one too few bases will be incorporated into the progeny strand, resulting in a one-base deletion. On the other hand, if a base in the progeny strand loops out, one too many bases will be incorporated, producing a one-base insertion.

Spontaneous Mutations Caused by Deamination

Spontaneous mutations can occur by mechanisms other than mistakes in DNA replication. For example, bases, especially cytosine, have a slight tendency to lose their amino groups in a process called **deamination.** When cytosine is deaminated, it receives a carbonyl oxygen in place of its amino group; this converts it to uracil, which base-pairs

FIGURE 11.12

Hypothetical mechanism for frameshift mutation. A parental DNA duplex (top) contains a string of five A–T pairs. During replication of this DNA, transient looping out of bases within this A–T stretch can occur. On the left, one of the T's (pink) in the parental strand loops out. This leaves only four T's to pair with A's in the growing progeny strand (red). After one more round of replication, these four A's serve as the template for the incorporation of four T's in the new progeny strand (blue). Now we have four A–T pairs instead of five; a one-base deletion has occurred. On the right, one of the newly incorporated A's (gray) in the progeny strand loops out, allowing six A's to be inserted instead of five. When this DNA replicates again, the strand with six A's will dictate the incorporation of six T's in the new progeny strand (light blue). A one-base insertion results.

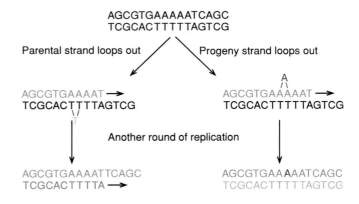

FIGURE 11.13

Deamination of cytosine and adenine. After deamination, the bases uracil and hypoxanthine are formed. These pair with adenine and cytosine, respectively, so transitions occur in each case.

with adenine instead of guanine. Adenine can also be deaminated, yielding the base hypoxanthine, which base-pairs with cytosine instead of thymine. In both of these cases, deamination can potentially cause a mutation because a new base, with new pairing properties, is created. Figure 11.13 shows the changes and new base pairing caused by deamination of cytosine and adenine.

The most common kind of deamination, the conversion of cytosine to uracil, does not usually lead to a mutation, because cells have a mechanism for removing uracils that find their way into DNA by mistake. It involves an enzyme called uracil-DNA-glycosidase (fig. 11.14), which cuts the bond between the uracil and its deoxyribose, thus removing the uracil and leaving behind a DNA strand containing one sugar without a base. Another enzyme soon supplies a cytosine to pair with the guanine on the opposite strand.

The DNA of some organisms contains a small number of modified bases in addition to the usual four. The most common of these is 5-methylcytosine (fig. 11.15a), which base-pairs with guanine exactly as ordinary cytosine does.

However, 5-methylcytosine does not behave in every respect like cytosine. In particular, the sites in DNA that contain 5-methylcytosine can become **hot spots** for spontaneous mutation via deamination.

If deamination of cytosine is usually not mutagenic, why is deamination of 5-methylcytosine so likely to cause mutations? The answer is simple: Although the deaminated product of cytosine is uracil, which is easily recognized and removed, the deaminated product of 5-methylcytosine is thymine, one of the ordinary DNA bases and therefore not recognizable as foreign. This means that once 5-methylcytosine deaminates, a GT mismatch appears. After this DNA replicates, one duplex will have the wild-type GC pair, but the other will have a mutant AT pair (fig. 11.15b).

The pattern of methylation of cytosines in bacteria and in higher eukaryotes is specific. That is, a certain few cytosines are selected for methylation, while all the others remain unmodified. In mammals, for example, cytosines in certain CG sequences are targets for methylation. This

FIGURE 11.14

Repair of deaminated cytosine. A deamination event (top) has converted a C (red) to a U (blue). The enzyme uracil-DNA-glycosidase recognizes the U as foreign and removes it (middle), leaving a sugar without a base in the bottom DNA strand. In the second step of the repair process (bottom), a C (red) is added to pair with the G in the top strand.

specificity has predictable consequences. The sequence CG has become much rarer than any other dinucleotide sequence, presumably because deamination has led to conversion of most of the CG's to TG's and then TA's during evolution. In bacteria, too, methylated C's become hot spots for mutation.

◆ Spontaneous mutations can occur by several mechanisms. First, the replication machinery can simply make a mistake and insert the wrong base. This kind of error becomes more likely if the DNA bases tautomerize, or change molecular form during replication. If normal keto groups on DNA bases tautomerize to the enol form, or if normal amino groups tautomerize to the imino form, the base-pairing properties of the bases change and the wrong base will be inserted into the replicating DNA. Spontaneous frameshift mutations occur when either the parental or progeny DNA strand loops out during replication, resulting in too few or too many bases being incorporated, respectively. Spontaneous deaminations of the DNA bases cytosine and adenine also occur. Each of these deaminations changes the base-pairing properties of the bases, which leads to subsequent incorporation of the wrong base during DNA replication.

Measuring Spontaneous Mutation Rates

How frequently do spontaneous mutations occur? This is a difficult determination to make. To begin with, many mutations are undetectable by ordinary means, so they are not counted. We will consider such silent mutations later. Furthermore, mutations are rare, so we must examine many

FIGURE 11.15

Mutation by deamination of 5-methylcytosine. (*a*) The deamination step converts 5-methylcytosine to thymine. (*b*) Making the change permanent. In step 1, deamination converts the 5-methylcytosine (red) to thymine (blue). Since thymine cannot be recognized as foreign, it is not removed. In step 2, DNA replication yields one duplex with a wild-type GC pair (red) and one with a mutant AT pair (blue).

organisms to find a significant number of mutants. Once we have detected a given number of mutants, we must determine the rate at which they are produced—that is, the number of mutations per unit of time. But geneticists measure time in different ways than physicists do. Instead of mutations per minute, day, or year, we speak of mutations per generation or per cell division. This is the **mutation rate.** In higher eukaryotes, including humans, the number of generations is usually taken as the number of gametes in the population under study. For example, if we are considering a population of 50,000 people, the number of gametes that produced this population is 100,000.

Table 11.1 presents mutation rates for several organisms. We do not fully understand why these numbers vary so much from one organism to another, but one contributing factor is likely to be the efficiency of proofreading during DNA synthesis (chapter 7). Another consideration may be the efficiency of mismatch repair.

As an example of how these data were collected, consider human achondroplasia, the dominant form of dwarfism mentioned in chapter 2. The records of 94,075 children at Lying-In Hospital in Copenhagen show that ten were afflicted with achondroplasia. Two of the ten came from families with a history of such dwarfism, so only

Table 11.1 A Sample of Spontaneous Mutation Rates in Different Organisms

Organism	Character	Rate	Units
Bacteriophage: T2	Lysis inhibition, $r \to r^+$ Host range, $h^+ \to h$	1×10^{-8} 3×10^{-9}	} Per gene per replication
Bacteria: *Escherichia coli*	Lactose fermentation, $lac^- \to lac^+$ Phage T1 sensitivity, T1-$s \to$ T1-r Histidine requirement, $his^- \to his^+$ $his^+ \to his^-$ Streptomycin sensitivity, str-$s \to str$-d str-$d \to str$-s	2×10^{-7} 2×10^{-8} 4×10^{-8} 2×10^{-6} 1×10^{-9} 1×10^{-8}	} Per cell per division
Algae: *Chlamydomonas reinhardi*	Streptomycin sensitivity, str-$s \to str$-r	1×10^{-6}	
Fungi: *Neurospora crassa*	Inositol requirement, $inos^- \to inos^+$ Adenine requirement, $ade^- \to ade^+$	8×10^{-8} 4×10^{-8}	} Mutant frequency among asexual spores
Corn: *Zea mays*	Shrunken seeds, $Sh \to sh$ Purple, $P \to p$	1×10^{-5} 1×10^{-6}	
Fruit fly: *Drosophila melanogaster*	Yellow body, $Y \to y$, in males $Y \to y$, in females White eye, $W \to w$ Brown eye, $Bw \to bw$	1×10^{-4} 1×10^{-5} 4×10^{-5} 3×10^{-5}	} Mutant frequency per gamete per sexual generation
Mouse: *Mus musculus*	Piebald coat color, $S \to s$ Dilute coat color, $D \to d$	3×10^{-5} 3×10^{-5}	
Man: *Homo sapiens*	Normal $\to$ hemophilic Normal $\to$ albino	3×10^{-5} 3×10^{-5}	
Human bone marrow cells in tissue culture	Normal $\to$ 8-azoguanine resistant Normal $\to$ 8-azoguanosine resistant	7×10^{-4} 1×10^{-6}	} Per cell per division

Data from R. Sager and F. J. Ryan, *Cell Heredity*, 1961, John Wiley & Sons, Inc., New York, NY.

eight represent new mutations. Since the total number of gametes involved is twice the total number of individuals, the mutation rate is $8/(2)(94,075) = 4.2 \times 10^{-5}$.

 Spontaneous mutation rates—the number of mutations arising per gamete each generation—vary considerably from one organism to another. The spontaneous mutation rate for a given human mutation can be calculated by dividing the number of individuals in a population having that mutation by twice the total number of individuals in the population.

CHEMICAL MUTAGENESIS

We have just seen that mutations can occur spontaneously, but they occur much more frequently in response to environmental agents: chemicals and radiation. We will consider chemical agents first.

Some chemicals act by accelerating the rate of mutations that occur spontaneously. We have already discussed several mechanisms of spontaneous mutation, including deamination, frameshifting, and tautomerization. Deamination can be greatly enhanced by agents such as nitrous acid or bisulfite, or even by heat. Frameshift mutations can be accelerated by planar molecules like the acridine dyes

FIGURE 11.16
Proflavin (hydrochloride), one of the acridine dyes.

(fig. 11.16), which **intercalate,** or insert themselves between the flat base pairs of DNA. This disruption may stabilize the looping out process that seems to be important in frameshift mutations (see fig. 11.12).

Mutations Caused by Base Analogues
Some synthetic compounds can enhance the frequency of tautomerization and thereby induce mutations. The classic example of such a mutagen is **5-bromodeoxyuridine (BrdU),** which resembles thymidine except for the substitution of a bromine atom for a methyl group in the 5-position (fig. 11.17). However, once BrdU is incorporated in place of thymidine, it can cause trouble. The trouble derives from the enhanced tendency of BrdU to switch occasionally to the enol tautomer that base-pairs as C instead of T. As we

FIGURE 11.17
Comparison of thymidine and bromodeoxyuridine.

FIGURE 11.18
Electron-rich centers in DNA. The targets most commonly attacked by electrophiles are the phosphate groups, N^7 of guanine and N^3 of adenine (red); other targets are in blue.

have seen, natural thymidine does this upon rare occasion; BrdU still switches to the enol tautomer only rarely, but it does so more frequently than thymidine.

Mutations Caused by Alkylation of Bases

Some substances in our environment, both natural and synthetic, are **electrophilic.** This name means electron (or negative charge)-loving; thus, electrophiles seek centers of negative charge in other molecules and bind to them. Many other environmental substances are metabolized in the body to electrophilic compounds. One of the most obvious centers of negative charge in biology is the DNA molecule. Every nucleotide contains one full negative charge on the phosphate and partial negative charges on the bases. When **electrophiles** encounter these negative centers, they attack them, usually adding carbon-containing groups called **alkyl groups.** Thus, we refer to this process as **alkylation.**

Aside from the phosphodiester bonds, the favorite sites of attack by alkylating agents are the N^7 of guanine and the N^3 of adenine (fig. 11.18). Since neither of these positions is involved in base pairing, such alkylations do not lead immediately to mispairing. However, they do make the bond between sugar and base more labile, or more apt to break. When this break occurs, it leaves an apurinic site, a sugar without its purine. This obviously cannot be replicated properly unless it is first repaired, and this repair itself can sometimes be mutagenic. Alkylation can also enhance the tendency of a base to form the rare (wrong) tautomer. As we have seen, this can lead to mutations. Moreover, all of the nitrogen and oxygen atoms involved in base pairing are also subject to alkylation, which can directly disrupt base pairing and lead to mutation (fig. 11.19).

Most environmental carcinogens, or cancer-causing agents, are electrophiles that seem to act by attacking DNA and alkylating it as we will see in chapter 12. Many of the favorite mutagens used in the laboratory for the express purpose of creating mutations are also alkylating agents. One example is ethylmethane sulfonate (EMS), which transfers ethyl (CH_3—CH_2) groups to DNA (fig. 11.19).

> Different chemicals induce different kinds of DNA damage. Nitrous acid and bisulfite cause deamination of bases, especially cytosine, causing them to base-pair differently. Alkylating agents like ethylmethane sulfonate add bulky alkyl groups to bases, either disrupting base pairing directly or causing loss of bases, either of which can lead to faulty DNA replication or repair. Base analogues such as BrdU are incorporated into DNA and then can lead to abnormal base pairing. Certain planar molecules, such as the acridine dyes, cause addition of extra bases or too few bases during DNA replication, which can alter the reading frame of a gene.

RADIATION-INDUCED MUTATIONS

Ultraviolet, gamma-, and X-radiation are the common types of mutagenic radiation found in nature and used in mutagenesis experiments. These kinds of radiation differ greatly in energy; therefore, they differ greatly in the kinds of DNA damage they cause. We will examine each in turn.

Ultraviolet Radiation

Ultraviolet radiation is relatively weak, so the damage it causes is relatively modest; it cross-links adjacent pyrimidines on the same DNA strand, forming dimers, usually **thymine dimers.** Figure 11.20 shows the structure of a thymine dimer and illustrates how it interrupts base pairing between the two DNA strands. This blocks DNA replication because the replication machinery cannot tell which bases to insert opposite the dimer. As we will see, replication sometimes proceeds anyway, and bases are inserted at random. If these are the wrong bases (and they usually are), a mutation results.

FIGURE 11.19

Alkylation of guanine by EMS. At the left is a normal guanine-cytosine base pair. Note the free O^6 oxygen (red) on the guanine. Ethylmethane sulfonate (EMS) donates an ethyl group (blue) to the O^6 oxygen, creating O^6-ethylguanine (right), which base-pairs with thymine instead of cytosine.

The kind of ultraviolet radiation that is most damaging to DNA has a wavelength of about 260 nm, which is not surprising, since this is the wavelength of radiation that is absorbed most strongly by DNA. Such radiation has great biological significance; it is abundant in sunlight, so most forms of life are exposed to it to some extent. The mutagenicity of ultraviolet light explains why sunlight causes skin cancer: Its ultraviolet component damages the DNA in skin cells, which sometimes causes those cells to lose control over their division.

Given the dangers of ultraviolet radiation, we are fortunate to have a shield—the ozone layer—in the earth's upper atmosphere to absorb the bulk of such radiation. However, scientists have recently noticed alarming holes in this protective shield—the most prominent one located over Antarctica. The causes of this ozone depletion are still somewhat controversial, but they probably include chlorofluorocarbon compounds. These are used as coolants in refrigerators and air conditioners (freon) and in the manufacture of plastics. Unless we can arrest the destruction of the ozone layer, we are destined to suffer more of the effects of ultraviolet radiation, including skin cancer.

Gamma and X Rays

The much more energetic **gamma rays** and **X rays,** like ultraviolet rays, can interact directly with the DNA molecule. However, they cause most of their damage by ionizing the molecules surrounding the DNA, especially water. This forms **free radicals,** chemical substances with an unpaired electron. These free radicals, especially those containing oxygen, are extremely reactive, and they immediately attack neighboring molecules. When such a free radical attacks a DNA molecule, it can change a base, but it frequently causes a single- or double-stranded break. Single-stranded breaks are ordinarily not serious because they are easily repaired, but double-stranded breaks are very difficult to repair properly, so they frequently cause a lasting mutation. Because ionizing radiation can break chromosomes, it is referred to not only as a mutagen, or mutation-causing substance, but also as a **clastogen,** which means "breaker."

FIGURE 11.20

Thymine dimers. (*a*) Ultraviolet light cross-links the two thymine bases on the top strand. This distorts the DNA so that these two bases no longer pair with their adenine partners. (*b*) The two bonds joining the two thymines form a four-membered cyclobutane ring (red).

Different kinds of radiation cause different kinds of mutations. Ultraviolet rays have comparatively low energy, and they cause a moderate type of damage: thymine dimers. Gamma and X rays are much more energetic. They ionize the molecules around DNA and form highly reactive free radicals that can attack DNA, altering bases or breaking strands.

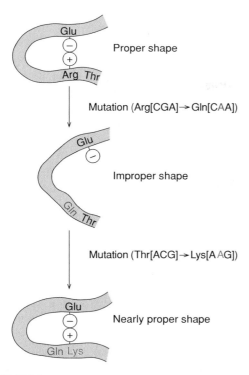

FIGURE 11.21
Second-site reversion. The original G → A mutation (blue) changed an arginine, which bears a positive charge, to a glutamine (blue), which is uncharged. This destroyed an important interaction with a negatively charged glutamic acid and resulted in an inactive protein. The C → A reversion (red) changed a neighboring threonine to a positively charged lysine (red), restoring the vital interaction with the glutamic acid and therefore restoring the protein's function.

SILENT MUTATIONS

Most mutations are harmful, leading to the production of faulty proteins or RNAs. Still, life is not perfect; there is always room for improvement. Accordingly, mutations can sometimes actually improve genes—change their products so that they work better or allow their hosts to survive better. These mutations allow fitter species to evolve.

Many mutations are not detectable by ordinary genetic means. We call them **silent mutations.** For example, consider the change of a codon from UCA to UCG. This is a base change, so it is a real mutation. But both triplets code for the same amino acid (serine), so no change occurs in the protein product of the gene. The mutation is silent. The only way we would ever know for sure that it had happened would be to clone the gene and determine its exact base sequence. Such mutations in the third base of a codon (the **wobble position**) are more likely to be silent since they frequently do not change the sense of the codon. Therefore, these mutations are tolerated and we find them much more frequently than mutations in the first two positions of a codon. Other types of silent mutations occur within introns, where they usually have no effect on the gene's function, or entirely outside of genes—in **intergenic regions**—where they are also usually without phenotypic effect.

REVERSION

Point mutations, especially frameshift mutations, may have drastic effects, but they are reversible. **Reversion,** or **back-mutation,** to the original phenotype can occur in two ways: A **true reversion** is the alteration of the mutated base back to its original identity; alternatively, a change can occur elsewhere in the same gene to compensate for the original mutation. For example, consider a gene that codes for a protein having a positively charged amino acid (arginine) that interacts with a negatively charged amino acid (glutamic acid), as in figure 11.21. This interaction helps hold part of the protein in the proper shape—a fold that is vital to the protein's function.

Suppose that we cause a mutation (a **forward mutation**—away from wild-type) in the codon for the arginine. This will prevent proper folding of the protein, so the protein will not function. Now suppose we cause another mutation that creates a new positively charged amino acid (lysine) near the position of the lost arginine. If this lysine can substitute for the missing arginine in holding the protein in the right shape, it will restore its function. The mutation that created the new lysine codon will therefore be a reversion. It is not a true reversion, because the reverted gene is different from the original. Instead, we call it a **second-site reversion.**

Second-site reversion can also be considered **suppression,** in that it is a compensation by one mutation for the effects of another. In particular, it is an **intragenic suppression,** since both mutations occur in the same gene. **Intergenic suppression**—suppression of a mutation in one gene by a mutation in another—also happens. Because two genes are involved, intergenic suppression is not the same as reversion. A familiar example of intergenic suppression is suppression of nonsense codons, such as amber codons, by mutant tRNAs (e.g., amber suppressors) that recognize these codons (chapter 10).

Frameshift mutations can revert in a number of ways (fig. 11.22): adding a base to compensate for a deleted base; losing a base to balance an extra one that has been inserted; losing two more bases if one has been lost, bringing the total lost to three and eliminating the frameshift; or gaining extra bases to bring the total number gained to three. However, such gains and losses are not likely to happen naturally, so further mutagenesis with frameshifting agents is generally required to obtain reversion of frameshift mutations.

DETECTING MUTAGENS

We have seen that many chemicals are now recognized as mutagens, and that many mutations are deleterious. In fact, as we will see in chapter 12, environmental mutagens are thought to be responsible for the majority of human cancers. It is important, therefore, to identify mutagenic chemicals so we can limit our exposure to them. But because there are so many potential mutagens to test and because

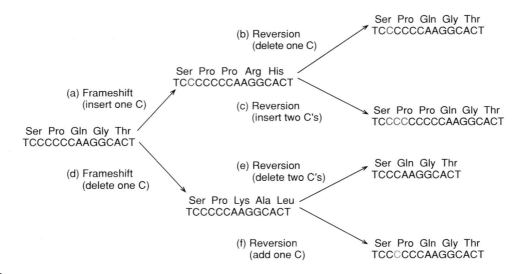

FIGURE 11.22

Reversion of frameshift mutation. The original coding strand is presented at left. It contains a string of six C's. (*a*) A frameshift occurs, involving the insertion of an extra C (blue). This shifts the reading frame at the point of the insertion and alters the coding, starting with the third codon. (*b*) This frameshift can revert simply by deleting one of the C's, returning the reading frame and coding to their exact original state. (*c*) The frameshift can also revert by inserting two more C's (red). Now the total number of inserted bases is three, so one whole new codon (CCC) has been created and the original reading frame has been restored. This results in a slightly altered protein with one extra proline. (*d*) A frameshift occurs, involving the deletion of one C. This shifts the reading frame at the point of the deletion and alters the coding, starting with the third codon. (*e*) This frameshift can revert by deleting two more C's, restoring the original reading frame but resulting in the loss of one codon (CCC). The corresponding protein will lack one proline. (*f*) The frameshift can also revert simply by inserting one C (pink) to compensate for the one that was lost. This restores the exact original reading frame and coding.

massive numbers of organisms must be examined to find significant numbers of mutants, using test animals such as mice is extremely expensive and terribly time-consuming.

Fortunately, excellent mutagen testing systems have been developed using bacteria. As a test organism, bacteria have tremendous advantages: They are cheap and easy to culture, and billions of cells can be tested in a day or two. They are also haploid, allowing mutants to be detected immediately. The best known of the bacterial assays was created by Bruce Ames and is known as the **Ames test.** Ames started with auxotrophic histidine mutants—strains of *Salmonella typhimurium* with mutations in the *his* operon that rendered them incapable of growing in the absence of the amino acid histidine. The basic idea of the assay is to observe the frequency of reversions to wild-type in the *his* operon (fig. 11.23). This is very simple to do. One just looks for colonies that can grow on plates lacking histidine. Without any mutagen, these reversions occur only rarely, so few colonies can grow. On the other hand, a powerful mutagen can greatly accelerate the mutation rate—including back-mutations, or reversions—and many colonies can be observed.

This concept sometimes gives students trouble because the mutation we are looking for is a beneficial back-mutation to wild-type, rather than a deleterious forward mutation. Nevertheless, it is a mutation, and it is much easier to observe than the forward mutation. To understand this, consider the ease of counting the colonies produced by a few bacteria that can grow, among a vast excess of bacteria that cannot; then compare this with the hopelessness of finding a few cells that cannot grow, among countless cells that can.

Ames had to make some additions and modifications to his bacteria in order to make them work well in detecting mutagens. First, he took note of the fact that most chemical mutagens must be metabolized by the body before they become mutagenic. This metabolism, carried out by liver enzymes, is part of the normal method by which we excrete toxic materials. Since bacteria do not perform such metabolism, Ames added a rat liver extract to his bacteria to mimic the metabolism a potential mutagen would experience in the body. In addition, Ames mutated his bacteria to make them leaky to exogenous chemicals and deficient in excision repair. He also added a plasmid that produces enzymes that convert most kinds of DNA damage to mutations.

How valid are the results of this kind of mutagenicity testing? Are these strains too sensitive to mutation? Are compounds that cause mutations in bacteria also mutagens in humans? It is true that we have much better defenses against mutagens than these severely weakened bacteria possess, but the Ames test does give a good idea about the relative mutagenicities of different chemical agents. And since bacteria and humans share the same kind of genetic material, agents that mutate their genes are likely to do the same to ours.

(a)

his⁻ tester strain 1 or 2
+ liver enzymes

his⁻ tester strain 1
+ mutagen
+ liver enzymes

his⁻ tester strain 2
+ mutagen
+ liver enzymes

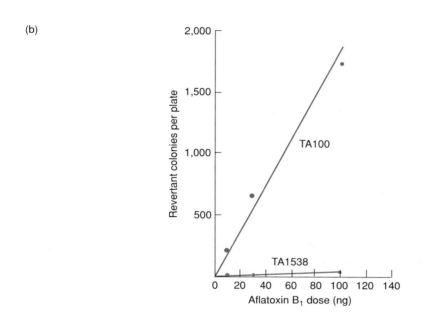

(b)

FIGURE 11.23

The Ames test. (*a*) Outline of the procedure. In this example, two *his⁻* tester strains are used. Strain 1 contains a missense mutation in the *his* operon and so will detect mutagens that can cause missense back-mutations, or reversions. Strain 2 contains a frameshift mutation in the *his* operon and so is sensitive to frameshift reversions. These strains are mixed with liver enzymes, which can metabolize compounds to their mutagenic forms. In the control on the left, no mutagen is added, thus only a few spontaneous reversions occur. In the middle, the missense mutation tester strain has experienced multiple reversions, showing that the compound under study can cause missense mutations. On the right, the frameshift mutation tester strain shows no significant difference from the control, so this compound does not cause frameshift mutations. (*b*) Data from a test of the mutagen aflatoxin B_1. The tester strain TA100, like the hypothetical strain 1 above, tests for missense mutations. Clearly, aflatoxin is a powerful mutagen in this strain. By contrast, the frameshift tester strain, TA1538, detects little if any tendency of aflatoxin to cause frameshift mutations.

> The Ames test uses a set of bacterial tester strains to screen for chemical mutagens. These *Salmonella* cells are histidine auxotrophs and are exquisitely sensitive to mutation by chemicals. To perform the Ames test, the researcher adds a suspected mutagen to these cells (plus liver enzymes to metabolize the chemical to its mutagenic form) and observes the appearance of histidine prototrophic colonies. The number of revertant colonies is a measure of the potency of the mutagen.

DNA REPAIR

We have discussed many different ways that DNA can be damaged. Since much of this damage is serious enough to threaten life, it is not surprising that living things have evolved mechanisms for dealing with it. One way to cope with a genetic lesion is to repair it, or restore it to its original, undamaged state. There are two basic ways to do this, and they can be illustrated by comparing damage in a strand of DNA to a knot in a piece of wire: (1) Directly undo the damage (untie the knot in the wire), or (2) remove the damaged section of DNA and fill it in with new, undamaged DNA (cut out the knot in the wire and solder in a fresh piece of straight wire).

DIRECTLY UNDOING DNA DAMAGE

In the late 1940s, Albert Kelner was trying to measure the effect of temperature on repair of ultraviolet damage to DNA in the bacterium *Streptomyces*. However, he noticed that damage was repaired much faster in some bacterial spores than in others kept at the same temperature. Obviously, some factor other than temperature was operating. Finally, Kelner noticed that the spores whose damage was repaired fastest were the ones kept most directly exposed to light from a laboratory window. When he performed control experiments with spores kept in the dark, he could detect no repair at all. Renato Dulbecco soon observed the same effect in bacteria infected with ultraviolet-damaged phages, and it now appears that most, if not all, forms of life share this important mechanism of repair, which is termed **photoreactivation,** or **light repair.**

It was discovered in the late 1950s that photoreactivation is catalyzed by an enzyme called **photoreactivating enzyme** or **DNA photolyase.** This enzyme operates by the mechanism sketched in figure 11.24. First, the enzyme detects and binds to the ultraviolet-damaged DNA site (a pyrimidine dimer). Then the enzyme absorbs visible light, which activates it so it can break the bonds holding the pyrimidine dimer together. This restores the pyrimidines to their original independent state. Finally, with the damage repaired, the enzyme dissociates from the DNA.

Organisms ranging from *E. coli* to human beings exhibit another mechanism for the direct reversal of a mutation, this time an alkylation of the O^6 of guanine. After DNA is methylated or ethylated, an enzyme called **O^6-methylguanine methyl transferase** comes on the scene to repair the damage. It does this by accepting the methyl or ethyl group itself, as outlined in figure 11.25. The acceptor site on the enzyme for the alkyl group is the sulfur atom of an amino acid called cysteine. Strictly speaking, this means that the methyl transferase does not fulfill one part of the definition of an enzyme—that it be regenerated unchanged after the reaction. Instead, this protein seems to be irreversibly inactivated, so we call it a suicide enzyme to denote the fact that it "dies" in performing its function. The repair process is therefore expensive; each repair event costs one protein molecule.

One more property of the O^6-methylguanine methyl transferase is worth noting. The enzyme, at least in *E. coli,* is induced by DNA alkylation. This means bacterial cells that have been exposed to alkylating agents for a period of time become much more resistant to DNA damage than cells that have just been exposed to such mutagens for the first time.

FIGURE 11.24

Model for photoreactivation. (*a*) and (*b*) Ultraviolet radiation causes a thymine dimer to form. (*c*) The photoreactivating enzyme (red) binds to this region of the DNA. (*d*) The enzyme absorbs visible light, breaks the dimer, and (*e*) finally dissociates.

 Ultraviolet damage to DNA (pyrimidine dimers) can be directly repaired by a photoreactivating enzyme that uses energy from visible light to break the bonds holding the two pyrimidines together. O^6-alkylations on guanine residues can be directly reversed by the suicide enzyme O^6-methylguanine methyl transferase, which accepts the alkyl group onto one of its amino acids, a cysteine.

EXCISION REPAIR

The percentage of mutations that can be handled by direct reversal is necessarily small. Most mutations involve neither pyrimidine dimers nor O^6-alkylguanine, so they must be corrected by a different mechanism. Most are removed by a process called **excision repair.** Excision repair is like cutting out a knotted wire segment and replacing it with fresh wire. The damaged DNA is first removed, then replaced with fresh DNA. This occurs by one of several mechanisms; some notable examples will be described here.

Using DNA Glycosylase
Certain mutations are recognized by an enzyme called **DNA glycosylase,** which breaks the **glycosidic bond** between the damaged base and its sugar (fig. 11.26). This leaves an **apurinic** or **apyrimidinic site (AP site),** which is a sugar without its purine or pyrimidine base. Once the AP site is created, it is recognized by a 5′ AP endonuclease that cuts, or **nicks,** the DNA strand on the 5′-side of the AP site.

(The "endo" in endonuclease means the enzyme cuts *inside* a DNA strand, not at a free end; Greek: *endo* = within.) This step, known as **incision,** creates free ends within the DNA strand that signal other enzymes to complete the excision process. One possibility is for a **5′ → 3′ exonuclease** to start at the nick and move left to right (5′ to 3′), removing nucleotides, including the damaged one. (The "exo" in exonuclease means the enzyme must start at a free end and degrade the DNA one nucleotide at a time, working toward the other end; Greek: *exo* = outside.) Another possibility is for a **3′ AP endonuclease** simply to cut just after the AP site, releasing the AP deoxyribose phosphate. Both of these mechanisms leave a gap in the DNA that must be repaired, using the opposite strand as the template.

Pyrimidine dimers can also be removed by a DNA glycosylase in a variation on the theme described above. This time, a DNA glycosylase specific for pyrimidine dimers cuts the glycosidic bond between the first pyrimidine and its sugar, leaving an AP site (fig. 11.27). However, the enzyme does not release a free base because the two pyrimidines are covalently bonded together. In the incision step, a 3′ AP endonuclease cuts the phosphodiester bond just after the AP site. Then either a 5′ AP endonuclease or a

FIGURE 11.25
Mechanism of O^6-methylguanine methyl transferase. A sulfhydryl group of the enzyme accepts the methyl group (blue) from a guanine on the DNA, thus inactivating the enzyme.

O^6-methylguanine methyl transferase

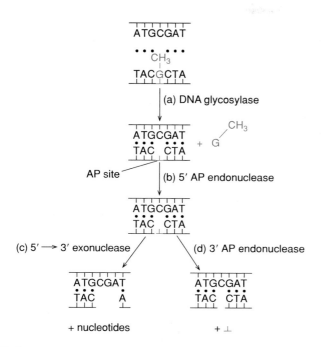

FIGURE 11.26
Removal of methylated base with a DNA glycosylase. (*a*) DNA glycosylase removes the methylguanine base, leaving an apurinic (AP) site. (*b*) A 5′ AP endonuclease cuts the DNA strand just upstream from the AP site. From this point on, two different pathways are possible: (*c*) A 5′ → 3′ exonuclease removes nucleotides, including the AP site, or (*d*) A 3′ AP endonuclease cuts the DNA on the other side of the AP site, removing the apurinic sugar. Both procedures leave a gap that must be filled.

FIGURE 11.27
Removal of pyrimidine dimer with a DNA glycosylase. (*a*) The pyrimidine dimer DNA glycosylase cuts the glycosidic bond between one thymine and its sugar, creating an AP site and leaving the dimer still attached to the DNA. (*b*) A 3′ AP endonuclease cuts just downstream from the AP site. (*c*) A 5′ AP endonuclease or a 3′ → 5′ exonuclease removes the AP sugar. (*d*) A 5′ → 3′ exonuclease removes several nucleotides, including the thymine dimer. As in figure 11.26, a gap remains that must be filled.

3′ → 5′ exonuclease excises the AP sugar. Finally, a 5′ → 3′ exonuclease removes a few nucleotides, including the one that contains the pyrimidine dimer, leaving a gap in the DNA just like the one in figure 11.26.

Repairing Damage Without Creating AP Sites
Bulky base damage, such as thymine dimers, can also be removed directly, without help from a DNA glycosylase. In principle, there are two ways to do this; both are illustrated in figure 11.28. In the one-nick pathway (*a*), a damage-specific DNA incising enzyme recognizes the bulky damage and cuts the DNA strand upstream from the damage. Then a 5′ → 3′ exonuclease removes the damaged DNA as described previously. In the two-nick pathway (*b*), the incising enzyme makes cuts on either side of the mutation, removing an oligonucleotide with the damage. *E. coli* cells seem to use the latter mechanism primarily, at least for repair of pyrimidine dimers and other bulky damage. The key enzyme in this process is called the uvrABC endonuclease

because it contains three polypeptides, the products of the *uvrA, uvrB,* and *uvrC* genes. This enzyme generates an oligonucleotide that is twelve to thirteen bases long, rather than the four shown for convenience here. Other repair systems may use the one-nick pathway.

In all cases just described, we wind up with a gapped DNA strand across from an intact strand. Recall from chapter 7 that this is exactly the same situation the DNA replication apparatus faces after an RNA primer is removed. And the same mechanism for filling in the gap is probably used here: DNA polymerase I synthesizes new DNA, using the intact strand as the template, and DNA

FIGURE 11.28

DNA incision without AP sites. A bulky, damage-dependent DNA incising enzyme attacks the damaged DNA strand (*a*) at one site or (*b*) at two sites. In the latter case, the bulky damage is removed as part of an oligonucleotide (twelve to thirteen bases long in *E. coli*). (*c*) In the former case, a 5′ → 3′ exonuclease is needed to remove the damaged DNA. Both procedures leave a gap that must be filled.

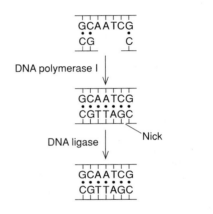

ligase forms the last phosphodiester bond to complete the job (fig. 11.29). In fact, DNA polymerase I contains both a 5′ → 3′ exonuclease and a DNA polymerase activity. Therefore, in principle at least, it can remove damage and fill in the gap at the same time. In that case, the long gaps shown in figure 11.29 would not occur. In fact, fairly long gaps do appear in *E. coli* DNA undergoing repair, so simultaneous DNA degradation and resynthesis apparently do not always take place.

 Most DNA damage is corrected by excision repair, in which the mutated DNA is removed and replaced with normal DNA. This can occur by a number of different mechanisms. (1) The damaged base can be clipped out by a DNA glycosylase, leaving an apurinic or apyrimidinic site that attracts the DNA cutting enzymes that remove the damaged region. (2) A pyrimidine dimer can also attract a specific DNA glycosylase to initiate the same kind of repair pathway. (3) The damaged DNA can be clipped out directly either by cutting on one side with an endonuclease and then removing the mutated DNA with an exonuclease, or by cutting on both sides with an endonuclease to remove the damaged DNA as part of an oligonucleotide.

MISMATCH REPAIR

So far, we have been dealing with repair of DNA damaged by external agents. What about DNA that simply has a mismatch due to incorporation of the wrong base and failure of the proofreading system (chapter 7)? At first glance, it seems tricky to repair such a mistake, because of the apparent difficulty in determining which strand is the newly synthesized one that has the mistake and which is the parental one that should be left alone. At least in *E. coli* this is not a problem, because the parental strand has identification tags that distinguish it from the progeny strand. These tags are methylated adenines, created by a methylating enzyme that recognizes the sequence GATC and places a methyl group

FIGURE 11.29

Filling in the gaps left by damage excision. First, DNA polymerase I links together nucleotides to fill the gap, leaving a nick (an unformed phosphodiester bond). Finally, DNA ligase seals the nick by forming the last phosphodiester bond.

on the A. Since this four-base sequence occurs approximately every 250 base pairs, there is usually one not far from a newly created mismatch.

Moreover, GATC is a symmetrical sequence, in that the opposite strand also reads GATC in the 5′ → 3′ direction. This means that a newly synthesized strand across from a methylated GATC is also destined to become methylated, but a little time elapses before this can happen. The **mismatch repair** system takes advantage of this delay; it uses the methylation on the parental strand as a signal to leave that strand alone and correct the nearby mismatch in the unmethylated progeny strand. This process must occur fairly soon after the mismatch is created, or both strands will be methylated and no distinction between them will be possible. It is not clear how eukaryotes carry out mismatch repair; some of them, such as yeast and *Drosophila,* do not seem to methylate their DNA.

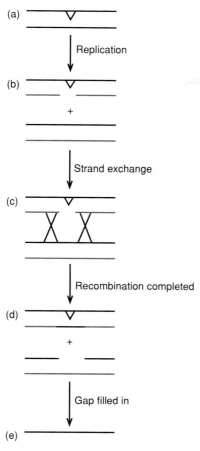

FIGURE 11.30

Recombination repair. (*a*) We begin with DNA with a pyrimidine dimer, represented by a V shape. (*b*) During replication, the replication machinery skips over the region with the dimer, leaving a gap; the complementary strand is replicated normally. The two newly synthesized strands are shown in red. (*c*) Strand exchange between homologous strands occurs. (*d*) Recombination is completed, filling in the gap opposite the pyrimidine dimer, but leaving a gap in the other daughter duplex. (*e*) This last gap is easily filled, using the normal complementary strand as template.

FIGURE 11.31

Error-prone (SOS) repair. Ultraviolet light activates the RecA co-protease, which stimulates the LexA protein (green) to cleave itself, releasing it from the *umuDC* operon. This results in synthesis of UmuC and UmuD proteins, which somehow allow DNA synthesis across from a thymine dimer, even though mistakes (blue) will usually be made.

COPING WITH DNA DAMAGE WITHOUT REPAIRING IT

The direct reversal and excision repair mechanisms described so far are all true repair processes. They eliminate the defective DNA entirely. However, cells have other means of coping with mutations that do not remove the damage but simply skirt around it. These are usually called repair mechanisms, even though they really are not.

Recombination Repair

So-called **recombination repair** is the most important of these mechanisms. Because it requires DNA replication before it can operate, it is also sometimes called **post-replication repair.**

Figure 11.30 shows how recombination repair works. First, the DNA must be replicated. This creates a problem for DNA with pyrimidine dimers because the dimers stop the replication machinery. Nevertheless, after a pause, replication continues, leaving a gap across from the dimer. (A new primer is presumably required to restart DNA synthesis.) Next, recombination occurs between the gapped strand and its homologue on the other daughter DNA duplex. This recombination depends on the *recA* gene product, which exchanges the homologous DNA strands (chapter 7). The net effect of this recombination is to fill in the gap across from the pyrimidine dimer but create a new gap in the other DNA duplex. However, since the other duplex has no dimer, the gap can easily be filled in by DNA polymerase and ligase. Note that the DNA damage still exists, but the cell has at least managed to replicate its DNA. Sooner or later, true DNA repair could presumably occur.

Error-Prone Repair

So-called **error-prone repair** is another way of dealing with damage without really repairing it. This pathway is induced by DNA damage, including ultraviolet damage, and depends on the product of the *recA* gene. We have encountered *recA* before in our discussion of the induction of a λ prophage during the SOS response (chapter 8). Indeed, error-prone repair is also part of the SOS response.

The chain of events seems to be as follows (fig. 11.31): Ultraviolet light or another mutagenic treatment somehow activates the RecA co-protease activity. This co-protease has several targets. One we have studied already is the λ repressor; another is the product of the **lexA** gene. This product, **LexA,** is a repressor for many genes, including repair genes; when the RecA co-protease stimulates it to cleave itself, all these genes are induced.

Two of the newly induced genes are *umuC* and *umuD,* which make up a single operon (*umuDC*). Somehow the products of these genes promote error-prone repair. We know this is so because mutations in either *umuC* or *umuD* prevent error-prone repair, but we do not know how these genes work. One thing seems clear: Error-prone repair requires DNA replication. When a gene for the replicating enzyme DNA polymerase III is mutated, no error-prone repair occurs. Our best guess at this time is that error-prone repair involves replication of DNA across from the pyrimidine dimer even though correct "reading" of the defective strand is impossible. This avoids leaving a gap, but it usually puts the wrong bases into the new DNA strand. When the DNA replicates again, these errors will be perpetuated.

> ◆◆ Cells can employ nonrepair methods to circumvent DNA damage. One of these is recombination repair, in which the gapped DNA strand across from a damaged strand recombines with a normal strand in the other daughter DNA duplex after replication. This solves the gap problem but leaves the original mutation unrepaired. Another mechanism to deal with DNA damage is to induce the SOS response. The effect of this on DNA damage seems to be to replicate the DNA, even though the damaged region cannot be read correctly. This results in errors in the newly made DNA, so the process is called error-prone repair.

SEVERE CONSEQUENCES OF DEFECTS IN DNA REPAIR MECHANISMS

Wild-type *E. coli* cells can tolerate as many as fifty pyrimidine dimers in their genome without ill effect because of their active repair mechanisms. Bacteria lacking one of the *uvr* genes cannot carry out excision repair, so their susceptibility to ultraviolet damage is greater. However, they are still somewhat resistant to DNA damage. On the other hand, double mutants in *uvr* and *recA* can perform neither excision repair nor recombination repair, and they are very sensitive to ultraviolet damage. Only one to two pyrimidine dimers per genome is a lethal dose.

A similar situation is found in humans. Congenital defects in DNA repair cause a group of human diseases, including xeroderma pigmentosum, ataxia telangiectasia, Fanconi's anemia, and Bloom's syndrome. The latter three conditions are characterized by increased incidence of chromosomal abnormalities, especially gaps and breaks.

This is understandable in the case of Bloom's syndrome, because that condition involves a deficiency in DNA ligase, the enzyme that repairs DNA breaks. Whether this defect explains the failure to repair damaged chromosomes is still unclear, but the result of this failure is that patients with any of these diseases are more susceptible to cancer than is the general population.

Xeroderma pigmentosum is better understood. Patients with this disease are thousands of times more likely to develop skin cancer than normal people if they are exposed to sunlight. In fact, their skin can become literally freckled with skin cancers. This reflects the fact that xeroderma pigmentosum cells are defective in excision repair and therefore cannot repair pyrimidine dimers effectively. In addition, patients with xeroderma pigmentosum have a somewhat higher than average incidence of internal cancers. This probably means that these people are also defective in repairing DNA damage caused by chemical mutagens. Notice the underlying assumption here that unrepaired genetic damage can lead to cancer. We will expand on this theme in chapter 12.

What repair steps are defective in xeroderma pigmentosum cells? There is no simple answer to this question because the defect varies from one patient to the next. The problem has been investigated by fusing cells from different patients to see if the fused cells still show the defect. Frequently they do not; instead, the genes from two different patients complement each other. Of course, this probably means that a different gene was defective in each patient. So far, eight different complementation groups have been identified this way, suggesting that the defect can lie in any of eight different genes. Most often, the first step in excision repair (incision) seems to be defective. There is even a variant of the disease in which excision repair occurs normally; in these patients, about 10% of the total, some other repair process must be blocked.

> ◆◆ Failure to correct genetic damage is very harmful to an organism. *E. coli* cells that are blocked in both excision repair and recombination repair can tolerate only a small fraction of the ultraviolet radiation that would be innocuous to wild-type bacteria. Humans defective in DNA repair suffer from a variety of conditions, the best understood of which is xeroderma pigmentosum. Most patients with this disease cannot carry out excision repair. As a result, ultraviolet damage persists in their skin cells and leads to multiple skin cancers.

WORKED-OUT PROBLEMS

Problem 1

Here is the double-stranded DNA sequence of a portion of a bacterial gene:

5′CACTATGCTTGCGTGGACGCATTAAC3′
3′GTGATACGAACGCACCTGCGTAATTG5′

The top strand is the coding strand.

(a) Assuming that transcription starts with the first A in the coding strand and continues to the end, what would be the sequence of the mRNA transcribed from this gene fragment?

(b) What is the amino acid sequence encoded by this mRNA?

What effects would the following events have on this DNA sequence, and on its protein product, assuming that any mutations go unrepaired and finally affect *both* DNA strands?

(c) A tautomerization to the enol form of T in the first T of the coding strand during DNA replication.

(d) A tautomerization to the imino form of A in the second A of the anticoding strand during DNA replication.

(e) Deamination of the third C of the coding strand prior to DNA replication.

(f) O^6-alkylation of the fourth G of the anticoding strand prior to DNA replication.

(g) Deletion of the third C in the coding strand.

Solution

(a) Since the top strand is the coding strand, it is the one that has the same polarity and sense as the mRNA. Therefore, this is the strand you should use as your model for the mRNA transcript of this gene. The only changes you need to bear in mind are that T's in the DNA coding strand will be replaced by U's in the mRNA, and that since transcription starts with the first A, you should ignore the first C. Thus, the mRNA will read:

ACUAUGCUUGCGUGGACGCAUUAAC

(b) Using the genetic code in chapter 10, and the strategy in the second worked-out problem in that same chapter, we can find the first AUG, beginning with the fourth base in the mRNA. The reading frame and coding from there on is:

fMet	Leu	Ala	Trp	Thr	His	Stop
AUG	CUU	GCG	UGG	ACG	CAU	UAA

Thus, the protein product will be fMet-Leu-Ala-Trp-Thr-His.

(c) A tautomerization to the enol form of the second T of the coding strand will make it pair like C, resulting in a T → C transition at that point. This will change the mRNA sequence from ACU etc., to ACC etc.

However, this will have no effect on the coding properties of the mRNA, since the mutation occurred outside the coding region of the gene.

(d) A tautomerization to the imino form of A in the second A in the anticoding strand will make it pair like G, resulting in an A → G transition at that point. This will in turn cause a T → C transition in the coding strand. The change in the gene can be summarized as follows:

CACTATG--- → CACTACG---
GTGATAC---   GTGATGC---

where the boldface denotes the altered base pair. This will change the mRNA sequence from ACUAUG etc., to ACUACG etc. The change of the initiation codon from AUG to ACG will destroy its function, so the ribosome will search for another AUG (corresponding to ATG in DNA). Since there is none, no protein product will be made.

(e) Deamination of the third C of the coding strand will convert it to a U. If this change goes unrepaired, it will result in a C → T transition, and the new mRNA sequence and coding properties will be:

	fMet	Phe
ACU	AUG	UUU	etc.,

where the boldface indicates the altered base. Thus, the protein product will now be fMet-Phe-Ala-Trp-Thr-His, with a phenylalanine (Phe) in place of the original leucine (Leu).

(f) O^6-alkylation of the fourth G of the anticoding strand will make it pair like A, resulting in a C → T transition in the coding strand. The change in the gene can be summarized as follows:

5′CACTATGCTTGCG--- → CACTATGCTTGTG---
3′GTGATACGAACGC---   GTGATACGAACAC---

where boldface denotes the altered base pair. The new mRNA sequence and coding properties will be:

	fMet	Leu	Val
ACU	AUG	CUU	GUG	etc.

Thus, the protein product will now be fMet-Leu-Val-Trp-Thr-His, with a valine (Val) in place of the original alanine (Ala).

(g) Deletion of the third C in the coding strand will result in the following DNA sequence:

CACTATGTTGCGTGGACGCATTAAC
GTGATACAACGCACCTGCGTAATTG

where the boldfaced bases flank the deleted base. The new mRNA sequence and coding properties will be:

	fMet	Leu	Arg	Gly	Arg	Ile	Asn
ACU	AUG	UUG	CGU	GGA	CGC	AUU	AAC

Thus, the protein product will now be fMet-Leu-Arg-Gly-Arg-Ile-Asn. Note that the frameshift has caused a drastic difference in the gene product.

Problem 2

You have isolated a protein from a wild-type organism and the corresponding, defective protein from a mutant. You find that the wild-type and mutant proteins differ in only one amino acid, as follows:

Wild-type: Arg Met Ser

Mutant: Arg Ile Ser

Three different missense mutations could have occurred. Using the genetic code, show the coding strand of the wild-type gene and the three different missense mutant genes. Show only the codon that is involved in the mutation and present your answer in a table as follows:

Organism	Codon	Transition or Transversion?
Wild-type		
Mutant #1		
Mutant #2		
Mutant #3		

Solution

We need to account for the change from Met to Ile in the mutant protein. The genetic code shows that the only codon for Met (methionine) is AUG (ATG in DNA), so that is the codon we need to use for the wild-type gene. Ile (isoleucine) has three codons: AUU, AUC, and AUA (ATT, ATC, and ATA in DNA). Thus, the mutant gene has one of these latter three Ile codons. If the Ile codon is ATA, then a G → A transition has occurred. A purine (G) has been replaced by another purine (A). If it is one of the others, then a transversion has occurred, because a purine (G) has been replaced by a pyrimidine (C or T). Thus, the table can be filled in as follows:

Organism	Codon	Transition or Transversion?
Wild-type	ATG	—
Mutant #1	ATA	Transition
Mutant #2	ATC	Transversion
Mutant #3	ATT	Transversion

Problem 3

Here is the sequence of a small piece of an mRNA and the protein it encodes.

Ser   Thr   Ala

AGU   ACG   GCU

A point mutation within its DNA coding sequence can change the sequence of amino acids in the protein to:

Arg   Tyr   Gly

What was the mutation? For the purposes of this problem, a point mutation is defined as changing, inserting, or deleting one base.

Solution

Since three codons in a row are changed, the mutation must be a frameshift rather than a missense mutation. The original amino acid sequence Ser Thr Ala was changed to Arg Tyr Gly after the mutation. Since the first codon was changed, and a frameshift mutation affects only the codon in which it occurs and codons downstream (toward the 3′-end), the mutation must have occurred in the first codon.

Next we need to determine the direction of the frameshift by looking at the genetic code. Since arginine (Arg) has six codons, it would be a bit time-consuming to examine all its codons; furthermore, the mutation occurred in the first codon, so that confuses the issue somewhat. It is easier to look downstream and see how changing the reading frame one base to the left (−1) or right (+1) could have given codons for tyrosine (Tyr) and glycine (Gly). Tyrosine has only two codons (UAU and UAC). You should look at the −1 and +1 reading frames for one of these codons. There is a UAC codon in the −1 reading frame, beginning with the third base of the wild-type mRNA. In order to shift to the −1 reading frame, the gene must have acquired a one-base insertion in the first codon, before the U. Notice that this will make the third codon GGC, which codes for glycine (Gly), as expected.

Now where was the base inserted, after the A, or after the G in the AGU codon? By consulting the genetic code, we discover that the only way inserting a base can convert the serine codon AGU to an arginine codon is by placing either an A or a G between the second and third bases to give AGA or AGG, both of which are arginine codons. This must have been the mutation in the DNA. Therefore, the sequence of the mutant mRNA is either/or AGA UAC GGC U AGG UAC GGC U.

Problem 4

The records of 200,000 children at a hospital show that 15 were afflicted with a disease caused by a dominant mutation. Three of the 15 came from families with a history of this disease, so only 12 represent new mutations. What is the mutation rate for the gene that causes this disease?

Solution

Since the number of new mutations is 12, and the total number of gametes involved is twice the total number of individuals, the mutation rate is $12/(2)(200,000) = 3 \times 10^{-5}$.

Summary

Mutations are heritable changes in the genetic material. They can be categorized according to their effects on the mutant organism's phenotype. Germ-line mutations affect the gametes, so the mutation can be transmitted to the next generation. Somatic mutations affect the nonsex, somatic cells, and thus are not passed on. Morphological mutations cause a visible change in the affected organism. Nutritional mutations render an organism dependent on added nutrients that a wild-type organism does not require. Lethal mutations occur in indispensable genes. Haploid organisms with lethal mutations die immediately, but lethal mutations are usually recessive, so diploid organisms can tolerate them in heterozygotes. Conditional mutations affect the mutant organism only under certain conditions, such as high temperature. If a conditional mutation causes death under nonpermissive conditions, it is a conditional lethal.

A number of different kinds of agents can cause genetic mutations. Radiation can either damage DNA directly or produce free radicals that do the damage. Ultraviolet light causes pyrimidine dimers, while the more energetic X rays and gamma rays usually cause breaks in DNA. Chemicals can injure DNA in a variety of ways, including deamination and alkylation of bases or by forcing the incorporation of the wrong base, or too many or too few bases, into a replicating DNA. Chemicals usually cause small lesions called point mutations that can revert readily to wild-type.

All living things have means of reversing genetic damage. Some mutations can be directly undone. For example, the photoreactivating enzyme can break pyrimidine dimers, and the suicide enzyme O^6-methylguanine methyl transferase can remove methyl and ethyl groups from guanine. However, most DNA damage is corrected by excision repair, in which the mutated DNA is removed and replaced with normal DNA. Cells can also employ nonrepair methods to replicate their damaged DNA without repairing it. Two examples are recombination repair and error-prone (SOS) repair.

PROBLEMS AND QUESTIONS ❓

1. You notice a rose that is all red except for a small patch of white. Is this likely to be due to a somatic or a germ-line mutation? Why?

2. Two mice, heterozygous for the yellow trait, mate. The progeny are ⅔ yellow and ⅓ wild-type color. The yellow mice are heterozygotes; the homozygous mutants died in utero. Is the yellow color dominant or recessive? Is the lethality of the yellow gene dominant or recessive?

3. You are mapping a DNA polymerase III gene in a new strain of bacteria. Can you use nonconditional mutations that inactivate the gene? Why, or why not?

4. What kind of mutations would be useful in the mapping studies in question 3?

5. Is a temperature-sensitive gene more easily melted (denatured) than a wild-type gene?

6. A Siamese cat was wounded and had to have a patch of white fur shaved. The fur grew in dark instead of white. With time, this dark fur was gradually replaced with white. Present a plausible explanation for these observations.

7. Does tautomerization of thymine cause transitions or transversions?

8. (a) What kind of DNA damage does ultraviolet light cause? (b) Why is this damage harmful to a cell?

9. (a) What kind of DNA damage do X rays cause? (b) Why is this damage harmful to a cell?

10. A hypothetical mRNA, AUGCGCCUAAAGAGG, codes for fMet-Arg-Leu-Lys-Arg. What happens to the coding if you delete the first C?

11. What would be the simplest way to cause a second-site reversion of the mutation in question 10?

12. Which two enzymes catalyze direct reversal of DNA damage?

13. What is the difference between an exonuclease and an endonuclease?

14. In excision repair in E. coli, two mechanisms could be used for removing damaged DNA and filling in the gap: (1) The old DNA could be removed and replaced in a single pass of the DNA polymerase I, or (2) the old DNA could be removed and then, in a separate step, replaced. What DNA feature would occur in the second case but not in the first? Which mechanism does the evidence favor?

15. Why are recombination repair and error-prone repair not real repair processes?

16. At a hospital, 150,000 patients were screened for a dominant genetic disease. Twenty were found to have the disorder. Fifteen of these had no family history of the disease and so represented new mutations. What is the mutation rate in the gene that causes this disease?

Answers appear at end of book.

CHAPTER 12: GENE MUTATION II: TRANSPOSABLE ELEMENTS AND CANCER

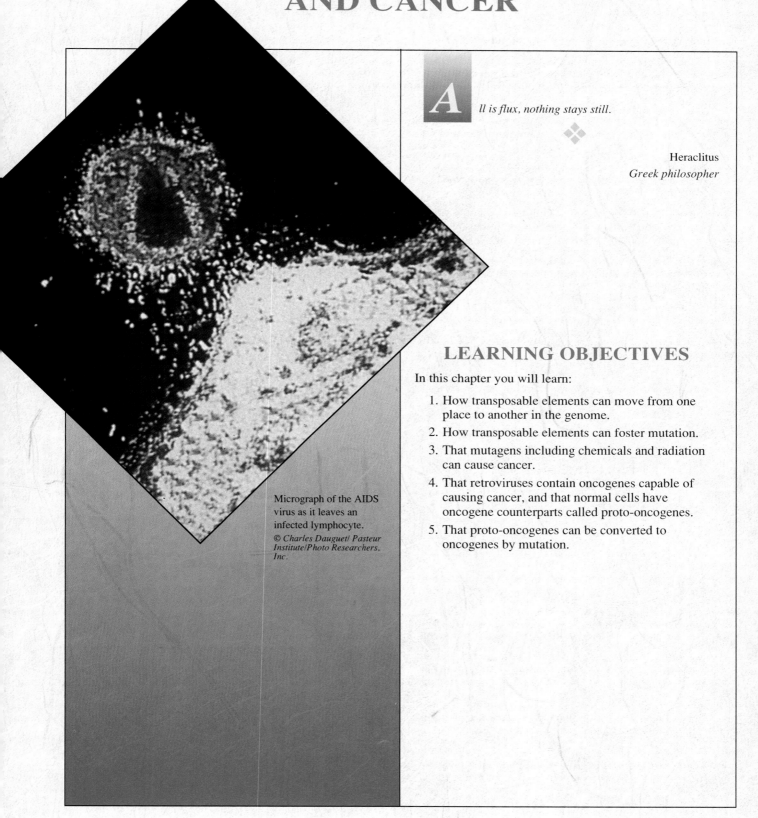

All is flux, nothing stays still.

Heraclitus
Greek philosopher

Micrograph of the AIDS virus as it leaves an infected lymphocyte.
© *Charles Dauguet/ Pasteur Institute/Photo Researchers, Inc.*

LEARNING OBJECTIVES

In this chapter you will learn:

1. How transposable elements can move from one place to another in the genome.

2. How transposable elements can foster mutation.

3. That mutagens including chemicals and radiation can cause cancer.

4. That retroviruses contain oncogenes capable of causing cancer, and that normal cells have oncogene counterparts called proto-oncogenes.

5. That proto-oncogenes can be converted to oncogenes by mutation.

e have already learned that an organism's genome is not absolutely fixed from the beginning to the end of its life. In addition to the point mutations already discussed, rearrangements of genetic material take place. This can be useful, as in the rearrangement of immunoglobulin gene segments to form a functioning gene, or in the activation of a trypanosome surface protein gene by moving it to an expression site (chapter 14). But pieces of DNA with no apparent function besides their own replication also move around the genomes of both prokaryotes and eukaryotes. These **transposable elements, or transposons,** are mobile in the sense that they can insert copies of themselves throughout the genome. They can also cause rearrangements of host DNA, giving rise to chromosome mutations. In the first part of this chapter, we will discuss the characteristics of bacterial and eukaryotic transposons in turn, showing how they can cause mutations, including chromosomal mutations.

The consequences of gene mutation can be destructive and often fatal for an organism. Mutations caused by mutagenic chemicals, radiation, and viruses can cause cancer. In the remaining portion of this chapter, we will discuss the role these environmental mutagens play in turning a normal cell into a cancer cell.

BACTERIAL TRANSPOSONS

Bacterial transposons were originally noticed by James Shapiro and others in the late 1960s as mutations that did not behave normally. For example, they did not revert readily the way point mutations do, and the mutant genes contained long stretches of extra DNA. In the rare cases when such mutants did revert, they lost the extra DNA. These extra DNAs that could inactivate a gene by inserting into it were the first transposons discovered in bacteria. They are called **insertion sequences (ISs).**

INSERTION SEQUENCES: THE SIMPLEST TRANSPOSONS

Bacterial insertion sequences contain only the elements necessary for transposition. The first of these elements is a set of special sequences at a transposon's ends, one of which is the inverted repeat of the other. The second element is the set of genes that code for enzymes that catalyze transposition.

Since the ends of an insertion sequence are inverted repeats, then if one end of an insertion sequence is 5′-ACCGTAG, the other end of that strand will be the reverse complement: CTACGGT-3′. This is akin to the palindromes, or inverted repeats, discussed in chapter 11, but the two parts of the palindrome are interrupted by hundreds of base pairs, corresponding to the body of the insertion sequence. The seven-nucleotide inverted repeats given here are hypothetical and are presented to illustrate the point. Typical insertion sequences have somewhat longer inverted repeats, from fifteen to twenty-five base pairs

long. IS1, for example, has inverted repeats twenty-three base pairs long. Larger transposons can have inverted repeats hundreds of base pairs long.

The main body of an insertion sequence encodes at least one protein, **transposase,** that mediates transposition. Mutations in the transposase gene can render the insertion sequence immobile.

One other feature of an insertion sequence, shared with more complex transposons, is found just outside the transposon itself. This is a pair of short direct repeats in the DNA immediately surrounding the transposon. These repeats did not exist before the transposon inserted; they result from the insertion process itself and tell us that the transposase cuts the target DNA in a staggered fashion rather than with two cuts right across from one another. Figure 12.1 shows how staggered cuts in the two strands of the target DNA at the site of insertion lead automatically to direct repeats. The length of these direct repeats depends on the distance between the two cuts in the target DNA strands. This distance depends in turn on the nature of the insertion sequence. The transposase of IS1 makes cuts nine base pairs apart and therefore generates direct repeats that are nine base pairs long.

Insertion sequences are the simplest of the transposons. They contain only the elements necessary for their own transposition: short inverted repeats at their ends and at least two genes coding for an enzyme called transposase that carries out transposition. Transposition involves duplication of a short sequence in the target DNA; one copy of this short sequence flanks the insertion sequence on each side after transposition.

MORE COMPLEX TRANSPOSONS

Insertion sequences are sometimes called "selfish DNA." You should not interpret this anthropomorphic term to mean that ISs think about what they do; of course they cannot. Rather, it means that an IS replicates at the expense of its bacterial host and apparently provides nothing useful in return. Other transposons do carry genes that are valuable to their hosts, the most familiar being genes for antibiotic resistance. Not only is this a clear benefit to the bacterial host, it is also valuable to geneticists, because it makes the transposon much easier to track. No longer is the transposon detectable only by the genes it inactivates; it now carries a conspicuous marker of its own.

For example, consider the situation in figure 12.2, in which we start with a donor plasmid containing a gene for kanamycin resistance (Km) and harboring a transposon **(Tn3)** with a gene for ampicillin resistance (Ap); in addition, we have a target plasmid with a gene for tetracycline resistance (Tc). After transposition, Tn3 has replicated, and a copy has moved to the target plasmid. Now the target plasmid confers both tetracycline and ampicillin resistance,

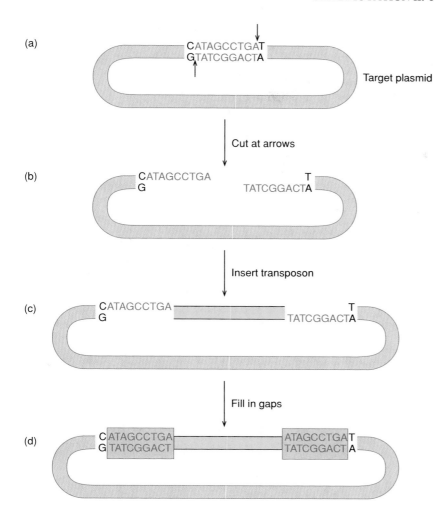

FIGURE 12.1
Generation of direct repeats in host DNA flanking a transposon. (*a*) The arrows indicate where the two strands of host DNA will be cut in a staggered fashion, nine base pairs apart. (*b*) After cutting. (*c*) The transposon (yellow) has been ligated to one strand of host DNA at each end, leaving two nine-base gaps.
(*d*) After the gaps are filled in, there are nine base-pair repeats of host DNA (red boxes) at each end of the transposon.

properties that we can easily monitor by transforming antibiotic-sensitive bacteria with the target plasmid and growing these host bacteria in medium containing both antibiotics. If the bacteria survive, they must have taken up both antibiotic resistance genes; therefore, Tn3 must have transposed to the target plasmid.

MECHANISMS OF TRANSPOSITION

One sometimes hears transposons referred to as "jumping genes," because of their ability to move from one place to another. However, the term is a little misleading, since it implies that the DNA always leaves one place and jumps to the other. This mode of transposition does occur and is called **conservative transposition,** because both strands of the original DNA are conserved as they move together from one place to the other. However, transposition frequently involves DNA replication, so one copy of the transposon remains at its original site as another copy inserts at the new site. This is called **replicative transposition** because when a transposon moves by this route, it also replicates itself.

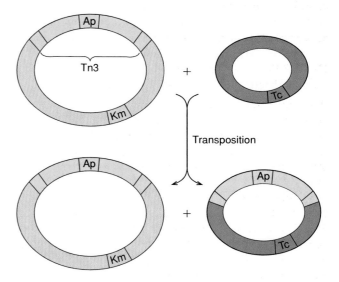

FIGURE 12.2
Tracking transposition with antibiotic resistance markers. We begin with two plasmids: The larger (blue) encodes kanamycin resistance (Km) and bears the transposon Tn3 (yellow), which codes for ampicillin resistance (Ap); the smaller (green) encodes tetracycline resistance (Tc). After transposition, the smaller plasmid bears both the Tc and Ap markers.

FIGURE 12.3

Structure of Tn3. The *tnpA* and *tnpR* genes are necessary for transposition; *res* is the site of the recombination that occurs during the resolution step in transposition; the *bla* gene encodes β-lactamase, which protects bacteria against the antibiotic ampicillin. This gene is also called Ap and Amp^r. Inverted repeats (IR) are found on each end. The arrows indicate the direction of transcription of each gene.

Tn3, whose structure is shown in figure 12.3, illustrates the replicative mechanism of transposition. In addition to the *bla* gene, which encodes ampicillin-inactivating **β-lactamase,** Tn3 contains two genes that are instrumental in transposition. Tn3 transposes by a two-step process, each step of which requires one of the Tn3 gene products. Figure 12.4 shows this sequence of events. We begin with two plasmids: the donor, which harbors Tn3, and the target. In the first step, the two plasmids fuse, with Tn3 replication, to form a **cointegrate** in which they are coupled through a pair of Tn3 copies. This step requires recombination between the two plasmids, which is catalyzed by the product of the Tn3 transposase gene ***tnpA.***

Most of the recombinations we have encountered so far required homology between the two recombining DNAs, but the recombination catalyzed by the Tn3 transposase requires little if any homology between donor and target DNAs. This means that transposition is relatively independent of DNA sequence, so Tn3 can transpose to a wide variety of target loci. This kind of **illegitimate recombination,** which shows little dependence on DNA sequence, is a common feature of transposons. Figure 12.4 illustrates transposition between two plasmids, but the donor and target DNAs can be other DNAs, including phage DNAs or the bacterial chromosome itself.

The second step in Tn3 transposition is a **resolution** of the cointegrate, in which the cointegrate breaks down into two independent plasmids, each bearing one copy of Tn3. This step, catalyzed by the product of the **resolvase** gene ***tnpR,*** is a recombination between homologous sites on Tn3 itself, called ***res*** sites. Several lines of evidence show that Tn3 transposition is a two-step process. First, mutants in the *tnpR* gene cannot resolve cointegrates, so they cause formation of cointegrates as the final product of transposition. This demonstrates that the cointegrate is normally an intermediate in the reaction. Second, even if the *tnpR* gene is defective, cointegrates can be resolved if a functional *tnpR* gene is provided *in trans* by another DNA molecule— the host chromosome or another plasmid, for example.

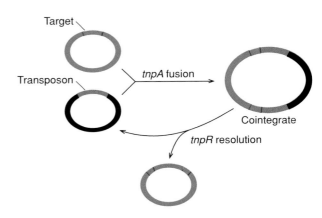

FIGURE 12.4

Simplified scheme of the two-step Tn3 transposition. In the first step, catalyzed by the *tnpA* gene product, the plasmid (black) bearing the transposon (blue) fuses with the target plasmid (green, target in red) to form a cointegrate. During cointegrate formation, the transposon replicates. In the second step, catalyzed by the *tnpR* gene product, the cointegrate resolves into the target plasmid, with the transposon inserted, plus the original transposon-bearing plasmid.

 Many transposons contain genes aside from the ones necessary for transposition. These are commonly antibiotic resistance genes. For example, Tn3 contains a gene that confers ampicillin resistance. Tn3 and its relatives transpose by a two-step process (replicative transposition). First the transposon replicates and the donor DNA fuses to the target DNA, forming a cointegrate. In the second step, the cointegrate is resolved into two DNA circles, each of which bears a copy of the transposon. An alternate pathway, not used by Tn3, is conservative transposition, in which no replication of the transposon occurs.

TRANSPOSONS AS MUTAGENIC AGENTS

A bacteriophage called **Mu** that infects *E. coli* also behaves as a transposon. In fact, its replication depends on transposition. Phage Mu particles contain a linear DNA genome 37 kb long that can replicate either lytically or lysogenically (fig. 12.5). (Recall the lytic and lysogenic replication of phage λ described in chapter 8.) In the lysogenic phase, the Mu genome inserts into the host chromosome, and a phage repressor prevents most phage gene expression, just as in the λ lysogenic phase. However, the lytic phase of Mu is very different from that of λ. Instead of remaining free, the phage DNA integrates into the host chromosome. More strikingly, the phage DNA remains integrated throughout the lytic phase and replicates by transposing.

Box 12.1 Transposition Impacts on Public Health

Even more complex than Tn3 and its relatives are **composite transposons.** These consist of two parts: a central region and two arms. The central region comprises the genes for transposition plus one or more antibiotic resistance genes. The arms are insertion sequences or IS-like elements. **Tn9** is an example of a composite transposon. Its central region carries a chloramphenicol resistance gene and is flanked by two IS1 elements. The ISs can cooperate to transpose the whole Tn9 element, or they can transpose on their own.

Figure B12.1 shows an antibiotic resistance plasmid that has been the target for a variety of different transposons, including the composite transposon **Tn10.** Such a plasmid, with multiple antibiotic resistance genes, is obviously a great advantage to the bacteria that harbor it, as long as antibiotic use is widespread. In fact, these plasmids may help explain how some pathogenic (disease-causing) bacteria are becoming resistant to the

antibiotics we use to kill them. Once these plasmids are assembled, they can be shared with other bacteria, instantly conferring multiple antibiotic resistance.

As an example of the potential severity of the antibiotic resistance problem, consider the case of a huge, modern dairy in Illinois that became contaminated with the bacterium *Salmonella typhimurium* in 1985. As a result, over 14,000 people in six states suffered salmonellosis, or salmonella poisoning. This thoroughly unpleasant condition, one of the diseases commonly known as "food poisoning," is characterized by nausea, abdominal cramps, vomiting, diarrhea, fever, and even death. In this outbreak, five people died. Fortunately, the bacteria involved in this epidemic were not antibiotic-resistant. If they had been, the death toll would probably have been much higher.

Another outbreak of salmonellosis had occurred two years earlier in the Midwest, mostly in Minnesota. It was much more limited in scope—only eighteen patients—but was potentially much more dangerous, because the

bacteria, *Salmonella newport* in this case, were antibiotic-resistant. Where did these bacteria come from? Scott A. Holmberg of the Centers for Disease Control (CDC) in Atlanta used a nice bit of detective work to identify the culprit.

Holmberg's only clues at the outset were that *Salmonella newport* is unusual in the north, and that most of the victims had taken antibiotics shortly before they fell ill. This pointed to the likelihood of a contaminated antibiotic, but Holmberg eliminated that possibility by discovering that different antibiotics were involved, and that some of the patients had not even taken antibiotics. The big break came when Holmberg learned that four other cases of drug-resistant salmonellosis had occurred somewhat earlier in South Dakota. Three of the patients were related, and one of them was a farmer who raised beef cattle.

The farmer told Holmberg that his herd had been infected by an epidemic of diarrhea several months before, and that several calves had died. He mentioned in passing that he thought *Salmonella* was involved. Was it the same strain of

FIGURE B12.1

Hypothetical antibiotic resistance plasmid. Several different transposons, each bearing its own resistance marker, have clustered in a single plasmid. Tn4, which includes Tn3, confers resistance to streptomycin (Sm), sulfonamide (Su), ampicillin (Ap), and mercury (Hg). Tn55 confers resistance to kanamycin (Km), and Tn10 confers resistance to tetracycline (Tc). All but the last transposon comprise the resistance-determining segment, which is bounded by two IS1 elements and also contains a gene for chloramphenicol resistance (Cm). The two IS1 elements allow the resistance-determining segment to transpose independently, carrying all its valuable resistance genes with it.

Box 12.1 (continued)

Salmonella implicated in the human cases? To find out, the plasmids in the *Salmonella* from one of the dead calves and those from the human patients in Minnesota were compared by electrophoresis. The pattern from the calf matched those of ten of the eleven patients, and was clearly different from ninety-one other samples of *Salmonella newport* from unrelated cases.

This suggested strongly that there was a relationship between the South Dakota cattle herd and all the cases of drug-resistant salmonellosis. Moreover, the beef from this herd could be traced directly to a meat processor in Nebraska

who sold boxed meat to two meat brokers in Minnesota and Iowa. These brokers in turn sold the meat to supermarkets in Minneapolis and St. Paul at which all the Minnesota victims had bought hamburger.

Why did antibiotics contribute to most of the cases? By taking antibiotics, the victims eliminated most of the natural bacteria in their bodies, allowing the drug-resistant bacteria to proliferate without competition. More importantly, how did this strain of *Salmonella* come to be antibiotic-resistant? Here the scenario becomes somewhat more controversial, but most epidemiologists now believe

that the use of low levels of antibiotics in livestock feed selects for antibiotic-resistant bacteria. It is probably significant that the South Dakota farmer, following a widespread practice dating back more than thirty years, had added the antibiotic tetracycline to his cattle feed to promote growth and prevent disease. The result seems to be that his cattle bred a strain of *Salmonella* resistant not only to tetracycline, but to penicillin and its derivative, amoxycillin. They probably obtained this multiple drug resistance by acquiring a resistance plasmid such as the one in figure B12.1.

(a) Free phage DNA

C   A   B   Head and tail genes   *gin*

Host DNA   G segment   Host DNA

(b) Prophage DNA

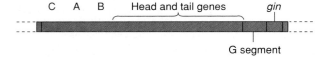

C   A   B   Head and tail genes   *gin*

G segment

FIGURE 12.5

Map of phage Mu. (*a*) The DNA found in phage particles is linear and contains unequal lengths of host DNA (red) attached to each end of the phage DNA (purple). (*b*) The prophage DNA has the same linear map and includes a 5-base-pair direct repeat of host DNA at each end (red). The host DNA present in the free phage DNA has been lost upon integration.

Austin Taylor discovered phage Mu because of the mutations it caused when it transposed. In fact, the name Mu stands for mutator. Obviously, the phage can mutate a gene simply by inserting into it and interrupting it, but Taylor found that Mu could cause many other types of mutations beyond simple insertions. It could cause inversions, deletions, translocations, and fusions of independent DNAs. We have already seen that other transposons can direct the formation of cointegrates (DNA fusion) and transposition (translocation). Figure 12.6 illustrates how Mu and other transposons can provide sites for ordinary homologous recombination (general recombination) to operate. Depending on the orientation of the two transposons, this can result in deletions or inversions.

◆ Bacteriophage Mu can behave as a transposable element. By moving from one site to another and simply providing sites of homology for recombination, Mu can cause rearrangements of host DNA, including replicon fusion, translocation, deletion, and inversion. These mutations in host DNA can also be directed by other transposons.

EUKARYOTIC TRANSPOSABLE ELEMENTS

It would be surprising if prokaryotes were the only organisms to harbor transposable elements, especially since these elements have powerful selective forces on their side. First, many transposons carry genes that are an advantage to their hosts. Therefore, their hosts can multiply at the expense of competing organisms, and can multiply the transposons along with the rest of their DNA. Second, even if transposons are not advantageous to their hosts, they can replicate themselves within their hosts in a "selfish" way.

THE FIRST EXAMPLES OF TRANSPOSABLE ELEMENTS

Barbara McClintock discovered the first transposable elements in a study of maize (corn) in the late 1940s. It had been known for some time that the variegation in color observed in the kernels of so-called Indian corn was caused by an unstable mutation. For example, in figure 12.7*a*, we see a kernel that is colored. This color is due to a factor encoded by a maize *C* locus. Figure 12.7*b* shows what happens when the *C* gene is mutated; no purple pigment is made, and the kernel appears almost white. The spotted kernel in figure 12.7*c* shows the results of reversion in some of the kernel's cells. Wherever the mutation has reverted, the revertant cell and its progeny will be able to make pigment, giving rise to a dark spot on the kernel.

(a) Deletion:

Recombination

(b) Inversion:

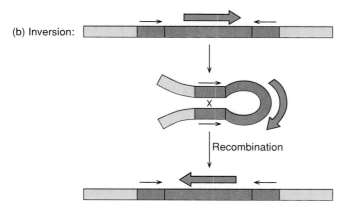

Recombination

FIGURE 12.6

Deletion and inversion promoted by transposons. (*a*) When two identical transposons (purple) are in the same orientation (arrows) on a chromosome, they can pair as shown in the middle figure. Recombination as indicated by the **X** releases the DNA between the two transposons (red) plus one of the transposons (purple). This leaves the original DNA with one transposon and a deletion. (*b*) When two identical transposons are in opposite orientations, they can pair as shown in the middle figure. Recombination as shown by the **X** inverts the DNA between the transposons (large green arrow).

It is striking that there are so many spots in this kernel. That means the mutation is very unstable. It reverts at a rate much higher than we would expect of an ordinary mutation.

In this case, McClintock discovered that the original mutation resulted from an insertion of a transposable element, called **Ds** for "dissociation," into the *C* gene (fig. 12.8*a* and *b*). Another transposable element, **Ac** for "activator," could cause *Ds* to transpose out of *C*, causing reversion (fig. 12.8*c*). In other words, *Ds* can transpose, but only with the help of *Ac. Ac,* on the other hand, is an autonomous transposon. It can transpose itself and therefore inactivate other genes without help from other elements.

(a)

(b)

(c)

FIGURE 12.7

Effects of mutations and reversions on maize kernel color. (*a*) Wild-type kernel has an active *C* locus that causes synthesis of purple pigment. (*b*) The *C* locus has mutated, preventing pigment synthesis, so the kernel is colorless. (*c*) The spots correspond to patches of cells in which the mutation in *C* has reverted, again allowing pigment synthesis.

From "Transposable Genetic Elements in Maize" by Nina V. Fedoroff. Copyright © 1984 by SCIENTIFIC AMERICAN, Inc. June, p. 86, figs. A, B, C. All rights reserved.

Now that molecular genetics tools are available, we can isolate and characterize these genetic elements decades after McClintock found them. Nina Fedoroff and her collaborators have obtained the structure of *Ac* and three different forms of *Ds. Ac* resembles the bacterial transposons we have already studied (fig. 12.9). It is about 4,500 base pairs long and is bounded by short, imperfect inverted repeats. It contains a transposase gene. The various forms of *Ds* are derived from *Ac* by deletion. *Ds-a* is very similar to *Ac*, except that a piece of the transposase gene has been deleted. This explains why *Ds* is unable to transpose itself. *Ds-b* is more severely shortened, retaining only a small fragment of the transposase gene, and *Ds-c* retains only the inverted repeats in common with *Ac.* These inverted repeats are all that *Ds-c* needs to be a target for transposition directed by *Ac.*

FIGURE 12.8

Transposable elements cause mutations and reversions in maize.
(*a*) A wild-type maize kernel has an uninterrupted, active *C* locus (blue) that causes synthesis of purple pigment. (*b*) A *Ds* element (red) has inserted into *C*, inactivating it and preventing pigment synthesis. The kernel is therefore colorless. (*c*) *Ac* (green) is present, as well as *Ds*. This allows *Ds* to transpose out of *C* in many cells, giving rise to groups of cells that make pigment. Such groups of pigmented cells account for the purple spots on the kernel. Of course, *Ds* must have transposed into *C* before it (*Ds*) became defective, or else it had help from an *Ac* element.

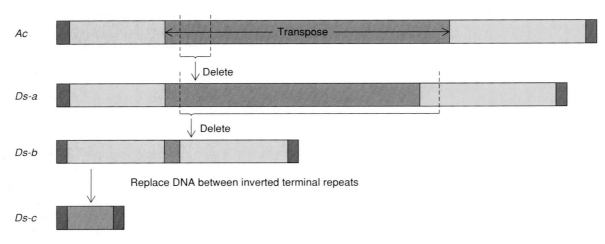

FIGURE 12.9

Structures of *Ac* and *Ds*. *Ac* contains the transposase gene (purple) and two imperfect inverted terminal repeats (blue). *Ds-a* is missing a 194-base-pair region from the transposase gene (dotted lines); otherwise, it is almost identical to *Ac*. *Ds-b* is missing a much larger segment of *Ac*. *Ds-c* has no similarity to *Ac* except for the inverted terminal repeats.

The variegation in the color of Indian corn kernels is caused by multiple reversions of an unstable mutation in the *C* locus, which is responsible for the kernel's coat color. The mutation and its reversion result from a *Ds* (dissociator) element, which transposes into the *C* gene, mutating it, and then transposes out again, causing it to revert to wild-type. *Ds* cannot transpose on its own; it must have help from an autonomous transposon called *Ac* (for activator), which supplies the transposase. *Ds* is an *Ac* element with more or less of its middle removed. All *Ds* needs in order to be transposed is a pair of inverted terminal repeats that the *Ac* transposase can recognize.

YEAST AND *Drosophila* TRANSPOSONS

In spite of the early discovery of the transposable elements in maize, the best-studied eukaryotic transposons are still those of yeast (*Saccharomyces cerevisiae*) and the fruit fly (*Drosophila melanogaster*). The prototype *Drosophila* transposon is called ***copia,*** because it is present in copious quantity. In fact, *copia* and related transposons called ***copia*-like elements** account for about 1% of the total fruit fly genome. Similar transposable elements in yeast are called **Ty** for "transposon yeast."

FIGURE 12.10

Retrovirus replication cycle. The viral genome is an RNA, with long terminal repeats (LTRs, green) at each end. Reverse transcriptase makes a linear, double-stranded DNA copy of the RNA, which then integrates into the host DNA (black), creating the provirus form. The host RNA polymerase II transcribes the provirus, forming genomic RNA. The viral RNA is packaged into a virus particle, which buds out of the cell and infects another cell, starting the cycle over again.

FIGURE 12.11

Retroviruses. (*a*) B-type retrovirus budding from a mouse mammary tumor cell. (*b*) Later in the budding process. (*c*) Free, but still immature B-type retrovirus. (*d*) Intracellular, A-type particles in a mouse mammary tumor.

Maugh, T. H. "RNA Virus: The Age of Innocence Ends." Science, March 22, 1974, Vol. 183, pp. 1181–85, fig. 1. Copyright 1974 by AAAS.

The physical structures of yeast and fruit fly transposons are somewhat related to those of bacterial transposons. They contain genes that are necessary for their own transposition and they cause duplication of host DNA at either end. For example, Ty is a 6,300-base-pair element flanked by 5-base-pair repeats of host DNA. However, there are major differences: Ty contains 330-base-pair direct terminal repeats called **delta repeats**. More importantly, Ty transposes by a pathway very different from the one taken by bacterial transposons. This pathway involves an RNA intermediate and is very similar to the replication scheme of eukaryotic tumor viruses called **retroviruses**. Let us first examine retrovirus replication, and then return to Ty.

RETROVIRUSES

In order to understand the mode of transposition of Ty, it is first necessary to introduce the retroviruses, since transposition of Ty resembles so closely the replication of this class of viruses. In fact, retroviruses can be considered transposons in that their genomes (or more precisely, DNA copies of their genomes) are found inserted randomly into host DNA. Retroviruses have a profound impact on public health, since they can cause both cancer and AIDS.

The most salient feature of a retrovirus, indeed the feature that gives this class of viruses its name, is its ability to make a DNA copy of its RNA genome. This reaction,

RNA → DNA, is the reverse of the transcription reaction, so it is commonly called **reverse transcription.** In 1970, Howard Temin and, simultaneously, David Baltimore convinced a skeptical scientific community that this reaction takes place. They did so by finding that the virus particles contain an enzyme that catalyzes the reverse transcription reaction. Inevitably, this enzyme has been dubbed **reverse transcriptase.**

Figure 12.10 illustrates the retrovirus replication cycle. We start with a virus infecting a cell. The virus contains two copies of its RNA genome, linked together by base pairing at their 5´-ends. When the virus enters a cell, its reverse transcriptase makes a double-stranded DNA copy of the viral RNA. This DNA fuses through its **long terminal repeats,** or **LTRs,** with the host genome. This integrated form of the viral genome is called the **provirus.** To complete virus replication, host RNA polymerase II makes viral mRNAs, which are then translated to viral proteins. RNA polymerase II also makes full-length RNA copies of the provirus, which of course are new viral genomes. These RNAs are packaged into virus particles (fig. 12.11) that bud out of the infected cell and go on to infect other cells.

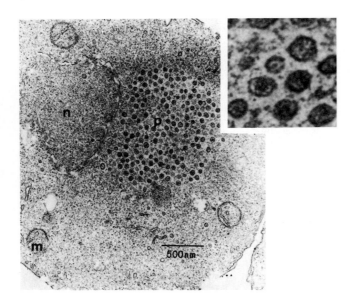

FIGURE 12.12

Retrovirus-like particles in a yeast cell induced for Ty transposition. The particles in this electron micrograph are labeled *p*. The inset shows the particles at higher magnification. The nucleus and a mitochondrion are labeled *n* and *m*, respectively.

Garfinkle and Boeke. "Retrovirus-like Particles in a Yeast Cell Induced for Ty Transposition," Cell 42:513, 1985. Reprinted with permission from Cell Press.

Ty Transposition by a Retrovirus-like Mechanism

Several lines of evidence indicate that Ty transposition resembles the replication of a retrovirus:

1. Ty encodes a reverse transcriptase. This seems likely because the *tyb* gene in Ty codes for a protein with an amino acid sequence closely resembling that of the reverse transcriptases encoded in the *pol* genes of retroviruses. If the Ty element really codes for a reverse transcriptase, this enzyme should appear when Ty is induced to transpose; moreover, mutations in *tyb* should block the appearance of reverse transcriptase. Gerald Fink and his colleagues have performed experiments that bear out both of these predictions.

2. Full-length Ty RNA and reverse transcriptase activity are both associated with particles that closely resemble retrovirus particles (fig. 12.12). These particles only appear in yeast cells that are induced for Ty transposition.

3. Finally, in a very clever experiment, Fink inserted an intron into a Ty element and then analyzed the element again after transposition. The intron was gone! This finding is incompatible with the kind of transposition prokaryotes employ, in which the transposed DNA looks just like its parent. But it is consistent with the following mechanism (fig. 12.13): The Ty element is first transcribed, intron and all; then the RNA is spliced to remove the intron; and finally, the spliced

FIGURE 12.13

Model for transposition of Ty. The Ty element has been experimentally supplied with an intron (yellow). The Ty element is transcribed to yield an RNA copy containing the intron. This transcript is spliced, and then the processed RNA is reverse transcribed, possibly in a viruslike particle. The resulting double-stranded DNA then reinserts into the yeast genome. (LTR = long terminal repeat.)

RNA is reverse transcribed within a viruslike particle, and the resulting DNA is inserted back into the yeast genome at a new location.

The similarity between this proposed mechanism of Ty transposition and retrovirus replication leads to the term **retrotransposon** (or even **retroposon**), which describes this and similar transposable elements. *Copia* and its relatives share many of the characteristics we have described for Ty, and they are also retrotransposons. Vertebrates too have agents that probably qualify as retrotransposons. Unlike Ty and *copia,* these are not elements that were first discovered because of their transposition. Rather, they were originally noticed as particles that looked and behaved like retroviruses, except that they were not infectious. These are called **A-type particles.** Their proviruses constitute as much as 0.3% of vertebrate genomes, and they sometimes transpose to interrupt vertebrate genes. The similarity between retrotransposons and retroviruses suggests that these genetic elements share a common evolutionary origin. Does it also mean that the viruslike retrotransposon particles are infectious? Apparently not; all attempts to make them infect other cells have failed.

A retrovirus replicates by reverse transcribing its RNA genome to double-stranded DNA, inserting this DNA into its host's genome, then using the host RNA polymerase II to make new RNA genomes that are packaged into virus particles. Several eukaryotic transposons, including Ty of yeast, *copia* of *Drosophila,* and probably the proviruses of A-type particles of vertebrates, apparently transpose in a similar way. They start with DNA in the host genome, make an RNA copy, then reverse transcribe it—probably within a viruslike particle—to DNA that can insert in a new location.

TRANSPOSONS AND THE MUTABILITY OF CHROMOSOMES

In our discussion of bacteriophage Mu, we pointed out that this and other prokaryotic transposons could cause re-arrangements of their host genomes. Such mutations can come as a by-product of the act of transposition itself, or simply because transposons provide multiple homologous sites on a genome that can serve as targets for general re-combination. This mutation enhancement is also a property of eukaryotic transposons, of course, and we have already seen one example: the unstable mutations caused by McClintock's maize transposons.

The phenomenon called **hybrid dysgenesis** illustrates one obvious kind of mutation enhancement caused by a eu-karyotic transposon. In hybrid dysgenesis, one strain of *Drosophila* mates with another to produce hybrid offspring that suffer so much chromosomal damage that they are dys-genic, or sterile. Hybrid dysgenesis requires a contribution from both parents; for example, in the **P-M system** the fa-ther must be from strain **P (paternal contributing)** and the mother must be from strain **M (maternal contributing).** The reverse cross, with an M father and a P mother, produces normal offspring, as do crosses within a strain (P × P or M × M).

What makes us suspect that a transposon is involved in this phenomenon? First, any P male chromosome can cause dysgenesis in a cross with an M female. Moreover, recom-binant male chromosomes derived in part from P males and in part from M males usually can cause dysgenesis, show-ing that the P trait is carried on multiple sites on the chro-mosomes. This is characteristic of transposable elements. Margaret Kidwell and her colleagues have investigated the **P elements** that inserted into the *white* locus of dysgenic flies. They found that these elements had great similarities in base sequence but differed considerably in size (from about 500 to about 2,900 base pairs). Furthermore, the P el-ements had direct terminal repeats and were flanked by short direct repeats of host DNA, both signatures of trans-posons. Finally, the *white* mutations reverted at a high rate by losing the entire P element—again a property of a trans-poson. Geneticists now introduce P elements intentionally into fruit flies to cause mutations or to carry extra genes into the flies.

Haig Kazazian has now identified transposition as the cause of at least one human genetic disease: hemophilia. In two of 240 hemophiliacs screened, Kazazian found a transposable element in place of the gene for a blood clot-ting protein, factor VIII. This destroyed the factor VIII gene, which in turn impaired the clotting process in the patients' blood. Now that transposons have been observed in humans, it seems certain that they will be linked to other mutations and therefore to other human genetic diseases.

Transposons enhance the mutability of eukaryotic as well as prokaryotic chromosomes. For example, the P-M system of hybrid dysgenesis in *Drosophila* is caused by the conjunction of two factors: (1) a transposable element (P) contributed by the male and (2) an environment contributed by the female, which allows transposition of P. Hybrid offspring of P males and M females therefore suffer multiple transpositions of the P element. This causes damaging chromosomal mutations that render the hybrids sterile. Transposons are also responsible for at least some human genetic disorders.

GENE MUTATIONS AND CANCER

We have examined some of the causes and effects of gene mutations, but it is becoming clear that these mutations can have another serious consequence: They can cause cancer. What changes turn a well-behaved normal cell into a renegade cancer cell? We certainly do not have the full answer to that question, but examinations of normal and tumor cells in culture show that malignant cells differ in three important respects from their normal counterparts.

First, malignant cells are immortal; they will go on di-viding indefinitely. Consider, for example, the HeLa cells cultured from a human cervical carcinoma in 1952 (see chapter 9). These malignant cells have already far outlived their donor, and we have every reason to believe they will continue to prosper as long as people are interested in growing them. This immortality stands in stark contrast to the behavior of normal vertebrate cells. Just as complex multicellular organisms have finite lifetimes, so their cells can divide only a limited number of times in culture before they too "grow old" and die.

In addition to immortality, cancer cells exhibit a group of characteristics that contrast with those of normal cells, which we lump together under the name **transformed.** Normal cells need hard surfaces on which to grow and serum to supply growth factors. Also, when they sense that they are crowded, they cease dividing. But transformed cells typically do not show these limitations. They can usu-ally grow in soft agar, as well as on hard surfaces; they usu-ally do not need growth factors; and they continue to divide when they are crowded, even piling up on top of one another until their medium is exhausted.

Besides immortality and the transformed phenotype, cancer cells usually exhibit a progressive loss of the characteristics of the normal cells from which they derived. Thus, liver tumor cells may at first produce many of the myriad enzymes that normal liver cells do, but they gradually lose those capabilities. Also, cancer cells are usually aneuploid, and this aneuploidy becomes more and more pronounced with time. Thus, tumor cells come to resemble their tissue of origin less and less, both biochemically and genetically.

It is fairly clear that the aneuploidy of cancer cells is an effect, rather than a cause, of malignancy. However, it is also obvious that genes are intimately involved in the malignant transformation of cells. Sometimes the genes themselves change; other times only gene expression seems to change. In this section we will explore the relationship between genes and **carcinogenesis** (the creation of a cancer cell).

Cancer cells differ from normal cells in several respects: They are immortal; they are transformed—capable of growing in physical and chemical environments where normal cells could not survive; and they progressively lose their biochemical and genetic resemblance to their normal ancestors.

CARCINOGENS AS MUTAGENS

One strong indication of the relationship between genes and cancer is the fact that many agents that cause cancer (**carcinogens**) are also mutagens. Since the late eighteenth century, we have suspected that chemicals can cause cancer. It was then that Sir Percival Pott noticed a high incidence of otherwise rare scrotal cancer in chimney sweeps. Pott concluded that these boys' constant exposure to soot had something to do with the disease. In fact, soot has the same kinds of mutagenic combustion by-products found in cigarette tar, and these very likely caused the trouble.

In the previous chapter we examined the mechanism behind most chemical carcinogens. They are electrophilic compounds that attack centers of negative charge in DNA, causing mutations. These mutations, in turn, apparently start the chain of events that leads to cancer. For example, consider benzo[a]pyrene, a major carcinogenic culprit in cigarette tar. Benzo[a]pyrene (fig. 12.14) belongs to a class of organic compounds called hydrocarbons, which are composed of hydrogen and carbon. Gasoline is primarily a mixture of hydrocarbons, although simpler ones than benzo[a]pyrene. Chemically, hydrocarbons are not very reactive, and they certainly are not electrophilic, so it is at first a little difficult to understand how benzo[a]pyrene could be carcinogenic. But hydrocarbons are also very insoluble in water, thus the body needs to change them chemically in order to excrete them in the urine. This chemical change is what converts the unreactive hydrocarbon to a dangerous carcinogen.

FIGURE 12.14
Conversion of benzo[a]pyrene to the ultimate carcinogenic form.

Figure 12.14 shows the enzymatic pathway whereby benzo[a]pyrene is converted to an electrophilic carcinogenic product called the **ultimate carcinogen.** (This is not the most carcinogenic compound in the world, just the most carcinogenic compound to which benzo[a]pyrene is metabolized.) The first step in the pathway is oxidation of the benzo[a]pyrene to an epoxide by an oxygenase. The epoxide then adds the elements of water under the direction of an epoxide hydrase. The product, a dihydrodiol, is oxidized again by the oxygenase to the ultimate carcinogen, a diol epoxide. Figure 12.15 demonstrates how this diol epoxide can attack an electron-rich center on a DNA base, altering the DNA.

Another environmental agent known to cause cancer is radiation. We have seen that radiation is mutagenic; several lines of evidence also show that it is carcinogenic. For example, consider the fact that people with light complexions,

Guanine

Diol epoxide

Adduct

FIGURE 12.15
Interaction between a benzo[a]pyrene diol epoxide (red) and the DNA base guanine.

who have relatively little defense against ultraviolet radiation from the sun, are much more likely than dark-skinned people to develop skin cancer. Also, as mentioned in chapter 11, some people, born with xeroderma pigmentosum, have defective DNA repair machinery. This suggests that the unrepaired DNA lesions caused by the ultraviolet radiation are responsible for their skin cancers. However, the situation may be a bit more complex than that.

Recall that bacteria have an error-prone DNA repair system. If this system is blocked, the bacteria have a much lower mutation rate. In other words, the improper repair of DNA damage, rather than the damage itself, seems to cause many permanent mutations. If eukaryotes also behave this way, the real mutagenic (and carcinogenic) event in xeroderma pigmentosum patients may not be unrepaired damage, but misrepaired damage. The same may be true of the DNA damaged by chemical carcinogens. It may be the cell's inaccurate attempts to repair bases damaged by these chemical agents that cause the permanent mutations leading to cancer.

A related argument may apply to the damage caused by more energetic radiation such as X rays. Remember that X rays cause double-stranded DNA breaks (chromosome breaks). This is obviously a mutagenic event, but it may become a carcinogenic event when the cell attempts to repair the damage by sticking together the severed ends of chromosomes. If more than one chromosome break has occurred, the wrong chromosome segments may be rejoined,

resulting in translocations. Such translocations could activate or inactive genes inappropriately, giving rise to a cancer cell.

 Mutagenic chemicals and radiation can lead to cancer, probably because of the DNA damage they cause. Chemicals can sometimes act directly, but usually they need to be metabolized to an electrophilic ultimate carcinogen that can attack DNA. Direct damage due to chemicals or radiation appears to be carcinogenic, but the inaccurate repair of such damage may actually be the root cause.

ONCOGENES

Viruses, especially retroviruses, represent a third kind of environmental agent capable of causing cancer. By studying **oncogenic** (tumor-causing) viruses, molecular virologists have revealed the identities of many of the viral genes (**oncogenes**) that cause tumors. More surprisingly, they have also discovered cellular counterparts of these viral oncogenes known as cellular oncogenes, or more accurately, cellular **proto-oncogenes.** We now have abundant evidence that these harmless cellular proto-oncogenes can convert to true oncogenes, which at least contribute to **oncogenesis,** the formation of tumors. We even know some of the conditions under which this conversion takes place.

Conversion Caused by Mutation

The most straightforward mechanism for converting a cellular proto-oncogene to an oncogene is mutation. Since viral oncogenes resemble their cellular cousins so closely, it is not difficult to imagine a few mutations causing the conversion. In fact, in at least one case—the **Ha-ras** oncogene—a single point mutation is sufficient.

The c-Ha-*ras* proto-oncogene is the cellular version of the Harvey rat sarcoma virus oncogene (v-Ha-*ras*). Here is the story of how this gene was implicated in human cancer: First, Michael Wigler and his colleagues demonstrated that human bladder carcinoma cells contain an oncogene capable of transforming NIH 3T3 cells. Specifically, this team of scientists introduced DNA from a line of human bladder cancer cells into 3T3 cells. The 3T3 cells are mouse cells that are immortal and apparently poised on the brink of transformation. Something in the bladder cancer cell DNA pushed them over into the transformed state (fig. 12.16). DNA from normal human cells does not transform them.

Which gene has the transforming activity? Robert Weinberg and his coworkers used radioactive oncogenes to investigate the nature of the DNA fragments from 3T3 cells transformed by human bladder carcinoma DNA. These experiments revealed that the transformed 3T3 cells have a new DNA fragment not found in ordinary 3T3 cells, which hybridizes with Ha-*ras* DNA. Therefore, the human cancer DNA has introduced a Ha-*ras* oncogene into the 3T3 cells. Furthermore, we know that this new gene is active in the

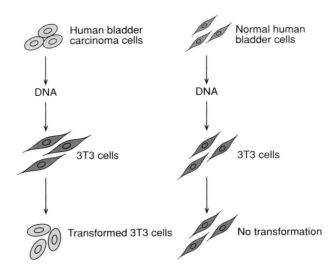

FIGURE 12.16

Human bladder carcinoma cells have DNA with transforming potential. Starting at the top of this figure, DNA is extracted from either human bladder carcinoma cells (left) or normal human bladder cells (right) and introduced into NIH 3T3 cells. The cells that receive the carcinoma DNA become transformed, but those that receive normal cell DNA remain normal.

transformed cells because these cells, but not untransformed 3T3 cells, make an RNA that hybridizes with the Ha-*ras* probe.

Since human bladder cancer cells, but not ordinary bladder cells, contain a transforming agent, and the agent seems to be the Ha-*ras* oncogene, we would like to know how this oncogene differs from the harmless Ha-*ras* proto-oncogene in normal bladder cells. To answer this question, Mariano Barbacid and his colleagues narrowed down the important region to a 350-base-pair segment of the gene. A single base change (a G → T transversion) is the only difference between this region in the oncogene and the same region in the proto-oncogene. Therefore, it appears that this change is what converts the proto-oncogene to the oncogene.

The product of the Ha-*ras* oncogene is a protein with a molecular weight of 21,000, called **p21** or **Ras.** The G → T transversion that converts the proto-oncogene to the oncogene in the human bladder carcinoma changes a glycine (the twelfth amino acid in Ras) to a valine. When does the change occur? It seems to be a somatic mutation, since it appears in human lung cancer cells, but not in normal trachea cells from the same patient. Is this the only cause of lung cancer? That seems very doubtful, because oncogenesis appears to be a multistep process that takes years to complete. The mutation in c-Ha-*ras* is probably just one step in that process.

Since the pioneering work of Wigler, Weinberg, and others, mutated *ras* genes have been found in a variety of human tumors, including 40% of colon cancers. These

mutations are not confined to codon 12, but almost all of them are clustered in three regions of the *ras* gene: codons 12–16; codons 59–63; and codons 116–119.

Certain human tumor cells contain an oncogene called Ha-*ras* that can transform mouse 3T3 cells in culture. Normal human DNA lacks this transforming activity. The cellular Ha-*ras* proto-oncogene in the first tumor cells studied had suffered a one-base mutation, a G → T transversion in the twelfth codon, which changed a glycine to a valine in the protein product. This was enough to convert the proto-oncogene to an oncogene. Almost all the *ras* mutations found in human tumor cells are clustered in three small regions of the gene.

THE c-Ha-*ras* GENE PRODUCT: A G PROTEIN

Why are these subtle changes so important? Clues come from the base sequence of the Ha-*ras* gene, which predicts the amino acid sequence of the protein product Ras. This amino acid sequence is very similar to those of proteins belonging to a class called **G proteins.**

The G proteins all bind GTP, enabling them to carry out their functions. They also have GTPase activity, so they can cleave the bound GTP to GDP, whereupon they become inactive until they bind a new molecule of GTP. The functions of the G proteins are quite varied. Some reside in cell membranes and respond to hormones and other substances. When activated by these signaling agents plus GTP, the G proteins stimulate the cell to appropriate activity. Another G protein responds to light and converts the light signal into a chemical signal inside the cell. Still another substance with G proteinlike characteristics is EF-Tu, which participates in protein synthesis. This protein binds GTP, which enables it to bind aminoacyl-tRNA and carry it to the ribosome where the amino acid can be joined to a growing protein chain (chapter 10).

The three-dimensional structure of normal human Ras has been solved by X-ray crystallography. Figure 12.17 shows this structure, along with a molecule of GDP bound to Ras. Although most G proteins have considerable GTPase activity, wild-type Ras has little on its own. However, when it interacts with another protein called GTPase activator protein (GAP), its low inherent GTPase activity increases about 20-fold. By contrast, the GTPase activity of the mutant Ras cannot be stimulated by GAP. The net effect is that the oncogene product loses much of its ability to cleave GTP to GDP. Knowing this, we might predict that the center of GTPase activity (the GTP or GDP binding site), and therefore the site of interaction with GAP, would lie close to the amino acids that are altered in the mutant proteins. This expectation is borne out beautifully. Recall, for example, that many of the altered amino acids in the mutant Ras proteins are clustered between amino acids 12 and 16. This region is highlighted in pink in figure 12.17, and we can see that it forms a loop that wraps around the phosphates of GDP (or GTP). It is not too surprising, therefore, that alterations in these amino acids affect the reaction with GTP (and/or GAP). Since GTP stimulates the activity of G proteins, including Ras, the inability of the protein to remove the GTP locks it in the activated state. In order to understand why this predisposes a cell to become malignant, we need to know one more key piece of the puzzle: the exact function of Ras. This is a function typical of G proteins: intracellular signaling. And it appears that this signal has something to do with stimulating cell division.

 The c-Ha-*ras* gene codes for a G protein called Ras, which binds GTP, carries out a signaling function, and then, with the help of GAP, hydrolyzes GTP turning off the signal. Conversion of the proto-oncogene to the oncogene causes an amino acid substitution in a key position near the presumed GTP and GAP binding site of Ras. This probably accounts for the reduction in the GTPase activity of the protein, which in turn means the protein spends a higher proportion of its time than normal in the activated state. This may predispose the cell to lose its control over growth.

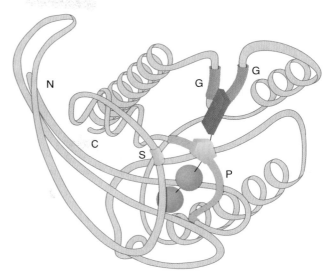

FIGURE 12.17

Interaction between the *ras* proto-oncogene product and GDP. The protein is shown as a blue ribbon, interacting with GDP (dark purple, yellow, and orange). The dark purple rectangle represents the guanosine; the dark yellow pentagon, the ribose; and the dark orange circles, the phosphates of GDP. The domains of the protein interacting with each of these parts of GDP are represented by sleeves on the protein ribbon, color coded to match the corresponding part of GDP. Thus, the purple sleeves (G) interact with the guanosine, the yellow sleeve (S) interacts with the ribose, and the orange sleeve (P) interacts with the phosphates of GDP. The P domain of the protein contains glycine-12, so this amino acid is in a critical position next to the phosphates of the nucleotide. N and C represent the N- and C-termini of the protein.

From L. Tong, et al., Science, 245:244. *Copyright © by the American Association for the Advancement of Science, Washington, DC. Reprinted by permission of the publisher and authors.*

WORKED-OUT PROBLEMS

Problem 1

You are interested in measuring the rate of transfer of transposon Tn3 from one plasmid, carrying no antibiotic resistance genes, to another, which carries a gene for chloramphenicol resistance. (Tn3 itself confers resistance to ampicillin.) Describe an experiment you would perform to assay for this transposition.

Solution

After allowing a certain time for transposition to occur, isolate the target plasmid and use it to transform cells with no inherent antibiotic resistance. Plate the transformed cells on medium containing both chloramphenicol and ampicillin. Only those cells that received a plasmid with both antibiotic resistance genes will survive; therefore, the number of survivors is related to the rate at which Tn3 transposed to the target plasmid, bringing the two antibiotic resistance genes together.

Problem 2

Assume you have two cell-free transposition systems that have all the enzymes necessary for transposition of Tn3 and Ty, respectively. What effect would the following inhibitors have on these two systems, and why?

(*a*) Inhibitors of double-stranded DNA replication; (*b*) inhibitors of transcription; (*c*) inhibitors of reverse transcription; (*d*) inhibitors of translation.

Solution

(*a*) Inhibitors of double-stranded DNA replication would block transposition of Tn3, but not Ty, because Tn3 transposition depends on this process, but the retrotransposon Ty does not. (*b*) and (*c*) Inhibitors of transcription and reverse transcription would block transposition of Ty, but not Tn3, because Ty transposition, just like retrovirus replication, requires these processes. Tn3 does not have to be transcribed, because all the gene products needed for its transposition are already present in the extract. (*d*) Inhibitors of translation will have no effect on either transposon; all proteins needed for transposition are supplied in the extract.

Summary

Transposable elements, or transposons, are pieces of DNA that can move from one site to another. Some transposable elements replicate, leaving one copy at the original location and placing one copy at a new site; others transpose conservatively, leaving the original location altogether. Bacterial transposons include the following types: (1) insertion sequences such as IS1 that contain only the genes necessary for transposition, flanked by inverted terminal repeats; (2) transposons such as Tn3 that are like insertion sequences but contain at least one extra gene, usually a gene that confers antibiotic resistance; (3) composite transposons such as Tn9 that have the added complexity of insertion sequences at each end; and (4) bacteriophage Mu, which can transpose in its prophage form.

Some examples of eukaryotic transposons are: (1) *Ds* and *Ac* elements of maize; (2) Ty of yeast; (3) *copia* of *Drosophila;* and (4) retroviruses. The latter three apparently transpose by a mechanism very different from the one used by prokaryotic transposons. RNA polymerase II from the host cell transcribes the transposon; this RNA is then reverse transcribed back to double-stranded DNA, probably within a viruslike particle, and finally inserted into the host genome at a new location.

Transposons enhance the mutability of both prokaryotic and eukaryotic genomes. They can directly disrupt genes by inserting into them during transposition; they can promote chromosome breakage, translocation, and inversion during transposition; and they can provide sites of homology for homologous recombination, even without transposing.

Mutagenic chemicals and radiation can lead to cancer, probably because of the DNA damage they cause. In addition, viruses—both DNA and RNA types—can cause tumors. All known RNA tumor viruses are retroviruses. Their RNA genomes usually contain an oncogene that can transform host cells, making them behave as tumor cells rather than as normal cells. The viral oncogenes appear to have descended from closely related cellular proto-oncogenes.

Various processes can convert a cellular proto-oncogene to a tumor-inducing oncogene. For example, the cellular Ha-*ras* proto-oncogene in certain human tumor cells has suffered a one-base mutation that is enough to convert the proto-oncogene to an oncogene.

The Ha-*ras* gene codes for a G protein called Ras, which binds GTP, sends a signal for the cell to divide, and then binds GAP protein, which stimulates hydrolysis of GTP, turning off the protein's function. Conversion of the proto-oncogene to the oncogene causes an amino acid substitution in a key position near the presumed GTP (and GAP) binding sites of Ras, leading to a reduction in the GTPase activity of the protein, which in turn forces the protein to spend a higher proportion of its time than normal in the activated state. This seems to predispose the cell to lose its control over growth.

PROBLEMS AND QUESTIONS ❓

1. (*a*) What elements, at a minimum, must a transposon contain? (*b*) What are the minimum elements that a transposable element must have in order to be transposed in a cell that harbors a separate, complete transposon?

2. Mutations in the transposase genes of IS1 render the transposon immobile (unable to transpose). What can we conclude from this finding?

3. Why is "jumping gene" a misleading nickname for some transposons?

4. (*a*) A certain transposon's transposase creates staggered cuts in the host DNA five base pairs apart. What consequence does this have for the host DNA surrounding the inserted transposon? (*b*) Draw a diagram to explain how the staggered cuts affect the host DNA.

5. What is the meaning of the term "selfish" or "parasitic" DNA?

6. Would a transposon that requires homologous recombination with its target DNA be able to transpose to more or fewer sites than one that uses illegitimate recombination? Why?

7. (*a*) What are the two steps in Tn3 transposition? (*b*) What is the intermediate called?

8. Identify the end product of abortive transposition carried out by Tn3 transposons with mutations in the following genes: (*a*) transposase; (*b*) resolvase.

9. What is the relationship between bacterial transposons and public health?

10. Transposon TnT in plasmid A transposes to plasmid B. How many copies of TnT are in the cointegrate? Where are they with respect to the two plasmids?

11. If the transposable element *Ds* of maize transposed by the same mechanism as Tn3, would we see the speckled kernels with the same high frequency? Why, or why not?

12. Assume that Ty transposes by a mechanism analogous to that used by Tn3. (Of course, it really does not.) Assume further that you have inserted an intron into a Ty element. What would the element look like after transposing?

13. Suspecting that Ty transposes by a mechanism analogous to the replication of a retrovirus, you have introduced an intron into a Ty element. What would be the fate of this intron upon transposition?

14. How do we know that unrepaired DNA damage can lead to cancer?

15. List three classes of environmental agents that are carcinogenic.

16. The term "transformation" has two meanings in biology. What is its meaning in the context of tumor biology?

17. Oncogenesis is a multistep process, yet the Ha-*ras* oncogene from human tumor cells is capable of transforming NIH 3T3 cells in a single step. Why is this statement not contradictory? *Hint:* Consider the nature of 3T3 cells.

18. Is the mutation in Ha-*ras* that contributes to human cancer a germ-line or a somatic mutation? How do we know?

Answers appear at end of book.

CHAPTER 13: GENETICS OF BACTERIA AND PHAGES

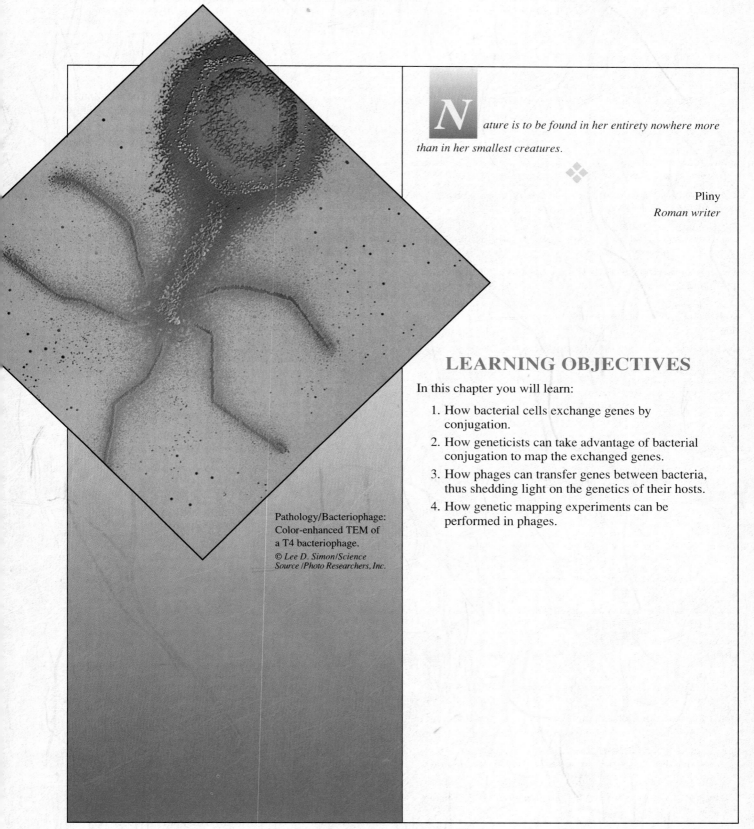

Pathology/Bacteriophage:
Color-enhanced TEM of
a T4 bacteriophage.
© *Lee D. Simon/Science
Source /Photo Researchers, Inc.*

*N*ature is to be found in her entirety nowhere more
than in her smallest creatures.

❖

Pliny
Roman writer

LEARNING OBJECTIVES

In this chapter you will learn:

1. How bacterial cells exchange genes by
   conjugation.
2. How geneticists can take advantage of bacterial
   conjugation to map the exchanged genes.
3. How phages can transfer genes between bacteria,
   thus shedding light on the genetics of their hosts.
4. How genetic mapping experiments can be
   performed in phages.

ur repeated references to *Escherichia coli,* the favorite model of bacterial geneticists, point to the importance of bacterial genetics. Why should this simple intestinal bacterium loom so large as a tool for understanding genetics? One reason is its very simplicity. It contains only about three thousand genes, compared to as many as one hundred thousand in humans. Another advantage is the fact that it exists for its whole life cycle in the haploid state, so mutations that might be masked by a wild-type allele in a diploid cell show up clearly in bacteria. But the most important qualification of bacteria for genetic studies is their extremely rapid rate of growth.

Given ideal growth conditions, *E. coli* cells divide every twenty minutes. Consider what that means: One ounce (30 grams) of bacteria growing constantly at that rate would surpass the mass of the earth in less than a day and a half! On a more practical level, a single *E. coli* cell will grow overnight into a visible colony containing millions of cells, even under relatively poor growth conditions. Thus, genetic experiments on *E. coli* usually last one day, whereas experiments on corn, for example, take months. It is no wonder that we know so much more about the genetics of *E. coli* than about the genetics of corn, even though we have been studying corn much longer.

In this chapter, we will examine the mapping techniques that laid the foundation for our present understanding of bacterial and phage genetics.

BACTERIAL CONJUGATION

In spite of the inherent advantages of bacteria, they are worthless as genetic tools unless we can perform genetic experiments on them. In other words, we must be able to cross them somehow. In 1946, Joshua Lederberg and E. L. Tatum figured out a way to do this. Lederberg reasoned that if bacteria exchange DNA, he might be able to detect this by mixing two different mutant strains and looking for wild-type recombinants.

BACTERIAL GENETIC NOTATION

In order to understand the experiments Lederberg and others performed, we should begin by defining some terms: Wild-type bacteria can grow on **minimal medium,** which contains only salts and the sugar glucose. These bacteria can make all their own amino acids and vitamins; they are called **prototrophs** (Greek: *proto* = first + *trophe* = nourishment). Mutants that lack the ability to make one or more nutrients are called **auxotrophs** (Latin: *auxilium* = help).

In the following discussion, you will also encounter the system of notation used by bacterial geneticists. According to this system, the names of genes or operons are rendered in italics and begin with lowercase letters (e.g., the *ara* operon). When we refer to the genotype of a bacterium, we again use this convention, with a + to denote the wild-type or a – to denote the mutant genotype. For

example, a *leu⁻* bacterium has a mutation in its *leu* operon and an *ara⁺* bacterium has a wild-type *ara* operon. When we write the phenotype of a bacterium, we capitalize and use standard letters. For example, Leu⁻ bacteria cannot make the amino acid leucine, while Ara⁺ bacteria can break down the sugar arabinose.

BACTERIAL EXCHANGE OF DNA

Lederberg foresaw difficulty with an experiment using auxotrophs having only a single mutated gene. These mutants reverted to wild-type with a high enough frequency (about one in a million) that the revertants might have been mistaken for recombinants, which occur at about the same rate. For instance, if Lederberg started with two strains (A⁺ B⁻ and A⁻ B⁺) and looked for A⁺ B⁺ recombinants, revertants of A⁻ B⁺ to A⁺ B⁺ would appear the same as recombinants and would therefore confuse the analysis. Accordingly, Lederberg decided to use double mutants in his crosses. The chance of both mutations reverting simultaneously would be negligible. The mutations used in these and other genetic experiments are sometimes called **markers** because they "mark" an organism's genome with a change that shows up clearly in the phenotype.

The decision to use two markers instead of one led Lederberg to a collaboration with Tatum, because the latter already had a variety of double mutants of *E. coli* available. Fortunately, Tatum's bacteria were *E. coli* strain *K12,* which just happened to be a strain that does exchange DNA. Almost all the other *E. coli* strains in common use at that time would not have worked.

Figure 13.1 shows the plan of the first experiment. One auxotrophic strain (A) required the amino acid methionine and the vitamin biotin; the other (B) required the amino acids threonine and leucine. Both strains grew on rich medium containing amino acids and vitamins, but not on minimal medium. However, if one strain could transfer DNA to the other, recombinants could presumably form that would bring together the wild-type genes for threonine and leucine synthesis (*thr⁺* and *leu⁺*) from strain A and the genes for methionine and biotin synthesis (*met⁺* and *bio⁺*) from strain B. These prototrophic recombinants would survive on minimal medium. In fact, that is just what Lederberg and Tatum found. One in ten million cells in the mixture of the two auxotrophs was a prototroph. On the other hand, if the two auxotrophs were grown separately, they produced no prototrophs.

This experiment showed that the DNAs from the two auxotrophs mixed and recombined, but it did not demonstrate conclusively that DNA passes directly from one cell to the other. Instead, one auxotroph could have burst open and its DNA could have transformed the other, as in Griffith's and Avery's experiments with *Streptococcus pneumoniae* (chapter 6). To rule this out, Lederberg and Tatum lysed one strain and filtered it to remove whole cells. Then they used the cell-free filtrate, which contained DNA, to try

(a)

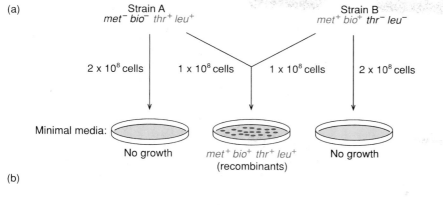

Strain A
met⁻ bio⁻ thr⁺ leu⁺

Strain B
met⁺ bio⁺ thr⁻ leu⁻

2 x 10⁸ cells   1 x 10⁸ cells   1 x 10⁸ cells   2 x 10⁸ cells

Minimal media:

No growth

met⁺ bio⁺ thr⁺ leu⁺
(recombinants)

No growth

(b)

Strain A: *met⁻ bio⁻ thr⁺ leu⁺*

Strain B: *met⁺ bio⁺ thr⁻ leu⁻*

met⁺ bio⁺ thr⁺ leu⁺
+
met⁻ bio⁻ thr⁻ leu⁻
(recombinants)

FIGURE 13.1

The classic bacterial conjugation experiment. (*a*) Two auxotrophic strains, A and B, were mixed. Strain A had the genotype *met⁻ bio⁻ thr⁺ leu⁺*; strain B had the genotype *met⁺ bio⁺ thr⁻ leu⁻*. Neither of these auxotrophic strains could grow by itself on minimal medium; however, when the two strains were mixed, prototrophic recombinants (*met⁺ bio⁺ thr⁺ leu⁺*) arose, which could grow on minimal medium. These gave rise to the colonies growing on the petri dish at center. (*b*) The recombination event that produced the prototrophs. Crossing over occurred between the *bio* and the *thr* genes, bringing together all the wild-type alleles on one chromosome and the mutant alleles on the other.

had the F pilus and, presumably, some fertility factor in its genome, so it was designated **F⁺**. Strain B had no F pilus, lacked the fertility factor, and was designated **F⁻**. Figure 13.3 shows two conjugating bacteria, joined by the F pilus. The pilus is hollow and could, in principle, allow DNA to pass through, but it is retracted during conjugation. Thus, it appears that this structure serves not as a direct conduit for passing DNA, but as a contractile element that draws the two cells close enough together to allow DNA to pass through their membranes by an alternate route.

> Two mutant (auxotrophic) bacterial strains can share DNA by direct contact, producing wild-type (prototrophic) recombinants. This process is called conjugation.

One-way DNA Transfer During Conjugation

Lederberg and Tatum interpreted their experiments in terms of classical crosses between haploid gametes, and therefore assumed that the recombinant prototrophs were true diploids. However, William Hayes and others produced evidence that this could not be the case. For one reason, the F⁺ partner could be killed with the antibiotic streptomycin and still function in conjugation, but if the F⁻ partner was treated this way, no conjugation occurred. This suggested that the F⁺ partner served as the donor and could continue to do so after streptomycin treatment, even though it was no longer viable. By contrast, a dead F⁻ recipient would hardly provide fertile ground for the donated genes. Moreover, if the genotype of an F⁺ donor differed in several markers from that of the F⁻ recipient, the recombinant usually wound up with mostly recipient markers. This indicated that the donor usually transferred only a portion of its genome to the recipient. Therefore, in spite of all the sexual nomenclature associated with fertility in bacteria, it appears that these organisms rarely share their whole genomes to form true diploid zygotes. Instead, they usually form partial diploids called **merodiploids** or **merozygotes** (Greek: *meros* = part).

FIGURE 13.2

E. coli with sex pili. Three kinds of filamentous structures project from the surface of the cell: common pili (C), flagella (F), and sex pili (S).
Elinor Meynell, G. G. Meynell, and Naomi Datta, Bacteriological Reviews *32:58, 1968.*

to transform the other strain. It did not. Therefore, they concluded that the two strains of bacteria in the original experiment had shared DNA by direct contact.

Direct contact between bacterial cells is called **conjugation;** it is made possible by a tubular structure, called a sex pilus, on the surface of one of the strains (fig. 13.2). The sex pilus is now usually called the **F pilus,** where F stands for "fertility." Strain A in the original experiment

The F Factor: A Plasmid

Hayes also found that an F$^+$ cell could convert an F$^-$ to F$^+$ very efficiently, with a frequency of about one in ten, compared to one in ten million for transferring other genes. This means the F factor can be passed easily, probably by a different mechanism than that used with other genes. How do we know the F factor consists of genes? In one experiment, *E. coli* F$^+$ cells were mixed with another species of bacteria (*Serratia marcescens*) containing a DNA of different density. After conjugation, a new species of DNA appeared in the recipient cells, and its density matched that of *E. coli*, not that of the recipient bacterium. This new DNA must be the F factor, since it is the component that is transferred readily during conjugation. As we will see, this is not true conjugation because only a small amount of DNA, the F factor, is transferred from the *E. coli* donor to the *S. marcescens* recipient. Subsequent work has shown that the F factor is a **plasmid,** a circular extra-chromosomal DNA (fig. 13.4) that replicates independently of the host chromosome. It is about 2% as large as the host chromosome, or 94,500 base pairs (94.5 kb). Now that we know what it is, we refer to the F factor as the **F plasmid,** and transfer of fertility can be seen as simply transferring the F plasmid from an F$^+$ cell to an F$^-$ cell.

 The fertility of F$^+$ *E. coli* is due to an F plasmid of 94.5 kb that can transfer at high frequency into F$^-$ bacteria.

Hfr STRAINS

In the 1950s, several investigators discovered a new type of *E. coli* in addition to the F$^+$ and F$^-$ already under study. The new type was called **Hfr,** which stands for "high-frequency recombinant." Hfr bacteria are mutant F$^+$ strains, but they have lost the ability to transfer fertility at high frequency. Instead, they transfer host chromosomal genes at high frequency.

For example, consider an Hfr strain isolated by Hayes in 1953. When Hayes conjugated this Hfr with auxotrophic recipient cells, it generated prototrophic recombinants a thousand times more efficiently than did its F$^+$ parent. Hayes also found that this transfer of host genes closely resembled the transfer by F$^+$ cells in two respects: It was not inhibited by treating the Hfr donor with streptomycin, and the recombinant genotype came mostly from the recipient, not from the donor. We will see that an F$^+$ cell converts to an Hfr when its F plasmid inserts into the host chromosome.

Hfr Cell Transfer of Host Genes

The fact that F$^+$ strains contain a few Hfr cells raises the question: Does the F plasmid in every F$^+$ cell have a small but significant potential to transfer host genes, or does this ability belong only to the few Hfr cells contaminating an

FIGURE 13.3

Conjugating *E. coli.* The two cells, joined by a connecting bridge, are an elongated donor (left) and a round recipient (right).

E. L. Wollman, F. Jacob, and W. Hayes. Symposia on Quantitative Biology *(Cold Spring Harbor), vol. 21, p. 153, 1956.*

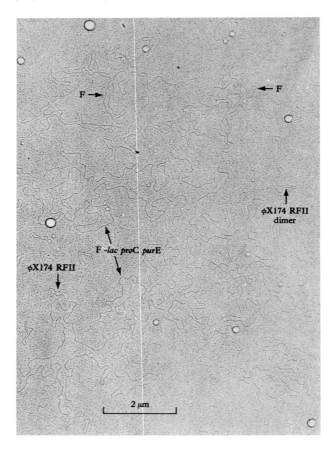

FIGURE 13.4

The F plasmid. This electron micrograph shows two circular molecules of F plasmid DNA among other circular DNAs of different sizes.

© *Dr. Richard Deonier.*

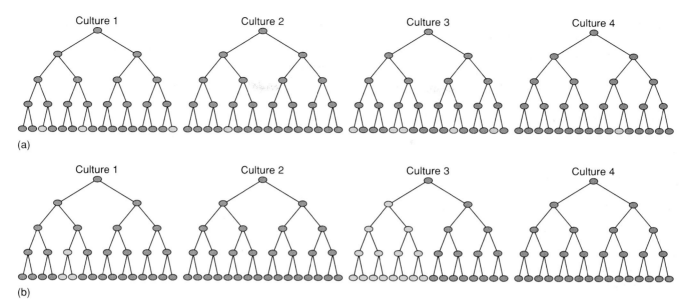

(a)

(b)

FIGURE 13.5

The fluctuation test. Two possibilities are illustrated: (*a*) Each culture has the same number of F$^+$ cells, and each cell has a small but equal chance of transferring host genes during conjugation (represented by the yellow cells). Therefore, each culture should give roughly the same amount of host gene transfer. (*b*) Only Hfr cells (mutant F$^+$ cells, yellow) are competent to transfer host genes. Therefore, cultures in which an F$^+ \rightarrow$ Hfr conversion occurs early in their growth (culture 3) will wind up with many competent cells, whereas other cultures can easily wind up with few (if the F$^+ \rightarrow$ Hfr conversion occurs late in the growth phase; culture 1) or none (if no conversion occurs; culture 4). In other words, the competence of the cells in host gene transfer will fluctuate widely. That is what Wollman and Jacob demonstrated.

F$^+$ population? By the mid-1950s, Elie Wollman and François Jacob had resolved this question using the **fluctuation test** devised by Max Delbrück and Salvador Luria. Here is how this test works: The investigators took a very dilute F$^+$ culture (only 250 cells per milliliter), split it in half, and then split the first half into fifty equal parts. The second half was not divided. After the bacteria in the cultures had multiplied to a million times their original concentration, each was tested for the ability to transfer host genes, thereby forming recombinant prototrophs.

The premise of the experiment is this: If each F$^+$ cell has an equal probability of transferring host genes, each of the fifty subcultures should show an equal ability to do this. On the other hand, if the occasional Hfr cells in the F$^+$ population are the only ones that can transfer genes, the fifty cultures should vary widely in this ability. This is because the original dilute cultures contained so few cells that most of them included no Hfr cells at all. Of course, as the cells multiplied, there was a chance of Hfr cells appearing in the cultures by mutation, and once an Hfr appears, it divides to yield Hfr progeny. Thus, a culture in which an F$^+ \rightarrow$ Hfr conversion occurred early in its growth would wind up with the most cells competent to transfer host genes. This is represented on a small scale in figure 13.5*b*, where culture 2 experiences an F$^+ \rightarrow$ Hfr conversion in generation 2 and therefore accumulates eight Hfr cells after four generations, while culture 4, with no F$^+ \rightarrow$ Hfr conversions, finishes with no Hfr cells.

When the experiment was performed, the small cultures did indeed show great variation—from one prototrophic recombinant in one culture to 116 in another. As a control, the half of the culture that was not split before the cells multiplied was finally divided into fifty equal parts and each of these was tested immediately. They varied much less: from a low of ten prototrophic recombinants to a high of twenty-three. Taken together, these results demonstrate that a genetic change in a small proportion of the total F$^+$ population renders them capable of transferring genes to auxotrophic F$^-$ cells to convert them to prototrophs. This small subpopulation is composed of Hfr cells.

F Plasmid Insertion into the Host Chromosome

How can Hfr cells transfer host genes so efficiently? A clue comes from the similarity between transfer of the F plasmid and transfer of the host genes. It is the F plasmid itself that causes both. An F$^+$ cell becomes an Hfr cell when the F plasmid ceases its solitary existence and inserts into the host chromosome. When this happens, the F plasmid loses the ability to transfer itself directly into a recipient cell, but confers this ability on the host chromosome. This leads to the hypothesis of F$^+ \rightarrow$ Hfr conversion shown in figure 13.6. The F plasmid recombines with the host chromosome at any of a number of different locations. This mobilizes the chromosome, which is presumably nicked within the F plasmid part, initiating replication by the rolling circle

(a) F-plasmid transfer:

(b) F$^+$ → Hfr conversion:

(c) Hfr transfer of host genes:

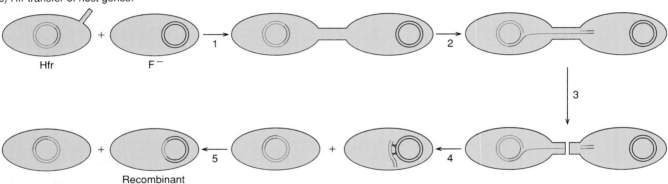

FIGURE 13.6

Model for F$^+$ → Hfr conversion. (*a*) The F plasmid (blue-green) normally passes from cell to cell, leaving a copy in the donor cell. This presumably involves rolling circle-style replication of the plasmid, as pictured. The F plasmid and host chromosome are not drawn to scale; the host chromosome is actually about fifty times larger than the F plasmid. (*b*) In F$^+$ → Hfr conversion, the F plasmid (blue-green) recombines with the host chromosome (red), mobilizing the whole chromosome. (*c*) Hfr transfer of host genes: 1. The F pilus establishes a bridge between the conjugating cells.

2. The Hfr chromosome begins replicating, with one daughter DNA molecule passing across the bridge to the recipient cell. 3. The bridge breaks during conjugation, leaving a piece of donor DNA in the recipient cell. 4. The donor DNA recombines with the recipient chromosome. 5. After the recipient cell replicates, it contains a recombinant DNA, with part (red) derived from the donor and part (black) from the recipient.

mechanism (see chapter 7). During replication, part of the integrated F plasmid acts as a locomotive and leads the newly synthesized chromosomal DNA into the F$^-$ cell.

The bridge between the two cells established by the F pilus is fragile and usually breaks before DNA transfer is complete. The part of the Hfr chromosome that has entered the recipient cell then recombines with the F$^-$ chromosome, changing its genotype. Occasionally, the whole Hfr chromosome makes it across before the bridge breaks. When this happens, the recipient cell is converted to Hfr, because the entire integrated F plasmid has been transferred. Part of it led the chromosome across the bridge; the rest of it came across last, like a caboose.

❖ Occasionally, the F plasmid recombines with its host's chromosome, converting the cell to an Hfr. This mobilizes the Hfr chromosome to transfer to recipient cells, presumably by the same mechanism the independent F plasmid would normally use. The favorite hypothesis for this mechanism involves a rolling circle-style replication of the Hfr chromosome, with one of the daughter DNA molecules entering the recipient cell and recombining with it to change its genotype.

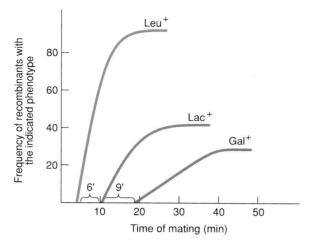

FIGURE 13.7

Timing of gene transfer in a bacterial conjugation experiment. An Hfr Leu$^+$ Lac$^+$ Gal$^+$ was mated with an F$^-$ Leu$^-$ Lac$^-$ Gal$^-$ recipient and the conjugation was interrupted at various times by blending. The phenotypes of the recipients were tested by growing in medium lacking leucine or containing lactose or galactose, instead of glucose, as the sole carbon source. Some Leu$^+$ recombinants (red line) began to appear at four minutes after the onset of conjugation, whereas Lac$^+$ and Gal$^+$ recombinants began to appear at ten and nineteen minutes, respectively (blue and green lines). This means that the time between transfer of the *leu*$^+$ and *lac*$^+$ markers was six minutes, while that between *lac*$^+$ and *gal*$^+$ was nine minutes. This tells us the relative positions and spacing of these three markers.

MAPPING BY INTERRUPTED CONJUGATION

The natural interruption of conjugation comes from random movements of the conjugating partners that can rupture the fragile link connecting them. Elie Wollman and François Jacob (pronounced "Jackobe") learned that they could stop conjugation at a defined time simply by placing the bacteria in a blender. This vigorous treatment guaranteed that all the conjugating partners would be separated. Furthermore, Wollman and Jacob took advantage of this procedure to build genetic maps.

Here is the basic strategy used by Wollman and Jacob: They conjugated two auxotrophs for a defined length of time, interrupted their conjugation by placing them in a Waring blender, then checked to see whether a given gene had been transferred. For example, consider an F$^-$ Leu$^-$ Lac$^-$ Gal$^-$ Str^r auxotroph (an F$^-$ incapable of growing on media lacking the amino acid leucine or on media containing the sugars lactose or galactose, but resistant to streptomycin). If it is conjugated with an Hfr Leu$^+$ Lac$^+$ Gal$^+$ Str^s (streptomycin-sensitive) auxotroph for two minutes, no Str^r recombinants that are Leu$^+$ will be formed. This tells us that in two minutes, transfer of the *leu* genes from the streptomycin-sensitive Hfr cell to the streptomycin-resistant F$^-$ cell has not yet occurred. On the other hand, if conjugation is

allowed for four minutes, a significant number of these recombinants will arise. Therefore, transfer of the *leu* genes begins about four minutes after conjugation begins. Notice the importance of the streptomycin resistance marker. It allows the experimenter to eliminate the donor bacteria simply by adding streptomycin after conjugation. Then, all the remaining bacteria should be recipients.

Taking the procedure a step further, if conjugation proceeds for ten minutes, some Leu$^+$ Lac$^+$ Str^r recombinants appear. This means that the Hfr cell begins to transfer its *lac* genes about ten minutes after the onset of conjugation. Furthermore, if conjugation proceeds for nineteen minutes, some Leu$^+$ Lac$^+$ Gal$^+$ Str^r recombinants appear. Therefore, the Hfr cell has begun to transfer its *gal* genes by this time. Figure 13.7 presents an idealized picture of the results of such an experiment. Why does the recipient cell remain streptomycin-resistant? There are two reasons: (1) The *str*^{s} locus lies beyond gal on the donor chromosome, and (2) the recipient's *str*^{r} allele is dominant.

These time values tell us something very important about the *E. coli* genetic map. If we assume that the rate of transfer is constant along the whole length of the bacterial chromosome, then the difference in the times it takes to transfer a given pair of genes is proportional to their separation on the genetic map. For example, the difference in the transfer times of the *leu* and *lac* genes is ten minutes minus four minutes, or six minutes; we therefore say that the *leu* and *lac* loci lie about "6 **minutes**" apart on the genetic map. Similarly, the difference in times of transfer of the *gal* and *lac* genes is 19–10, or 9 minutes. Thus these two loci lie 9 minutes apart, and the *leu* and *gal* loci are 9 + 6, or 15 minutes apart. It takes just one hundred minutes for the entire *E. coli* chromosome to be transferred, including its F "caboose." Therefore, the *E. coli* map is composed of 100 minutes.

In his Nobel address, Jacob made the following tongue-in-cheek observation about this long conjugation time: "Marvelous organism, in which conjugal bliss can last for nearly three times the life span of the individual." Actually, this is an underestimate; *E. coli* cells can divide five times in the time it takes for complete conjugation to occur. However, Jacob forgot to note that this "conjugal bliss" is almost always interrupted.

The Circular *E. coli* Genetic Map

Not all Hfr strains transfer the *leu* genes first; each strain has its own characteristic starting point. Moreover, some strains transfer genes in an order opposite to what we have just described. Thus, *gal* would be transferred before *lac*, and *leu* would be transferred last of all. Table 13.1 lists eleven different Hfr strains and the order in which each transfers host genes. Even though the orders vary, you can see that they can all be made the same by considering them not as open-ended lines, but as closed circles. For example, *thr* is the first and *thi* is the last gene transferred by

TABLE 13.1 Order of Gene Transfer During Conjugation with a Variety of Hfr Strains

Hfr Strain	Order of Gene Transfer
Hayes	*O-thr-leu-azi-ton-pro-lac-pur-gal-trp-his-gly-str-mal-xyl-mtl-ile-met-thi*
Hfr 1	*O-ieu-thr-thi-met-ile-mtl-xyl-mal-str-gly-his-trp-gal-pur-lac-pro-ton-azi*
Hfr 2	*O-pro-ton-azi-leu-thr-thi-met-ile-mtl-xyl-mal-str-gly-his-trp-gal-pur-lac*
Hfr 3	*O-pur-lac-pro-ton-azi-leu-thr-thi-met-ile-mtl-xyl-mal-str-gly-his-trp-gal*
Hfr 4	*O-thi-met-ile-mtl-xyl-mal-str-gly-his-trp-gal-pur-lac-pro-ton-azi-leu-thr*
Hfr 5	*O-met-thi-thr-leu-azi-ton-pro-lac-pur-gal-trp-his-gly-str-mal-xyl-mtl-ile*
Hfr 6	*O-ile-met-thi-thr-leu-azi-ton-pro-lac-pur-gal-trp-his-gly-str-mal-xyl-mtl*
Hfr 7	*O-ton-azi-leu-thr-thi-met-ile-mtl-xyl-mal-str-gly-his-trp-gal-pur-lac-pro*
AB311	*O-his-trp-gal-pur-lac-pro-ton-azi-leu-thr-thi-met-ile-mtl-xyl-mal-str-gly*
AB312	*O-str-mal-xyl-mtl-ile-met-thi-thr-leu-azi-ton-pro-lac-pur-gal-trp-his-gly*
AB313	*O-mtl-xyl-mal-str-gly-his-trp-gal-pur-lac-pro-ton-azi-leu-thr-thi-met-ile*

From F. Jacob and E. L. Wollman, *Sexuality and the Genetics of Bacteria.* Copyright © 1961 Academic Press, Inc., Orlando, FL. Reprinted by permission of the publisher and authors.

Hfr Hayes, whereas these two genes are transferred one after the other (*thr,* then *thi*) in Hfr 1. All we must do to make these two orders identical is to form a circle of each one. Then the only difference will be the location of *O,* the origin of replication, which determines which gene will be transferred first. Now look at the order of Hfr 5. You will see that the order of transfer of these two genes is reversed: *thi,* then *thr.* Again, the order of gene transfer by these two strains can be reconciled by forming a circle of each one and turning one of them over as you would flip a coin.

What does all this mean in physical terms? First, it tells us that the *E. coli* chromosome is circular and that the F plasmid inserts in different places in different Hfr strains. This accounts for the fact that the *order* of transfer is the same in different strains, but that the starting points (defined by the site of F plasmid insertion) vary. Second, the F plasmid can insert in either orientation, thus inverting the order of gene transfer. These phenomena are illustrated in figure 13.8. In the 1960s, John Cairns used electron micrographs (chapter 7) to show that *E. coli* DNA is circular, so we accept the proposition of a circular map naturally. Nevertheless, in 1957, when Jacob and Wollman proposed the idea, it was a radical one. A partial circular map of the *E. coli* chromosome is presented in figure 13.9.

By interrupting conjugation between an Hfr donor and an F⁻ recipient, we can determine the length of time it takes to transfer a given gene. The differences in the transfer times of different genes tell us their relative positions on the circular genetic map of *E. coli.*

COMPLETION OF CONJUGATION

Just because a given gene enters an F⁻ cell during conjugation does not mean that the gene will be expressed indefinitely in the recipient. In order to become a stable part of the recipient cell's genetic makeup, the newly arrived gene must recombine with the recipient cell's chromosome. If it does not, it has no way of replicating and will be lost. J. Tomizawa demonstrated this phenomenon clearly in a conjugation experiment with an Hfr Lac⁺ donor and an F⁻ Lac⁻ recipient. After various times, he interrupted conjugation and tested the merodiploids for the Lac⁺ phenotype by growing them on indicator plates containing EMB-lactose agar. Lac⁺ cells metabolize the lactose, causing chemical changes that turn the indicators in the medium red; Lac⁻ cells cannot metabolize the lactose and so remain white.

At the earliest times that Lac⁺ cells appear, they produce colonies that are not pure red, but red and white **variegated,** a phenotype called **Lac^v.** This is because the *lac* genes in the merodiploids are not yet stably integrated into their new hosts' chromosomes. Thus, they can be lost, giving rise to Lac⁻ cells that will be white. On the other hand, when Tomizawa tested the merodiploids for longer periods of time, he observed a drop in the Lac^v phenotype, beginning about fifty minutes after the onset of conjugation, and a corresponding rise in the number of pure Lac⁺ colonies (fig. 13.10). This tells us that integration of the *lac* genes from the Hfr cell into the F⁻ chromosome begins about fifty minutes after conjugation begins, or about forty minutes after the *lac* genes entered the recipient cell. This integration of the *lac* genes into the *lac⁻* genome stabilizes the Lac⁺ phenotype. The name of the integration process, of course, is recombination.

(a)   (b)   (c)

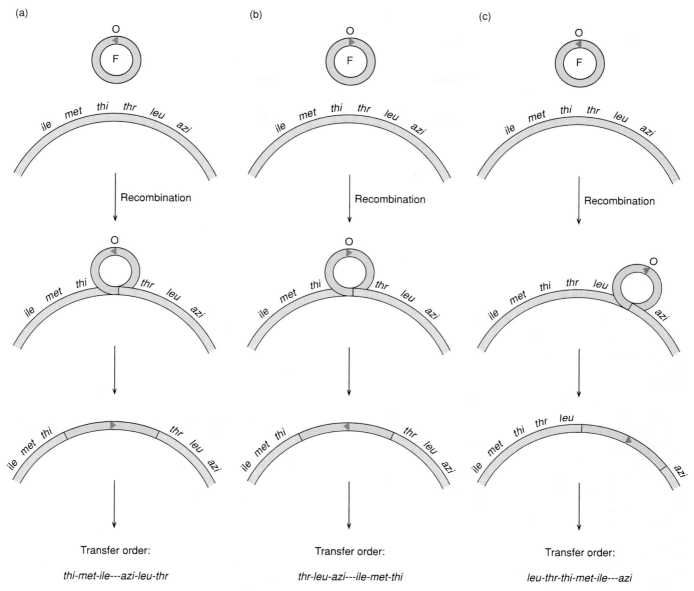

Transfer order:

thi-met-ile---azi-leu-thr

Transfer order:

thr-leu-azi---ile-met-thi

Transfer order:

leu-thr-thi-met-ile---azi

FIGURE 13.8

The F plasmid portion of an Hfr chromosome determines the direction and order of host gene transfer. (*a*) The F plasmid (yellow circle with red arrowhead showing orientation) inserts between *thi* and *thr* in a rightward orientation; thus, the host genes will be transferred in the order *thi-met-ile - - - azi-leu-thr,* where the three dashes in the middle represent the entire remainder of the Hfr chromosome (not shown in this picture).

The last three genes are therefore transferred only in those rare conjugations that last throughout the entire one hundred minutes. (*b*) The F plasmid inserts between *thi* and *thr* in a leftward orientation; thus, the host genes will be transferred in the order *thr-leu-azi- - - ile-met-thi.* (*c*) The F plasmid inserts between *leu* and *azi* in a rightward orientation; thus, the order of transfer is *leu-thr-thi-met-ile- - - azi.*

FIGURE 13.9
Circular map of the *E. coli* genome. This map was prepared in 1976 and contains only a fraction of the genes that have now been mapped.

FIGURE 13.10
Recombination is the final step in conjugation. The Lac^V phenotype, indicative of unstable association between donor and recipient chromosomes, decays, beginning about fifty minutes after the onset of conjugation. At the same time, the Lac^+ phenotype, indicative of a stable association that results from recombination between the two chromosomes, begins a steady rise.

❖ The completion of conjugation in *E. coli* requires recombination between the donated piece of Hfr chromosome and the F^− chromosome. Otherwise, the donated DNA cannot replicate and will be lost. This recombination requires extensive sequence similarity between the recombining DNAs and therefore follows the homologous recombination pathway.

GENETIC MAPPING WITH *E. coli*

We have seen that conjugation occurs in *E. coli* and that this sharing of DNA allows us to construct genetic maps by following the progressive entry of one gene after another into the recipient F^− cell. This technique works fairly well when the distance separating the markers is great, but it is too crude a method for mapping closely spaced genes. For such fine mapping, more delicate genetic tools have been developed. We will examine the most important of these.

F′ PLASMIDS

In 1959, E. H. Adelberg discovered that one of his Hfr strains was contaminated with an F^+ revertant. At first this seemed like an impediment, but it turned out that the F^+ revertant had very interesting properties. In fact, this was one of those occasions when an annoyance turned out to be a boon.

Adelberg's F^+ revertant did not behave normally. Whereas an F plasmid usually inserts randomly into the host chromosome to generate a new Hfr, the F plasmid in this revertant always went back to the same locus at which it had resided in the original Hfr. And it generated Hfrs

with a much higher frequency than normal. Adelberg recognized that the F factor in his F^+ revertant had suffered a genetic change, so he gave it a new name: **F′**. He also guessed the reason for the unusual behavior of the F′ plasmid: It had picked up a piece of the host chromosome during its sojourn there in the Hfr. This host DNA gave the F′ plasmid a "memory" of its former location, since it provided homologous sequences on the plasmid and the host chromosome that could serve as sites for homologous recombination. This ensured that recombination between F′ plasmid and host chromosome would be rapid and would always occur at the same locus on the chromosome (fig. 13.11).

Apart from its theoretical interest, the F′ plasmid also provides a handy genetic tool. We can use it to introduce a defined small segment of the *E. coli* genome into an F^− cell very efficiently. This process, sometimes called **F-duction,** or even **sex-duction,** is so efficient because the segment of host DNA is now part of the F plasmid and transfers with the same high frequency as the F plasmid itself. Conjugation between an F′ cell and an F^− therefore produces a merodiploid for genetic analysis.

For example, consider an F′ called **F-*lac*** that carries part of the *lac* operon. When an F^+ cell bearing F-*lac* is conjugated with an F^− Lac^− cell, it converts the recipient to F^+ Lac^+ (fig. 13.12). Thus, even though the recipient retains a mutant *lac* gene in its chromosome, the wild-type gene on the F-*lac* plasmid provides the missing enzyme. Note that no recombination is necessary; the F-*lac* plasmid reproduces along with the host cell. This cross tells us that the wild-type *lac* gene is dominant over the *lac*^− gene in the F^− chromosome. Of course, merodiploids make possible

FIGURE 13.11
Formation of F-*lac*. The integrated F plasmid (red) in an Hfr chromosome is clipped out along with some host DNA, including the nearby *lac* operon (blue). This produces a recombinant DNA called F-*lac*, part F plasmid and part host DNA, including *lac*.

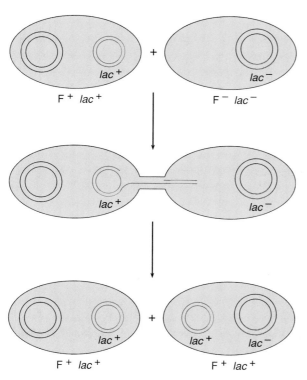

FIGURE 13.12
Transfer of F-*lac* to an F⁻ *lac*⁻ cell. The F-*lac* plasmid (blue, with *lac*⁺ genes in red) crosses the connecting bridge, just as an F⁺ normally would. This converts the F⁻ *lac*⁻ recipient to F⁺ *lac*⁺. The host chromosomes are black, with *lac*⁻ genes in gray. Again, the host chromosome and F-*lac* plasmid are not drawn to scale.

many other, more sophisticated genetic experiments, which we will also investigate. These experiments have been aided considerably by a collection or F′ plasmids that represent the entire *E. coli* genome (fig. 13.13).

> Sometimes a recombination event within an Hfr chromosome clips out the F plasmid along with some of the host chromosome. This produces an F′ plasmid that includes the clipped out piece of host DNA. The host genes thus become part of a highly mobile element that can deliver them to another cell. We can take advantage of this efficient sharing of limited pieces of DNA to perform genetic experiments.

FIGURE 13.13
F′ plasmids cover the whole *E. coli* map.

Modified from Barbara J. Bachmann and K. B. Low, in Microbiological Reviews, *44:1–56, 1980. Copyright © 1980 American Society for Microbiology, Washington, DC. Used with permission.*

THE *CIS-TRANS* COMPLEMENTATION TEST

Suppose we have isolated two strains of *E. coli* that cannot grow on the sugar lactose. Two genes (*lacZ* and *lacY*) code for the two enzymes a cell needs to metabolize this sugar. Therefore, an important question about these two Lac⁻ mutants is: Do both strains have mutations in the same gene, or are two different genes affected? We can answer this by **complementation** as follows. We conjugate cells to form a merodiploid, thus placing both mutant *lac* operons together in the same cell. If the two mutations are in different genes, they will **complement** each other in the merodiploid to give a Lac⁺ cell. In other words, the donor provides the gene product that the recipient is missing, and vice versa. In the case illustrated in figure 13.14*a*, the donor (F-*lac*) cell contained a wild-type *lacZ* gene, but a mutant *lacY* gene. Thus, the wild-type *lacZ* gene product from the donor cooperated with the wild-type *lacY* gene product from the recipient to allow the merodiploid to use lactose.

On the other hand, if the mutations are in the same gene in the two cells, they will usually not complement each other, so the merodiploid will be Lac⁻. Figure 13.14*b*

FIGURE 13.14

The *cis-trans* complementation test. (*a*) Mutations, represented by small red x's, are in different genes—*lacZ* and *lacY*, respectively. Therefore, their products can complement each other, and the cell will be *lac*⁺. (*b*) The mutations are in the same gene (*lacY*). Now both chromosomes will produce active *lacZ* product, but inactive *lacY* product, so the cell will be *lac*⁻. (*c*) and (*d*) are complications that must be considered when running the *cis-trans* test. (*c*) Sometimes recombination can occur between markers in the same gene on homologous chromosomes. This produces a wild-type gene and a double-mutant gene. The cell is therefore *lac*⁺, which makes it look as if the two mutations were in different genes. (*d*) Sometimes the products of two homologous genes with different mutations (two different *lacZ* mutants in our example) can get together to make an active, multi-subunit protein, in this case an active β-galactosidase. This makes it look as if the two mutations were in different genes.

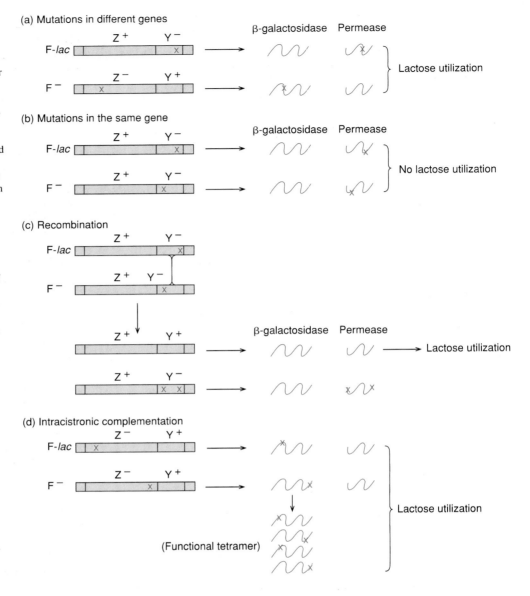

illustrates what happens when both mutations are in the same gene—the *lacY* gene, in this case. Both the donor plasmid and the recipient chromosome provide wild-type *lacZ* products, but mutant *lacY* products. They cannot complement because neither can provide a functional *lacY* product. Notice the emphasis here on gene *products*. In order for complementation to work, the gene products have to be **diffusible**. That is, they must be capable of moving within the cell so that they can cooperate.

As an aid to understanding the complementation concept, consider the following analogy: Two factories in the same town make the same kind of automobile. Each has two assembly lines, one for the engines and one for the bodies. One day the engine line of one factory and the body line of the other factory break down. Nevertheless, the two factories are able to continue producing cars by cooperating—workers put the engines from the second factory together with the bodies from the first. This is complementation, and

you can see how it depends on the portable nature of the product. Of course, if the engine assembly lines of both factories break down at the same time, the two defective factories cannot complement each other and there will be no way to continue making cars.

Several qualifications must be made about the complementation test. First, it must be performed in a situation where recombination does not occur to a significant extent. If recombination is common, it can take place between two homologous mutant genes, creating one gene with no mutations and one with two (fig. 13.14c). Thus, the two mutations would seem to complement each other and would be scored as existing in different genes even though they actually started out in homologous genes in the two chromosomes. One way to eliminate this problem is to use a *recA*⁻ recipient that cannot carry out homologous recombination.

Another kind of problem can arise if one of the mutations affects the expression of several other genes. Such a

pleiotropic mutation behaves as if it were mutations in several genes instead of just one, which complicates the complementation test. A final problem is that mutations in homologous genes in two different chromosomes can sometimes complement each other. This occurs when the gene product is a polypeptide that aggregates with one or more identical polypeptides to form an **oligomeric protein.** β-galactosidase, a **tetramer** (composed of four identical polypeptides, or monomers), is a good example. Each *lacZ* gene in figure 13.14d is nonfunctional; each gives rise to an inactive tetramer of four identical defective β-galactosidase polypeptides. However, when the two different *lacZ* genes are present in the same cell, they produce two different defective polypeptides that somehow cooperate to form an active tetramer. Therefore, these two *lacZ* mutations complement each other; if we did not know better, we would say they were on different genes. Instead, they are on the same gene, or **cistron,** and they therefore engage in **intracistronic complementation.**

We can use the complementation examples just given to illustrate some important genetic terms. The mutant genes in these cases were supplied *in trans*—on separate chromosomes (Latin: *trans* = across). Alternatively, we could supply the mutant genes *in cis*—on the same chromosome (Latin: *cis* = here). These two possible configurations of the mutant genes—*cis* and *trans*—have given the complementation test another name: the **cis-trans** test. That in turn has provided the term **cistron,** which Seymour Benzer coined to describe a genetic unit that could be distinguished by the *cis-trans* test from another genetic unit. Of course, this is a unit of genetic function—a piece of chromosome that gives rise to a protein (or RNA) product. As such, it is synonymous with our current understanding of the word "gene."

We also use the terms *cis* and *trans* to describe the way two genetic loci interact. Promoters, the binding sites for RNA polymerase (chapters 8 and 9), are ***cis*-acting** elements: They must be on the same DNA molecule as the gene in question in order to influence that gene's activity. In our factory analogy, they are like the main power switch. If it breaks down, the factory cannot operate and the defect cannot be complemented by a product of the other factory. By contrast, proteins are ***trans*-acting.** Each is a product of one gene and can interact with other genes or their products. The mobility of proteins allows them to act at a distance.

 Two different mutations that affect the same function can be subjected to the *cis-trans* test, or complementation test, to determine whether they occur in the same or in different genes. In this test, the two mutant genes are supplied to the same cell *in trans* (on separate chromosomes). If the mutations complement each other to give wild-type function, they are probably in separate genes. If the mutations fail to complement, they affect the same gene.

TRANSDUCTION

The term **transduction** refers to the use of phages to carry bacterial genes into new host cells. Two different transduction mechanisms are widely used: **specialized transduction,** in which λ or a similar phage DNA carries the bacterial genes, and **generalized transduction,** in which **P1** or other phage heads carry the bacterial genes instead of phage genes.

Phages as Carriers of Host Genes

Lambda (λ) is a temperate phage that infects *E. coli* cells. Temperate phages, unlike T2, T4, and other virulent phages, do not necessarily kill their hosts. Instead, as we saw in chapter 8, λ DNA can incorporate itself into the host genome. This gives rise to a lysogenic infection in which the phage DNA becomes a passenger and replicates along with the host DNA. On the other hand, the λ phage can initiate a *lytic* infection in which the phage DNA remains independent of the host chromosome. Under these circumstances, host cells fill up with progeny phage particles and finally burst, or lyse, releasing the new phages to infect other cells. The spreading infection started by a single phage replicating lytically on a lawn of bacteria can produce a hole, or *plaque,* devoid of living cells.

During lysogenic infection, λ DNA inserts into a site called ***att B,*** which lies between the *gal* and *bio* genes in the *E. coli* DNA. We also saw that mutagens can induce the λ lytic cycle, causing the phage DNA to be excised from the host chromosome. Usually, the λ DNA excises perfectly, but sometimes it comes out carrying some excess baggage: a piece of host DNA.

Since the λ DNA usually inserts between the host *gal* and *bio* genes, these are the genes that λ DNA picks up when it excises improperly (fig. 13.15). And since the room available inside a λ phage head is limited, the amount of extra DNA that can be accommodated is only about 10% of the size of the λ DNA itself, or about 5 kb. In practice, more host DNA than this is usually included, which means that some phage DNA must be excluded. If a large person sits down on one end of a crowded bench, someone on the other end gets pushed off. By the same token, if the λ DNA picks up the *E. coli bio* locus at one end, it loses some of its regulatory region from the other. The result is a **transducing phage,** a particle that contains a recombinant DNA: part *E. coli bio* and part λ. The λ part is missing a regulatory region required for lysogeny but not for the λ lytic cycle, so the resulting transducing phage can replicate lytically to form plaques. It is therefore called λ*pbio,* where the *p* stands for "plaque-forming."

On the other hand, if the λ DNA picks up the *gal* locus on one end, it loses genes on the other end that are essential for the lytic cycle, and the resulting transducing phage (λ*dgal*) is defective. The *d* stands for "defective," of course. Such a defective transducing phage needs a nondefective **helper phage** to supply the functions it requires to replicate.

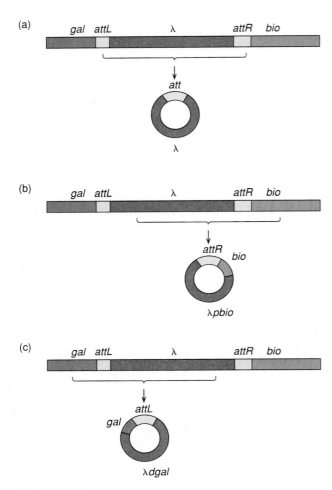

FIGURE 13.15

Creation of λ transducing phages. (*a*) Normal λ excision: The λ DNA (red) is correctly cut out of the host genome (colored purple on the left and green on the right) by clipping in the *att* sites (yellow) on either side of the λ DNA. (*b*) The cutting occurs too far to the right, thus including the host *bio* operon (green), but missing the left end of the λ genome. The resulting phage will still be able to replicate and can transduce the *bio* genes; it is called λ*pbio*. (*c*) The cutting occurs too far to the left, thus including the host *gal* genes (purple), but missing the right end of the λ genome, which is essential for replication. The resulting transducing phage is called λ*dgal*.

What can we do with specialized transduction? We can use it instead of conjugation to form merodiploids for complementation tests. It is somewhat more limited than conjugation in that the *gal* and *bio* loci are the only ones that usually participate. However, λ DNA sometimes goes to alternate sites in the host genome, so it can pick up and transfer a variety of genes. This makes specialized transduction more versatile than it seems at first glance.

There is a significant difference between the physical relationship of the two genes in merodiploids resulting from F-duction and specialized transduction. After F-duction, the two complementing genes are on different DNA molecules: one on the host chromosome and the other on the F′ plasmid. After specialized transduction, both

complementing genes are part of the host chromosome, because the λ DNA, with its piece of host DNA attached, inserts into the recipient genome. Nevertheless, both genes are presumably active in each case, so complementation can occur equally well in both.

When λ DNA escapes from the host chromosome after lysogenic infection, it sometimes carries a piece of host DNA along with it. A phage carrying such a recombinant DNA is called a transducing phage because it can be used to "transduce" a piece of host DNA into a new host cell to form a merodiploid. Geneticists use such merodiploids for complementation tests. λ transduction is called specialized transduction because it usually carries only the *gal* or *bio* genes that flank the λ integration site.

P1 Phage Mediation

In specialized transduction, a temperate phage such as λ carries host genes as passengers on its genome. In generalized transduction, little if any phage DNA is involved. Instead, the phage head forms around a piece of host DNA, producing a transducing particle that can deliver this DNA to a new host. Of course, the resulting infection cannot produce new phages in the absence of phage genes, but the donated host DNA can recombine with the recipient's chromosome and change its genotype. In fact, recombination is necessary for successful generalized transduction, just as it is for mating between Hfr and F⁻ strains. In both cases, linear pieces of DNA enter new host cells and must recombine with the host chromosome or be lost.

The favorite phage for generalized transduction is P1, whose genome normally contains 91.5 kb of DNA. This means that a P1 phage head can hold up to 91.5 kb of donor DNA if it does not have to accommodate any phage DNA. This is about 2.4% of the whole *E. coli* genome, enough room for about seventy-five average-size genes—many times more DNA than can be transferred by specialized transduction.

About 0.3% of the phages from a given P1 infection are transducing phages; the rest are normal. Since a given transducing phage contains at most 2.4% of the host genome, picked up more or less at random, the chance of finding a given host locus in a given phage particle is at most 0.024×0.003, or 7.2×10^{-5}. This is less than one in ten thousand; in practice, it is only about one in one hundred thousand, so the experimenter must perform an efficient selection to find this "needle in a haystack."

Consider an experiment in which we want to transduce a *leu⁺* gene into a *leu⁻* recipient (fig. 13.16). This is relatively straightforward. We simply infect the *leu⁺* donor with P1, collect the progeny phage, including the transducing phage, then infect the *leu⁻* recipient and transfer the cells to medium lacking the amino acid leucine. Cells infected by normal P1 phage will lyse; cells infected by transducing phages carrying genes other than *leu⁺*, and cells not infected at all, will fail to grow in the absence of leucine.

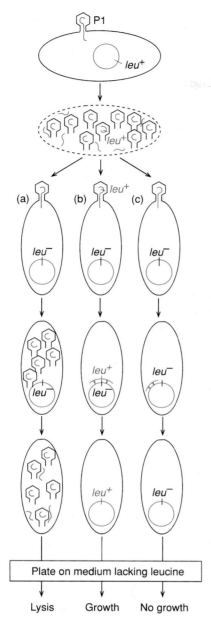

FIGURE 13.16

Generalized transduction. Infection of a *leu*⁺ bacterium by the phage P1 generates three kinds of particles: (*a*) a normal P1 particle containing only phage DNA (gray); (*b*) a transducing particle containing host DNA (red) that includes the *leu* operon (blue); (*c*) a transducing particle containing non-*leu* host genes only (red). These three kinds of particles can be tested by using them to infect *leu*⁻ hosts, then plating on medium lacking leucine. The normal P1 particles (left) will infect and destroy these cells. The transducing particles with the *leu* genes (middle) will render the host cells *leu*⁺, so they will grow in the leucine-free medium. The transducing particles lacking the *leu* genes (right) cannot confer the *leu*⁺ genotype, so the infected cells will not grow on the leucine-free medium.

FIGURE 13.17

Crossing over between two mutations. The donor, transducing DNA, contains the *ara*⁺ and *leu*⁻ alleles; the chromosome of the recipient cell (purple) contains the *ara*⁻ and *leu*⁺ alleles. Recombination occurs between these two DNAs (dashed lines), resulting in an Ara⁺ Leu⁺ cell. Both mutations are now on the short, linear piece of DNA (bottom), which is lost. This leaves the chromosomal DNA free of mutations in both *ara* and *leu*. The frequency with which this recombination occurs is a measure of the distance between the two markers, *ara* and *leu*.

Only those rare cells infected by transducing phages carrying the *leu*⁺ gene will survive. These will be easy to spot as opaque colonies on petri dishes, surrounded by transparent regions of nongrowing cells. This type of procedure is called a **selection.** It removes all undesirable cells—in this case, by preventing them from reproducing. Only the desired cells (*leu*⁺ in this example) can grow.

P1 phage can mediate generalized transduction by incorporating host DNA instead of phage DNA. Such particles can deliver up to 91.5 kb of host DNA to a new cell, allowing recombination between the donor DNA and the recipient chromosome. Since only a small fraction of the phages from a P1 infection will carry a given gene, a selection must be performed to locate the transductant of interest. This is easiest if transduction creates a prototrophic recombinant.

Mapping Mutations

Suppose we want to measure the distance between two markers, *ara* and *leu*. Since the *leu* locus is very close to the *ara* locus, which codes for the enzymes that metabolize the sugar arabinose, chances are that a transducing phage containing *leu* will also contain *ara*. We learned in chapter 5 that the farther apart two genes are, the better the chance that a recombination event will occur between them. This applies equally well to the *ara* and *leu* mutations in figure 13.17. They are very far apart in the figure, so the chance of a crossover between them is relatively high. As the figure shows, the crossover results in an *ara*⁺ *leu*⁺

recombinant. Generalized transduction makes this type of mapping by **cotransduction** relatively convenient. The linkage between two mutations within the same gene can also be examined by this same kind of cotransduction analysis. Again, the farther apart the mutations, the greater the frequency of recombination between them.

A Direct Mapping Scheme Based on Cotransduction

We have seen that cotransduction is a useful way to "piggy-back" a gene we want to study with another that is selectable. Cotransduction also gives us another mapping method. First, if two genes cotransduce, we immediately know that they lie within 91.5 kb of each other and are therefore fairly closely linked. Furthermore, the closer the linkage between two genes, the more likely they are to cotransduce. Notice that this is just the opposite of the recombination test. There, the closer the linkage between two genes, the less likely they are to recombine. T. T. Wu has proposed a formula for calculating the frequency (x) of cotransduction of two markers:

$$x = (1 - d/L)^3$$

where d is the distance between the two markers, and L is the length of the DNA that can be accommodated by the transducing particle. Using this expression, we find that markers separated by 60 kb (kilobase pairs), 30 kb, and 10 kb are expected to cotransduce with frequencies of about 4%, 30%, and 70%, respectively. An average bacterial gene is only about 1 kb long, which means that two markers within such a gene cotransduce with a frequency higher than 96%.

We can also use transduction to deduce the order of genes. For example, consider three genes: A, B, and C. We know from previous experiments that genes B and C are close together and A is farther away. The question is this: Is the order of these genes A B C, as shown in figure 13.18a, or A C B, as shown in figure 13.18b? We can answer this by doing a three-factor cross as illustrated in figure 13.18.

We start with a recipient cell with the phenotype $A^+ B^+ C^-$ and transduce DNA from a cell with the phenotype $A^- B^- C^+$. Then we screen for transductants that are $B^+ C^+$. In order to produce this phenotype, crossovers 1 and 2, at least, must occur. Among these $B^+ C^+$ transductants, we then test for the A alleles. We ask: How many transductants are $A^+ B^+ C^+$, and how many are $A^- B^+ C^+$? Figure 13.18a shows us what to expect if the gene order is A B C: Crossovers 1 and 2 alone will produce $A^+ B^+ C^+$, but two additional crossovers (3 and 4) must occur to bring the A^-, B^+, and C^+ alleles together. Clearly, two crossovers are much more likely to occur than four, so if the gene order is A B C, we would expect to find many more $A^+ B^+ C^+$ transductants than $A^- B^+ C^+$.

On the other hand, if the gene order is A C B, figure 13.18b shows us what to expect. Here, crossovers 1 and 2 or 1 and 3 will produce the $B^+ C^+$ phenotype we first screen for. The difference is that crossovers 1 and 2 will produce $A^+ B^+ C^+$, whereas 1 and 3 will give $A^- B^+ C^+$. And crossover 3 is much more likely to occur than 2 because

FIGURE 13.18

Three-factor cross to determine the order of genes A, B, and C. We know that genes B and C are closer together than either is to gene A. We want to know if the order is A B C as shown in (a), or A C B as shown in (b). We perform a transduction with an $A^- B^- C^+$ donor and an $A^+ B^+ C^-$ recipient, and screen for $B^+ C^+$ transductants. Then we check to see whether among these transductants we get more A^+ or A^-. In (a) we know that crossovers 1 and 2 have occurred to bring together the B^+ and C^+ alleles on the recipient chromosome. If those are the only crossovers, the transductant will also be A^+; if crossovers 3 and 4 also occur, the transductant will be A^-. But two crossovers will occur much more frequently than four, so A^+ will be much more common if (a) is correct. In (b) we again know that crossovers 1 and 2 or 1 and 3 have occurred to bring together B^+ and C^+. But 1 and 3 are more likely than 1 and 2, because there is much more space for 3 to the left (not fully shown) than for 2, which must occur between A and C. And 1 and 3 give A^-, while 1 and 2 give A^+. Therefore, A^- will be more common if (b) is correct.

there is much more room at the left end of both DNAs (not fully shown in the figure) than between genes A and C. Therefore, if the gene order is A C B, we would expect to find more $A^- B^+ C^+$ transductants than $A^+ B^+ C^+$. This is the opposite of what we expect for the A B C gene order.

> Transduction allows us to do several types of genetic mapping. We can transduce two genes, or two markers within a gene, and observe the frequency of recombination between them. The closer together they are, the lower their frequency of recombination. Alternatively, we can observe the frequency of cotransduction of two markers. The closer together the markers, the higher their frequency of cotransduction. We can also deduce the order of the markers on the genome by performing three-factor crosses in which we transduce three markers.

GENETICS OF T-EVEN PHAGES

Scientists began to work on bacterial viruses as early as 1915, when F. W. Twort reported that very small, invisible agents could destroy bacteria. He proposed that these agents were viruses that parasitized their bacterial hosts. Twort's work, like other genetic studies we have

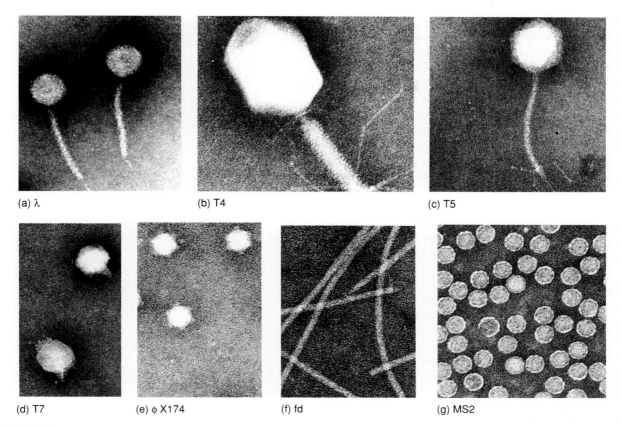

(a) λ   (b) T4   (c) T5

(d) T7   (e) φ X174   (f) fd   (g) MS2

FIGURE 13.19

Structures of some phages of *E. coli*. All are shown at approximately equal magnification.

From Robley C. Williams and Harold W. Fisher. An Electron Micrographic Atlas of Viruses, pp. 26, 70, 104, 112, 116, 120; 1974. Courtesy of Charles C. Thomas, Publisher, Springfield, Illinois.

encountered, escaped recognition, allowing F. d'Herelle to rediscover the same phenomenon two years later. D'Herelle popularized the notion of a bacterial virus and gave this agent the name **bacteriophage** (Greek: *phagein* = to devour; hence, bacteriophage = bacteria-devourer). Delbrück and his colleagues initiated the modern era of phage investigation in 1938 when they began to focus on the genetics of a small group of phages called **T1–T7,** where the T stands for "type." This focusing of attention was important, especially since it was directed mostly on the closely related **T-even** phages (T2, T4, and T6). Thus, new findings could be compared with one another and organized into a coherent picture, rather than remaining isolated fragments of information about a bewildering variety of very different phages. Figure 13.19 illustrates some of the various forms of *E. coli* phages.

STRUCTURE AND GROWTH OF T-EVEN PHAGES

Figure 13.20 is a schematic diagram of the structure of T4, one of the T-even phages. The most obvious features are a polyhedral **head,** which contains the phage genome—175 kb of double-stranded DNA tightly coiled inside; a **tail,**

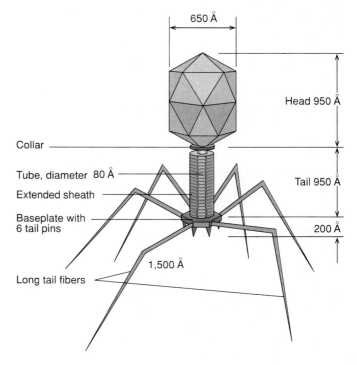

650 Å

Head 950 Å

Collar

Tube, diameter 80 Å

Extended sheath

Tail 950 Å

Baseplate with 6 tail pins

200 Å

1,500 Å

Long tail fibers

FIGURE 13.20

Schematic diagram of a T-even phage particle.

which the phage uses to inject its DNA into the host bacterium *E. coli;* and **tail fibers** and a **baseplate,** by which the phage attaches to the surface of the host cell. Figure 13.21 dramatically illustrates the tight coiling of phage DNA inside the head by showing that when the head bursts upon osmotic shock, the DNA spews out in all directions. Indeed, it is difficult to believe all that DNA could fit into the phage head until you remember that the DNA is heavily shadowed with metal to make it show up; it is really far thinner than it appears here.

Even though phages are invisible to all but the electron microscopist, they can be easily counted by observing their effects on their hosts. In this procedure, called a **plaque assay,** an experimenter mixes phage particles with bacterial cells, suspends the mixture in molten agar containing nutrients for the bacteria, and then pours the mixture into a petri dish. The agar quickly solidifies, and the bacteria grow overnight to form a uniform, turbid "lawn." However, as figure 13.22 shows, phages make their presence known by creating clear holes, or **plaques,** in the bacterial lawn.

Here is how phages create plaques: Each phage that infects a bacterial cell in the original mixture takes over the host machinery for DNA, RNA, and protein synthesis, forcing the cell to make phage products instead of its own. Ultimately, the phage DNAs and proteins assemble to form complete progeny phages, and about twenty-five minutes after the start of infection, the cell lyses, releasing hundreds of progeny phages. By that time, however, the agar has hardened, so the progeny phages cannot float away; instead, they are restricted to infecting adjacent cells. This process repeats itself over and over, until the phages, by lysing one host cell after another, succeed in clearing out a visible plaque in the bacterial lawn.

Each infectious phage in the original suspension will, in principle, cause one plaque to form. Therefore, by counting plaques, we get an idea of the number of infectious phages, or **plaque-forming units (pfu),** in a given volume. For example, suppose we start with a phage suspension, dilute it 1:100,000 (a 10^5-fold dilution), and spread ¼ ml on a plate of bacteria. If we obtain one hundred plaques, that means we had $100 \times 10^5 \times 4 = 4 \times 10^7$ pfu, or infectious phages, per milliliter of the original suspension. Why did we do the dilution? If we had plated the concentrated suspension, there would have been so many plaques that they all would have blended into one, and there would have been no way to count them. In practice, we perform several different dilutions so that some of the plates in every experiment have a reasonable number of plaques.

Our description of phage replication contains a flavor of discontinuity. One phage infects a cell, and after a period in which no new phages appear, gives rise to a burst of hundreds of progeny phages. This idea comes from the "one-step growth experiment" performed by Delbrück and E. Ellis at the beginning of their investigation into phage genetics. In this study, Delbrück and Ellis infected *E. coli* cells with T4 phages and then measured the number of

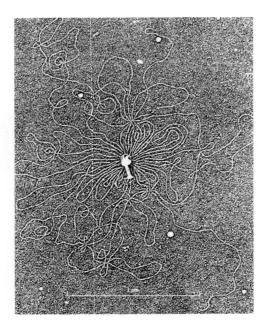

FIGURE 13.21
Lysed T4 phage. The phage ghost is at center, with escaped phage DNA all around.

A. K. Kleinschmidt et al. "Darstellung und Längenmessungen des gesamten Desoxyribosenucleinsäure-Inhaltes von T2-Bacteriophagen," Biochemica et Biophysica Acta *61(1962):857–64. Reproduced by permission.*

phages (pfu) per milliliter at various times thereafter. Figure 13.23 shows that there was no increase in the phage concentration for almost twenty-five minutes. This is the **latent period** (or **lag period**). Then, very rapidly, the concentration rose from about 10^5 per milliliter to more than 10^7, an increase of more than 100-fold. This is the **rise period.** The difference between the final and initial phage concentrations (over 100) is the **burst size.**

The actual production of infectious particles within the host cell may be somewhat more gradual than suggested by the rise period, but each infected cell will be counted as only one pfu in the plaque assay until it lyses. In other words, the rise period really reflects **lysis,** or rupture (Greek: *lysis* = a loosening), of infected cells, which allows each progeny phage to produce a separate infection. Before lysis, all the infectious particles are contained in one cell; thus that cell and all the infectious particles within still behave as only one infectious unit.

 T-even phages have polyhedral heads, containing the double-stranded DNA genome of about 175 kb, and tails through which they inject the DNA into their host, *E. coli.* The phages take over their hosts' DNA, RNA, and protein synthesizing machinery, reproduce themselves, lyse their hosts, and go on to infect other cells. Since T-even phages lyse their host, they can be conveniently counted by observing the number of plaques, or areas devoid of living cells, they produce on a lawn of bacteria.

FIGURE 13.22
Plaque assay for T2 phage. The phages were mixed with *E. coli* cells and plated on nutrient agar in petri dishes. Each hole, or plaque, in the lawn of bacteria started from a single phage infecting a single cell. The infection spread to surrounding cells to form the plaque.
From Molecular Biology of Bacterial Viruses *by Gunther S. Stent. Copyright © 1963 by W. H. Freeman and Company. Reprinted with permission.*

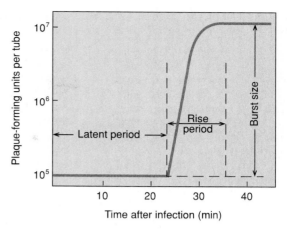

FIGURE 13.23
The one-step growth experiment. Plaque-forming units per tube were recorded as a function of time after infection. No change occurred for almost twenty-five minutes (the latent period). At the end of this time, a rapid rise in phage numbers began (the rise period). The difference in phage number at the beginning and end of the rise period is the burst size.

CROSSING MUTANT T-EVEN PHAGES

Since phages are detected by the plaques they form, it is perhaps not surprising that some of the first T-even phage mutants discovered were those that formed abnormal plaques. Among these was the *r* mutant, which, for reasons we still do not fully understand, forms extra-large plaques. A. D. Hershey (the same Hershey who later participated in the famous Hershey-Chase experiment, chapter 6) discovered this mutant in 1946 and called it *r*, for "rapid lysis," because it seemed that rapid lysis of the host cells was what allowed the plaques to grow so large. The wild-type allele of the *r* gene is r^+. Hershey also demonstrated that the *r*-mutant phenotype was indeed due to a mutated gene, since phages from an *r*-mutant plaque gave rise only to *r* plaques, not to wild-type plaques. In other words, the *r* phenotype was heritable.

Another early T-even mutant, also discovered in 1946, was T2*h*, where the *h* stands for **host-range.** This is a T2 phage that can infect *E. coli* cells (Tto^r) that cannot be infected by wild-type T2 phage. Tto is simply a genetic way of writing T2, so Tto^r is T2-resistant. (Similarly, Ton^r means T1-resistant.) Since wild-type T2 phages (h^+) cannot infect Tto^r cells, they will not produce plaques on this strain of cells. Nevertheless, we can get them to form plaques by plating them on a mixture of Tto^r and Tto^s cells. The T2 phages will infect the Tto^s cells in the lawn and therefore form turbid plaques that still contain unlysed Tto^r cells.

The existence of two different T2 mutants allowed phage geneticists to do crosses and look for recombinants. Here is how the experiments were performed: A mixture of Tto^r and Tto^s cells was infected with a high concentration of two different phages, T2*h* r^+ and T2h^+ *r*. The high concentration of phages gave a high ratio of phages to bacteria

FIGURE 13.24
Distinguishing four phenotypes of phage T2. The phages were plated on a mixture of Tto^s and Tto^r *E. coli*. Four different kinds of plaques are visible, corresponding to $h^+ r^+$, $h^+ r$, $h r^+$, and $h r$ phages.
From Molecular Biology of Bacterial Viruses *by Gunther S. Stent. Copyright © 1963 by W. H. Freeman and Company. Reprinted with permission.*

(a high **multiplicity of infection,** or **m.o.i.**) and ensured that most bacteria would be infected by more than one phage. This allowed the two different phage DNAs to recombine to produce the nonparental types, T2h^+ r^+ and T2*h r,* in the progeny.

The mixed lawn of Tto^r and Tto^s cells offers a neat way to distinguish all four possible combinations of the two pairs of genotypes in the progeny phages (fig. 13.24): T2h^+ r^+ phages will produce small, turbid plaques (small

Box 13.1 Inadvertent Sharing of a Phage

Many stories circulate in the scientific world, some apocryphal, some true. Here is one concerning phages that shows how cleverness can triumph over selfishness. It is probably true, but no names will be used.

It is traditional for biologists to share materials with each other; when one scientist reads about a new mutant, plasmid, or virus that another has isolated or made, all he or she usually has to do is call or write the first scientist requesting a sample, and it will be delivered.

This system has obvious advantages: It greatly accelerates the pace of research because it puts new materials into many hands, instead of bottling them up in one lab. However, nowadays these biological materials, especially clones of bacteria harboring useful eukaryotic genes, can be exploited commercially. This makes them potentially very valuable and, regrettably, leads to less sharing than there used to be.

Our story predates the era of genetic engineering, so no immediate profit motive was involved. Instead, it simply reflects the quite understandable desire to avoid the intense competition that characterizes much genetic research.

It seems that a certain phage geneticist (let us call him A) discovered a novel phage, and a second scientist (B) got wind of the discovery. B wrote to A and asked for a sample of the new phage. A wrote back and declined to cooperate, claiming the right to exploit the phage himself, without competition from B.

Nevertheless, A had already unwittingly cooperated, simply by writing back, because B had a clever idea. He cut the letter into pieces and dropped it into a culture of *E. coli*. Soon, the bacteria were infected and producing the new phage. How can a letter be infectious? Since A was already actively growing the phage in his lab, the air in the lab contained significant numbers of phage particles. Some of these settled on the letter A sent to B, so without intending to, A actually did send B the phage!

---

because of the r^+ gene, turbid because of the h^+ gene); T2h^+ r phages will produce large, turbid plaques; T2h r^+ phages will produce small, clear plaques; and T2h r phages will produce large, clear plaques. In fact, all four combinations were observed. Does this prove that recombination has occurred? Since the progeny phages went on producing nonparental plaques, this new plaque morphology bred true. Thus, recombination, not just complementation, has indeed occurred (fig. 13.25).

Recombination between two different strains of a T-even phage makes it possible to map phage genes by methods analogous to those used for bacterial genes. For example, Hershey and his colleagues isolated several different r mutants of phage T2, numbered $r1$, $r2$, etc., and studied the recombination frequency of each one with an h mutant. Let us consider two r mutants whose recombination frequencies with h differ significantly: $r7$ and $r13$. In fact, recombinants between $r7$ and a particular h mutation are found in 12.3% of the progeny phages, but recombinants between $r13$ and this same h mutation are found in only 1.7% of the progeny. This means that the $r13$ and h mutations lie much closer on the T2 genetic map than do the $r7$ and h mutations. This can be explained by two possible arrangements of $r7$, $r13$, and h, as illustrated in figure 13.26.

To distinguish between these two possibilities, we need to examine the recombination frequency between $r13$ and $r7$ and see if it is greater or less than that between h and $r7$. If it is greater, $r13$ and $r7$ are farther apart than h and $r7$, and figure 13.26b is correct. If it is less, $r13$ and $r7$ are closer than h and $r7$, and figure 13.26a is correct. In fact, it is less, so diagram (a) is the right one. You may be thinking that this was a foregone conclusion, since the two r mutations surely lie in the same gene and that gene is not likely to be interrupted by another gene, h. However, it turns out that $r7$ and $r13$ lie in *different* genes, but mutations in either gene lead to the same mutant plaques. So the experiment was not so trivial after all.

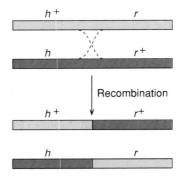

FIGURE 13.25

Recombination between h and r genes of phage T2. Bacteria were coinfected with phages of genotypes h^+ r and h r^+. Recombination between these two markers produced progeny phage of genotype h^+ r^+ and the double-mutant h r.

FIGURE 13.26

Two possible arrangements of the markers: h, $r13$, and $r7$.

T-even phages can be crossed by coinfecting host cells with phages of two different genotypes that differ at two distinct loci. Recombination occurs, producing nonparental genotypes. Recombination frequencies in these experiments can be measured to determine the order and linkage of phage genes.

FINE STRUCTURE MAPPING

We have already examined some mapping of T-even phage *r* mutants. Now we are going to focus on a class of T4 *r* mutants that Hershey mapped to a region he named the **rII** locus. These *rII* mutants were especially interesting because they exhibited a phenotype in addition to the *r*-mutant plaque morphology. They could grow on *E. coli* strain B, but not on strain K(λ), which harbors a λ phage DNA in its chromosome. Thus, *rII* mutants are a kind of conditional lethal—the lethality in this case depends on the host strain—and Benzer exploited this trait to perform an exhaustive study of the *rII* locus, mapping a large number of very closely linked *rII* mutants.

Benzer's *rII* Mapping

This kind of detailed mapping, or **fine structure mapping,** required that Benzer isolate many mutants in a single locus and map them relative to one another. Obviously, if all these mutations are in one gene, they will all be relatively close together. In fact, some will be only one or a few nucleotides apart. Mutations that close together will recombine with a very low frequency. Thus, Benzer had to hunt through thousands of unrecombined, parental (T4*rII*) phages to find one recombinant phage. This is where the conditional lethality of *rII* mutants came in handy. The unrecombined parental phages could not grow on *E. coli* strain K(λ); only the rare, recombinant, wild-type (T4*r*$^+$) phages could do so. This simplified the search tremendously. It was much easier to see a wild-type plaque against a smooth lawn of bacteria than it would have been to pick one wild-type plaque out of a sea of mutant plaques.

Benzer's conditional lethal screening method was so powerful that he could have detected a recombination frequency of one in a million. In fact, the lowest real frequency of *r*$^+$ recombinants he observed was one in ten thousand, or 0.01%. This seems to represent the closest that two mutations can be and still recombine.

How close are two mutations that give *r*$^+$ recombinants with a frequency of 0.01%? We can make an estimate as follows: First, we must realize that if *r*$^+$ recombinants appeared with a frequency of 0.01%, another 0.01% of double-mutant recombinants went undetected. So the real rate of recombination was 0.02%. We know from other mapping data that the circular phage map consists of about 1,500 **map units,** where a map unit is the distance separating two loci that recombine with a frequency of 1% (chapter 5). Thus, two loci that recombine with a frequency of 0.02% are separated by a distance equal to 0.02/1,500, or 1.3×10^{-5} of the total phage genome, assuming that the recombination frequency is constant throughout the genome. Since the T4 phage genome contains about 1.75×10^5 base pairs, the distance between recombinable sites appeared to be one or two base pairs.

The mechanisms we envision for recombination (chapter 7) predict that it should be able to occur between mutations on adjacent nucleotides. Benzer's experiments were not quite precise enough to demonstrate that, but later experiments nailed down the point: Recombination can indeed occur between mutations on adjacent nucleotides.

The finding that recombination could occur *within* a gene was an important one. It destroyed the classical notion of the gene as an indivisible unit. Instead of the gene itself being the unit that recombines, a subgenic unit that Benzer called the "recon" was recognized as the real indivisible unit of recombination. When it became obvious in later years that a recon is really just a base pair, Benzer's term fell into disuse.

Figure 13.27 plots the locations of hundreds of Benzer's spontaneous mutations in the *rII* region. One fact is immediately obvious. The mutations are not evenly distributed throughout the region. Instead, some sites are not mutated at all, whereas others are mutated repeatedly. One site was affected in more than five hundred of the 1,612 mutants screened. Benzer called these highly mutable sites **hot spots.** The same phenomenon has been observed with mutagen-induced mutations.

Seymour Benzer collected thousands of mutants in the *rII* region of phage T4 and mapped their mutations exhaustively. This fine structure mapping revealed that recombination can occur between mutations as close as two (and probably only one) nucleotide(s) apart. Benzer also showed that his mutations were not evenly distributed throughout the *rII* region. Instead, many were clustered in hot spots.

APPLYING THE *CIS-TRANS* TEST TO THE *rII* REGION

We have already observed, in the context of bacterial genetics, the usefulness of the *cis-trans* complementation test in determining whether two mutations affect the same or different genes. The idea is to bring the two mutant genomes together in one organism and see whether a wild-type phenotype results. If so, the two mutant genomes have complemented one another, demonstrating that their mutations lie in different genes. On the other hand, if they cannot complement each other, the mutations must be in the same gene. You may already have recognized that this is really the same thing that geneticists do when they mate two mutant eukaryotic organisms. They place two mutant genomes in the same offspring and observe the phenotype of that offspring.

In 1955, Benzer applied this approach to phage genetics—in particular to the *rII* region of phage T4. We have already seen that recombination between some *rII* mutants is quite frequent, which tells us that the mutated loci are relatively far apart, and therefore suggests that more than one gene is involved in the *rII* phenotype. Benzer proved this point as follows: He coinfected bacteria with pairs of *rII* mutants at a high ratio of phages to bacteria to ensure that each host cell would receive both mutant phage genomes. (Notice that this is the equivalent of producing a diploid by crossing bacteria or higher organisms.) The question was: Would the two *rII* mutants complement each other? The answer was that many of them would. Thus, more than one gene was indeed involved. How many genes? Benzer found

FIGURE 13.27

Hot spots. The distribution of over 1,000 separate mutations in the *rII* locus. Each mutation is represented by a square. Obviously, many of the mutations are clustered in hot spots.

From: Seymour Benzer, Proceedings of the National Academy of Sciences, USA, *47:403–16, 1961. Copyright © 1961 by Seymour Benzer. Reprinted by permission of the author.*

that all of his *rII* mutants fell into only two **complementation groups,** or cistrons. That is, members of one group would complement members of the other, but not members of their own group. These two cistrons represent the two genes of the *rII* region, which Benzer called *rII*A and *rII* B.

Figure 13.28 shows the arrangement of these two genes and some of the mutations that lie in each one. At the time Benzer carried out these studies his term cistron became popular as a substitute for the word gene, because the latter term was tied to the outdated notion of an indivisible unit. However, *gene* has proven to be a very durable word and is now widely used by geneticists of all stripes.

 Benzer applied the *cis-trans* test to the *rII* region of phage T4 and identified two complementation groups, or cistrons, that are responsible for the *rII*-mutant phenotype. Thus, the *rII* region consists of two genes.

FIGURE 13.28

The two cistrons of the *rII* locus. The *rII*A and B genes and a few of their mutations are depicted.

From: Seymour Benzer, Proceedings of the National Academy of Sciences, USA, 41:344, 1955. Copyright © 1955 by Seymour Benzer. Reprinted by permission of the author.

WORKED-OUT PROBLEMS

Problem 1

Consider an experiment in which we do not know anything about the order of three genes, *X*, *Y*, and *Z*, but we do know that they are close enough together to cotransduce. We use an $X^+ Y^+ Z^+$ donor and an $X^- Y^- Z^-$ recipient. We select for X^+ transductants and find the following frequencies of transduction:

103 are $X^+ Y^- Z^-$

315 are $X^+ Y^+ Z^-$

57 are $X^+ Y^+ Z^+$

0 are $X^+ Y^- Z^+$

We can ask the following questions:

1. What is the order of these markers?
2. What is the cotransduction frequency of *X* and *Y*? Of *X* and *Z*?
3. What are the DNA distances between *X* and *Y*? Between *X* and *Z*?

Solution

First, what is the marker order? We will see that all of the combinations but one can be achieved by two crossovers between donor and recipient DNA. The fourth can only be produced by four crossovers and will therefore be the rarest. The rarest combination is $X^+ Y^- Z^+$, which appeared in none of the transductants. How can four crossovers between donor and recipient produce this combination? There is only one way, as shown in figure 13.29, where *Y* is in the middle. Therefore, the order of markers is *XYZ*.

Here is another way of looking at this: If there are four crossovers, there must be one before and after each of the three markers. This will automatically give the transductant the donor alleles of the two end markers, and the recipient allele of the middle marker. Thus, a shortcut is to find the rarest combination ($X^+ Y^- Z^+$ in this case) and determine which of the three loci has the recipient phenotype (Y^- in this case). Therefore, *Y* must be in the middle.

What are the cotransduction frequencies? For *X* and *Y*: (315 + 57)/475 = 0.78; for *X* and *Z*: 57/475 = 0.12. Figure 13.29 shows the crossovers that create each of the four marker combinations. Notice again that the first three involve only two crossovers, while the fourth involves four. Also note that a crossover before *X* is required in all cases, since only X^+ transductants were selected.

Finally, what are the DNA distances among the markers? Rearrangement of the expression $x = (1 - d/L)^3$ for cotransduction frequency (*x*) gives:

$$d = L(1 - \sqrt[3]{x})$$

Given a length (L) of 91.5 kb, and a cotransduction frequency of 0.78 for *X* and *Y,* the distance between *X* and *Y* would be 91.5 kb $(1 - \sqrt[3]{.78})$, or 7.3 kb. Similarly, the distance between *X* and *Z* would be 91.5 kb $(1 - \sqrt[3]{.12})$, or 46.7 kb.

Problem 2

An exotic bacterium takes just 60 minutes to conjugate. You perform a conjugation (mating) experiment with this bacterium and find that the following markers are transferred at the respective times: *pud,* 10 minutes; *spa,* 15 minutes; *ara,* 25 minutes; *dip,* 40 minutes; and *lip,* 50 minutes. Draw a genetic map of this bacterium, assuming a linear rate of DNA transfer during conjugation, and ignoring any lag at the beginning of conjugation.

Solution

Since the total conjugation time is 60 minutes, the genetic map of this organism looks just like a clock face. Each gene goes on the clock at the position corresponding to the time of transfer. For example, *pud* took 10 minutes to be transferred, so it goes at the 10 minute position (two

FIGURE 13.29
Three-factor cross to determine gene order and cotransduction frequency.

	Crossovers	Resulting phenotype	Number of transductants
(a)		$X^+Y^-Z^-$	103
(b)		$X^+Y^+Z^-$	315
(c)		$X^+Y^+Z^+$	57
(d)		$X^+Y^-Z^+$	0

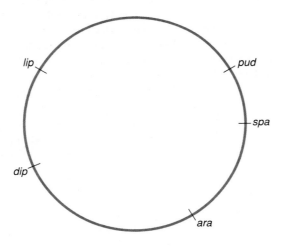

o'clock). Similarly, *spa* took 15 minutes to be transferred, so it goes at the 15 minute position (three o'clock), and so forth. If the conjugation time were not 60 minutes, the process would not be quite as simple, but the principle would be the same. For example, if the conjugation time were 150 minutes, a gene transferred in 50 minutes would map at a position 50/150, or ⅓ of the way around the circle. This would correspond to 20 minutes (or four o'clock) on a 60-minute map.

Problem 3
You want to determine the titre (plaque-forming units [pfu] per milliliter) of phage particles in a phage suspension. You dilute the suspension by a factor of 10^7 and plate 0.2 ml. You observe 120 plaques. What is the titre of the original phage suspension?

Solution
Simply multiply the number of plaques observed by the dilution factor and divide by the volume in milliliters:

$$\frac{120 \text{ pfu} \times 10^7}{0.2 \text{ ml}} = 6 \times 10^9 \text{ pfu /ml}$$

Summary

Two mutant (auxotrophic) bacterial strains can share DNA by direct contact, producing wild-type (prototrophic) recombinants. This process, called conjugation, is mediated in F$^+$ *E. coli* by an F plasmid of 94.5 kb that can transfer at high frequency into F$^-$ bacteria. Occasionally, the F plasmid recombines with its host's chromosome, converting the cell to an Hfr and mobilizing the Hfr chromosome to transfer to recipient cells. By interrupting mating between an Hfr donor and an F$^-$ recipient, we can determine the length of time it takes to transfer a given gene. The differences in the transfer times of different genes tell us their relative positions on the circular genetic map of *E. coli*. The completion of mating in *E. coli* requires recombination between the donated piece of Hfr chromosome and the F⁻ chromosome.

Sometimes a recombination event brings together a host DNA and an F plasmid, creating an F′ plasmid. The host genes thus become part of a highly mobile element that can deliver them to another cell. We can take advantage of this efficient sharing of limited pieces of DNA to perform genetic experiments. For example, two different bacterial mutations that affect the same function can be subjected to the *cis-trans* test, or complementation test. If they complement each other to restore wild-type function, the mutations are usually in separate genes.

Phages can also be used to carry, or transduce, DNA from one bacterium to another. Transduction by phage λ is called specialized transduction because it usually carries only the *gal* or *bio* genes that flank the λ integration site. P1 phage heads can mediate generalized transduction by incorporating up to 91.5 kb of host DNA instead of phage DNA. Transduction allows us to do several types of genetic mapping. We can transduce two genes, or two markers within a gene, and observe the frequency of recombination between them. The closer together they are, the lower their frequency of recombination. Alternatively, we can observe the frequency of cotransduction of two markers. The closer together the markers, the higher their frequency of cotransduction. We can also perform three-factor crosses, transducing three markers and then deducing the order of the markers on the genome.

T-even phages can be crossed by coinfecting host cells with phages of two different genotypes differing at two different loci. Recombination occurs, producing nonparental genotypes. Recombination frequencies in these experiments can be measured to determine the order and linkage of phage genes. Conditional lethal mutations are used in essential genes to avoid destroying the phage's viability. The two common kinds of conditional lethal mutations are temperature-sensitive and amber.

Benzer's fine structure mapping of the *rII* region of phage T4 revealed that recombination can occur between mutations as close as two (and probably only one) nucleotide(s) apart, and that the mutations were clustered in hot spot regions. The *cis-trans* test, applied to the *rII* region of phage T4, identified two complementation groups, or cistrons, that are responsible for the *rII* phenotype. Thus, the *rII* region consists of two genes.

PROBLEMS AND QUESTIONS

1. Write the genotypes and phenotypes of bacteria having the following characteristics:

   (*a*) Unable to metabolize the sugar lactose (lac).
   (*b*) Able to make the amino acid leucine (leu).
   (*c*) Unable to make the vitamin thiamine (thi).
   (*d*) Resistant to the antibiotic streptomycin (str).

(*e*) Sensitive to the antibiotic puromycin (pur).
(*f*) Able to make the vitamin biotin (bio).

2. Which of the strains in problem 1 are prototrophs, and which are auxotrophs, assuming that they are wild-type in all genes but the given one?

3. Given a reversion rate of one in a million and a recombination rate of one in ten million in Lederberg and Tatum's first experiment, was Lederberg justified in his use of double mutants? Why, or why not?

4. Why is a prototrophic recombinant in a bacterial conjugation experiment rarely a true zygote? What do we call it instead?

5. You have discovered that certain *E. coli* cells are resistant to infection by phage T1 (*ton*ʳ). You would like to know whether this characteristic is caused by a mutation in the bacterial genome or by an adaptation of the cells to the presence of the phage. To answer this question, you perform a fluctuation test and find that twenty individual cultures vary quite a bit in their resistance to T1—from a low of zero to a high of 107 *ton*ʳ colonies per culture. Is the characteristic caused by mutation or adaptation? Why?

6. What event converts an F⁺ cell to an Hfr? How does this work?

7. Why does an Hfr transfer its whole genome only rarely?

8. The giant sandworms of the planet Arrakis harbor a gut bacterium (*Xenobacterium gigantii*) that undergoes conjugation. However, like everything in these worms, the circular bacterial chromosome is large, and conjugation takes 200 minutes. You perform a conjugation experiment with this organism and find that the following markers are transferred at the respective times: *ple,* 25 minutes; *bar,* 75 minutes; *gap,* 100 minutes; *zap,* 150 minutes; *chr,* 200 minutes. Draw a genetic map of this exotic bacterium, assuming a linear rate of transfer of DNA during conjugation, and ignoring any lag at the beginning of conjugation.

9. In the experiment in problem 8, you find that the efficiency of transfer of the *ple* marker is very much greater than that of the *chr* marker. What is the probable explanation for this phenomenon?

10. In the experiment in problem 8, if you used a different Hfr that begins its transfer just after *zap,* what difference in transfer times would you expect between *chr* and *ple?*

11. In another Hfr of *X. gigantii*, the fertility plasmid inserts between *gap* and *zap* in the opposite orientation to that in problem 8. What would be the order of transfer of markers by this Hfr?

12. One strain of *X. gigantii* harbors an F′ plasmid, F-*zap,* that contains the *zap* operon and is composed of two genes: *zapA* and *zapB.* You have two Zap⁻ mutants and you want to determine whether their mutations lie in

the same or different genes. Assuming that you can readily place one of these mutant genes on the F-*zap* plasmid, outline an experiment to answer your question.

13. You perform the experiment described in the answer to problem 12 (see answer section at end of book) and obtain only a few Zap$^+$ colonies. What are two possibilities for the cause of these Zap$^+$ cells?

14. Why should we not expect to find both *bio* and *gal* markers in a λ transducing phage?

15. In *E. coli*, the markers *thr* and *leu* are separated by about 2% of the total bacterial genome. Would you expect these genes to cotransduce in P1 phage? Why, or why not? If so, at what frequency?

16. You know that a gene called *bop* is tightly linked to both *zip* and *zap* in the *X. gigantii* genome and readily cotransduces with these other markers in a generalized transducing phage called Q2. You want to know whether the gene order is *bop zip zap* or *zip zap bop*. Therefore, you perform a three-factor cross by transduction as follows: You transduce DNA from a Zip$^+$ Zap$^-$ Bop$^+$ strain into a Zip$^-$ Zap$^+$ Bop$^-$ recipient and screen for Bop$^+$ transductants. The lowest frequency transductant is Zip$^+$ Zap$^+$ Bop$^+$. What is the order of these three genes? What results would you expect if the other gene order were correct? (Drawing diagrams of the two possibilities should help you figure out the answer.)

17. You perform a three-factor cross experiment in which you use Q2 phage to transduce DNA with these three markers: Awk$^+$ Kat$^+$ and Nrd$^+$ into an Awk$^-$ Kat$^-$ Nrd$^-$ recipient. You select for Awk$^+$ transductants and observe the following:

Awk$^+$ Kat$^-$ Nrd$^-$: 638
Awk$^+$ Kat$^-$ Nrd$^+$: 309
Awk$^+$ Kat$^+$ Nrd$^+$: 115
Awk$^+$ Kat$^+$ Nrd$^-$:   1

What is the order of these markers? What are the cotransduction frequencies between *awk* and *kat* and between *awk* and *nrd*?

18. You have a suspension of phages and you want to know its titer (plaque-forming units per milliliter). You dilute the suspension by a factor of 10^6 and plate 0.05 ml. You observe 250 plaques. What was the titre of the original suspension?

19. You perform a phage growth experiment with phage T2. Instead of waiting for the phage to lyse the host cells, you break the cells open at various times after infection and assay for plaque-forming units. How will your growth curve differ from the one-step growth curve? Why?

20. You infect *X. gigantii* cells with different strains of Q2 phage and observe the following frequencies of recombination among the three markers K, L, and M: between K and L, 18%; between K and M, 2.4%; between L and M, 15%. What is the order of these three markers?

Answers appear at end of book.

CHAPTER 14:
DEVELOPMENTAL GENETICS

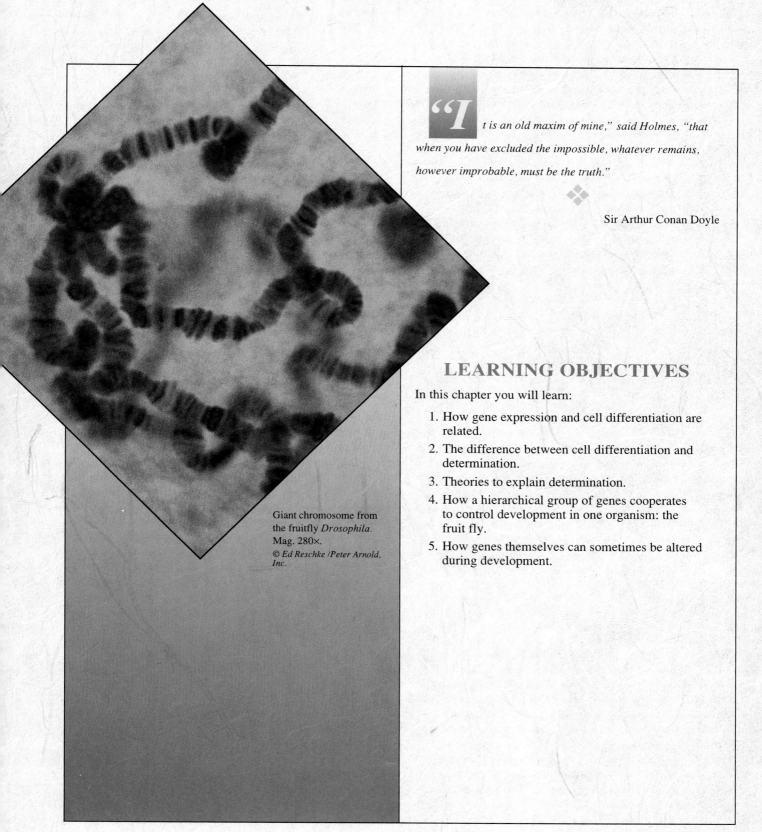

Giant chromosome from
the fruitfly *Drosophila*.
Mag. 280×.
© Ed Reschke /Peter Arnold,
Inc.

*"**I**t is an old maxim of mine," said Holmes, "that when you have excluded the impossible, whatever remains, however improbable, must be the truth."*

Sir Arthur Conan Doyle

LEARNING OBJECTIVES

In this chapter you will learn:

1. How gene expression and cell differentiation are related.
2. The difference between cell differentiation and determination.
3. Theories to explain determination.
4. How a hierarchical group of genes cooperates to control development in one organism: the fruit fly.
5. How genes themselves can sometimes be altered during development.

ne of the greatest mysteries, if not the central mystery, of modern biology is this: How does a fertilized egg develop into a human being? The question is phrased in human terms because we tend to be most curious about ourselves, but it is just as mystifying when we ask about the development of any other complex organism. Through the ages, scientists have offered explanations of increasing sophistication, but we are still a long way from solving the mystery.

In the seventeenth century, biologists believed that a tiny, preformed person, or homunculus (fig. 14.1), resided in each sperm cell. They assumed that the homunculus simply grew in the fertilized egg until it reached the proportions of a newborn baby. Of course there were serious theoretical problems with this idea—one being that the homunculus itself would also have to contain miniature sperm cells enclosing even tinier homunculi, and so on, to include all possible future generations.

Scientists studying the embryology of chickens found no support for the homunculus theory. In fact, as they looked farther and farther back in the development of the chick, they noticed that the embryo not only became smaller, it also became simpler. This gave rise to the idea of **epigenesis,** which holds that tissues do not exist in the earliest embryo, but arise from simpler structures by developmental change. Taken to its logical conclusion, this means that a single fertilized egg cell develops into an adult human composed of trillions of cells that are organized into dozens of different organs, each of which performs a different, highly specialized function.

THE GENETIC BASIS OF CELL DIFFERENTIATION

We have already seen that specialized cellular functions depend on proteins: structural proteins to give cells their shapes; enzymes to catalyze specific chemical reactions; hormones to carry signals from cell to cell; and so forth. For example, consider two different specialized cells in the human body: a red blood cell and a β cell from the pancreas (fig. 14.2). The blood cell is highly specialized to carry oxygen through the bloodstream. Its smooth disk shape allows it to glide easily through tiny capillaries, and it contains a large concentration of a single protein, hemoglobin. As blood flows through the lungs, this protein binds oxygen and turns bright red. Later, it releases the oxygen to peripheral tissues, sustaining their lives.

By contrast, β cells do not move throughout the body; they are stuck in one place—in small conglomerations of cells called islets of Langerhans in the pancreas. The main job of the β cells is to produce a hormone called insulin. The cells secrete this small protein into the bloodstream, after which it travels to remote areas of the body, signaling cells to alter their metabolism. For example, insulin responds to a high level of sugar (glucose) in the blood by promoting the uptake of glucose by cells and depressing the

FIGURE 14.1
Homunculus.

breakdown of fats in adipose (fat) tissue that would normally produce more sugar. The result of these and other actions of insulin is a lowering of the blood glucose level.

Obviously, the functions of red blood cells and β cells are vastly different. These different functions are clearly reflected in the proteins the two types of cells make. Immature red blood cells make huge amounts of hemoglobin and very little else; in particular, they do not make insulin. On the other hand, β cells make lots of insulin, but no hemoglobin. Yet both of these cell types descended directly from the same fertilized egg cell. The early embryo contained many cells, all of which looked more or less alike, and none of which made either hemoglobin or insulin. However, at some point during development, the cells **differentiated** so that they began to manufacture these specialized products. Some became blood cells and made hemoglobin; others became β cells and made insulin.

All cells, even though they begin to synthesize different specialized products, continue to make the key proteins essential for life. For example, they all make RNA polymerase and ribosomal proteins. Because of their maintenance function, the genes for these fundamental proteins are sometimes called **housekeeping genes.** By contrast, genes that code for specialized cell products like hemoglobin or insulin are sometimes called **luxury genes.** They are a "luxury" only in that their products are not immediately vital to the survival of the cells that make them; they are certainly not a luxury to the whole organism.

In short, different specialized cells make different specialized proteins, and those proteins determine what the cells can do. Moreover, as we have already learned,

(a)

(b)

FIGURE 14.2

Contrasting cell morphologies. (*a*) Mature red blood cells; (*b*) an islet of Langerhans (light area at center) composed of insulin-producing β cells. *(a) © MEOLA/ARS-USDA/Science Source/Photo Researchers, Inc. (b) © Edwin A. Reschke.*

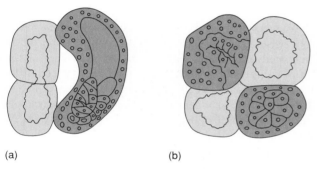

(a)   (b)

FIGURE 14.3

Demonstration of mosaic development. *Styela* embryos were forced through a pipette at the four-cell stage to kill some of the blastomeres, then allowed to develop. (*a*) Both cells on the far left (yellow) were killed, but the two remaining cells (red) continued to divide and differentiate. (*b*) The lower left and upper right cells (yellow) were killed, but again the remaining cells (red) developed into partial embryos.

proteins are products of genes. This means that the genes for globins, the proteins in hemoglobin, are turned on in immature red blood cells, and the gene for insulin is turned on in β cells of the pancreas. Cell differentiation is therefore a reflection of differential gene activity, and those who seek to understand differentiation and development must ultimately ask: What allows expression of certain genes in one kind of cell and other genes in a second kind of cell? We will examine two theories that attempt to answer this question—mosaic development and regulative development.

MOSAIC DEVELOPMENT VERSUS REGULATIVE DEVELOPMENT

One simple explanation of gene differentiation would be that a differentiating cell loses all its genes, except the ones

it needs for housekeeping and for its specialized function. This would be a true genetic change, since the genetic material itself would be altered or lost. A version of this theory, presented by Wilhelm Roux, held that embryonic cells begin losing genetic potential from the very beginning, so that each cell division during development involves the unequal partitioning of genetic material between the daughter cells. This would produce an embryo whose genetically dissimilar cells resemble a mosaic of many different pieces. Hence the term **mosaic embryo.**

Mosaic Development

The mosaic development theory found support in the fact that several invertebrate embryos, even at a very early stage, are composed of cells that behave quite differently; in fact, they seem to be irreversibly programmed to develop along preset lines. A favorite example of such a mosaic embryo is the ascidian *Styela.* Even at the two-cell stage, the two embryonic cells (**blastomeres**) in this organism are different. These differences continue at the four-cell stage, as shown by an experiment performed by E. G. Conklin in 1905, which is depicted in figure 14.3. Conklin forced four-cell *Styela* embryos through a pipette to destroy some of the blastomeres, then looked to see what would happen to the remaining live cells.

In figure 14.3*a,* the two cells on the left (anterior cells) were killed, but the two on the right (posterior cells) went on to differentiate as they would have in a complete embryo, giving rise to chorda cells and muscle cells. In figure 14.3*b,* the bottom left (anterior-vegetal) and top right (posterior-animal) cells were killed, but again the two remaining cells carried out their own developmental programs. The top left cell made neural plate cells, and the bottom right cell again made muscle cells. In principle, this could be caused by the embryo's dividing up its genes with each cell division, so that the top left cell received the genes for making neural plate cells and the bottom right cell got the genes for muscle. No evidence exists for such a partitioning of genes among embryonic cells, and it is increasingly unlikely that any will be found.

Regulative Development

Even if we someday find an unusual organism that does divide its genes among embryonic cells, this cannot be a general phenomenon. Embryos from higher eukaryotes, such as amphibians, can be separated into individual cells as late as the eight-cell stage, and all of the resulting cells are **totipotent,** which means they can go on to produce whole, normal organisms. Clearly, these individual embryonic cells have not lost any of their genetic potential. Even in humans, we know that embryonic cells retain all of their genes at least through the first three cell divisions, because identical quintuplets have occurred. Humans and frogs are examples of organisms that have **regulative** embryos. The cells in these embryos look and behave identically much longer than those of mosaic embryos. Of course, other organisms, such as sea urchins, fall in between these extremes, and sometimes it is difficult to label an embryo either mosaic or regulative.

We have established the fact that early regulative embryos contain totipotent cells. But what about differentiated cells? Are they still totipotent, or have they lost some of their genes? A number of experiments have shown that they are sometimes totipotent. Figure 14.4 illustrates John Gurdon's classic nuclear transplantation experiment, which worked as follows: First, Gurdon used ultraviolet light to kill the nuclei in several eggs from the South African clawed frog (*Xenopus laevis*), producing **enucleate eggs** (eggs without nuclei). Then he extracted nuclei from the intestinal epithelium of a *Xenopus* tadpole and transplanted these nuclei into the enucleate eggs by **microinjection** with a very fine glass pipette.

The transplanted nuclei were diploid, just like **zygote** (fertilized egg) nuclei, and many of the eggs began dividing just like zygotes, to produce early "hollow ball" embryos called **blastulas.** Gurdon took nuclei from these blastulas and transplanted them to new enucleate eggs. Some of the resulting eggs actually developed into tadpoles, and a few made it all the way to the adult frog stage. This demonstrated that even a differentiated intestinal cell nucleus retained the genetic power to produce a whole adult frog. In other words, the nucleus was still totipotent. It could not have lost all the superfluous genes it did not need for housekeeping and intestinal function. In fact, since it gave rise to every cell type in a whole frog, it seems not to have lost any genes at all.

A potential objection to this experiment could be that the ultraviolet light failed to kill all of the original egg nuclei, and these nuclei provided the genes that caused **embryogenesis** (embryonic development). To counter this argument, Gurdon used transplanted nuclei that were genetically different from the original nuclei in that they

had only one nucleolus, whereas the original nuclei had two. The resulting tadpoles and frogs consisted of cells with only one nucleolus, proving that the transplanted nuclei really had directed their development.

The nuclear transplantation experiment has been remarkably difficult to duplicate in other animal species. Even before Gurdon's experiment on *Xenopus,* Robert Briggs and Thomas King showed that another frog species, *Rana pipiens,* does not contain totipotent nuclei after gastrulation, which is a relatively early phase of embryogenesis. However, totipotent adult cells are abundant in plants. F. C. Steward provided the first such example in the early 1950s by regenerating a whole carrot plant from a single carrot root cell. Since then, several different kinds of plants, including tobacco, petunias, and peas, have been grown from single cells. Notice that this furnishes us with a way to clone (produce many identical copies of) a plant. In fact, the word clone comes from the greek *klon,* meaning "twig," and refers to a centuries-old method of propagating plants by taking cuttings.

We are left with the somewhat messy conclusion that, whereas some animals and plants do contain totipotent differentiated cells, many seem not to. At least we can state that loss of genetic material is not necessary for cell differentiation. Specialized cells that are not totipotent may have suffered genetic changes, and we will see some examples of these later in this chapter, but there is certainly no evidence for wholesale loss of genes. Instead, most cell differentiation seems to be caused by **epigenetic** changes. These are changes in gene expression, rather than changes in the genes themselves.

 Multicellular organisms with specialized tissues develop from a single cell, or zygote, by a process called embryogenesis. During embryogenesis, cells differentiate to produce tissues with specialized functions. Differentiation seems to require little if any change in a cell's genetic makeup, since differentiated cells or nuclei from some plants and animals are still totipotent. Instead, the pattern of expression of a common set of genes varies from one kind of differentiated cell to another.

DETERMINANTS

Even before a cell becomes differentiated and starts expressing its specialized functions, it undergoes a process called **determination.** This is the stage at which the cell passes the point of no return in its developmental program. It may not yet look or behave any differently than its neighbors, because it has not differentiated, but it has become irreversibly committed to its fate. Indeed, determination is a prerequisite for differentiation. We will look in detail at a possible mechanism for determination later in the chapter; for now, let us consider some examples of determination that imply the existence of cytoplasmic effector substances called **determinants.**

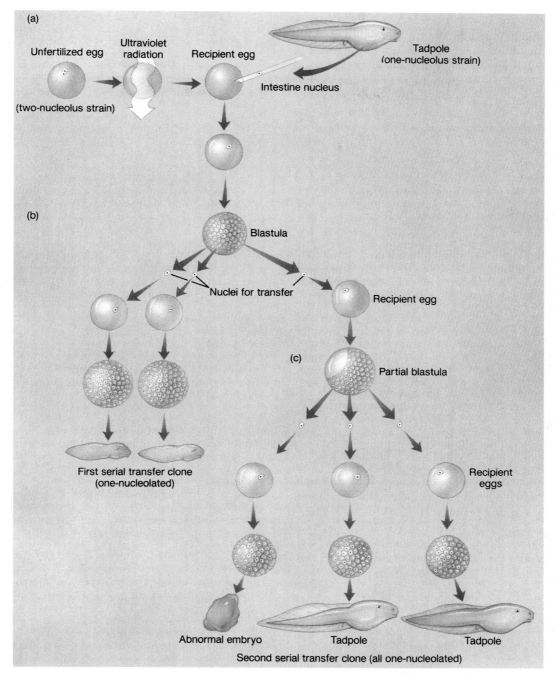

FIGURE 14.4

Demonstration of regulative development. (*a*) An unfertilized frog egg from a two-nucleolus strain was enucleated by ultraviolet radiation, then given a replacement nucleus from the intestine of a one-nucleolus strain. The artificially fertilized egg subsequently developed into a blastula. (*b*) Nuclei from this blastula were transplanted to new enucleate eggs from the two-nucleolus strain. These developed into partial embryos or tadpoles. (*c*) Nuclei from a partial embryo (a blastula) were transplanted to new enucleate eggs, some of which developed into normal, one-nucleolated tadpoles. The first transfer rarely produces normal tadpoles because the differentiated nuclei cannot divide fast enough to support normal development. After the first transfer, the nuclei seem to be geared up to divide more rapidly.

The very existence of mosaic embryos, in which the position of a blastomere in the embryo determines absolutely the fate of that cell, suggests that determinants are present in the egg and that they are unequally distributed. The simplest form of unequal distribution of determinants would be a gradient from the top (**animal pole**) to the bottom (**vegetal pole**) of the egg. But a second gradient, from one side to the other, could coexist with the vertical one (fig. 14.5).

Styela provides a good example of unequal distribution of determinants. The top left blastomere in figure 14.3*b* must have contained the determinants for neural plate development, while the bottom right blastomere must have held muscle determinants. This point was made even more powerfully by J. R. Whittaker, who pushed some of the cytoplasm from the posterior-vegetal (bottom right) blastomere into the posterior-animal (top right) blastomere by squeezing the embryo during the third cleavage. He then blocked further cell division with the drug cytochalasin B and looked for the appearance of the muscle protein acetylcholinesterase (AChE). It appeared in a new site, the posterior-animal part of the embryo, in addition to the posterior-vegetal part, where it is normally made. Only the cytoplasm moved in this experiment, not the nuclei, yet this drastically altered the fate of the cells containing those nuclei. This means that the determinants at work here are cytoplasmic, not nuclear.

What is the chemical nature of determinants? Some are no doubt **maternal messages**—mRNAs made and stored in large quantity in the developing egg. However, embryonic development can be blocked at an early stage by inhibitors of RNA synthesis, proving that there are not nearly enough maternal messages to direct all of the embryogenic processes. For instance, synthesis of AChE in *Styela* embryos depends on mRNA synthesis. In short, maternal messages may account for some of the determinants, but others must be something else, such as transcription factors. Many of these are probably not found in the egg, but are made by the embryonic cells themselves.

 Determinants are substances, some of which are present and unequally distributed in the egg. Determinants cause cells to become determined, or irreversibly committed, to a given fate.

INDUCERS

Substances that one cell makes to influence the development of another cell are called **inducers.** This property of signaling from one cell to another sets the inducers apart from the determinants we have just discussed. An inducer can produce rather startling effects if its signal goes to the wrong address. For example, consider the development of feathers in an embryonic chicken. Feathers are products of one of the three classical embryonic tissue layers, the

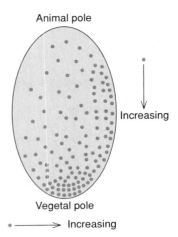

FIGURE 14.5
Hypothetical 2-fold gradient of determinants in an egg. One type of molecule (red dots) is arranged in a gradient from top to bottom, where the highest concentration is at the bottom. The other type of molecule (blue dots) is arranged in a horizontal gradient, where the highest concentration is on the right.

ectoderm. But the underlying **mesoderm** plays an inductive role; that is, it produces substances (inducers) that dictate what kind of feathers the ectoderm will make. We know this because we can transplant embryonic thigh mesoderm to the wing of a developing chick and produce thigh feathers on that wing. The reverse experiment yields wing feathers on the thigh.

Another example of induction sheds additional light on determination. We can transplant embryonic cells of the amphibian *Triturus* to new locations in the embryo and follow their fates. Up to the gastrula stage, the transplanted cells are still totipotent; they will differentiate just like their neighbors in their new location, not like the cells in their old location. Clearly, the transplanted cells are responding to inducers they would not have experienced in their original home. After gastrulation, something quite different happens. The transplanted cells have become determined in their original locations, so they are no longer responsive to external influences; they follow their predetermined program in spite of the inducers they encounter after transplantation.

 Embryonic inducers are substances that diffuse from one embryonic site to another and cause differentiation of the target cells.

CONTROL OF GENE EXPRESSION

Given that cell differentiation occurs via the turn-on of selected genes, differentiation is largely a question of how gene expression is controlled. Chapters 8, 9, and 10 demonstrated that structural gene expression involves three main

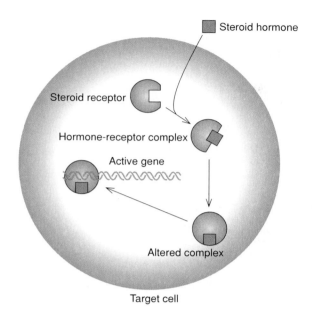

Steroid hormone

Steroid receptor

Hormone-receptor complex

Active gene

Altered complex

Target cell

FIGURE 14.6

Steroid hormone action. A steroid hormone enters a cell and binds to a steroid receptor. The hormone-receptor complex changes shape, binds to a target gene, and activates it.

levels: transcription, translation, and, especially in eukaryotes, an intervening, **posttranscriptional** phase during which transcripts are processed (by splicing, clipping, and modification) into mature form. In addition, many proteins experience posttranslational modifications of various sorts.

Prokaryotes undergo radical changes in gene expression and cellular activity that resemble cell differentiation in higher organisms. A prime example, discussed in chapter 8, is sporulation in bacteria. In that case, changes in RNA polymerase σ factors allow recognition of new promoters, shifting transcription from vegetative genes to sporulation genes. Other examples of massive gene control are found in the replication cycles of bacteriophages, which effect complex, time-dependent shifts in gene expression. Again, these changes in gene expression take place at the transcription level.

MECHANISMS OF TRANSCRIPTION CONTROL

We have also seen that eukaryotic genes, at least the luxury genes that are expressed actively, are controlled primarily at the transcription level. For this reason, and because we understand much more about transcription control, a few examples of such control will be given here. However, it is important to remember that eukaryotes also use other mechanisms to regulate their genes. Even though a gene is subject to transcriptional control, these other mechanisms may also play an important role.

Hormone Activation of Genes

Many hormones, notably the steroids, act by turning on transcription of specific genes in certain target cells. They

do this by a complex process: First, the steroid hormone penetrates the cell membrane; then it couples with a receptor protein, forming a hormone-receptor complex and changing the structure of the receptor in the process. The hormone-receptor complex then enters the nucleus and binds to a set of sites on the chromatin and activates the corresponding genes (fig. 14.6). A key element missing from this scheme is the nature of the chromatin sites that recognize the hormone-receptor complex.

In chapter 9 we looked briefly at a well-studied example of hormone activation, the activation by estrogen of the ovalbumin gene in the chick oviduct. Pierre Chambon and his coworkers have identified a site about 250 base pairs upstream from the TATA box that binds the estrogen-receptor complex.

The hormone-receptor binding site, or **hormone-response element (HRE),** is actually an enhancer. It now seems clear that steroid hormone-receptor complexes are really transcription factors that activate genes by binding to and stimulating these genes' enhancers. In fact, the steroid hormone receptor is just one of a family of hormone receptors that act in this same way. This family also includes (1) the glucocorticoid receptor, which mediates control of genes by hormones such as cortisol; (2) the vitamin D_3 receptor; (3) the thyroid hormone receptor; and (4) the retinoic acid receptor. The latter three are especially relevant to the discussion in this chapter, since vitamin D_3 thyroid hormone (thyroxin), and retinoic acid all play crucial roles in development.

It is important to note that not all cells are responsive to the same hormones. The key to whether or not a cell will respond is the presence or absence of the specific hormone receptor in that cell. Most hormones interact with receptors in only a few types of cells. The receptors for these various hormones are themselves proteins—the products of genes—and most of them are made only in a small subset of differentiating cells. In other words, the receptor is itself the product of cell differentiation; further differentiation, with activation of new genes, will occur when the cell encounters the hormone.

 Hormones such as steroids, thyroid hormone, and retinoic acid control genes by binding to receptor proteins. The receptor proteins interact with hormone-response elements (enhancers) attached to the genes controlled by these hormones.

Light-Induction of Plant Genes

When plant seedlings grow in the dark, they do not make the products needed for photosynthesis. This is economical, because photosynthesis is impossible in the dark and it would be a waste of energy to make the photosynthetic apparatus. Since one of the key components of this apparatus is the green pigment *chlorophyll,* an obvious consequence of this behavior is that the seedlings remain pale yellow. When the experimenter turns the light on, the seedlings

FIGURE 14.7
Light activation of a plant gene. (*a*) The rubisco
gene from the pea was cloned and introduced
into petunia cells. The wild-type gene (top) was
still light-inducible in its new host, though
activity was weak. When all but the thirty-five
base pairs upstream from the transcription start
site were deleted (middle), the gene was still
light-inducible. However, when only thirteen
base pairs remained upstream from the
transcription start site (bottom), light
inducibility was lost. The red color denotes
deleted DNA. (*b*) The cloned rubisco gene was
introduced into a whole tobacco plant (see
chapter 15 for details). The wild-type gene (top)
was strongly active and light-inducible. When
all but the forty-eight base pairs upstream from
the transcription start site were deleted (middle),
the gene was only weakly active. When the
region between −46 and +8 was replaced with a
non-light-responsive cauliflower mosaic virus
(CaMV) promoter (yellow, bottom), strong,
light-inducible transcription was restored. Thus,
the upstream sequence contains a light-
responsive enhancer.

rapidly turn green, reflecting the activation of genes.
Chlorophyll is not the direct product of a gene, since it is
not a protein. It is a small molecule whose synthesis depends on enzymes that are, of course, gene products.
These genes are obviously among those activated by light.
Another light-activated gene is one that codes for the
small polypeptide subunit of an enzyme called **ribulose
bisphosphate carboxylase,** or **rubisco,** a vital element in
photosynthesis.

How does the plant control this rubisco gene in response to light? We are not sure of all the details, but at
least we know that the 5′-flanking region of the gene is important to the light response. Nam-Hai Chua and his colleagues cloned the rubisco gene from the pea and introduced it into petunia cells. They found that, although the
transplanted gene was expressed only at a low level in its
new surroundings, it was still light-inducible. Next, they
pared away various amounts of the 5′-flanking region
of the gene to see what would affect light inducibility
(fig. 14.7*a*). As long as at least 35 base pairs of DNA were
preserved upstream from the start of transcription, the gene
remained light-inducible. However, as soon as the experimenters encroached on this 35-base-pair region, which contains the gene's TATA box as well as a sequence that is
conserved in other light-inducible genes, they destroyed
light inducibility. This indicates that a factor that carries
light's message to the rubisco gene must interact with this
small DNA region.

In further, more sophisticated experiments (fig. 14.7*b*),
Chua introduced the pea rubisco gene into a whole tobacco
plant, forming what we call a **transgenic** plant. In this case,
the transplanted gene was expressed just as actively as in

the pea, its natural home, and was responsive to light.
Moreover, the expression of the gene in tobacco was tissue-
specific, being most active in leaves, weakly active in
stems, and essentially inactive in roots. This is the same
pattern of gene expression observed in the pea.

In contrast to the previous experiment, deletions of the
region upstream from the −35 position in the transplanted
gene greatly reduced the gene's expression in transgenic
plants. When this 5′-flanking region was added back, the
gene's activity was restored and it was light-dependent.
Further probing revealed that the region from −327 to −48
could be inverted or moved and still exert its effect. Therefore, it appears that this region contains an enhancer that
stimulates rubisco transcription.

Is this enhancer responsible only for stimulation of
gene transcription, or is it also responsive to light? To answer this question, Chua removed the region from position
−46 to +8, previously shown to be light-responsive, and
substituted a promoter from cauliflower mosaic virus,
which is not light-responsive. He knew that if the gene still
responded to light, the region upstream from position −46
must be responsible—and indeed it did respond to light.
Apparently this enhancer also helps mediate the stimulation
of the rubisco gene by light.

Thus in plants, as well as in animals, the 5′-flanking
regions of genes contain *cis*-acting DNA elements that
modulate expression of the nearby genes. Recall from
chapter 9 the abundant other *cis*-acting elements (TATA
boxes, CCAAT boxes, GC boxes, for example) and the
trans-acting proteins (TFIID, CTF, and Spl) that interact
with them. These also play major roles in controlling gene
expression, and therefore in controlling cell differentiation.

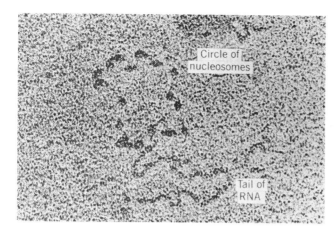

FIGURE 14.8

Transcription of SV40 minichromosome. The SV40 DNA, complexed with histones to form a circle of nucleosomes, is at center. A transcript trails away to the bottom of the figure.

Benjamin Lewin, GENES III, *John Wiley and Sons, p. 543, 1987.*

 The gene for the small subunit of the rubisco enzyme is responsive to light. Two regions in the 5´-flank of this gene mediate this light stimulation: (1) a short region immediately upstream from the cap site and (2) an enhancer located further upstream.

THE NATURE OF DETERMINATION

Determination, as we previously defined it, is the process by which a cell makes the irreversible "decision" to follow a certain developmental path. Much of the process of determination is still unclear, but we are beginning to appreciate that determination involves a change in the state of the chromatin. In this section we will examine two examples of the developmental activation of genes that is accompanied by a more or less permanent alteration in the physical state of the chromatin. In particular, the genes in question seem to become determined by factors that prevent histones from covering the genes' control regions.

DEVELOPMENTAL ACTIVATION OF GENES AND DNase HYPERSENSITIVITY

Gary Felsenfeld and his colleagues have provided us with some insight into this development process with their studies of the expression of the adult β-globin gene (β^A) in the chick embryo. In order to understand this work, we need to be acquainted with the fact that active genes are very sensitive to degradation by **DNase I.** One explanation for this heightened sensitivity to DNase could be that active genes are devoid of nucleosomes. However, we now have ample evidence that this is not the case. First of all, another DNase called micrococcal nuclease, which is known to degrade DNA that is not involved in nucleosomes, does not break down the active genes that are sensitive to DNase I any more than it breaks down inactive genes. Furthermore, electron micrographs such as the one in figure 14.8 show

FIGURE 14.9

SV40 minichromosomes late in infection. These minichromosomes show the nucleosome gap most clearly; sometimes it is less obvious, and sometimes it is not visible at all.

S. Saragosti, G. Moyne, and M. Yaniv, Cell, *Vol. 20, p. 67, May 1980.*

directly that active chromatin contains nucleosomes. We know this SV40 minichromosome is being transcribed because we can see an RNA transcript trailing away from it. It is also clear that the entire circular chromosome contains nucleosomes. Therefore, active genes contain nucleosomes, and DNase sensitivity reflects something subtler than a mere absence of these structures.

Certain sites on active chromatin are especially sensitive to degradation by DNase I. These **DNase hypersensitive sites** give us added insight into the mechanisms by which eukaryotic genes are turned on. In contrast to the ordinary DNase sensitive sites we have discussed, the hypersensitive sites are devoid of nucleosomes. Figure 14.9 shows the pattern of nucleosomes on the SV40 minichromosome late in viral infection. There is a 350-base-pair gap in the nucleosome pattern, and this corresponds to the DNase hypersensitive region.

Further experiments have revealed something else interesting about the DNase hypersensitive region: It corresponds to the 5´-flanking region of both the late and early genes of the virus, where the promoters and enhancer lie. In fact, this is a general finding among genes transcribed by RNA polymerase II. DNase hypersensitive sites in such genes are usually located in the 5´-flanking regions, where the controls are. You should not infer from this that enhancers, as well as promoters, always lie in the 5´-flanking regions of polymerase II genes. In fact, there are several examples of enhancers *within* such genes.

FIGURE 14.10

Activation of the chicken adult β-globin gene. (*a*) When the gene is active in vivo, it contains a nucleosome-free, DNase-hypersensitive region (blue) upstream from the coding sequence of the gene. (*b*) When the gene is mixed with histones and an extract from early (five-day) embryo erythrocytes in vitro, nucleosomes (red) cover the entire locus and no DNase-hypersensitive site results. However, when the extract is from fourteen-day embryo erythrocytes, where the adult β-globin gene is active,

the same DNase-hypersensitive site as seen in vivo appears. Factors in late embryos, not found in earlier embryos, can generate the DNase hypersensitivity. The discontinuities in the DNA in the β^A genes on the right in (*b*) denote DNA not drawn. This was done to avoid having to draw the gene six times longer on the right. (Remember that nucleosomes cause a 6-fold compacting of DNA.)

Now we can return to the studies by Felsenfeld and colleagues. In erythrocytes (red blood cells) from a five-day chick embryo, the β^A gene is not yet turned on, but by nine days, it has become active. Felsenfeld's group found that the β^A gene in erythrocyte chromatin from nine-day embryos is DNase-hypersensitive, whereas the same gene in five-day embryos is not. The hypersensitive site in this case, as in most such cases, is in the 5′-flanking region of the gene; it extends approximately 200 base pairs, covering the region between 60 and 260 base pairs upstream from the start of transcription. Furthermore, the DNase-hypersensitive region is devoid of nucleosomes (fig. 14.10*a*).

What causes the DNase hypersensitivity in this gene? Whatever it is, we strongly suspect that it also plays a major role in turning the gene on. If the hormone model holds true, we would expect the factor to be a protein that binds to the 5′-flanking region of the gene. This is just what Felsenfeld found: a factor in fourteen-day embryos, in which the gene is active, that confers DNase hypersensitivity on the globin gene.

The assay for the factor works as follows (fig. 14.10*b*): We mix a cloned chicken β^A globin gene with a frog oocyte extract to assemble nucleosomes on the bare, cloned DNA. If no other substances are added, nucleosomes form and the globin gene does not become DNase-hypersensitive. If we add an extract from cells that do not express the globin gene (oviduct cells or five-day embryonic erythrocytes), we see the same result—no hypersensitivity. But if we add an extract from fourteen-day embryonic erythrocytes, along with the frog oocyte extract, the hypersensitive region appears just in the right spot in the gene's 5′-flank. The same

extract has no effect if it is added after the nucleosomes have already had a chance to form. This tells us that the factor acts by preventing nucleosomes from forming, thereby preserving a small, nucleosome-free, DNase-hypersensitive region in the gene. We assume that this helps activate the gene and that these experiments have therefore discovered a new transcription factor, or factors. The factor behaves like a protein: Its activity is abolished by proteinase, an enzyme that destroys protein.

The adult β-globin gene (β^A) is inactive in red blood cells from the five-day chick embryo; neither is it DNase-hypersensitive. The same gene is both active and DNase-hypersensitive in red blood cells from nine-day and older embryos. The DNase-hypersensitive region is found in a nucleosome-free zone in the 5′-flanking region of the gene. The hypersensitivity, and presumably the gene activity, are caused by a factor that is found only in cells where the adult β-globin gene is active. The factor seems to be a protein, and it apparently acts by preventing nucleosomes from forming just upstream from the transcription start site.

CONTROL OF THE 5S rRNA GENES
IN *Xenopus laevis*

Donald Brown and Robert Roeder and their colleagues have performed detailed studies on the control of transcription of the 5S rRNA genes in the frog *Xenopus laevis*. Brown believes this simple system gives us important insights into the more general phenomenon of determination.

(a) Transcription factors win:

(b) Histones win:

FIGURE 14.11

The race between transcription factors and histones for the 5S rRNA control region. (*a*) Transcription factors (blue, red, and green) win the race, associate with the control region (purple), and prevent histones from binding; the gene is determined. (*b*) Histones (yellow) win the race, associate with the control region, and tie it up in a nucleosome; the gene is unavailable to transcription factors and remains inactive.

First, let us examine the details of the control of the 5S rRNA genes; then we will analyze Brown's hypothesis of determination.

As we learned in chapter 9, three proteins, TFIIIA, TFIIIB, and TFIIIC, cooperate to activate the 5S rRNA genes so they can be transcribed by RNA polymerase III. The haploid number of 5S rRNA genes is quite large, about 20,000, and these genes compose two different families. The first family includes approximately 98% of the total; since the transcription of these genes occurs only in oocytes, they are called the **oocyte 5S rRNA genes.** The smaller family contains only about four hundred genes, which are transcribed in both oocytes and somatic cells and are therefore called the **somatic 5S rRNA genes.** These observations raise a crucial question: What causes the oocyte genes to be active in oocytes but inactive in somatic cells?

If this system operates the same way as the chicken globin model, we would predict that nucleosomes play a role in gene inactivation. In particular, we would expect the oocyte gene, or at least its internal control region, to be complexed with nucleosomes in somatic cells and therefore repressed, but to be relatively free of nucleosomes in oocytes and therefore active. In fact, Brown's group has found this to be true.

Evidence for this assertion comes from the following in vitro transcription experiments. First, Brown transcribed purified DNA from oocyte and somatic 5S genes and found that RNA polymerase III plus the three transcription factors could transcribe both genes very well. Next, he gently extracted chromatin from oocytes and from somatic cells, and tried to transcribe both in vitro. Here he obtained a much different result: The oocyte 5S gene in oocyte chromatin was active in vitro, but the same gene in somatic cell chromatin was not. Something in the somatic cell chromatin was repressing the oocyte gene.

As we have just suggested, that "something" involves histones and, therefore, nucleosomes. When Brown mixed purified 5S genes with histones and then added transcription factors and RNA polymerase, no transcription occurred. But when he added the factors before the histones, RNA polymerase III could transcribe the gene very well. This suggested a sort of race between factors and histones (fig. 14.11). If the factors get to the gene first, they form a stable complex that prevents nucleosomes from forming and repressing the gene. Thus the gene stays on. But if histones win the race, they form nucleosomes that bind stably to the gene's control region and therefore prevent the factors from binding to activate the gene. In that case, the gene stays off.

Brown also identified histone H1 as a key player in this mechanism. Any procedure that removed histone H1 from oocyte chromatin activated the repressed gene. How might histone H1 exert such profound influence? We have already discussed (chapter 9) how histone H1 does not itself form part of the nucleosomal core, but cross-links nucleosomes in chromatin. This cross-linking fixes the nucleosomes in place, and nucleosomes lacking histone H1 are therefore presumably free to slide along the DNA. This means the 5S gene's control region will sometimes be uncovered as the histone H1-deficient nucleosomes slide away from it; at those moments, transcription factors can bind, turning the gene on.

The real developmental question is: How do the somatic 5S rRNA genes escape the repression that befalls the oocyte 5S rRNA genes in the same somatic cells? Clues to this mechanism come from a comparison of the activity of oocyte 5S genes in embryonic somatic cells and the concentration of TFIIIA in those same cells. We find that the amount of factor available to each 5S rRNA gene drops dramatically during embryogenesis. When it falls below a threshold level, the oocyte 5S genes become stably repressed, while the somatic 5S genes remain active.

More specifically, we see that immature oocytes contain an excess of factors and make tremendous quantities of oocyte 5S rRNA, which is saved for use during embryogenesis. Later in the oocyte's development, all 5S rRNA synthesis ceases, then switches back on again at a time when the developing embryo, a blastula, contains about four thousand cells. In the meantime, the total concentration of TFIIIA has remained constant, but the number of 5S genes this factor must service has increased 4,000-fold. Still, there is an approximate 10-fold excess of factor to 5S genes, and the oocyte genes remain somewhat active. After three more cell divisions, however, only about one molecule of TFIIIA exists per 5S gene in the embryo, now a gastrula, and the oocyte genes have become stably repressed.

The simplest explanation for this behavior is that there is no longer enough factor to go around and the somatic 5S genes compete successfully with the oocyte genes for the limited supply. Without transcription factors, the oocyte genes are vulnerable to histones, which form nucleosomes, repressing the genes.

Brown believes that the important difference between the somatic and oocyte 5S rRNA genes lies in their relative ability to form stable complexes with transcription factors. The somatic genes form more stable complexes and are therefore able to resist being repressed. By contrast, the oocyte genes form relatively unstable complexes that allow histones to tie up and repress the genes.

We can see that this behavior of the 5S rRNA genes can be generalized to give us a hypothesis of determination: Every time a cell divides, the DNA replication machinery strips all proteins—histones and transcription factors alike—off the DNA. This leaves a clean slate, and the histones and factors then compete for re-binding. If the transcription factors are in low concentration, or if they form a weak complex with the DNA, the histones will tie up the gene in nucleosomes and it will not be determined. On the other hand, if the transcription factors are plentiful, or if they bind tightly to the DNA, they prevent nucleosomes from tying up the gene's control region and the gene is determined. Therefore, a cell can become determined by increasing its transcription factor concentration and replicating. This hypothesis might explain most determination, but it now appears that repressed genes can sometimes be activated without DNA replication. Brown's model does not explain how this happens.

 The two families of 5S rRNA genes in the frog *Xenopus laevis* are oocyte genes and somatic genes. The former are expressed only in oocytes; the latter are turned on in both oocytes and somatic cells. The apparent reason for this difference is that the somatic genes form more stable complexes with transcription factors, including TFIIIA. The transcription factors seem to keep the somatic genes active by preventing nucleosomes from forming a stable complex with the internal control region. This stable complex requires the participation of histone H1. Once it forms, transcription factors are excluded and the gene is repressed. Genes can therefore become determined by binding tightly to transcription factors, and one or more determined genes can determine a cell.

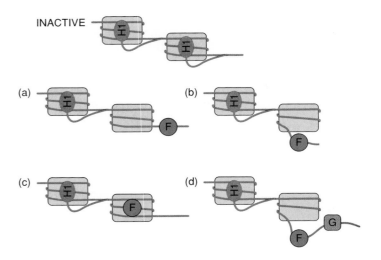

FIGURE 14.12

Four models to explain how transcription factors displace nucleosomes from promoters by removing histone H1. At top, inactive chromatin with cross-linked nucleosomes is shown. (*a*) A transcription factor (F) has bound to the linker DNA between nucleosomes, displacing the resident histone H1, which was bound there through its carboxyl terminal tail. (*b*) The transcription factor has bound to an accessible site on the DNA just as it exits the nucleosome. Again, this displaces histone H1. (*c*) The transcription factor has bound to a site on the surface of the nucleosome, removing histone H1. (*d*) A second factor (G) has bound to the DNA linker, removing histone H1 and opening up a site to which the transcription factor (F) can bind.

DETERMINATION OF POLYMERASE II GENES

Brown's hypothesis of 5S rRNA gene determination can also be called a pre-emptive mechanism, in that transcription factors can pre-empt histones by beating them to the punch at promoters. Similarly, histones can pre-empt the transcription factors by binding first. This neatly explains how genes in dividing cells, with replicating chromosomes, can be determined, and it can be generalized to include genes transcribed by RNA polymerase II. But we already know of several examples of genes transcribed by RNA polymerase II that become determined without benefit of DNA replication. How does this happen? Felsenfeld presents four variations on a theme of dynamic competition between transcription factors and histones. Each of these mechanisms (fig. 14.12) could cause removal of nucleosomes, or at least of histone H1, from a gene's control region without the necessity for DNA replication:

(*a*) One or more transcription factors may bind to the linker region between nucleosomes, where the basic histone H1 tail binds, thus loosening the binding between histone and DNA.

(*b*) Transcription factors may bind to a site on the DNA just at its exit point from the nucleosome. Again, this would disrupt H1-DNA binding.

(*c*) Transcription factors may even bind to DNA within the nucleosome if the site is on the outside and positioned appropriately.

FIGURE 14.13
Structures of (*a*) 5-methylcytosine and (*b*) 5-azacytosine.

(*d*) Some other factor may bind outside the nucleosome, displacing H1 and allowing transcription factors to bind to a previously hidden DNA site.

Remember that removing only histone H1 should be sufficient, since that should break the cross-links between nucleosomes so they can dissociate from the promoter by sliding, jumping, or falling completely off the DNA.

The pre-emptive model makes it seem that histones are just a nuisance, as Felsenfeld puts it, that must be swept aside and then can be forgotten. By contrast, the dynamic competition models require that histones play an active role, interacting with transcription factors to turn genes off and on. This is likely to be the case most often in vivo. Nevertheless, most of the experiments performed so far on gene activation in vitro (chapter 9) have ignored histones, because naked DNA is much easier to deal with than chromatin. We have learned a great deal from these experiments, but it is increasingly clear that they are telling us only part of the story.

Polymerase II genes appear to be determined by two different mechanisms: (1) a pre-emptive model in which histones and transcription factors race to promoters after DNA replication cleans the slate; and (2) a dynamic competition model in which transcription factors remove the histones that had previously covered up a promoter and blocked transcription. This could presumably happen if histone H1 is removed, which would break the cross-links between nucleosomes and allow them to leave the promoter.

THE ROLE OF DNA METHYLATION

Rollin Hotchkiss first discovered methylated DNA in 1948. He found that DNA from certain sources contained, in addition to the standard four bases, a fifth: 5-methylcytosine (fig. 14.13*a*). Watson and Crick ignored this base when they built their DNA model; fortunately, the model is still valid, because the methyl group does not change the base-pairing characteristics of cytosine. It took almost thirty years after Hotchkiss discovered DNA methylation for molecular biologists to begin to find a role for it. That role is in development.

In the mid-1970s, Harold Weintraub and his colleagues noticed that active genes were low in methyl groups, or **undermethylated.** Therefore, a relationship between undermethylation and gene activity seemed likely, as if methylation helped repress genes. This would be a valuable means of keeping genes inactive if methylation is passed on from parent to daughter cells during cell division. And, in fact, it turned out that methylation is inherited from cell to cell: Each parental strand retains its methyl groups, which serve as signals to the methylating apparatus to place methyl groups on the newly made progeny strands (fig. 14.14). Notice that the methyl groups are on the cytosines of 5′-CG-3′ pairs—that is, cytosines followed by guanines. This is the general rule for methylation in higher eukaryotes.

Thus, methylation has two of the requirements for a mechanism of determination: (1) It represses gene activity, and (2) it is permanent. Once a gene is turned off by methylation, it remains that way. This is an epigenetic mechanism. Strictly speaking, the DNA is altered, since methyl groups are attached, but because methyl cytosine behaves the same as ordinary cytosine, the genetic coding remains the same. Thus, methylation cannot qualify as a true genetic change. A striking example of such a role for methylation is seen in the inactivation of the X chromosome in female mammals. One of the two X chromosomes in each cell of a female placental mammal is turned off early in development. The inactive X chromosome becomes heterochromatic and appears as a dark fleck when viewed with a microscope; this chromosome is said to be **Lyonized,** in honor of Mary Lyon, who first postulated the effect in mice. The dark fleck is called a **Barr body,** for Murray Barr, its discoverer. The Lyonized X chromosome can be either the maternal or paternal one; the selection process is completely random. One consequence of this random inactivation of X chromosomes is the patchy coat pattern in calico cats.

Of course, we can explain the heterochromatization of the X chromosome by the formation of solenoids and higher-order chromatin structures discussed in chapter 9. But the repression of the genes in this chromosome goes beyond merely covering up with protein; something permanent has happened to the DNA itself. This conclusion comes from the following type of experiment: DNA is taken from the inactive X chromosome and stripped of all protein, then introduced into another cell and checked for activity. It is one hundred times less active than DNA taken from active X chromosomes. Again, this difference cannot be caused by protein, since all protein has been removed. It has to be due to the DNA itself.

An obvious explanation is that the DNA in the Lyonized X chromosome is methylated, whereas the DNA in the active X chromosome is not. To check this hypothesis, Peter Jones and Lawrence Shapiro and their colleagues grew cells in the presence of the drug 5-azacytosine (fig. 14.13*b*), which prevents DNA methylation.

FIGURE 14.14
Methylation persists after DNA replication. After replication, only one methyl group remains on each daughter molecule. This attracts the methylation machinery, which methylates the opposite strand.

This reactivated the Lyonized X chromosome. Furthermore, Shapiro showed that these reactivated chromosomes could be transferred to other cells and still remain active. It appears that methylation can be a potent and lasting inhibitor of gene activity, and this has led some to speculate about methylation as a mechanism of determination.

As attractive as this idea may be, it cannot be a general explanation. While 80% of DNA in higher vertebrates is methylated, only 20% of that in lower vertebrates is methylated. The situation in lower eukaryotes is even more depressing for methylation enthusiasts. Fruit flies and yeast seem to lack methylated DNA altogether. We are left with the conclusion that methylation can be a mode of inactivating genes, at least in mammals, but that it is probably a mechanism of recent origin in evolutionary history. As such, its role is most likely supplementary rather than fundamental. Methylation also seems to play a crucial role in a phenomenon called genomic imprinting (see box 14.1).

 Methylation of cytosines in CpG sequences in mammals is associated with inactive genes and can indeed turn genes off. The inactivation of methylated genes is perpetuated by methylation of progeny DNA strands across from a methylated parental strand after every round of DNA replication. Methylation can therefore play a part in determination of mammalian cells during embryogenesis; however, it is unlikely that this mechanism is a general one, since lower eukaryotes have little methylated DNA and sometimes none at all.

DETERMINATION IN THE FRUIT FLY *Drosophila*

We have discussed some genes that become determined during embryogenesis, but a more important, and more difficult question is this: What are the genes that *cause* determination? The best clues to this mystery so far have come from one of the standard subjects of genetics research, the fruit fly *Drosophila*. In fact, the genes that are causing so much excitement today—called **homeotic** genes—came to light in mutants in the late nineteenth century. William Bateson coined the term **homoeosis** in 1894 to describe these mutants. He defined the term as a transformation in which "something has been changed into the likeness of something else." That is what seems to happen in a homeotic mutant. A body part appears in a location normally occupied by another part. For example, in **Antennapedia (Antp)**

FIGURE 14.15
The *Antennapedia* phenotype. Legs appear on the head where antennae would normally be.
Courtesy of Walter J. Gehring, University of Basel.

mutants, legs appear on the head in place of antennae (fig. 14.15). It seems that antennae have been changed into the likeness of legs, in Bateson's words. Actually, cells that would normally have developed into antennae have been programmed to produce legs instead.

In other *Drosophila* mutants, different transformations occur. In some, wings replace eyes; in others, legs appear where mouth parts ought to be. The drastic transformations seen in homeotic mutants must involve the activity of many genes, yet only one gene is mutated in each case. This suggests that the mutated genes normally control many other genes, which act during embryogenesis to build body parts. The homeotic genes are therefore "master genes" that are involved in determination. Mutating these genes diverts determination to a new path.

Since these homeotic genes play such a central role in development, we would like to know what kinds of products they encode and how they function. The powerful tools of molecular genetics give us a means of attacking this problem. Several laboratory groups have cloned homeotic genes and have found that some of them fall into two very large clusters: One of these, the **Antennapedia complex (ANT-C)**, spans more than 100 kb of DNA and includes the *Antp* gene, among at least four others. The other is the **Bithorax complex (BX-C)**, which controls the segmentation pattern in the fly's thorax and abdomen. Other homeotic genes exist outside these two major complexes.

Box 14.1 Genomic Imprinting, Genetic Disease, and the Battle of the Sexes

In our discussion of transmission genetics, we have assumed that it made no difference which allele came from the female parent and which came from the male. Thus, in garden peas, crossing a homozygous green-seeded and a homozygous yellow-seeded parent will produce all yellow-seeded hybrid offspring, regardless of which parent contributed the gamete that specified yellow seeds. This is indeed generally true, but there are some important exceptions. These fall into three categories, the first two of which can be explained easily. First, as we saw in chapters 3 and 4, some genes are sex-linked. For example, the outcome of a cross between white-eyed and red-eyed fruit flies is different if the female parent is white-eyed than if the male parent is white-eyed. However, since the eye color gene resides on the X chromosome in fruit flies, and since females have two X chromosomes and males have one X and one Y chromosome, these results still conform to Mendelian principles. The second category of exceptions includes traits that are carried on genes that lie outside the nucleus—in the mitochondria or chloroplasts of a cell. For instance, the gene for a human neuromuscular disease called encephalomyopathy resides on a mitochondrial gene. Since mitochondria come exclusively from the mother, in the egg cytoplasm, this disease can be inherited only from the mother. As in the sex-linked case, this exception turns out to be readily explainable.

On the other hand, we now recognize a third class of exceptions that are not so easily explained. The genes in this case are not necessarily sex-linked, and they inhabit the nucleus, not a cytoplasmic organelle like the mitochondrion. Still, traits governed by these genes are inherited in a non-Mendelian fashion: that is, it matters whether the gene involved comes from the mother or from the father.

The first evidence for the existence of this phenomenon came from studies with mouse eggs just after fertilization, in which the maternal and paternal nuclei had not yet fused. At this stage, the maternal nucleus can be removed and replaced with a second paternal nucleus. Similarly, the paternal nucleus can be removed and replaced with a second maternal nucleus. In either case, the embryo will have chromosomes contributed by only one parent. In principle, that should not have made a big difference, because the parental mice were from an inbred strain in which all the individuals are genetically identical (except, of course, for the XY versus XX difference between males and females). In fact, however, it made a tremendous difference. All of these embryos died during development, most at a very early stage. Those that made it the longest before dying showed an interesting difference, depending on whether their genes came from the mother or the father. Those with genes derived only from the mother had few abnormalities in the embryo itself but had abnormal and stunted placentas and yolk sacs. Embryos with genes derived only from the father were small and poorly formed, but had relatively normal placentas and yolk sacs. How can we account for this difference if the genes contributed by the mother and father are identical? One explanation for this phenomenon is that the genes—that is, the base sequences of the genes—are identical, but they are somehow modified, or **imprinted,** differently in males and females.

Bruce Cattanach provided more evidence for **imprinting** with his studies on mice with fused chromosomes. For example, in some mice, chromosome 11 is fused, so it cannot separate during mitosis. This means that some gametes produced by such a mouse will have two copies of chromosome 11, while others will have none. These mice made it possible for Cattanach to produce offspring with both chromosomes 11 from the father, or both from the mother (using sperm with a double dose of chromosome 11 and eggs with no

chromosome 11 in the former case, and sperm with no chromosome 11 and eggs with a double dose in the latter). Again, if inheritance were strictly Mendelian, it should not matter which parent contributes the chromosomes 11, as long as the embryo gets two copies. But it did matter. In cases where both chromosomes came from the mother, the pups were abnormally small; if both chromosomes came from the father, the pups were giants.

Furthermore, these experiments demonstrated that the imprint is erased at each generation. That is, a runty male mouse whose chromosomes 11 came from his mother generally would produce normal size offspring himself. The production of male gametes somehow erased the maternal imprint.

Genomic imprinting also occurs in humans, sometimes with tragic results. Inheritance of chromosome 15 only from the mother is associated with Prader-Willi syndrome, in which the patient is typically mentally retarded, short, and obese, with very small hands and feet. By contrast, inheritance of chromosome 15 only from the father is connected with Angelman syndrome, characterized by a large mouth and abnormally red cheeks, as well as by mental retardation with inappropriate laughter and jerky movements. Because of this appearance and behavior, such patients used to be called "happy puppets."

What is the mechanism of genomic imprinting? How is the DNA modified in a reversible way so that the imprint can be erased? The evidence points to DNA methylation as the most probable answer. To begin with, experiments show that genes derived from males and females are methylated differently, and that this methylation can be reversed. More tellingly, several geneticists have inserted foreign genes (transgenes) into developing mice and observed differences in their behavior in the resulting male and female transgenic mice. In general, these investigations have found that transgenes inherited from the mother are methylated, while those inherited from the father are not. In one strain, a paternally inherited

Box 14.1 (continued)

transgene is poorly methylated and is active, whereas the same transgene inherited from the mother is methylated and is inactive. Thus, methylation of a gene can indeed depend on which parent provides it, and the methylation state of a gene can indeed determine whether or not the gene will be active.

But what about erasure? Philip Leder and his colleagues have used transgenic mice to follow the methylated state of a transgene as it moves through gametogenesis (the production of sperm or eggs in male or female mice) and into the developing embryo. These experiments reveal that the methyl groups on the transgene are removed in the early stages of gametogenesis in both males and females. The developing egg then establishes the maternal methylation pattern before the oocyte is completely mature. In the male, some methylation occurs during sperm development, but this methylation pattern is further modified in the developing embryo. Thus, methylation has all the characteristics we expect of an imprinting mechanism: it occurs differently in male and female gametes;

it is correlated with gene activity; and it is erased after each generation.

Do any benefits derive from genomic imprinting, or is it just another cause of genetic disorders? David Haig believes that certain imprinting patterns have evolved in response to environmental demands. His favorite example is the insulin-like growth factor (IGF) and its receptor in the mouse. The growth factor tends to make baby mice bigger, but it must interact with its receptor (the type-1 IGF receptor) in order to do so. To complicate the problem, mice have an alternate receptor (a type-2 IGF receptor) that binds IGF but does not pass the growth-promoting signal along. Thus, expression of the IGF gene in developing mice will produce bigger offspring, but expression of the type-2 IGF receptor will sop up the IGF and keep it away from the type-1 receptor, and therefore produce smaller offspring.

Haig points to an inherent biological conflict between the interests of a mother and those of a father of a baby mammal. If the benefits to the mother and father are viewed simply in terms of getting their own genes passed on to their offspring, then the father should favor large

offspring, and the mother should favor small ones. The reason is that a large baby is more likely to survive and therefore perpetuate the father's genes. On the other hand, a large baby saps the mother's strength and leaves her fewer resources to provide to other offspring, which could be sired by a different father but still would perpetuate her genes. This is a coldhearted way of looking at parenthood, but it is the sort of thing that can influence evolution. Viewed in this context, it is very interesting that imprinting of male and female gametes in the mouse dictate that the IGF gene provided by a mother mouse is repressed, while that provided by the father is active. On the other hand, the type-2 IGF receptor gene from the father is turned off, whereas that from the mother is active. Both of these phenomena fit with the premise that a male should favor large offspring and a female should favor small ones. We seem to have a battle of the sexes going on at the molecular level, but neither side is winning, because the strategies of each side are canceled by those of the other!

The Homeobox

DNA sequencing studies on these homeotic genes have uncovered a short (about 180-base-pair) sequence that most of them have in common. Several other genes that are, strictly speaking, not homeotic also contain these sequences, called **homeoboxes.** Even though these other genes do not share the homeotic genes' dramatic effect of transforming one body part into another, all of them are just as important as homeotic genes in the physical organization of the developing embryo. An example is the *engrailed (en)* gene, which participates in the division of the embryo's segments into anterior and posterior compartments.

Another gene with a homeobox is the *ftz* gene, which maps to the ANT-C region. The ftz stands for *fushi tarazu,* meaning "not enough segments" in Japanese; homozygous *ftz* mutations are lethal, but the mutant embryos live long enough to reveal that they have only half the normal number of segments. Thus, the *ftz* gene apparently helps establish the correct segmentation pattern in the embryo. Although this function is not characteristic of a homeotic gene, *ftz* could be considered homeotic because mutants in *ftz* can also have a homeotic phenotype.

Several features of the homeobox stand out. First, it is very well conserved. The homeoboxes of *ftz, Antp,* and **Ubx** (one of the genes in the BX-C) are 75% to 79% conserved,

which means that their base sequences correspond in almost eighty out of one hundred positions. The conservation of amino acids is even higher, approaching 90%. This indicates all the homeobox genes share a common function that depends on the homeobox itself. That function is to control other genes so as to orchestrate development.

In this discussion, bear in mind that a homeobox makes up only part of a gene, so it encodes only a part, or domain, of the protein encoded by the whole gene. In fact, the homeoboxes code for the DNA-binding domains of transcription factors. Of course, a transcription factor is strategically placed to influence the activities of other genes, which is precisely the function postulated for the homeobox genes.

 Homeotic genes of *Drosophila* appear to control other genes that determine the nature of segments in the animal. Mutations in these genes cause dramatic transformations of one body part to another. These, and related genes that influence the insect's body plan, frequently contain regions of about 180 base pairs called homeoboxes. The homeoboxes in different *Drosophila* genes are very well conserved and code for the DNA-binding domains of transcription factors.

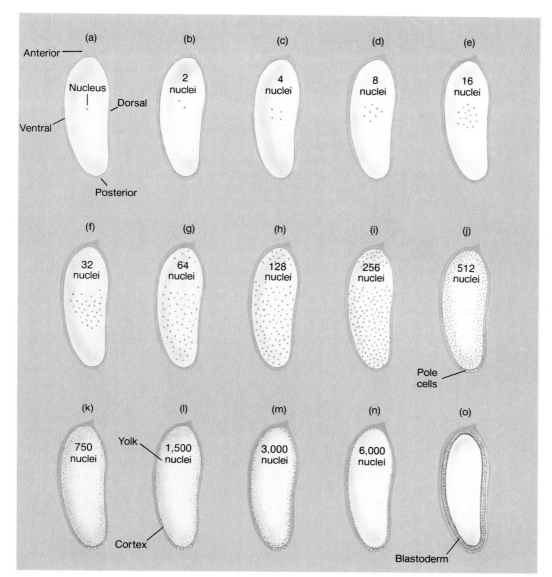

FIGURE 14.16
Early embryogenesis of *Drosophila*. (*a–i*) Nuclei have divided eight times without cell division, producing 256 free nuclei. After the seventh division, the nuclei begin to migrate toward the periphery of the embryo (the cortex, brown). (*j*) A group of nuclei are surrounded by cell membranes and isolated in the posterior end of the embryo. These pole cells will become the germ cells. (*k–n*) The nuclei continue to divide in the cortex. (*o*) When there are about 6,000 nuclei in the cortex, cell membranes form, producing blastoderm cells.

Expression of Homeobox Genes During Embryogenesis

We do not yet know the precise functions of the homeobox genes. However, the simplest explanation is that they are master control genes that switch on batteries of other genes whose products specify certain body parts. To understand this, it will help to look at an outline of early *Drosophila* development (fig. 14.16). One characteristic of this process worth noting is that the nuclei of the early embryo divide several times without cell division. This continues until the 512-nucleus stage, when membranes form around some of the nuclei in the posterior end of the embryo, giving rise to pole cells that will eventually differentiate into the animal's germ cells. The majority of the nuclei are still free at this stage, but they have begun to migrate toward the periphery, or cortex, of the embryo. The nuclei go on dividing in the cortex until their number reaches about six thousand. At this point, cell membranes finally form to segregate the nuclei into individual **blastoderm** cells.

Later in embryogenesis (fig. 14.17), the embryo extends, and the posterior region, which will develop into the abdomen, folds back on itself. Later still, the embryo shortens again, and overt segmentation occurs. After one day of development, the embryo hatches into a larva that feeds, molts twice, then pupates. The pupa metamorphoses into an adult fly. The external adult structures (legs, wings, eyes, antennae, and the like) derive from groups of larval cells called **imaginal disks** (Latin: *imago* = copy or likeness). One disk gives rise to a leg, another to a wing, and so forth. The disks are further divided into regions, each responsible for part of an adult structure, such as a leg segment. The imaginal disks find expression as adult structures only when their cells differentiate under the influence of the appropriate hormone. It is important to note that imaginal disks and body segments are not the same structures. A given body segment may contain several imaginal disks.

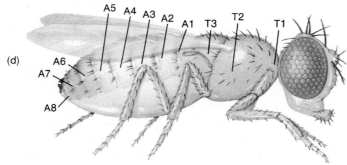

FIGURE 14.17
Late embryogenesis of *Drosophila*. (*a*) "Fate map" of the embryo, in which vertical dashed lines delineate the regions of the embryo that will develop into segments. (*b*) The embryo has extended and folded back on itself, with most of the abdominal segments (A4–A8) lying on the dorsal surface. Md, Mx, and Lb will develop into the head, T1–T3 will become thoracic segments, and A1–A8 will form abdominal segments.
(*c*) The embryo has shortened again and segments have become obvious.
(*d*) Adult fly, showing the ultimate positions of the segments.

The genes that contribute to the fruit fly's body plan can be grouped into three classes. First, and most fundamental, are the **maternal effect genes,** which are active in the mother during oogenesis. Thus, only the mother contributes the products of this first class of genes to the egg. These gene products establish the egg's overall spatial characteristics, such as the anterior-posterior axis. The second group is comprised of the **segmentation genes,** which establish the segmentation pattern of the animal. The homeotic genes, responsible for the identity of each segment, comprise the third class of genes.

Homeobox genes are master control genes that encode transcription factors. These factors switch on the groups of genes that ultimately determine the insect's body plan. These genes can be divided into three hierarchical groups: the maternal effect genes, the segmentation genes, and the homeotic genes.

Maternal Effect Genes

If the maternal effect genes establish polarity in the egg, we might expect to find a polarized gradient of products of these genes: high concentration in one part of the egg changing continuously to low concentration in another. This is precisely what Christiane Nüsslein-Volhard, her colleagues, and others observed with the maternal effect *bicoid (bcd)* gene. *Bicoid* mutant mothers produce mutant embryos that have only abdomens—no heads or thoraxes. However, by injecting such a mutant with cytoplasm from a wild-type embryo, we can induce an almost complete head and thorax at the site of injection. Cytoplasm from the anterior pole of the donor embryo works best; the activity drops off rapidly as one moves toward the posterior pole. Furthermore, *bcd* mRNA works the same way as anterior cytoplasm in these experiments. This suggests that the product of the *bcd* gene is responsible for organizing the production of anterior body parts and is polarized from the anterior to the posterior end of the wild-type embryo.

Markus Noll and his colleagues demonstrated that *bcd* gene products are indeed polarized in *Drosophila* oocytes and early embryos. These researchers constructed a radioactive piece of the *bcd* gene that could be used to hybridize to *bcd* RNA *in situ* (meaning in a whole embryo, its natural situation, rather than in a test tube; Latin: *situs* = position or situation). After the radioactive *bcd* probe hybridized to the mRNA in a thin, longitudinal section of the embryo, autoradiography of the section revealed where the radioactivity had gone. Figure 14.18 shows the results: The radioactivity, and therefore the *bcd* mRNA, was found in highest concentration at the anterior end of the oocyte or embryo.

(a)

(b)

(c)

(d)

(e)

(f)

FIGURE 14.18

Location of *bicoid (bcd)* mRNA during oogenesis and early embryogenesis in *Drosophila*. Thin sections of an adult female abdomen (*a*) and (*b*), a cleavage stage embryo (*c*) and (*d*), and an early stage 14 syncytial blastoderm stage embryo, before cell membranes have formed (*e*) and (*f*), were hybridized *in situ* to a radioactive cDNA probe specific for *bcd* mRNA, then autoradiographed to locate the radioactivity. In the micrographs on the left (*a, c,* and *e*), phase-contrast microscopy was used, so the silver grains from the autoradiography show up as dark spots. These pictures are included primarily to give an overview of the whole oocyte or embryo. Dark-field illumination was used in the micrographs on the right (*b, d,* and *f*), so the silver grains appear as bright spots on a dark background and are much easier to see. Embryos are oriented in all cases with their anterior ends to the left and their dorsal sides up. In the oocytes, designated "o" in micrograph (*a*), *bcd* mRNA is deposited primarily at the anterior end, as can be seen in (*b*). This anterior localization of *bcd* mRNA persists in the cleavage embryo (*d*) and in the stage 14 embryo (*f*).

Gabriella Frigero, Maya Burri, Daniel Bopp, Stefan Baumgartner, and Markus Noll, Cell *47:740, 1986. Reprinted with permission from Cell Press.*

How is this *bcd* mRNA gradient established? It depends on the activity of three other maternal effect genes called *exuperantia (exu), swallow (swa),* and *staufen (stau).* Nüsslein-Volhard showed that mutations in any of these genes result in the spread of *bcd* mRNA to more posterior positions. We are not sure of the detailed molecular mechanism by which the products of these genes operate, but we do know the following: Nüsslein-Volhard's *in situ* hybridization experiments have demonstrated that the *bcd* mRNA is specifically made in nurse cells surrounding the developing oocyte, but only in those nurse cells near the anterior end of the oocyte. Thus, the localization of the *bcd* mRNA begins even before it enters the oocyte; this localization in the nurse cells depends on the function of *exu.* Once in the oocyte, the *bcd* mRNA binds to the cortex at the anterior end. This binding depends on *swa.* Finally, the *bcd* mRNA migrates from the anterior cortex of the oocyte into the anterior cytoplasm, where it remains. This migration relies on *stau.* Struhl and colleagues have also shown that all these localizations require base sequences in the 3′ untranslated region of the *bcd* mRNA. Perhaps these sequences bind to one or more of the products of *exu, swa,* and *stau.*

Upon fertilization, the *bcd* mRNA, located primarily in the anterior egg cytoplasm, is translated to bicoid protein, which forms a concentration gradient, with the highest concentration at the anterior pole of the embryo. This makes sense of course; the bicoid protein is made at the anterior pole, since that is where its mRNA is located, and it then spreads out from this site, forming the gradient. The bicoid protein then acts as a transcription factor, turning on target genes that determine head and thorax formation.

 Maternal effect genes are responsible for the general spatial organization of the fly. For example, *bicoid* is responsible for anterior structures, and its product is distributed in a gradient, with the highest concentration at the anterior end. The bicoid protein acts as a transcription factor, which turns on anterior-determining genes.

Segmentation Genes

The segmentation genes, as their name implies, direct the organization of the fly into discrete segments. *Fushi tarazu* is a good example. As mentioned previously, *ftz* mutants have only half the normal number of segments; this is because every second segment is missing. Since the studies performed so far on homeobox gene expression have revealed a very satisfying correspondence between the site of gene expression and the phenotype of mutants in that gene, we could predict that the expression of *ftz* in wild-type embryos would occur in every second segment. Walter Gehring performed *in situ* hybridization experiments with a radioactive *ftz* probe to detect *ftz* mRNA; figure 14.19 shows that the radioactivity, and therefore the expression of the *ftz* gene, was confined to seven neat bands in the early wild-type embryo. Moreover, these bands correspond precisely to the parts of the embryo that are missing in *ftz* mutants, which is just what we would expect if expression of this gene is required for those parts to form. (*Fushi tarazu* is just one of a class of segmentation genes called **pair-rule** genes. Pair-rule mutants lack body parts that are roughly segment-sized, and these deletions occur at two-segment intervals. *Engrailed* is another pair-rule gene.)

It is also interesting that the *ftz* gene is expressed only early in embryogenesis, from the early blastoderm stage to gastrulation. In fact, expression begins even before cell membranes have formed between the embryonic nuclei. In spite of this, *ftz* expression at this early stage has already assumed a characteristic banded pattern. This means the *ftz* genes are responding to determinants in the early embryo that must also be present in a banded pattern. If a nucleus finds itself in a band of high determinant concentration, it will express the *ftz* gene; but a short distance away is a band of low determinant concentration where little *ftz* expression will occur.

What are these determinants? Gehring and his coworkers performed an experiment that suggests they are transcription factors that act on the 5′-flanking region of the *ftz* gene. Here is how the experiment worked: Gehring coupled the 5′-flanking region of the *ftz* gene to a "reporter" gene—the coding region of the bacterial *lacZ* gene, which codes for β-galactosidase. The product of this gene can be measured easily, since β-galactosidase cleaves a colorless dye into two parts, one of which is bright blue. Next, Gehring injected the fused gene into an early *Drosophila* embryo. When the indicator dye was added, blue bands appeared in the same locations that *ftz* mRNAs would normally occupy (fig. 14.20c). This showed that transcription factors were already arranged in these bands, poised to activate the *ftz* gene.

 One of the first segmentation genes to be expressed in *Drosophila* is *fushi tarazu*. This gene responds to determinants, probably transcription factors, arranged in a banded pattern in the embryo even before cell membranes have formed around the nuclei. The expression of *fushi tarazu* therefore assumes the same banded pattern and probably participates in establishing the segmentation of the embryo.

(a)

(b)

FIGURE 14.19

Expression of *fushi tarazu* in *Drosophila*. A radioactive *ftz* cDNA was hybridized to a thin section of an early embryo, then autoradiographed to detect *ftz* transcripts. (*a*) Autoradiograph showing seven bands of dark silver grains that indicate seven zones of transcription of the *ftz* gene. (*b*) Dark-field micrograph showing the bands more clearly. The bands do not extend across the embryo because by this stage the nuclei are all located in the cortex of the embryo.
From Dr. Walter J. Gehring, University of Basel, Cover of Cell *37:838, July 1984. Reprinted by permission of Cell Press.*

Homeotic Genes

The true homeotic gene, *Ubx,* is also expressed only in the segments affected by *Ubx* mutations, although this situation is more complex than those just described. *Ubx* mutations cause obvious transformations in four contiguous compartments of the fruit fly: the posterior compartment of the second thoracic segment (T2p); the anterior and posterior compartments of the third thoracic segment (T3a and T3p); and the anterior compartment of the first abdominal segment (A1a). In addition, these mutations cause more subtle changes of the larval cuticle, or skin, in the anterior compartments of abdominal segments 2–7. We would therefore predict that *Ubx* would be expressed most strongly in T2p–A1a, and weakly in the posterior regions of A2–A7. Using fluorescent antibodies, David Hogness and his coworkers have confirmed this hypothesis.

Similar experiments with other homeotic genes have shown a good correlation between the locus of expression of a gene and the sites affected by mutants in the same gene. Figure 14.21 summarizes these data on a **fate map** of a *Drosophila* embryo at the blastoderm stage. A fate map shows the parts of the embryo that are "fated" to differentiate into each segment of the adult. This particular fate map

(a)

(b)

(c)

FIGURE 14.20

Activity of the *ftz* promoter in early *Drosophila* embryos.
(*a*) A fluorescent antibody was used to detect the *ftz* gene product in the typical seven-banded pattern. (*b*) At a later stage, the embryo has extended, and the posterior abdominal segments are doubled back on top of the embryo, as shown in figure 14.17*b*. We can still see the seven bands of *ftz* gene product. (*c*) The *ftz* promoter was fused to the β-galactosidase coding region and introduced into a fly embryo. Such a transformed embryo at the same stage as in (*b*) was stained for β-galactosidase activity. The usual seven-band pattern was observed, demonstrating that these seven-band regions contain transcription factors capable of activating the *ftz* promoter.

Courtesy of Dr. Walter J. Gehring. Figures 5, 6, and 7, Science, Vol. 236, pp. 1245–52, 5 June 1987, "Homeo Boxes in the Study of Development." Copyright 1987 by the AAAS.

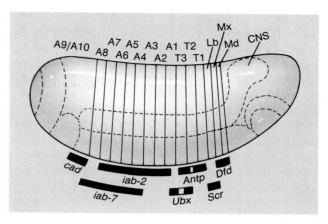

FIGURE 14.21

Patterns of expression of various homeobox genes. The black bars represent the sites on the fate map where each gene is expressed.

 The true homeotic genes, such as *Antennapedia,* switch on after the segmentation genes and determine the nature of each segment. Although we do not understand the details of these genes' operations, one of their functions seems to be to repress the homeotic gene that operates in the neighboring anterior section of the embryo.

A Model for Genetic Control of *Drosophila* Development

The phenomenon of control of one gene by another probably extends to all the developmental control genes, not just the homeotic genes. In fact, we can envision the following model: The maternal effect genes, such as *bcd,* are expressed very early, beginning during oogenesis, and they give products (transcription factors) that become nonuniformly distributed in the embryo. This distribution reflects the function of these genes: to establish the general structure of the embryo. The segmentation genes, such as *ftz,* are expressed later, during early embryogenesis, and help establish the segments of the embryo. These genes respond to signals put in place by the earlier-expressed maternal effect genes. The homeotic genes, such as *Ubx* and *Antp,* are expressed later still and determine the identities of the segments they control. These genes are in turn controlled by products of the segmentation genes. Thus, we see a hierarchy of control genes: Those that are expressed first accomplish the most fundamental developmental steps in embryogenesis, while the more specialized genes are turned on later and control finer steps.

HOMEOBOXES IN OTHER SPECIES

Homeobox genes clearly play a crucial role in *Drosophila* development, but what relevance do they have for higher organisms? It is likely that similar genes play similar roles in vertebrates, including humans. One of the strongest lines of evidence for this hypothesis is the very great similarity in

also shows the parts of the embryo that will express each homeotic gene. We can see that the sites of expression of *Antp* and *Ubx* overlap somewhat, but that *Antp* will generally be expressed more anteriorly than *Ubx*. This is satisfying, because recessive loss-of-function mutants in *Antp* affect more anterior segments than do similar mutations in *Ubx*.

Furthermore, inactivation of the *Ubx* gene leads to the turn-on of *Antp* in segments where *Ubx*, not *Antp*, would normally be expressed. This suggests that one of the functions of *Ubx* is to keep *Antp* turned off. Other studies suggest that *Ubx* is repressed by *iab-2*, the homeotic gene expressed just posterior to *Ubx*. These findings lead to the hypothesis that there is a sequential pattern of control among the homeotic genes, with the posteriorly expressed genes repressing the more anteriorly expressed ones.

sequence between homeoboxes in *Drosophila* and in vertebrates. For example, there is 95% similarity between the
amino acid sequences encoded in certain homeoboxes in
the mouse and in the *Antp* gene of the fruit fly. The similarity extends for a considerable distance outside the homeobox itself, especially toward the 3´-end of the gene. Scientists have now found homeobox genes in a wide variety of
organisms, ranging from those as simple as hydras to those
as complex as humans. There is a startling similarity among
these genes from such different species, and it suggests an
important function for these genes. This function is apparently so vital that changes in the genes are lethal and are
therefore not perpetuated.

What can this vital function be? It is tempting to guess
that it is the same as the apparent function of the homeobox
genes in fruit flies: determination, especially determination
of segments or compartments of the animal. However, segmentation cannot always be involved, since hydras have no
segments.

GENE REARRANGEMENT DURING DEVELOPMENT

We have been assuming that an organism's genome remains constant during development as gene expression
changes. This is generally true, but there are conspicuous
exceptions.

REARRANGEMENT OF IMMUNOGLOBULIN GENES

One example of rearrangement involves the mammalian
genes that produce **antibodies,** or **immunoglobulins,** the
proteins that help us fight infections. An antibody is composed of four polypeptides: two heavy chains and two light
chains. Figure 14.22 illustrates an antibody schematically
and shows the sites that combine with an invading antigen.
These sites, called **variable regions,** vary from one antibody to the next and give these proteins their specificities;
the rest of the molecule (the **constant region**) does not vary
from one antibody to another within an antibody class,
though there is some variation between the few classes of
antibodies. Any given immune cell can make antibody with
only one kind of specificity. Remarkably enough, humans
have immune cells capable of producing antibody to react
with virtually any foreign substance we would ever encounter. That means we can make literally millions,
perhaps even billions, of different antibodies.

Does this imply that we have millions or billions of
different antibody genes? That is an untenable hypothesis;
it would place an impossible burden on our genomes to
carry all the necessary genes. So how do we solve the antibody diversity problem? As unlikely as it may seem, a cell
that is destined to make antibody rearranges its genome to
bring together separate parts of its antibody genes. The machinery that puts together the gene selects these parts at
random from heterogeneous groups of parts, rather like
ordering from a Chinese menu ("Choose one from column

FIGURE 14.22
The antibody is composed of two light chains (blue) bound through
disulfide bridges to two heavy chains (red), which are themselves held
together by a disulfide bridge. The antigen binding sites are at the ends of
the protein chains, where the variable regions lie.

A and one from column B"). This arrangement greatly increases the variability of the genes. For instance, if there
are 200 possibilities in "column A" and 5 in "column B,"
the total number of combinations of A + B is 200 × 5, or
1,000. Thus, from 205 gene fragments, we can assemble
1,000 genes. And this is just for one of the antibody
polypeptides. If a similar situation exists for the other, the
total number of antibodies will be 1,000 × 1,000, or one
million. This description, though correct in principle, is actually an oversimplification of the situation in the antibody
genes; as we will see, they have somewhat more complex
mechanisms for introducing diversity, which lead to an
even greater number of possible antibody products.

Antibody genes have been studied most thoroughly in
the mouse because of very handy mouse tumors called
myelomas. A myeloma is a clone of malignant immune
cells that overproduce one kind of antibody. Studies on the
antibodies made by these myelomas have revealed two
families of mouse antibody light chains called kappa (κ)
and lambda (λ). Figure 14.23a illustrates the arrangement
of the gene parts for a κ light chain. "Column A" of this
Chinese menu contains 90–300 variable region parts (V);
"Column B" contains 5 **joining region** parts (J). The J segments actually encode the last thirteen amino acids of the
variable region, but they are located far away from the rest
of the V region and close to a single constant region part.
This is the situation in the animal's germ cells, before the
antibody-producing cells differentiate and before rearrangement brings the two unlinked regions together. The rearrangement and expression events have been worked out
by several groups, including those led by Susumu Tonegawa, Philip Leder, and Leroy Hood; these events are
depicted in figure 14.23b.

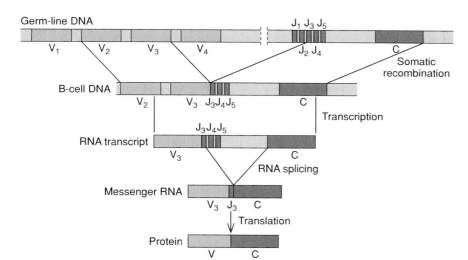

Kappa light chain

90–300 V

5 J   C

(a)

Germo-line DNA

V₁   V₂   V₃   V₄   J₁ J₃ J₅   J₂ J₄   C

Somatic recombination

B-cell DNA

V₂   V₃ J₃J₄J₅   C

Transcription

RNA transcript

J₃J₄J₅   V₃   C

RNA splicing

Messenger RNA

V₃   J₃   C

Translation

Protein

V   C

(b)

Heavy chain

100–200 V   20 D   4 J   C

FIGURE 14.24
Structure of an antibody heavy chain gene. The heavy chain is encoded in
100–200 variable segments (V; light green), 20 diversity segments
(D; purple), four joining segments (J; red), and one constant
segment (C; blue).
*From "The Molecules of the Immune System" by Susumu Tonegawa. Copyright
© 1985 by Scientific American, Inc. All rights reserved.*

FIGURE 14.23
Rearrangement of an antibody light chain gene.
(*a*) The kappa antibody light chain is encoded in
90–300 variable gene segments (V; light green),
five joining segments (J; red), and one constant
segment (C; blue). (*b*) During maturation of an
antibody-producing cell, a DNA segment is deleted,
bringing a V segment (V₃, in this case) together
with a J segment (J₃, in this case). The gene can
now be transcribed to produce the mRNA precursor
shown here, with extra J segments and intervening
sequences. The material between J₃ and C is then
spliced out, yielding the mature mRNA, which is
translated to the antibody protein shown at the
bottom. The J segment of the mRNA is translated
into part of the variable region of the antibody.

First, a recombination event brings one of the V regions together with one of the J regions. In this case, V₃ and J₃ fuse together, but it could just as easily have been V₁ and J₄; the selection is random. After the two parts of the gene assemble, transcription occurs, starting at the beginning of V₂ and continuing until the end of C. Next, the splicing machinery joins the J₃ region of the transcript to C, removing the extra J regions and the intervening sequence between the J regions and C. It is important to remember that the rearrangement step took place at the DNA level, but this splicing step occurs at the RNA level by mechanisms we studied in chapter 9. The messenger RNA thus assembled moves into the cytoplasm to be translated into an antibody light chain with a variable region (encoded in both V and J) and a constant region (encoded in C).

Why does transcription begin at the beginning of V₃ and not further upstream? The answer seems to be that an enhancer in the intron between the J regions and the C region activates the promoter closest to it: the V₃ promoter in this case. This also provides a convenient way of activating the gene after it rearranges; only then is the enhancer close enough to turn on the promoter.

The rearrangement of the heavy chain gene is even more complex, because there is an extra set of gene parts in between the V's and J's. These gene fragments are called D, for "diversity," and they represent a third column on our Chinese menu. Figure 14.24 shows that the heavy chain is assembled from 100 to 200 V regions, approximately 20 D regions, and 4 J regions. On this basis alone, we could put together up to 200 × 20 × 4, or 16,000 different heavy chain genes. Furthermore, 16,000 heavy chains combined with 1,000 light chains yield more than ten million different antibodies.

But there are even more sources of diversity. The first derives from the fact that the mechanism joining V, D, and J segments is not precise. It can delete a few bases on either side of the joining site, or even add bases that are not found in the original segments. This leads to extra differences in antibodies' amino acid sequences.

FIGURE 14.25
The 8;14 translocation in Burkitt's lymphoma as a corruption of a normal antibody gene rearrangement. (*a*) Normal antibody gene rearrangement. The variable and constant regions (V and C, respectively) start out as separate entities; recombination brings them together as a functioning antibody heavy chain gene. D and J regions and introns have been omitted throughout this figure for simplicity. (*b*) One of a kind abnormal rearrangement involving *myc* from chromosome 8 and an antibody heavy chain constant region from chromosome 14. The *myc* gene, minus its first exon, becomes attached to part of the antibody constant region gene, with the two gene fragments in opposite orientations.

Another source of antibody diversity is somatic mutation (chapter 11). In this case, the mutations occur in antibody genes, probably at the time that a clone of antibody-producing cells proliferates to meet the challenge of an invader. This offers an even wider variety of antibodies and, therefore, a better response to the challenge. Assuming that the extra diversity introduced by imprecise joining of antibody gene segments and somatic mutation is a factor of about 100, this brings the total diversity to one hundred times ten million, or about one billion, enough different antibodies to match almost any attacker.

 The immune systems of vertebrates can produce millions, if not billions, of different antibodies to react with virtually any foreign substance. These immune systems generate such enormous diversity by three basic mechanisms: (1) assembling genes for antibody light chains and heavy chains from two or three component parts, respectively, each part selected from heterogeneous pools of parts; (2) joining the gene parts by an imprecise mechanism that can delete bases or even add extra bases, thus changing the gene; and (3) causing a high rate of somatic mutations, probably during proliferation of a clone of immune cells, thus creating slightly different genes.

REARRANGEMENT OF AN ONCOGENE

As we have just seen, antibody genes normally undergo a translocation in which a variable region is juxtaposed with a constant region. In certain cells, this process seems to be corrupted, resulting in cancer. For example, in Burkitt's lymphoma cells, an 8;14 translocation occurs (fig. 14.25). However, instead of a variable region, it is the *myc* oncogene that winds up next to an H-chain constant region. In the process, two things happen to the *myc* gene that may disturb its normal function. First, the gene loses the first of its three exons. This has no effect on the protein product of the gene, since the first exon does not encode any amino acids, but it may affect the regulation of the gene. Second, the translocated gene has moved close to the very active

antibody H-chain genes. This could have an activating effect on *myc*, which is indeed frequently overactive in Burkitt's lymphoma cells.

Thus, the *myc* story provides an example, not only of altering a proto-oncogene by translocation, but of activating an oncogene. And that is not all. The translocated *myc* gene may also escape from the mechanisms that normally keep it under control. Which of these factors is important in the transformation process? We do not know the answer yet; it may be that all are important.

Some proto-oncogenes suffer translocations that convert them to oncogenes. In Burkitt's lymphoma, the *myc* oncogene is translocated next to a cluster of antibody H-chain genes. In the process, it loses its first (noncoding) exon, becomes activated, and loses its normal control pattern.

REARRANGEMENT OF TRYPANOSOME COAT GENES

Do the mechanisms for introducing diversity during development of the immune system apply to development in general? Probably not. In most cases, a gene in the germ cells is the same as the corresponding one in somatic cells in the adult organism. However, changes during an animal's lifetime are not unique to antibody genes. Another well-known example is the rearrangement of surface protein genes during the life of the **trypanosomes**, the protozoa that cause African sleeping sickness. These parasites have protein coats that are easily recognized as foreign, yet they manage to evade human immune defenses by changing the nature of their coat proteins just as our bodies are mounting an effective response to the previous coat. The change in coat comes from switching on new coat protein genes, but in an unorthodox way: The genome rearranges periodically to bring new coat protein genes into an **expression site** at the end of a chromosome where they can be transcribed efficiently. This is rather like taking a lightbulb out of a dead socket and plugging it into a live one.

FIGURE 14.26
The human globin gene families. The α gene family, arranged on the genome in order of expression, consists of a ζ gene, two pseudogenes, and two α genes. The pseudogenes are not expressed. The β family, also arranged in order of expression, consists of an ε gene, two γ genes, a β pseudogene, a δ gene, and a β gene. Again, the pseudogene is not expressed.

SEQUENTIAL ACTIVATION OF MEMBERS OF A GENE FAMILY

Most eukaryotic genes, even those we call unique, or single-copy genes, are actually represented by a family of closely related DNA sequences. Such a family is usually modest in size. For example, the globin genes in vertebrates exist in two families, one for the α-like globins and one for the β-like globins. The members of each of these families are arranged together in **gene clusters.** In the human, the α-globin gene cluster is located on chromosome 16 and contains five members. The β-globin cluster, with six members, lies on chromosome 11. Both of these clusters probably arose by repeated duplication of a single ancestral globin gene. After duplication, the extra gene copies were free to mutate and take on somewhat different functions. At present, we are interested in the way the cell activates these genes in turn during development.

Figure 14.26 shows the structures of the α- and β-globin gene clusters. All the genes are transcribed from left to right, as drawn here. On the far left of the α cluster is a zeta gene (ζ), which is active only very early in embryogenesis. This is followed by two **pseudogenes,** which have collected mutations that render them nonfunctional. Finally, on the right, lie two α genes that code for exactly the same protein. Such genes are obviously not alleles, since they exist on the same chromosome; they are called nonallelic copies. One or both of these genes are active from embryonic through adult life.

The β-globin genes are subject to even finer control. On the left is an epsilon gene (ε) that turns on only during embryonic life. Next come two gamma genes (γ) whose products differ from each other in only one amino acid. These genes are active from embryonic through fetal life. A pseudogene comes next, followed by the delta (δ)- and β-globin genes, which are active only during adult life (β being about fifty times more active than δ).

What is the developmental advantage of this gene switching? Remember that the purpose of hemoglobin is to carry oxygen through the bloodstream to the various tissues of the body. If an embryo or fetus's hemoglobin had the same affinity for oxygen as its mother's hemoglobin, it would not capture enough oxygen to support life. Fortunately, embryonic and fetal hemoglobins have a higher affinity for oxygen than does adult hemoglobin, so they can accept oxygen from the mother's blood in the placenta.

 The globin genes exist in two clusters on separate chromosomes. During embryonic life, the embryonic and adult α-like genes (ζ and α), as well as the embryonic β-like gene (ε), are active; during the fetal stage, the fetal β-like genes (γ) and the adult α gene are active; finally, just before birth, expression switches over almost exclusively to the adult genes (α and β plus δ).

WORKED-OUT PROBLEMS

Problem 1
Suppose human identical nonuplets (nine identical human children) occurred naturally. If this were true, what would be the minimum number of cell divisions after which all the human embryonic cells remained totipotent?

Solution
The existence of identical nonuplets implies that enough cell divisions had occurred to produce at least nine totipotent embryonic cells. Thus, at least four cell divisions must have occurred, because there are two cells after one division, four cells after two divisions, and eight after three divisions. Therefore, one more division, a fourth, would be necessary to produce more than eight cells.

Problem 2
(*a*) How could you use the *lacZ* reporter gene to demonstrate the existence of localized transcription factors that activate the *bicoid (bcd)* gene during early development of

Drosophila embryos? (*b*) If such localized transcription factors exist, how would your experimental result look?

Solution

(*a*) Attach the *bcd* promoter/enhancer region to the *lacZ* reporter gene and inject this recombinant DNA construct, along with the β-galactosidase indicator X-gal into very early *Drosophila* embryos. Then, if transcription factors for *bcd* are localized somewhere in the embryo, they should interact with the *bcd* promoter/enhancer on the *bcd* promoter/enhancer-*lacZ* construct, and activate the *lacZ* gene, producing β-galactosidase. The β-galactosidase should in turn cleave the X-gal to a blue dye, which we can detect microscopically. The location of the dye should tell us the location of the transcription factors.

(*b*) Since activity of the *bcd* gene in the early embryo is normally confined to the extreme anterior end, this is where we should find the transcription factors, and therefore the blue dye.

Summary

Multicellular organisms with specialized tissues develop from a single cell, or zygote, by a process called embryogenesis. During embryogenesis, cells differentiate to produce tissues with specialized functions. Differentiation seems to require little if any change in a cell's genetic makeup, since differentiated cells or nuclei from some plants and animals are still totipotent. Instead, the pattern of expression of a common set of genes varies from one kind of differentiated cell to another. Even before they differentiate, cells become determined, or irreversibly committed to a given fate.

Cell determination in turn depends on the determination of one or more genes. The pre-emptive model of gene determination invokes tissue-specific transcription factors that associate with the control region of a gene and keep it free of nucleosomes. This allows RNA polymerase and any other effector molecules, such as hormones and their receptors, to locate and transcribe the gene, completing the differentiation process. The dynamic competition model involves the same elements, but the transcription factors can sometimes actively remove nucleosomes from a gene and turn it on. Methylation of cytosines in CG sequences in mammals is associated with inactive genes and represents another potential mechanism for controlling determination, at least in mammals.

In the fruit fly *Drosophila,* determination is controlled by a cascade of gene activities, with the product of one gene influencing the activities of others farther down in the cascade. The most fundamental of these genes—at the top of the cascade—are the maternal effect genes, which are active in the mother during oogenesis. The products of these maternal effect genes, or determinants, are placed asymmetrically into the maturing egg, so they can exert their influence more strongly in one part of the early embryo than in another. Some of these determinants are maternal mRNAs that code for transcription factors that turn on other genes. Other determinants act less directly. The targets of the maternal effect gene products are the segmentation genes that determine the segmented pattern of the embryo. The targets of these genes, in turn, are the homeotic genes that determine the identity of each segment. Most of the genes in this cascade code for transcription factors that bind to DNA via a common protein motif called a homeobox.

The immune systems of vertebrates can produce millions, if not billions, of different antibodies to react with virtually any foreign substance. These immune systems generate this enormous diversity by three basic mechanisms: (1) assembling genes for antibody light chains and heavy chains from three or four component parts, respectively, each part selected from heterogeneous pools of parts; (2) joining the gene parts by an imprecise mechanism that can delete bases or even add extra bases, thus changing the gene; (3) causing a high rate of somatic mutations, probably during proliferation of a clone of immune cells, thus creating slightly different genes.

PROBLEMS AND QUESTIONS

1. In what sense is the vertebrate hemoglobin gene a "luxury" to the red blood cell? Why is it not a luxury to the whole organism?

2. Is a chicken oviduct cell determined, differentiated, or both in the following instances? (*a*) When it is competent to produce ovalbumin in response to estrogen, but the chick does not yet make estrogen. (*b*) After the chicken matures and begins producing estrogen and ovalbumin.

3. Why does the existence of identical human quintuplets prove that human embryonic cells remain totipotent at least through three cell divisions?

4. Based on the results of experiments with the 5S rRNA gene, what would happen to activity and DNase hypersensitivity if the chicken β^A gene were mixed first with histones, then with extract from fourteen-day embryos?

5. The experiment cited in figure 14.7 examined the location of the control region of the rubisco gene that is light-sensitive. How does having a cloned gene greatly simplify this kind of experiment?

6. In the experiment cited in figure 14.7, why was it important to move the mutated rubisco gene to a new species, petunia, rather than to reinsert it into a pea plant?

7. If the cloned *Xenopus* 5S rRNA gene is mixed with histones H2A, H2B, H3, and H4, and finally with transcription factors, would you predict that the gene would be active or inactive? Why?

8. Why do you think it is important for an early embryo to have a good supply of 5S rRNA?

9. Which genes are determined in the liver of an immature chick that has not begun producing estrogen—the ovalbumin gene and/or the vitellogenin gene?

10. In what kinds of *Drosophila* genes are homeoboxes found? How do you think this is related to the fact that a homeobox encodes a DNA-binding domain of a protein?

11. How does an *in situ* hybridization experiment show us where a particular gene is expressed in a *Drosophila* embryo?

12. How did the *ftz* promoter-β-galactosidase construct help Gehring demonstrate the presence of bands of transcription factors in the *Drosophila* embryo?

13. How do mutations in homeotic genes and other homeobox genes show that these genes are important in development?

Answers appear at end of book.

CHAPTER 15: GENE CLONING AND MANIPULATION

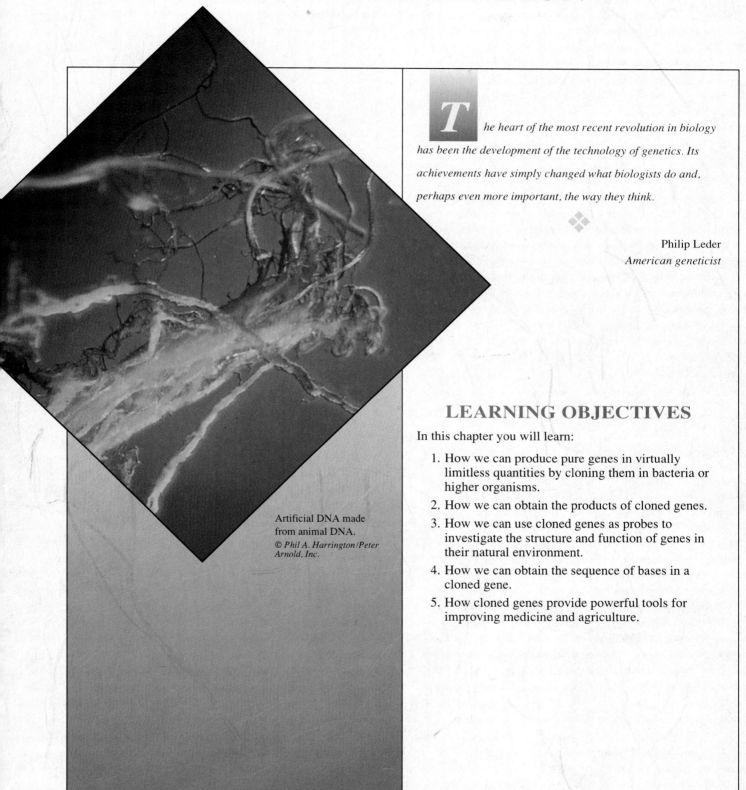

Artificial DNA made from animal DNA.
© Phil A. Harrington/Peter Arnold, Inc.

The heart of the most recent revolution in biology has been the development of the technology of genetics. Its achievements have simply changed what biologists do and, perhaps even more important, the way they think.

❖

Philip Leder
American geneticist

LEARNING OBJECTIVES

In this chapter you will learn:

1. How we can produce pure genes in virtually limitless quantities by cloning them in bacteria or higher organisms.

2. How we can obtain the products of cloned genes.

3. How we can use cloned genes as probes to investigate the structure and function of genes in their natural environment.

4. How we can obtain the sequence of bases in a cloned gene.

5. How cloned genes provide powerful tools for improving medicine and agriculture.

*I*magine that you are a geneticist in the year 1972. You want to investigate the function of eukaryotic genes at the molecular level. In particular, you are curious about the molecular structure and function of the human globin genes: What are the base sequences of these genes? What do their promoters look like? How does RNA polymerase interact with these genes? What changes occur in these genes to cause diseases like thalassemia, where one of the globin polypeptides is not produced at all?

These questions cannot be answered unless you can purify enough of your gene to study—probably about a milligram's worth. A milligram does not sound like much, but it is an overwhelming amount when you imagine purifying it from whole human DNA. Consider that the DNA involved in one globin gene is less than one part per million in the human genome. This means you will need kilograms of human DNA to start with, which is a very depressing prospect. And even if you could round up that much material somehow, you would not know how to separate the one gene you are interested in from all the rest of the DNA.

Gene cloning neatly solves these problems. By linking eukaryotic genes to small bacterial or phage DNAs and inserting these recombinant molecules into bacterial hosts, we can purify large quantities of these genes in pure form. In this chapter we will see how to clone genes in bacteria and in eukaryotes. We will also learn the basic manipulations molecular geneticists use to discover their cloned genes' secrets.

GENE CLONING

The purpose of any cloning experiment is to produce a **clone,** a group of identical organisms. In chapter 14 we saw that some plants can be cloned simply by taking cuttings (Greek: *klon* = twig), and that others can be cloned by growing whole plants from single cells collected from one plant. Even vertebrates can be cloned. John Gurdon produced clones of identical frogs by transplanting many nuclei from one frog embryo to many enucleate eggs (chapter 14). Identical twins also constitute a clone.

The procedure we usually follow in a gene cloning experiment is to place a foreign gene into a bacterial cell, then to grow a clone of these modified bacteria, with each cell in the clone containing the foreign gene. Thus, as long as we ensure that the foreign gene can replicate, we can clone the gene by cloning its bacterial host. Stanley Cohen, Herbert Boyer, and their colleagues performed the first cloning experiment in 1973.

THE ROLE OF RESTRICTION ENDONUCLEASES

Cohen and Boyer's elegant plan depended on invaluable enzymes called **restriction endonucleases.** Stewart Linn and Werner Arber discovered restriction endonucleases in

E. coli in the late 1960s. These enzymes get their name from the fact that they "restrict" invasion by foreign DNA, such as viral DNA, by cutting it up. Furthermore, they cut at sites within the foreign DNA, rather than chewing it away from the ends, so we call them endonucleases (Greek: *endo* = within), rather than exonucleases (Greek: *exo* = outside). Linn and Arber hoped that their enzymes would cut DNA at specific sites, giving them finely honed molecular knives with which to slice DNA. Unfortunately, these particular enzymes did not fulfill that hope.

However, another such enzyme, discovered in *Haemophilus influenzae* strain R_d by Hamilton Smith, did show specificity. This enzyme is called *Hin*dII (pronounced Hin-dee-2). All restriction enzymes derive the first three letters of their names from the Latin name of the microorganism that produces them. The first letter is the first letter of the genus and the next two letters are the first two letters of the species (hence: *Haemophilus influenzae* yields *Hin*). In addition, the strain designation is sometimes included; in this case, the "d" from R_d is used. Finally, if the strain of microorganism produces just one restriction enzyme, the name ends with the Roman numeral I. If more than one enzyme is produced, they are numbered I, II, III, etc.

*Hin*dII recognizes this sequence:

$$\downarrow$$

GTPyPuAC

CAPuPyTG

$$\uparrow$$

and cuts in the middle as shown by the arrows. Py stands for either of the pyrimidines (T or C), and Pu stands for either purine (A or G). Wherever this sequence occurs, and *only* when this sequence occurs, *Hin*dII will make a cut. Happily for molecular geneticists, *Hin*dII turned out to be only one of hundreds of restriction enzymes, each with its own specific recognition site. Table 15.1 lists the sources and cutting sites for several popular restriction enzymes. Note that some of these enzymes recognize four-base sequences instead of the more common six-base sequences. As a result, they cut much more frequently. This is because a given sequence of four bases will occur about once in every $4^4 = 256$ bases, while a sequence of six bases will occur only about once in every $4^6 = 4,096$ bases. Some restriction enzymes, such as *Not*I, recognize eight-base sequences, so they cut much less frequently (once in $4^8 \cong 65,000$ bases).

Notice also the three possible positions of the cutting sites, indicated by the arrows on the table: in the middle of the sequence, therefore leaving blunt ends (*Alu*I, *Sma*I); at the 5′-end, leaving protruding 5′-ends (*Bam*HI, *Hpa*II); or at the 3′-end, leaving protruding 3′-ends (*Pst*I, *Pvu*I). Finally, notice that the recognition sequences for *Sma*I and *Xma*I are identical, although the cutting sites are different. We call such enzymes that recognize identical sequences **isoschizomers** (Greek: *iso* = equal; *schizo* = split).

·Table 15.1 Recognition Sequences and Cutting Sites of Selected Restriction Enzymes

Enzyme	Recognition Sequence
*Alu*I	A G′C T
*Bam*HI	G′G A T C C
*Bgl*II	A′G A T C T
*Cla*I	A T′C G A T
*Eco*RI	G′A A T T C
*Hae*III	G G′C C
*Hind*II	G T Py′Pu A C
*Hind*III	A′A G C T T
*Hpa*II	C′C G G
*Kpn*I	G G T A C′C
*Mbo*I	′G A T C
*Pst*I	C T G C A′G
*Pvu*I	C G A T′C G
*Sal*I	G′T C G A C
*Sma*I	C C C′G G G
*Xma*I	C′C C G G G
*Not*I	G C′G G C C G C

Only the top strand of each recognition sequence is shown. Thus, the actual double-stranded recognition sequence of *Pst*I, for example, is

C T G C A′G
G‚A C G T C

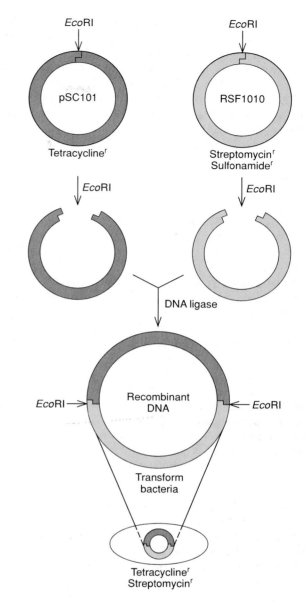

FIGURE 15.1

The first cloning experiment involving a recombinant DNA assembled in vitro. Boyer and Cohen cut two plasmids, pSC101 and RSF1010, with the same restriction endonuclease, *Eco*RI. This gave the two linear DNAs the same sticky ends, which were then linked in vitro using DNA ligase. The investigators reintroduced the recombinant DNA into *E. coli* cells by transformation, and screened for clones that were resistant to both tetracycline and streptomycin. These clones were therefore harboring the recombinant plasmid.

The main advantage of restriction enzymes is their ability to cut a DNA reproducibly in the same places. This property is the basis of many techniques used to analyze genes and their expression. We have seen examples of some of these in previous chapters; in this chapter we will see several more. But this is not the only advantage. Many restriction enzymes make staggered cuts in the two DNA strands (table 15.1), leaving single-stranded overhangs, or **sticky ends,** that can base-pair together briefly. Note, for example, the complementarity between the ends created by *Eco*RI (pronounced Eeko R-1 or Echo R-1):

---GAATTC---   ---G   AATTC---
                → +
---CTTAAG---   ---CTTAA   G---

Restriction enzymes can make staggered cuts because the sites they recognize are usually symmetrical. That is, they read the same forwards and backwards. Thus, *Eco*RI cuts between the G and the A in the top strand (on the left), and between the G and the A in the bottom strand (on the right).

These symmetrical sites are also called **palindromes.** In ordinary language, palindromes are sentences that read the same forwards and backwards. Examples are Napoleon's lament: "Able was I ere I saw Elba," or a wart remedy: "Straw? No, too stupid a fad; I put soot on warts." DNA palindromes also read the same forwards and backwards, but you have to be careful to read in the same sense ($5' \rightarrow 3'$) in both directions. This means that you read the top strand left to right and the bottom strand right to left.

Cohen and Boyer took advantage of the sticky ends created by a restriction enzyme in their cloning experiment (fig. 15.1). They cut two different DNAs with the same restriction enzyme, *Eco*RI. Both DNAs were plasmids. The first, called pSC101, carried a gene that conferred resistance to the antibiotic tetracycline; the other, RSF1010, gave resistance to both streptomycin and sulfonamide. Both plasmids had just one *Eco*RI **restriction site,** or cutting site for *Eco*RI. Therefore, when *Eco*RI cut these circular DNAs, it converted them to linear molecules and left them with the same sticky ends. These sticky ends then base-paired with one another, at least briefly. Finally, DNA ligase completed the task of joining the two DNAs.

The result was a **recombinant DNA,** two previously separate pieces of DNA linked together. This new, recombinant plasmid was easy to detect. When introduced into bacterial cells, it conferred resistance to both tetracycline, a property of pSC101, and to streptomycin, a property of RSF1010. We have seen several examples of recombinant DNA in this text, but this one differs from the others in that it was not created naturally in a cell. Instead geneticists put it together in a test tube.

 Restriction endonucleases recognize specific sequences in DNA molecules and make cuts in both strands. This allows very specific cutting of DNAs. Also, because the cuts in the two strands are frequently staggered, restriction enzymes create sticky ends that help link together two DNAs to form a recombinant DNA in vitro.

VECTORS

Both plasmids in the Cohen and Boyer experiment are capable of replicating in *E. coli*. Thus, both can serve as carriers to allow replication of recombinant DNAs. All gene cloning experiments require such carriers, which we call **vectors,** but a typical experiment involves only one vector, plus a piece of foreign DNA that depends on the vector for its replication. Since the mid-1970s, many vectors have been developed; these fall into two classes: plasmids and phages.

Plasmids as Vectors

In the early years of the cloning era, Boyer and his colleagues developed a set of very popular vectors known as the pBR plasmid series. One of these plasmids, **pBR322** (fig. 15.2), contains genes that confer resistance to two antibiotics, ampicillin and tetracycline. Between these two genes lies the origin of replication. The plasmid was engineered to contain only one cutting site for several of the common restriction enzymes, including *Eco*RI, *Bam*HI, *Pst*I, *Hin*dIII, and *Sal*I. This is convenient because it allows each enzyme to be used to create a site for inserting foreign DNA without losing any of the plasmid DNA.

For example, let us consider cloning a foreign DNA fragment into the *Pst*I site of pBR322 (fig. 15.3). First, we would cut the vector with *Pst*I to generate the sticky ends characteristic of that enzyme. In this example, we also cut the foreign DNA with *Pst*I, thus giving it *Pst*I sticky ends. Next, we combine the cut vector with the foreign DNA and incubate them with DNA ligase. As the sticky ends on the vector and on the foreign DNA base-pair momentarily, DNA ligase seals the nicks, attaching the two DNAs

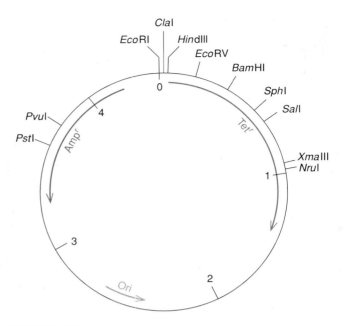

FIGURE 15.2

The plasmid pBR322, showing the locations of eleven unique restriction sites that can be used to insert foreign DNA. The locations of the two antibiotic resistance genes (Amp^r = ampicillin resistance; Tet^r = tetracycline resistance) and the origin of replication (Ori) are also shown. Numbers refer to kilobase pairs (kb) from the *Eco*RI site.

together covalently. Once this is done, they cannot come apart again unless they are recut with *Pst*I.

In the following step, we transform *E. coli* with our DNA mixture. It would be nice if all the cut DNA had been ligated to plasmids to form recombinant DNAs, but that never happens. Instead, we get a mixture of re-ligated plasmids and re-ligated inserts, along with the recombinants. How do we sort these out? This is where the antibiotic resistance genes of the vector come into play. First we grow cells in the presence of tetracycline, which selects for cells that have taken up either the vector or the vector with inserted DNA. Cells that received no DNA, or that received insert DNA only, will not be tetracycline-resistant and will fail to grow.

Next, we want to find the clones that have received recombinant DNAs. To do this, we screen for clones that are both tetracycline-resistant and ampicillin-sensitive. Figure 15.3 shows that the *Pst*I site, where we are inserting DNA in this experiment, lies within the ampicillin resistance gene. Therefore, inserting foreign DNA into the *Pst*I site inactivates this gene and leaves the host cell vulnerable to ampicillin.

How do we do the screening? One way is to transfer copies of the clones from the original tetracycline plate to an ampicillin plate. This can be accomplished with a felt transfer tool as illustrated in figure 15.4. We touch the tool lightly to the surface of the tetracycline plate to pick up cells from each clone, then touch the tool to a fresh

ditto

(a)   Original   Replica

(b)   Original (tetracycline)   Replica (ampicillin)

FIGURE 15.4

Screening bacteria by replica plating. (*a*) The replica plating process. We touch a felt-covered circular tool to the surface of the first dish containing colonies of bacteria. Cells from each of these colonies stick to the felt and can be transferred to the replica plate in the same positions relative to one another. (*b*) Screening for inserts in the pBR322 ampicillin resistance gene by replica plating. The original plate contains tetracycline, so all colonies containing pBR322 will grow. The replica plate contains ampicillin, so colonies bearing pBR322 with inserts in the ampicillin resistance gene will not grow (these colonies are depicted by dotted circles). The corresponding colonies from the original plate can then be picked.

FIGURE 15.3

Cloning foreign DNA using the *Pst*I site of pBR322. We cut both the plasmid and the insert (yellow) with *Pst*I, then join them through these sticky ends with DNA ligase. Next, we transform bacteria with the recombinant DNA and screen for tetracycline-resistant, ampicillin-sensitive cells. The plasmid no longer confers ampicillin resistance because the foreign DNA interrupts that resistance gene (blue).

ampicillin plate. This deposits cells from each original clone in the same relative positions as on the original plate. We look for colonies that do *not* grow on the ampicillin plate and then find the corresponding growing clone on the tetracycline plate. Using that most sophisticated of scientific tools—a sterile toothpick—we transfer cells from this positive clone to fresh medium for storage or immediate use. Notice that we did not call this procedure a selection, because it does not remove unwanted clones automatically. Instead, we had to examine each clone individually. We call this more laborious process a **screen.**

Nowadays we can choose from many plasmid cloning vectors besides the pBR plasmids. One useful class of plasmids is the **pUC** series (fig. 15.5). These plasmids are based on pBR322, from which almost half the DNA, including the tetracycline resistance gene, has been deleted. Furthermore, the pUC vectors have their cloning sites clustered into one small area called a **multiple cloning site (MCS).** The pUC vectors contain the pBR322 ampicillin resistance gene to allow selection for bacteria that have received a copy of the vector. Moreover, to compensate for the loss of the other antibiotic resistance gene, they have genetic elements that permit a very convenient screen for clones with inserts in their multiple cloning sites.

Figure 15.5*b* shows the multiple cloning sites of pUC18 and pUC19. Notice that they lie within a DNA sequence coding for the amino portion of β-galactosidase (denoted *lacZ´*). The host bacteria used with the pUC vectors carry a gene fragment that encodes the carboxyl portion of β-galactosidase. By themselves, the β-galactosidase fragments made by these partial genes have no activity. But they can complement each other by intracistronic

(a)

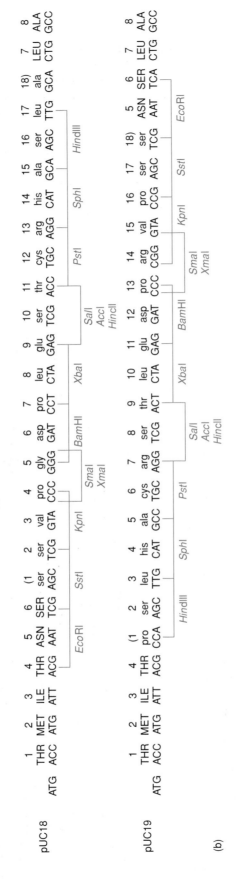

(b)

FIGURE 15.5

Architecture of a pUC plasmid. (*a*) The pUC plasmids retain the ampicillin resistance gene and the origin of replication of pBR322. In addition, they include a multiple cloning site (MCS) inserted into a gene encoding the amino part of β-galactosidase (*lacZ′*). (*b*) The MCSs of pUC18 and pUC19. In pUC18, the MCS, containing thirteen restriction sites, is inserted after the sixth codon of the *lacZ′* gene. In pUC19, it comes after the fourth codon and is reversed. In addition to containing the restriction sites, the MCSs preserve the reading frame of the *lacZ′* gene, so plasmids with no inserts still support active expression of this gene. Therefore, clones harboring the vector alone will turn blue in the presence of the synthetic β-galactosidase substrate X-gal, while clones harboring the vector plus an insert will remain white.

Courtesy of Life Technologies, Inc.

368

complementation, which we encountered in chapter 13. In other words, the two partial gene products can cooperate to form an active enzyme. Thus, when pUC18 by itself transforms a bacterial cell carrying the partial β-galactosidase gene, active β-galactosidase is produced. If we plate these clones on medium containing a β-galactosidase indicator, colonies with the pUC plasmid will turn color. The indicator X-gal, for instance, is a synthetic, colorless galactoside; when β-galactosidase cleaves X-gal, it releases galactose plus a dye that stains the bacterial colony blue.

On the other hand, if we have interrupted the plasmid's partial β-galactosidase gene by placing an insert into the multiple cloning site, the gene is usually inactivated. It can no longer make a product that complements the host cell's β-galactosidase fragment, so the X-gal remains colorless. Thus, it is a simple matter to pick the clones with inserts. They are the white ones; all the rest are blue. Notice that, in contrast to screening with pBR322, this is a one-step process. We look simultaneously for a clone that (1) grows on ampicillin and (2) is white in the presence of X-gal. The multiple cloning sites have been carefully constructed to preserve the reading frame of β-galactosidase. Thus, even though the gene is interrupted by eighteen codons (placed in parentheses, with amino acid names in lowercase), a functional protein still results. Further interruption by large inserts, especially those that shift the reading frame, is usually enough to destroy the gene's function.

The multiple cloning site also allows us to cut it with two different restriction enzymes (say, *Eco*RI and *Bam*HI) and then to clone a piece of DNA with one *Eco*RI end and one *Bam*HI end. This is called **directional cloning,** or **forced cloning,** because we force the insert DNA into the vector in only one orientation. (Obviously, the *Eco*RI and *Bam*HI ends of the insert have to match their counterparts in the vector.) Knowing the orientation of an insert has certain benefits, which we will explore later in this chapter. Forced cloning also has the advantage of preventing the vector's simply re-ligating by itself, since its two restriction sites are incompatible.

Notice that the multiple cloning site of pUC18 is just the reverse of that of pUC19. That is, the restriction sites are in opposite order. This means we can clone our fragment in either orientation simply by shifting from one pUC plasmid to the other.

> Among the commonly used plasmid cloning vectors are pBR322 and the pUC plasmids. The former has two antibiotic resistance genes and a variety of unique restriction sites into which we can introduce foreign DNA. Most of these sites interrupt one of the antibiotic resistance genes, making screening straightforward. Screening is even easier with the pUC plasmids. These have an ampicillin resistance gene and a multiple cloning site that interrupts a β-galactosidase gene. We select for ampicillin-resistant clones that do not make β-galactosidase and therefore do not turn the indicator, X-gal, blue. The multiple cloning site also makes it convenient to carry out forced cloning into two different restriction sites.

Phages as Vectors

We have already seen how phages serve as natural vectors in transducing bacterial DNA from one cell to another. It was only natural, then, to engineer phages to do the same thing for *all* kinds of DNA.

λ *phage vectors*

Fred Blattner and his colleagues constructed the first phage vectors by modifying the well-known λ phage. They took out the region in the middle of the phage DNA, which codes for proteins needed for lysogeny, but retained the genes needed for lytic infection. These phages are no longer capable of lysogenic infection, but their missing genes can be replaced with foreign DNA. Blattner named these vectors **Charon phages** after Charon, the boatman on the river Styx in classical mythology. Just as Charon carried souls to the underworld, the Charon phages carry foreign DNA into bacterial cells. Charon the boatman is pronounced "Karen," but Charon the phage is usually pronounced "Sharon."

One clear advantage of the Charon phages over plasmid vectors is that they can accommodate much more foreign DNA. For example, Charon 4 can accept up to about 20 kb of DNA, a limit imposed by the capacity of the λ phage head. When would we need such high capacity? The most common use for Charon phage vectors is in constructing **genomic libraries.** Suppose we wanted to clone the entire human genome. This would obviously require a great many clones, but the larger the insert in each clone, the fewer total clones would be needed. In fact, such genomic libraries have been constructed for humans and for a variety of other organisms, and the Charon phages have been popular vectors for this purpose.

Aside from their high capacity, some of the Charon phages have the advantage of minimum size requirement for their inserts. Figure 15.6 illustrates the reason for this requirement: To get the Charon 4 vector ready to accept an insert, we cut it with *Eco*RI. This cuts at two sites near the middle of the phage DNA, yielding two "arms" and two "stuffer" fragments. Next, we purify the arms by ultracentrifugation and throw away the stuffers. The final step is to ligate the arms to our insert, which then takes the place of the discarded stuffers.

At first glance, it may appear that the two arms could simply ligate together without accepting an insert. Indeed, this may happen, but it will not produce a clone, because the two arms constitute too little DNA and will not be packaged into a phage head. The packaging is done in vitro; we simply mix the ligated arms plus inserts with all the components needed to put together a phage particle. This extract has rather stringent requirements as to the size of DNA it will package. It must have at least 12 kb of DNA in addition to Charon 4 arms, but no more than 20 kb, or the phage head will overflow.

Since we can be sure that each clone in our genomic library has at least 12 kb, we know we are not wasting space in our library with clones that contain insignificant amounts

FIGURE 15.6
Cloning in Charon 4. (*a*) Forming the recombinant DNA. We cut the vector (yellow) with *Eco*RI to remove the stuffer fragments, and save the arms. Next, we ligate partially digested insert DNA (blue) to the arms. (*b*) Packaging and cloning the recombinant DNA. We mix the recombinant DNA from (*a*) with an in vitro packaging extract that contains λ phage head and tail components and all other factors needed to package the recombinant DNA into functional phage particles. Finally, we plate these particles on *E. coli* and collect the plaques that form.

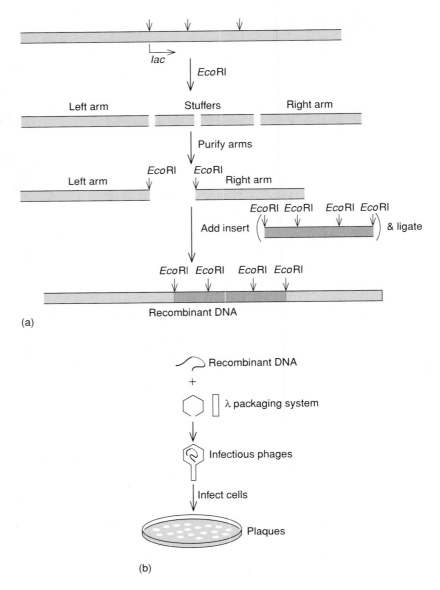

of DNA. This is an important consideration since, even at 12–20 kb per clone, we need about half a million clones to be sure of having each human gene represented at least once. It would be much more difficult to make a human genomic library in pBR322, since plasmids tend to ligate selectively to small inserts, and bacteria selectively reproduce small plasmids. Therefore, most of the clones would contain inserts of a few thousand, or even just a few hundred base pairs. Such a library would have to contain many millions of clones to be complete.

Since *Eco*RI produces fragments whose average size is about 4 kb and yet the vector will not accept any inserts smaller than 12 kb, it is obvious that we cannot cut our DNA completely with *Eco*RI, or most of the fragments will be too small to clone. Furthermore, *Eco*RI, or any other restriction enzyme, cuts in the middle of most eukaryotic genes one or more times, so a complete digest would contain only fragments of most genes. We can avoid these problems by performing an incomplete digestion with

*Eco*RI; if the enzyme cuts only about every fourth or fifth site, the average length of the resulting fragments will be about 16–20 kb, just the size the vector will accept and big enough to include most eukaryotic genes, introns and all.

Another vector designed especially for cloning large DNA fragments is called a **cosmid.** Cosmids resist classification because they behave both as plasmids and as phages. They contain the cos sites, or cohesive ends, of λ phage DNA, which allow the DNA to be packaged into λ phage heads (hence the "cos" part of the name cosmid). They also contain a plasmid origin of replication, so they can replicate as plasmids in bacteria (hence the "mid" part of the name).

Because almost the entire λ genome, except for the cos sites, has been removed from the cosmids, they have room for very large inserts (40–50 kb). Once these inserts are in place, the recombinant cosmids are packaged into phage particles. These particles cannot replicate as phages because they have almost no phage DNA, but they are

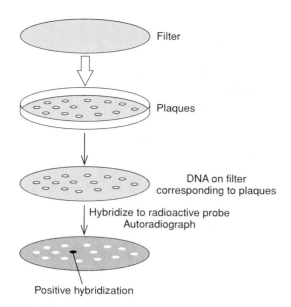

Filter

Plaques

DNA on filter
corresponding to plaques

Hybridize to radioactive probe
Autoradiograph

Positive hybridization

FIGURE 15.7

Selection of positive genomic clones by plaque hybridization. First, we touch a nitrocellulose or similar filter to the surface of the dish containing the Charon 4 plaques from figure 15.6. Phage DNA released naturally from each plaque will stick to the filter. Next, we denature the DNA with alkali and hybridize the filter to a radioactive probe for the gene we are studying, then autoradiograph to reveal the position of any radioactivity. Cloned DNA from one plaque near the center of the filter has hybridized, as shown by the dark spot on the autoradiograph.

infectious, so they carry their recombinant DNA into bacterial cells. Once inside, the DNA replicates as a plasmid, using its plasmid origin of replication.

A genomic library is very handy. Once it is established, we can search it for any gene we want. The only problem is that there is no catalog for such a library, so we need some kind of probe to tell us which clone contains the gene of interest. An ideal probe would be a labeled nucleic acid whose sequence matches that of the gene we are trying to find. We would then carry out a **plaque hybridization** (or **colony hybridization**) procedure in which the DNA from each of the thousands of λ phage clones (or cosmid clones) from our library is hybridized to the labeled probe. The DNA that forms a labeled hybrid is the right one.

We have encountered hybridization before in several chapters, and we will discuss it again later in this chapter. Figure 15.7 shows how plaque hybridization works. We grow thousands of plaques on each of several petri dishes (only a few plaques are shown here for simplicity). Next, we touch a filter made of a DNA-binding material such as **nitrocellulose** to the surface of the petri dish. This transfers phage DNA from each plaque to the filter. The DNA is then denatured with alkali and hybridized to the radioactive probe. When the probe encounters complementary DNA,

which should only be the DNA from the clone of interest, it will hybridize, making that DNA spot radioactive. This radioactive spot is then detected by autoradiography. The black spot on the autoradiograph shows us where to look on the original petri dish for the plaque containing our gene. In practice, the original plate may have been so crowded with plaques that it is impossible to pick out the right one, so we pick several plaques from that area, replate at a much lower phage density, and rehybridize to find the positive clone.

M13 phage vectors

Another phage frequently used as a cloning vector is the filamentous phage M13. Joachim Messing and his coworkers endowed the phage DNA with the same β-galactosidase gene fragment and multiple cloning sites found in the pUC family of vectors. In fact, the M13 vectors were engineered first; then the useful cloning sites were simply transferred to the pUC plasmids.

What is the advantage of the M13 vectors? The main factor is that the genome of this phage is a single-stranded DNA, so DNA fragments cloned into this vector can be recovered in single-stranded form. As we will see later in this chapter, single-stranded DNA is an invaluable aid to site-directed mutagenesis, by which we can introduce specific, premeditated alterations into a gene.

Figure 15.8 illustrates how we can clone a double-stranded piece of DNA into M13 and harvest a single-stranded DNA product. The DNA in the phage particle itself is single-stranded, but after infecting an *E. coli* cell, it is converted to a double-stranded replicative form (RF). This double-stranded replicative form of the phage DNA is what we use for cloning. After it is cut by one or two restriction enzymes at its multiple cloning site, foreign DNA with compatible ends can be inserted. This recombinant DNA is then used to transform host cells, giving rise to progeny phages that bear single-stranded recombinant DNA. The phage particles are secreted from the transformed cells and can be collected from the growth medium.

Two kinds of phages have been especially popular as cloning vectors. The first of these is λ, from which certain nonessential genes have been removed to make room for inserts. In some of these engineered phages, called Charon phages, inserts up to 20 kb can be accommodated. This makes the Charon phages useful for building genomic libraries, in which it is important to have large pieces of genomic DNA in each clone. Cosmids can accept even larger inserts—up to 50 kb—making them a favorite choice for genomic libraries. The second major class of phage vectors is composed of the M13 phages. These vectors have the convenience of a multiple cloning site and the further advantage of producing single-stranded recombinant DNA, which can be used for DNA sequencing and for site-directed mutagenesis.

IDENTIFYING A SPECIFIC CLONE WITH A SPECIFIC PROBE

We have already mentioned the need for a probe to identify the clone we want among the thousands we do not want. What sort of probe could we employ? Two different kinds are widely used: polynucleotides and antibodies. Both are molecules able to bind very specifically to other molecules. We will discuss oligonucleotide probes here and antibody probes later in this chapter.

Polynucleotide Probes

If the gene we are trying to locate is a very active one, appropriate cells may make its corresponding mRNA in large enough quantities to purify. In principle, we could use this mRNA itself as a probe, but it is more practical to make a cDNA copy by reverse transcription as described later in this chapter. Once we have made the cDNA, we can clone it or use it directly as a probe.

More often than not, the gene we want is not active enough to allow this direct approach. In that case, we might use the homologous gene from another organism if someone has already managed to clone it. For example, if we were after the human insulin gene and another research group had already cloned the rat insulin gene, we could ask them for their clone to use as a probe. We would hope the two genes have enough similarity in sequence that the rat probe could hybridize to the human gene. This hope is usually fulfilled. However, we generally have to lower the **stringency** of the hybridization conditions so that the hybridization reaction can tolerate some mismatches in base sequence between the probe and the cloned gene.

Researchers use several means to control stringency. High temperature, high organic solvent concentration, and low salt concentration all tend to promote the separation of the strands of a DNA double helix. We can therefore adjust these conditions until only perfectly matched DNA strands will form a duplex; this is high stringency. By relaxing these conditions (lowering the temperature, for example), we lower the stringency until DNA strands with a few mismatches can hybridize.

Without cDNA or homologous DNA from another organism, what could we use? There is still a way out if we know at least part of the sequence of the protein product of the gene. We faced a problem just like this in our lab when we cloned the gene for a plant toxin known as ricin. Fortunately, the entire amino acid sequences of both polypeptides of ricin were known. That meant we could examine the amino acid sequence and, using the genetic code, deduce the nucleotide sequence that would code for these amino acids. Then we could construct the nucleotide sequence chemically and use this synthetic probe to find the ricin gene. This sounds easy, but there is a hitch. The genetic code is degenerate, so for most amino acids we would have to consider several different nucleotide sequences.

Fortunately, we were spared much inconvenience because one of the polypeptides of ricin includes this amino acid sequence: Trp-Met-Phe-Lys-Asn-Glu. The first two

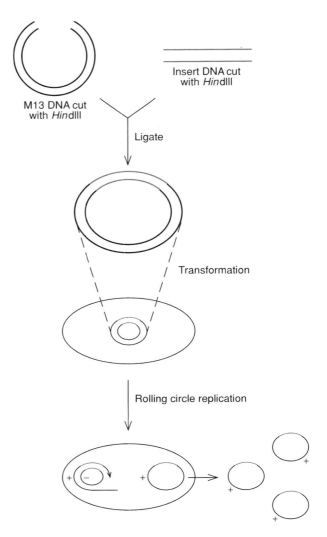

FIGURE 15.8

Obtaining single-stranded DNA by cloning in M13 phage. Foreign DNA (red), cut with *Hin*dIII, is inserted into the *Hin*dIII site of the double-stranded phage DNA. The resulting recombinant DNA is used to transform *E. coli* cells, whereupon the DNA replicates by a rolling circle mechanism, producing many single-stranded product DNAs. The product DNAs are called positive (+) strands, by convention. The template DNA is therefore the negative (−) strand.

amino acids in this sequence have only one codon each, and the next three only two each. The sixth gives us two free bases because the degeneracy occurs only in the third base. Thus, we had to make only eight 17-base oligonucleotides (*17-mers*) to be sure of getting the exact coding sequence for this string of amino acids. This degenerate sequence can be expressed as follows:

```
              U     G    U
UGG  AUG  UUC  AAA  AAC  GA
Trp  Met  Phe  Lys  Asn  Glu
```

Using this mixture of eight 17-mers (UGGAUGUU-CAAAAACGA, UGGAUGUUUAAAAACGA, etc.), we quickly identified several ricin-specific clones.

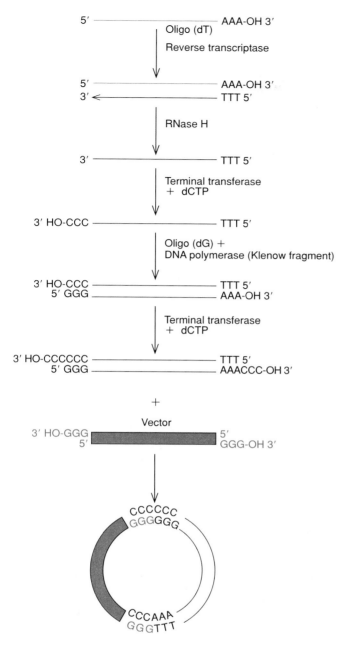

FIGURE 15.9

cDNA cloning. We start with a eukaryotic messenger RNA (blue) having poly(A) at its 3′-end. Oligo(dT) hybridizes to the poly(A) and primes reverse transcription, forming the first cDNA strand. We remove the mRNA template, using RNase H, then add an oligo(dC) tail to the 3′-end of the cDNA, using terminal transferase. Oligo(dG) hybridizes to the oligo(dC) and primes second-strand cDNA synthesis by the Klenow fragment of DNA polymerase I. In order to give the double-stranded cDNA sticky ends, we add oligo(dC) with terminal transferase, then anneal these ends to complementary oligo(dG) ends of a suitable vector (red). The recombinant DNA can then be used to transform bacterial cells.

 Specific clones can be identified using polynucleotide probes that bind to the gene itself. Knowing the amino acid sequence of a gene product, one can design an oligonucleotide—a short polynucleotide that encodes part of this amino acid sequence. This can be one of the quickest and most accurate means of identifying a particular clone.

cDNA CLONING

A **cDNA** is a DNA copy of an mRNA. Thus the "c" in cDNA stands for "copy." Molecular geneticists use a variety of techniques to clone cDNAs; here we will consider a fairly simple, yet effective strategy, which is illustrated in figure 15.9. The central part of any cDNA cloning procedure is synthesis of the cDNA from an mRNA template using **reverse transcriptase.** This reverse transcriptase is like any other DNA-synthesizing enzyme in that it cannot initiate DNA synthesis without a primer; give it only mRNA and we get nothing in return. To get around this problem, we take advantage of the poly(A) tail at the end of most mRNAs and use oligo(dT) as the primer. The oligo(dT) is complementary to poly(A), so it binds to the poly(A) at the 3′-end of the mRNA and primes DNA synthesis, using the mRNA as the template.

After the mRNA has been copied, yielding a single-stranded DNA (the "first strand"), we remove the mRNA with alkali or **ribonuclease H** (RNase H). This enzyme degrades the RNA part of an RNA/DNA hybrid—just what we need to remove the RNA from our first strand cDNA. Next, we must make a second DNA strand, using the first as the template. Again, we need a primer, and this time we do not have a convenient poly(A) to which to hybridize the primer. Instead, we build an oligo(dC) tail at the 3′-end of the first strand, using the enzyme **terminal transferase** and the substrate dCTP. The enzyme adds dCs, one at a time, to the 3′-end of the first strand. To this tail, we hybridize a short oligo(dG), which primes second strand synthesis. We can use reverse transcriptase again to make the second strand, but DNA polymerase works better. Actually, the most successful enzyme is a fragment of DNA polymerase called the **Klenow fragment.** This piece of enzyme is generated by cleaving *E. coli* DNA polymerase I with a proteolytic enzyme. The Klenow fragment contains the DNA polymerase activity and the 3′ → 5′ exonuclease activity, (the proofreading activity), but it lacks the 5′ → 3′ exonuclease activity normally associated with DNA polymerase I. The latter activity is undesirable because it degrades DNA from the 5′-end, which is damage that DNA polymerase cannot repair.

Once we have a double-stranded cDNA, we must ligate it to a vector. This was easy with our pieces of genomic DNA, since they had sticky ends, but the cDNA has no sticky ends. This problem is easily solved. We simply tack sticky ends (oligo[dC]) onto the cDNA, again using terminal transferase and dCTP. In the same way, we attach oligo(dG) ends to our vector and allow the oligo(dC)s to anneal to the oligo(dG)s. This brings the vector and cDNA together in a recombinant DNA that can be used directly for transformation. The base pairing between the oligonucleotide tails is strong enough that no ligation is required before transformation. The DNA ligase inside the transformed cells finally performs this task.

What kind of vector should we use? Several choices are available, depending on the way we wish to detect positive clones (those that bear the cDNA we want). We can use a simple vector such as one of the pUC plasmids; if we do, we usually identify positive clones by colony hybridization with a radioactive DNA probe. This procedure is analogous to the plaque hybridization described previously. Or we can use a λ phage, such as λgt11, as a vector. This vector places the cloned cDNA under the control of the *lac* promoter, so that transcription and translation of the cloned gene can occur. We can then use an antibody to screen directly for the protein product of the correct gene. We will describe this procedure in more detail later in this chapter.

 A cDNA can be synthesized one strand at a time, using mRNA as template for the first DNA strand and this first strand as template for the second. The double-stranded cDNA can be endowed with oligonucleotide tails that basepair with complementary tails on a cloning vector. This recombinant DNA can then be used to transform bacteria. Positive clones can be detected by colony hybridization with radioactive DNA probes, or with antibodies if an expression vector such as λgt11 is used.

METHODS OF EXPRESSING CLONED GENES

Why would we want to clone a gene? An obvious reason, suggested at the beginning of this chapter, is that cloning allows us to produce large quantities of pure eukaryotic (or prokaryotic) genes so that we can study them in detail. Several specific studies using cloned genes were described in chapter 14. For example: Cloned genes gave us labeled probes to localize the expression of homeobox genes in fruit flies. Cloned genes were manipulated to discover the important control regions of the pea rubisco gene and the frog 5S rRNA genes. Thus, the gene itself can be a valuable product of gene cloning. Another goal of gene cloning is to make a large quantity of the gene's product, either for investigative purposes or for profit.

EXPRESSION VECTORS

The vectors we have examined so far are meant to be used primarily in the first stage of cloning—when we first put a foreign DNA into a bacterium and get it to replicate. By

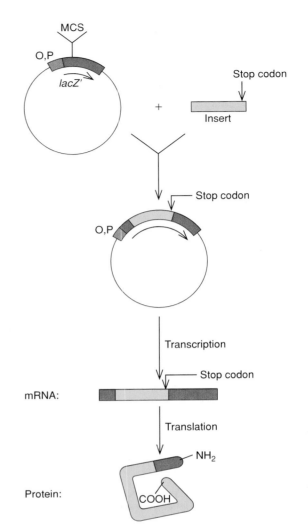

FIGURE 15.10
Formation of a fusion protein by cloning in a pUC plasmid. We insert foreign DNA (yellow) into the multiple cloning site (MCS); transcription from the *lac* promoter (purple) gives a hybrid mRNA beginning with a few *lacZ* codons, changing to insert sequence, then back to *lacZ* (red). This mRNA will be translated to a fusion protein containing a few β-galactosidase amino acids at the beginning (amino end), followed by the insert amino acids for the remainder of the protein. Since the insert contains a translation stop codon, the remaining *lacZ* codons will not be translated.

and large, they work well for that purpose, growing readily in *E. coli* and producing high yields of recombinant DNA. Some of them even work as **expression vectors** that can yield the protein products of the cloned genes. For example, the pUC vectors place inserted DNA under the control of the *lac* promoter, which lies upstream from the multiple cloning site. If an inserted DNA happens to be in the same reading frame as the *lac* gene it interrupts, a **fusion protein** will result. It will have a partial β-galactosidase protein sequence at its amino end and another protein sequence, encoded in the inserted DNA, at its carboxyl end (fig. 15.10).

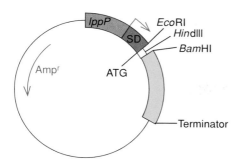

FIGURE 15.11

The pIN-I-A1 expression vector. The coding region of a gene to be expressed is inserted into one of the three unique restriction sites (*Eco*RI, *Hin*dIII, or *Bam*HI). Transcription starts at the arrow and is directed by the strong lipoprotein promoter (*lppP*) and terminated by the transcription terminator. Translation, abetted by the Shine-Dalgarno ribosome binding site (SD), begins at the ATG just upstream from the inserted DNA, so the ultimate product will be a fusion protein containing just a few extra amino acids at the amino end. The Amp^r gene confers resistance to ampicillin for selection purposes.

However, if we are interested in high expression of our cloned gene, specialized expression vectors usually work better. These typically have two elements that are required for active gene expression: a strong promoter and a ribosome binding site that includes a Shine-Dalgarno sequence near an initiating ATG codon.

Expression Vectors with Strong Promoters

The main function of an expression vector is to yield the product of a gene—usually, the more product the better. Therefore, expression vectors are ordinarily equipped with very strong promoters; the rationale is that the more mRNA is produced, the more protein product will be made.

One of the strongest of all *E. coli* promoters is the *lpp* promoter, which directs transcription of the lipoprotein gene. A consequence of the strength of this promoter is that the product of the gene, the major membrane lipoprotein, is one of the most abundant proteins in the bacterial cell. Thus, this promoter is a natural for an expression vector, and Masayori Inouye and his colleagues have constructed a set of such vectors, called the *pIN* plasmids. Figure 15.11 shows the architecture of a typical pIN vector. Just behind the *lpp* promoter lie three cloning sites—restriction sites for the enzymes *Eco*RI, *Bam*HI, and *Hin*dIII. If we ligate a gene's coding region into any of these three sites, being careful to preserve the reading frame of the *lpp* coding region, we can get a protein with just a few extra amino acids at the amino end.

Another strong promoter, though not as strong as the one from the *lpp* gene, is the *trp* (tryptophan operon) promoter. It forms the basis for several expression vectors, including p*trp*L1. Figure 15.12 shows how this vector works. It has the *trp* promoter/operator region, followed by a Shine-Dalgarno sequence, and can be used directly as an

expression vector by inserting a foreign gene into the *Cla*I site. Alternatively, the *trp* control region can be made "portable" by cutting it out with *Cla*I and *Hin*dIII and inserting it in front of a gene to be expressed in another vector.

Inducible Expression Vectors

It is usually advantageous to keep a cloned gene turned off until we are ready to express it. One reason is that eukaryotic proteins produced in large quantities in bacteria can be toxic. In such a case, if the cloned gene were allowed to remain turned on constantly, the bacteria bearing the gene would never grow to a great enough concentration to produce meaningful quantities of protein product. The solution is to keep the cloned gene turned off by placing it behind an inducible promoter.

The *lac* promoter is inducible to a certain extent, presumably remaining off until stimulated by the inducer allolactose or by its synthetic analog IPTG. However, the repression wrought by the *lac* repressor is incomplete, and some expression of the cloned gene will be observed even in the absence of inducer. One way around this problem is to express our gene in a host cell that has an overactive *lac* repressor gene, I^q. The excess repressor produced by such a cell keeps our cloned gene turned off until we are ready to induce it.

Another strategy is to use a very tightly controlled promoter such as the λ phage promoter P_L. Expression vectors with this promoter/operator system are cloned into host cells bearing a temperature-sensitive λ repressor gene (*cI*857). As long as the temperature of these cells is kept relatively low (32° C), the repressor functions, and no expression takes place. However, when we raise the temperature to the nonpermissive level (42° C), the temperature-sensitive repressor can no longer function and the cloned gene is induced. Figure 15.13 illustrates this mechanism for the expression vector pKC30.

Expression vectors are designed to yield the protein product of a cloned gene, usually in the greatest amount possible. To optimize expression, these vectors provide strong bacterial promoters and bacterial ribosome binding sites that would be missing on cloned eukaryotic genes. Most cloning vectors are inducible, to avoid premature overproduction of a foreign product that could poison the bacterial host cells.

Expression Vectors that Produce Fusion Proteins

When most expression vectors operate, they produce fusion proteins. This might at first seem a disadvantage because the natural product of the inserted gene is not made. However, the extra amino acids at the amino terminus of the fusion protein can be useful. Consider the expression vector pβ-gal13C. It has the *lac* operator and promoter, and the

FIGURE 15.12

Two uses of the *ptrpL*1 expression vector. The vector contains a *Cla*I cloning site, preceded by a Shine-Dalgarno ribosome binding site (SD) and the *trp* operator/promoter region (*trpO,P*). Transcription occurs in a counterclockwise direction as shown by the arrow (top). The vector can be used as a traditional expression vector (left) simply by inserting a foreign coding region (X, green) into the unique *Cla*I site. Alternatively (right), the *trp* control region (purple) can be cut out with *Cla*I and *Hin*dIII and inserted into another plasmid bearing the coding region (Y, yellow) to be expressed.

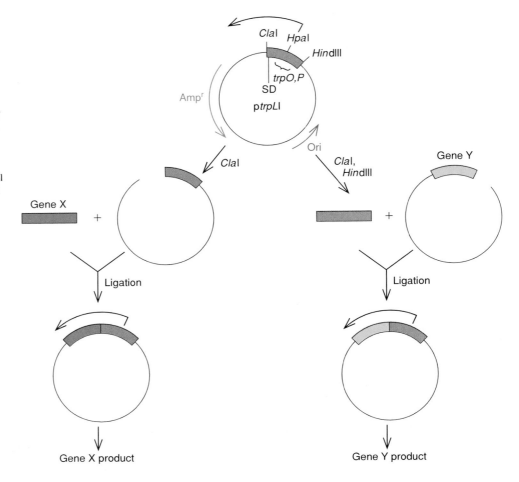

*Eco*RI cloning site lies near the end of the *lacZ* gene (fig. 15.14). This means the product will be a fusion protein with a large piece of β-galactosidase at its amino end.

How can fusion proteins be useful? One reason is that they are sometimes more stable in bacterial cells than normal eukaryotic proteins are. Furthermore, if we do not have a good way of purifying the protein product of our cloned gene, the long β-galactosidase addition can be a big help. For example, at a molecular weight of 116,000, the β-galactosidase monomer is already among the largest polypeptides in an *E. coli* cell and is correspondingly easy to purify by techniques that separate molecules by size. A fusion protein, with another polypeptide tacked onto the carboxyl terminus, will be even larger, and it is usually a simple matter to identify and purify such a protein.

λ phages have also served as the basis for expression vectors. One of these, λgt11 (fig. 15.15), contains the *lac* control region followed by the *lacZ* gene. The cloning sites are located within the *lacZ* gene, so products of a gene inserted into this vector will be fusion proteins with a leader of β-galactosidase.

The expression vector λgt11 has become a popular vehicle for making and screening cDNA libraries. In the examples of screening presented earlier, we looked for the proper DNA sequence by probing with a labeled oligonucleotide or polynucleotide. By contrast, λgt11 allows us to screen a group of clones directly for the expression of the right protein. The main ingredients required for this procedure are a cDNA library in λgt11 and an antiserum directed against the protein of interest.

Figure 15.16 shows how this works. We plate our λ phages with various cDNA inserts and blot the proteins released by each clone onto a support such as nitrocellulose. The nice thing about a λ clone in this regard is that the host cells lyse, forming plaques. In so doing, they release their products, making it easy to transfer proteins from thousands of clones simultaneously, simply by touching a nitrocellulose filter to the surface of a petri dish containing the plaques.

Once we have transferred the proteins from each plaque to nitrocellulose, we probe with our antiserum. Next, we probe for antibody bound to protein from a particular plaque, using radioactive protein A from *Staphylococcus*

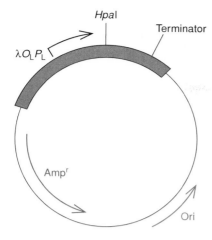

FIGURE 15.13

The inducible expression vector pKC30. A gene to be expressed is inserted into the unique *Hpa*I site, downstream from the $\lambda\,O_L P_L$ operator/promoter region. The host cell is a λ lysogen bearing a temperature-sensitive λ repressor gene (*cI*857). To induce expression of the cloned gene, we raise the temperature from 32° to 42° C, which inactivates the temperature-sensitive λ repressor, removing it from O_L and allowing transcription to occur.

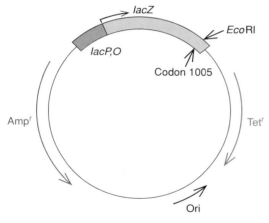

FIGURE 15.14

The pβ-gal13C expression vector. The unique *Eco*RI cloning site lies near the end of the *lacZ* coding region, just after codon number 1005. Therefore, products of genes cloned into this site will contain a 1,005-amino acid tag of β-galactosidase at their amino ends.

aureus. This protein binds tightly to antibody and makes the corresponding spot on the nitrocellulose radioactive. We detect this radioactivity by autoradiography, then go back to our master plate and pick the corresponding plaque. Note that we are detecting a fusion protein, not the cloned protein by itself. Furthermore, it does not matter if we have cloned a whole cDNA or not. Our antiserum is a mixture of antibodies that will react with several different parts of our protein, so even a partial gene will do, as long as its coding region is cloned in the same orientation and reading frame as the leading β-galactosidase coding region.

FIGURE 15.15

Forming a fusion protein in λgt11. The gene to be expressed (green) is inserted into the *Eco*RI site near the end of the *lacZ* coding region (red) just before the transcription terminator. Thus, upon induction of the *lacZ* gene by IPTG, a fused mRNA results, containing the inserted coding region just downstream from that of β-galactosidase. This mRNA is translated by the host cell to produce a fusion protein.

> Expression vectors frequently produce fusion proteins, with the first part of the protein coming from coding sequences in the vector and the latter part from sequences in the cloned gene itself. Fusion proteins have certain advantages: They may be more stable in prokaryotic cells than unmodified eukaryotic proteins, and they are much simpler to isolate.

Eukaryotic Expression Systems

Eukaryotic genes are not really "at home" in prokaryotic cells, even when they are expressed under the control of their prokaryotic vectors. One reason is that *E. coli* cells frequently recognize the protein products of cloned eukaryotic genes as outsiders and destroy them. Another is that prokaryotes do not carry out the same kinds of post-translational modifications as do eukaryotes. For example, a protein that would ordinarily be coupled to sugars in a eukaryotic cell will be expressed as a bare protein when cloned in bacteria. This can affect a protein's activity or stability, or at least its response to antibodies. A more serious problem is that eukaryotic proteins frequently do not fold into their native shapes in bacterial cells. The result can be improperly folded, inactive products of cloned genes. This phenomenon explains the fact that we can sometimes express a cloned gene at a stupendously high level in bacteria, but the product forms highly insoluble, inactive granules that are useless unless we can somehow get the protein to dissolve and regain its activity by folding properly. Finally, it is obviously hopeless to try to express genes with introns in prokaryotes, since only eukaryotes have the machinery to splice introns out.

In order to avoid the incompatibility between a cloned gene and its host, we can express our gene in a eukaryotic cell. In such cases, we usually do the initial cloning in *E. coli*, using a **shuttle vector** that can replicate in both bacterial and eukaryotic cells. We then transfer the recombinant DNA to the eukaryote of choice by transformation.

One class of eukaryotic vectors that has been remarkably successful is derived from the baculovirus that infects the caterpillar known as the alfalfa looper. Viruses in this class have a rather large circular DNA genome, approximately 130 kb in length. The major viral structural protein, polyhedrin, is made in copious quantities in infected cells. In fact, it has been estimated that when a caterpillar dies of a baculovirus infection, up to 10% of the dry mass of the dead insect is this one protein. This indicates that the polyhedrin gene must be very active, and indeed it is—apparently due to its powerful promoter. The baculovirus vectors ultimately place cloned genes under the control of this promoter, so they can be expressed abundantly. This system has produced yields of over 100 mg of protein per liter of insect cell culture.

 Eukaryotic expression systems have the advantage that they produce eukaryotic proteins that are usually folded and modified properly. Baculovirus systems allow us to express foreign genes in insect cells to produce high yields of protein.

MANIPULATING CLONED GENES

Besides simply obtaining the products of cloned genes, we can put them to many uses, as shown in the following three examples: (1) We do not have to be satisfied with the natural product of a cloned gene; once the gene is cloned, we can change it any way we want and collect the correspondingly changed gene product. (2) We can make cloned genes radioactive and use them as probes for many purposes. For example, we can find out how many similar genes exist in an organism's genome or how actively the gene is expressed in living cells. (3) We can determine the exact base sequence of a cloned gene.

PROTEIN ENGINEERING WITH CLONED GENES

Traditionally, protein biochemists have relied on chemical methods to alter certain amino acids in the proteins they study, so that they can then observe the effects of these

FIGURE 15.16

Detecting positive λgt11 clones by measuring expression. A filter is used to blot proteins from phage plaques on a petri dish. One of the clones (red) has produced a plaque containing a fusion protein including β-galactosidase and a part of the protein we are interested in. The filter with its blotted proteins is incubated with an antibody directed against our protein of interest, then with radioactive *Staphylococcus* protein A, which binds specifically to antibodies. It will therefore bind only to the antibody-antigen complexes at the spot corresponding to our positive clone. A dark spot on the autoradiograph of the filter reveals the location of our positive clone.

changes on protein activities. But chemicals are rather crude tools for manipulating proteins; it is difficult to be sure that only one amino acid, or even one kind of amino acid, has been altered. Cloned genes make this sort of investigation much more precise, allowing us to perform microsurgery on a protein. By replacing specific bases in a gene, we also replace amino acids at selected spots in the protein product and observe the effects of those changes on that protein's function.

How do we perform such **site-directed mutagenesis?** First, we need a cloned gene whose base sequence is known. (We will see later in this chapter how DNA sequencing is done; for the moment, let us assume that we have already determined the sequence of the gene we want to manipulate.) Then, we need to obtain our gene in single-stranded form. This can be done by cloning it into M13 phage and collecting the single-stranded progeny phage DNA (plus strands) as previously described in figure 15.8.

Our next task is to change a single codon in this gene. Let us suppose the gene contains the sequence of bases given in figure 15.17, which codes for a sequence of amino acids that includes a tyrosine. The amino acid tyrosine contains a phenolic group:

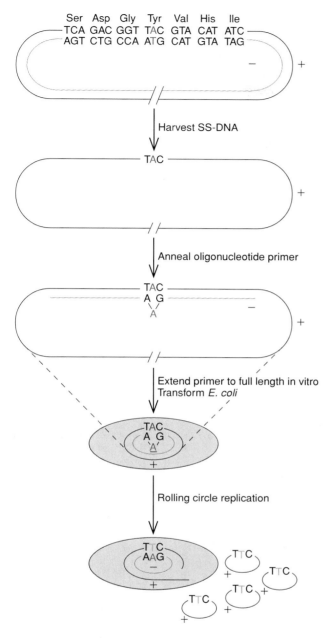

FIGURE 15.17
Site-directed mutagenesis using an oligonucleotide. We begin with the gene we want to mutagenize, cloned into double-stranded phage M13 DNA. Our goal is to change the tyrosine codon TAC (blue) to the phenylalanine codon TTC. We transform *E. coli* cells with the recombinant phage DNA and harvest single-stranded (plus strand) progeny phage DNA. To this plus strand, we anneal a synthetic primer that is complementary to the region of interest, except for a single base change, an A (red) for a T. This creates an A–A mismatch (blue and red). We extend the primer to form a full-length minus strand in vitro and use this double-stranded DNA to transform *E. coli* cells. The minus strand with the single base change then serves as the template for making many copies of the plus strand, each bearing a TTC phenylalanine codon (pink) in place of the original TAC tyrosine codon.

To investigate the importance of this phenolic group, we can change the tyrosine codon to a phenylalanine codon. Phenylalanine is just like tyrosine except that it lacks the phenolic group; instead, it has a simple phenyl group:

If the tyrosine phenolic group is important to a protein's activity, replacing it with phenylalanine's phenyl group should diminish that activity.

We want to change the DNA codon TAC (Tyr) to TTC (Phe). The simplest way is to use an automated DNA synthesizer to make an oligonucleotide (a "21-mer") with the following sequence:

3′-AGTCTGCCAAAGCATGTATAG-5′

This has the same sequence as a piece of the original minus strand, except that the central triplet has been changed from ATG to AAG. Thus, this oligonucleotide will hybridize to the plus strand we harvested from the M13 phage, except for the one base we changed, which will cause an A–A mismatch. The oligonucleotide in this hybrid can serve as a primer for DNA polymerase to complete the negative strand in vitro; then we can transform *E. coli* with the resulting double-stranded replicative form (RF), which still contains the one-base mismatch.

Once inside the host bacterium, the RF replicates by a rolling circle mechanism, producing many copies of the positive strand and using the mutated negative strand as template. The first positive strand released will be the original, with its TAC tyrosine codon. However, subsequent positive strands will be made using the altered negative strand as template, so they will contain a TTC phenylalanine codon instead. These mutated positive strands will be packaged into phage particles (not shown in fig. 15.17), which can be used to infect more bacterial cells. In the final step of the process, we can collect the mutant RF molecules produced by these cells, cut out the mutant gene with restriction enzymes, and clone it into an appropriate expression vector. The protein product will be identical to the wild-type product except for the single change of a tyrosine to a phenylalanine.

Note the precision of this technique compared to traditional chemical mutagenesis. With chemical mutagenesis, we assault organisms with an often fatal dose of mutagen, then examine progeny organisms for the desired mutant characteristics. Before DNA sequencing was possible, we could not be certain of the molecular nature of such a mutation without laborious protein sequencing. Even with modern DNA sequencing techniques, we would have to determine the sequence of an entire mutant gene before we could be sure what specific damage a chemical had caused.

By contrast, site-directed mutagenesis enables us to decide in advance what parts of a protein we want to change and to tailor our mutants accordingly. Nevertheless, in spite of its lack of precision, traditional mutagenesis still plays an important role; one such experiment can quickly and

FIGURE 15.18
Southern blotting. First, we electrophorese DNA fragments on an agarose gel. Next, we denature the DNA with base, and transfer the single-stranded DNA fragments from the gel (yellow) to a sheet of nitrocellulose (red) or similar material. This can be done in two ways: by diffusion, in which buffer passes through the gel, carrying the DNA with it (left), or by electrophoresis (right). The blot can then be hybridized to a radioactive probe and autoradiographed to visualize bands corresponding to the probe.

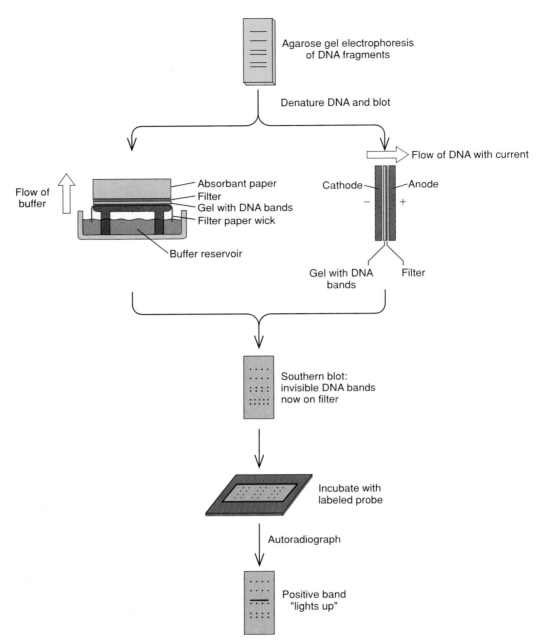

easily create a rich variety of mutants, some of which involve changes we might never think of—or never get around to making by site-directed mutagenesis.

Using cloned genes, we can introduce changes at will, thus altering the amino acid sequences of the protein products. This is conveniently done with single-stranded cloned DNA from a bacteriophage vector and synthetic oligonucleotide primers containing the desired base change.

USING CLONED GENES AS PROBES

The phenomenon of hybridization—the ability of one single-stranded nucleic acid to form a double helix with another single strand of complementary base sequence—

is one of the backbones of modern molecular genetics. So pervasive are techniques using hybridization that it would be difficult to overestimate its importance. We encountered plaque and colony hybridization earlier in this chapter; here we will illustrate two further examples of hybridization techniques.

Using DNA Probes to Count Genes

In chapter 14 we learned that many eukaryotic genes are not found just once per genome, but are parts of families of closely related genes. Suppose we want to know how many genes of a certain type exist in a given organism. If we have cloned a member of that gene family—even a partial cDNA—we can estimate this number.

We begin by using a restriction enzyme to cut genomic DNA that we have isolated from the organism (fig. 15.18).

It is best to use a restriction enzyme such as *Eco*RI or *Hind*III that recognizes a 6-base-pair cutting site. These enzymes will produce thousands of fragments of genomic DNA, with an average size of about 4,000 base pairs. Next, we electrophorese these fragments on an agarose gel. The result, if we visualize the bands by staining, will be a blurred streak of thousands of bands, none distinguishable from the others. Eventually, we will want to hybridize a radioactive probe to these bands to see how many of them contain coding sequences for the gene of interest. First, however, we must transfer the bands to a medium on which we can conveniently perform our hybridization.

E. Southern was the pioneer of this technique; he transferred, or blotted, DNA fragments from an agarose gel to nitrocellulose by diffusion as depicted in figure 15.18. This process has been called **Southern blotting** ever since. Nowadays, blotting is frequently done by electrophoresing the DNA bands out of the gel and onto the blot. Figure 15.18 also illustrates this process. Before blotting, the DNA fragments are denatured with base so that the resulting single-stranded DNA can bind to the nitrocellulose, forming the Southern blot. Media superior to nitrocellulose are now available; some of them use nylon supports and have clever trade names like Gene Screen. Next, we apply DNA polymerase to our cloned gene in the presence of radioactive DNA precursors to make it radioactive. Then we hybridize this labeled probe to the Southern blot. Wherever the probe encounters a DNA fragment bearing the same sequence (actually the complementary sequence), it hybridizes, forming a band of radioactivity corresponding to the band of DNA containing our gene of interest. Finally, we visualize these bands by autoradiography.

If we find only one band, the interpretation is relatively easy; there is probably only one gene corresponding to our cDNA. If we find multiple bands, there are probably multiple genes, but it is difficult to tell exactly how many. One gene can give more than one band if it contains one or more cutting sites for the restriction enzyme we used. We can minimize this problem by using a short probe, such as a 100–200-base-pair restriction fragment of the cDNA, for example. Chances are, a restriction enzyme that cuts on average only every 4,000 base pairs will not cut within the 100–200-base-pair regions of the genes that hybridize to such a probe (unless these regions contain introns, of course).

Using DNA Probes to "Fingerprint" DNA

In 1985, Alec Jeffreys and his colleagues were investigating a DNA fragment from the gene for a human blood protein, α-globin, when they discovered that this fragment contained a sequence of bases repeated several times. For reasons that are not important here, this repeated DNA is called a **minisatellite.** More interestingly, this same minisatellite sequence was found in other places in the human genome, again repeated several times. This simple finding

turned out to have far-reaching consequences, because individuals differ in the pattern of repeats of the basic sequence. In fact, they differ enough that two individuals have only a remote chance of having exactly the same pattern. That means that these patterns are like fingerprints; indeed, they are called **DNA fingerprints.**

How do we make a DNA fingerprint? We first cut the DNA under study with a restriction enzyme such as *Hae*III. Jeffreys chose this enzyme because the repeated sequence he had found did not contain a *Hae*III recognition site. That means that *Hae*III will cut on either side of the minisatellite regions, but not inside, as shown in figure 15.19*a*. In this case, the DNA has three sets of repeated regions, containing four, three, and two repeats, respectively. Thus, we will get three different-sized fragments bearing these repeated regions.

Next, we electrophorese the fragments we have produced. This separates all the fragments into bands according to their sizes, including the three bearing the minisatellites. Then we separate the two strands of the DNA fragments and blot the single-stranded DNA as just described. At this point, if we have used human DNA, we have millions of fragments, but none of them is visible on the blot; all we can see is a smear of DNA.

We can find the bands with the minisatellites, among millions of irrelevant bands, by hybridizing the blot to a radioactive single-stranded DNA containing the minisatellite sequence. This labeled probe DNA will form a radioactive double helix with any single-stranded fragments on the blot containing the complementary minisatellite sequence, making them radioactive. Finally, we can detect the radioactive bands by autoradiography. In this case, there are three radioactive bands, so we will see three dark bands on the film (fig. 15.19*d*).

Of course, real animals have a much more complex genome than the simple piece of DNA in this example, so they will have many more than three fragments that react with the probe. Figure 15.20 shows an example of the DNA fingerprints of several unrelated people and a set of monozygotic twins. As we have already mentioned, this is such a complex pattern of fragments that the patterns for two individuals are extremely unlikely to be identical, unless they come from monozygotic twins. This complexity makes DNA fingerprinting a very powerful identification technique (see box 15.1).

Using DNA Probes to Measure Gene Activity

Suppose we have cloned a cDNA and want to know how actively the corresponding gene (gene X) is transcribed in a number of different tissues of organism Y. We could answer that question in several ways, but the method we describe here will also tell us the size of the mRNA the gene produces.

We begin by collecting RNA from several tissues of the organism in question. Then we electrophorese these RNAs in an agarose gel and blot them to a suitable support.

FIGURE 15.19

DNA fingerprinting. (*a*) First, we cut DNA with a restriction enzyme. In this case, the enzyme *Hae*III cuts the DNA in seven places (arrows), generating eight fragments. Only three of these fragments (labeled A, B, and C according to size) contain the minisatellites, represented by blue boxes. The other fragments (yellow) contain unrelated DNA sequences. (*b*) We electrophorese the fragments from part (*a*), which separates them according to their sizes. All eight fragments are present in the electrophoresis gel, but we cannot see any of them. The positions of all fragments, including the three (A, B, and C) with minisatellites are indicated by dotted lines. (*c*) We denature the DNA fragments and Southern blot them. (*d*) We hybridize the DNA fragments on the Southern blot to a radioactive DNA with several copies of the minisatellite. This probe will bind to the three fragments containing the minisatellites, but with no others. Finally, we autoradiograph the blot, which reveals the three positive, radioactive bands.

FIGURE 15.20

A DNA fingerprint. (*a*) The nine parallel lanes contain DNA from nine unrelated Caucasians. Note that no two patterns are identical, especially at the upper end. (*b*) The two lanes contain DNA from monozygotic twins, so the patterns are identical.

From G. Vassart, et al. "A Sequence in M13 Phage Detects Hypervariable Minisatellites in Human and Animal DNA." Science, February 6, 1987, p. 683, fig. 1A,B. Copyright by the AAAS.

Box 15.1 Uses of DNA Fingerprinting

A valuable feature of DNA fingerprinting is the fact that, although almost all individuals have different patterns, parts of the pattern (sets of bands) are inherited in a Mendelian fashion. Thus, fingerprints can be used to establish parentage. A recent immigration case in England illustrates the power of this technique. A Ghanaian boy born in England had moved to Ghana to live with his father. When he wanted to return to England to be with his mother, British authorities questioned whether he was a son or a nephew of the woman. Information from blood group genes was inconclusive, but DNA fingerprinting of the boy demonstrated that he was indeed her son.

Scientists studying family relationships in other vertebrate species are finding DNA fingerprinting a very valuable tool. Alec Jeffreys' human probe works surprisingly well with a wide variety of vertebrate species, and probes for other repeated sequences have also been developed.

Aside from testing parentage, DNA fingerprinting has already become a valuable tool for identifying criminals. One dramatic case involved a man who murdered another man and a woman as they slept in a pickup truck, then about forty minutes later went back and raped the woman. This act not only compounded the crime, it also provided forensic scientists with the means to convict the perpetrator. They obtained DNA from the sperm he had left behind, fingerprinted it, and showed that the fingerprint matched that of the suspect's DNA. The chance that another person's DNA fingerprint would be identical was claimed to be about one in 9.34 billion, more than the entire population of the earth. Thus, this procedure can in principle eliminate all other suspects and therefore offer something very close to absolute identification. In this case, it helped convince the jury to convict the defendant and recommend severe punishment.

In a related DNA-identifying procedure, the probe hybridizes to specific variable number tandem repeat (VNTR) sequences, thus this method is called VNTR typing. These sequences resemble minisatellites in that they are composed of repeats of a core sequence and different individuals have different numbers of repeats. However, unlike minisatellites, there is usually only one VNTR of a given type in an individual. Thus, VNTR analysis gives patterns that are simpler and easier to analyze than DNA fingerprinting, but several different probes must be used to be convincing. Figure B15.1.1 presents an example of DNA typing that was used to identify a rape suspect. The pattern from the suspect clearly matches that from the sperm DNA.

One advantage of DNA typing of the kind described here is its extreme sensitivity. Only a few drops of blood or semen are sufficient to perform a test. However, sometimes forensic scientists have even less to go on—a hair pulled out by the victim, for example. While this is not enough for fingerprinting, it still contains a considerable amount of information, especially if accompanied by hair follicle cells. Selected segments of DNA from these cells can be amplified by a technique called *polymerase chain reaction* (PCR) invented by a team of scientists at Cetus Corporation led by Kary Mullis. As figure B15.1.2 explains, PCR uses the enzyme DNA polymerase to make a copy of a defined region of the DNA. We can select the part of the DNA we want to amplify by putting in short pieces of DNA called primers that hybridize to DNA sequences on either side of a given DNA region and cause initiation (priming) of DNA synthesis through it.

The copies of both strands of the given DNA region, as well as the original DNA strands, then serve as templates for the next round of amplification. In this way, the amount of the selected DNA region doubles over and over—up to at least a million times the starting amount—until there is enough to analyze. Of course, it helps to choose a variable DNA region, such as an HLA region, to amplify, since this increases the information we can

FIGURE B15.1.1
A VNTR analysis used to identify a rape suspect. Note that the banding pattern of the suspect's DNA is identical to that of the DNA in the rapist's semen, but clearly different from the pattern of the victim's DNA. This very simple pattern was produced by only one probe; to get a more precise identification, three or four probes would be used.
© *Lifecodes Corporation.*

Box 15.1 *(continued)*

obtain. Using a variety of probes on such amplified DNA may in the future allow us to say that a given pattern would only be found once in 10,000 to 100,000 people. This is not as good as the one in ten billion possible with DNA fingerprinting, but coupled with other evidence it could still be convincing.

In spite of its potential accuracy, DNA fingerprinting has recently been effectively challenged in court. One problem is that the technique is tricky and must be performed very carefully to give reliable results. In two celebrated cases, commercial forensic laboratories did not exercise enough care, and the DNA fingerprinting evidence had to be thrown out. The legal profession will have to establish standards for acceptable DNA fingerprint evidence. Recently, population geneticists, including Daniel Hartl and Richard Lewontin, have further argued that the DNA fingerprints of members

of a given ethnic group may resemble each other somewhat. To the extent that such similarities do exist, they reduce the confidence we can have in fingerprinting as an absolute means of

identification. Nevertheless, even with these qualifications, DNA fingerprinting and similar DNA typing methods are powerful tools for demonstrating guilt or innocence.

FIGURE B15.1.2

Amplifying DNA by polymerase chain reaction (PCR). Cycle 1: We start with a DNA duplex (top), and heat it to separate its two strands (red and blue). We then add short, single-stranded DNA primers (pink and blue) complementary to sequences on either side of the region we want to amplify. The primers hybridize to the appropriate sites on the separated DNA strands; now DNA polymerase can use these primers to start synthesis of complementary DNA strands. The arrows represent newly made DNA. At the end of cycle 1, we have two DNA duplexes including the region we want to amplify, while we started with only one. Cycle 2: We repeat the process, heating to separate DNA strands, adding primers (or simply relying on the primers we already added), and adding DNA polymerase (or relying on a heat-stable DNA polymerase added at the beginning). Now each of the four DNA strands, including the two newly made ones, can serve as templates for complementary DNA synthesis. The result is four DNA duplexes that have the region we want to amplify. Notice that each cycle doubles the number of molecules of DNA, because the products of each cycle join the parental molecules in serving as templates for the next cycle. This exponential increase leads to very high numbers in only a short time.

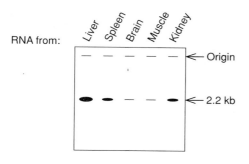

RNA from:

Liver Spleen Brain Muscle Kidney

— Origin

— 2.2 kb

FIGURE 15.21

A hypothetical Northern blot. mRNAs from vertebrate liver, spleen, brain, muscle, and kidney were electrophoresed and then blotted to nitrocellulose. The blot was hybridized to a radioactive probe for gene X, then autoradiographed. A 2.2-kb mRNA is observed in all tissues, but it is most abundant in the liver.

Since a similar blot of DNA is called a Southern blot, it was natural to name a blot of RNA a **Northern blot.** (By analogy, a blot of protein is called a Western blot.)

Next, we hybridize the Northern blot to a radioactive cDNA probe. Wherever an mRNA complementary to the probe exists on the blot, hybridization will occur, resulting in a radioactive band that we can detect by autoradiography. If we run marker RNAs of known size next to the unknown RNAs, we can tell the sizes of the RNA bands that "light up" when hybridized to the probe.

Furthermore, the autoradiograph tells us how actively gene X is transcribed. The more RNA the band contains, the more probe it will bind and the darker the band will be on the autoradiograph. We can quantify this darkness by measuring the amount of light it absorbs in an instrument called a densitometer. Figure 15.21 shows a hypothetical Northern blot of RNA from the liver, spleen, brain, muscle, and kidney of a vertebrate, hybridized to a probe for gene X. Clearly, gene X is most actively expressed in the liver.

Cloned DNAs make excellent probes because of their ability to hybridize to DNAs or RNAs with the same, or very similar, sequences. These cloned DNAs can be made radioactive and hybridized to DNAs on a Southern blot to reveal the sizes of DNA fragments containing similar sequences. Or they can be hybridized to Northern blots to tell us the sizes and amounts of RNAs containing similar sequences.

DETERMINING THE BASE SEQUENCE OF A GENE

In 1975, Alan Maxam and Walter Gilbert, working as a team, and Frederick Sanger and his colleagues developed two methods for determining the exact base sequence of a defined piece of DNA. These were spectacular breakthroughs. They allowed us to dissect the fine structure of a gene much faster than Benzer could using his traditional genetic tools (chapter 13). Moreover, although Benzer's fine structure mapping placed many mutants close together on a genetic map, it did not reveal anything about base sequence. As a result, Benzer's elegant experiments are unlikely to be applied to any other genetic systems; they are already obsolete.

Let us consider the modern Sanger method that has evolved from the original. We focus on this method because it is easier to understand than the Maxam-Gilbert procedure, and it is also more widely used. To sequence a piece of DNA by the Sanger method (fig. 15.22), we begin the same way we did for site-directed mutagenesis. We clone the DNA into a vector, such as M13 phage, that can give us the cloned DNA in single-stranded form. To the single-stranded DNA we hybridize an oligonucleotide primer about twenty bases long. This primer is designed to hybridize right beside the multiple cloning region of the vector and is oriented with its 3′-end pointing toward the insert in the multiple cloning region.

If we extend the primer using the Klenow fragment of DNA polymerase, we will produce DNA complementary to the insert. The trick to Sanger's method is to carry out such DNA synthesis reactions in four separate tubes, but to include in each tube a different chain terminator. The chain terminator is a **dideoxy ribonucleotide** such as **dideoxy ATP (ddATP).** Not only is this terminator 2′-deoxy, like a normal DNA precursor, it is 3′-deoxy as well. Thus, it cannot form a phosphodiester bond, since it lacks the necessary 3′-hydroxyl group. That is why we call it a chain terminator; whenever a dideoxy nucleotide is incorporated into a growing DNA chain, that growth stops.

If we add only dideoxy nucleotides, the chains will not grow at all, so we must put in an excess of normal deoxy nucleotides, with just enough dideoxy nucleotide to stop chain growth once in a while. In each tube we place a different dideoxy nucleotide; ddATP in tube 1, so chain termination will occur with A's; ddCTP in tube 2, so chain termination will occur with C's; and so forth. We also include radioactive dATP in all the tubes so the DNA products will be radioactive.

The result is a series of fragments of different lengths in each tube. In tube 1, all the fragments end in A; in tube 2, all end in C; in tube 3, all end in G; and in tube 4, all end in T. Next, we electrophorese all four reaction mixtures in parallel lanes in a high-resolution polyacrylamide gel. Finally, we perform autoradiography to visualize the DNA fragments, which appear as horizontal bands on an X-ray film.

FIGURE 15.22

The Sanger dideoxy method of DNA sequencing.
(*a*) The primer extension (replication) reaction.
A primer, 21 bases long in this case, is hybridized to
the single-stranded DNA to be sequenced, then
placed with DNA polymerase Klenow fragment and
dNTPs to allow replication. One dideoxy NTP is
included to terminate replication after certain bases;
in this case, ddTTP is used, and it has caused
termination at the second position where dTTP was
called for. (*b*) Products of the four reactions
(rxn's). In each case, the template strand is shown at
the top, with the various products underneath. Each
product will begin with the 21-base primer and will
have one or more nucleotides added to the 3′-end.
The last nucleotide is always a dideoxy nucleotide
(color) that terminated the chain. The total length of
each product is given in parentheses at the left end
of the fragment. Thus, we wind up with fragments
ranging from 22 to 33 nucleotides long.
(*c*) Electrophoresis of the products. The products of
the four reactions are loaded into parallel lanes of a
high-resolution electrophoresis gel and
electrophoresed to separate them according to size.
By starting at the bottom and finding the shortest
fragment (22 bases in the A lane), then the next
shortest (23 bases in the T lane), and so forth, we
can read the sequence of the product DNA. Of
course, this is the complement of the template
strand.

(a) Primer extension reaction:

—/ /————————————————————————TACTATGCCAGA————————/ /—
 ————————————————
        21-base primer

                          Replication with ddTTP

—/ /————————————————————————TACTATGCCAGA————————/ /—
 ————————————————————————ATGAT
            (26 bases)

(b) Products of the four reactions:

Products of ddA rxn

Template: _____ TACTATGCCAGA ____
(22) _____ A
(25) _____ ATGA
(27) _____ ATGATA

Products of ddC rxn

Template: _____ TACTATGCCAGA ____
(28) _____ ATGATAC
(32) _____ ATGATACGGTC

Products of ddG rxn

Template: _____ TACTATGCCAGA ____
(24) _____ ATG
(29) _____ ATGATACG
(30) _____ ATGATACGG

Products of ddT rxn

Template: _____ TACTATGCCAGA ____
(23) _____ AT
(26) _____ ATGAT
(31) _____ ATGATACGGT
(33) _____ ATGATACGGTCT

(c) Electrophoresis of the products:

Figure 15.23 shows a typical sequencing film. To begin
reading the sequence, we start at the bottom and find the
first band. In this case, it is in the C lane, so we know that
this short fragment ends in C. Now we move to the next
longer fragment, one step up on the film; the gel electrophoresis used to separate these fragments has such good
resolution that it can separate fragments differing by only
one base in length. And the next fragment, one base longer
than the first, is found in the A lane, so it must end in A.
Thus, so far, we have found the sequence C–A. We simply
continue reading the sequence in this way as we work up the
film. At first we will just be reading the sequence of part of
the multiple cloning region of the vector. However, before
very long, the DNA chains will extend into the insert—and
unknown territory. An experienced sequencer can continue
to read sequence from one film for hundreds of bases.

 Cloned genes give us a homogeneous population of DNA
molecules whose base sequence can be determined by either
of two methods. The Sanger method uses dideoxy
nucleotides to terminate DNA synthesis, yielding a series of
DNA fragments whose sizes can be measured by
electrophoresis. The last base in each of these fragments is
known, since we know which dideoxy nucleotide was used
to terminate each reaction. Therefore, ordering these
fragments by size—each fragment one (known) base longer
than the next—tells us the base sequence of the DNA.

FIGURE 15.23

A typical sequencing film. The sequence begins CAAAAAACGG. How much more of the DNA sequence can you read?

Courtesy of Bethesda Research Laboratories, Life Technologies, Inc. Gaithersburg, MD.

DNA SEQUENCING: THE ULTIMATE IN GENETIC MAPPING

We have already studied the traditional methods for mapping genes. However, the ultimate in fine structure genetic mapping is to obtain the exact base sequence of a gene, a goal that has already been realized for hundreds of genes. By the same token, the ultimate in genetic mapping of a phage would be to learn the base sequence of its whole genome. This has been accomplished for several viruses, including phages.

The first phage to have its entire base sequence revealed was the *E. coli* phage φX174. Frederick Sanger applied his DNA sequencing technique to this single-stranded, circular phage DNA and determined the total sequence of its 5,375 nucleotides.

What kind of information can we glean from this sequence? First, we can locate exactly the coding regions for all the genes. This tells us the spatial relationships among genes and the distances between them (to the exact nucleotide) without any guesswork about recombination frequency. How do we recognize a coding region? It contains an **open reading frame,** a sequence of bases that, if translated in one frame, contains no stop codons for a relatively long distance—long enough to code for one of the phage proteins. Furthermore, the open reading frame must start with an ATG (or occasionally a GTG) triplet, corresponding to an AUG (or GUG) translation initiation codon. In other words, an open reading frame is the same as a gene's coding region.

The base sequence of the phage DNA also tells us the amino acid sequence of all the phage proteins, in case we did not have this information before. All we have to do is use the genetic code to translate the DNA base sequence of each open reading frame into the corresponding amino acid sequence. This sounds like a laborious process, but we now have computer programs that allow a personal computer to perform this task in a split second.

Sanger's analysis of the open reading frames of the φX174 DNA revealed something unexpected and fascinating: Some of the phage genes overlap. Figure 15.24 shows that the coding region for gene B lies within gene A and the coding region for gene E lies within gene D. Furthermore, genes D and J overlap by one base pair. How can two genes occupy the same space and code for different proteins? The answer is that the two genes are translated in different reading frames (fig. 15.25). Since entirely different sets of codons will be encountered in these two frames, the two protein products will also be quite different.

This was certainly an interesting finding, and it raised the question of how common this phenomenon would be. So far, major overlaps seem to be confined to viruses, which is not surprising since these simple genetic systems have small genomes and there is a premium on efficient use of the genetic material. Moreover, viruses have prodigious power to replicate, so enormous numbers of generations have passed during which evolution has honed the viral genomes.

 The base sequence of the simple phage φX174 reveals all its genes' coding regions, or open reading frames, and shows that two of the phage genes lie within other genes. These genes-within-genes still code for different products because they are translated in different frames.

PRACTICAL APPLICATIONS OF GENE CLONING

Our discussion of expression vectors demonstrated that it is possible to engineer bacteria or other cells to produce the products of cloned genes. Obviously, some of these protein products may be very useful, or very scarce, or both, and are therefore potentially valuable. In this section we will see how cloned gene products, and the genes themselves, have the potential to reshape medicine and agriculture.

PROTEINS: THE NEW PHARMACEUTICAL FRONTIER

What sorts of valuable proteins would we want to produce? The first cloned pharmaceutical product to be sold was human insulin, which is valuable for two reasons: First,

FIGURE 15.24
The genetic map of phage φX174. Each letter stands for a phage gene.

FIGURE 15.25
Overlapping reading frames of φX174. Gene D (red) begins with the base numbered 1 in this diagram and continues through base number 459. This corresponds to amino acids 1–152 plus the stop signal TAA. Dots represent bases or amino acids not shown. Only the coding strand is shown. Gene E (blue) begins at base number 179 and continues through base number 454, corresponding to amino acids 1–90 plus the stop signal TGA. This gene uses the reading frame one space to the right, relative to the reading frame of gene D. Gene J (gray) begins at base number 459 and uses the reading frame one space to the left, relative to gene D.

some diabetics become allergic to the pig and bovine insulin traditionally used. Since pig insulin is a little different from the human variety, our bodies can detect that difference and mount an immune response, or even an allergic response, to the foreign protein. In principle, human insulin should not cause a problem, because the body does not recognize it as foreign (although a few patients have suffered unexplained reactions). Furthermore, the cloned gene assures an unlimited supply of insulin, so we do not have to depend on the vagaries of the livestock market.

Another example of a useful protein is **human growth hormone (hGH).** Some children inherit a condition called **hypopituitary dwarfism.** They do not grow normally because their bodies make too little growth hormone. They are destined to be dwarves unless they are treated with hGH. Growth hormone from other animals will not help. In the past, these children were treated with growth hormone extracted from the pituitary glands of human cadavers. But this was very expensive; hormone from more than seventy cadavers was required in order to treat one child for one year. Moreover, a few early preparations of hGH were apparently contaminated with the agent that causes a lethal brain condition called Creutzfeld-Jacob disease. The solution was to obtain hGH by cloning its gene. This has been accomplished, and the product is effective.

IMPROVED VACCINES

Novel, safer vaccines may be another valuable product of cloned genes. Usually, vaccines are made from whole viruses that have been either killed or weakened so that they can no longer cause disease but can still arouse the body's immune defenses. The problem is that sometimes these viruses are not killed or are not weakened enough, thus they end up causing the disease, and sometimes even death, after all. For example, smallpox vaccine is now considered more dangerous than the almost nonexistent chance of catching smallpox, so it is no longer given routinely in the United States.

One of the most fearful viral diseases in the United States is AIDS (acquired immune deficiency syndrome). The most promising long-term solution to the AIDS problem is vaccination, but consider the potential dangers of an AIDS vaccine containing live virus. No pharmaceutical company would risk selling such a product to the general public.

Gene cloning may provide a solution to this problem. Several research groups have already cloned the genes for the surface proteins of the AIDS virus (**human immunodeficiency virus,** or **HIV**). They are now harvesting the proteins from the bacteria harboring those genes and using them to trigger the body's immune system in much the same way a whole virus would. In other words, by immunizing people against a viral surface protein, we ought to be able to immunize them against the virus and therefore against AIDS. Notice that these proteins do not come directly from the virus but from relatively harmless bacteria. Thus, the vaccine should be safe.

A related strategy would be to insert a cloned AIDS surface protein gene into a harmless virus such as vaccinia. This virus, otherwise known as the cowpox virus, has been used since the end of the eighteenth century to immunize people against smallpox. With an added gene for HIV surface protein, it could become the basis for an effective AIDS vaccine. However, some observers have expressed concern that the HIV surface protein might direct the recombinant vaccinia virus to the very helper T-cells that HIV normally attacks. This could disrupt those cells' functions and possibly cause an AIDS-like condition.

In addition to this reservation, there are other reasons to doubt that a preventive for AIDS is just around the corner. Vaccines are tricky to develop, even in the best of circumstances, and researchers are finding many different forms of AIDS virus, each of which would require a different vaccine. Furthermore, the virus seems capable of rapid genetic change, so it may change its coat to frustrate vaccines, the way flu virus does.

The most obvious use of cloned genes is to express them and harvest their products. This has already yielded proteins that have pharmaceutical value, including human insulin and human growth hormone. In the future, it may provide purified viral proteins that can serve as safe, effective vaccines against deadly diseases. We may even be able to engineer harmless viruses that can immunize us against unrelated dangerous viruses.

INTERVENING IN HUMAN GENETIC DISEASE

Ultimately, we may be able to use cloned genes to treat patients suffering from genetic diseases. The idea would be to implant a normal gene to correct the defect.

The classic example of a genetic disorder is sickle-cell disease (chapter 11). Recall that sickle-cell patients are homozygous for a defective β-globin gene; as a result, they make abnormal hemoglobin, which causes the trouble. This suggests that a genetic engineering solution is possible. Since blood cells are made in the bone marrow, we should be able to remove all of a patient's abnormal marrow and replace it with marrow that has a normal gene for hemoglobin. It would be best if we could use the patient's own marrow, to avoid a rejection problem. But how can we get a normal gene into abnormal marrow cells? Again, cloning is the key. We have already cloned human globin genes, and these can be added to cells in much the same way that we add recombinant DNAs to bacterial cells.

An experimental procedure might work like this: (1) Take out some of the patient's bone marrow cells. (2) Add a cloned normal hemoglobin gene to the extracted cells. (3) Grow these engineered cells in culture. (4) Irradiate the patient with a high enough dose of X rays to kill all the remaining marrow cells. (5) Rescue the patient, who is destined to die without marrow cells, by implanting the engineered cells; these will grow and repopulate the bone marrow. (6) Presumably, the normal gene that was added will function well enough to prevent sickling.

Do we know this procedure will work? Actually, we don't, but we do know that it is possible to cure sickle-cell disease with a marrow transplant. An eight-year-old girl in Boston suffered from both leukemia and sickle-cell disease. Doctors treated the leukemia by destroying the patient's cancerous bone marrow and replacing it with normal marrow from another person. Not only did this seem to cure the girl's leukemia, but it had the side effect of curing her sickle-cell disease as well, because of course the donor cells had come from a person who didn't suffer from that disease. So we know the transplant part of the procedure can work.

The current roadblock is getting a cloned gene to function normally in the patient's cells. This has been a problem, not only for globin genes, but for most genes transplanted into animals. For some reason, they are not usually expressed and regulated as they would be in their natural state. This might be especially serious for the globin genes, which are normally regulated so that their α- and β-globin products are produced in roughly equal amounts. Moreover, this procedure requires that the gene be expressed in bone marrow stem cells. This is a problem because no one has yet achieved long-term expression of transplanted genes in such cells in large mammals. For these reasons, sickle-cell disease was not the first genetic disorder attacked with cloned genes.

More attractive targets for cloned genes are diseases caused by failure of a gene to function at all, where a donor gene producing any amount of product will help and overproduction will not hurt. In fact, the first actual trials of human gene therapy have been on two young patients with severe combined immunodeficiency, caused by an adenosine deaminase deficiency. The lack of this enzyme, which normally degrades 2′-deoxyadenosine, leads to a build-up of this nucleotide in the circulatory system, which kills the white blood cells that normally fight disease. The result is an immunodeficiency so severe that some of these unfortunate youngsters must live in sterile "bubbles" in order to avoid fatal infections that they could not combat. In the trials, these patients received infusions of their own white blood cells to which cloned adenosine deaminase genes had been added. As of June 1992, both patients showed significantly improved immunity. One patient exhibited an increase in adenosine deaminase from less than 2% of normal to 20% of normal, and an increase in white blood cell count from well below normal to within the normal range. Both patients attained sufficient immunity to live normal lives, including attending school.

The second clinical trial of gene therapy concerns a condition called familial hypercholesterolemia. This syndrome is caused by a defect in the gene for the low-density lipoprotein (LDL) receptor, which leads to a deficiency in the LDL receptor on cell surfaces, which in turn results in the inability to transport LDL into cells. Since LDL is rich in cholesterol, the failure to get it into cells means that cholesterol builds up to abnormally high levels in the bloodstream. This leads to formation of plaques in the arteries of young patients, and heart attacks that kill most patients at an early age, many before the age of 20. The idea will be to place a wild-type LDL receptor gene in some of the patients' liver cells and see whether this will lower the deadly level of cholesterol in their blood. This procedure has already given a 30% serum cholesterol reduction in hyperlipidemic rabbits, which augurs well for human trials.

Another genetic disease that might yield to gene therapy is Lesch-Nyhan syndrome, which is characterized by mental retardation and bizarre self-mutilation such that the patient may literally bite off his or her own fingers and lips. This genetic disease results from a defective gene for the enzyme hypoxanthine-guanine phosphoribosyl transferase (HPRT). Nobody knows why lack of this enzyme causes such dramatic effects, but placement of a normal HPRT gene in transplanted bone marrow might yield at least some of the enzyme and therefore partially alleviate the symptoms.

It is becoming possible to treat genetic defects by transplanting cloned, normal genes into patients whose own genes are damaged.

USING CLONED GENES IN AGRICULTURE

In 1983, Richard Palmiter and Ralph Brinster placed the gene for human growth hormone in an early mouse embryo. The result, dubbed "supermouse" by the popular press, was a giant mouse—twice normal size. We call such animals **transgenic** because they contain a transplanted gene from another organism. It is important to realize that a true transgenic animal receives the foreign gene (the *transgene*) at a very early stage in development so that the transgene becomes incorporated into the host animal's chromosomes and can be passed on to progeny as any normal gene would be. This is different from adding a gene to one tissue in an adult animal, as would be done in the genetic disease cases we just discussed. In those situations, the transplanted genes would not enter the germ cells, and so would not be passed on to the next generation.

As soon as people heard about "supermouse," they began wondering, "Why not 'supercow,' or 'superpig?' " (Never mind "superman.") Palmiter, Brinster, and their colleagues have now attempted to create a "superpig," and so far their experiment is only a qualified success. These workers created transgenic pigs by injecting genes for human growth hormone (hGH) and bovine growth hormone (bGH) into pig eggs. The bGH worked best, but the transgenic pigs containing bGH probably do not deserve the name "superpig." They did not grow to supernormal size, but they did show a greater than normal feed efficiency—that is, weight gain per unit of feed. This is important, since feed accounts for up to 70% of the day-to-day cost of raising pigs. Furthermore, the transgenic pigs had a much smaller amount of subcutaneous fat, the layer of fat just beneath the skin. This is a benefit during the current era of concern about the adverse effects of animal fat in our diet.

Offsetting these benefits, however, are some serious health problems in the transgenic pigs. They have an abnormally high incidence of stomach ulcers, arthritis, enlarged heart, dermatitis, and kidney disease. These conditions depress appetite and lead to higher mortality, already a significant problem in domestic pigs. Clearly, more work is required to produce a real superpig.

PROTEIN PRODUCTS FROM TRANSGENIC ORGANISMS

Geneticists have recently begun using transgenic farm animals for a more mundane purpose: producing valuable proteins from cloned genes. Consider, for example, the gene for a human blood clotting protein called factor IX. People homozygous for a defect in this gene suffer the blood clotting disorder hemophilia B. They have traditionally been treated with human plasma or partially purified factor IX, but these human blood products can be contaminated with hepatitis virus, or even HIV, the AIDS virus, making that method of treatment somewhat risky. Thus, a cloned source of the protein would be a big advantage.

However, factor IX is produced by human cells as a precursor that requires extensive modification to make it active. Bacteria, the usual cells in which we express cloned genes, do not perform these key alterations. Therefore, expression in mammalian cells is called for, but cultured mammalian cells are expensive to maintain. An alternative is to make a transgenic animal do the job, and A. J. Clark and his colleagues have done just that. They linked the human factor IX gene to the control sequences of a sheep milk protein gene called β-lactoglobulin (BLG). Then they injected this fused gene into sheep eggs and implanted the eggs in ewes. The rationale for this, of course, is that the factor IX gene will be expressed in the mammary gland, where BLG is normally produced, and the factor IX can therefore be conveniently recovered in the milk of transgenic ewes. Two transgenic ewes were born; after they were mated, they did indeed produce human factor IX in their milk. Admittedly, the amounts of factor IX produced were very modest, but there is good reason to hope that this performance will improve. Geneticists have now made transgenic sheep carrying the human α_1 antitrypsin gene, and one ewe gives milk containing 35 grams per liter of the protein!

This idea is not confined to transgenic animals. Recent experiments have shown that transgenic tobacco plants can also be used as protein factories. So far, some of the favorite proteins to make in plants have been antibodies. These "plantibodies" are not exactly like their human counterparts, but they function well enough to be extremely valuable. For example, they might find use as supplements for infant formula. If so, they will be needed in large quantities. This brings up another great advantage of transgenic plants: They can produce prodigious amounts of protein. We expect to harvest tens to hundreds of pounds of foreign protein per acre from these green factories.

GENETICALLY ENGINEERED CROPS

The ability to place cloned genes into plants raises the possibility of engineering new, improved strains of crops. Of course, humankind has been improving crops by selective breeding for millenia, but it has been a time-consuming, hit-and-miss process in which many undesirable genes are transferred along with the advantageous ones. Now we can identify useful genes, isolate them by cloning, and insert them directly into a new plant host. If these genes behave normally in their new environment (a big "if," indeed), they may be able to effect dramatic improvements in the recipient plants.

One of the first useful genes to be transferred this way was one that confers herbicide resistance. This trait is useful because an herbicide-resistant plant can survive treatment while weeds all around it are dying. The herbicide in question was **glyphosate,** the active ingredient in Monsanto's weed killer Roundup. Glyphosate kills by inhibiting an

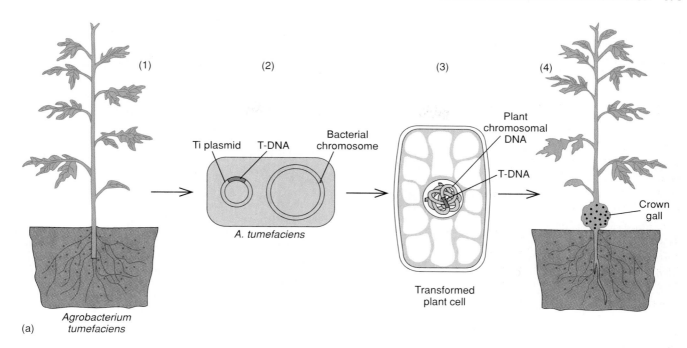

(a)

Agrobacterium
tumefaciens

(1) (2) (3) (4)

Ti plasmid   T-DNA   Bacterial chromosome

A. tumefaciens

Plant chromosomal DNA

T-DNA

Transformed plant cell

Crown gall

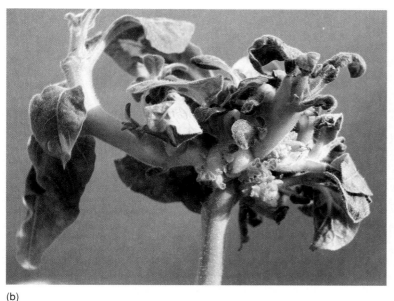

(b)

FIGURE 15.26

(*a*) Formation of a crown gall. 1. *Agrobacterium* cells enter a wound in the plant, usually at the crown, or the junction of root and stem. 2. The *Agrobacterium* contains a Ti plasmid in addition to the much larger bacterial chromosome. The Ti plasmid has a segment (the T-DNA, red) that promotes tumor formation in infected plants. 3. The bacterium contributes its Ti plasmid to the plant cell, and the T-DNA from the Ti plasmid integrates into the plant's chromosomal DNA. 4. The genes in the T-DNA direct the formation of a crown gall, which nourishes the invading bacteria. (*b*) Photograph of a crown gall tumor generated by cutting off the top of a tobacco plant and inoculating with *Agrobacterium*. This crown gall tumor is a teratoma, which generates normal as well as tumorous tissue. Note the disorganized, more or less normal plant tissues springing from the tumor.

enzyme called EPSP synthase, which is necessary for making the essential amino acids phenylalanine, tyrosine, and tryptophan. Without these amino acids, plants cannot live, making glyphosate a very effective herbicide. But one would hardly want to spray it on fields planted with ordinary crops, since it kills all kinds of plants—weeds and crops alike. The aim, therefore, was to make crops resistant to glyphosate.

David Stalker and his coworkers managed to engineer glyphosate-resistant tobacco plants. First, they grew *Salmonella* bacteria in the presence of glyphosate and isolated a resistant mutant that could still grow. The EPSP synthase of the mutant has suffered a single amino acid change, which accounts for its glyphosate resistance. Next, these workers cloned the mutant gene and transferred it to a tobacco plant.

The transfer is not as easy as it sounds. In order to make the foreign gene function in a plant, genetic engineers had to clone it in a vector that could function effectively in plant cells. The common bacterial vectors, such as the pUC vectors, would not serve this purpose, because plant cells cannot recognize their prokaryotic promoters and replication origins. Instead, the research team placed the glyphosate resistance gene into a plasmid containing so-called **T-DNA.** This is a piece of DNA from a plasmid known as **Ti** (tumor-inducing).

The Ti plasmid inhabits the bacterium *Agrobacterium tumefaciens,* which causes tumors called **crown galls** (fig. 15.26) in dicotyledonous plants. When this bacterium infects a plant, it transfers its Ti plasmid to the host cells, whereupon the T-DNA integrates into the plant DNA, causing the abnormal proliferation of plant cells that gives rise to a crown gall. This is advantageous for the invading bacterium, because the T-DNA has genes directing the synthesis of unusual organic acids called **opines;** these opines are worthless to the plant, but the bacterium has enzymes that can break down opines so that they can serve as an exclusive energy source for the bacterium.

The T-DNA genes coding for the enzymes that make opines (e.g., mannopine synthase) have strong promoters. Plant molecular geneticists take advantage of them by placing T-DNA into small plasmids, then putting foreign genes under the control of one of these promoters. Stalker and his colleagues used the mannopine synthetase promoter. After they had formed their recombinant DNA containing the herbicide resistance gene, they placed it back into *Agrobacterium,* which they then used to infect plant cells and thereby transfer the cloned gene.

Figure 15.27 outlines the process these investigators used to transfer the herbicide resistance gene to a tobacco plant. They punched out a 2-mm diameter disk from a tobacco leaf and placed it in a dish with nutrient medium. Under these conditions, tobacco tissue will grow around the edge of the disk. Next, they added *Agrobacterium* containing the cloned glyphosate resistance gene; these bacteria infected the growing tobacco cells and introduced the cloned gene.

When the tobacco tissue grew roots around the edge, Stalker and his coworkers transplanted those roots to medium that encouraged shoots to form. These plantlets gave rise to full-sized tobacco plants that, if they contained a functioning glyphosate resistance gene, should be resistant to the herbicide. The next step was to spray the plants with herbicide to find out if they really were resistant. Figure 15.28 shows the results of the test. The wild-type tobacco plants at the top were obviously sensitive, even to intermediate levels of the herbicide, while the engineered plants at the bottom were quite resistant to intermediate levels and somewhat resistant to high levels.

This is only one example of plant genetic engineering. Other plant geneticists have made additional strides, including the following: (1) conferring glyphosate resistance on petunias by inserting extra copies of a normal EPSP gene; (2) conferring virus resistance on tobacco plants by inserting a gene for the viral coat protein; (3) conferring pest resistance on tobacco plants by inserting a bacterial gene for an insecticidal toxin; and (4) inserting the gene for firefly luciferase into tobacco plants. The latter experiment has no practical value, but it does have the arresting effect of making the plant glow in the dark (fig. 15.29).

Why all this emphasis on tobacco? Would it not be more productive to work on an important food crop such as wheat, corn, or rice? It certainly would, but tobacco happens to be a very convenient plant to manipulate, for two

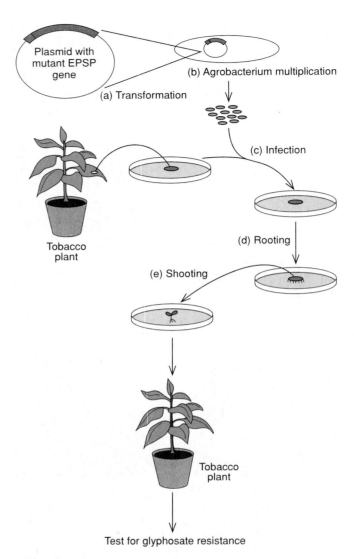

FIGURE 15.27
Using a T-DNA plasmid to introduce a gene into tobacco plants. (*a*) A plasmid is formed with the mutant EPSP gene (red) under the control of the mannopine synthetase promoter (blue). This plasmid is used to transform *Agrobacterium* cells. (*b*) The transformed bacterial cells divide repeatedly. (*c*) A disk of tobacco leaf tissue is removed and incubated in nutrient medium, along with the transformed *Agrobacterium* cells. These cells infect the tobacco tissue, transferring the plasmid bearing the mutant EPSP gene. (*d*) The disk of tobacco tissue sends out roots into the surrounding medium. (*e*) One of these roots is transplanted to another kind of medium, where it forms a shoot. This plantlet grows into a tobacco plant that can be tested for glyphosate resistance.

important reasons: First, because it is a *dicot,* it can be infected with *Agrobacterium,* providing an easy way to insert foreign DNA. (By contrast, all cereal plants are *monocots* that, unfortunately, cannot be infected with *Agrobacterium.*) Furthermore, in the experiments just described, it was necessary to regrow a whole plant from individual plant cells that had been transformed with the cloned gene. This is easy with tobacco and its relative, the petunia, but could not be done with cereals until the mid-1980s.

FIGURE 15.28

Testing transformed tobacco plants for herbicide resistance. The top nine plants are wild-type; the bottom nine have been transformed with a glyphosate-resistant EPSP gene. The left vertical row received no glyphosate. The middle vertical row received the equivalent of 0.6 kg glyphosate per hectare. The right vertical row received the equivalent of 1.0 kg glyphosate per hectare. Note the resistance of the transformed plants, especially to the lower concentration of glyphosate.

From L. Comai, D. Facciotti, W. R. Hiatt, G. Thompson, R. E. Rose, and D. M. Stalker. Reprinted by permission from Nature *317:744. Copyright © 1985 Macmillan Magazines Limited.*

 It is possible to transfer cloned genes to plants, thereby creating transgenic plants with altered properties. In dicots, this is generally done using a gene under control of a promoter from a T-DNA plasmid derived from *Agrobacterium.* The recombinant DNA is then placed back in the bacterium, which is used to infect plant cells. Finally, the infected cells, with their transplanted genes, are grown into whole plants and tested for new traits, such as herbicide or disease resistance.

MAPPING GENETIC DEFECTS IN HUMANS

As we have seen, certain debilitating diseases are heritable, indicating that they are caused by genetic defects. In chapter 11, for example, we saw that a defective gene for the

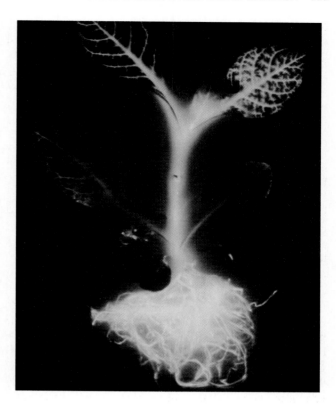

FIGURE 15.29

A tobacco plant that has been transformed with the firefly luciferase gene. The green light emitted by the plant indicates the activity of luciferase. Note the concentration of the enzyme in the roots and stem.

D. W. Ow, Keith Wood, Marlene DeLuca, Jeffrey R. DeWet, Donald R. Helinski, Steven Howell. "Transient and Stable Expression of the Firefly Luciferase Gene in Plant Cells and Transgenic Plants." Science *234:856–59, fig. 5, November 1986. Copyright 1986 by AAAS.*

β-globin blood protein can cause sickle-cell disease. There is great interest in locating the defective genes responsible for such human genetic diseases because knowing the nature of the defect may lead us to the cure. But this is more difficult than it sounds. Finding the defective gene was relatively simple in the case of sickle-cell disease, since we knew the identity of the defective protein. But in most cases, we know only the symptoms of the disease, not the cause.

Consider **Huntington disease,** the degenerative nerve disorder that killed folksinger Woody Guthrie; **Duchenne muscular dystrophy,** a sex-linked muscle-wasting disease; or **cystic fibrosis,** a common disease in which thick mucous secretions build up in the lungs of young patients. In none of these disorders did we know at the outset what gene products were defective, so we could not rely on that kind of information to lead us to the mutant genes. But in each case, imagination, hard work, and a measure of luck have brought us the answer after all.

HUNTINGTON DISEASE

As mentioned, Huntington disease is a progressive nerve disorder. It begins almost imperceptibly with small tics and clumsiness. Over a period of years, these symptoms

FIGURE 15.30

Detecting a RFLP. Two individuals are polymorphic with respect to a *Hin*dIII restriction site (blue). The first individual contains the site, so cutting the DNA with *Hin*dIII yields two fragments, 2 and 4 kb long, that can hybridize with the probe, whose extent is shown at top. The second individual lacks this site, so cutting that DNA with *Hin*dIII yields only one fragment, 6 kb long, which can hybridize with the probe. The results from electrophoresis of these fragments, followed by blotting, hybridization to the radioactive probe, and autoradiography, are shown at right. The fragments at either end, represented by dotted lines, do not show up because they cannot hybridize to the probe.

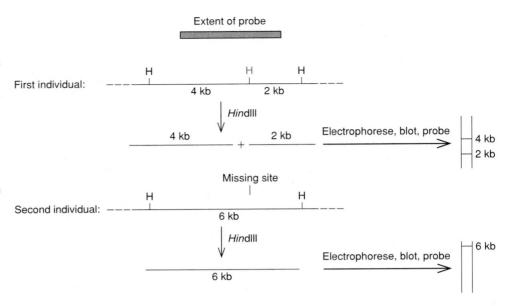

intensify and are accompanied by emotional disturbances. Nancy Wexler describes the advanced disease as follows: "The entire body is encompassed by adventitious movements. The trunk is writhing and the face is twisting. The full-fledged Huntington patient is very dramatic to look at." Finally, after ten to twenty years, the patient dies.

Huntington disease is controlled by a single dominant gene. A child of a Huntington patient has a 50:50 chance of being afflicted with the disease. People who have the disease could avoid passing it on by not having children, except that the first indications of trouble usually do not appear until after the child-bearing years.

RESTRICTION FRAGMENT LENGTH POLYMORPHISMS

Because we do not know the nature of the product of the Huntington disease gene, we cannot look for the gene directly. The next best approach is to look for a gene that is tightly linked to the Huntington disease gene (*HD*). Michael Conneally and his colleagues spent more than a decade trying to find such a linked gene, but with no success.

Another approach, which does not depend on finding linkage with a known gene, is to establish linkage with an "anonymous" stretch of DNA that may not even contain any genes. We recognize such a piece of DNA by its pattern of cleavage by restriction enzymes.

Since one person differs genetically from another, the sequences of their DNAs will differ somewhat, as will the pattern of cutting by restriction enzymes. Consider the restriction enzyme *Hin*dIII, which recognizes the sequence AAGCTT. One individual may have three such sites separated by 4 and 2 kilobases (kb), respectively, in a given region of a chromosome (fig. 15.30). Another individual may lack the middle site but have the other two 6 kb apart. This means that if we cut the first person's DNA with *Hin*dIII, we will produce two fragments 2 kb and 4 kb long. The

second person's DNA will yield a 6-kb fragment instead. In other words, we are dealing with a **restriction fragment length polymorphism** (see chapter 5). This very clumsy term simply means that cutting the DNA from any two individuals with a restriction enzyme may yield fragments of different lengths. The abbreviated term, **RFLP**, is usually pronounced "rifflip."

Because geneticists failed to find a standard genetic marker, or gene, linked to *HD*, they looked for a RFLP that is linked. But how do we go about looking for a RFLP? Clearly, we cannot analyze the whole human genome at once. It contains hundreds of thousands of cut sites for a typical restriction enzyme, so each time we cut the whole genome with such an enzyme, we release hundreds of thousands of fragments. No one would relish sorting through that morass for subtle differences between individuals.

Fortunately, there is an easier way. We can cut the DNA with a restriction enzyme, electrophorese the hundreds of thousands of fragments to separate them by size, then transfer them to a piece of nitrocellulose or other DNA-binding medium by Southern blotting. Now, how can we identify the one or two bands we are interested in among thousands of irrelevant bands? We isolate a piece of DNA that spans at least part of our region of interest (about 4 kb in fig. 15.30), make it radioactive, denature it, and use it to probe the Southern blot. The 4-kb and 2-kb bands, but no others, from the first individual should hybridize to the probe. Similarly, only the 6-kb band from the second individual should hybridize. Since the probe is radioactive, we can easily detect it by placing a piece of X-ray film over the Southern blot and allowing the radioactive emissions to expose the film by autoradiography (chapter 7).

Using RFLPs to Map the Huntington Disease Gene

The approach just described was used by a team of scientists, including Wexler, Conneally, and James Gusella, to find a RFLP linked to *HD*. Of course, there is a catch. Each

Box 15.2 Ethical Issues

Discussion of gene cloning raises ethical questions about genetic engineering, especially as it applies to changing human genes. These questions have been with us ever since the first gene cloning experiments. Actually, the early debate about cloning revolved around the safety issue. Some scientists were so alarmed about the potential dangers of introducing new genes into bacteria that they convinced the molecular biology community to stop this kind of research until it could be studied more fully. The moratorium lasted for eighteen months, at which time it began to be clear that dangerous germs have evolved along with their hosts for thousands of years and that

dozens of genes are probably needed to make a bacterium pathogenic. It is extremely improbable that we could create a dangerous new germ simply by adding one new gene to an innocuous microbe. Solid evidence now indicates that those early fears were unfounded; thousands of scientists have been cloning genes during the last two decades, with no untoward effects.

But as the safety debate has subsided, a new one has heated up regarding the ethics of altering human genes. It is not too hard to identify sickle-cell disease as a genetic defect, but where do you draw the line? Is a below-average IQ a genetic defect? Is left-handedness? Who would make these decisions? Fortunately, we are probably at least a century away from

knowing the genetic basis of intelligence, let alone being able to change it genetically, so worries of this sort are probably premature.

What about creating transgenic humans—changing not only the genes of patients, but the genes in their germ cells, so that their descendants' genes will be changed as well? This is tantamount to interfering in human evolution, and some would argue that it is too much like playing God. Others point out that we have a duty to minimize human suffering, and thus, that eliminating a defective allele from the population could only be good. Whatever your feelings on these issues, it is clear that they raise questions for which there may never be easy answers.

FIGURE 15.31

The RFLP associated with the Huntington disease gene. The *Hin*dIII sites in the region covered by the G8 probe are shown. The families studied show polymorphisms in two of these sites, marked with an asterisk and numbered 1 (blue) and 2 (red). Presence of site 1 results in a 15-kb fragment plus a 2.5-kb fragment that is not detected because it lies outside the region that hybridizes to the G8 probe. Absence of this site results in a 17.5-kb fragment. Presence of site 2 results in two fragments of 3.7 and 1.2 kb. Absence of this site results in a 4.9-kb fragment. Four haplotypes

(A–D) result from the four combinations of presence or absence of these two sites. These are listed at right, beside a list of polymorphic *Hin*dIII sites and a diagram of the *Hin*dIII restriction fragments detected by the G8 probe for each haplotype. For example, haplotype A lacks site 1 but has site 2. As a result, *Hin*dIII fragments of 17.5, 3.7, and 1.2 are produced. The 2.3- and 8.4-kb fragments are also detected by the probe, but we ignore them because they are common to all four haplotypes.

radioactive probe hybridizes only to a small fraction of the total human DNA, so hundreds of probes may have to be screened before the right one is found. In this case, the investigators were lucky; they found a promising probe (called G8) among the first dozen they tried.

They were also lucky to have a very large family to study. Living around Lake Maracaibo in Venezuela is a family whose members have suffered from Huntington disease since the early nineteenth century. The first member of the family to be so afflicted was a woman whose father, presumably a European, carried the defective gene. So the

pedigree of this family can be traced through seven generations, and the number of individuals is unusually large, a typical nuclear family having fifteen to eighteen children.

The Gusella et al. group used their probe to examine RFLPs in the members of this Venezuelan family, then checked to see if the RFLPs correlated with the disease. Indeed, the probe located a RFLP that is very tightly linked to HD. Figure 15.31 shows the locations of *Hin*dIII sites in the stretch of DNA that hybridizes to the probe. We can see seven sites in all, but only five of these are found in all

FIGURE 15.32
Southern blots of *Hind*III fragments from members of two families, hybridized to the G8 probe. The bands in the autoradiographs represent DNA fragments whose sizes are listed at the right. The genotypes of all family members.

the children and three of the parents are shown at the top. The fourth parent was deceased, so his genotype could not be determined.

From James F. Gusella, et al. Reprinted by permission from Nature *306:236. Copyright © 1983 Macmillan Magazines Limited.*

The other two, marked with asterisks and numbered 1 and 2, may or may not be present. These latter two sites are therefore **polymorphic,** or variable.

Let us see how the presence or absence of these two restriction sites gives rise to a RFLP. If site 1 is absent, a single fragment 17.5 kb long will be produced. However, if site 1 is present, the 17.5-kb fragment will be cut into two pieces having lengths of 15 kb and 2.5 kb. Only the 15-kb band will show up on the autoradiograph, since the 2.5-kb fragment lies outside the region that hybridizes to the G8 probe. If site 2 is absent, a 4.9-kb fragment will be produced. On the other hand, if site 2 is present, the 4.9-kb fragment will be subdivided into a 3.7-kb fragment and a 1.2-kb fragment.

There are four possible haplotypes (sets of linked genetic loci) with respect to these two polymorphic *Hind*III sites, and they have been labeled *A–D*:

Haplotype	Site 1	Site 2	Fragments observed
A	Absent	Present	17.5; 3.7; 1.2
B	Absent	Absent	17.5; 4.9
C	Present	Present	15.0; 3.7; 1.2
D	Present	Absent	15.0; 4.9

Of course, each member of the family will inherit two haplotypes, one from each parent. This means that a person with the *AD* genotype will have the same RFLP pattern as one with the *BC* genotype, since all five fragments will be present in both cases. However, the true genotype can be deduced by examining the parents' genotypes. Figure 15.32 shows autoradiographs of Southern blots of two families, using the radioactive G8 probe. The 17.5- and 15-kb fragments migrate very close together, so they are difficult to

distinguish when both are present, as in the *AC* genotype; nevertheless, the *AA* genotype with only the 17.5-kb fragment is relatively easy to distinguish from the *CC* genotype with only the 15-kb fragment. The *B* haplotype in the first family is obvious because of the presence of the 4.9-kb fragment.

Now we would like to know which haplotype is associated with the disease in the Venezuelan family; figure 15.33 demonstrates that it is *C*. Nearly all individuals with this haplotype have the disease. Those who do not will almost certainly develop it later. Equally telling is the fact that no individual lacking the *C* haplotype has the disease. Thus, this is a very accurate way of predicting whether a member of this family is carrying a defective *HD* gene. A similar study of an American family showed that the *A* haplotype was linked with the disease. Therefore, each family varies in the haplotype associated with the disease, but within a family, the linkage between the RFLP site and *HD* is so close that recombination between these sites is very rare.

Aside from its use as a genetic counseling tool, what does this analysis reveal? For one thing, it has allowed us to locate *HD* on chromosome 4. Gusella et al. did this by making mouse-human hybrid cell lines, each containing only a few human chromosomes, as described in chapter 5. They then prepared DNA from each of these lines and hybridized it to the radioactive G8 probe. Only the cell lines having chromosome 4 hybridized; the presence or absence of all other chromosomes was irrelevant. Therefore, human chromosome 4 carries *HD*.

At this point, the *HD* mapping team's luck ran out. One long detour arose from a mapping study that indicated the gene lay far out at the end of chromosome 4. This made

FIGURE 15.33
Pedigree of the large Venezuelan Huntington disease family. Family members with confirmed disease are represented by purple symbols. Notice that most of the individuals with the *C* haplotype already have the disease, and that no sufferers of the disease lack the *C* haplotype. Thus, the *C* haplotype is strongly associated with the disease, and the corresponding RFLP is tightly linked to the Huntington disease gene.

the search much more difficult because the tip of the chromosome is a genetic wasteland, full of repetitive sequences and apparently devoid of genes. Finally, after wandering for years in what he calls a genetic junkyard, Gusella and his group turned their attention to a more promising region. Some mapping work suggested that *HD* resided, not at the tip of the chromosome, but in a 2.2-Mb region several Mb removed from the tip. Over 2 Mb is a tremendous amount of DNA to sift through to find a gene, so Gusella decided to focus on a 500-kb region that was highly conserved among about one-third of HD patients, who seem to have a common ancestor.

To find genes in the 500-kb region, Gusella used an **exon trapping** strategy that selects only DNA sequences bordered by splicing signals, which are usually exons. The beauty of this procedure is that it eliminates intergenic and intron DNA and focuses attention on introns, where the coding regions lie. Using this method, he identified a handful of exon clones. He then used these exons to probe a cDNA library to find the whole coding regions of the corresponding genes. One of the clones, called IT15, for "interesting transcript number 15," hybridized to cDNAs that identified a large (10,366 nucleotide) transcript that codes for a large (3,144 amino acid) protein. The presumed protein product did not resemble any known proteins, so that did not provide any evidence that this is indeed *HD*. However, the gene had an intriguing repeat of 23 copies of the triplet CAG (one copy is actually CAA). If, as appears likely, this repeated sequence lies in the coding region of the gene, it encodes a stretch of 23 glutamines.

Is this really *HD?* Gusella's comparison of the gene in affected and unaffected individuals in 75 HD families demonstrates that it is. In all unaffected individuals, the number of CAG repeats ranged from 11 to 34, and 98% of these unaffected people had 24 or fewer CAG repeats. In

all affected individuals, the number of CAG repeats had expanded to at least 42, up to a high of about 100. Thus, we can predict whether an individual will be affected by the disease by looking at the number of CAG repeats in this gene. Furthermore, the severity or age of onset of the disease correlates at least roughly with the number of CAG repeats. People with a number of repeats at the low end of the affected range (42) generally survive well into adulthood before symptoms appear, whereas people with a number of repeats at the high end of the range tend to show symptoms in childhood. In one extreme example, an individual with the highest number of repeats detected (about 100) started showing disease symptoms at the extraordinarily early age of two. Finally, two individuals were affected, even though their parents were not. In both cases, the affected patients had expanded CAG repeats, whereas their parents did not. New mutations (expanded CAG repeats) apparently caused both these cases of disease.

How can we put this new knowledge to work? One obvious way is to perform accurate genetic screening to detect people who will be affected by the disease. In fact, by counting the CAG repeats, we may even be able to predict the age of onset of the disease. However, that kind of information is a mixed blessing; it can be psychologically devastating (see box 15.3). What we really need, of course, is a cure, but that may be a long way off.

In order to find a cure for HD, we need to answer at least two questions. First, what is the normal function of the *HD* product (now called *huntingtin*)? To get at that question, geneticists are constructing transgenic mice carrying the mutant form of *HD*. This should shed light on the function of the gene in the brain and in other tissues, where it is also active. Second, we need to know how the expansion of the CAG repeat causes disease. We know it does not simply inactivate the gene, because the mutant allele is

Box 15.3 Problems in Genetic Screening

Now that we have identified RFLPs and other genetic markers closely associated with genes for Huntington disease, cystic fibrosis, and other serious maladies, how will we use this information? We are in a position to identify carriers of genetic diseases or people who are at risk for developing these debilitating conditions. This sounds like a good thing; in general, it undoubtedly will be, but there are some ethical problems.

Consider, for example, the actual case of an adoption agency that was trying to place a baby girl whose mother suffered from Huntington disease. Understandably, potential adoptive parents wanted to know if the baby had inherited the gene. However, Michael Conneally, the geneticist consulted in the case, decided not to test the child for the following reason: If the results were positive and were later shared with the girl, she would know that she was doomed to get the disease. Many people at risk of developing this terrible disease feel that they cannot cope with knowing for sure that they have the gene, so they elect not to take the test. In this case, the child was too young to choose for herself.

Another dilemma would occur in the following situation: A woman is at risk of developing Huntington disease because her father has it. She has decided she cannot bear to know that she has inherited the gene, so she will not submit to a test. Then she gets pregnant. She and her husband want the child, but not if it carries the defective gene. They could check the fetus's genotype by amniocentesis. A negative result would end their worries about the child and allow the mother to cling to hope for her own future, but a positive result would confirm her worst fears, not only for the baby but for herself.

In the case of the autosomal recessive condition cystic fibrosis, a different set of problems arises. Cystic fibrosis is the most common lethal genetic disease afflicting American children. One in twenty-five Caucasians is a CF carrier, and therefore, one in 2,500 children will be homozygous for CF and suffer the

disease. Now that the CF gene has been identified, one would predict that millions of people would want to be tested to help them with their family planning. Nevertheless, in 1989 the American Society for Human Genetics called for a moratorium on widespread screening.

The immediate reason for this reluctance was that the mutation found by Tsui, Riordan, Collins, and colleagues is present in only 70% of the CF carriers. The other 30% have a variety of other mutations and would therefore test negative. This means that only about 50% ($70\% \times 70\% = 49\%$) of couples at risk would be identified, and that was deemed too low to warrant mass screening.

However, the equation is changing rapidly, since dozens of labs are working together, identifying the remaining mutations. At what point will it be appropriate to begin widespread testing? This is a difficult enough question, but even if the test were perfect, problems would remain. The major difficulty arises from the sheer magnitude of the problem. In the United States alone, there are 100–200 million people in the reproductive age group. Even assuming we could test that many people, someone has to explain to them what the results mean, and there are not nearly enough professional genetic counselors to do that.

Our experience with screening African-Americans for sickle-cell disease teaches us what can go wrong with a genetic screening program. In the early 1970s, many states passed laws requiring testing of African-Americans for sickle-cell disease, but did not provide enough counseling to make the program successful. This had several unfortunate outcomes. For one, many people who merely carried the sickle-cell gene came to the mistaken and distressing conclusion that they actually had the disease. Also, the suggestion to African-American couples who were both carriers that they forego having children, coming as it frequently did from whites, provoked charges of racism. Moreover, in some cases confidentiality was not respected. This led to the denial of health insurance to some carriers. The result of all this was a loss of confidence in the screening

process among those it was designed to help. Thus, few people took part, and even fewer used the information they received to help in their family planning.

These cases exemplify the tough decisions that people must make about their own genetic testing and the psychological impact such tests may have on them. Quite a different kind of question arises concerning the mandatory testing of others. Would it be ethical, for example, for insurance companies to demand genetic tests? Those who test positive for such conditions as Huntington disease or Alzheimer's disease would either be denied health insurance or forced to pay exorbitant premiums. We have already seen that this happened to carriers of the sickle-cell gene. If it is not fair to burden people because of their genes, over which they have no control, would it be fair to require them to submit to tests before obtaining life insurance? Keep in mind that life insurance companies have traditionally required proof of good health before issuing insurance and that genetic screening could simply be interpreted as an extension of this practice.

Let us take the problem one step further and assume that a life insurance company has paid for genetic testing of an individual and found him or her likely to develop a debilitating disease such as Alzheimer's. Would it be ethical for the insurance company to share such information with others—the person's employer or prospective employer, for instance? This information would certainly give an indication of the person's long-term fitness for employment, but it would just as certainly place an enormous burden on that person and maybe even render him or her unemployable. The problem becomes more difficult if the insurer is also the employer, as is the case with some large corporations.

These thorny questions have no easy answers, but the development of more and more tests for genetic diseases will force us to deal with them.

dominant. Gusella postulates that the extra glutamines in the protein give it a new function that somehow disrupts brain activity. This new function could be something as simple as binding to the wild-type protein and blocking its operation. Given that the sequence of huntingtin does not provide a clue to its function, and given the technical difficulty of working with unstable repetitive DNAs, these questions will not be easy to answer. But at least now we know what the questions are.

Using RFLPs, geneticists mapped the Huntington disease gene (*HD*) to a region near the end of chromosome 4. Then they used an exon trap to identify the gene itself. The mutation that causes the disease is an expansion of a CAG repeat from the normal range of 11–34 copies, to the abnormal range of at least 42 copies.

Mapping the Cystic Fibrosis Gene

Investigators have used RFLPs and a variety of other methods to identify the cystic fibrosis (CF) gene. One of the techniques used in this fine mapping takes advantage of the fact that active human genes tend to be associated with unmethylated CG sequences, whereas the CG's in inactive regions are almost always methylated (chapter 14). Furthermore, the restriction enzyme *Hpa*II cuts at the sequence CCGG, but only if the second C is unmethylated. In other words, it will cut active genes that have unmethylated CG's within CCGG sites, releasing small DNA fragments, but it will leave inactive genes (with methylated CCGG's) alone. Thus, Robert Williamson and colleagues scanned a relatively large region of DNA linked to the CF gene for "islands" of sites that could be cut with *Hpa*II in a "sea" of other DNA sequences that could not be cut. They found such a DNA region extremely tightly linked to the CF gene. (This region is called an **HTF island** because it yields *Hpa*II tiny fragments.)

Once the gene had been located within a 1.5 megabase (million base pairs) region of the q31 band on the long arm of chromosome 7, a large team of researchers, headed by Lap-Chee Tsui, Jack Riordan, and Francis Collins, obtained the base sequence of more than 250 kb of this DNA. Within this sequence they found the CF gene. Its relationship with the disease is quite clear. First, it codes for a membrane-spanning protein that mediates the flow of chloride ions across the membrane. (Abnormal chloride ion transport in lung and sweat gland tissue is a hallmark of cystic fibrosis.) Second, the normal allele of this gene is active in sweat glands. Finally, CF patients show a mutation in this gene. In fact, 68% of them show exactly the same mutation: the loss of one codon out of a total of 1,480 in the whole gene.

This results in the loss of one amino acid (a phenylalanine) from the protein, but it is in a key position—a part of the protein that binds ATP. Apparently this disrupts ATP binding, preventing the protein from obtaining energy from ATP to drive chloride ion transport in the lung and other tissues.

Identifying the Duchenne Muscular Dystrophy Gene

Geneticists have also found the exact gene responsible for Duchenne muscular dystrophy (DMD)—without using RFLPs. Several methods were employed, including classical mapping, but a key piece of the puzzle came from studying the rare cases of girls afflicted with the disease. Ronald Worton and coworkers examined the X chromosomes of twelve such patients and found that all had X chromosomes broken in about the same place—band Xp21 near the middle of the chromosome's short arm. This strongly suggests that the break in every case occurred in the DMD gene, inactivating it and thereby causing the disease.

L. M. Kunkel and his colleagues have cloned the DMD gene and have found that it is enormous. It contains at least sixty exons and spans 2 megabases of DNA. The mature mRNA produced from this gene comprises more than 14 kilobases and encodes a gigantic polypeptide having a molecular weight of about 400,000. This protein, called dystrophin, is about ten times the size of an average polypeptide.

Dystrophin has been located in human muscle cells using antibodies directed against the product of the very similar mouse DMD gene. It is satisfying that these antibodies react with a protein in normal muscle cells but not in muscle cells from DMD patients. It is also gratifying that this protein has a relative molecular mass of about 400,000, as predicted. The geneticists studying DMD have thus completed an important part of their job. The next step will be for cell biologists to determine the role dystrophin plays in normal muscle cells. This, in turn, may give clues about how to intervene in the disease. It may even be possible someday to perform gene therapy by providing DMD patients with a functional dystrophin gene.

We can map genes for human diseases even if we do not know the genes' functions. A common way to do this is by identifying a pattern of cutting by a restriction enzyme (a restriction fragment length polymorphism, or RFLP) that is found in people afflicted by the disease but not in other members of the same family. Once linkage between the disease gene and such a RFLP has been established, the gene can be localized to a chromosome, and more detailed mapping can begin.

WORKED-OUT PROBLEMS

Problem 1

Here is the amino acid sequence of part of a hypothetical protein whose gene you want to clone:

Phe-Leu-Pro-Trp-Ile-Gln-Met-Arg-His

(a) What sequence of five codons for amino acids from the sequence above would give a 15-mer probe with the least degeneracy?

(b) How many different 15-mers would you have to make in order to be sure that one of the 15-mers in your probe perfectly matches the corresponding sequence in your cloned gene?

(c) If you started your probe one codon to the left relative to the one you chose, how many different 15-mers would you have to make?

(d) If you started your probe one codon to the right relative to the one you chose, how many different 15-mers would you have to make?

Solution

(a) We begin by consulting the genetic code to determine the coding degeneracy of each amino acid in the sequence. This yields:

$$2 \quad 6 \quad 4 \quad 1 \quad 3 \quad 2 \quad 1 \quad 6 \quad 2$$
Phe-Leu-Pro-Trp-Ile-Gln-Met-Arg-His

where the numbers above the amino acids represent the coding degeneracy for each. In other words, there are 2 codons for phenylalanine (Phe), 6 for leucine (Leu), 4 for proline (Pro), etc. Now the task is to find the contiguous set of five codons with the lowest degeneracy. This is Pro-Trp-Ile-Gln-Met.

(b) To compute the number of different 15-mers we have to make, we multiply the degeneracies at all positions within the region covered by the probe. For the five amino acids we have chosen this is $4 \times 1 \times 3 \times 2 \times 1 = 24$.

(c) If we start our probe one codon to the left, we will add Leu and lose Met. This substitutes an amino acid with sixfold degeneracy (Leu) for one with no degeneracy (Met), which is clearly undesirable. To calculate the number of different 15-mers we would need to construct, we perform the same calculation as in part (b), but leave off the Met degeneracy (1) and add the Leu degeneracy (6): $6 \times 4 \times 1 \times 3 \times 2 = 144$. This dramatically illustrates the negative impact of adding an amino acid with sixfold degeneracy. It multiplies the number of oligonucleotides we need by a factor of 6.

(d) If we start our probe one codon to the right of the original, we add Arg and lose Pro. This substitutes an amino acid with sixfold degeneracy (Arg) for one with fourfold degeneracy (Pro). To perform the new calculation, we leave off the Pro degeneracy (4) and add the Arg degeneracy (6): $1 \times 3 \times 2 \times 1 \times 6 = 36$. This is only slightly worse than the original, but still worth avoiding.

Problem 2

Here is the base and amino acid sequence of a portion of a gene and its protein product:

Leu Pro Cys Gly Asn
TTA CCG TGC GGA AAC

You want to probe the effect of the SH group on the Cys by changing this amino acid to Ser, which has an OH group instead. What 15-mer would you prepare to serve as a primer in a site-directed mutagenesis experiment to make this change? Consult the genetic code for the appropriate Ser codons.

Solution

There are six Ser codons, but one of them (TCC) differs from the Cys codon TGC in only one base. This is the one to use. Therefore, the sequence of the whole 15-mer would be

TTACCGT**C**CGGAAAC

The boldfaced C is the only altered nucleotide. In practice, we would add more nucleotides on either side of the altered one to stabilize hybridization between the primer and the DNA template.

Summary

To clone a gene, we must insert it into a vector that can carry the gene into a host cell and ensure that it will replicate there. The insertion is usually carried out by cutting the vector and the DNA to be inserted with the same restriction nucleases to endow them with the same "sticky ends." Vectors for cloning in bacteria come in two types: plasmids and phages.

Among the plasmid cloning vectors are pBR322 and the pUC plasmids. pBR322 has two antibiotic resistance genes and a variety of unique restriction sites into which we can introduce foreign DNA. Most of these sites interrupt one of the antibiotic resistance genes, making selection straightforward. Selection is even easier with the pUC plasmids. These have an ampicillin resistance gene and a multiple cloning site that interrupts a β-galactosidase gene whose product is easily detected with a color test. We select for ampicillin-resistant clones that do not make β-galactosidase.

Two kinds of phage have been especially popular as cloning vectors. The first is λ, which has had certain nonessential genes removed to make room for inserts. In some of these engineered phages, called Charon phages, inserts up to 20 kb in length can be accommodated. Cosmids, a cross between phage and plasmid vectors, can accept inserts as large as 50 kb. This makes these vectors very useful for building genomic libraries. The second major class of phage vectors is the M13 phages. These vectors have the convenience of a multiple cloning region and the further advantage of producing single-stranded recombinant DNA, which can be used for DNA sequencing and for site-directed mutagenesis.

Expression vectors are designed to yield the protein product of a cloned gene, usually in the greatest amount possible. To optimize expression, these vectors provide strong bacterial promoters and bacterial ribosome binding sites that would be missing on cloned eukaryotic genes. Most cloning vectors are inducible, to avoid premature overproduction of a foreign product that could poison the bacterial host cells. Expression vectors frequently produce fusion proteins, which often have the advantages of stability and ease of isolation. We can introduce changes into cloned genes at will, thus changing the amino acid sequences of the protein products.

Cloned DNAs have many uses. They make excellent probes because of their ability to hybridize to DNAs or RNAs with the same, or very similar, sequences. Because of their homogeneity, they also provide ideal material for determining a DNA's base sequence. Some products of cloned genes (human insulin, human growth hormone) are already finding uses as pharmaceuticals. In the near future, they may also provide safe, effective vaccines. In the more distant future, it may be possible to treat genetic defects by transplanting cloned, normal genes into patients whose own genes are damaged. Genes are already being transferred to plants, and there is great potential for using this technique to improve crops.

Genes for human diseases can be mapped even if we do not know their functions. A common tool for performing such mapping is a restriction fragment length polymorphism (RFLP). A particular pattern of cutting by a given restriction endonuclease can be treated as a genetic marker; if it is tightly linked to the disease gene, it can be used as a first step in locating the gene.

PROBLEMS AND QUESTIONS ❓

1. The hypothetical bacterium *Xenobacterium giganticus* produces one restriction endonuclease. According to the convention for naming these enzymes, what would you call it?

2. Here is the sequence of a small piece of DNA that contains a 6-base palindromic recognition site for a restriction endonuclease: GACGATATCAACT. What is the sequence of this recognition site?

3. Some restriction endonucleases recognize 5-base sites in target DNA. Approximately how frequent are these 5-base sites?

4. Here are the recognition sites of four restriction enzymes:

*Bam*HI GGATCC
CCTAGG

*Eco*RI GAATTC
CTTAAG

*Hin*dIII AAGCTT
TTCGAA

*Bgl*II AGATCT
TCTAGA

Each cuts between the first two bases on the top strand and between the last two bases on the bottom strand, as indicated by the arrows. Which of these enzymes will create compatible (complementary) sticky ends? What is the sequence of this compatible sticky end?

5. Suppose you cut two different DNAs, one with *Bam*HI and one with *Bgl*II, then ligate them together through their compatible sticky ends. Once joined, could you separate these two DNAs again with either restriction enzyme? Why, or why not?

6. Suppose you transform bacteria with a recombinant DNA and, by mistake, plate the transformed cells on medium lacking any antibiotic. What result would you observe? Why?

7. You want to insert a gene into the *Bam*HI site of pBR322. Which antibiotic would you use to distinguish between cells that have taken up plasmid without insert and those that have taken up plasmid with insert? Consult the map of pBR322 (fig. 15.2).

8. Is it possible to use antibiotics to determine whether pBR322 has received an insert in its *Eco*RI site? Why, or why not?

9. Would a plasmid like pUC8 be a good choice as a vector with which to construct a genomic library of a complex organism? Why, or why not?

10. Would a plasmid like pUC8 be a good choice as a vector with which to construct a genomic library of a relatively simple viral genome? Why, or why not?

11. You have constructed a cDNA library in λgt11 using mRNAs from a pea plant. You want to identify a rubisco clone among the 50,000 clones in the library. You have a rubisco cDNA from tobacco to use as a probe. When hybridizing the probe to the plaques in the library, would you use a higher, lower, or the same temperature as you would use for hybridizing the tobacco DNA to a tobacco rubisco clone? Why?

12. Here is the amino acid sequence of part of a hypothetical protein whose gene you want to clone:

Pro-Arg-Tyr-Met-Cys-Trp-Ile-Leu-Met-Ser

(a) What sequence of five amino acids would give a 15-mer probe with the least degeneracy? (b) How many different 15-mers would you have to make in order to be sure that your probe matches the corresponding sequence in your cloned gene perfectly? (c) If you started your probe one amino acid to the left of the one you chose in (a), how many different 15-mers would you have to make? Use the genetic code to determine degeneracy.

13. When and why can you ignore the ligation step in forming a recombinant DNA for transforming host cells?

14. What is the advantage of inducibility in an expression vector?

15. When we use λgt11 as a cDNA cloning vector, we can screen for expression of our inserted cDNA as a fusion protein, using antibody as described in figure 15.16. In order for this to work, the cDNA must be in the same orientation and in the same reading frame as the λ gene to which it is fused. From a population of λgt11 cDNA clones containing fragments of our gene of interest, what is the theoretical fraction that would satisfy these two conditions? Why?

16. Here is the base and amino acid sequence of a portion of a gene and its protein product:

Thr Asp Ala Val Gly
ACC GAT GCA GTG GGC

You are performing a site-directed mutagenesis experiment to change the gene so that it encodes Ser instead of Ala in this region. What 15-mer would you construct to serve as primer for making this change? Use the genetic code to find the Ser codons.

17. A certain animal virus contains a circular DNA that has five recognition sites for the restriction enzyme EcoRI. Infected cells make a large quantity of a viral surface protein. A cDNA corresponding to the mRNA for this surface protein has been cloned. How would you use a Southern blot to determine which EcoRI fragment(s) contain the gene for the viral surface protein?

18. You have cloned a gene from the petunia and want to know which tissue (leaves, stems, roots, flowers, or seeds) transcribes this gene to the greatest extent. How would you use a Northern blot to answer this question?

19. What will the Northern blot in question 18 tell you besides the extent of transcription of the petunia gene in each tissue?

20. You find a family that is polymorphic for the second HindIII site from the right in figure 15.31. What would be the size of the new restriction fragment observed in family members lacking this site?

21. If the family in problem 20 is polymorphic for the two other HindIII sites, as well as the new one, how many haplotypes could there be with respect to this region of the genome?

22. If the rightmost HindIII site lay just to the right of the edge of the G8 probe, would polymorphisms with respect to this site be detectable by this probe? Why, or why not?

23. Imagine the situation in problem 22 and still another HindIII site even further to the right. Would polymorphisms in this site be detectable by the G8 probe? Why, or why not?

24. Find the longest open reading frame in this sequence of bases:

CCAGATGCCTAAATGAGTTGGCCAGCAGAGCGAGCATGGATGTAATCAG

Answers appear at end of book.

Female gametes (frequency)

A_1(p)　　　　A_2(q)

Male gametes (frequency)

A_1(p)

$A_1A_1(p^2)$　　　$A_1A_2(pq)$

A_2(q)

$A_1A_2(pq)$　　　$A_2A_2(q^2)$

Δq

0.00
−0.01
−0.02
−0.03
−0.04

0.0　0.2　0.4　0.6　0.8　1.0

Chapter 16:
*An Introduction to
Population
Genetics*

Chapter 17:
*Extensions and
Applications of
Population Genetics*

CHAPTER 16: AN INTRODUCTION TO POPULATION GENETICS

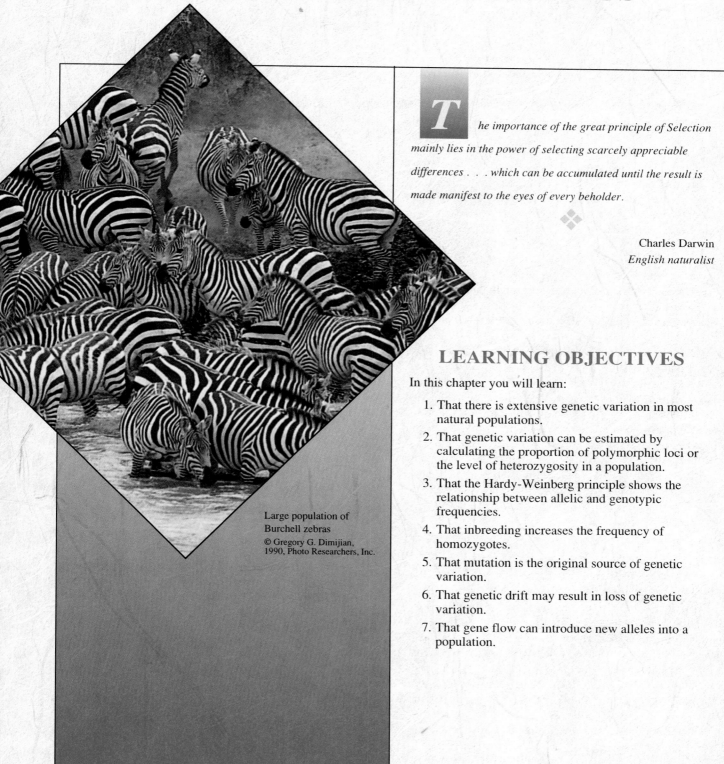

Large population of
Burchell zebras
© Gregory G. Dimijian,
1990, Photo Researchers, Inc.

*T*he importance of the great principle of Selection
mainly lies in the power of selecting scarcely appreciable
differences . . . which can be accumulated until the result is
made manifest to the eyes of every beholder.

Charles Darwin
English naturalist

LEARNING OBJECTIVES

In this chapter you will learn:

1. That there is extensive genetic variation in most natural populations.

2. That genetic variation can be estimated by calculating the proportion of polymorphic loci or the level of heterozygosity in a population.

3. That the Hardy-Weinberg principle shows the relationship between allelic and genotypic frequencies.

4. That inbreeding increases the frequency of homozygotes.

5. That mutation is the original source of genetic variation.

6. That genetic drift may result in loss of genetic variation.

7. That gene flow can introduce new alleles into a population.

U ntil now, we have focused on genetic variants that occur rarely in a given group of individuals—for example, mutations that result in inborn errors of metabolism, such as albinism in humans. However, natural populations of most species have alleles in substantial frequencies that affect such traits as body color, red-blood-cell types, or pesticide resistance. Why do these different genetic types exist within a population, and what factors determine their frequencies? Answering these questions necessitates an understanding of **population genetics,** the field that focuses on the extent and pattern of genetic variation in natural populations and the explanations for these observations.

Natural selection has often been considered the major force determining the extent and pattern of genetic variation, an idea that goes back to Charles Darwin's *On the Origin of Species by Means of Natural Selection,* first published in 1859. However, other factors also appear to be important in given situations or in certain species. In this chapter we will introduce those factors: inbreeding, mutation, genetic drift, and gene flow. In chapter 17 we will look at the combined effects of several of these factors in order to understand specific topics in population genetics: namely, the incidence of genetic disease in humans, the evolution of molecules, conservation genetics, agricultural genetics, and pesticide resistance.

We will find Mendelian genetic principles essential to understanding the genetic composition of populations. For example, the principle of segregation (that two alleles in a heterozygous individual are equally represented in the gametes produced by that individual), when applied to all individuals in a population, allows us to understand the overall genetic variation in that population. The genetic constitution of a population is sometimes referred to as its **gene pool.** Let us first describe some examples of genetic variation before discussing the principles of population genetics that explain the basis of this variation and how it may change.

TYPES OF GENETIC VARIATION

Genetic variants can directly affect the morphology or color of an organism in a natural or wild population. A classic example of genetic variation that has fascinated evolutionary biologists in England and France for decades is the color and banding pattern of *Cepaea nemoralis,* a European land snail common in gardens and uncultivated fields. The morphological variation is determined by the alleles at a group of closely linked genes, so that the snails vary strikingly in color and pattern from yellow, unbanded types to yellow, banded types to brown, unbanded ones (fig. 16.1). The frequencies of the alleles that determine yellow color and bandedness vary on a large scale from one part of Europe to another and on a small scale from one field to another.

In one detailed study, the frequency of yellow, unbanded types over a 200-meter stretch on an earthen dam in the Netherlands varied from 60% at one end to 40% at the other. Furthermore, these populations remained at the same frequencies over a twelve-year period, suggesting that evolutionary forces were maintaining the differences. For example, it appears from other studies that the different forms confer protective camouflage from the predation of birds, depending upon the habitat in which they are found. When the snails are resting on grass blades, the bands are vertical, so the banded snails blend in with the grass. But when the snails are in the ground litter, the unbanded brown snails are the most difficult to detect.

In a number of species, variation in chromosome structure has been observed among different individuals. Recently, new techniques developed to examine morphological details of chromosomes (see chapter 4) have uncovered many new chromosomal variants, and it is likely that numerous other small variations in chromosome structure, such as deletions, replications, and inversions, are still to be discovered. One of the most thoroughly studied examples of chromosomal variation was initiated by Theodosius Dobzhansky, a Russian geneticist who moved to the United States in the 1920s. He and his students analyzed the inversion variants on the third chromosome of the fruit fly *Drosophila pseudoobscura,* a widespread species in the mountains of western North America (see fig. 4.25).

In the 1930s, Dobzhansky began a long-term study of variation in this inversion. Some of these data, showing the frequencies of four inversion types over thirty years of sampling at a site in New Mexico, are given in figure 16.2. Initially, the inversion type AR (Arrowhead) was most common, but in 1965 type PP (Pikes Peak) increased in frequency, only to decline in later years. Two other inversion types, CH (Chiricahua) and ST (Standard), were always present, but at low frequencies in the samples from this population.

These inversions, although inherited as single units like alleles (see chapter 4), generally contain hundreds of genes. Because different inversions may have different alleles at these loci, they can substantially affect the survival of individual flies. In fact, some of the changes observed over time may have been caused by differential survival of various inversions to DDT, an insecticide that was first applied in the 1940s.

Most alleles that cause genetic diseases in humans occur in very low frequencies. However, some disease alleles have higher frequencies in certain human populations than in others. For example, the frequency of the sickle-cell allele at the β-globin locus is quite high in many African populations and in populations with African ancestry, such as African-Americans. The sickle-cell allele also occurs in some Mediterranean and Asian populations (fig. 16.3a).

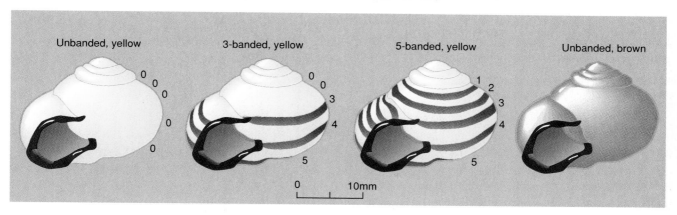

FIGURE 16.1

Several shell banding patterns observed in the snail *Cepaea nemoralis*, where the non-zero numbers indicate the presence of particular bands.

Reproduced, with permission, from the Annual Review of Ecology and Systematics, *Vol. 8, © 1977 by Annual Reviews, Inc.*

FIGURE 16.2

Frequencies of inversions on the third chromosome in *Drosophila pseudoobscura* that have taken place over a period of three decades in the Capitan area of New Mexico. (AR = Arrowhead; PP = Pikes Peak; CH = Chiricahua; ST = Standard.)

The presence of the sickle-cell allele can be ascertained by biochemical examination of the β-globin molecule (see a photo of normal and sickle cells, figure 11.5). Variant forms of a protein can be detected by **gel electrophoresis** (see also chapter 6), a technique that allows the separation of different proteins extracted from blood, tissues, or whole organisms. The process is carried out by imposing an electric field across a gelatinous supporting medium (such as a polyacrylamide or starch gel) into which the protein has been placed. The proteins are allowed to migrate for a specific amount of time and are then stained with various protein-specific chemicals, resulting in bands on the gel that permit the relative mobility of a specific protein to be determined (fig. 16.4). The relative mobility is generally a function of the charge, size, and shape of the molecule. If two proteins that are the products of different alleles have different amino acid sequences, they often have different mobilities as well, because the differences in sequence result in a change in charge, size, or shape of the molecule.

The three β-globin genotypes (homozygotes for the normal and sickle-cell allele plus a heterozygote) have different banding patterns when they are run on an electrophoretic gel as shown in figure 16.5. The normal A_1A_1 genotype has only one band that migrates relatively fast as in the first column, while individuals with sickle-cell disease, genotype A_2A_2 in column 3, also have only one band, but it migrates more slowly. The heterozygous individuals, A_1A_2, who have the sickle-cell trait (they are carriers) have two bands (see column 2), each band representing the protein synthesized from the respective allele. Sometimes these genotypes at the β-globin locus are represented as *AA*, *AS*, and *SS*, instead of A_1A_1, A_1A_2, A_2A_2.

The reason for the high frequency of this disease allele is that heterozygotes (those who carry this allele and the normal allele) have a relatively high resistance to malaria. The relative immunity of sickle-cell carriers to malaria is

FIGURE 16.3

(*a*) The geographic distribution of the sickle-cell allele, indicating the regions of low frequency (white) and higher frequency (dark green) and (*b*) The geographic distribution of malaria (blue-green).

From "The Genetics of Human Populations" by L. L. Cavalli-Sforza. Copyright © 1974 by Scientific American, Inc. All rights reserved.

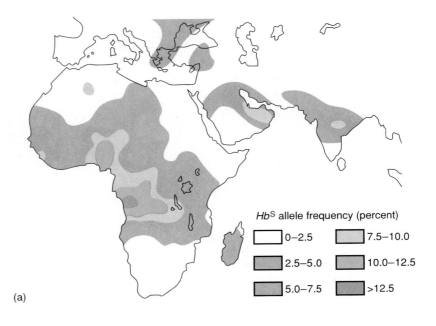

Hb^S allele frequency (percent)

0–2.5	7.5–10.0
2.5–5.0	10.0–12.5
5.0–7.5	>12.5

(a)

(b)

FIGURE 16.4

Diagram of a gel electrophoresis apparatus. The buffers are used to conduct electricity and ensure a given pH.

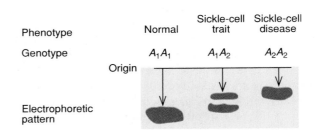

FIGURE 16.5

The phenotypes, genotypes, and electrophoretic patterns of the β-globin gene.

thought to occur because the red blood cells of heterozygotes are sickle-shaped, a state that inhibits infection by the malarial parasite. When we compare the frequency of the sickle-cell allele and the past incidence of malaria on a geographic scale (before the extensive eradication of the mosquito that carries the malarial parasite), we find an extremely high concordance (see fig. 16.3).

 Extensive genetic variation occurs in most populations at several different levels, including variants affecting color, chromosomal structure, and protein characteristics.

MEASURING GENETIC VARIATION

How much genetic variation exists in a given population? How can we quantify this variation in a standardized manner? To properly answer these questions, we need to introduce some quantitative measures and the symbols and statistics that accompany them. However, let us first define a **population** as a group of interbreeding individuals of the same species that exist together in time and space. For example, from a genetic viewpoint, all the breeding largemouth bass in a small lake in a given year constitute a population.

To determine the frequencies of the genotypes or alleles at a given locus in a population, we must first count the number of individuals having different genotypes. For example, at the β-globin gene, let N_{11}, N_{12}, and N_{22} indicate the number of individuals counted with genotypes A_1A_1, A_1A_2, and A_2A_2, respectively. If there are N total individuals in the sample, the estimated frequencies of these three genotypes in the population are:

$$P = \frac{N_{11}}{N} \quad \text{for} \quad A_1A_1$$

$$H = \frac{N_{12}}{N} \quad \text{for} \quad A_1A_2$$

$$Q = \frac{N_{22}}{N} \quad \text{for} \quad A_2A_2$$

Because there are three genotypes, the sum of the frequencies of the three genotypes must be one; that is, $P + H + Q = 1$.

Notice that all the alleles in the homozygote A_1A_1 are A_1 and that all the alleles in the homozygote A_2A_2 are A_2. In addition, half of the alleles in the heterozygotes are A_1 and half are A_2. Therefore, we can calculate the frequency of allele A_2 (designated by q) as the sum of half the frequency of the heterozygote plus the frequency of the A_2A_2 homozygote. In other words:

$$q = \tfrac{1}{2}H + Q$$

If we substitute the numbers of different genotypes given in the first formulas, then:

$$q = \frac{\tfrac{1}{2}N_{12}}{N} + \frac{N_{22}}{N}$$
$$= \frac{\tfrac{1}{2}N_{12} + N_{22}}{N}$$

If we designate the frequency of the other allele, A_1, as p, then using the same logic:

$$p = P + \tfrac{1}{2}H$$

and

$$p = \frac{N_{11} + \tfrac{1}{2}N_{12}}{N}$$

Because there are only two alleles, the sum of their frequencies must be one, or $p + q = 1$. Furthermore, if we subtract q from both sides of this equation, we get the useful relationship $p = 1 - q$.

Let us illustrate the use of these formulas by a numerical example. In parts of Africa, the frequency of the sickle-cell allele is quite high, considering that it causes a severe disease when homozygous. In one sample of 400 individuals from western Africa, 320 were normal (genotype A_1A_1), 72 were carriers of the sickle-cell trait (A_1A_2), and 8 (mostly young individuals) actually had sickle-cell disease (A_2A_2). In other words, N_{11}, N_{12}, and N_{22} were 320, 72, and 8, respectively. The frequencies of the three genotypes in this population would be expressed as:

$$P = \,^{320}\!/_{400} = 0.80$$
$$H = \,^{72}\!/_{400} = 0.18$$
$$Q = \,^{8}\!/_{400} = 0.02$$

This means that 2% of the individuals have sickle-cell disease and 18% are carriers of the sickle-cell allele. Using the formula given for the frequency of allele A_2, the frequency of the sickle-cell allele is:

$$q = \tfrac{1}{2}(0.18) + 0.02$$
$$= 0.11$$

The frequency of the normal allele, A_1, is $p = 1 - q = 0.89$.

Now that we have introduced these formulas, we can discuss two useful ways to quantify the extent of genetic variation in a population. Using the allelic frequencies calculated for a given gene, we can define a gene as **polymorphic** (having many forms; that is, having two or more alleles in substantial frequency) or **monomorphic** (having only one allele in high frequency). The simplest measure of genetic variation is to categorize loci as polymorphic or monomorphic and then to calculate the proportion of all polymorphic loci. By general (although somewhat arbitrary) definition, if the frequency of the most common allele is 0.99 or greater (or if all other alleles combined have a frequency of 0.01 or less), the gene is considered monomorphic; otherwise, it is polymorphic. For example, the sickle-cell gene in the example just given is polymorphic for the two alleles A_1 and A_2 because the frequency of A_1 is less than 0.99.

A second approach for measuring genetic variation in a population is to calculate the average heterozygosity over all loci. The measure of heterozygosity is useful because it takes into account all levels of genetic variation rather than just classifying loci into two categories as does the first method. The average heterozygosity for a sample of n different loci is:

$$H = \,^1\!/_n(H_A + H_B + H_C \ldots)$$

where H_A is the heterozygosity for gene A, H_B for gene B, and so on. As a simple example, if the heterozygosities at five loci are $H_A = 0.5$, $H_B = 0.0$, $H_C = 0.1$, $H_D = 0.0$, and $H_E = 0.2$, then n = 5, and the average heterozygosity is:

$$H = (^1/_5)(0.5 + 0.0 + 0.1 + 0.0 + 0.2)$$
$$= 0.16$$

Note that in this case three loci, A, C, and E, are polymorphic, making the proportion of polymorphic loci $^3/_5 = 0.6$.

The first attempts to quantify the overall extent of genetic variation in a given species were made in the mid-1960s by Richard Lewontin in *D. pseudoobscura* and by Harry Harris in humans. Both surveys used protein electrophoresis and demonstrated substantial genetic variation from a sample of genes. As an illustration, some of Harris's data from a Caucasian population in England is given in table 16.1 for seventy-one different genes. Twenty of these loci are polymorphic; that is, they have the most common allele with a frequency of 0.99 or less. Therefore, the proportion of polymorphic loci in this sample is $^{20}/_{71} = 0.282$. Notice that the level of heterozygosity ranges from 0.00 for the monomorphic loci to around 0.5 for five polymorphic loci, most of which have multiple alleles. The average heterozygosity over all seventy-one loci is 0.067. Another way to interpret this value is that, on the average, about 6.7% of the loci in a given individual are heterozygous for such biochemical variants.

A relatively recent discovery in population genetics (and one that caused a radical reexamination of its foundation) is the high degree of polymorphism and heterozygosity that was found in nearly all natural populations once modern biochemical techniques were applied. At first thought, this level of genetic variation is surprising, because little variation appears in the wild-type morphology of most populations. However, the success of plant and animal breeding programs in changing morphological traits such as size (see chapters 3 and 17) suggests that there is extensive background genetic variation.

Surveys of the genetic variation in various species demonstrate that up to 20% of the genes examined by electrophoresis are polymorphic and that 10% of these genes are heterozygous in a given individual. Table 16.2 lists a number of species with different levels of polymorphism. Generally, the highest levels of genetic variation occur in invertebrates and in some plants, while vertebrates generally have lower levels of variation. Several organisms are of particular interest. The horseshoe crab, an animal that has not changed morphologically in millions of years according to fossil records, has a high level of genetic variation. Thus, **stasis** (no change) in morphology over geological time does not necessarily indicate lack of genetic variation on a biochemical level. On the other hand, the Northern elephant seal, which was nearly hunted to extinction in the nineteenth century, and the cheetah, also an endangered species, show no genetic variation in these electrophoretic surveys. In both cases, small population

Table 16.1 Heterozygosity for Seventy-one Human Loci

Locus	Heterozygosity (H)
51 monomorphic loci	0.00
Peptidase C	0.02
Peptidase D	0.02
Glutamate-oxaloacetate transaminase	0.03
Leukocyte hexokinase	0.05
6-phosphogluconate dehydrogenase	0.05
Alcohol dehydrogenase-2	0.07
Adenylate kinase	0.09
Pancreatic amylase	0.09
Adenosine deaminase	0.11
Galactose-1-phosphate uridyl tranferase	0.11
Acetylcholinesterase	0.23
Mitochondrial malic enzyme	0.30
Phosphoglucomutase-1	0.36
Peptidase A	0.37
Phosphoglucomutase-3	0.38
Pepsinogen	0.47
Alcohol dehydrogenase-3	0.48
Glutamate-pyruvate transaminase	0.50
RBC acid phosphatase	0.52
Placental alkaline phosphatase	0.53

From Harris and Hopkinson, *Annals of Human Genetics*, 36:9–20. Copyright © 1972 Cambridge University Press, New York, NY. Reprinted with the permission of Cambridge University Press.

numbers appear to have caused a loss of genetic variation via genetic drift, a phenomenon that will be discussed later in this chapter and in chapter 17. Finally, barley is a plant with approximately 99% self-fertilization, and thus it has almost no heterozygotes (see discussion of inbreeding), yet it has a substantial number of polymorphic loci when plants are examined over several populations.

 By counting the numbers of different genotypes in a population, we can calculate genotypic frequencies, allelic frequencies, and heterozygosity. Two measures of genetic variation, the proportion of polymorphic loci and the level of heterozygosity, are used to quantify the extent of genetic variation. In many species, the amount of genetic variation observed for genes coding for proteins is substantial.

THE HARDY-WEINBERG PRINCIPLE

The description of the genetic variation in a population is often greatly simplified when allelic frequencies, rather than genotypic frequencies, can be used. In 1908, shortly after the rediscovery of Mendel's work, an English mathematician, Godfrey Hardy, and a German physician, Wilhelm Weinberg, showed that a simple relationship often exists between allelic frequencies and genotypic frequencies.

Table 16.2 Average Proportions of Polymorphic Loci and Average Heterozygosity for Various Organisms

Organism	Number of Genes Examined	Proportion Polymorphic	Heterozygosity
Humans	71	0.28	0.067
Northern elephant seal	24	0.0	0.0
Elephant	32	0.29	0.089
Cheetah	52	0.0	0.0
Tree frog	27	0.41	0.074
Drosophila pseudoobscura	24	0.42	0.12
Horseshoe crab	25	0.25	0.057
Barley	28	0.30	0.003

FIGURE 16.6

Hardy-Weinberg proportions as generated from the random union of gametes, using a unit square.

(Actually, William Castle, an American geneticist, had published a special case of this relationship in 1903.) This relationship, generally known today as the **Hardy-Weinberg principle,** states that the genotypic frequencies for a gene with two different alleles are a binomial function of the allelic frequencies.

To understand the Hardy-Weinberg principle, assume that a population is segregating for two alleles, A_1 and A_2, at gene A, with frequencies of p and q (remember p + q = 1), respectively. We will assume that the male and female gametes unite at random to form zygotes; that is, the product rule of probability discussed in chapter 2 is applicable. In addition, we should assume that the factors that we will discuss in some detail later—mutation, genetic drift, gene flow, and selection—are not affecting genetic variation.

The random union of gametes can be illustrated graphically, as in figure 16.6. The unit length across the top of the square in this figure is divided into proportions p and q, representing the frequencies of the female gametes, A_1 and A_2, respectively. Likewise, the side of the square is divided into proportions representing the frequencies of the male gametes. (The frequencies of the alleles in female and male gametes are assumed to be identical.) To envision random union of gametes, one must imagine a pool composed of both female and male gametes, p with A_1 alleles and q with A_2 alleles—and then assume that zygote formation occurs by chance collision in the gametic pool.

The areas within the unit square represent the probabilities (or proportions) of different progeny zygotes. For example, the upper left square has an area of p^2, the frequency of genotype A_1A_1, and the lower right square has an area of q^2, the frequency of A_2A_2. The other two gametic combinations both have frequencies of pq and result in heterozygotes, making the total proportion of heterozygotes 2pq. Overall, therefore, the three progeny zygotes (A_1A_1, A_1A_2, A_2A_2) are formed in proportions (p^2, 2pq, q^2). If we square the sum of the allelic frequencies (that is, $[p + q]^2 = p^2 + 2pq + q^2$), we can see that the Hardy-Weinberg proportions are really binomial proportions.

Of course, different populations may have different allelic frequencies. To illustrate the effect of different allelic frequencies on genotypic frequencies, figure 16.7 gives the frequencies of the three different genotypes, assuming Hardy-Weinberg proportions for the total range of allelic frequencies. Notice that the heterozygote is the most common genotype for intermediate allelic frequencies, while one of the homozygotes is the most common genotype for nonintermediate frequencies. The maximum frequency of the heterozygote occurs when p = q = 0.5. In this case, half of the individuals are heterozygotes. When the locus is monomorphic (p > 0.99 or q > 0.99), the amount of heterozygosity is very low.

In most organisms, random union of gametes does not occur. Instead, the parental genotypes pair and mate; then these mated pairs produce gametes that unite independently of the alleles they contain. Let us illustrate this situation

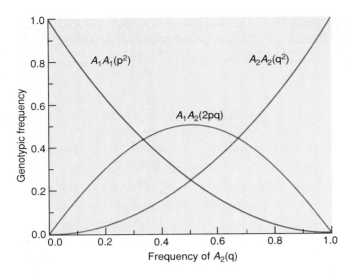

FIGURE 16.7
The allelic frequencies of three different genotypes for a population in Hardy-Weinberg proportions.

Table 16.3 Frequencies of Different Mating Types for Two Alleles, with Random Mating

Female Genotypes (Frequencies)	Male Genotypes (Frequencies)		
	$A_1A_1(P)$	$A_1A_2(H)$	$A_2A_2(Q)$
$A_1A_1(P)$	P^2	PH	PQ
$A_1A_2(H)$	PH	H^2	HQ
$A_2A_2(Q)$	PQ	HQ	Q^2

when **random mating** occurs between different female and male genotypes. We will consider a diploid organism and assume that the three possible genotypes, A_1A_1, A_1A_2, and A_2A_2, are present in the population in frequencies P, H, and Q, respectively (remember that P + H + Q = 1). Given the three genotypes in each sex, nine combinations of matings between the male and female genotypes are possible, as shown in table 16.3. The frequency of a particular mating type, given random mating and using the product rule of probability again, is equal to the product of the frequencies of the genotypes that constitute a particular mating type. For example, the frequency of mating type $A_1A_1 \times A_1A_1$ is P^2. Only six mating types will be distinguished here, because reciprocal matings—for example, $A_1A_1 \times A_1A_2$ and $A_1A_2 \times A_1A_1$, where the first genotype is the female—have the same genetic consequences.

The mating types, their frequencies, and the expected frequencies of their offspring genotypes, using the principle of segregation are given in table 16.4. For example, the mating $A_1A_1 \times A_1A_1$ produces only A_1A_1 progeny; the mating $A_1A_1 \times A_1A_2$ produces ½ A_1A_1 and ½ A_1A_2; and so on. If the frequency of A_1A_1 progeny contributed by each mating type is summed (adding down the first progeny column), we find that $P^2 + PH + ¼H^2 = (P + ½H)^2$ of the progeny are A_1A_1. From the expression for the frequency of A_1 given earlier, we know that p = P + ½H and, therefore, that p^2 of the progeny are genotype A_1A_1. Similarly, the frequencies of A_1A_2 and A_2A_2 progeny are 2(P + ½H)(Q + ½H) = 2pq and $(Q + ½H)^2 = q^2$, respectively. In other words, the Hardy-Weinberg principle is also true for random mating in the population, an assumption that is entirely reasonable for most genes.

The crucial point of the Hardy-Weinberg principle is that, given any set of initial genotypic frequencies (P, H, Q)

after one generation of random mating, the genotypic frequencies in the progeny are in the proportions p^2, 2pq, and q^2. For example, given the initial genotypic frequencies (0.2, 0.4, 0.4) in which p = 0.4 and q = 0.6, after one generation, the genotypic frequencies become $[(0.4)^2$, 2(0.4)(0.6), $(0.6)^2 = 0.16, 0.48, 0.36]$. Furthermore, the genotypic frequencies will stay in these exact proportions generation after generation, given continued random mating and the absence of factors that change allelic or genotypic frequencies. (After all this discussion of p's and q's, one might think that it is the source of the phrase "Mind your p's and q's." A more probable origin, however, is the tradition of British bartenders who tell their customers, "Mind your pints and quarts"—p's and q's for short.)

In the allelic frequency estimates given earlier in this chapter, all alleles were assumed to be expressed (observable) in heterozygotes. However, in many cases—for example, metabolic disorders in humans and color polymorphisms in various other organisms—the heterozygote is generally indistinguishable from one of the homozygotes because one of the alleles is completely dominant. In these instances, the population is usually assumed to be in Hardy-Weinberg proportions, to allow the estimation of allelic frequencies. For example, with two alleles and complete dominance, it is assumed that the proportion of A_2A_2 genotypes in the population is:

$$q^2 = \frac{N_{22}}{N}$$

By taking the square root of both sides of this equation, an estimate of the frequency of A_2 is:

$$q = \left(\frac{N_{22}}{N}\right)^{½}$$

(The superscript ½ means square root.)

This approach is applied in many human metabolic diseases where it is often important to know the proportion of heterozygotes (carriers) in a population—individuals who usually cannot be distinguished from homozygotes for the normal allele. An estimate of the proportion of carriers in a population is:

H = 2pq

Table 16.4 The Hardy-Weinberg Principle, Assuming Random Mating and Normal Segregation

Parents		Progeny		
Mating Type	Frequency	$A_1 A_1$	$A_1 A_2$	$A_2 A_2$
$A_1 A_1 \times A_1 A_1$	P^2	P^2	—	—
$A_1 A_1 \times A_1 A_2$	$2PH$	PH	PH	—
$A_1 A_1 \times A_2 A_2$	$2PQ$	—	$2PQ$	—
$A_1 A_2 \times A_1 A_2$	H^2	$\frac{1}{4}H^2$	$\frac{1}{2}H^2$	$\frac{1}{4}H^2$
$A_1 A_2 \times A_2 A_2$	$2HQ$	—	HQ	HQ
$A_2 A_2 \times A_2 A_2$	Q^2	—	—	Q^2
Total	1	$(P + \frac{1}{2}H)^2 = p^2$	$2(P + \frac{1}{2}H)(Q + \frac{1}{2}H) = 2pq$	$(Q + \frac{1}{2}H)^2 = q^2$

where q is the estimate of allelic frequency and $p = 1 - q$. For example, assuming that only 1 in 40,000 individuals at birth has a particular rare recessive disease such as albinism (or is an $A_2 A_2$ homozygote), then from the equation, $q = (^1/_{40,000})^{1/2} = 0.005$. Therefore, $p = 0.995$, and the frequency of carriers is $H = 2(0.005)(0.995) = 0.00995$. In other words, about 1% of such a population carries the disease allele. People who are not geneticists are often surprised at how many individuals are carriers of recessive disease alleles when the actual frequency of the disease itself is so low.

THE CHI-SQUARE TEST

The observed genotypic numbers are often quite close to the expected genotypic numbers calculated from the Hardy-Weinberg principle. But in some organisms, such as self-fertilizing plants, the observed and expected genotypic frequencies in a population may differ greatly. Therefore a statistical test is necessary to determine whether the fit is sufficiently close to expected Hardy-Weinberg proportions. The common test employed for this purpose is the chi-square (χ^2) test used in chapter 2 to examine Mendel's principles. To determine whether the observed numbers are consistent with Hardy-Weinberg predictions, a chi-square value can be calculated as:

$$\chi^2 = \Sigma \frac{(O - E)^2}{E}$$

O and E are the observed and expected *numbers* of a particular genotype (remember to use numbers and not proportions), and Σ indicates that it is summed over all genotypic classes. In other words, this chi-square value is the squared difference of the observed and expected genotypic numbers, divided by the expected genotypic numbers, and summed over all genotypic classes. Knowing the calculated value of chi-square and the degrees of freedom, the probability that the observed numbers will deviate from the expected numbers by chance can be obtained from a chi-square table (see table 2.5).

Using the Hardy-Weinberg expected frequencies for the two-allele case, where all genotypes are distinguishable, the chi-square expression becomes:

$$\chi^2 = \frac{(N_{11} - p^2N)^2}{p^2N} + \frac{(N_{12} - 2pqN)^2}{2pqN} + \frac{(N_{22} - q^2N)^2}{q^2N}$$

The three terms represent the squared, standardized deviations from the expected numbers for genotypes A_1A_1, A_1A_2, and A_2A_2, respectively. For example, N_{11} is the observed number of genotype A_1A_1, and p^2N is the expected number of A_1A_1.

As an example of this procedure, let us use the hemoglobin data given previously. The observed numbers of the three genotypes were: $N_{11} = 320$, $N_{12} = 72$, and $N_{22} = 8$, giving $p = 0.89$ and $q = 0.11$. Therefore, the expected numbers of the three genotypes are $p^2N = (0.89)^2400 = 316.8$, $2pqN = 2(0.89)(0.11)400 = 78.3$, and $q^2N = (0.11)^2400 = 4.8$. Using the chi-square formula:

$$\chi^2 = \frac{(320 - 316.8)^2}{316.8} + \frac{(72 - 78.3)^2}{78.3} + \frac{(8 - 4.8)^2}{4.8}$$
$$= 0.03 + 0.51 + 2.13$$
$$= 2.67$$

To determine whether this value is statistically significant, we first need to know the degrees of freedom. As in chapter 2, we start by counting the number of categories (or genotypes)—three in this case. Next, we subtract 1 from this total because the numbers of all genotypes sum to N. Furthermore, in this application of the chi-square, we subtract an additional 1 because we had to estimate the allelic frequency (in this case only q because $p = 1 - q$) in order to calculate the expected numbers of the genotypes. Overall then, the remaining degrees of freedom are $3 - 1 - 1 = 1$.

Table 16.5 The Hardy-Weinberg Principle Applied to the ABO Blood Group System

Female Gametes (Frequencies)	Male Gametes (Frequencies)		
	I^A(p)	I^B(q)	I^O(r)
I^A(p)	$I^A I^A$(p)^2 A	$I^A I^B$(pq) AB	$I^A I^O$(pr) A
I^B(q)	$I^A I^B$(pq) AB	$I^B I^B$(q^2) B	$I^B I^O$(qr) B
I^O(r)	$I^A I^O$(pr) A	$I^B I^O$(qr) B	$I^O I^O$(r^2) O

Table 16.6 Frequencies of Gametes and Genotypes for an X-linked Gene

Female Gametes (Frequencies)	Male Gametes (Frequencies)		
	Female Progeny		*Male Progeny*
	A_1(p)	A_2(q)	Y(1)
A_1(p)	$A_1 A_1$(p^2)	$A_1 A_2$(pq)	A_1Y(p)
A_2(q)	$A_1 A_2$(pq)	$A_2 A_2$(q^2)	A_2Y(q)

Now we can look at table 2.5 to determine whether a chi-square value of 2.67 with 1 degree of freedom is significant. Because 2.67 is less than 3.84 (the lowest level of statistical significance for 1 degree of freedom), we can conclude that the observed numbers in the sample are consistent with those expected under the Hardy-Weinberg principle.

MORE THAN TWO ALLELES

Many genes have more than two alleles, as we discussed in chapter 3. For example, the β-globin gene containing the sickle-cell allele actually has more than one hundred variants, most of which are rare. Also, many genes examined by protein electrophoresis have more than two alleles. Further, when DNA sequences for genes are considered, many alleles are generally present in a population.

In the case of multiple alleles, the frequency of an allele can be calculated as the sum of the frequency of its homozygote plus half the frequency of each heterozygote that carries the allele in question. To illustrate, assume that a sample of N individuals is categorized for the genotypes from three alleles. As discussed in chapter 3, six genotypes are possible: three homozygotes ($A_1 A_1$, $A_2 A_2$, and $A_3 A_3$) and three heterozygotes ($A_1 A_2$, $A_1 A_3$, and $A_2 A_3$). The frequency of allele A_1 is calculated in the same manner as for two alleles, except that A_1 alleles are in two different heterozygotes, $A_1 A_2$ and $A_1 A_3$. Therefore, the frequency of A_1 is:

$$p = \frac{N_{11} + \frac{1}{2}N_{12} + \frac{1}{2}N_{13}}{N}$$

where N_{13} is the number of $A_1 A_3$ heterozygotes.

The Hardy-Weinberg principle can also be extended to more than two alleles. Let us illustrate this with the human ABO blood group system, which has three alleles: I^A, I^B, and I^O. Table 16.5 gives the frequencies of these alleles in gametes p, q, and r, respectively, and the frequencies of the resulting genotypes and phenotypes. As for two alleles, the genotypic frequencies can be calculated as the product of particular male and female gametic frequencies. For example, the frequency of genotype $I^A I^A$ is (p)(p) = p^2. Overall, the genotypic frequencies are equal to the square of the following trinomial (because there are three alleles):

$$(p + q + r)^2 = p^2 + 2pq + q^2 + 2pr + 2qr + r^2$$

Check table 16.5 to see that you understand the genotypes associated with these frequencies.

Note that because of the recessivity of allele I^O, two genotypes have the A phenotype ($I^A I^A$ and $I^A I^O$) and two have the B phenotype ($I^B I^B$ and $I^B I^O$). The frequency of the O phenotype, using the Hardy-Weinberg principle, is the square of the frequency of the recessive allele I^O, that is r^2; the frequencies of the A and B phenotypes are p^2 + 2pr and q^2 + 2qr, respectively; and the frequency of the AB phenotype is 2pq. In the United States Caucasian population, the frequencies of alleles I^A, I^B, and I^O are approximately 0.28, 0.06, and 0.66, resulting in phenotypic frequencies of 0.45, 0.08, 0.03, and 0.44 for ABO blood group types A, B, AB, and O, respectively. Luckily, the rarest phenotype, type AB, is also the universal recipient, so people with this phenotype can receive transfusions from individuals of any ABO blood group type.

An important consequence of having many alleles at a locus is that, in general, most individuals in a population are heterozygotes. For example, at the *HLA-A* and *HLA-B* loci, over 90% of the individuals are heterozygotes and less than 10% are homozygotes (see chapter 3). To illustrate this calculation, the Hardy-Weinberg heterozygosity for the ABO system is:

$$\begin{aligned} H &= 2pq + 2pr + 2qr \\ &= 2(0.28)(0.06) + 2(0.28)(0.66) + 2(0.06)(0.66) \\ &= 0.482 \end{aligned}$$

When there are many alleles (which means many more heterozygotes than homozygotes), the simplest approach is to calculate the Hardy-Weinberg homozygosity and subtract it from unity. The remainder is the Hardy-Weinberg heterozygosity.

X-LINKED GENES

A number of human diseases, such as hemophilia and muscular dystrophy, are determined by recessive genes on the X chromosome. These diseases are much more common in males than in females, a fact that can be explained by the Hardy-Weinberg principle. For example, table 16.6 gives

Table 16.7 Positive Assortative Mating in Studies of Five Human Traits

Trait	Number of Studies Showing Correlation Coefficients of:			
	<0.1	0.1–0.2	0.2–0.3	>0.3
Height	7	8	7	5
Weight	1	2	3	1
Cephalic index	14	5	3	—
Hair color	—	2	2	1
Eye color	2	1	1	1

Table 16.8 Frequency of Parents and Progeny with Complete Self-Fertilization

Parent		Progeny		
	Frequency	A_1A_1	A_1A_2	A_2A_2
A_1A_1	P	P	—	—
A_1A_2	H	¼H	½H	¼H
A_2A_2	Q	—	—	Q
	1.0	P + ¼H	½H	Q + ¼H

the expected frequencies of different genotypes, assuming random union of gametes for an X-linked trait. The frequency of affected males, A_2Y, is q, while the frequency of affected females, A_2A_2, is q^2. If q is low, as it is for most disease alleles (say, 0.005), the frequencies of affected males and females are 0.005 and 0.000025, respectively. In other words, the ratio of affected males to females is $q:q^2$ or 1:q. Thus, if q = 0.005, the ratio is 200:1. Because of the high number of males affected by X-linked traits and the relatively high frequency of the hemophilia allele for a disease allele, a high proportion of individuals with a recessive genetic disease (nearly one individual in four in some populations) are male hemophiliacs. On the other hand, very few females have hemophilia.

 The Hardy-Weinberg principle states that genotypic frequencies are a binomial function of allelic frequencies after one generation of random mating. Genotypic numbers can be examined statistically, using the chi-square test to determine whether they are consistent with Hardy-Weinberg expectations. The Hardy-Weinberg principle and estimation of allelic frequencies can also be applied to multiple alleles and X-linked genes.

INBREEDING

So far in this chapter, we have considered populations in which mating is random, but there are obvious exceptions to this assumption in some species, such as self-fertilizing plants or invertebrates. **Inbreeding,** nonrandom mating that results from mating of close relatives, is of particular importance in humans, because many individuals who have recessive diseases are the products of matings between relatives. Furthermore, inbreeding is used in both plant and animal breeding to develop varieties or lines that have certain desired characteristics. In chapter 17 we will discuss the joint effects of inbreeding and selection that result in the phenomenon of inbreeding depression.

Nonrandom mating with respect to genotype occurs in populations where the mating individuals are either more closely or less closely related than mates drawn by chance (at random) from the population. The results of these two

types of matings are called inbreeding and **outbreeding,** respectively. As we will see, neither inbreeding nor outbreeding can by itself cause a change in allelic frequency, but both do cause a change in genotypic frequencies. In an inbred population, the frequency of homozygotes is increased and the frequency of heterozygotes is reduced relative to random mating (or Hardy-Weinberg) proportions. In outbreeding, the opposite occurs, with the frequency of heterozygotes increased and the frequency of homozygotes reduced relative to random mating proportions. These genotypic changes affect all loci in the genome.

Nonrandom mating based on phenotypes rather than on genotypes may also occur. If the mated pairs in a population are composed of individuals with the same phenotype more often than would be expected by chance, **positive assortative mating** has occurred. By the same token, if mated pairs share the same phenotype less often than expected, we say that **negative assortative mating** has occurred. For example, positive assortative mating occurs in humans for traits such as height and eye color (table 16.7), although even for these traits it is absent in some populations. From these studies, it is obvious that cephalic index, a measure of head shape, is unimportant in selecting mates! Assortative mating generally affects the genotypic frequencies of only those loci involved in determining the phenotypes used in mate selection. Because positive assortative mating has effects similar to those of inbreeding, except that it generally affects only one or a few loci rather than all loci, we will consider only the effects of inbreeding in the following discussion.

SELF-FERTILIZATION

In general, the most extreme type of inbreeding is self-fertilization, or "selfing," where both pollen and egg (or sperm and egg) are produced by the same individual. With complete self-fertilization, a population is divided into a series of inbred lines that quickly become highly homozygous. For example, the lines of pea plants used by Mendel were homozygous for different alleles that affected morphology or color because they were allowed to self-fertilize.

To illustrate how such high homozygosity occurs, assume that the frequencies of the three genotypes in the parents are P, H, and Q for A_1A_1, A_1A_2, and A_2A_2, respectively (table 16.8). Notice that A_1A_1 and A_2A_2 are true-breeding genotypes; that is, all selfed progeny produced by them are A_1A_1 and A_2A_2, respectively. However, only half the progeny of A_1A_2 are A_1A_2, with the other half split equally between homozygotes A_1A_1 and A_2A_2 due to Mendelian segregation.

If we sum the columns of table 16.8 for the different progeny types, we find that the frequency of heterozygotes is decreased by one-half compared to the parents, while the frequencies of the two homozygotes are each increased. For example, if the initial frequencies were 0.25, 0.5, and 0.25 for genotypes A_1A_1, A_1A_2, and A_2A_2, these frequencies would be 0.375, 0.25, and 0.375 after one generation of complete selfing.

To understand how inbreeding changes the genotypic proportions, let us examine the frequency of heterozygotes when complete self-fertilization occurs over several generations. The relationship for heterozygosity in two succeeding generations is a general one; thus, after one generation, the heterozygosity is halved and after two generations, it is halved again, so that the result is one-fourth the initial value. After t generations of complete selfing, heterozygosity is halved t times, so that it is $(\frac{1}{2})^t$ of its initial value. For example, if H was initially 0.5, after five generations the expected heterozygosity would be $(\frac{1}{2})^5 \, 0.5 = 0.016$. This is a reduction of heterozygosity from 50% to less than 2% in five generations, showing that the reduction in heterozygosity through selfing is very rapid. The change in heterozygosity and homozygosity $(1 - H)$ is given in figure 16.8.

One of the most important aspects of inbreeding is that, even though genotypic frequencies may be greatly altered, allelic frequencies remain unchanged overall. To illustrate this, let us indicate frequencies of the alleles and genotypes in the progeny generation by primes ($'$). Therefore, the frequency of allele A_1 in the next generation, p', is:

$$p' = P' + \frac{1}{2}H'$$

where P' and H' are the frequencies of genotypes A_1A_1 and A_1A_2 in the next generation. By substituting the values of P' and H' from table 16.8 (the sums of columns 3 and 4), then:

$$p' = P + \frac{1}{4}H + \frac{1}{4}H$$
$$= p$$

In other words, as stated previously, no change in allelic frequency occurs from generation to generation due to self-fertilization.

THE INBREEDING COEFFICIENT

In order to describe the effect of inbreeding on genotypic frequencies in general, we will use a measure called the **inbreeding coefficient** (f). This value is defined as the probability that the two alleles at a gene in an individual are identical by descent. (Alleles are identical by descent when the two alleles in a diploid individual are derived from one particular allele in their ancestry, a situation we will illus-

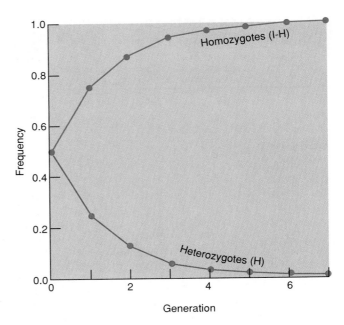

FIGURE 16.8
Change in heterozygosity and homozygosity over time with self-fertilization.

trate later using pedigrees.) The largest value f can take is 1, when all genotypes in a population are homozygotes containing alleles identical by descent, as in the pure-breeding lines of Mendel's peas. The lowest value f can take is 0, when no alleles are identical by descent. In large, random mating populations, it is generally assumed that f is close to 0, because any inbreeding that may occur is between very distant relatives and thus will have little effect on the inbreeding coefficient.

A general formulation of the proportion of the three genotypes A_1A_1, A_1A_2, and A_2A_2, using the inbreeding coefficient and the allelic frequencies, can be derived as shown in table 16.9. Here it is assumed that the population can be divided into an inbred proportion (f) and a random mated proportion $(1 - f)$. In the inbred proportion, the frequencies of A_1A_1, A_1A_2, and A_2A_2 are p, 0, and q, respectively. These frequencies are the proportions of lines expected for each of the genotypes if, for example, complete selfing continues. When the inbred and random mated proportions are added together (using the relationship $q = 1 - p$), the genotypic frequencies become:

$$P = p^2 + fpq$$
$$H = 2pq - 2fpq$$
$$Q = q^2 + fpq$$

The first term in each equation is the Hardy-Weinberg proportion of the genotypic frequencies, and the second term is the deviation from that value. Note that an individual may be homozygous either because the two alleles are identical by descent (the fpq terms) or because the two alleles are identical in kind through random mating

Table 16.9 Frequency of Different Genotypes with Inbreeding

Genotype	Inbred (f)	Random mated $(1-f)$	Total	
A_1A_1	fp	$(1-f)p^2$	$fp + (1-f)p^2$	$= p^2 + fpq$
A_1A_2	—	$(1-f)2pq$	$(1-f)2pq$	$= 2pq - 2fpq$
A_2A_2	fq	$(1-f)q^2$	$fq + (1-f)q^2$	$= q^2 + fpq$
	f	$1-f$	1	1

(the p^2 and q^2 terms in the equations for P and Q, respectively). For example, A_2A_2 homozygotes that occur unrelated to inbreeding contain alleles identical in kind, while A_2A_2 homozygotes resulting from the same ancestral allele are termed identical by descent. The size of the coefficient of inbreeding reflects the genotypes' deviation from Hardy-Weinberg proportions such that, when f is 0, the zygotes are in Hardy-Weinberg proportions, and when f is positive, as from inbreeding, a deficiency of heterozygotes and an excess of homozygotes occur.

CALCULATING THE INBREEDING COEFFICIENT

We will discuss two ways the inbreeding coefficient can be estimated—from genotypic frequencies or from pedigrees. By the first method, we estimate the inbreeding coefficient in a natural population using the expression for the frequency of heterozygotes given previously. Thus, we can solve the expression for f as:

$$H = 2pq - 2fpq$$

$$1 - f = \frac{H}{2pq}$$

$$f = 1 - \frac{H}{2pq}$$

From the bottom equation, it is clear that the inbreeding coefficient (f) is a function of the ratio of the observed heterozygosity (H) over the Hardy-Weinberg heterozygosity (2pq). With inbreeding, H is less than 2pq, so f is greater than 0. As expected, if no heterozygotes are observed (H = 0), the breeding coefficient is 1.

Many species of plants have mating systems that include both self-fertilization and random mating (often called outcrossing) with other individuals. If the proportion of self-fertilization is high, nearly all individuals in the population should be homozygotes. An example is *Avena fatua*, a highly self-fertilizing species of wild oats, which is widespread in many areas around the Mediterranean Sea and was introduced into California where it is responsible for the "golden" hills in summer. A sample from the plant revealed that the frequencies of the three genotypes at an enzyme locus were P = 0.67, H = 0.06, and Q = 0.27. In

order to estimate the inbreeding coefficient, we must first calculate the frequencies of the alleles. As before, the frequency of A_2 (q) is:

$$q = \frac{1}{2}H + Q$$
$$= (\frac{1}{2})(0.06) + 0.27$$
$$= 0.03$$

the value of p is therefore $1 - q = 0.7$. Using the expression given previously:

$$f = 1 - \frac{0.06}{2(0.3)(0.7)}$$
$$= 0.86$$

This is an extremely high inbreeding coefficient, not far from 1, the maximum value possible. This high value suggests that most of the population reproduces by self-fertilization and that very little mating is random.

As stated earlier, the inbreeding coefficient is known as the probability of identity by descent, which implies that homozygosity is the result of the two alleles in an individual having descended from the same ancestral allele. The probability of identity by descent varies, depending upon the relationship of the parents of the individual being examined. For example, if the parents are unrelated (or not closely related), there is no possibility (or a very low one) that the individual will be homozygous by descent. The other extreme (given one generation of sexual reproduction) occurs when the same individual produces both gametes; that is, when self-fertilization occurs. In this case, the probability of an offspring having identical alleles by descent is 0.5. To illustrate, assume that the parent has the genotype A_1A_2. Progeny are produced in the proportions $\frac{1}{4} A_1A_1$, $\frac{1}{2} A_1A_2$, and $\frac{1}{4} A_2A_2$; thus, half the progeny, the A_1A_1 and the A_2A_2, will have alleles identical by descent. We assumed the parent was heterozygous so that we could identify the alleles in the progeny. However, even if the parent were homozygous for alleles identical in kind, we could arbitrarily designate them as different in order to calculate the probability of identity by descent.

A second way to obtain the inbreeding coefficient for a progeny is from a pedigree in which a **consanguineous mating** (a mating between relatives) has occurred. We use the pedigree to calculate the probability of identity by descent in the progeny. As an example, let us calculate the

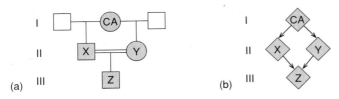

FIGURE 16.9
A pedigree illustrating a mating between two half sibs, X and Y. (*a*) With all individuals; (*b*) without the father. (CA = common ancestor, and the double line indicates a consanguineous mating.)

inbreeding coefficient for an offspring of two half sibs, individuals who share one common parent. Figure 16.9*a* gives the pedigree for this type of mating, where X and Y are two half sibs having the same mother but different fathers. The two fathers are indicated as open squares because they do not contribute to the inbreeding coefficient. The mother of X and Y is indicated as the common ancestor (CA). In figure 16.9*b*, the same pedigree appears in a different form, with the fathers omitted and diamonds symbolizing all individuals, because sex is not important in determining the inbreeding coefficient here. The arrows in this figure indicate the direction of parent to offspring transmission.

Let us assume that the mother (CA) has the genotype A_1A_2. To calculate the inbreeding coefficient, we need to know the probability that her grandchild (Z) is either A_1A_1 or A_2A_2—that is, identical by descent for either of her alleles. The first alternative, that Z is A_1A_1, can occur only if X and Y each contribute to Z a gamete containing A_1. The probability of an A_1 allele in X is the probability that an A_1 allele came from CA, or 0.5. Because the probability of transmission of A_1 to Z from X is also 0.5, the joint probability of both events (that is, A_1 from CA to X and then to Z) is $0.5 \times 0.5 = 0.25$, using the product rule. Likewise, the probability of A_1 to Z through Y is 0.25. Therefore, the probability of an A_1A_1 progeny, who has received an A_1 from X and Y each, is $(0.25)(0.25) = 0.0625$. Using the same approach, the probability of an A_2A_2 progeny is 0.0625. The overall probability of identity by descent in Z is then $0.0625 + 0.0625 = 0.125$.

A more straightforward way to calculate the inbreeding coefficient from a pedigree is the **chain-counting technique.** A chain for a given common ancestor starts with one parent of the inbred individual, goes up the pedigree to the common ancestor, and comes back down to the other parent. For the example in figure 16.9, the chain is quite simple and is expressed X-CA-Y. The number of individuals in the chain (N) can be used in the following formula to calculate the inbreeding coefficient as:

$$f = (\tfrac{1}{2})^N$$

For example, for the pedigree in figure 16.9, the inbreeding coefficient would be $(\tfrac{1}{2})^3 = 0.125$, the same result we obtained by the longer approach, using the definition of the inbreeding coefficient.

> Inbreeding increases the frequency of homozygotes and reduces the frequency of heterozygotes, without changing allelic frequencies. The inbreeding coefficient can be measured either from genotypic frequencies or from pedigrees.

MUTATION

Mutation is a particularly important process in population genetics and evolution because it is the major source of genetic variation in a species or population. Mutation can take many forms (see chapter 11). For example, it may involve only one DNA base, several bases, the major part of a chromosome, whole chromosomes, or sets of chromosomes. The immediate cause of a particular mutation might be a mistake in DNA replication, the physical breakage of the chromosome by a mutagenic agent, the insertion of a transposable element, or a failure in disjunction of meiosis. Our discussion will center on **spontaneous mutations,** which appear without apparent explanation, rather than on **induced mutations,** which are caused by some mutagenic agent.

In order to examine the effect of mutation on genetic variation in a population, let us assume that the rate of mutation from wild-type alleles (A_1) to deleterious (or detrimental) alleles (A_2) for a given gamete per generation is u. This is known as the **forward mutation rate.** We will call the rate of mutation from deleterious alleles to wild-type alleles v, or the **backward mutation rate.** Because only A_1 alleles can mutate to A_2 and the A_1 allele has a frequency of p, the increase in the frequency of A_2 from forward mutation is $u \times p$. Using the same logic, the frequency of A_2 alleles can be decreased by backward mutation by the amount $v \times q$. Overall then, the change per generation in the frequency of A_2 (Δq) due to mutation is:

$$\Delta q = up - vq$$

The maximum positive value for this change is u, when p = 1 and q = 0 (all alleles are wild-type). The maximum negative value is v, when p = 0 and q = 1. However, because the mutation rates u and v are generally small (see chapter 11), the expected change due to mutation is also quite small. For example, if $u = 10^{-5}$, $v = 10^{-6}$, and q = 0.0, then:

$$\Delta q = (0.00001)(1.0) - (0.000001)(0.0)$$
$$= 0.00001$$

This is obviously quite a small change in allelic frequency.

Even so, mutation is of fundamental importance in determining the incidence of rare genetic diseases (see chapter 17). In fact, the balance between mutation (increasing the frequency of the disease allele) and selection (reducing the frequency of the disease allele) can explain the observed incidence of diseases such as albinism. In addition,

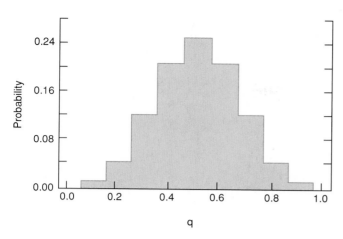

q

FIGURE 16.10

The probability that a population with ten gametes will have a given allelic frequency after one generation of genetic drift.

mutation in concert with genetic drift provides a reasonable explanation for the large amounts of molecular variation recently observed in many species (see chapter 17).

  Mutation is the original source of genetic variation. It can cause only a small change in allelic frequency per generation.

GENETIC DRIFT

We have assumed until now that the population under study is large. But in fact, many populations are small, such as the number of individuals in an endangered species or in a species colonizing a new habitat. As a result of small population numbers, chance effects, called **genetic drift,** which are introduced by sampling gametes from generation to generation in a small population, can change allelic frequencies. In a large population, usually only a small chance change in the allelic frequency will be due to genetic drift. But in a small population, allelic frequency can undergo large fluctuations in different generations in a seemingly unpredictable pattern. Genetic drift is fundamental to efforts toward genetic conservation and understanding molecular evolution (see chapter 17).

An example will illustrate the type of allelic frequency change that can be expected in a small population. Assume that a diploid population has five individuals (ten copies of the gene, or ten gametes) and that initially, half of these are A_1 and half are A_2. The ten gametes for the next generation are randomly chosen from the gametic pool, a process analogous to flipping a coin ten times and counting the number of heads and tails. Sometimes the proportion of A_1 alleles is greater than the proportion of A_2,

sometimes the proportions are equal, and sometimes the proportion of A_2 is greater than the proportion of A_1. If many small populations, each containing ten gametes, are randomly chosen from a population with equal numbers of A_1 and A_2, the proportions of populations with different frequencies of A_2 will be as shown in figure 16.10. For example, a frequency of 0.4 indicates that four of the ten gametes were A_2. Notice that in most of the populations, the frequencies are close to the initial frequency of 0.5, but that significant variation occurs around that value. For example, there is a greater than 1% chance that the allele frequency will change from 0.5 to 0.1 (the far left end of the histogram).

To illustrate the potential for cumulative genetic change over time due to genetic drift, let us follow four particular populations of size twenty (forty gametes) over a number of generations (fig. 16.11). These values were obtained using computer simulation in which random numbers represented the chance segregation that took place from generation to generation. The solid lines indicate the four replicates, and the broken line is the mean frequency of allele A_2 over the four replicates. All of the replicates were initiated with the frequency of A_2 equal to 0.5. After one generation of genetic drift, the frequencies were 0.625, 0.55, 0.55, and 0.475. By generation 15, they were even more widely dispersed—0.8, 0.5, 0.475, and 0.3. In one of the replicates, the frequency of A_2 went to 1.0 in generation 19, meaning that all of the alleles in the population were A_2. In another replicate, the A_2 allele frequency went to 0.0 in generation 28. The other two replicates were still segregating for both alleles at the end of thirty generations. The mean of the four replicates varied from 0.625 in generation 19 to 0.475 in generation 30, but was generally near the initial frequency of 0.5. (In cases where there are enough replicate populations, no change occurs in the mean allelic frequency over all populations, because increases in allelic frequency in some populations are cancelled by reductions in allelic frequency in others.)

From this example, it is obvious that genetic drift can cause large and erratic changes in the allelic frequencies of individual populations in a rather short time and can lead to differentiation among populations. Within a small population, there is a high probability that a gene will become fixed for one allele or the other over a period of time. But in a large population over the same time span, there would be little change in allelic frequency due to genetic drift, and consequently, little cumulative differentiation among populations.

One prediction of genetic drift—that a small population will become highly homozygous over time—is supported by the findings in several endangered species. For example, the Northern elephant seal was hunted to near-extinction at the turn of the century, until fewer than one

FIGURE 16.11
The allelic frequencies over thirty generations for four
replicate populations of size twenty. The mean is
represented by the broken line.

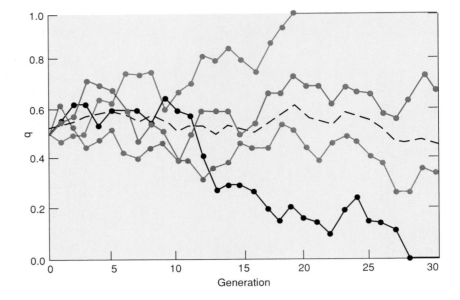

hundred individuals survived (fig. 16.12). Although the population size has rebounded to over 70,000 individuals in recent years, it was low for a number of generations. Apparently due to this "bottleneck," genetic variation was lost, resulting in an estimated zero heterozygosity for the twenty-four genes mentioned in table 16.2.

Another example of a seemingly genetically impoverished species is the Pere David's deer. This species was discovered in the hunting compound of a Chinese emperor in 1856, presumably having been sequestered there for three hundred years. Approximately four hundred of the deer are living today, all descended from eleven animals taken from China. A survey of this species has also shown no heterozygosity for the enzyme genes examined.

We can identify the effects of genetic drift in several ways. First, populations eventually go either to fixation for $A_2(q = 1)$ or to loss of $A_2(q = 0.0)$. The probability that a population will eventually become fixed for A_2 is equal to the initial frequency of A_2. For example, if q is initially 0.2, the probability that the population will be fixed for A_2 is 0.2, and the probability that it will lose the A_2 allele is 0.8. We can understand this point by realizing that, for A_2 to become fixed, it must change by chance from 0.2 to 1.0, a much bigger change than for the loss of A_2 (from 0.2 to 0.0).

Second, because the mean frequency over replicate populations does not change and the distribution of the allelic frequencies over replicate populations does, the overall effect of drift can be understood by examining the variation of the allelic frequency over replicate populations. One statistic used to measure the extent of variation is the variance (see chapter 3). The expected variance (V) in allelic frequency over replicate populations, due to one generation of genetic drift, is:

$$V = \frac{pq}{2N}$$

FIGURE 16.12
The Northern elephant seal.
© *Richard Humbert/Biological Photo Service.*

Where N is the population size in a large population (large N), V is small, indicating only a minor effect from genetic drift. The impact of genetic drift can also be understood by examining the rate of loss of heterozygosity. In chapter 17, we will use the decay of heterozygosity to measure the effectiveness of genetic conservation in endangered species.

To establish a general yardstick for measuring the effect of genetic drift on allelic frequency relative to other factors, we note that the square root of the variance (the standard deviation, see chapter 3) is approximately equal to the average absolute value (indicated by parallel vertical lines) of the allelic frequency change, or:

$$|\Delta q| = V^{1/2}$$

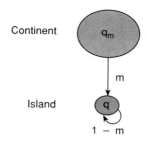

FIGURE 16.13

A model of the gene flow from a continental population to an island population, where m is the rate of gene flow and q_m is the allelic frequency in the migrants.

For example, if $p = q = 0.5$ and $N = 50$, then:

$$V = \frac{0.5 \times 0.5}{2 \times 50} = 0.0025$$

Taking the positive square root of V, then:

$$|\Delta q| = 0.05$$

Obviously, if N is relatively small, the effect of genetic drift on allelic frequency may be substantial, much larger than that of mutation and comparable to that of selection or gene flow, as we will see later.

When N is large, as in most species, the expected per-generation change in allelic frequency from genetic drift is small. However, the cumulative change over many generations may be quite important, as we will find out when we discuss molecular evolution in chapter 17.

 Genetic drift in small populations can result in large chance changes in allelic frequency. Genetic drift may be the cause of the low heterozygosity observed in some endangered species.

GENE FLOW

Previously, we have considered only a single large population in which random mating occurred throughout. Many species, however, have smaller populations that are divided because of ecological or behavioral factors. For example, the populations of fish in tide pools, deciduous trees in farmlands, and insects on host plants are separated from each other because the suitable habitat for the species is not continuous. With this type of population structure, differing amounts of genetic connectedness can exist among the separated populations. The extent of this genetic connection depends primarily on the amount of **gene flow**—that is, genetically effective migration—among the populations. As a result of gene flow, genetic variants can be introduced from one population to another.

For example, the genetic variants that confer resistance to insecticides like DDT generally are initially present in only a limited number of populations (see chapter 17).

However, the movement of individuals between populations introduces resistant alleles to populations that did not have them previously. As a result, DDT resistance can spread rapidly throughout mobile species such as mosquitoes or houseflies.

The simplest type of gene flow takes place when an allele enters a single population from an outside source. For example, such unidirectional gene flow occurs in an island population that receives migrants from a continental population. This description basically applies to species having populations on islands and nearby large landmasses, to aquatic species that live in ponds and have lakes as a source of migration, and to peripheral populations that are constantly replenished by individuals from the main part of the species range.

To examine the effect of this type of population structure, let us assume that an island population receives migrants from a large source (continent) population, as shown in figure 16.13. Although reciprocal gene flow may occur, we will assume it is so low that it has a negligible effect on the allelic frequency in the source population. Let the proportion of migrants moving into the island population each generation be m, so that the proportion of nonmigrants (or residents) in the island population is $1 - m$. If the frequency of A_2 in the migrants is q_m and the frequency of A_2 on the island before migration is q, the allelic frequency of A_2 after migration, q', is:

$$q' = (1 - m)q + mq_m$$
$$= q - m(q - q_m)$$

The change in allelic frequency after one generation of gene flow is the difference between the frequencies after gene flow and before gene flow, or:

$$\Delta q = q' - q$$

If we substitute q' from above into this expression, then:

$$\Delta q = -m(q - q_m)$$

From these expressions, it is obvious that no change in allelic frequency will occur (that is, $\Delta q = 0$) if $m = 0$ or if $q = q_m$. Only the second alternative is of interest here, because a value of $m = 0$ indicates no gene flow. Note that q_m and m are assumed to be constant over time and that they have values between zero and unity. If q is smaller than q_m, the frequency of A_2 increases on the island because of the gene flow; that is, Δq is positive. If q is greater than q_m, the frequency of A_2 will decrease, and Δq is negative. As a result of these two effects, there is a **stable equilibrium frequency** of A_2 on the island at $q_m = q$, as indicated by the solid circles in figure 16.14. (A stable equilibrium occurs where the population frequency returns to a given value after being perturbed either above or below that value.)

The effect of different allelic frequencies in the migrants on Δq when $m = 0.1$ (10% of the individuals of each generation are migrants) can be seen in figure 16.14. Note that the allelic frequency changes linearly as the frequency moves away from the equilibrium value and that it reaches an absolute maximum at either zero or unity, depending

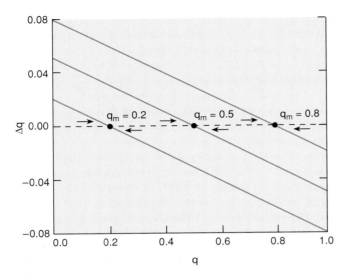

FIGURE 16.14
The expected change in allelic frequency in one generation due to gene flow when m = 0.1 and q_m = 0.2, 0.5, or 0.8. The arrows indicate the direction of changes in allelic frequency.

upon the value of q_m. From the values in this figure, it is obvious that the change in allelic frequency due to gene flow can be important. It may equal the size of changes caused by genetic drift in small populations and may be much larger than changes caused by mutation.

Many populations are composed of individuals who have descended from different populations. If we assume that only two ancestral populations have contributed to the gene pool in a given population (for clarity, we will call this population the **hybrid population**), we can measure the amount of **admixture,** the proportion of the gene pool descended from one of the ancestral populations. We achieve this estimate by rearranging the first equation of this section so that it reads:

$$q' = (1 - m)q + mq_m$$

Then, we solve for m so that:

$$m = \frac{q - q'}{q - q_m}$$

Now, so that we won't confuse the estimate for the amount of admixture (M) with the rate of gene flow (m), let us substitute M for m in the equation above, so that:

$$M = \frac{q - q'}{q - q_m}$$

Such admixture estimates have been made for human populations that have descended from different racial groups. For example, the African-American population is primarily of African origin, mixed with some more recent Caucasian ancestry. How can we estimate the proportion of Caucasian ancestry in African-American populations? One approach is to use a gene that has quite different allelic frequencies in the two ancestral populations, such as an allele

at the Duffy blood group locus, another gene that codes for red-blood-cell antigens. Table 16.10 gives estimates of the ancestral frequencies of this allele and estimates of admixture in two United States cities. For example, to calculate M in the Oakland sample:

$$M = \frac{0.0 - 0.094}{0.0 - 0.429}$$

$$= 0.220$$

Although the estimate of admixture is relatively low in both samples, notice that the estimate of admixture is much higher (0.22) in the northern locality (Oakland) and only 0.037 in Charleston, indicating that certain southern African-American populations may have very little Caucasian admixture.

Table 16.10 Estimates of Caucasian Admixture (M) in African-Americans

Locality	African-Americans (q')	African (q)	Caucasian (q_m)	M
Oakland, CA	0.094	0.0	0.429	0.220
Charleston, SC	0.016	0.0	0.429	0.037

Gene flow between populations can introduce new alleles and, in some cases, can rapidly change allelic frequencies. The extent of past gene flow between populations can be estimated.

NATURAL SELECTION

Charles Darwin is often considered the chief architect of the theory of natural selection. He provided strong evidence for it from his collection of plants and animals gathered on an around-the-world voyage on the H.M.S. *Beagle.* Darwin's theory held that variation among individuals of a species was the raw material through which selection could eventually produce individuals better adapted to their environment. Darwin was primarily concerned with the evolution of quantitative traits (discussed in chapter 3), but the same general principles operate regarding traits determined by single genes.

Selection can occur in several different ways (affecting fertility or mating success, for example), but here we will assume for simplicity that it results from differential survival of genotypes. To illustrate, assume 50% of the haploid genotype A_1 survive to adulthood, while only 40% of A_2 survive to adulthood, making the survival rate of A_1 higher than that of A_2. Intuitively, one would expect the frequency of A_1 to increase and the frequency of A_2 to decrease over time in a population initially composed of these two genotypes.

Table 16.11 Frequency of Genotypes Before and After Selection, Assuming Hardy-Weinberg Proportions Before Selection

	Genotype			Total
	A_1A_1	A_1A_2	A_2A_2	
Relative fitness	1	1	$1 - s$	—
Frequency before selection	p^2	$2pq$	q^2	1.0
Weighted contribution	p^2	$2pq$	$q^2(1 - s)$	$1 - sq^2$
Frequency after selection	$\dfrac{p^2}{1 - sq^2}$	$\dfrac{2pq}{1 - sq^2}$	$\dfrac{q^2(1 - s)}{1 - sq^2}$	1.0

Natural selection at work

Selection has two basic effects on the amount of genetic variation. First, selection favoring a particular allele may lead to the reduction of genetic variation and, consequently, to homozygosity for the favored allele, as for DDT resistance alleles in mosquitoes exposed to DDT for many generations. Second, selection may result in the maintenance of two or more alleles in a population, such as for the banding polymorphism in snails or for the sickle-cell allele in humans.

In order to understand the effects of selection on genetic variation, we must consider the relative fitnesses of the different genotypes. **Relative fitness** can be defined as the relative ability of different genotypes to pass on their alleles to future generations. If, as in the example given previously, the relative fitnesses of two genotypes are based solely on their viabilities, there values can be used to calculate relative fitness values. Generally, the highest relative fitness value is assumed to be 1.0, and the others are standardized by dividing by this value. (As we will see, this is done to simplify the algebra.) Therefore, in the previous example, the relative fitness of genotype A_1 is $^{0.5}/_{0.5} = 1.0$ and that of genotype A_2 is $^{0.4}/_{0.5} = 0.8$.

SELECTION AGAINST A HOMOZYGOTE

We will begin by considering the situation in which selection operates against a homozygote, with the other homozygote and the heterozygote having equal relative fitnesses. This is sometimes called purifying selection because it purifies the gene pool of detrimental alleles. Let the relative fitnesses of the three possible genotypes, A_1A_1, A_1A_2, and A_2A_2, be designated as 1, 1, and $1 - s$, respectively, where the fitness of genotype A_2A_2 is smaller by an amount s, called the **selective disadvantage** (or **selection coefficient**) of the homozygote. If s = 1, then A_2 is a recessive lethal like a number of disease alleles in humans. If the relative fitness of A_2A_2 is 0.8, the selective disadvantage is s = $1 - 0.8 = 0.2$.

Because we are considering only selection resulting from different viabilities of the genotypes, the relative fitness of a genotype is its relative probability of surviving from the newly formed zygote stage until reproduction. We can give the genotypic frequencies before selection in terms of allelic frequencies, assuming Hardy-Weinberg proportions. The weighted contribution of the three genotypes to the next generation is the product of the frequency of the genotypes before selection and their relative fitnesses (table 16.11). The **mean fitness** of the population (w) is the sum of the weighted contributions of the different genotypes, or:

$$w = p^2(1) + 2pq(1) + q^2(1 - s)$$
$$= p^2 + 2pq + q^2 - sq^2$$

Because $p^2 + 2pq + q^2 = 1$ (that is, the sum of the frequencies of all the genotypes is 1), the mean fitness is:

$$w = 1 - sq^2$$

If the relative contributions are standardized by the mean fitness, the frequencies of the genotypes after selection can be obtained as shown in the bottom line of table 16.11. The frequency of A_2 after selection (q´) is equal to half the frequency of the heterozygote A_1A_2 (because half the genes in A_1A_2 are A_2 alleles) plus the frequency of the homozygote A_2A_2, or:

$$q' = (\tfrac{1}{2}) \left(\frac{2pq}{1 - sq^2} \right) + \frac{q^2(1-s)}{1 - sq^2}$$

$$= \frac{pq + q^2(1-s)}{1 - sq^2}$$

$$= \frac{q(p + q - sq)}{1 - sq^2}$$

$$= \frac{q(1 - sq)}{1 - sq^2}$$

Therefore, the frequency of A_2 after selection is a function of the frequency before selection and of the selection coefficient.

As an example, let us assume that the homozygote A_2A_2 is inviable; in other words, that A_2 is a recessive lethal. In order to make the fitness of A_2A_2 zero, then $1 - s = 0$, or $s = 1.0$. Let us now calculate the change in the frequency of A_2 over several generations. If the initial frequency of A_2 is 0.5, then from the formula just given:

$$q' = \frac{0.5[1 - (1.0)(0.5)]}{1 - (1.0)(0.5)^2}$$

$$= 0.333$$

The same formula can be used to calculate the frequency in the next generation, assuming that $q = 0.333$ before selection, so that:

$$q' = \frac{0.333[1 - (1.0)(0.333)]}{1 - (1.0)(0.333)^2}$$

$$= 0.250$$

This process can be repeated over subsequent generations to calculate the allelic frequency. After twenty generations of selection against A_2A_2, the frequency of A_2 has been reduced from 0.5 to approximately 0.045 (see box 16.1).

Many of the principles of population genetics have been investigated using *Drosophila* laboratory populations as a model. For example, the *Glued* mutant is a lethal homozygote that also reduces eye size and affects eye appearance in heterozygotes. Michael Clegg and several coworkers followed the decline in the frequency of the *Glued* mutant in artificial populations initially composed of all heterozygotes, that is, those having an initial frequency of 0.5. Figure 16.15 gives the expected change (broken line) just as we calculated above and the observed change in two replicates, A and B. Notice that the general trend observed is close to that expected for the first few generations but that there is a difference in the later generations. Further investigation also showed selection against the heterozygotes, explaining why the observed frequency declined faster than expected.

The amount of allelic frequency change in one generation due to selection is defined as:

$$\Delta q = q' - q$$

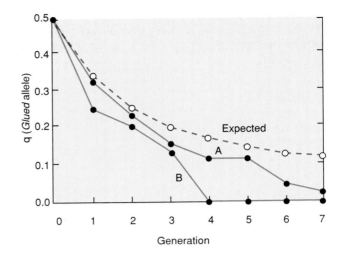

FIGURE 16.15

The expected change in the frequency of the lethal allele *Glued* (blue line) and the observed change in the replicate populations (red lines).

Source: Data from Clegg, et al., Genetics, 83:793–810, 1976.

By substituting q' given previously, then:

$$\Delta q = \frac{q(1 - sq)}{1 - sq^2} - q$$

$$= \frac{q(1 - sq) - q(1 - sq^2)}{1 - sq^2}$$

$$= \frac{q(1 - sq - 1 + sq^2)}{1 - sq^2}$$

$$= \frac{-spq^2}{1 - sq^2}$$

Notice that the change in allelic frequency is negative, indicating that selection in this case is always reducing the frequency of A_2.

To illustrate the connection between this formula and the one given for q', let us assume again that $q = 0.5$ and that $s = 1.0$. Putting these values in the formula for change in allelic frequency:

$$\Delta q = \frac{-(1.0)(0.5)(0.5)^2}{1 - (1.0)(0.5)^2}$$

$$= -0.167$$

In other words, the allelic frequency has been reduced from 0.5 to 0.333 as we found earlier.

The change in allelic frequency is a function of allelic frequency and selective disadvantage. Figure 16.16 illustrates how Δq changes for different initial frequencies when the selective disadvantage is 0.2. The change in frequency is 0.0 when either $p = 0$ or $q = 0$, because the population is monomorphic for either the A_2 or the A_1 alleles. The change in frequency is greatest for an intermediate allelic frequency and becomes quite low when q approaches 0.0. This occurs because most of the A_2 alleles are in heterozygotes and are not subject to selection.

Box 16.1 Industrial Melanism

T he peppered moth, *Biston betularia,* is generally a mottled light color, called typical, but in a number of areas of England, a dark, melanic form is common (fig. B16.1). The melanics were quite rare 150 years ago, but now they constitute nearly 100% of some urban populations. The melanic moths have increased because their dark color serves as protective camouflage from bird predation in the polluted areas they inhabit. This phenomenon, which has occurred in a number of other species, is called **industrial melanism.** A further confirmation of this hypothesis is the increasing frequency of the nonmelanic forms in areas where the air has become cleaner due to pollution control. In other words, there are both spatial and temporal associations between air pollution and the frequency of melanic peppered moths.

To measure the extent of selection between the two moth phenotypes, British biologists Cedric Clarke and Phillip Sheppard pinned dead moths of the two types on dark and light backgrounds and measured their survival from bird predation. On the dark background, fifty-eight of seventy melanic moths survived, while only thirty-nine of seventy typical moths survived (box table 16.1*a*). As a result, the relative fitness of the typical moths was only 0.67, compared to 1.0 for the melanics in this environment. On the other hand, on the light background (box table 16.1*b*), the relative fitness of typicals was highest (or 1) and that of melanics was only 0.75.

Let us use the relative fitnesses estimated on the dark background to calculate the expected change in the typical allele. In this case, the selective disadvantage against typical homozygotes is s = 0.33, because 1 − s = 0.67. If we assume that the frequency of the typical allele is 0.2, then using the formula:

$$q' = \frac{0.2[1 - (0.33)(0.2)]}{1 - (0.33)(0.2)^2}$$
$$= 0.189$$

In other words, the allelic frequency will decline from 0.2 to 0.189, a significant but relatively small change.

(a)

(b)

FIGURE B16.1
Melanic and typical forms of the peppered moth *Biston betularia* (*a*) on a soot-covered trunk and (*b*) on a lichen-coated tree.
Both photos © Michael Tweedie/Photo Researchers.

Box Table 16.1 Survival of Melanic and Typical Moths on Dark and Light Backgrounds

(a) Dark Background

	Melanic		Typical	
	Exposed	*Survived*	*Exposed*	*Survived*
Number	70	58	70	39
Survival	$\frac{58}{70} = 0.83$		$\frac{39}{70} = 0.56$	
Relative fitness	$\frac{0.83}{0.83} = 1$		$\frac{0.56}{0.83} = 0.67$	

(b) Light Background

	Melanic		Typical	
	Exposed	*Survived*	*Exposed*	*Survived*
Number	40	24	40	32
Survival	$\frac{24}{40} = 0.6$		$\frac{32}{40} = 0.8$	
Relative fitness	$\frac{0.6}{0.8} = 0.75$		$\frac{0.8}{0.8} = 1$	

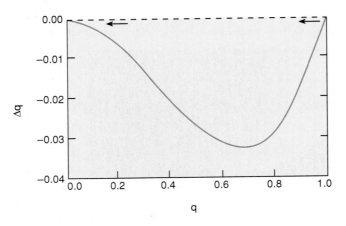

FIGURE 16.16

The change in the frequency of A_2 in one generation for different initial frequencies (q) when there is selection against the recessive genotype A_2A_2. The arrows at top indicate the direction of allele frequency change.

HETEROZYGOUS ADVANTAGE

The mode of selection just described, that is, selection against homozygotes, leads to the eventual fixation of one allele and, as a result, reduces the genetic variation in the population. However, when the heterozygote has a higher fitness than both homozygotes, two alleles can be maintained in the population. A classic example of this type of selection is sickle-cell disease, mentioned earlier; another example is resistance to the rodenticide warfarin in rats, which will be discussed in chapter 17.

To investigate the heterozygous advantage model, we will use a fitness array where the heterozygote, A_1A_2, has the maximum relative fitness (1) and where the homozygotes' fitnesses are $1 - s_1$ and $1 - s_2$ (s_1 and s_2 being the selective disadvantages, or selection coefficients, of the homozygotes A_1A_1 and A_2A_2, respectively). If we follow the same procedure used before, the change in allelic frequency due to selection is:

$$\Delta q = \frac{pq(s_1p - s_2q)}{w}$$

In order to maintain both alleles in the population, Δq must be zero for some value of q between 0 and 1. The point where this occurs is called the stable equilibrium frequency of allele A_2—that is, the allelic frequency for which the change is zero. We found earlier that gene flow could also cause a stable equilibrium.

The value of q, when this expression equals zero and both alleles are present, occurs when the term inside the parentheses in the equation is zero, or:

$$s_1p - s_2q = 0$$

By solving this equation for q, one can show that an equilibrium occurs when:

$$q_e = \frac{s_1}{s_1 + s_2}$$

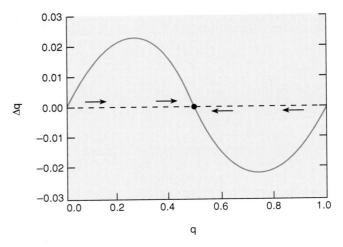

FIGURE 16.17

The change in the frequency of A_2 in one generation for different initial frequencies when there is a symmetric heterozygote advantage ($s_1 = s_2$). The arrows indicate the direction of allele frequency change, and the solid circle indicates the equilibrium.

where q_e signifies the equilibrium frequency of allele A_2. In other words, for this allelic frequency, there is no change, and both alleles are maintained in the population. Notice that the equilibrium is a function only of the selection coefficients for the two homozygotes.

The properties of the equilibrium can be seen in another way by rewriting Δq using the equilibrium frequency so that:

$$\Delta q = \frac{-pq(s_1 + s_2)(q - q_e)}{w}$$

From this formulation, one can see that when q is greater than q_e, Δq is negative; thus, the allelic frequency will decrease toward the equilibrium. Likewise, when q is less than q_e, Δq is positive and the allelic frequency will increase toward the equilibrium. In other words, as in the gene flow example discussed earlier, the equilibrium is stable because after a perturbation away from it, the allelic frequency will change back toward the equilibrium frequency again.

The change in allelic frequency is a function of the allelic frequency and the selection coefficients. An example is given in figure 16.17, where the fitnesses of the three genotypes are 0.8, 1.0, and 0.8, so that $s_1 = s_2 = 0.2$. As a result of these fitness values, the equilibrium frequency is $q_e = {}^{0.2}/_{(0.2 + 0.2)} = 0.5$. If q is smaller than 0.5, the change in allelic frequency is positive, with a maximum value just above q = 0.2. When the value of q is greater than 0.5, the change in allelic frequency is negative, with a maximum value just below 0.8.

We mentioned earlier that heterozygote advantage is the basis of the maintenance of the sickle-cell allele at the β-globin gene in humans. A number of studies have shown that heterozygotes have a higher viability than normal homozygotes in areas where malaria is prevalent.

Table 16.12 Number of Eggs Laid and Hatched by Pigeons of Three Different Transferrin Genotypes

	Female Genotype		
	$Tf^A Tf^A$	$Tf^A Tf^B$	$Tf^B Tf^B$
Number of eggs laid	128	267	144
Number of eggs hatched	59	180	75
Proportion hatched	0.46	0.67	0.52
Relative fitness	0.69	1.0	0.78

Heterozygotes are apparently resistant to malaria because their abnormal red blood cells, sickle-shaped rather than donut-shaped, are hard for the parasite to penetrate. If we let the relative fitness of heterozygotes be 1, that of normal homozygotes should be about 0.8, making $s_1 = 0.2$. In malarial areas, sickle-cell disease is generally fatal before individuals can reproduce, so that the fitness of genotype A_2A_2 is 0, making $s_2 = 1$. Putting these selection coefficients into the formula given previously, the predicted equilibrium is $q_e = {}^{0.2}/_{(0.2 + 1.0)} = 0.17$. In fact, the observed frequency of the sickle-cell allele is between 0.15 and 0.2 in many regions of Africa where malaria is endemic, making these calculations consistent with observations in human populations (see fig. 16.3).

Transferrin is an iron-binding protein that is polymorphic in most vertebrate species. In pigeons there are two common alleles, called Tf^A and Tf^B, which vary in frequency from about 0.38 to 0.59 in different breeds. Jeffrey Frelinger found that the eggs from female pigeons heterozygous for these two alleles were better able to inhibit microbial growth than were the eggs from homozygous females. His data, given in table 16.12, show that 67% of the eggs from heterozygous females survived infection while only 46% and 52% of the two homozygotes survived. If we standardize these values to calculate relative fitnesses, then $s_1 = 0.31$ and $s_2 = 0.22$, and the predicted equilibrium is 0.584, a value that is at the upper end of the range observed in different pigeon breeds.

 Natural selection can result in the elimination of detrimental alleles by selection against homozygotes or in the maintenance of genetic variation by selection for heterozygotes.

WORKED-OUT PROBLEMS

Problem 1

One of the most common types of consanguineous matings in humans is that between two first cousins. In this case, the mating individuals have the same set of grandparents. Draw a pedigree showing this type of mating. In this pedigree there are two common ancestors, the two grandparents. Write out the chains for the two different common ancestors and calculate the inbreeding coefficient (f) that results from each. The total inbreeding coefficient is the total of these two values. What is it?

Solution

The pedigree for this mating is given below. CA_1 and CA_2 are the two grandparents who are common ancestors, and X and Y are the mating individuals.

The chains are Z-X-V-CA_1-W-Y-Z and Z-X-V-CA_2-W-Y-Z. The f value for CA_1 is $(\frac{1}{2})^7$, and the f value for CA_2 is also $(\frac{1}{2})^7 = 0.03125$. Therefore, the total inbreeding coefficient is the sum of these or $0.03125 + 0.03125 = 0.0625$ or $\frac{1}{16}$.

Problem 2

A certain plant population has 289 individuals, some bearing white flowers, others red flowers. The red allele is dominant over the white allele. Of the 289 plants, 246 have red flowers. Assuming the population is in equilibrium, (a) What is the frequency of the white allele? (b) How many of the red plants are expected to be heterozygous?

Solution

(a) Several important relationships must be considered to solve this problem. First, the sum of the frequency of both alleles is unity: $p + q = 1$ (where p is the frequency of one allele and q is the frequency of the other allele). Second, the frequency of each of the genotypes also adds up to

or

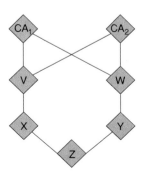

unity: $p^2 + 2pq + q^2 = 1$ (where p^2 is the frequency of one homozygous genotype, q^2 is the frequency of the other homozygous genotype, and $2pq$ is the frequency of the heterozygotes).

In problems of this sort, it is best to make q^2 the frequency of the homozygous recessive genotype. Hence, q is equal to the frequency of the recessive allele. For this problem, you know that 246 plants have red flowers. The remaining 43 plants have white flowers (289 − 246 = 43). To determine the frequency of white flowered plants, divide 43 by 289. This gives 0.149. Thus, $q^2 = 0.149$. The square root of q^2 is q. So, $q = \sqrt{0.149} = 0.386$. The frequency of the white allele is 0.386.

To find the number of heterozygous red plants in the population (b), you need the frequency of the heterozygotes. First, find p (the frequency of the red allele), since the frequency of the heterozygotes is $2pq$. To find p, substitute into the equation $p + q = 1$. Thus, $p = 1 - q = 1 - 0.386 = 0.614$. The frequency of heterozygotes is determined by substituting values in $2pq = 2(0.614)(0.386) = 0.474$. The frequency of heterozygotes is therefore 0.474. To find the number of heterozygotes, multiply the total number of plants in the population by the frequency. In this case, the number of heterozygous red flowered plants is $(289)(0.474) = 136.986$ or 137.

Summary

Extensive genetic variation occurs in most populations at several different levels, including variants affecting color, chromosomal structure, and protein characteristics. By counting the numbers of different genotypes in a population, genotypic frequencies, allelic frequencies, and heterozygosity can be calculated. Two measures of genetic variation, the proportion of polymorphic loci and the level of heterozygosity, can be used to quantify the extent of genetic variation. In many species, the amount of genetic variation observed for genes coding for proteins is substantial. The Hardy-Weinberg principle states that genotypic frequencies are a binomial function of allelic frequencies after one generation of random mating. Genotypic numbers can be examined statistically using the chi-square test to determine whether they are consistent with Hardy-Weinberg expectations. The Hardy-Weinberg principle and the methods of estimating allelic frequencies also apply to multiple alleles and X-linked genes.

Inbreeding increases the frequency of homozygotes and reduces the frequency of heterozygotes, without changing allelic frequencies. The inbreeding coefficient can be measured either from genotypic frequencies or from pedigrees. Mutation is the original source of genetic variation, but it can cause only a small change in allelic frequency per generation. Genetic drift in small populations can result in larger chance changes in allelic frequency. Genetic drift may be the cause of the low heterozygosity observed in endangered species. Gene flow between populations introduces new alleles, and in some cases rapidly changes allelic frequencies. The extent of past gene flow between populations can be estimated. Natural selection eliminates detrimental alleles by selection against homozygotes, or maintains genetic variation by selection for heterozygotes.

PROBLEMS AND QUESTIONS

1. Calculate the frequency of alleles A_1 and A_2 for the following populations:

Population	N_{11}	N_{12}	N_{22}
a	20	40	40
b	30	8	12
c	72	122	91

What are the Hardy-Weinberg frequencies and expected genotypic numbers for the three populations? Calculate the chi-square values for these populations. Are these populations statistically different from Hardy-Weinberg proportions?

2. Assume that the frequencies of genotypes A_1A_1, A_1A_2, and A_2A_2 are 0.1, 0.4, and 0.5, respectively. Use these values and table 16.3 to illustrate that random mating will produce Hardy-Weinberg proportions in one generation. Using a chi-square test and assuming that there are 200 individuals, is this population statistically different from Hardy-Weinberg proportions?

3. If $q = 0.3$ and there are Hardy-Weinberg proportions, what is the most common genotype and what is its frequency? What is the least frequent genotype and its frequency?

4. If A_2 is recessive and 4 out of 400 individuals are A_2A_2 (the rest are the dominant phenotype), what is the estimated frequency of A_2? What proportion of the population is expected to be heterozygous?

5. A survey of 50 *Drosophila* by Dobzhansky and his colleagues in New Mexico (see fig. 16.2) found the following numbers of the various inversion genotypes: AR AR, 18; AR PP, 19; AR ST, 5; PP PP, 5; PP ST, 3; and ST ST, 0. (a) Treating the inversions AR, PP, and ST as if they are three different alleles, calculate the frequency of the three inversion types. (b) Using these frequencies, calculate the expected numbers of the six genotypes if they were in Hardy-Weinberg proportions.

6. Calculate the frequency of alleles A_1, A_2, and A_3 in the following populations:

Population	N_{11}	N_{12}	N_{13}	N_{22}	N_{23}	N_{33}
a	10	20	30	10	20	10
b	40	—	—	40	—	120

What are the expected genotypic frequencies for these two populations? Calculate the chi-square values for these populations. Are these populations statistically different from Hardy-Weinberg proportions? (There are 3 degrees of freedom in this case.)

7. In some populations, approximately 8% of the males have a particular type of color blindness (an X-linked recessive trait). What proportion of females would you expect to have this trait? What proportion of females would you expect to be carriers for this allele?

8. If the initial genotypic frequencies are 0.04, 0.32, and 0.64 for A_1A_1, A_1A_2, and A_2A_2, what are the frequencies after one generation of complete self-fertilization? What is the frequency of heterozygotes after five generations of complete self-fertilization?

9. Calculate the frequency of the three genotypes, A_1A_1, A_1A_2, and A_2A_2, given f = 0.2 and q = 0.4. Assuming that the frequencies of the three genotypes are 0.35, 0.1, and 0.55, what is f?

10. A survey of three populations of wild oats in northern California found that the observed frequencies of heterozygotes were 0.05, 0.07, and 0.03, while the expected frequencies of heterozygotes were 0.41, 0.44, and 0.32, respectively. (a) What are the estimated inbreeding coefficients in these three populations? (b) What do you think accounts for this level of inbreeding in this species?

11. (a) Draw the pedigree for a mating between two half-first cousins who share only one grandparent. What is the expected inbreeding coefficient? (b) Draw a pedigree for a mating between full sibs. What is the expected inbreeding coefficient? (There are two chains in this case because there are two common ancestors. To calculate the overall inbreeding coefficient, you need to add the two f values together.)

12. If u = 10^{-6} and v = 10^{-7}, what is Δq for q = 0.0, 0.25, 0.5, 0.75, and 1.0? When is Δq largest? Explain why.

13. What is the effect of a small population size on the amount of genetic variation in that population over a period of time? Discuss the effects on fixation of alleles and the level of homozygosity in the population.

14. If the initial frequency in a small population is 0.3 for A_2, what is the probability that the population will become fixed for A_2? For A_1? Explain your answers.

15. In a population of 100 individuals with q = 0.4, what are V and the approximate value of $|\Delta q|$ after one generation? How does this value of Δq compare to that for mutation? (For example, see problem 12.)

16. What is the effect of gene flow on the amount of genetic variation in a population? How does it depend upon the level of gene flow and the allele frequency in the migrants?

17. Assume that m = 0.05 and q_m = 0.1. (a) Draw a Δq curve (as in fig. 16.14) with these values for different levels of q. (b) What is the expected equilibrium frequency?

18. In two ancestral populations, the frequency of an allele is 0.0 and 0.6. In a hybrid population, the frequency of the allele is 0.08. What is the estimate of admixture? What kinds of assumptions need to be made for such estimates of gene flow?

19. (a) For a recessive with s = 0.4, what is Δq when q = 0.0, 0.25, 0.5, 0.75, and 1.0? (b) Graph these results on a Δq curve. (c) For a recessive with s = 0.1, what is the allelic frequency after one and two generations when the initial frequency is 0.2?

20. In another experiment like that described in box table 16.1a, 120 melanic moths were exposed and 96 survived, while only 74 of 120 typicals survived. Estimate the relative fitness of the two morphs. What is the expected frequency of the melanic allele after one generation, given that its initial frequency is 0.4?

21. If the relative fitnesses of genotypes A_1A_1, A_1A_2, and A_2A_2, are 0.8, 1.0, and 0.2, respectively, what is the expected equilibrium frequency of A_2? Calculate Δq for q = 0.1 and q = 0.8. What does this tell you about the expected change in frequency?

22. As discussed in the text, it appears that the two common transferrin alleles are maintained in pigeon populations by heterozygote advantage. Three other breeds of pigeons have frequencies of 0.32, 0.81, and 1.0, respectively. After a number of generations, what would you expect the frequencies in these breeds to become? Why do you think they were different from these values initially?

23. Compare in general the expected change in allelic frequencies that result from mutation, gene flow, genetic drift, and selection. Which factors will have the largest effect on Δq, and when? In what ways is the effect on allelic frequency from genetic drift different from that due to mutation, selection, and gene flow?

24. Inbreeding affects genetic variation in a different way than do other factors. Describe its influence, and tell how you could detect whether high inbreeding is present in a population. How might inbeeding be used in plant or animal breeding?

25. In the study of snails mentioned at the beginning of this chapter, it was suggested that differential selection via protective camouflage is important in determining the frequencies of different snail types. Design an experiment that would illustrate this directly. Some people suggest that genetic drift is important in determining snail types. How would you experimentally measure the importance of genetic drift? (In this experiment, you can remove snails, add snails, look at different sites etc.)

Answers appear at end of book.

CHAPTER 17: EXTENSIONS AND APPLICATIONS OF POPULATION GENETICS

Smokeybrown cockroach,
Periplaneta fuliginosa
© *Jack K. Clark,*
1990/Comstock

I *t is the triumph of scientific men . . . to desire*
tests by which the value of beliefs may be ascertained, and to
feel sufficiently masters of themselves to discard
contemptuously whatever may be found untrue.

❖

Sir Francis Galton
British scientist

LEARNING OBJECTIVES

In this chapter you will learn:

1. That the incidence of many human genetic diseases can be explained by a mutation-selection balance.

2. That selection, inbreeding, and genetic drift can also influence the incidence of genetic diseases.

3. That the theory of neutrality is consistent with many findings in molecular evolution.

4. How the molecular clock and phylogenetic trees are used to understand the relationships between species.

5. That conservation genetics is concerned with the preservation of genetic variation in agricultural plants and animals as well as in rare and endangered species.

6. That the primary goals of conservation genetics are to avoid inbreeding depression and loss of genetic variation.

7. That pesticide resistance occurs in many pest species.

8. That pesticide resistance can provide excellent case studies of evolutionary change.

*I*n the introduction to population genetics in chapter 16, we discussed each of the important population genetic factors separately. However, in reality, more than one factor is often at work in a given situation. In this chapter we will illustrate the combined effects of two or more population genetic factors as they operate in several important areas of biological research.

The principles of population genetics can provide insight into a number of areas of biology, medicine, and agriculture, including human genetic disease, molecular evolution, genetic conservation, animal and plant breeding, and pesticide resistance. Here we will use these areas to illustrate the applications of population genetics and to show why it is important to consider several population genetic factors simultaneously. For example, understanding the incidence of human genetic disease necessitates considering both mutation and selection, while studying molecular evolution involves discussing both mutation and genetic drift. Furthermore, a broad comprehension of conservation genetics, animal or plant breeding, or pesticide resistance requires understanding and integrating nearly all the population genetic factors.

FREQUENCY OF HUMAN GENETIC DISEASE

Generally, human genetic diseases are quite rare, with some occurring only in a few families, such as Mal de Meleda and Ellis-van Crevald syndrome, which will be discussed later in this chapter. Such diseases have sometimes been referred to as **orphan diseases.** Because of their rarity, they do not have advocates for research, as do more common diseases such as cancer or AIDS. However, other inherited diseases, such as sickle-cell disease and cystic fibrosis, reach relatively high incidences in certain populations (table 17.1). Here we will consider some of the factors that determine the incidence of both the rare and the more common genetic diseases. Overall, the combined effect of genetic disease is substantial—nearly 2% of newborns in the United States have a single-gene defect. We will begin this discussion by examining the joint effects of mutation and selection in determining the frequency of recessive diseases. In addition, we will mention the influences of inbreeding, genetic drift, and limited gene flow on the incidence of genetic disease in various populations.

MUTATION-SELECTION BALANCE

Selection is the major force that keeps detrimental alleles from increasing in frequency in a population. For example, individuals with cystic fibrosis, an autosomal recessive condition, usually do not survive to adulthood without extensive medical intervention and very seldom reproduce (although recent advances in treatment are delaying their mortality). As a result of this strong selection, the

Table 17.1 Frequency of Alleles Causing Diseases in Different Human Populations

Disease	Population	Allelic Frequency
Tay-Sachs	Israel	0.014
	U.S. (Jewish)	0.013
	U.S. (non-Jewish)	0.0013
Cystic fibrosis	United Kingdom	0.022
	Ohio	0.013
	Africa	0.0036
	Hawaii (Asian)	0.0033
Sickle-cell disease	West Africa	0.10–0.20
	African-American	0.08
	U.S. Caucasian	<0.001
Phenylketonuria	Ireland	0.0141
	U.S. Caucasian	0.0079
	Israel	0.0063
	African-American	0.0023

frequency of the detrimental alleles is kept quite low. On the other hand, new mutations are constantly occurring, so that the cystic fibrosis allele is never completely eliminated. The counterbalancing effects of selection and mutation provide a reasonable explanation for the frequency of many genetic diseases.

To understand this situation, let us assume that there are two alleles and that one of them, the recessive mutant A_2, causes a reduction in viability. If the detrimental allele is recessive, the reduction in allelic frequency due to selection is:

$$\Delta q_s = \frac{-spq^2}{1 - sq^2}$$

where s is the selective disadvantage of homozygote A_2A_2 (see chapter 16). The increase in allelic frequency due to mutation is approximately:

$$\Delta q_{mu} = up$$

if we assume that v, the backward mutation rate (A_2 to A_1), is small compared with u, the forward mutation rate (A_1 to A_2) (see chapter 16). Because these two forces have opposite effects on allelic frequency, they balance each other, so that at some point, the increase in allelic frequency from mutation is the same magnitude as the decrease from selection, or:

$$\Delta q_{mu} + \Delta q_s = 0$$

If we substitute the expressions just given into this equation, then:

$$up - \frac{spq^2}{1 - sq^2} = 0$$

If we assume q^2 is small, because the frequency of the detrimental allele is low (if q = 0.01, then $q^2 = 0.0001$), the

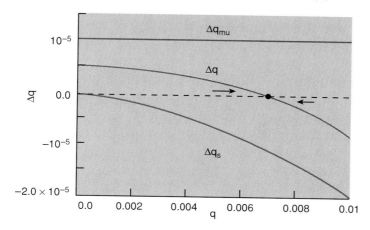

FIGURE 17.1

The variation in allelic frequencies due to both mutation to and selection against recessive alleles. The line labeled Δq is the summation of Δq_{mu} and Δq_s.

denominator of the expression giving the change due to selection will be approximately unity ($sq^2 = 0$). Thus:

$$up - spq^2 = 0$$

This expression can then be solved for the frequency of the recessive genotype as:

$$q_e^2 = \frac{u}{s}$$

The equilibrium allelic frequency is then:

$$q_e = \left(\frac{u}{s} \right)^{1/2}$$

The allelic frequency is obviously increased with a higher mutation rate and decreased with a higher selective disadvantage. Of course, if the allele is a recessive lethal, then $s = 1.0$, and the expected genotypic and allelic frequencies are the mutation rates, u and $(u)^{1/2}$, respectively. Because the mutation rate is generally quite low (see the estimates in table 11.1), we would expect the frequency of lethal disease alleles to be quite low also.

The balance (or equilibrium) between selection and mutation can be understood by examining Δq_{mu} and Δq_s at low frequencies of A_2. Figure 17.1 gives these values when $u = 10^{-5}$ and $s = 0.2$. Notice that Δq_{mu} is about 10^{-5} throughout this low range in allelic frequencies, while Δq_s is 0.0 when $q = 0.0$ and becomes increasingly negative as q increases. The equilibrium for this example is:

$$q_e = \left(\frac{10^{-5}}{0.2} \right)^{1/2}$$

$$= 0.00707$$

In other words, this is the point at which $\Delta q_{mu} + \Delta q_s = 0$. The two curves are summarized in the resultant Δq curve, and the solid circle indicates the equilibrium.

As we mentioned in chapter 2, albinism is a recessive condition caused by a biochemical error that results in a lack of the pigment melanin. Individuals with albinism often have poor eyesight and are sensitive to sunlight, but their relative fitness is probably not greatly reduced. One estimate (based on relative viabilities) is that the fitness of albinos is about 0.9 that of unaffected individuals, making $s = 0.1$. Estimation of the mutation rate per gene is generally

between 10^{-5} and 10^{-6} (see chapter 11). If we use 5×10^{-6} as the mutation rate, the expected frequency of albino individuals from mutation-selection balance, using the equation just given, is:

$$q_e^2 = \frac{5 \times 10^{-6}}{0.1}$$

$$= 0.00005$$

or about 1 in 20,000 individuals. In most populations, the incidence of albinism ranges between 1 in 10,000 and 1 in 40,000, and is thus consistent with our mutation-selection balance explanation.

We can also use the mutation-selection balance to predict the future incidence of genetic disease. For example, the frequency of a genetic disease may rise either because of improved medical care that enables affected individuals to survive and reproduce or because of an increased mutation rate that results in the production of more mutant alleles. When the selective disadvantage against the disease decreases due to improved medical care so that s' is the new selective disadvantage, the new equilibrium becomes:

$$q_e^2 = \frac{u}{s'}$$

For example, if a disease was formerly lethal before the age of reproduction ($s = 1.0$) and now, with better medical care, is only slightly disadvantageous (say, $s' = 0.1$), its equilibrium frequency should increase 10-fold. Phenylketonuria is a disease for which the fitness has increased from almost zero when untreated (because the severely retarded individuals with PKU are generally institutionalized) to nearly that of unaffected individuals after diet therapy. As a result, we might expect the frequency of individuals with PKU to increase over time, with a possible consequent long-term increase in medical costs.

Likewise, an increase in the mutation rate should also increase the equilibrium value. Assume that because of environmental pollutants, food additives, or some other mutagenic factor, there is an elevated mutation rate, u'. The new equilibrium is then:

$$q_e^2 = \frac{u'}{s}$$

If $u' = 10u$, a 10-fold increase in mutation rate, the equilibrium is also increased 10-fold. However, an increase in mutation rate may influence a number of loci simultaneously. We should caution that higher incidence of disease as a result of both improved medical care and increased mutation rate would generally occur slowly over many, many generations.

OTHER FACTORS AFFECTING DISEASE FREQUENCY

Although mutation-selection balance appears to be a good general explanation for the frequency of many genetic diseases, other evolutionary factors can also be important. For example, the frequency of the sickle-cell allele is increased by the advantage of heterozygotes (the homozygotes are still at a disadvantage) in areas with endemic malaria (see chapter 16). However, the sickle-cell allele is affected by different evolutionary factors in contemporary U.S. populations. Although malaria was once endemic in parts of the southern United States, the present rare incidence of the disease has resulted in a different set of fitness values here than in Africa; that is, there is no heterozygous advantage, and the homozygote is often no longer lethal.

Genetic counseling may lead to lower reproduction of sickle-cell carriers and selective abortion of sickle-cell homozygotes. In addition, past and present gene flow into African-American populations from Caucasians, in which the frequency of the sickle-cell allele is near zero, further dilutes its frequency. Overall then, one would predict that the frequency of the sickle-cell allele will eventually decline to a very low level among African-Americans in the United States.

Examining the DNA of sickle-cell alleles suggests that these favorable mutants have been incorporated into populations only a few times. Support for this conclusion comes from the fact that in most individuals the β-globin gene is contained in a 7.6-kb restriction fragment when the enzyme HpaI is used (fig. 17.2). However, in African-American populations, a longer type of restriction fragment is almost always associated with the sickle-cell allele. In fact, the S allele is really associated with absence of a restriction site that is present on the 3′-end of the fragment with the normal allele.

The association of particular alleles at different loci, as in the example of allele S lacking a restriction site at a linked region, is called **linkage disequilibrium.** Linkage disequilibrium between disease alleles and linked restriction sites is used to locate disease genes on particular DNA fragments. Furthermore, examining other sickle-cell alleles in Africa and checking linkage disequilibria with various DNA markers suggests that nearly all of the African sickle-cell alleles descend from only three independent mutants. In other words, mutation (or lack of mutation) may actually limit the evolutionary response in this case.

Another factor influencing the incidence of genetic disease is inbreeding. In isolated human populations,

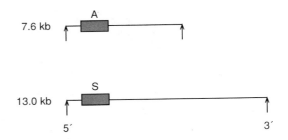

FIGURE 17.2

A diagram of the restriction fragments for the normal β-globin allele and the sickle-cell allele. The arrows indicate the cleavage sites caused by restriction enzyme HpaI.

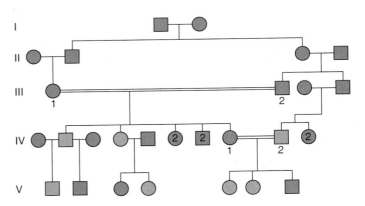

FIGURE 17.3

A pedigree in which two first-cousin matings (double lines) have taken place, III-1 with III-2 and IV-1 with IV-2. Notice the high incidence of Mal de Meleda, a rare recessive disease (purple).

consanguineous matings often occur because of the limited number of available mates, Such inbreeding will increase the incidence of recessive diseases in these populations (see further discussion later in this chapter). For example, in the Kel Kummer Taureg tribe of the remote South Sahara, which is composed of only three hundred people, approximately 30% of marriages are between first cousins. Excellent documentation is also available in Italy, where the Catholic church requires special permission for consanguineous marriages. Records show that consanguineous marriages are three to four times more frequent in the remote, less densely populated mountain villages than in the more densely populated areas.

To further illustrate inbreeding, figure 17.3 gives a pedigree of an isolated island population off the coast of the former Yugoslavia. The disease Mal de Meleda is a rare, autosomal recessive disorder that results in thickened skin on the hands and feet. In this pedigree, two first-cousin matings are indicated by the double lines connecting individuals 1 and 2 in generation III and individuals 1 and 2 in generation IV. Individuals III-1 and III-2 must both have been heterozygotes, because two of their seven offspring were affected. The other first-cousin mating, between heterozygote IV-1 and diseased homozygote IV-2, resulted in

FIGURE 17.4
(*a*) A child with Ellis-van Crevald syndrome, a form of dwarfism found in the Amish of Pennsylvania. Affected individuals have short limbs, malformed hearts, and six fingers on each hand (see X ray in *b*).
Both photos © Victor McKusick.

(a)   (b)

two out of three progeny being affected. (Although this disease is rare elsewhere, its commonness in this population results in many heterozygotes.) The high proportion of consanguinity in this population has resulted in a much higher incidence of Mal de Meleda than would occur in a random mating population (or on the mainland nearby).

Some isolated religious groups have relatively high incidences of particular rare diseases. In fact, a number of diseases were originally described in populations such as the Amish, an Anabaptist, farming population, in the United States. Often, groups like this were originally established by a small number of founders and have subsequently kept their community closed to immigrants. As a result, genetic drift, or **founder effect,** is a primary factor accounting for the abnormally high incidence of certain genetic diseases in their populations. Of course, in the case of the Amish, the high incidence is relatively temporary, and selection will probably reduce the frequency over a number of generations.

The Amish population of Lancaster County, Pennsylvania, has a high incidence of a recessive disorder known as six-fingered dwarfism, or Ellis-van Crevald (EvC) syndrome (fig. 17.4). In this population of about thirteen thousand, eighty-two affected individuals in forty affected sibships have been diagnosed. Pedigrees of the eighty parents in these sibships showed that all trace their ancestry to Samuel King and his wife, early members of the community. Thus, it appears quite certain that the high incidence of EvC syndrome is due primarily to the founder effect; that is, either Samuel King or his wife carried the recessive allele, and because many individuals in the population are their descendants, the incidence of the disease is high.

The incidence of human genetic disease can be generally explained by a mutation-selection balance in which mutation increases the frequency of the mutant and selection reduces it. In addition to mutation and selection, the frequency of disease alleles is sometimes influenced by genetic drift and gene flow. The incidence of recessive genetic diseases is also affected by inbreeding.

MOLECULAR EVOLUTION

One of the most exciting and potentially useful applications of population genetics is in understanding the process of **molecular evolution**—the factors that have caused changes in the nucleotide sequence or amount of DNA and in the amino acid sequence of proteins over evolutionary time. In addition, the differences between these macromolecules can explain the relationship between species. In the ensuing discussion, we will assume that most differences in DNA sequences and many changes in protein sequences are not under strong selective constraints. As a result, other evolutionary factors that affect genetic variation, such as genetic drift and mutation, appear to have predominant roles in influencing the variation and evolution of these molecular sequences.

NEUTRALITY

Perhaps the best perspective for understanding molecular evolution is the theory based on mutation and genetic drift that was developed by Japanese geneticist Motoo Kimura and his colleagues. Although genetic drift has only a small effect in a large population each generation, its cumulative effect may be substantial if there are many generations. Likewise, the mutation rate in any one generation is small, but the likelihood of a mutation becomes substantial over a number of generations. Thus, the joint effects of genetic drift and mutation may be important over long periods of time and may be an important influence on molecular evolution.

In order to examine mutation in a finite population, let us assume selective differences are not important—that different alleles (or sequences) have no (or a very small) selective advantage or disadvantage. This concept is called **neutrality.** Generally, neutrality refers to a situation where the effects of genetic drift on allelic frequency are larger than the effects of selection. For example, alleles A_1 and A_2 are neutral with respect to each other when the relative fitnesses of A_1A_1, A_1A_2, and A_2A_2 are all equal. Another approach is to say that the alleles are all equally important.

FIGURE 17.5

The frequency of four new mutants over time in a finite population. In this example, only the first mutant eventually became fixed. The rest were lost due to genetic drift.

Let us assume that a new mutant, A_2, occurs in a population of size N that otherwise consists only of A_1 alleles. The initial frequency of the mutant is:

$$q = \frac{1}{2N}$$

where 2N is the number of alleles in a diploid population. Assuming the two alleles are neutral with respect to each other, the probability of fixation of the new mutant is equal to its initial frequency, or $1/(2N)$, as discussed in chapter 16. The probability that the new mutant will be lost (equal to the fixation of the original allele) is then $1 - 1/(2N)$. In other words, unless the population is quite small, a new neutral mutant will nearly always be lost from the population. If N = 50, the probability of fixation of a new mutant is only 0.01.

The two alternatives, fixation or loss of the new mutant, take quite different amounts of time to occur. Because the change of frequency necessary when loss occurs is small, from $1/(2N)$ to zero, the average amount of time for loss to occur is short. On the other hand, fixation of the new mutant requires a substantial change in allelic frequency, from $1/(2N)$ to unity, and the time required for fixation is much longer. Figure 17.5 illustrates the time necessary for fixation and loss to occur for four separate mutants. In this example, the first mutant eventually becomes fixed, a quite unlikely event, while the others are lost from the population due to genetic drift.

If a large number of potential alleles exist at a locus, mutation will increase the number of alleles, and genetic drift will reduce the number of alleles. The equilibrium heterozygosity (H_e) resulting from the balance of these two factors is:

$$H_e = \frac{4Nu}{4Nu + 1}$$

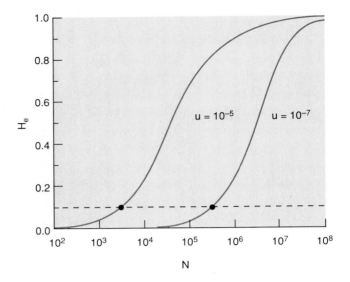

FIGURE 17.6

The equilibrium heterozygosity H_e expected from a balance of mutation with a rate u and finite population of size N.

where N is the population size and u is the mutation rate. If it is assumed that the mutation rate is between 10^{-5} and 10^{-7}, then the expected range of heterozygosity at equilibrium can be calculated for different population sizes (fig. 17.6). For example, if N is 10,000 or 10^4, then equilibrium heterozygosity would be nearly 0.0 when $u = 10^{-7}$ and 0.286 when $u = 10^{-5}$. In general, if Nu is much greater than one and the population size is very large, then the mutation rate determines for the most part the amount of heterozygosity, so that H_e is quite high. On the other hand, if Nu is much less than one, then genetic drift becomes the major factor and H_e is low.

One can look at the predictions from this neutrality expression in another way. If we use the average observed electrophoretic heterozygosity found in surveys of 0.10 (see discussion in chapter 16), and assume u is between 10^{-5} and 10^{-7}, then we can obtain an idea of what N must have been. In this case, N should be between 3,000 and 300,000 (the intersection of the vertical broken lines with the horizontal axis in figure 17.6, given as solid circles).

In fact, the heterozygosity resulting from the balance of mutation and genetic drift may be rather large if the population is big. As we discussed in chapter 16, extensive variation occurs for electrophoretically determined alleles in many species. The neutrality theory suggests that much molecular variation is primarily due to the joint effects of genetic drift and mutation.

However, some electrophoretic variation observed in natural populations is probably maintained by selection. The actual proportion of loci affected by selection or by the forces of neutrality has been a controversial topic. One of the most thoroughly studied biochemical polymorphisms is the variation at the alcohol dehydrogenase (*Adh*) locus in

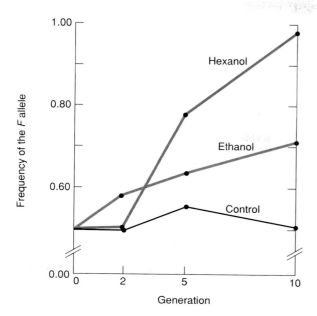

FIGURE 17.7

The change in frequency of the Adh^F allele in *D. melanogaster* when the flies are exposed to different alcohols.

Table 17.2 Hypothetical Array of Amino Acids from Two Species[1]

Species	Position				
	1	*2*	*3*	*4*	*5*
a	arg	asp	val	glu	lys
b	arg	asp	arg	glu	leu

[1] Abbreviations indicate different amino acids.

Drosophila melanogaster. The two common alleles, *F* and *S,* differ only in one out of 255 amino acids; that is, in position 192, the *F* allele has a threonine and the *S* allele has a lysine. Because alcohols are an important component of the rotting fruit and vegetables in which *Drosophila* live, breakdown of alcohols by alcohol dehydrogenase is probably necessary for fly survival. A number of researchers have found that the frequency of the two alleles is affected by alcohols. Furthermore, the rate of change appears to be affected by the type of alcohol present, as illustrated in figure 17.7 for the change in allele *F* for different alcohols. For nearly all the alcohols tested, the *F* allele increased in frequency. In other words, strong selection seems to favor the *F* allele under these conditions, making neutrality an unlikely explanation for the maintenance of this variation.

MOLECULAR CLOCK

The amino acid sequences for a number of proteins from different organisms are known. Paleontological information dating the splits between the ancestors of various living organisms indicates that substitutions between these organisms accumulate at a regular rate over evolutionary time for a given protein. Of course, this apparent regular replacement is an average of a number of events over a long period of time. Assuming that such a regular replacement occurs, the differences in amino acid sequence between two species with an unknown common ancestor may serve as a **molecular clock,** indicating the time since the two species diverged. In addition, by comparing the amino acid sequences of homologous proteins (or nucleotide sequences)

from different species, we can estimate the rate of substitution of molecular variants, given the approximate time since the species diverged.

As an example, let us consider the hypothetical array of homologous amino acids from two species in table 17.2. The five amino acids are identical at positions 1, 2, and 4 and different at positions 3 and 5. Therefore, the proportion of sites that are different is $^2/_5 = 0.4$. A neutrality explanation of this difference would state that a mutation occurred in one of the species at these positions and has subsequently become fixed by genetic drift. If the two species were very closely related, they would have the same amino acid at nearly every site, whereas if they were distantly related, they would be different at many different sites.

The molecular clock hypothesis is consistent with the neutrality theory. In other words, if the replacement of molecular variants over time is a function of genetic drift and the mutation rate, relatively regular turnover (replacement) of molecular variants will take place over a long period of time. To illustrate, assume that genetic drift and mutation are the only factors that change the frequencies of molecular variants. Let the mutation rate to a new molecular variant be u, so that in a diploid population of size N, there are 2Nu new mutants per generation. Remember that the probability of fixation of a new neutral mutant is equal to its initial frequency, $^1/_{(2N)}$. Assuming an equilibrium situation between mutation and genetic drift, the **rate of allelic substitution** per generation (k) is the product of the number of mutants generated per generation and their probability of fixation, or:

$$k = 2Nu \frac{1}{2N}$$

$$= u$$

In other words, the average rate of allelic substitution in this simple model is equal to the mutation rate at the locus.

If we assume that the mutation rate per gene is about 10^{-6} and that about 100 amino acids exist in a typical protein, the mutation rate per amino acid should be about 10^{-8}. When there are no selective constraints anywhere in the amino acid sequence, this should be an approximate estimate of the rate of amino acid substitution. However, the rate of amino acid substitution appears to differ for different protein molecules. For example, the amount of divergence for three molecules that have been sequenced in a number of organisms is given in figure 17.8. From these comparisons, it appears that the rate of

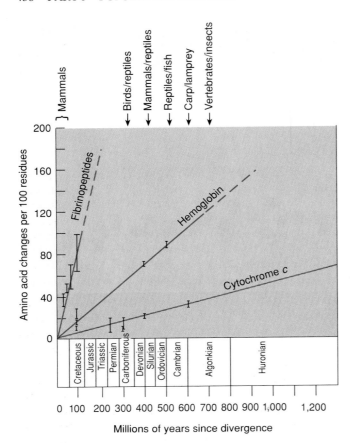

FIGURE 17.8

The amino acid differences for three separate macromolecules between divergent lines of organismic evolution.

Source: Data from R. C. Dickerson, "The Structure of Cytochrome C and the Rates of Molecular Evolution" in J. Molec. Evol. 1:26–45, 1971.

amino acid substitution is highest for fibrinopeptides, intermediate for hemoglobin, and slowest for cytochrome c. The estimated rates of amino acid substitutions for fibrinopeptides, hemoglobin, and cytochrome c are 0.9×10^{-8}, 0.14×10^{-8}, and 0.03×10^{-8}, respectively.

These differences may result from the relative importance of the three-dimensional structure of these molecules and the proportion of amino acids involved in active enzymatic sites. For example, histone IV, which has the very low substitution rate of 0.0006×10^{-8}, fits in tightly with the DNA molecule; thus nearly the whole amino acid sequence is critical for the three-dimensional structure. At the other extreme, the fibrinopeptide molecule appears to be hardly more than a spacer molecule, with no enzymatic sites. One explanation is that, where constraints are greatest, the rate of amino acid substitution is reduced. In other words, the rate for fibrinopeptides is close to that expected from neutrality, while the rate for histones is much lower, presumably because of the functional constraints on a large part of the histone molecule.

PHYLOGENETIC TREES

The differences among related species may be measured for morphological, biochemical, immunological, or other traits, with more closely related species presumably being more

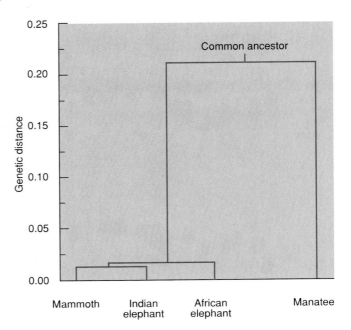

FIGURE 17.9

A phylogenetic tree constructed from immunological genetic distance data, indicating a close relationship between the mammoth, the Indian elephant, and the African elephant.

Source: Data from J. M. Lowenstein, et. al., "Albumin Systematics of the Extinct Mammoth and Tasmanian Wolf" in Nature 291:409–411, 1981.

similar. Thus, one important use of genetic or phenotypic measures of differences is in organizing related populations or species in a biologically meaningful way—that is, arranging these groups to illustrate ancestral relationships. Such relationships are usually shown on a **phylogenetic tree,** in which the root or ancestral type is at the top of the tree and the branches, or evolutionarily derived types, are at the bottom. The similarities and differences in molecular structure among different organisms can help us understand such evolutionary relationships. Presumably, organisms having similar molecules are closely related, and those having quite different molecules are only distantly related.

To show how to construct a simple phylogenetic tree, let us examine a number of groups that are either different populations or species. Assume that the genetic differences between these groups have been measured, using such genetic data as allelic frequencies, amino acid differences, or restriction map information, so that a matrix of values is available.

These differences can be organized in a way that reflects biological relationships. For example, proteins are still intact in specimens of some extinct animals. Comparing the immunological properties of these proteins to those of related extant animals gives a measure, called the **genetic distance,** of the molecular differences among these organisms. In one case, the remains of a 40,000-year-old baby mammoth were discovered in the Soviet Union in 1977. Comparison of the immunological properties of the protein albumin from the mammoth, two species of elephants, and a manatee yielded the phylogenetic tree shown in figure 17.9. Here the differences in the properties of

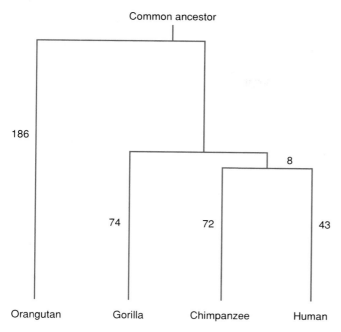

Common ancestor

FIGURE 17.10

A phylogeny of humans, chimpanzees, gorillas, and orangutans, based on their base sequences in the β-globin region. The numbers indicate the number of changes that have occurred along each lineage.

Table 17.3 Proportion of Difference for Each Species Pair at the 7,100-Base β-globin Sequence

	Chimpanzee	Gorilla	Orangutan
Human	0.016	0.017	0.034
Chimpanzee	—	0.021	0.038
Gorilla	—	—	0.037

panzees share eight changes that are not present in gorillas, supporting the hypothesis that humans are most closely related to chimpanzees. In the near future, other comparative nucleotide sequences from unlinked regions should be available to help further resolve the ancestry of these species.

USE OF mtDNA IN STUDYING GENETIC RELATIONSHIPS

The mitochondria, large cell organelles, have their own DNA, called mtDNA, which is inherited independently of nuclear DNA because it is generally transmitted almost entirely maternally, that is, mother to daughter. Further, because mitochondrial DNA can be physically separated from nuclear DNA, many of the recent studies of DNA variation have used mtDNA. Since mtDNA is generally maternally inherited and there appears to be little genetic exchange between mitochondria, one can assume that individuals sharing the same mtDNA type have a common female ancestor. In addition, the rate of substitution for mtDNA in mammals is approximately 5- to 10-fold higher than that of nuclear DNA, possibly due to inefficient DNA repair in mitochondria that results in a high mutation rate.

Because of the maternal inheritance of mitochondria, mtDNA and nuclear DNA may show quite different patterns of variation. For example, two species of house mice form a hybrid zone about 50 kilometers wide in Denmark, with a marked anatomical difference between the species. A number of nuclear genes identified by protein electrophoresis have different frequencies above and below this zone; that is, they are concordant with the known anatomical differences. However, mtDNA shows a different pattern; mtDNA frequencies above and below the hybrid zone are quite similar.

The power of examining mtDNA is illustrated in the recent study of green sea turtles by Jon Avise and his colleagues. Each year female turtles migrate 2,000 kilometers from their feeding grounds near Brazil to Ascension Island in the mid-Atlantic. It has been hypothesized that this migration has endured over forty million years since a time when South America and the mid-Atlantic were near each other (they have since moved apart by continental drift). The female turtles are believed to migrate to these beaches to lay eggs because of *natal homing,* a behavior in which adults return to their birthplace. If in fact the Ascension Island population has been isolated from other Atlantic populations for millions of years, it should be genetically quite distinct.

albumin between the mammoth, the Indian elephant, and the African elephant are small, corresponding to the current belief that the three species are closely related. The molecular similarity of the three species suggests that they diverged from each other at about the same time.

In figure 17.9, we cannot really tell whether the mammoth is more closely related to the Indian or the African elephant. In fact, we would need more detailed data to determine which of the three species first branched off from the other two. The exact sequence of branching regarding the relationship of humans, chimpanzees, and gorillas is a major controversy. As we saw in chapter 4, the detailed chromosomal patterns of humans, chimpanzees, and gorillas show high similarity. Recently, the DNA sequences of these three species and of the orangutan were compared for a 7,100-base-pair sequence in the β-globin region. The proportion of divergence between all of them is small, but the orangutan has the greatest difference from the others, at over 3% (table 17.3). Notice that the least different pair, the human-chimpanzee, is slightly more similar than the human-gorilla pair.

These data can be organized into a phylogeny to illustrate the ancestral relationships between the species (fig. 17.10). The numbers in this most parsimonious phylogeny (there are other arrangements that do not explain the data quite as well) indicate the number of DNA changes along each branch of the tree. For example, humans have accumulated forty-three changes and chimpanzees seventy-two since they diverged. Interestingly, humans and chim-

When the mtDNA of turtles from Hawaii, Florida, Venezuela (as a sample of South American turtles), and Ascension Island was examined, it was clear that the Ascension Island population mtDNA was identical to the most common type found in Venezuela. For most restriction sites (ninety-five were assayed), the mtDNA from all populations was identical. Figure 17.11 gives an example of some of the variable sites that show the individuals from Venezuela and Ascension Island were identical, while individuals from Florida and Hawaii were different. Therefore, it appears that the Ascension Island population has not been isolated for long, since it is identical to the Venezuelan sample, and is probably recently descended from other Atlantic-breeding populations.

Recent studies using phylogenetic techniques in human mtDNA have suggested that the current mtDNA descended from a female some 200,000 years ago, prompting speculation about a human Eve. This suggestion is based on the observation that less variation in mtDNA occurs among humans than would be expected, given the high mtDNA mutation rate. First, mtDNA consists of only 16 kilobases, a small amount of DNA compared to that found in the nucleus. Second, as we just discussed, the patterns of variation observed for mtDNA and nuclear DNA may be quite different. Finally, remember that genetic variation is reduced by genetic drift. A small population size 50,000 years ago could have the same effect as the "bottleneck" of a single woman 200,000 years ago. When similar information is available about nuclear DNA, we will have a better picture of recent human evolution.

MULTIGENE FAMILIES

A number of genes, such as the immunoglobins, HLA genes, the histones, rRNA genes, and tRNA genes, are now known to exist as multigene families—multiple copies that have apparently arisen as a series of tandem duplications (see chapter 4) on a chromosome. The number of copies in such a multigene family may vary in different individuals, and unequal crossing over may lead to either the contraction or expansion of the gene number in a family.

In humans, two copies of the α-globin gene ordinarily occur 3.7 kilobases apart on chromosome 16. If one or both of these loci are deleted, the individual has thalassemia, a blood disorder that causes anemia. The loss of the α-globin genes from unequal crossing over can also produce gametes with three copies of the α-globin gene. A survey using endonuclease mapping on three populations with high incidences of thalassemia for triplicated α-globins found seven individuals heterozygous for a triplicated chromosome and a normal chromosome (table 17.4). None of the individuals had clinical symptoms related to the triplication.

Understanding the reasons for the number and types of genes in a multigene family requires knowing about the relevant genetic processes—unequal crossing over, gene

FIGURE 17.11

mtDNA digested by restriction endonuclease SpeI from green sea turtles at four different rookeries. The turtles from Venezuela and Ascension are identical, while the mtDNA from the turtles in Florida and Hawaii is different.

From Brian W. Bowen, Anne B. Meylan, and John C. Avise. AN ODYSSEY OF THE GREEN SEA TURTLE: Ascension Island Revisited. Proceedings of the National Academy of Science 86:573–76, January 1989, Evolution fig. 2A.

Table 17.4 Frequencies of Single and Triplicated Copies of the α-globin Gene

| | Number of α-globin Genes | | |
Population	1	2	3
African-American	0.16	0.836	0.004
Sardinian	0.18	0.816	0.004
Greek Cypriot	0.07	0.88	0.05

conversion, and transposition. In addition, the size and makeup of a multigene family appear to be affected by genetic drift, mutation, and selection. The overall scenario, therefore, becomes quite complex, but the combination of genetic processes and population genetic factors should explain the evolution of these important sequences.

 The evolution of molecules is generally consistent with the neutrality model, a theory based on the combined effects of genetic drift and mutation. Changes in DNA and proteins generally occur at a regular rate, resulting in a molecular clock. Molecular differences in different species can be used to construct phylogenetic trees that indicate the evolutionary relationships between the species. Mitochondrial DNA has been used to determine phylogenetic relationships.

FIGURE 17.12
Corn suffering from southern corn leaf blight, a disease that severely damaged crops until a blight-resistant strain of corn was developed.
Courtesy of U.S. Department of Agriculture.

CONSERVATION GENETICS

In recent decades, habitat destruction, hunting, and other factors have resulted in the extinction of a number of species and the near-extinction of many others. In fact, some animals, such as the California condor and the black-footed ferret, now exist only in small numbers in captivity. In other parts of the world, such as the Amazonian river basin, whole ecosystems containing many species of plants and animals are in danger. Furthermore, the number of animal breeds and crop varieties used in agriculture has precipitously declined in recent years. All of these factors have stimulated attempts to maintain genetic variation in many different species, a discipline known as **conservation genetics.**

CROP SPECIES

In the past, a number of different varieties of each crop plant were grown, many of which were adapted to local environmental conditions. Now, fewer and fewer varieties are being planted, resulting in a restriction of the gene pool of these crops. Part of the reason for this change is that in many areas of the world, newly developed high-yield strains have replaced the local plant varieties, or **land races.** For example, in some areas, a "green revolution" has caused new strains of rice and wheat to be substituted for the land races. Such high-yield varieties often require large-scale planting and mechanized spreading of fertilizer and pesticides. As a result, local varieties that were selected to yield well in a labor-intensive situation are no longer used. However, these local varieties often have alleles that are well adapted to local conditions, such as particular soil types, diseases, and pests, which could be valuable in developing future varieties. Unfortunately, many land races may already have been lost, and their wild relatives may now be the only source of advantageous alleles.

Two related phenomena can result from such developments. First, a crop species can become genetically vulnerable to a particular disease, because in a large planting, the varieties have the same alleles at the loci for pathogen or insect resistance. For example, in 1970, genetic uniformity in planted corn led to a loss of approximately 15% of the United States crop due to southern corn leaf blight (fig. 17.12). Recovery from this disaster only occurred when a blight-resistant strain of corn was developed.

The second possible result of such developments is that genetic variation in the total species, that is, the combination of land races, germ plasm (or seed) collections, and wild progenitors of the crop, may be reduced. Several different approaches have been advocated to ensure the maintenance of genetic variation in crop species. One system would conserve genetic variation by growing the land races in situ—that is, in the area where they developed. Obviously, this is preferred (if possible), so that further adaptive change may occur while present genetic types are maintained.

A second approach to genetic conservation is the accumulation of germ plasm collections of many cultivated varieties and their wild relatives. The general strategy of collection should consider three factors: the number of seeds to be collected per local population, the number of populations sampled, and the distribution of these populations over the species range. For example, in a species with high gene flow, local populations may be quite similar, while in species with high amounts of self-fertilization and low gene flow, local populations may differ greatly. Where knowledge of the species is inadequate, it is usually advisable to sample as many populations as possible, basing the distribution of the samples upon both the structure of the species and the heterogeneity of the environment.

When measuring genetic diversity in either land races or the wild relatives of crop plants, three major types of traits are considered: economic or other metric traits, disease or pest resistance genes, and Mendelian morphological markers. All of these traits, plus the adaptation of different varieties to a range of environments, would be important considerations in a sampling regime. In addition, electrophoretic variants may be useful, since the distribution of alleles over populations can indicate the population structure of the species and help identify local populations. Electrophoretic variants may also be important in identifying genotypes common in only one or a few populations.

One crop that exemplifies the importance of genetic variation in wild relatives is the tomato (fig. 17.13). The tomato was domesticated from a wild species present in the Andean region of South America. It is actually rather new as a crop, having been avoided even into the early twentieth century in Europe and the United States because of its alleged toxicity (the plant is a member of the night-shade family). However, today the tomato yields more tons of produce and has a greater value than any other vegetable grown for human consumption in the United States.

A number of important phenotypic qualities have been incorporated into tomato cultivars in recent decades. These include an allele that causes uniform ripening of the fruit (no green area near the stem) and another that prevents the plant from growing in an indeterminate fashion (not ending in a flower or bud). Certain tomato varieties also have traits that make them easily harvestable by machine; for instance, they are uniformly short and have tough-skinned fruit that ripens in a concentrated period of time.

The extensive phenotypic variation among the wild species of tomatoes also has potential breeding importance. One type of tomato lives only five meters from the high tide line in the Galápagos Islands. These plants can survive in the salty spray from seawater, suggesting that it might be feasible to develop a tomato variety capable of using water containing a high proportion of salt. A second species exists in very dry habitats in Peru, where it obtains most of its water from fog or mist. These plants appear to have extreme resistance to water loss, a trait of potential importance for irrigated tomatoes.

INBREEDING DEPRESSION

As we discussed in chapter 16, inbreeding increases the frequency of homozygotes and decreases the frequency of heterozygotes. Thus, if recessive genotypes have lower fitness, the incidence of recessive diseases or traits may increase, and consequently, the mean fitness of a population may be reduced. This, of course, may be particularly detrimental to endangered species where population numbers are already low.

To understand the joint effect of inbreeding and selection, let us first examine the incidence of recessive genetic diseases among individuals from consanguineous matings. Inbreeding increases the frequency of homozygous recessives, which in this case also increases the incidence of the recessive disease. The genetic basis for this effect becomes clear when the proportion of recessive homozygotes in a noninbred population ($Q = q^2$) is compared to that for a given inbreeding coefficient ($Q_f = q^2 + fpq$). The ratio of these two quantities is:

$$\frac{Q_f}{Q} = \frac{q^2 + fpq}{q^2}$$
$$= \frac{q + fp}{q}$$

which is greater than unity for any allelic frequency but very large for low allelic frequencies. For example, when $f = \frac{1}{16}$ (the inbreeding coefficient that results from a first-cousin mating) and $q = 0.01$, as it does for albinism, then:

$$\frac{Q_f}{Q} = \frac{0.01 + (0.0625)(0.99)}{0.01}$$
$$= 7.2$$

FIGURE 17.13
The variation in tomatoes showing both domesticated and wild fruits.
Cover of Science of Food and Agriculture, *Vol. 4, No. 1, January 1986. Photo by Dr. R. Clark, Regional Plant Introduction Station, Washington State University.*

In other words, there are over seven times as many affected individuals in the inbred group as in the random mating group. If $q = 0.001$, as it does for many rarer genetic diseases, this ratio becomes 63, illustrating why many individuals with very rare recessive diseases are the progeny of consanguineous matings.

Because individuals are more likely to be homozygous for recessive deleterious alleles in an inbred population, the mean fitness of the total population may decline. **Inbreeding depression**—the decline of population fitness due to inbreeding—seems to be a nearly universal phenomenon. However, the extent of inbreeding depression varies among different organisms. For example, three generations of mating full sibs in Japanese quail result in complete loss of reproductive fitness, while many plant species with partial self-fertilization show little decline in fitness when inbred.

Let us examine how inbreeding can affect the mean fitness of the population. Assume, as in chapter 16, that selection causes the relative fitness values of genotypes A_1A_1, A_1A_2, and A_2A_2 to be 1, 1, and $1 - s$, respectively. If we use the genotypic frequencies given for inbreeding in chapter 16, the mean fitness with inbreeding is:

$$w = (1)(p^2 + fpq) + (1)(2pq - 2fpq)$$
$$+ (1 - s)(q^2 + fpq)$$
$$= 1 - sq^2 - sfpq$$

using the fact that $p^2 + 2pq + q^2 = 1$. Remember from chapter 16 that the mean fitness for a recessive gene when there is random mating is $1 - sq^2$. In other words, the mean fitness is lowered from inbreeding by the amount $sfpq$. Obviously, this formula predicts that the fitness is lowered (inbreeding depression increased) linearly as the inbreeding

Table 17.5 Proportion of Juvenile Mortality in Noninbred and Inbred Progeny

	Juvenile Mortality (Sample Size)	
	Noninbred	Inbred
Degu	0.106	0.235*
Elephant shrew	0.153	0.233
Pygmy hippopotamus	0.245	0.549*
Cheetah	0.263	0.442*
Dorcas gazelle	0.280	0.595*
Japanese serow	0.288	0.565*
Greater galago	0.308	0.386
Golden lion tamarin	0.501	0.634*

*Indicates p < 0.05.

FIGURE 17.14

The golden lion tamarin. This species shows an increased mortality rate when inbred.

© Michael Dick/Animals Animals.

coefficient is increased. In fact, results of extensive studies in *Drosophila,* humans, and other species generally support this prediction.

Until recent years, some individuals breeding endangered species in captivity were not aware of, or thought unimportant, the extent to which fitness is reduced by inbreeding. One reason for this misconception was that, until recently, few records (stud books) were kept to show the relationships between breeding animals in zoos. Using pedigree information from a number of captive species, Kathy Ralls and Jon Ballou at the National Zoo documented that the overall mortality of inbred offspring is nearly double that of noninbred offspring. As an example, table 17.5 gives the infant mortality for eight mammalian species with relatively large numbers of both noninbred and inbred progeny. The species are ordered by their juvenile mortality in noninbreds, which ranges from 0.106 in the degu to 0.501 in the golden lion tamarin (fig. 17.14). For all the species, the mortality in inbreds is higher, and in six of the eight, the difference is significant, using the chi-square test.

The maintenance of endangered species may result in a number of detrimental evolutionary changes. In addition to the general reduction in fitness caused by inbreeding, inadvertent selection may produce animals more adapted to captive life than to natural conditions. That is, selection may result in tame and docile animals or individuals that are not resistant to the diseases or parasites encountered in nature. These animals might not be able to survive if reintroduced to nature or to nature preserves. Furthermore, some captive animals appear to have hereditary diseases, due either to their small founder number or to inbreeding. For example, the captive golden lion tamarin population, derived from fifty-one founder individuals, has a high frequency of hereditary diaphragmatic hernia.

LOSS OF GENETIC VARIATION

Because the population size of a species in captivity is by necessity small (about 100 Bengal tigers can be maintained by the zoos in the United States), future genetic drift may cause the loss of genetic variation or the possible fixation of detrimental alleles. The best way to understand the loss of genetic variation from genetic drift is to examine the change in heterozygosity from one generation to the next as the result of finite population size. A formula that predicts this change is:

$$H_1 = \left(1 - \frac{1}{2N}\right)H_0$$

where H_0 is the initial heterozygosity. H_1 is the heterozygosity after one generation, and N is the number of individuals. For example, if N = 50, then:

$$H_1 = 0.99H_0$$

Thus, the heterozygosity is reduced 1% in one generation because of genetic drift.

However, for captive animals, we are generally concerned about the reduction in genetic variation over a longer period of time, possibly several generations. In this case, we can use the formula:

$$H_t = \left(1 - \frac{1}{2N}\right)^t H_0$$

where t is the number of generations. For example, if the average generation length of an endangered antelope is two years, there would be about fifty generations in 100 years. If we assume again that the breeding population is composed of fifty animals, then after fifty generations (100 years):

$$H_{50} = \left(1 - \frac{1}{100}\right)^{50} H_0$$
$$= 0.61 H_0$$

In other words, the heterozygosity would be only 61% of the initial value.

Table 17.6 gives the proportions of heterozygosity retained for combinations of four population sizes and four generation lengths. To preserve 90% of the original heterozygosity after 100 years for an organism with a two-year generation length, the population size would have to be 250

Table 17.6 Proportion of Initial Heterozygosity Remaining After 100 Years

Population Size	Generation Length (in Years)			
N	1	2	5	10
10	0.01	0.08	0.36	0.60
50	0.37	0.61	0.82	0.90
250	0.82	0.90	0.96	0.98

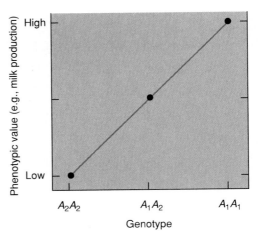

FIGURE 17.15

Effect of a gene on the phenotypic value of milk production. In this case, the A_1A_1 genotype has the highest milk production.

individuals. Obviously, the population size can be smaller for species with a longer generation length. However, these are optimistic predictions that don't include problems, such as disease, that may drastically reduce the population size.

In theory, the loss of genetic variation in a captive population can be counteracted by periodically introducing captured wild animals. However, often it is not possible to obtain wild animals or to induce them to breed in captivity. Another strategy for reducing the effect of genetic drift is to exchange individuals or their gametes (semen or eggs) between zoos. This artificial gene flow may delay the reduction of genetic variation, but it is not a long-term solution. Obviously, genetic conservation is based on a thorough understanding of population genetics fundamentals, including genetic drift, selection, and inbreeding.

 Conservation genetics is concerned with maintaining genetic variation in agricultural plants and animals and in rare and endangered species. The term *inbreeding depression* refers to a decline in fitness resulting from inbreeding. Loss of genetic variation occurs in captive populations when genetic drift has a large effect.

ANIMAL AND PLANT BREEDING

Population genetic principles are widely used to develop new breeds of livestock and new varieties of crop plants. They are also useful in understanding and predicting changes in the traits important in a breeding program. The phenotypic mean in a population is most frequently changed by animal and plant breeding, using selection for particular types. Some of the characteristics that are important in animal and plant breeding, such as resistance to diseases, are often controlled by only a few genes, while other traits, such as crop yield, appear to be influenced by literally hundreds of genes. In addition to selection, inbreeding is used to develop lines that have particular traits, and gene flow is used to bring in new alleles from other breeds or varieties. Genetic drift is often a problem in breeding programs because favorable alleles may be lost by chance and unfavorable alleles may be accidentally increased in frequency. In the following section, we will discuss artificial

selection, inbred lines, and heterosis to illustrate the applications of population genetic principles to animal and plant breeding.

ARTIFICIAL SELECTION

Generally, the increase of a favorable trait in a population is accomplished by **artificial selection**—that is, selecting as parents for the next generation those individuals that are closest to the ideal phenotype. As we discussed in chapter 3, the rate of progress from artificial selection is a result of the heritability of a trait and the selection differential, the difference between the mean of the selected parents and the mean of all the individuals in the parental population. If the heritability is high (most phenotypic differences are genetically determined) and the selection differential is large, progress can be made quickly.

The impact of directional selection is to increase the frequency of alleles that cause a change in the phenotypes being selected. To illustrate, let us assume that gene A is one of the genes influencing a given economically important trait, say milk production in dairy cattle. With two alleles, A_1 and A_2, the effect of this gene on the phenotypic value of milk production can be visualized as in figure 17.15. In this case, genotype A_1A_1 has the highest phenotypic value, so artificial selection should eventually increase the frequency of A_1 to fixation.

After many generations of selection, say for high milk yield, there often appears to be a **selection limit,** a point beyond which selection appears unable to change the phenotypic value. For example, the two-way (up and down) selection experiment for body weight in six-week-old mice conducted by Richard Roberts (fig. 17.16) reaches a selection limit in the down line between 10 and 15 grams. On the other hand, the up line reached a plateau at slightly

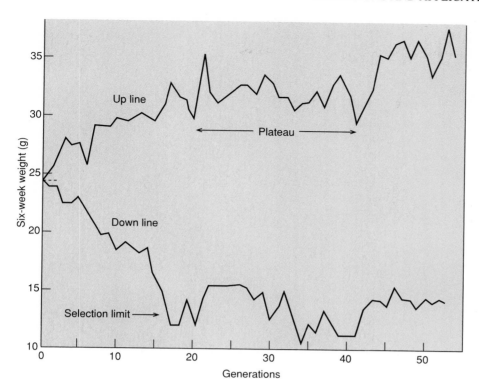

FIGURE 17.16

Two-way selection experiment for body weight in mice. The down line reached a limit of 13 g after seventeen generations, while the up line reached a temporary plateau of about 32 g after thirteen generations.

From F. W. Nicholas, Veterinary Genetics. *Copyright © Oxford University Press, Oxford, England. Reprinted by permission.*

above 30 grams for many generations and then unexpectedly responded again to selection in generations 40–45, reaching a new level at 35 grams.

There are a number of reasons why a selection limit occurs during artificial selection. First, there may be no more genetic variation in the population because of fixation for favorable alleles at all the loci that were originally variable in the population. Second, fertility or viability may be reduced in the extreme phenotypic individuals so that, for example, very large animals have fewer progeny.

The problem of reaching a selection limit may be overcome if new genetic variation is introduced into the population. For instance, mutation to favorable variants or recombination between favorable and unfavorable genes may allow selection to proceed further. In the selection experiment for mouse body weight in figure 17.16, the delayed response from 30 grams to 35 grams probably resulted either from a new mutation or from recombination.

Another approach is to cross new genes from other breeds into the selected populations. This artificial gene flow generally brings in genes that initially reduce the phenotypic value as well as genes that increase it. As a result, there may be an initial loss before the new and old favorable alleles together produce superior plants or animals. Molecular genetic techniques, such as those discussed in chapter 15, promise a way to introduce only the favorable alleles.

Often more than one trait is of economic importance in a selection program. For example, in pig breeding programs,

both growth rate per day and back fat thickness (an indicator of body fat) are important. Figure 17.17 gives the values for these two traits in 100 pigs. The selection scheme is to reduce back fat and increase growth rate, so the "best" pig is in the lower right with a back fat value of under 16 millimeters and a growth rate of 856 grams per day. To continue the breeding program, twenty pigs are needed out of the one hundred. One way to pick the appropriate number of animals is by **independent culling levels,** in which animals must exceed a given value for each trait or be culled (not allowed to breed). In our example, 780 grams per day is the culling level for growth rate while 21.4 millimeters is that for back fat. The twenty selected pigs are indicated by X's in the shaded part of figure 17.17.

When a particular combination of a number of traits is desired, selection response may be quite slow. An example of such a long-term selection experiment for multiple traits was supervised by Gordon Dickerson for eleven characteristics in leghorn chickens that together constitute *performance*. This flock was selected for twenty years for viability, egg production, egg weight, freedom from blood spots in the eggs, and seven other traits. Heritability estimates for many of these traits were above 0.2, suggesting that response to selection would be substantial over a period of time. However, the response to selection was much less than expected; for example, egg production actually declined during the experiment, and egg weight showed response to selection much smaller than expected (fig. 17.18).

FIGURE 17.17

The growth rate and back fat thickness for one hundred pigs. The twenty pigs with the combination of highest growth rate and lowest back fat thickness are indicated by X, while the rest are indicated by •.

From F. W. Nicholas, Veterinary Genetics. *Copyright © Oxford University Press, Oxford, England. Reprinted by permission.*

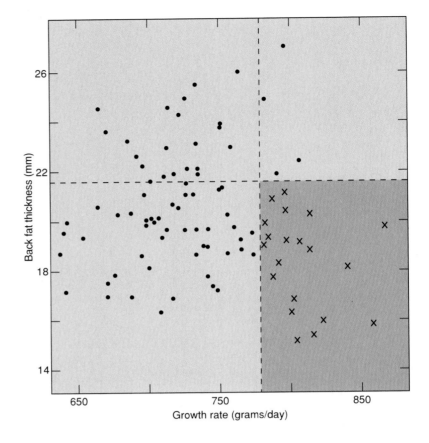

FIGURE 17.18

Changes in egg production and egg weight in long-term selection experiments for eleven traits in chickens.

From G. E. Dickerson, "Genetic Slippage in Response to Selection for Multiple Objectives" in Cold Spring Harbor Symposia on Quantitative Biology, 20:213–224. Copyright © 1955 Cold Spring Harbor Lab Press, Cold Spring Harbor, NY.

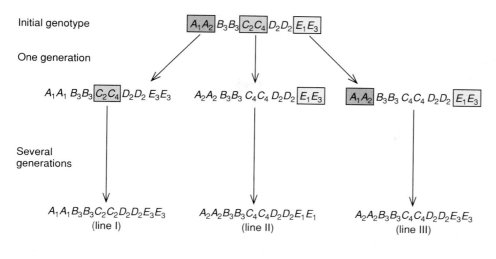

FIGURE 17.19
The differentiation of three inbred lines produced by self-fertilization from an initial genotype. Heterozygous loci are indicated in colored boxes.

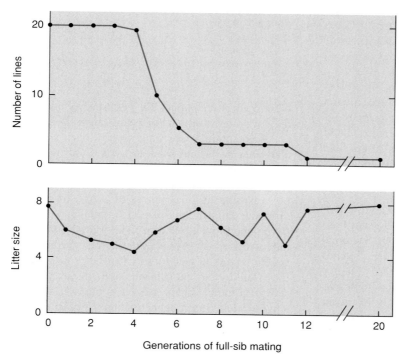

FIGURE 17.20
Results of an inbreeding experiment with twenty lines of mice, showing the number of lines surviving and their average litter size as a result of continuous full-sib mating.

INBRED LINES

In plant breeding, medical research, and to some extent animal breeding, inbred lines are useful because they are genetically uniform for known traits. Inbred lines of mice and rats (lines that have been brother-sister mated for over a hundred generations) are used to test new medical treatments because they provide a genetically uniform test animal.

What happens genetically when a plant is self-fertilized repeatedly or when an animal is brother-sister mated generation after generation? As we discussed in chapter 16, continuous inbreeding results in homozygosity for a given locus. As an example, let us assume that an individual genotype $A_1A_2B_3B_3C_2C_4D_2D_2E_1E_3$ . . . is self-fertilized and that we take three progeny to start inbred lines (fig. 17.19). These progeny after one generation are already more homozygous than the parent, and after several generations they are completely homozygous for the genes

we have shown. If we want to obtain a group of genetically identical individuals, all we have to do is raise individuals from a given inbred line, say line I, all of which would be $A_1A_1B_3B_3C_2C_2D_2D_2E_3E_3$. . . .

We showed that all three lines survived inbreeding and could be maintained. In most cases, many of the lines die out from inbreeding because of low fertility or viability resulting from fixation of lethal or deleterious alleles. For example, if the genotype C_2C_2 were lethal, then inbred line I would have died out. Douglas Falconer illustrated this phenomenon in an experiment in which he started with twenty lines of mice that he maintained by brother-sister mating. After five generations, only half of the lines survived (fig. 17.20), and after twelve generations, only one survived. This one line survived twenty generations (until the experiment was stopped) and still had a litter size of nearly eight, similar to the average before inbreeding began.

HETEROSIS

Long before Mendelism was understood, many researchers noticed the increase in vigor and productivity associated with crossing plant varieties. Even Darwin experimented briefly with corn and noted, "The first and most important conclusion . . . is that cross-fertilization is generally beneficial and self-fertilization injurious." Two early maize geneticists, George Shull and Edward East, documented the detrimental effects of inbreeding and the positive effects of crossbreeding on corn yield. The increase in yield of an F_1 cross over its parental inbred lines is termed **heterosis.**

The extent of heterosis, a term nearly synonymous with **hybrid vigor,** can be measured by the difference in the yield (or other traits) for an F_1 and the mean of the parents, or:

$$H = F_1 - \frac{1}{2}(P_1 + P_2)$$

For example, the yields in bushels per acre in two parental corn strains, P_1 and P_2, were 20, while the F_1 cross between them was 101. The heterotic effect is then 81 bushels per acre. Today, nearly all of the United States corn crop is produced from hybrid corn to capitalize upon this heterotic effect.

Often in recent years, the market animal is also crossbred rather than purebred. Although different breeds are not really inbred lines, they may be somewhat homozygous because of breed selection. If we examine the increase in calf weight weaned per cow for purebred Hereford, Angus, and Shorthorn cattle, and the crosses between them, the effect of heterosis is 8.5% (fig. 17.21), a small but economically important amount. In addition, if we have a three-breed cross in which the mother is an F_1 and is crossed to a bull of the third breed, say (Hereford × Angus) ♀ × Shorthorn ♂, an additional 14.8% increase is observed. The maternal heterotic effect is often substantial because large F_1 females produce large offspring.

What is the genetic basis of heterosis? This topic has been controversial for many decades, and only in recent years does there seem to be some consensus. Initially, many researchers thought heterosis was caused by heterozygous advantage at some loci in the F_1. However, it is now generally accepted that inbreeding (or selection) leads to divergence between lines or breeds and to chance fixation of different, somewhat deleterious, alleles. Because of the general dominance of the favorable or wild-type allele assumed to be present in the other line, this effect is covered up in the progeny when two lines are crossed, resulting in an increased yield.

PESTICIDE RESISTANCE

The rapid development of pesticide resistance in many organisms, including the yellow fever mosquito, the Colorado potato beetle, the Norway rat, and pigweed, is a lesson in population genetics in itself. The genetic variation that confers resistance must initially arise by some sort of mutation, then spread to other populations by selection and gene flow. Obviously, to control species with pesticides, one must understand basic population genetics—or expect eventual failure if resistance arises. (Perhaps we should learn to use

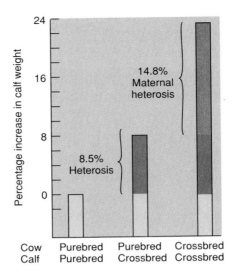

FIGURE 17.21

The percentage increase in calf weight due to heterosis in the calf and in the cow that produced the calf.

Source: Data from Cundiff, et al., 1982b, as appeared in F. W. Nicholas, Veterinary Genetics.

other control techniques for pest species and thereby save our environment from toxic chemicals as well.)

Agricultural crops have always been beset by insects and other pests, but modern agricultural techniques have changed the environment in a number of ways that have made diligent pest control even more necessary. For example, planting large expanses of land with the same crop (a monoculture) allows a pest or pathogen easier dispersal from plant to plant or from field to field. In addition, using only a few varieties of a particular plant reduces the natural genetic variation that confers resistance to pest infestation. In a similar vein, human infectious diseases passed by insect and animal vectors have also become more of a problem now that greater concentrations of people live in cities.

Following World War II, synthetic organic pesticides were introduced to control both agricultural and public health pests. These pesticides greatly increased crop production and reduced the incidence of many diseases. In addition, they were inexpensive, thus they quickly replaced other suppressive measures. By now, however, the use of pesticides has greatly altered the biological environment, reducing or eliminating predators or parasites that originally helped control the pest species.

Studies of evolutionary response to pesticides have shown that many pest organisms are capable of quickly evolving resistance to pesticides. Many pests, such as the mosquitoes that transmit malaria, are vectors of human diseases, but the problems of control are similar to those for agricultural pests. For example, in 1969, fifteen species of anopheline mosquitoes, the malarial vector, were resistant to DDT, and in 1976, forty-three species were resistant. A rather discouraging observation is that malaria is rapidly increasing in many areas where it had nearly been eradicated; for example, India reported twenty-seven million cases in 1977 compared with only a half million in 1969.

Box 17.1 Lysenko—Soviet Anti-Geneticist

Probably no person has had so great an effect on a country's science as Trofim Lysenko (1898–1976) in the former Soviet Union and other former and contemporary Communist countries. Lysenko, born in a peasant family, had a slow educational start but finally obtained a degree by correspondence from Kiev Agricultural Institute. This background fit perfectly with Stalin's intent to replace the old specialists of the tsar with newly trained persons of worker or peasant origin.

Lysenko's scientific training was minimal, and many of his experimental results showed an obvious lack of rigor. For example, his first experiment was to plant peas for pasture in the winter before a cotton crop. The year he tried this, 1925–26, was a mild winter and produced a good yield of peas to use for pasture. However, he did not replicate the experiment in other more typical years, and today this approach has proven completely impractical.

One of Lysenko's "discoveries" was vernalization, a practice of treating seeds before germination to cause them to mature more quickly. For example, he suggested soaking the seeds of wheat and other crops for several days before sowing in order to protect the plants from later drought. However, many technical problems and extra work resulted from this procedure, and in fact vernalization often actually lowered the yield.

Stalin created a personality cult and sought to destroy all those who did not agree with him. In keeping with this, Lysenko brought criticism of his controversial ideas into the polemics of

FIGURE B17.1
Trofim Lysenko acquainting collective farmers with a new product called branched wheat.
© SOVFOTO/EASTFOTO AGENCY.

Communism. For example, in a 1935 speech, he stated, "Tell me, comrades, was there not a class struggle on the vernalization front?" At the end of this speech, Stalin exclaimed, "Bravo, Comrade Lysenko, bravo!"

Lysenko went on to claim that the inheritance of characteristics in plants could be altered by changing their environment. Of course, this notion was completely at odds with modern genetics and with the opinions of the many outstanding Soviet geneticists in the 1930s. However, the idea was compelling to Stalin because it fit with how he wanted to change Soviet society and offered a solution to the perennial agricultural failure of the collective farms. As a result, Lysenko moved up the Soviet scientific hierarchy quickly,

denouncing and persecuting his critics as adherents of the old Mendel-Morgan genetic "heresies." As director of the Genetic Institute of the Academy of Sciences, he even ordered all the fruit flies to be destroyed by boiling!

Lysenko's impact was widespread until 1964 when Khrushchev fell. He had managed not only to extinguish Soviet genetics for nearly three decades but also to inhibit progress in many other scientific areas. In addition, other Communist countries, such as China, suffered similar suppression of scientific thought. The former Soviet Union, some other former Communist countries, and China today have experienced a resurgence of activity in genetics, but recovery from the Lysenko era has been slow and difficult.

This trend is apparently due to the increased use of pesticides associated with new agricultural practices and to the consequent increased pesticide resistance. The housefly seems to be the organism most broadly resistant to pesticides geographically, but *Tribolium,* a genus of flour beetle, and seven species of rodents, including two rats, are also widely resistant to various pesticides. Until the 1970s, no weeds were resistant to herbicides, but by 1985, forty-eight species had evolved resistance to at least one.

FACTORS IMPORTANT IN PESTICIDE RESISTANCE

A number of factors can affect the evolution of pesticide resistance. One basic requirement is that an organism have

some physiological, anatomical, or other mechanism that enables it to resist a pesticide. In numerous examples, pesticide resistance takes different forms in different strains of the same species. If the potential for resistance is present, a number of other factors help determine whether a population actually becomes resistant.

First, the new genetically advantageous variant or type must already be present in the population, be generated by mutation, or migrate in from other populations. Once the resistant form is established in the population, its increase is dependent upon such factors as level of dominance, selective advantage, and initial frequency. For example, a variant that is dominant (expressed in heterozygotes) and has a large selective advantage should increase quickly.

Table 17.7 Relative Fitness at the R_w Locus in the Norway Rat

	SS	RS	RR
Warfarin	Susceptible	Resistant	Resistant
Vitamin K dependence	No	Intermediate	Yes
Relative fitness	0.68	1.00	0.37

Source: Data from J. H. Greaves, "Warfarin Resistance: A Balanced Polymorphism in the Norway Rat" in *Genet. Res.* 30:257–263, 1977.

Also, many pest species have short generation lengths, multiple generations per year, and high fecundity, all of which are advantageous for the evolution of resistance. In addition, it is possible that chemical resistance genes sometimes reside in the DNA of chloroplasts, mitochondria, or plasmids (small nonnuclear pieces of DNA). Bacterial resistance to antibiotics is often present in plasmids, and transmission of plasmids may occur horizontally (at times other than reproduction) both between members of the same species and of different species; thus antibiotic-resistant bacteria have become a major medical problem.

To illustrate the potential for developing resistance, consider the use of juvenile hormone, once thought to be a panacea pesticide for insects. Because it is a necessary chemical for insects' normal development, researchers thought insects could not develop a mechanism of resistance to it. However, in the past few years, several pest insects have developed resistance, making juvenile hormone pesticides ineffective.

In addition, the spread of antibiotic resistance may have increased in many bacteria due to the presence of preadapted genotypes. Specimens of bacteria from the pre-antibiotic era, which have been stored since the 1930s, do have some resistant strains. Furthermore, collections from remote parts of the world where antibiotics have never been used also show resistant strains. These strains may occur because they confer resistance to certain factors in the environment.

EXAMPLES OF PESTICIDE RESISTANCE

To better understand the genetic basis of pesticide resistance, let us examine several case studies. The Norway or brown rat, *Rattus norvegicus,* a widespread agricultural and domestic pest, was controlled for about 15 years by the anticoagulant warfarin. However, resistance to this chemical evolved quickly, and resistant rats have now been found in the United States and in many European countries. The mechanism of resistance has been examined in some detail and appears to be the result of a dominant resistant allele, *R,* at an autosomal locus. Most of the resistant animals are heterozygotes for this allele and for the wild-type or susceptible allele, *S,* because homozygotes for the resistant allele have quite low viability (table 17.7). The lowered viability in *RR* individuals results from a 20-fold increase in vitamin K requirement.

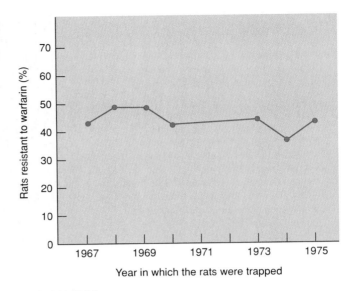

FIGURE 17.22
The frequency of rats resistant to warfarin over an eight-year period along the England-Wales border.

As a result, there appears to be a stable polymorphism when warfarin is applied, owing to the overall higher net survival advantage of the heterozygote. Further support for this conclusion comes from a survey of a population where the frequency of the resistant allele appeared to be stable over a number of years (fig. 17.22) and showed a large excess of heterozygotes compared to Hardy-Weinberg expectations. In this population, where warfarin was regularly applied, the overall relative fitnesses of the three genotypes were estimated to be 0.68, 1.0, and 0.37 (see table 17.7), respectively, showing a substantial advantage for the heterozygote. Furthermore, in another population, the frequency of the resistant allele declined very quickly when warfarin was no longer applied.

Let us mention two other instances of pesticide resistance in which detailed genetic information is known. First, a number of weed species have quickly developed resistance to the commonly used herbicide atrazine. The resistance is maternally inherited, indicating that the trait is coded for by a gene in the chloroplast (or mitochondrion). In fact, a protein called pbsA, encoded by a chloroplast gene, is involved in the part of photosynthesis affected by atrazine. The amino acid and nucleotide sequences of both resistant and susceptible *Amaranthus* plants (pigweed) were compared. Three differences in nucleotide sequence were found, two of which were silent and resulted in no amino acid change. However, the third, the replacement of an adenine by a guanine at position 682, caused the substitution of a glycine for a serine at amino acid position 228 (fig. 17.23). This simple change apparently resulted in a pbsA protein that has lost affinity for the herbicide atrazine. Further evidence that this is the critical change comes from two other species in which resistant and susceptible strains also differ at amino acid position 228.

Resistant	CAA	TAT	GCT	GGT	TTC	AAC
	Gln	Tyr	Ala	Gly	Phe	Asn
Susceptible	CAA	TAT	GCT	AGT	TTC	AAC
	Gln	Tyr	Ala	Ser	Phe	Asn
Amino acid position	225	226	227	228	229	230

FIGURE 17.23

The DNA and amino acid sequence of pbsA-resistant and pbsA-susceptible strains of pigweed. The boxed nucleotide and amino acid indicate the difference in the resistant strain. The amino acid position is shown at the bottom.

FIGURE 17.24

The amounts of esterase B1 produced in susceptible and resistant mosquitoes are shown in lanes 1 and 2, respectively. Lanes 3 through 9 give the amount when the resistant mosquito extract was subjected to serial dilution.

Instead of undergoing a qualitative change, some insects apparently become resistant to organophosphates by producing large amounts of detoxifying esterases. For example, the amount of esterase in susceptible and resistant mosquitoes was evaluated using immunological techniques (fig. 17.24). The amount of B1 esterase in the electrophoretic gel is given in the first column, while the amount of esterase under different dilutions is given in columns 2–9 for the resistant strain. It appears that the resistant strain has approximately a 500-fold increase in the level of the esterase; that is, a dilution of 1 to 512 gives about the same amount of esterase in the susceptible strain as in the resistant strain. It is not known whether the overproduction of esterase results from a change in regulation or from an increased number of genes.

CONTROL OF PESTICIDE RESISTANCE

To control pesticide resistance, all of the evolutionary factors just discussed must be considered. For example, to localize warfarin resistance in rats, efforts were made to eradicate every rat within a 5-kilometer diameter area; however, the gene flow of the rat is such that resistant rats escaped beyond this border. Atrazine resistance is hard to control because independent resistant types have arisen; this is due

Table 17.8 Resistance (R) or Susceptibility (S) of Two Flax Varieties to Flax Rust

Rust Strain	Flax		
	Ottawa	Bombay	F₁ Progeny
22	S	R	R
24	R	S	R

Source: Data from H. Flor, "The Complementary Genic Systems in Flax and Flax Rust" in *Adv. Genet.* 8:29–54.

either to a large number of independent mutations (some individual plants can produce 100,000 seeds, making this a realistic possibility) or to some variation in populations prior to exposure to atrazine.

One way to control insect pests on crops without using insecticides is to plant crop varieties that are themselves toxic or avoided by the pest. However, even with this strategy, pests have evolved virulence to resistant crop varieties. For example, the Hessian fly, a pest of wheat, has caused severe damage to susceptible varieties. Certain alleles in wheat can inhibit larvae development, but the flies may have corresponding virulence alleles. However, only when there is a virulence allele for each resistance allele can the flies develop normally.

The first definitive studies on the genetics of plant resistance to a rust were done in flax. Two varieties of flax, Ottawa and Bombay, were tested for their susceptibility (S) or resistance (R) to two rust strains, 22 and 24. As shown in table 17.8, Ottawa is resistant to rust strain 24 and Bombay is resistant to rust strain 22. When the flax varieties are crossed, the F₁ is resistant to both rust strains, demonstrating that the alleles conferring resistance are dominant over the alleles showing susceptibility.

Obviously, reducing applications of particular pesticides will slow the development of resistance. In fact, integrated pest management strategies, such as using pest-resistant strains of crops, introducing predators or parasites to control pests, managing the agroecosystem, and so on, should reduce reliance on chemical control. Pesticides could then be introduced effectively at critical times or places. In addition, resistant alleles often have associated negative pleiotropic effects (lowered survival), such as for the warfarin resistance allele, which should result in a natural decrease of resistance in the absence of the pesticide. In other words, prudent use may actually make a given pesticide more useful over a longer period of time.

Pesticide resistance is common in many agricultural and medical pests. Pesticide resistance spreads quickly in species with preadaptation, large population numbers, and high gene flow. Warfarin resistance in rats, atrazine resistance in pigweed, and organophosphate resistance in mosquitoes are case studies of evolutionary change as well as examples of pesticide resistance. Pesticide resistance can be controlled by using pest-resistant crops or other integrated pest management strategies.

WORKED-OUT PROBLEMS

Problem 1

Many populations of endangered species have been reduced to small numbers. Assume one species with only ten individuals had a heterozygosity of 0.2 at a given gene, how much heterozygosity would be expected to remain after one generation? After twenty generations? Discuss your answers.

Solution

Using the information in the question, we know that $H_0 = 0.2$ and $N = 10$. Using the relationship

$$H_1 = \left(1 - \frac{1}{2N}\right)H_0$$

then $H_1 = 0.19$. Likewise

$$H_{20} = \left(1 - \frac{1}{2N}\right)^{20} H_0$$

so the $H_{20} = 0.072$. In other words, after one generation, the heterozygosity has hardly changed, while after twenty generations, it is only 36% of the initial value.

Problem 2

The theory of neutrality assumes that the level of heterozygosity is a balance between the effects of genetic drift and mutation. If we assume that the estimated size of a population of grasshoppers on an island is 5,000 and the estimated mutation rate is 0.00001, what would be the expected level of heterozygosity under neutrality? In an effort to determine the actual level of heterozygosity in this population, a sample of individuals was scored for heterozygosity at 30 electrophoretic loci and it was found that the heterozygosity was 0.01. Is this consistent with the level expected by the neutrality theory? What factors could account for any difference?

Solution

The expected level is $H_e = (4 \times 5,000 \times 0.00001)/(4 \times 5,000 \times 0.00001 + 1) = 0.167$. The observed level is much below that expected from the neutrality theory. There are a number of possible explanations. For example, the estimate of the population size may be too high, or the estimate of the mutation rate may be too high. Another possibility is that the population is fairly young and is the result of a recent founder event with low initial genetic variation. As a result, there hasn't been enough time for mutation to regenerate heterozygosity so that the predicted equilibrium value is reached.

Summary

The incidence of human genetic disease can be generally explained by a mutation-selection balance in which mutation increases the frequency of the mutant and selection reduces it. In addition to mutation and selection, the frequency of disease alleles is sometimes influenced by genetic drift, gene flow, and inbreeding.

The evolution of molecules is generally consistent with the neutrality model, a theory based on the combined effects of genetic drift and mutation. Changes in DNA and proteins generally occur at a regular rate, resulting in a molecular clock. Molecular differences in different species can be used to construct phylogenetic trees that indicate the evolutionary relationships between the species. Mitochondrial DNA has been employed to determine phylogenetic relationships.

Conservation genetics is concerned with maintaining genetic variation in agricultural plants and animals as well as in rare and endangered species. Inbreeding depression refers to the decline in fitness that results from inbreeding. Loss of genetic variation occurs in captive populations when genetic drift has a large effect. Pesticide resistance is common in many agricultural and medical pests. Pesticide resistance spreads quickly in species with preadaptation, large population numbers, and high gene flow. Warfarin resistance in rats, atrazine resistance in pigweed, and organophosphate resistance in mosquitoes are case studies of evolutionary change as well as examples of pesticide resistance. Pesticide resistance can be controlled by using pest-resistant crops or other integrated pest management strategies.

PROBLEMS AND QUESTIONS

1. If $u = 10^{-5}$, $v = 0.0$, and the selection coefficient against a recessive (s) = 0.6, what are the expected equilibrium frequencies of genotype A_2A_2 and allele A_2? If u is doubled, what are the new frequencies? By how much are the equilibrium genotypic and allelic frequencies changed?

2. Assume that $u = 10^{-6}$ for a lethal recessive gene. What is the equilibrium genotypic frequency? Assume that with better medical care, homozygotes have only a slight disadvantage, so that s = 0.05. What is the new equilibrium genotypic frequency?

3. Discuss the factors that affected the frequency of the sickle-cell allele in African blacks in the past and those that are influencing its frequency in African-Americans today.

4. Cystic fibrosis is a recessive disease that has a high frequency in Caucasians (about 1 in 2,500 newborns). Until recently, it was lethal in childhood. Given this information, calculate the expected mutation rate. Compare this value to known mutation rates, and give a better explanation for the frequency of this disease.

5. The founder effect may be important in influencing the frequency of rare disease alleles in some populations. What type of populations do you think would be most affected by the founder effect?

6. Compare the predictions of the neutrality theory and the selection theory in influencing the amount of

genetic variation. Suggest an experiment that might differentiate between them as explanations for variation at a given polymorphic locus.

7. What is the molecular clock? What is the rate of amino acid substitution? Would you expect the rate of nucleotide substitution to be faster or slower than the rate of amino acid substitution? Why?

8. Let us assume that the heterozygosities for a survey of electrophoretic loci in two populations of a leopard frog are 0.1 and 0.05. (*a*) Since the same loci were analyzed in both populations (they have the same mutation rate, let us say it is $u = 10^{-5}$) and assuming the neutrality theory, what could be the cause of the difference? (*b*) What are the values of the population sizes that could account for the difference?

9. If two species of rodents differ at 40 out of 90 amino acid positions for one protein and at 20 out of 80 amino acid positions for another, what proportion of sites differ for the two proteins? Why might these values differ for the two proteins?

10. Assume that four closely related species of birds (a, b, c, and d) were analyzed and that the genetic distances (a measure of the biochemical differences) between them were:

	b	c	d
a	0.2	0.12	0.55
b		0.3	0.7
c			0.65

From these data, which pair of species is most closely related? Which species is least related to the other species?

11. Table 17.3 suggests that molecular data may be able to resolve the deviation of closely related species. How could this information be used to understand evolutionary changes, including differences in morphology or physiology between species?

12. How do you think the population genetics of mtDNA differs from that of nuclear DNA? (In answering, remember that mtDNA is maternally inherited and haploid.)

13. The amount of the inbreeding depression is equal to sfpq. What is the expected inbreeding depression when $f = 0.25$, $s = 0.5$, and $q = 0.01$? If $q = 0.5$? Why is one of these values larger than the other?

14. In a population of 100 individuals where $q = 0.4$ and there are initial Hardy-Weinberg proportions, what are H_1 and H_2? If $N = 10$, what would H_1 and H_2 be? Discuss your answers.

15. Let us say that the goal of a captive breeding program is to retain 75% of the initial heterozygosity $(H_t/H_0 = 0.75)$ after two hundred years of captive breeding. For a species that has a generation length of two years, what population size would be necessary?

16. How would you identify different land races or wild varieties of rice having genetic variation that might be useful in future plant breeding? How could you maintain these samples so that none of the genetic variation would be lost?

17. The progeny from a mating of half-first cousins have an inbreeding coefficient of $1/32$. For a recessive disease like PKU, with an allelic frequency of about 0.008, what is the expected frequency of PKU cases in the progeny of half-first cousin matings? Compare this to the frequency expected in the progeny of unrelated individuals.

18. Some animals, such as the black-footed ferret and the North American condor, now exist only in captivity. What kind of plan would you propose to conserve these species, considering such factors as inbreeding depression, loss of genetic variation, and reintroduction to the wild?

19. Calculate the expected equilibrium frequency of the allele that confers resistance to warfarin in rats, using the relative fitnesses in table 17.7. Some populations do not have this allele. Give two possible reasons why.

20. From table 17.8, it appears that resistance to two rust strains is dominant in flax. If the frequencies of the two alleles for resistance to rust strains 22 and 24 were 0.1 and 0.2, what proportions of the population would be susceptible to rust strain 22? To both rust strains?

21. In some species, pesticide resistance has spread very quickly. For example, the malaria mosquito is resistant to DDT nearly everywhere. Discuss the population genetic factors that may be important in causing such a speedy spread of resistance.

22. It is tempting to imagine using genetic engineering to produce new crops. However, in light of our discussion on genetic conservation and pesticide resistance, what problems may result from this approach?

Answers appear at end of book.

APPENDIX A:
ANSWERS TO END-OF-CHAPTER PROBLEMS AND QUESTIONS

CHAPTER 2

1. Mendel's first principle is segregation, which states that heterozygotes produce equal numbers of gametes having two different alleles. Mendel's second principle is independent assortment, which states that alleles at different genes assort independently of each other.

2. Garden peas have a number of contrasting characteristics, produce a large number of progeny, and are available commercially in a number of true-breeding varieties.

3. You would cross it to true-breeding r plants which would have the rr genotype. If the progeny all had round seeds (genotype of progeny would be Rr), then the genotype of the individual in question would be RR. If some of progeny had round (Rr) seeds and some had wrinkled (rr) seeds, then the genotype of the individual in question would be Rr.

4. (a) 0.5. (b) 0.5.

5. Recessive. Because the parents are not affected, the trait is recessive and not dominant.

6. 0.35, 0.61; with 1 degree of freedom, they are consistent.

7. 0.5; 0.33.

8. 0.0625; 0.00391; product rule.

9. (a)

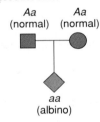

Aa (normal) × Aa (normal) → aa (albino)

   (b) 0.25; 0.5; 0.667.

10. 0.25; 0.240.

11. (a) 0.25 for each type. (b) 0.62; with 3 degrees of freedom, it is consistent.

12. 0.0625; 0.1875.

13. 0.0156; 0.1406.

14. Recessive; see table 2.7; Aa, Aa, aa; 0.5; 0.0.

15. Dominant; see table 2.7; bb, Bb, bb; 0.0; II-3 does not carry the dominant allele, so his child cannot be affected.

16. Her husband is heterozygous; 0.5.

17. 0.052; with 1 degree of freedom, it is consistent.

18. 0.429; with 1 degree of freedom, it is consistent.

19. 0.125; 0.75.

20. (a)

	AB	Ab	aB	ab
AB	AABB	AABb	AaBB	AaBb
Ab	AABb	AAbb	AaBb	Aabb
aB	AaBB	AaBb	aaBB	aaBb
ab	AaBb	Aabb	aaBb	aabb

   (b) 0.5625, 0.1875, 0.1875, 0.0625.

21. 0.667; 0.25.

22. 0.333; 0.5.

23. Dominant. See table 2.7 for a complete list.

24. 22.25.

25. 0.00694; 0.0833.

26. (a) Parents, AA (tall) × aa (short); F_1 Aa (tall); F_2 153 AA or Aa (tall), 48 aa (short); (b) 0.761 tall, 0.239 short.

27. Parents, Aa (tall) × aa (short); ½ Aa tall and ½ aa (short).

28. (a) Although it is hard to see all the kernels well, approximately 27 are purple and 12 are yellow. The observed ratio of purple to yellow is 2.25:1. The expected numbers would be ¾(39) = 29.25 purple and ¼(39) = 9.75 yellow. The chi-square is 0.69, consistent with expectations for one degree of freedom. (b) The number of normal and shrunken kernels are 25 and 14; the observed ratio of normal to shrunken kernels is 1.79:1. The expected are the same as in (a). In this case, the chi-square is 2.47, again consistent with expectations. (c) The observed ratio is 3.6 purple normal to 1.78 purple shrunken to 1.4 yellow normal to 1 yellow shrunken. The observed numbers are for purple, normal, 18; purple, shrunken, 9; yellow, normal, 7; yellow, shrunken, 5. The expected numbers are $\frac{9}{16}$ (39) = 21.94, $\frac{3}{16}$ (39) = 7.31, $\frac{3}{16}$ (39) = 7.31 and $\frac{1}{16}$ (39) = 2.44. The chi-square here (with three degrees of freedom) is 2.83, again consistent with expectations.

29. 1.0, 0.5, 0.25.

30. (a) $RY = \frac{1}{4}$, $Ry = \frac{1}{4}$, $rY = \frac{1}{4}$, $ry = \frac{1}{4}$; (b) ¼, ¼, 0.

CHAPTER 3

1. Traits can exhibit complete dominance as did Mendel's traits; incomplete dominance, in which the phenotype of the heterozygote is intermediate between the homozygotes, such as in the flower color of four-o'clocks; or codominance, in which the characteristics of both homozygotes are expressed in the heterozygotes, such as in the antigens in the MN blood group system.

2. The level of penetrance is the proportion of individuals of a given genotype that exhibit a given phenotype. The level of expressivity, or expression, of a trait is the extent to which an individual with a trait actually exhibits it. For a disease, expressivity can vary from mild to severe.

3. (a) 6. (b) 5, 10.

4. (a) Sex-linked traits are traits that are caused by genes on the sex chromosomes, such as the X and Y in mammals. Sex-limited traits occur only in one sex. The expression of sex-influenced traits is a function of the sex of the particular individual. Neither sex-limited nor sex-influenced traits are determined directly by genes on the sex chromosomes.
(b) If the trait is sex-linked and not sex-limited, crosses between individuals with different phenotypes should result in some progeny in which the wrong sex has the given trait. Further, if a comprehensive survey is done, then no individuals of the wrong sex for the phenotype would be observed for a sex-limited trait, while some should be observed for a sex-linked trait.

5. Mendel would have observed a variety of different phenotypes, probably most of which would have been intermediate between the parental types. He would not have seen two discrete phenotypes and probably would not have figured out particulate inheritance. Part of his genius was to avoid polygenic traits in these experiments.

6. 0.26; with 2 degrees of freedom, they are consistent.

7. 0.28; with 1 degree of freedom, it is consistent.

8. A × A, A × B, A × AB, A × O, B × B, B × AB, B × O, AB × AB, AB × O, O × O.
$I^A I^A \times I^B I^B$ (all AB); $I^A I^A \times I^B I^O$ (½ AB, ½ A); $I^A I^O \times I^B I^B$ (½ AB, ½ B); $I^A I^O \times I^B$ (¼ A, ¼ AB, ¼ B, ¼ O).

9. 5.24—statistically significant at 0.05 level with 1 degree of freedom; 0.21—consistent with a 2:1 hypothesis.

10. (a) ⅔. (b) ½.

11. To document the pleiotropic effects of a newly discovered genetic disease, assuming that the disease is due to a single gene, you would compare individuals with and without the disease for a number of different characteristics, especially those directly related to the genetic defect (as in PKU, where the basic defect is the inability to metabolize phenylalanine), and other characteristics that are possibly a consequence of the basic defect.

12.

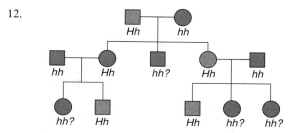

$$hh \quad Hh \quad hh? \quad Hh \quad hh$$
$$hh? \quad Hh \qquad Hh \quad hh? \quad hh?$$

13. (a) Compare the phenotypes of groups of flies raised at different levels of temperature and diet, measuring closely the phenotypic and environmental factors. (b) Cross the mutation into different stocks and observe the offspring to determine if different crosses result in different phenotypes.

14. Type O is the universal donor; type AB is the universal recipient. Type O has no antigens and therefore is accepted by all types. Type AB has no antibodies and therefore can accept all types of blood.

15. $MM' \times mm$.

16. (a) ¼ B_1B_2, ¼ B_2B_5, ¼ B_1B_4, ¼ B_4B_5. (b) $B_1B_4 \times B_2B_2$.

17. Woman: $I^A I^O$, man: $I^B I^O$, child: $I^O I^O$; 0.25; 0.25.

18. Man: $X^C Y$, woman: $X^C X^c$, female child: $X^C X^c$: cc 0.50.

19. Male, unaffected ($X^a Y$): 0.25; male, affected ($X^A Y$): 0.25; female, unaffected ($X^a X^a$): 0.25; female, affected ($X^A X^A$): 0.25.

20. (a) ½ $X^w X^{w+}$ (red-eyed females), ½ $X^w Y$ (white-eyed males). (b) 0.36—with 3 degrees of freedom, they are consistent.

21.

$$\frac{1}{4} A_1 A_1 \Rightarrow \begin{cases} \frac{1}{4} B_1 B_1 \\ \frac{1}{2} B_1 B_2 \\ \frac{1}{4} B_2 B_2 \end{cases} \quad \frac{1}{16} \text{ red } (\frac{1}{16} A_1 A_1 B_1 B_1)$$

$$\frac{1}{2} A_1 A_2 \Rightarrow \begin{cases} \frac{1}{4} B_1 B_1 \\ \frac{1}{2} B_1 B_2 \\ \frac{1}{4} B_2 B_2 \end{cases}$$

$\frac{1}{4}$ light red ($\frac{1}{8} A_1 A_1 B_1 B_2, \frac{1}{8} A_1 A_2 B_1 B_1$)

$\frac{3}{8}$ pink ($\frac{1}{16} A_1 A_1 B_2 B_2, \frac{1}{4} A_1 A_2 B_1 B_2, \frac{1}{16} A_2 A_2 B_1 B_1$)

$\frac{1}{4}$ light pink ($\frac{1}{8} A_1 A_2 B_2 B_2, \frac{1}{8} A_2 A_2 B_1 B_2$)

$$\frac{1}{4} A_2 A_2 \Rightarrow \begin{cases} \frac{1}{4} B_1 B_1 \\ \frac{1}{2} B_1 B_2 \\ \frac{1}{4} B_2 B_2 \end{cases}$$

$\frac{1}{16}$ white ($\frac{1}{16} A_2 A_2 B_2 B_2$)

22. (a)

Frequency	Category	Genotypes
$\frac{1}{64}$	0	$(\frac{1}{64}A_1A_1B_1B_1C_1C_1)$
$\frac{3}{32}$	1	$(\frac{1}{32}A_1A_1B_1B_1C_1C_2, \frac{1}{32}A_1A_1B_1B_2C_1C_1, \frac{1}{32}A_1A_2B_1B_1C_1C_1)$
$\frac{15}{64}$	2	$(\frac{1}{64}A_1A_1B_1B_1C_2C_2, \frac{1}{64}A_1A_1B_2B_2C_1C_1, \frac{1}{64}A_2A_2B_1B_1C_1C_1,$ $\frac{1}{16}A_1A_1B_1B_2C_1C_2, \frac{1}{16}A_1A_2B_1B_1C_1C_2, \frac{1}{16}A_1A_2B_1B_2C_1C_1)$
$\frac{5}{16}$	3	$(\frac{1}{32}A_1A_1B_1B_2C_2C_2, \frac{1}{32}A_1A_2B_1B_1C_2C_2, \frac{1}{32}A_1A_1B_2B_2C_1C_2,$ $\frac{1}{32}A_1A_2B_2B_2C_1C_1, \frac{1}{32}A_2A_2B_1B_1C_1C_2, \frac{1}{32}A_2A_2B_1B_2C_1C_1,$ $\frac{1}{8}A_1A_2B_1B_2C_1C_2)$
$\frac{15}{64}$	4	$(\frac{1}{64}A_2A_2B_2B_2C_1C_1, \frac{1}{64}A_2A_2B_1B_1C_2C_2, \frac{1}{64}A_1A_1B_2B_2C_2C_2,$ $\frac{1}{16}A_2A_2B_1B_2C_1C_2, \frac{1}{16}A_1A_2B_2B_2C_1C_2, \frac{1}{16}A_1A_2B_1B_2C_2C_2)$
$\frac{3}{32}$	5	$(\frac{1}{32}A_1A_2B_2B_2C_2C_2, \frac{1}{32}A_2A_2B_1B_2C_2C_2, \frac{1}{32}A_2A_2B_2B_2C_1C_2)$
$\frac{1}{64}$	6	$(\frac{1}{64}A_2A_2B_2B_2C_2C_2)$

(b)

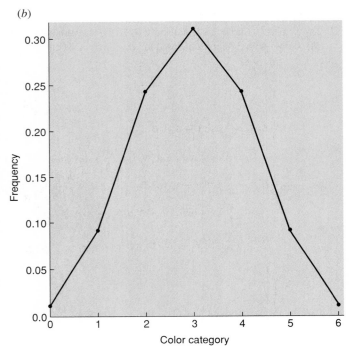

(c) Three genes have more classes and approach a bell-shaped curve.

23. Dominance concerns the interaction of two alleles at a given gene to determine phenotype. Epistasis involves the interaction of alleles at different genes. All of the traits examined by Mendel were determined at a single gene.

24. 1.56; with 1 degree of freedom, they are consistent.

25. 0.5.

26. (a) *BbCc*, black; (b) ⁹⁄₁₆ black: *BBCC*, *BBCc*, *BbCC*, and *BbCc*; ³⁄₁₆ brown: *bbCC* and *bbCc*; ¼ albino: *BBcc*, *Bbcc*, and *bbcc*.

27. Duplicate gene action could not result in white plants that differed only at one gene. In other words, a cross differing by two genes would produce all white progeny. Complementary gene action would give results as shown in figure 3.19.

28.

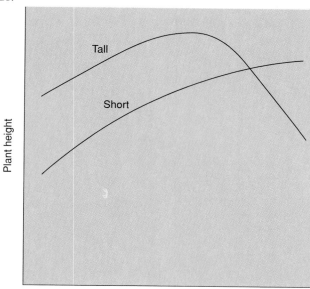

Temperature

Plant height

29. Polygenic means many genes affect the trait. Quantitative means that the trait is measured on a continuous scale. In some cases, polygenic traits may not have a continuous scale, while quantitative traits may not have a polygenic basis. These traits may be either single-gene or polygenic. Pedigree information would be needed in order to determine the genetic basis of such a trait.

30. The rate of response would be quite fast because nearly all the phenotypic variation is in the additive genetic component.

31. 0.4375; 0.0.

32. (a) No. (b) Yes. The child had neither of the alleles of the man at the ABO locus.

CHAPTER 4

1. Metacentric, which has the centromere in the center; acrocentric, which has the centromere toward one end; and telocentric, which has the centromere very near the end.

2. S, G_2, mitosis, G_1. DNA synthesis takes place during the S phase.

3. 0.04; with 1 degree of freedom, it is consistent. 2.19; with 2 degrees of freedom, it is consistent.

4. 0.00118.

5. The synthesis of DNA and chromosomal events are not synchronous. See figure 4.10 for details.

6. Oogenesis results in only one gamete (ovum) descended from each oogonium (and three polar bodies), while spermatogenesis produces four gametes (sperm) for each spermatogonium. Spermatogenesis generally occurs continuously and produces many millions of sperm, while oogenesis begins before birth and the cells stay in a resting stage until ovulation. A woman produces only about 500 mature eggs in her lifetime.

7. In mitosis, chromosomes are aligned so that sister chromatids are on opposite sides of the metaphase plate. In meiosis I, homologues are paired and are on opposite sides of the plate. Meiosis II is like mitosis except that only one homologue is in each cell.

8. (a) 0.5 with A, 0.5 with a.

   (b)

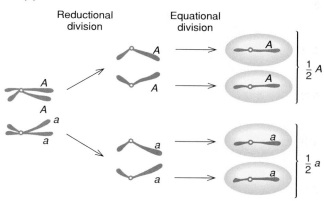

(c) 0.25 with AB, 0.25 with Ab, 0.25 with aB, and 0.25 with ab.

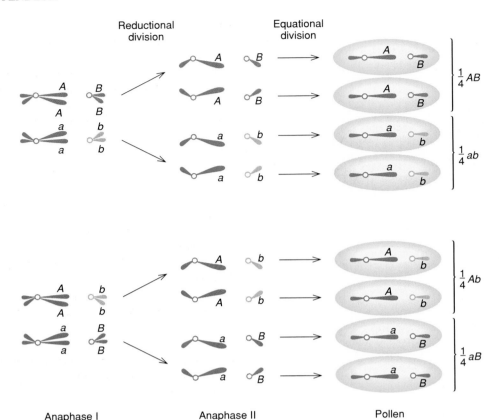

Anaphase I   Anaphase II   Pollen

9. Duplication: a region of a given chromosome is present more
   than once; deletion: a region of a given chromosome is
   missing; inversion: the sequence of a group of genes is
   reversed from the normal order; translocation: there is an
   exchange of chromosomal segments between two
   nonhomologous chromosomes.

10. (a)

(b)

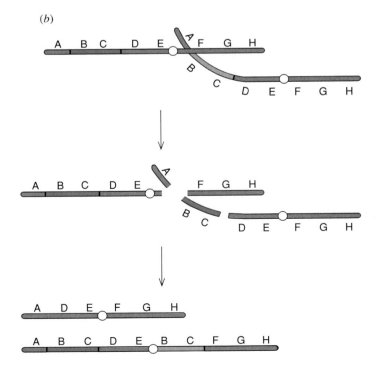

11. Genes *e* and *f*.

12. Pericentric inversions include the centromere, while paracentric inversions do not. Meiosis involving a paracentric results in an acentric chromosome and a dicentric bridge, while the pericentric does not. Both have chromosomes with combinations of duplications and deletions, but those from the paracentric may be more severe.

13. (*a*)

(*b*)

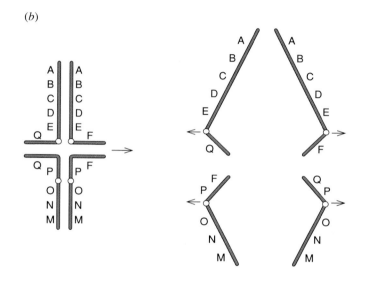

14. See table 4.4.

15. See figure 4.29. Many individuals with Down syndrome die before birth.

16. (*a*) See figure 4.33. (*b*) See table 4.5; XXY, XO. (*c*) The expected genotypes from male nondisjunction are: XXX, XYY, XO, XX, XY. The expected genotypes from female nondisjunction are: XXX, XXY, XO, YO (unviable), XY, and XX.

17. Chromosome 6.

18. 4 copies; *AAAA, AAAa, AAaa, Aaaa, aaaa*; 2 phenotypes.

19. 0.0625; 0.0625.

20. The smaller inversion is located just below the large one indicated by arrows (six bands long). The pericentric inversion starts below the third band from the top in humans and ends below the eighth band.

21. 0.25 $Z^A Z^A$ males, 0.25 $Z^A Z^a$ males, 0.25 $Z^A W$ females, 0.25 $Z^a W$ females.

22. Normal female; normal male. Equal numbers of X and autosomes result in females, while an X/A ratio of 0.5 results in males.

23. See figures 4.7 and 4.9.

24. The two parental genomes may cover up deleterious alleles in the other genome. Yes, normal meiosis.

25. 3 chromosomal fissions.

CHAPTER 5

1. Mendel's second principle, that of independent assortment. Alleles at different linked genes that are together in a parent are more likely to be together in the offspring than is predicted by the principle of independent assortment.

2. 139.0; with 3 degrees of freedom, not consistent with hypothesis at 0.01 level.

3. *PL* (0.443), *Pl* (0.057), *pL* (0.057), *pl* (0.443).

4. *P–L–* (0.696), *P–ll* (0.054), *ppL–* (0.054), *ppll* (0.196); 1.42— with 3 degrees of freedom, they are consistent. (See chapter 4.)

5. Apparently, through luck, the seven genes he studied were on different chromosomes or far apart on the same chromosome (thus showing no linkage). He did not use the only pair that were closely linked to study independent assortment.

6. 0.130. The estimate for the first testcross was 0.107, a lower estimate of recombination.

7.

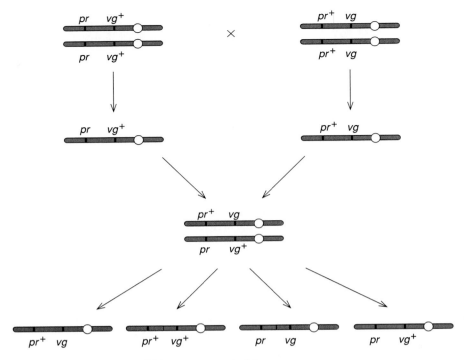

8. The gene order is *a–c–b*. 0.3 > 0.27 because some double crossovers have occurred.

9. The gene order would be *a–c–d–b*. *bcd* /+ + + × *bcd* / *bcd*.

10. The use of three-point crosses confirmed the linearity of genes on chromosomes predicted from two-locus crosses. In addition, one three-point cross can give as much information as three crosses involving two genes each.

11. *k*–0.141–*m*–0.083–*l*.

12. 0.0117. I = 1 − 0.00636/0.0117 = 0.46. There is some interference because the observed number of double recombinants is about half the proportion of doubles expected.

13.

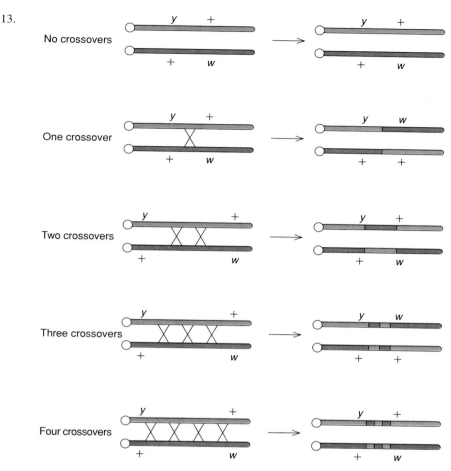

14. In humans, one cannot obviously make specific crosses planned to determine the linkage of two genes. Further, the progeny number in humans is much smaller than in these other organisms in which a single mating may produce over a hundred progeny.

15. 2 h C/Y, 2 H c/Y, 1 H C/Y; 0.2.

16. 0.214. Families having no recombinant male offspring are omitted here, causing a bias in the data.

17. 0.5 A_1A_2, 0.5 A_2—; all would be A_1A_2.

18. Chromosome 11.

19. In four-stranded crossing over, only two of the four present chromatids are involved in any recombination event, while in two-stranded crossing over all (or both) are involved. As a result, four-stranded crossing over generally will give both parental and nonparental gametes in the progeny, while two-stranded crossing over will give only parental (100%), when there is no crossing over, or only nonparental (100%), when there is crossing over.

20. X^+Y, X^wX^+Y; wild-type male and female; white is a recessive allele.

21. *DDDDdddd; DDddDDdd, DDddddDD, ddDDDDdd,* or *ddDDddDD.*

22. Centromere–F–E.

23. 0.028; 0.0045.

24. Recombination may be localized into certain regions.

25. Mendel might not have formulated the principle of independent assortment correctly at all.

CHAPTER 6

1. The transforming agent induced a heritable change in *Pneumococci.* Since genes control heredity, it was very likely that the transforming agent was itself a gene or genes.

2. Avery et al. used cell-free extracts of virulent (S) cells as the transforming agent. Since no virulent cells were mixed with the avirulent ones, there was no possibility of resurrecting any virulent cells.

3. The Avery et al. approach offers the stronger argument for DNA as the genetic material. These workers carefully purified their DNA before using it to transform cells. In the Hershey-Chase approach, 10% of the phage protein was not accounted for and therefore could have entered the host cells along with the phage DNA. In principle, this protein could have been the genetic material.

4. No. Both protein and DNA contain C and H.

5. 5′-CGAATGCT-3′; 3′-TCGTAAGC-5′.

6. (a) GTCGTCA. (b) CAGCAGT.

7. TCCGATTGAC
   AGGCTAACTG

8. Thymine, 30%; cytosine, 20%.

9. 50%.

10. (a) 12,000 helical turns; (b) 40.8 microns; (c) 79.2×10^6; (d) 240,000 phosphorous atoms (120,000 in each strand).

11. 0.54 kb.

12. 0.3 kb.

CHAPTER 7

1. (a) Conservative:

(b) Semiconservative:

2. (a) Conservative:

(b) Semiconservative:

3. (a)

Progeny strands: ······ = Unlabeled
               ―――― = Labeled

(b)

4.

5. There are two classes of replicons; one begins replicating before the other.

6.

―――― = Low specific activity
―――― = High specific activity

7. (a) The gradient centrifugation of the pulse-labeled DNA would show only a peak of large DNA. No Okazaki fragments would appear, no matter how short the labeling period. (b) The DNA polymerase must be able to make DNA in both the $5′ \rightarrow 3′$ and $3′ \rightarrow 5′$ directions.

8. Helicases separate the two DNA strands in advance of the replicating fork. This process requires energy, which is provided by the ATPase's cleavage of ATP.

9. A helicase unwinds DNA, while a topoisomerase changes the topology of the DNA by increasing or decreasing its number of supercoils. Helicases break no DNA strands, whereas transient strand breakage is a necessary part of a topoisomerase's function.

10. SSB binds cooperatively to single-stranded DNA, but not to double-stranded DNA. This means that once one molecule of SSB binds to single-stranded DNA, it greatly facilitates the binding of other molecules, but this facilitation does not operate with double-stranded DNA.

11. Assuming that the linear DNA is not constrained somehow by proteins or other attached molecules, it should have no need for a swivel; since it has free ends, it can readily rotate to compensate for the separation of parental strands at the fork. Therefore, no strain will develop in the molecule.

12. Positive.

13. The primosome binds to the template for the trailing strand. This makes sense because the trailing strand needs a new primer for each new Okazaki fragment, whereas the leading strand replicates more or less continuously and rarely needs a new primer.

14. A replisome with two polymerase centers would be able to replicate both strands more or less at the same time and do so processively in the direction of the replicating fork, without having to waste time dissociating from and reassociating with the DNA.

15. Primer synthesis cannot require a based-paired nucleotide to add to. If it did, the first nucleotide could never be incorporated. Therefore, the editing function cannot operate. If the wrong nucleotide is incorporated into a primer, it must remain there.

CHAPTER 8

1. RNA polymerase holoenzyme transcribes un-nicked T4 DNA actively in vitro. The transcription is specific (immediate early genes only) and asymmetric. The core enzyme transcribes such a template only weakly. What transcription occurs is nonspecific and symmetric.

2. (a) Up. A C $\rightarrow$ T mutation in the first C would create a −10 box with this sequence: TATAGT. This is closer to the consensus sequence. (b) Down. A T $\rightarrow$ A mutation in the last position would create CATAGA. This is less like the consensus −10 box.

3. The genes in the operon are all transcribed together, from a single promoter. Therefore, controlling transcription from that promoter controls all the genes simultaneously.

4. (*a*) The *lac* operon would always be turned on, because repressor cannot turn it off unless it can bind to the *lac* operator. (*b*) Same as (*a*). (*c*) The operon would be uninducible. Repressor would remain bound to operator even in the presence of inducer. (*d*) The operon would be transcribed only weakly. Even in the absence of glucose, CAP plus cAMP would be unable to facilitate polymerase binding to the *lac* promoter.

5. Phenotype for β-galactosidase Production

	No Inducer	Inducer	Reasoning
(*a.*)	–	+	Wild-type.
(*b.*)	–	–	Mutant Z gene, which is the β-galactosidase gene.
(*c.*)	+	+	Constitutive operator.
(*d.*)	+	+	No repressor made from I^- gene.
(*e.*)	–	–	No β-galactosidase can be made from the Z^- gene, even though the operator is constitutive.
(*f.*)	+	+	Both repressor and operator are constitutive.
(*g.*)	–	–	The mutant repressor cannot be removed from the operator by the inducer.
(*h.*)	+	+	The constitutive operator cannot bind repressor, including the product of the I^s gene.

6. (*a*) Since the mutant's phenotype is constitutive, its genotype must carry a mutation. Therefore, I^x must be I^-.

(*b*) Since the mutation in I prevents *lac* function, and is dominant, I^x must be I^s.

(*c*) Since the partial diploid behaves as wild-type, I^x is either I^+ or I^- (I^- is recessive).

(*d*) Since the mutant is constitutive, I^x must be I^-.

(*e*) Since the mutant is constitutive, O^x must be O^c.

(*f*) Since the phenotype of the partial diploid is wild-type, O^x must be O^+.

(*g*) Since the partial diploid fails to make β-galactosidase, even with wild-type forms of the control loci (I^+O^+), Z^x must be Z^-.

7.

	No Inducer	Inducer	
(*a*)	–	+	P^+ is dominant. The *lac* operon with the wild-type promoter will still function, even if the other does not. The phenotype is therefore wild-type.
(*b*)	–	–	With neither promoter able to bind RNA polymerase, neither *lac* operon can function.
(*c*)	–	–	The operon with the wild-type Z gene has a nonfunctional promoter, and the operon with a wild-type promoter has a nonfunctional Z gene.
(*d*)	–	–	Even in the absence of active repressor, neither *lac* operon can be productive without a functional promotor.

8. (*a*) $G^+ O^+ R^+$ (wild-type).

(*b*) $G^+ O^c R^+$; this operator-constitutive mutant will make the gene product even without inducer because the operator does not bind repressor well.

$G^+ O^+ R^-$; this genotype produces inactive repressor, so G is turned on even in the absence of inducer.

$G^+ O^c R^-$; this genotype includes both a defective operator and a defective repressor, so G is always on.

(*c*) $G^- O^+ R^+$; if G is defective, no product will be made, regardless of the condition of the other two genes, so all possible combinations of O^+, O^c, R^+, R^-, and R^s will give the same phenotype, as long as they are coupled with G^-. These are

$G^- O^c R^+$

$G^- O^+ R^-$

$G^- O^+ R^s$

$G^- O^c R^-$

$G^- O^c R^s$

$G^+ O^+ R^s$; even if G is wild-type, no product will be made if the repressor cannot bind to inducer.

9.

Genotype	Phenotype for G product +inducer	–inducer
$G^+ O^+ R^+/G^- O^+ R^+$	+	–
$G^+ O^c R^+/G^- O^+ R^+$	+	+
$G^+ O^+ R^-/G^- O^+ R^+$	+	–
$G^+ O^c R^-/G^- O^+ R^+$	+	+
$G^- O^+ R^+/G^- O^+ R^+$	–	–
$G^- O^c R^+/G^- O^+ R^+$	–	–
$G^- O^+ R^-/G^- O^+ R^+$	–	–
$G^- O^+ R^s/G^- O^+ R^+$	–	–
$G^- O^c R^-/G^- O^+ R^+$	–	–
$G^- O^c R^s/G^- O^+ R^+$	–	–
$G^+ O^+ R^s/G^- O^+ R^+$	+	+

10. (*a*) The early genes only, because the gene 28 product is needed for the switch to middle transcription. (*b*) and (*c*) The early and middle genes only, because the products of both genes 33 and 34 are needed for the switch to late transcription.

11. Only the early (class I) genes would be transcribed, since gene 1 encodes the T7 RNA polymerase that transcribes the later genes.

12. The cells would not be able to sporulate, since σ^{29} and σ^{32} are needed to switch the transcriptional specificity of *B. subtilis* RNA polymerase from vegetative to sporulation genes.

13. The repressor gene (*cI*).

14. All *but* the repressor gene.

15. Transcripts made in the delayed early phase of infection all initiate at the immediate early promoters. Therefore, immediate early products, including pN, continue to be made throughout the delayed early phase.

16. It cannot result in *cro* expression because the *cro* gene is transcribed backwards, and therefore an anticoding mRNA is produced.

17. Positive. The expression of the delayed early genes is turned off unless pN acts positively to turn it on.

CHAPTER 9

1. Because a polymerase III promoter lies within the coding region of the gene.

2. First, nucleosomes are found in all eukaryotes that have been studied. Second, most of the histones are extremely well conserved. If they were not so vitally important, they would have evolved more rapidly.

3. The 11-nm fiber is essentially a string of nucleosomes. The nucleosomes contain all histones, but H1 is only bound to the outside of each nucleosome. Since H1 is not necessary for forming nucleosomes, its absence would not perturb the 11-nm fiber, beyond reducing its average diameter slightly. On the other hand, histone H1 plays an active role in coiling the nucleosome chain into the 30-nm fiber, or solenoid. Removing H1 would prevent formation of this chromatin structure.

4. The euchromatic chromosome.

5. Both stimulate transcription of specific genes. σ causes holoenzyme to bind tightly to promoters, but does not bind to them by itself. Eukaryotic transcription factors, on the other hand, bind independently to promoters.

6.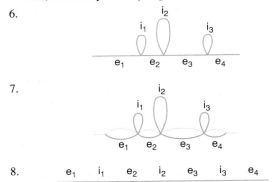

7.

8.   e₁   i₁   e₂   i₂   e₃   i₃   e₄

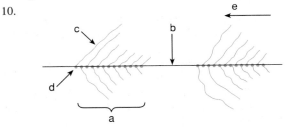

9. It would not hybridize because there is no tract of poly (T) in a gene that would encode poly (A).

10.

11. GpppXpY . . .

12. Messenger RNA levels in the cytoplasm reflect much more than just transcription. They are also potentially influenced by capping, polyadenylation, splicing, transport, and RNA degradation.

13. 28 bases downstream from the second A of the TATA box (TATAAAT) lies the sequence CCCGTTA, where the boldfaced G is 31 bases downstream from the middle of the TATA box. This G is the most likely cap site, although the A preceding and the A and G following it are also candidates.

CHAPTER 10

1. 100 proteins (120,000/1,200 = 100).

2. Test the mutant for growth on media containing the supplements x, a, b, c, or z, individually. Should the mutant grow on medium supplemented with c but not on medium supplemented with b, this would indicate that the enzyme that catalyzes the b → c reaction is defective. Therefore, the gene controlling the production of that enzyme is defective.

3.
    (a)
    5′ - - - - - - - - - - - - ⟶ 3′

    (b) 5′ ———————— 3′
    - - - - - - - - - - - ⟶

    (c)
    3′ ———————— 5′
    - - - - - - - - - - - ⟶

4. The proteins would be made with arginines where phenylalanines should have been. Most or all of these proteins, because of their altered primary structures, would function abnormally or not at all.

5. (a) 5′GTCAATCGTAGCGGCCATAT3′
     3′CAGTTAGCATCGCCGGTATA5′
   (b) 5′AUAUGGCCGCUACGAUUGAC3′

6. The tRNAs bind together much more tightly than free trinucleotides because their anticodons are bent into a helical shape that facilitates base pairing.

7. The alanine would go where a methionine is supposed to go because the anticodon, not the amino acid, determines the response to a given codon.

8. The aminoacyl-tRNA synthetases are crucial to fidelity of translation; if they make mistakes, the translation machinery will incorporate incorrect amino acids.

9. First reaction: aminoacyl-AMP + pyrophosphate. Second reaction: aminoacyl-tRNA.

10. (a) Overlapping, 10 (AUG, UGG, GGC, etc.); nonoverlapping, 4 (AUG, GCA, GUG, CCA).
    (b) Overlapping, 3 (UGC, GCC, CCA); nonoverlapping, 1 (CCA).

11. (a) The reading frame would shift by two bases, and the gene would therefore not function. (b) The reading frame would not change, and the gene would therefore be functional, unless the extra three bases coded for an amino acid that disrupted the function of the gene product. (c) The reading frame would not change, and the gene would therefore be functional, unless the changes in the two adjacent codons were deleterious.

12. A repeating dipeptide. For example, the two codons would be UUCGUU and CGUUCG if translation started at the beginning of the message, and they would be UCGUUC and GUUCGU if translation started with the second base.

13. It would eliminate initiation at that point. One of the AUGs in the middle of the message might then function as the start codon, but the protein product would be incomplete—truncated at the amino terminus.

14. (*a*) fMet-Phe-Lys-Met-Val-Thr-Trp. The next to last codon (UAA) causes termination, so the last codon (AUC) is not translated. (*b*) fMet-Val-Thr-Trp. Note that the base change destroyed the first start codon, so the second had to be used. (*c*) No change in amino acid sequence; UUU and UUC both code for Phe. (*d*) fMet-Leu-Lys-Met-Val-Thr-Trp. (*e*) fMet-Phe-Lys-Met-Val-Thr. The change converted the Trp codon to a stop signal.

15. (*a*) Leu, Ser, and Arg. (*b*) Trp and Met.

16. 256 (4^4).

17. CUU and CUC.

18. UAI.

19. Cloacin DF13 should profoundly inhibit translation, because it removes the part of the 16S rRNA that binds to Shine-Dalgarno sequences on mRNAs and therefore binds mRNA and ribosome together.

20. (*a*) Before translocation; (*b*) before aminoacyl-tRNA transfer to the ribosome; (*c*) before peptide bond formation.

21. fMet-tRNA. This aminoacyl-tRNA must move to the P site to make room in the A site for the next aminoacyl-tRNA.

22. 402.

23. All QT proteins made in this suppressor strain would have an extra tyrosine at the carboxyl end, because the suppressor tRNA would recognize the first stop codon (UAG) in every gene as a tyrosine codon.

24. Wild-type Gln anticodon: 3′ GUU5′ (recognizes the Gln codon 5′CAA3′). Ochre suppressor anticodon: 3′AUU5′ (recognizes the ochre codon 5′UAA3′). The change was a G → A transition in the 3′-end base of the anticodon of the tRNA.

CHAPTER 11

1. A somatic mutation. If it were a germ-line mutation, the color would be uniform, or at least in a pattern of some sort. The unique spot of white was probably caused by a somatic mutation in this particular blossom that inactivated a gene responsible for the red color.

2. We know the yellow color is dominant, because the heterozygotes are yellow; the lethality of the yellow gene must be recessive, because only the homozygotes die.

3. No. Mutations in DNA polymerase III would surely be lethal, since they would block a function crucial to life: replication of DNA.

4. Conditional lethal mutations would be appropriate. Bacteria could be tested for the trait by attempting to grow them at normal temperature, but they could be maintained alive at lower temperature.

5. No, the product of a temperature-sensitive gene, not the gene itself, is temperature-sensitive.

6. After the patch of fur was shaved, the skin was relatively cool, so the temperature-sensitive, color-producing enzyme was active. This led to a patch of dark fur. The new fur warmed the skin, inactivating the color-producing enzyme, so the dark patch gradually turned white.

7. Transitions.

8. (*a*) Pyrimidine (especially thymine) dimers. (*b*) If this damage goes unrepaired, it can stall DNA replication or cause error-prone repair, which can induce missense mutations.

9. (*a*) DNA strand breaks. (*b*) Double-strand breaks are difficult to repair properly and therefore constitute lasting damage.

10. The mRNA becomes AUGGCCUAAAGAGG; the frameshift mutation has shifted the reading frame one base to the right. Now the mRNA codes for fMet-Ala. The next codon, UAA, codes for stop.

11. Add a base, preferably early in the mRNA, to compensate for the lost base. This will restore the reading frame.

12. Photoreactivating enzyme and O^6-methylguanine methyl transferase.

13. An exonuclease degrades a DNA molecule from the end, one base at a time; an endonuclease cuts within a DNA molecule.

14. A long gap in one of the strands. Long gaps do occur during repair in *E. coli,* suggesting that this bacterium uses the second, stepwise method.

15. Because the damage remains, even though DNA replication has occurred.

16. $15/(2)(150,000) = 5 \times 10^{-5}$

CHAPTER 12

1. (*a*) Genes for transposase and terminal inverted repeats that the transposase recognizes. (*b*) Terminal inverted repeats that the other transposon's transposase recognizes.

2. Transposase is necessary for transposition.

3. The DNA does not always jump. In replicative transposition, it does not even leave its original location; instead, it replicates, and one of the copies goes to the new location.

4. (*a*) The inserted transposon will be surrounded by 5-bp direct repeats of host DNA. (*b*) See figure 12.1; imagine the cuts 5 bp apart, rather than 9 bp.

5. A "selfish" or "parasitic" DNA is a DNA element that replicates within a cell but seems to provide nothing useful to the cell in return.

6. Fewer. Homologous recombination requires some homology between the transposon and its target DNA, which automatically limits the number of target sites; illegitimate recombination requires little if any homology, so a transposon using this mechanism is free to go almost anywhere.

7. (*a*) Replicon fusion (cointegrate formation) and resolution. (*b*) A cointegrate.

8. (*a*) No product; the transposase is needed to form the cointegrate intermediate. (*b*) A cointegrate.

9. Several transposons conferring resistance to a variety of antibiotics can transpose to the same plasmid in a pathogenic bacterium. This bacterium will be resistant to all these antibiotics, and it can even share the plasmid with other dangerous bacteria, making them antibiotic-resistant. When these bacteria cause disease, they are very difficult to kill because they have built-in resistance to our drugs.

10. Two copies. At the boundaries between the two plasmids.

11. No. Tn3 transposes by replicating itself, so it would not leave the *C* gene during transposition, and reversion would not be nearly as frequent.

12. The intron would remain in the transposon.

13. The intron would be spliced out.

14. Xeroderma pigmentosum patients have defective DNA repair systems that cannot repair ultraviolet damage properly. As a result, these people respond to ultraviolet exposure by developing multiple skin cancers.

15. Chemicals, radiation, and viruses.

16. Transformation means the change in a cell's behavior from normal to tumor cell-like.

17. NIH 3T3 cells are not normal. They are already immortal and poised on the brink of transformation, needing just one final nudge to push them over. The mutant Ha-*ras* gene provides that nudge.

18. It is a somatic mutation. We know this because tumor cells have a mutant Ha-*ras* gene, while normal cells from the same patient have only the normal gene.

CHAPTER 13

1. Genotypes and phenotypes (respectively) are: (*a*) *lac$^-$*, Lac$^-$; (*b*) *leu$^+$*, Leu$^+$; (*c*) *thi$^-$*, Thi$^-$; (*d*) *str^r*, Str^r; (*e*) *pur^s*, Pur^s; (*f*) *bio$^+$*, Bio$^+$.

2. Prototrophs: (*a*), (*b*), (*d*), (*e*), and (*f*). Auxotroph: (*c*). This is the only one that cannot make a substance needed for growth on minimal medium containing glucose.

3. Yes, he was justified. If single mutants had been used, their reversion rate (one in a million) would have produced ten times as many prototrophs as their recombination rate (one in ten million). Therefore, the investigators would have observed almost as many prototrophs with the two mutant strains grown separately as when they were grown together; they would not have known whether they were observing recombination plus reversion, or just reversion.

4. It usually does not contain a full set of genes from the donor cell, so it is not a true diploid. We call it a merozygote or a merodiploid.

5. The Ton^r phenotype is caused by a mutation. If it were due to an adaptation, all cultures would be about equal in resistance to the phage. (These were the actual results obtained by Luria and Delbrück in 1943.)

6. Integration of the F plasmid into the host genome. This mobilizes the host genome so that it behaves as the F plasmid normally would and transfers the chromosome (usually only in part) to a recipient cell.

7. Usually, the connection between conjugating cells breaks before the whole chromosome has a chance to transfer.

8.

9. Conjugation is usually interrupted, just as it is in earthly bacteria. This lowers the likelihood of transfer of genes near the end of the chromosome; conjugation usually ceases before such transfer can take place.

10. 25 minutes.

11. *gap-bar-ple-chr-zap*

12. Place one mutant *zap* gene on an F-*zap* plasmid (by recombination) and conjugate this mutant with the other *zap$^-$* mutant. This places both mutant genes in the same cell— one in the chromosome and the other on the F′ plasmid. If the mutations are in different genes, they will complement and give rise to a large number of Zap$^+$ cells. If the mutations are in the same gene, very few Zap$^+$ cells will arise.

13. They could be due to recombination between two DNAs with mutations in the same gene (either *zapA* or *zapB*). Alternatively, they could be due to intracistronic complementation, such as that observed in the *E. coli lacZ* gene.

14. λ phage heads can accommodate only about 10% extra DNA. Thus, the λ DNA will be able to pick up host genes from one side (*gal*) or the other (*bio*) of the prophage, but not both.

15. Yes, they would cotransduce, because they are separated by less than 2.4% of the *E. coli* genome, and this is the amount of DNA that can fit into a P1 phage head. Using the expression: frequency = $(1-d/L)^3$, we would predict a cotransduction frequency of 2%, not too far from the actual frequency of 1%.

16. *zip zap bop*. If the order were *bop zip zap*, Bop$^+$ Zip$^-$ Zap$^-$ transductants would be less common than Bop$^+$ Zip$^+$ Zap$^+$.

17. The marker order is *awk nrd kat*. The cotransduction frequency between *awk* and *kat* is $(1 + 115)/1,063 = 0.11$. The cotransduction frequency between *awk* and *nrd* is $(309 + 115)/1,063 = 0.40$.

18. (250 plaques/0.05 ml) × 10^6 dilution factor = $5 × 10^9$ pfu/ml.

19. The curve will start to rise sooner and the rise will be more gradual. This is because infectious phages appear in infected cells before lysis occurs. These will be detected by breaking the cells open.

20. K M L (or L M K).

CHAPTER 14

1. The red blood cell itself can live without hemoglobin, but the whole organism would die without this product to carry oxygen to all its tissues.

2. (*a*) Determined only; (*b*) determined and differentiated.

3. After two cell divisions, there are only four cells, enough for identical quadruplets if the cells separate instead of staying together in one embryo, but not enough to produce quintuplets. One more cell division, or a total of three, would produce eight blastomeres. This is more than enough for quintuplets, but the cells must all be totipotent. If they have lost any of their genetic potential, they cannot give rise to a whole person.

4. You would observe neither DNase hypersensitivity nor activity. This is because the histones, which you added first, would bind throughout the gene and its 5′-flanking region, tying it up in nucleosomes. The factors in the fourteen-day extract would not be expected to remove these nucleosomes.

5. With a cloned gene, deletions of precisely defined length and position can be introduced, and the altered genes can then be tested for light sensitivity. Without a cloned gene, the deletions would have to be introduced by hit-and-miss genetic means, and their locations and extents would not be as well defined.

6. Expression of the gene with the mutated 5′-flanking region must be distinguishable from the unmutated gene. If the mutated pea rubisco gene were reintroduced into a pea plant, its expression would not be as easily distinguishable from expression of the resident pea rubisco gene.

7. The gene will be active because no histone H1 is present to cross-link the nucleosomes.

8. Early embryonic cells divide rapidly, thus they must produce a great deal of protein. 5S rRNA is one of the components of the ribosomes that must make this protein.

9. The vitellogenin gene (ovalbumin is not produced in the liver).

10. Homeoboxes are found in genes that determine the body plan of *Drosophila*. Since the products of these genes probably influence the activity of several other genes, it is likely that the products of the homeobox genes in *Drosophila* bind to the control regions of these other genes. This probably explains the importance of the DNA-binding domain.

11. A radioactive probe hybridizes to a given mRNA in a thin slice of an embryo. Wherever the gene for the mRNA is expressed, that mRNA will be available to hybridize to the probe. Hybridization is then detected by autoradiography.

12. When the fused *ftz-lacZ* DNA was injected into an embryo, β-galactosidase was expressed in a banded pattern, detected by cleavage of a colorless material to a blue dye by the enzyme. Since the only remaining element in the fused gene that was of *Drosophila* origin was the *ftz* promoter, it was very likely that the banded pattern of expression was due to a banded pattern of factors that could act on that promoter, namely transcription factors.

13. Mutants in these genes experience abnormal development.

CHAPTER 15

1. *Xgi*I.

2. GATATC

3. Every $4^5 = 1,024$ bases.

4. *Bam*HI and *Bgl*II; 5′-GATC-3′.

5. No; after ligating the two DNAs together, the site (GGATCT) is a hybrid—*Bam*HI-like on one end and *Bgl*II-like on the other. Therefore, neither enzyme will recognize this site.

6. You would see a uniform lawn of bacteria instead of individual colonies. This is because *all* cells will grow, not just those containing plasmids with antibiotic resistance genes.

7. Tetracycline. Inserts in the *Bam*HI site of pBR322 will inactivate the tetracycline gene.

8. No. The *Eco*RI site of pBR322 lies outside both the ampicillin and tetracycline resistance genes.

9. No. The pUC plasmids selectively accept small inserts. Therefore, it would take millions of pUC clones to build a complete genomic library from a complex genome.

10. Yes. The pUC plasmids are very convenient, and since the total amount of DNA we need to clone is not large, we do not have to worry about the fact that each clone will have only a limited amount of DNA.

11. Since the tobacco probe and the pea gene will almost certainly have different sequences, you should use a lower temperature (lower stringency) so that a stable hybrid can form between the two DNAs of similar, but not identical, sequence.

12. (a) Tyr-Met-Cys-Trp-Ile. (b) $2 \times 1 \times 2 \times 1 \times 3 = 12$. (c) $6 \times 2 \times 1 \times 2 \times 1 = 24$.

13. When the vector and insert have tails long enough to form stable hybrids (approximately 12 bases). Then the ligation step can be completed within the host cell after transformation.

14. It prevents premature synthesis of the protein product of the cloned gene, which may be harmful to the host cell.

15. One in six; one-half will be in the right orientation and one-third of these will be in the right reading frame, since there are three possible reading frames; $(\frac{1}{2} \times \frac{1}{3} = \frac{1}{6})$.

16. T h r A s p S e r V a l G l y
A C C G A T T C A G T G G G C

Boldface type denotes the base change that alters the Ala codon to a Ser codon.

17. Cut the viral DNA with *Eco*RI, electrophorese the resulting five fragments, and Southern blot the fragments. Hybridize the Southern blot to a radioactive probe made from the cloned cDNA, and autoradiograph to see which of the five DNA fragments "light(s) up."

18. Collect mRNAs from all five tissues, electrophorese them, then Northern blot. Hybridize a radioactive probe (made from the cloned petunia gene) to the Northern blot and autoradiograph. The tissue that contained the highest concentration of the specific mRNA will hybridize best and will give the most intense band on the autoradiogram.

19. The size(s) of the mRNA(s) made from this gene.

20. 10.7 kb.

21. $8(2 \times 2 \times 2)$.

22. Yes. Absence of this site would result in a longer fragment, but it would still be detected by the probe, because it would overlap the probe region by just as much.

23. No. The probe would not be able to hybridize to the right-hand fragment resulting from the presence of this site. Therefore, this fragment would not show up on the autoradiograph.

24. ATGAGTTGGCCAGCAGAGCGAGCATGGATGTAA

CHAPTER 16

1.

	p	q	p^2	2pq	q^2
a	0.4	0.6	0.16	0.48	0.36
b	0.68	0.32	0.462	0.435	0.102
c	0.467	0.533	0.218	0.498	0.284
	p^2N	2pqN	q^2N	χ^2	
a	16	48	36	2.78	
b	23.1	21.8	5.1	20.1	
c	62.1	141.9	80.9	5.63	

With 1 degree of freedom, *a* is not significant, *b* is significant at the 0.01 level, and *c* is significant at the 0.05 level.

2. The expected numbers are 18, 84, and 98 for genotypes A_1A_1, A_1A_2, and A_2A_2. 0.45—with 1 degree of freedom, it is consistent.

3. A_1A_1, 0.49; A_2A_2, 0.09.

4. 0.1; 0.18.

5. (*a*) AR: 0.6, PP: 0.32, and ST: 0.08. (*b*) AR AR: 18, AR PP: 19.2, AR ST: 4.8, PP PP: 5.12, PP ST: 2.56, and ST ST: 0.32.

6.

	A_1	A_2	A_3	χ^2
a	0.35	0.3	0.35	2.27
b	0.2	0.2	0.6	400

Expected genotypic frequencies

	A_1A_1	A_1A_2	A_1A_3	A_2A_2	A_2A_3	A_3A_3
a	0.1225	0.21	0.245	0.09	0.210	0.1225
b	0.04	0.08	0.24	0.04	0.24	0.36

With 3 degrees of freedom, *a* is not significant and *b* is significant at the 0.01 level.

7. 0.0064; 0.147.

8. 0.12, 0.16, 0.72; 0.01.

9. 0.408, 0.384, 0.208; f = 0.792.

10. (*a*) 0.878, 0.841, and 0.906. (*b*) This species has a very high level of self-fertilization and a low level of random mating so that the frequency of heterozygotes is quite low.

11. (*a*)

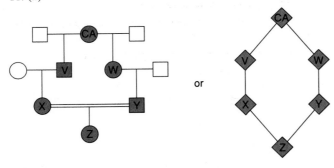

Inbreeding coefficient = 0.03125

(*b*)

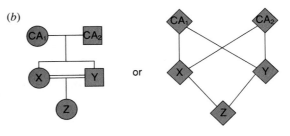

Inbreeding coefficient = 0.25

12. 10^{-6}, 7.25×10^{-7}, 4.5×10^{-7}, 1.75×10^{-7}, 10^{-7}. Δq is largest when p = 1 and q = 0 because all alleles are wild-type.

13. A small population size results in genetic drift, a factor that increases the fixation of alleles in a population and thereby reduces the heterozygosity.

14. 0.3; 0.7. The probability of fixation of an allele is based on its initial frequency. Hence, the probability of fixation is lower for the allele in lowest frequency because it is less likely to increase all the way to 1.0.

15. 0.0012, 0.035. The expected is much larger.

16. Gene flow may have several effects on the genetic variation in a population. For example, if an allele is missing from a population, gene flow from a polymorphic population can introduce it to the population and therefore increase the genetic variation. On the other hand, if the population is polymorphic, then gene flow from a monomorphic population can reduce the level of genetic variation. A high level of gene flow can speed up these changes, while a low level of gene flow may cause the changes to occur over many generations.

17. (*a*)

(*b*) 0.1.

18. 0.13. We assume that there has been no selection or genetic drift and that the allelic frequency estimates are correct.

19. (a) 0.0, −0.019, −0.056, −0.073, 0.0.

(b)

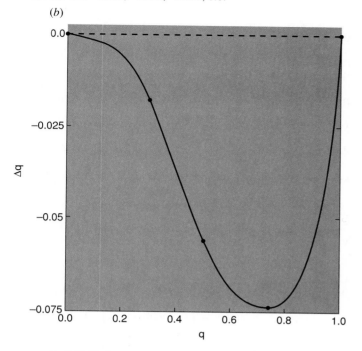

(c) 0.1968, 0.1937.

20. Melanic 1, typical 0.771; 0.4.

21. 0.2, 0.011, −0.2; there is a stable equilibrium at q = 0.2.

22. 0.585, 0.585, and 1.0; assuming that the level of selection is the same in the three breeds, the observed differences could be due to chance, i.e., genetic drift.

23. In general, selection and gene flow have the largest effects on Δq, and mutation has the smallest. Genetic drift can have a large effect in small populations. Genetic drift can either increase or decrease allelic frequency, while the other factors in particular situations will change allelic frequencies only in one direction.

24. Inbreeding increases the frequency of homozygotes and decreases the frequency of heterozygotes. Examine genotypic frequencies compared to Hardy-Weinberg expectations. Inbreeding in concert with selection may fix populations for favorable homozygotes.

25. Snails could be placed on different backgrounds to determine the rate of predation. Look at changes in allelic frequency from generation to generation to see if they vary more in small populations than in large.

CHAPTER 17

1. 0.0000167, 0.0041; new frequencies 0.0000333, 0.0058. When u is doubled, the equilibrium genotypic frequency is essentially doubled. This is not true of the equilibrium allelic frequencies as indicated by the value differences; .0000166 difference in genotypic frequency and 0.0017 difference in allelic frequency.

2. 0.000001; 0.00002.

3. Malaria increased the frequency to equilibrium in Africa; gene flow and selection are reducing the frequency today. Mutation may be a limiting factor.

4. $u = sq^2 = 0.0004$. This value is much higher than known mutation rates, which are about 10^{-5} to 10^{-7}. Genetic drift, heterozygote advantage, or past selection have been offered as explanations.

5. Populations that are started with a very few founders are most likely to have some disease alleles that are higher in frequency than in other populations. This is an extreme form of genetic drift in which in one generation the frequency of an allele changes dramatically because of the restricted number of individuals.

6. Neutrality considers genetic variation a function of mutation and genetic drift, predicting alleles to be transient and fewer in small populations. Selection predicts constant alleles and more variation in small populations. One experiment would involve studying the change in frequency of a polymorphic allele in *D. melanogaster* when the flies are exposed to different environmental or nutritional conditions (see pp. 436–437).

7. The molecular clock assumes a regular replacement of molecular variants over time. The rate of amino acid substitution is the change in amino acid sequence per unit time. Nucleotide substitution would be faster because there are fewer constraints and the third position changes fast.

8. (a) The population size in the first population is larger so there is less genetic drift reducing the amount of heterozygosity. (b) Population sizes (N) of 2,778 and 1,316.

9. 0.444, 0.25; there are more physical constraints on the second protein.

10. Species a and c; species d.

11. If we assume that the molecular information gives us the true relationship, we can examine morphology and physiology, knowing these patterns and the time since divergence.

12. Selection is haploid, no dominance. Genetic drift is more likely because of only one copy versus four for diploid (male and female).

13. 0.00124; 0.03125. The second is larger because more recessive homozygotes are exposed to selection.

14. 0.478, 0.475; 0.456, 0.433; the reduction in heterozygosity is very low for the larger population and not too great for the smaller one.

15. 167.

16. Look at morphological, economic, and disease and pest resistance traits. Grow separately in their natural habitats, avoiding unnatural selection if possible.

17. 0.000312; 0.000064; nearly five times as many progeny from half-first cousins are affected.

18. Avoid inbreeding, keep a large population, and avoid unnatural selection.

19. 0.337. Gene flow or mutation has not introduced the allele, and/or selection for resistance has not occurred. (See chapter 16.)

20. 0.81; 0.518. (See chapter 16.)

21. Mosquitoes have a high reproductive capacity, short generation length, and high dispersal. Furthermore, they adapt to DDT in several different ways.

22. All the new crops may be identical, resulting in loss of genetic variation and susceptibility to particular diseases.

APPENDIX B:
SELECTED READINGS

CHAPTER 1

Creighton, H. B., and B. McClintock. 1931. A correlation of cytological and genetical crossing-over in *Zea mays*. *Proceedings of the National Academy of Sciences* 17:492–97.

Mirsky, A. E. 1968. The discovery of DNA. *Scientific American* 218(June):78–88.

Morgan, T. H. 1910. Sex-limited inheritance in *Drosophila*. *Science* 32:120–22.

Sturtevant, A. H. 1913. The linear arrangement of six sex-linked factors in *Drosophila*, as shown by their mode of association. *Journal of Experimental Zoology* 14:43–59.

CHAPTER 2

Bodmer, W. F., and L. L. Cavalli-Sforza. 1986. *Genetics, evolution, and man*. San Francisco: W. H. Freeman.

Hartl, D. L., and V. Orel. 1992. What did Gregor Mendel think he discovered? *Genetics* 131:245–53.

McKusick, V. 1990. *Mendelian inheritance in man*, 9th ed. Baltimore: Johns Hopkins University Press.

Olby, R. C. 1966. *Origins of Mendelism*. London: Constable.

Peters, J. A. 1962. *Classic papers in genetics*. Englewood Cliffs, N.J.: Prentice-Hall.

Piergorsch, W. W. 1990. Fisher's contributions to genetics and heredity, with special emphasis on the Gregor Mendel controversy. *Biometrics* 46:915–24.

Stern, C. 1973. *Principles of human genetics*, 3rd ed. San Francisco: W. H. Freeman.

Stern, C., and E. R. Sherwood. 1966. *The origin of genetics: A Mendel source book*. San Francisco: W. H. Freeman and Co.

Sutton, H. E. 1988. *An introduction to human genetics*, 4th ed. New York: Harcourt, Brace, Jovanovich.

Todd, N. B. 1977. Cats and commerce. *Scientific American* 237 (November):100–107.

Vogel, F., and A. G. Motulsky. 1979. *Human genetics*. New York: Springer.

CHAPTER 3

Bodmer, W. F., and L. L. Cavalli-Sforza. 1970. Intelligence and race. *Scientific American* 223(October):19–29.

Bouchard, T. J., D. T. Lykken, M. McGue, N. L. Segal, and A. Tellegen. 1990. Sources of human psychological differences: The Minnesota study of twins reared apart. *Science* 250:223–28.

Clausen, J., D. D. Keck, and W. H. Heisey. 1941. Regional differences in differentiation in plant species. *American Naturalist* 75:231–50.

Falconer, D. S. *Quantitative genetics*, 3rd ed. London: Halstead Press.

Hedrick, P. W. 1984. *Population biology*. Boston: Jones and Bartlett Publishers.

Kettlewell, B. 1973. *The evolution of melanism*. Oxford: Oxford University Press.

Nakamura, Y., M. Leppert, P. O'Connell, R. Wolff, T. Holm, M. Culver, C. Martin, E. Fujimoto, M. Hoff, E. Kumlin, and R. White. 1987. Variable number of tandem repeat (VNTR) markers for human gene mapping. *Science* 235:1616–22.

Searle, A. G. 1968. *Comparative genetics of coat colour in mammals*. London: Logos Press.

Silvers, W. K. 1979. *The coat colors of mice: A model for mammalian gene action and interaction*. New York: Springer.

Singer, S. 1978. *Human genetics*. San Francisco: W. H. Freeman.

Sokal, R. R., and F. J. Rohlf. 1981. *Biometry*. San Francisco: W. H. Freeman.

CHAPTER 4

Borgaonkar, D. S. 1991. *Chromosomal variation in man*. New York: Wiley-Liss.

Bridges, C. 1931. Nondisjunction as proof of the chromosome theory of heredity. *Genetics* 1:1–52; 107–63.

Cherfas, J. 1991. Sex and the single gene. *Science* 252:782.

Hassold, T. J. 1986. Chromosome abnormalities in human reproductive wastage. *Trends in Genetics* 3(April):105–10.

Hassold, T. J., and P. A. Jacobs. 1984. Trisomy in man. *Annual Review of Genetics* 18:69–98.

Mcintosh, J. R., and K. L. McDonald. 1989. The mitotic spindle. *Scientific American* 261(October):48–56.

McLeish, J., and B. Snoad. 1958. *Looking at chromosomes*. New York: Macmillan.

Swanson, C. P., T. Merz, and W. J. Young. 1981. *Cytogenetics: The chromosome in division, inheritance, and evolution*. Englewood Cliffs, N.J.: Prentice-Hall.

Varma, R. S., and A. Babu. 1989. *Human chromosomes: A manual of basic techniques*. New York: Pergamon Press.

Von Wettstein, D., et al. 1984. The synaptonemal complex in genetic segregation. *Annual Review of Genetics* 18:331–414.

CHAPTER 5

Beeman, R. W. 1987. A homoeotic gene cluster in the red flour beetle. *Nature* 327:247–49.

Culliton, B. J. 1990. Mapping terra incognita (humani corporis). *Science* 250:210–12.

Finchman, J. R. S., P. R. Day, and A. Radford. 1979. *Fungal genetics*. London: Blackwell.

Lindsley, D. L., and G. G. Zimm. 1992. *The genome of Drosophila melanogaster*. New York: Academic Press.

Morton, N. E. 1991. Parameters of the human genome. *Proceedings of the National Academy of Sciences* 88:7474–76.

Ott, J. 1985. *Analysis of human genetic linkage*. Baltimore: Johns Hopkins University Press.

Ruddle, F. H. 1974. Hybrid cells and human genes. *Scientific American* 231(July):36–44.

Stephens, J. C., M. L. Cavanaugh, M. I. Gradie, M. L. Mador, and K. K. Kidd. 1990. Mapping the human genome: Current status. *Science* 250:237–44.

CHAPTER 6

Adams, R. L. P., R. H. Burdon, A. M. Campbell, and R. M. S. Smellie, eds. 1976. *Davidson's The Biochemistry of the Nucleic Acids*, 8th ed. The structure of DNA, chapter 5. New York: Academic Press.

Avery, O. T., C. M. McLeod, and M. McCarty. 1944. Studies on the chemical nature of the substance-inducing transformation of pneumococcal types. *Journal of Experimental Medicine* 79:137–58.

Chargaff, E. 1950. Chemical specificity of nucleic acids and their enzymatic degradation. *Experimentia* 6:201–209.

Dickerson, R. E. 1983. The DNA helix and how it is read. *Scientific American* 249(December):94–111.

Hershey, A. D., and M. Chase. 1952. Independent functions of viral protein and nucleic acid in growth of bacteriophage. *Journal of General Physiology* 36:39–56.

Watson, J. D., and F. H. C. Crick. 1953. Molecular structure of nucleic acids: A structure for deoxyribose nucleic acid. *Nature* 171:737–38.

———. 1953. Genetical implications of the structure of deoxyribonucleic acid. *Nature* 171:964–67.

CHAPTER 7

Blackburn, E. H. 1990. Telomeres: Structure and Synthesis. *Journal of Biological Chemistry* 265:5919–21.

Cairns, J. 1963. The chromosome of *Escherichia coli*. *Cold Spring Harbor Symposia on Quantitative Biology* 28:43–46.

Callan, H. G. 1973. DNA replication in the chromosomes of eukaryotes. *Cold Spring Harbor Symposia on Quantitative Biology* 38:195–203.

Cox, M. M., and I. R. Lehman. 1987. Enzymes of general recombination. *Annual Review of Biochemistry* 56:229–62.

Dressler, D., and H. Potter. 1982. Molecular mechanism of genetic recombination. *Annual Review of Biochemistry* 51:727–62.

Kornberg, A., and T. A. Baker. 1992. *DNA replication,* 2nd ed. New York: W. H. Freeman.

Kunkel, T. A. 1992. DNA replication fidelity. *Journal of Biological Chemistry* 267:18251–54.

McHenry, C. S. 1988. DNA polymerase III holoenzyme of *Escherichia coli. Annual Review of Biochemistry* 57:519–50.

Meselson, M., and F. W. Stahl. 1958. The replication of DNA in *Escherichia coli. Proceedings of the National Academy of Sciences* 44:671–82.

Meselson, M. S., and C. M. Radding. 1975. A general model for recombination. *Proceedings of the National Academy of Sciences* 72:358–61.

Radman, M., and R. Wagner. 1988. The high fidelity of DNA replication. *Scientific American* 259(August):24–30.

Smith, G. R. 1987. Mechanism and control of homologous recombination in *Escherichia coli. Annual Review of Genetics* 21:179–201.

Wang, J. C. 1982. DNA topoisomerases. *Scientific American* 247(July):94–109.

CHAPTER 8

Adhya, S., and S. Garges. 1990. Positive control. *Journal of Biological Chemistry* 265:10797–10800.

Beadle, G. W., and E. L. Tatum. 1941. Genetic control of biochemical reactions in *Neurospora. Proceedings of the National Academy of Sciences* 27:499–506.

Beckwith, J. R., and D. Zipser, eds. 1970. The lactose operon. Cold Spring Harbor Laboratory.

Conceptual foundations of genetics: Selected readings. 1976. Edited by H. O. Corwin and J. B. Jenkins. Boston: Houghton Mifflin Co.

Friedman, D. I., M. J. Imperiale, and S. L. Adhya. 1987. RNA 3′-end formation in the control of gene expression. *Annual Review of Genetics* 21:453–88.

Helmann, J. D., and M. J. Chamberlin. 1988. Structure and function of bacterial sigma factors. *Annual Review of Biochemistry* 57:839–72.

Hendrix, R. W., J. W. Roberts, F. W. Stahl, and R. A. Weisberg, eds. 1983. Lambda II. Cold Spring Harbor Laboratory.

Jacob, F., and J. Monod. 1961. Genetic regulatory mechanisms in the synthesis of proteins. *Journal of Molecular Biology* 3:318–56.

Jacob, F. 1966. Genetics of the bacterial cell (Nobel lecture). *Science* 152:1470–78.

Losick, R., and M. Chamberlin. 1976. RNA polymerase. Cold Spring Harbor Laboratory.

Miller, J. H., and W. S. Reznikoff, eds. 1978. The operon. Cold Spring Harbor Laboratory.

Monod, J. 1966. From enzymatic adaptation to allosteric transitions (Nobel lecture). *Science* 154:475–83.

Ptashne, M., and W. Gilbert. 1970. Genetic repressors. *Scientific American* 222(June):36–44.

Ptashne, M. 1989. How gene activators work. *Scientific American* 260(January):24–31.

Schleif, R. 1987. DNA binding proteins. *Science* 241:1182–87.

Zubay, G., D. Schwartz, and J. Beckwith. 1970. Mechanism of activation of catabolite-sensitive genes: A positive control system. *Proceedings of the National Academy of Sciences* 66:104–10.

CHAPTER 9

Borg, J. 1990. Zinc fingers and other metal-binding domains. Elements for interactions between macro-molecules. *Journal of Biological Chemistry* 265:6513–16.

Burlingame, R. W., W. E. Love, B.-C. Wang, R. Hamlin, N.-H. Xuong, and E. N. Moudrianskis. 1985. Crystallographic structure of the octameric histone core of the nucleosome at a resolution of 3.3A. *Science* 228:546–53.

Chambon, P. 1981. Split genes. *Scientific American* 244(May):60–71.

Conaway, J. W., and R. C. Conaway. 1991. Initiation of eukaryotic messenger RNA synthesis. *Journal of Biological Chemistry* 266:17721–24.

Friedman, D. E., M. J. Imperiale, and S. L. Adhya. 1987. RNA 3′-end formation in the control of gene expression. *Annual Review of Genetics* 21:453–88.

Geiduschek, E. P., and G. P. Tocchini-Valentini. 1988. Transcription by RNA polymerase III. *Annual Review of Biochemistry* 57:873–914.

Kornberg, R. D., and A. Klug. 1981. The nucleosome. *Scientific American* 244(February):52–64.

McKnight, S., and R. Tjian. 1986. Transcriptional selectivity of viral genes in mammalian cells. *Cell* 46:795–805.

McKnight, S. L. 1991. Molecular zippers in gene regulation. *Scientific American* 264(April):54.

Miller, O. L. 1973. The visualization of genes in action. *Scientific American* 228(March):34–42.

Proudfoot, N. 1991. Poly(A) signals. *Cell* 64:671.

Ptashne, M. 1989. How gene activators work. *Scientific American* 260(January):24–31.

Rhodes, D., and A. Klug. 1993. Zinc fingers. *Scientific American* 268(February):56–65.

Sharp, P. A. 1987. Splicing of messenger RNA precursors. *Science* 235:766–71.

Steitz, J. A. 1988. "Snurps." *Scientific American* 258(June):36–41.

CHAPTER 10

Bjork, G. R., et al. 1987. Transfer RNA modification. *Annual Review of Biochemistry* 56:263–87.

Brenner, S., F. Jacob, and M. Meselson. 1961. An unstable intermediate carrying information from genes to ribosomes for protein synthesis. *Nature* 190:576–81.

Crick, F. H. C. 1958. On protein synthesis. *Symposium of the Society for Experimental Biology* 12:138–63.

Crick, F. H. C. 1966. The genetic code. *Scientific American* 215(October):55–62.

Crick, F. H. C., L. Barnett, S. Brenner, and R. J. Watts-Tobin. 1962. General nature of the genetic code for proteins. *Nature* 192:1227–32.

Fox, T. D. 1987. Natural variation in the genetic code. *Annual Review of Genetics* 21:67–91.

Khorana, H. G. 1968. Synthesis in the study of nucleic acids. *Biochemical Journal* 109:709–25.

Kozak, M. 1991. Structural features in eukaryotic mRNAs that modulate the imitation of translation. *Journal of Biological Chemistry* 266:19867–70.

Lake, J. A. 1981. The ribosome. *Scientific American* 245(August):84–97.

Leder, P., and M. Nirenberg. 1964. RNA codewords and protein synthesis, II. Nucleotide sequence of a valine RNA codeword. *Proceedings of the National Academy of Sciences* 52:420–27.

Moore, P. B. 1988. The ribosome returns. *Nature* 331:223–27.

Nomura, M. 1984. The control of ribosome synthesis. *Scientific American* 250(January):102–14.

Quigley, G. J., and A. Rich. 1976. Structural domains of transfer RNA molecules. *Science* 194:796–806.

Rich, A., and S. H. Kim. 1978. The three-dimensional structure of transfer RNA. *Scientific American* 238(January):52–62.

CHAPTER 11

Brenner, S., L. Barnett, F. H. C. Crick, and A. Orgel. 1961. The theory of mutagenesis. *Journal of Molecular Biology* 3:121–24.

Little, J. W. 1982. The SOS regulatory system of *Escherichia coli. Cell* 29:11–12.

Loeb, L. A. 1985. Apurinic sites as mutagenic intermediates. *Cell* 40:483–84.

Miller, J. H. 1983. Mutational specificity in bacteria. *Annual Review of Genetics* 17:215–38.

Morell, V. 1993. Huntington's gene finally found. *Science* 260:28–30.

Radman, M., and R. Wagner. 1986. Mismatch repair in *Escherichia coli. Annual Review of Genetics* 20:523–38.

Rennie, J. 1993. DNA's new twists. *Scientific American* 266(March):122–32.

Sancar, A., and G. B. Sancar. 1988. DNA repair enzymes. *Annual Review of Biochemistry* 57:29–67.

Suzuki, D. T. 1970. Temperature-sensitive mutations in *Drosophila melanogaster. Science* 170:695–706.

Walker, G. C. 1985. Inducible DNA repair systems. *Annual Review of Biochemistry* 54:425–58.

CHAPTER 12

Ames, B. N., W. E. Durston, E. Yamasaki, and F. D. Lee. 1973. Carcinogens are mutagens: A simple test system combining liver homogenates for activation and bacteria for detection. *Proceedings of the National Academy of Sciences* 72:2423–27.

Baltimore, D. 1985. Retroviruses and retrotransposons: The role of reverse transcription in shaping the eukaryotic genome. *Cell* 40:481–82.

Barbacid, M. 1987. *ras* genes. *Annual Review of Biochemistry* 56:779–827.

Cohen, S. N., and J. A. Shapiro. 1980. Transposable genetic elements. *Scientific American* 242(February):40–49.

Doering, H.-P., and P. Starlinger. 1984. Barbara McClintock's controlling elements: Now at the DNA level. *Cell* 39:253–59.

Drake, J. W. 1970. *The molecular basis of mutation.* San Francisco: Holden-Day.

Engels, W. R. 1983. The P family of transposable elements in *Drosophila. Annual Review of Genetics* 17:315–44.

Fedoroff, N. V. 1984. Transposable genetic elements in maize. *Scientific American* 250(June):84–99.

Garfinkel, D. J., J. D. Boeke, and G. R. Fink. 1985. Ty element transposition: Reverse transcriptase and virus-like particles. *Cell* 42:507–17.

Mizuuchi K., and R. Craigie. 1986. Mechanism of bacteriophage Mu transposition. *Annual Review of Genetics* 20:385–429.

Shapiro, J. A. 1983. *Mobile genetic elements.* New York: Academic Press.

Weinberg, R. A. 1988. Finding the anti-oncogene. *Scientific American* 259(September):34–41.

Wessler, S. R. 1988. Phenotypic diversity mediated by the maize transposable elements *Ac* and *Spm. Science* 242:399–405.

CHAPTER 13

Benzer, S. 1961. Genetic fine structure. In *Harvey Lectures,* Series 56. New York: Academic Press.

Hayes, W. 1968. *The genetics of bacteria and their viruses,* 2nd ed. New York: Wiley.

Ippen-Ihler, K. A., and E. G. Minkley. 1986. The conjugation system of F, the fertility factor of *Escherichia coli. Annual Review of Genetics* 20:593–624.

Jacob, F., and E. L. Wollman. 1961. *Sexuality and the genetics of bacteria.* New York: Academic Press.

Morse, M. L., E. M. Lederberg, and J. Lederberg. 1956. Transduction in *Escherichia coli* K–12. *Genetics* 41:142–56.

Stent, G. S., and R. Calendar. 1978. *Molecular genetics: An introductory narrative,* chapters 9–13. San Francisco: W. H. Freeman.

Wollman, E. L., F. Jacob, and W. Hayes. 1956. Conjugation and genetic recombination in *Escherichia coli* K–12. *Cold Spring Harbor Symposia on Quantitative Biology* 21:141–62.

CHAPTER 14

Barinaga, M. 1991. Dimers direct development. *Science* 251:1176.

Borst, P., and D. R. Greaves. 1987. Programmed gene rearrangements altering gene expression. *Science* 235:658–67.

Brown, D. D. 1981. Gene expression in eukaryotes. *Science* 211:667–74.

Coulter, D., and E. Wieschaus. 1986. Segmentation genes and the distributions of transcripts. *Nature* 321:472–74.

DeRobertis, E. M., and J. B. Gurdon. 1979. Gene transplantation and the analysis of development. *Scientific American* 241(December):74–82.

DeRobertis, E. M., G. Oliver, and C. V. E. Wright. 1990. Homeobox genes and the vertebrate body plan. *Scientific American* 263(July):46.

Gehring, W. 1985. The homeobox: A key to the understanding of development? *Cell* 40:3–5.

————. 1985. The molecular basis of development. *Scientific American* 253(October): 152B–162.

Gehring, W. J., and Y. Hiromi. 1986. Homeotic genes and the homeobox. *Annual Review of Genetics* 20:147–73.

Gurdon, J. B. 1968. Transplanted nuclei and cellular differentiation. *Scientific American* 219(December):24–35.

Gurdon, J. B. 1992. The generation of diversity and pattern in animal development. *Cell* 68:185–99.

Hoffman, M. 1991. How parents make their mark on genes. *Science* 252:1250.

Holliday, R. 1987. The inheritance of epigenetic defects. *Science* 238:163–70.

————. 1989. A different kind of inheritance. *Scientific American* 260(June):40–48.

Levine, M., and T. Hoey. 1988. Homeobox proteins as sequence-specific transcription factors. *Cell* 55:537–40.

Moses, P. B., and N.-H. Chua. 1988. Light switches for plant genes. *Scientific American* 258(April):64–69.

Nüsslein-Volhard, C. 1991. Determination of the embryonic axes of *Drosophila. Development Supplement* 1:1–10.

Sapienza, C. 1990. Parental imprinting of genes. *Scientific American* 263(October):52–60.

Thomas, B. J., and R. Rothstein. 1991. Sex, maps, and imprinting. *Cell* 64:1.

Tonegawa, S. 1985. The molecules of the immune system. *Scientific American* 253(October):122–31.

Wolffe, A. P., and D. D. Brown. 1988. Developmental regulation of two 5S ribosomal RNA genes. *Science* 241:1626–32.

CHAPTER 15

Chilton, M.-D. 1983. A vector for introducing new genes into plants. *Scientific American* 248(June):50–59.

Cohen, S. 1975. The manipulation of genes. *Scientific American* 233(July):24–33.

Cohen, S., A. Chang, H. Boyer, and R. Helling. 1973. Construction of biologically functional bacterial plasmids *in vitro. Proceedings of the National Academy of Sciences* 70:3240–44.

Gasser, C. S., and R. T. Fraley. 1992. Transgenic crops. *Scientific American* 266(June):62–69.

Gilbert, W., and L. Villa-Komaroff. 1980. Useful proteins from recombinant bacteria. *Scientific American* 242(April):74–94.

Gusella, J. F., et al. 1983. A polymorphic DNA marker genetically linked to Huntington's disease. *Nature* 306:234–38.

Landegren, U., R. Kaiser, C. T. Caskey, and L. Hood. 1988. DNA diagnostics: Molecular techniques and automation. *Science* 242:229–37.

Miller, A. D. 1992. Human gene therapy comes of age. *Nature* 247:455–60.

Morell, V. 1993. Huntington's disease gene finally found. *Science* 260:28–30.

Murray, T. H. 1991. Ethical issues in human genome research. *FASEB Journal* 5:55–60.

Nathans, D., and H. O. Smith. 1975. Restriction endonucleases in the analysis and restructuring of DNA molecules. *Annual Review of Biochemistry* 44:273–93.

Neufeld, P. J., and N. Colman. 1990. When science takes the witness stand. *Scientific American* 262(May):18–25.

Sanger, F., G. M. Air, B. G. Barrell, N. L. Brown, A. R. Coulson, J. C. Fiddes, C. A. Hutchison, P. M. Slocombe, and M. Smith. 1977. Nucleotide sequence of the φX174 DNA. *Nature* 265:687–95.

Verma, I. M. 1990. Gene therapy. *Scientific American* 263(November):68.

Watson, J. D., J. Tooze, and D. T. Kurtz. 1983. *Recombinant DNA: A Short Course.* New York: W. H. Freeman.

White, R., M. Leppert, D. T. Bishop, D. Barker, J. Berkowitz, C. Brown, P. Callahan, T. Holm, and L. Jerominski. 1985. Construction of linkage maps with DNA markers for human chromosomes. *Nature* 313:101–104.

White, R., and J.-M. Lalouel. 1988. Chromosome mapping with DNA markers. *Scientific American* 258(February):20–29.

CHAPTER 16

Allison, A. C. 1955. Aspects of polymorphism in man. *Cold Spring Harbor Symposia on Quantitative Biology* 20:239–55.

Bowen, B. W., A. B. Meylan, and J. C. Avise. 1989. An odyssey of the green sea turtle: Ascension Island revisited. *Proceedings of the National Academy of Sciences* 86:573–76.

Darwin, C. 1859. *On the origin of species by means of natural selection.* London: Murray.

Dobzhansky, T. 1973. Nothing in biology makes sense except in the light of evolution. *American Biology Teacher* 35:125–29.

Falconer, D. S. 1989. *Introduction to quantitative genetics,* 3rd ed. London: Halstead.

Hartl, D., and A. G. Clark. *Principles of population genetics,* 2nd ed. Sunderland, Mass.: Sinauer Associates.

Hedrick, P. W. 1985. *Genetics of populations.* Boston: Jones and Bartlett Publishers.

Jones, J. S., B. H. Leith, and P. Rawlings. 1977. Polymorphism in *Cepaea:* A problem with too many solutions. *Annual Review of Ecological Systems* 8:109–43.

Lewontin, R. C. 1974. *The genetic basis of evolutionary change.* New York: Columbia University Press.

Maynard Smith, J. 1989. *Evolutionary genetics.* Oxford: Oxford University Press.

CHAPTER 17

Beadle, G. W. 1980. The ancestry of corn. *Scientific American* 242(January):112–19.

Clarke, B. 1975. The causes of biological diversity. *Scientific American* 233(August):50–60.

Hedrick, P. W. 1984. *Population biology.* Boston: Jones and Bartlett Publishers.

Hedrick, P. W., and P. S. Miller. 1992. Conservation genetics: Techniques and fundamentals. *Ecological Applications* 2:30–46.

Kimura, M. 1985. *The neutral theory of molecular evolution.* Cambridge: Cambridge University Press.

Li, W-H., and D. Graur. 1991. *Fundamentals of molecular evolution.* Sunderland, Mass.: Sinauer Associates.

Nicholas, F. W. 1987. *Veterinary genetics.* Oxford: Oxford University Press.

O'Brien, S. J., D. E. Widt, and Mitchell Bush. 1986. The cheetah in genetic peril. *Scientific American* 254(May):84–92.

Pesticide resistance. 1986. Washington, D.C.: National Academy Press.

Pirchner, F. 1983. *Population genetics in animal breeding.* New York: Plenum Press.

Rick, C. M. 1978. The tomato. *Scientific American* 239(August):76–87.

Soule, M. 1986. *Conservation biology: The science of scarcity and diversity.* Sunderland, Mass.: Sinauer Associates.

GLOSSARY

A

Ac A transposable element found in corn, capable of transposing itself or directing the transposition of a defective *Ds* element.

acceptor stem The part of a tRNA involving the 5'- and 3'-ends base-paired together, which binds, or accepts, an amino acid.

acrocentric Chromosome with the centromere toward one end.

adenine (A) The purine base that pairs with thymine in DNA.

adenosine A nucleoside containing the base adenine.

adenovirus A DNA tumor virus in rodents, having a linear genome of moderate size.

admixture The amount of gene flow that has occurred from one parental population into a hybrid population.

Agrobacterium tumefaciens A bacterium that invades plants and causes crown gall tumors.

albinism A recessive genetic disease in which the pigment melanin is not reproduced.

alcaptonuria A genetic disease characterized by an abnormal buildup in the urine of homogentisic acid, an intermediate in breakdown of the amino acid phenylalanine. When exposed to air, the urine turns black.

alkylation Addition of a carbon-containing alkyl group—to DNA, for example.

allele A particular form of a gene, presumably reflecting a certain DNA sequence, such as A_1 or A_2.

allelic frequency The proportion of alleles in the population that are of a given type.

allolactose The natural inducer of the *lac* operon.

allopolyploid *See* polyploid.

allosteric protein A protein that changes shape and function when it binds to a small molecule.

Alu sequence A sequence about 300 bp long that is repeated about 300,000 times in the human genome and contains the recognition site (AGCT) for the restriction enzyme *Alu*I.

amber codon UAG, coding for termination.

amber mutation *See* nonsense mutation.

amber suppressor A tRNA bearing an anticodon that can recognize the amber codon (UAG) and thereby suppress amber mutations.

Ames test A test for mutagens developed by Bruce Ames, in which the rate of reversion of auxotrophic bacterial strains (*Salmonella*) to prototrophy upon exposure to a chemical is taken as a measure of the chemical's mutagenicity.

amino acid The building block of proteins.

aminoacyl-tRNA synthetase The enzyme that links a tRNA to its cognate amino acid.

amino terminus The end of a polypeptide with a free amino group.

amplification Selective replication of a gene so that it is represented more than the normal once in a haploid genome.

anaphase A short stage of nuclear division in which the chromosomes move to the poles.

aneuploid Not having the normal diploid number of chromosomes.

annealing of DNA The process of bringing back together the two separate strands of denatured DNA.

***Antennapedia (Antp)* mutants** Mutants of *Drosophila* in which a leg appears where an antenna ought to be.

***Antennapedia* complex (ANT-C)** A large genetic locus in *Drosophila* that contains several homeotic genes, including *Antennapedia.*

antibody A protein with the ability to recognize and bind to a substance, usually another protein, with great specificity. Helps the body's immune system recognize and trigger an attack on invading agents.

anticoding strand The complement of the coding strand of a gene. Serves as template for an mRNA and therefore has polarity opposite that of the mRNA.

anticodon A three-base sequence in a tRNA that base-pairs with a specific codon.

antigen A substance recognized and bound by an antibody.

antiparallel The relative polarities of the two strands in a DNA double helix; if one strand goes 5'–3', top to bottom, the other goes 3'–5'. The same antiparallel relationship applies to any double-stranded polynucleotide or oligonucleotide, including the RNAs in a codon-anticodon pair.

antiterminator A protein, such as the λN gene product, that overrides a terminator and allows transcription to continue.

apurinic site (AP site) A deoxyribose in a DNA strand that has lost its purine base.

artificial selection *See* selection.

assembly map A scheme showing the order of addition of ribosomal proteins during self-assembly of a ribosomal particle in vitro.

assortative mating Mating that is nonrandom with respect to the phenotypes in a population—that is, where the probability of individuals of two given phenotypes mating is not equal to the product of their frequencies in the population.

asymmetric transcription Transcription of only one strand of a given region of a double-stranded polynucleotide.

ATPase An enzyme that cleaves ATP, releasing energy for other cellular activities.

***att* sites** Sites on phage and host DNA where recombination occurs, allowing integration of the phage DNA into the host genome as a prophage.

A-type particles Intracellular, noninfectious particles that resemble retroviruses. May be vehicles for transposition of retrotransposons.

autopolyploid *See* polyploid.

autoradiography A technique in which a radioactive sample is allowed to expose a photographic emulsion, thus "taking a picture of itself."

autosome Any chromosome except the sex chromosomes.

auxotroph An organism that requires one or more substances in addition to minimal medium.

B

backcross A cross between an F_1 and an individual from one of the parental lines.

back-mutation *See* reversion.

bacteriophage *See* phage.

Barr body Structure formed when the inactivated X chromosome is heterochromatinized.

base A cyclic, nitrogen-containing compound linked to deoxyribose in DNA and ribose in RNA.

base pair A pair of bases (A–T or G–C), one in each strand, that occur opposite one another in a double-stranded DNA or RNA.

B-DNA The standard Watson-Crick model of DNA favored at high relative humidity and in solution.

β-galactosidase An enzyme that breaks the bond between the two constituent sugars of lactose.

binomial formula The probability that in a group of a given size, a certain number of individuals will be one type and a certain number will be another type.

***Bithorax* complex (BX-C)** A large genetic locus in *Drosophila* containing several homeotic genes that control segmentation in the fly's thorax and abdomen.

bivalent The structure formed when two homologous chromosomes are paired in meiosis I.

blastomere An early embryonic cell formed by cleavage of the egg.

blastula The hollow ball stage of animal embryogenesis.

branch migration Lateral motion of the branch of a chi structure during recombination.

bromodeoxyuridine (BrdU) An analogue of thymidine that can be incorporated into DNA in place of thymidine, and can then cause mutations by forming an alternate tautomer that base-pairs with guanine instead of adenine.

burst size The difference between the final and initial phage concentrations in a cell; the number of progeny phage produced by a single cell.

C

cap A methylated guanosine bound through a 5′–5′ triphosphate linkage to the 5′-end of an mRNA or an snRNA.

cap binding protein (CBP) A protein that associates with the cap on a eukaryotic mRNA and allows the mRNA to bind to a ribosome.

CAP (catabolite activator protein) A protein that, together with cAMP, activates operons that are subject to catabolite repression.

carboxyl terminus The end of a polypeptide with a free carboxyl group.

carcinogen A substance that causes cancer.

carcinogenesis The sequence of events that convert a normal cell to a cancer cell.

catabolite repression The repression of a gene or operon by glucose, or more likely, by a catabolite, or breakdown product of glucose.

cDNA A DNA copy of an RNA, made by reverse transcription.

centimorgan (cM) *See* map unit.

centromere Constricted region on the chromosome where spindle fibers are attached during cell division.

charging Coupling a tRNA with its cognate amino acid.

Charon phages A set of cloning vectors based on λ phage.

chiasma (chiasmata, pl.) The structure formed between nonsister chromatids during meiosis; physical evidence of recombination.

chi-square test A statistical test to determine if the observed numbers deviate from those expected under a particular hypothesis. Often expressed as χ^2 test.

chi structure The branched DNA structure formed by strand crossover during recombination.

chloramphenicol An antibiotic that kills bacteria by inhibiting the peptidyl transferase reaction catalyzed by 50S ribosomes.

chloroplast The photosynthetic organelle of green plants and other photosynthetic eukaryotes.

chromatids Copies of a chromosome produced in cell division.

chromatin The material of chromosomes, composed of DNA and chromosomal proteins.

***cis*-acting** A term that describes a genetic element, such as an enhancer, a promoter, or an operator, that must be on the same chromosome in order to influence a gene's activity.

***cis*-dominant** A mutation that is dominant, but only with respect to genes on the same piece of DNA (*in cis*).

***cis-trans* test** A test to see whether two mutations are true alleles (located in the same gene) or pseudoalleles (located in different genes). When the mutations are in *cis* position, wild-type behavior is always observed. When the mutations are in *trans* position, wild-type behavior is observed if the mutations are pseudoalleles, but mutant behavior is observed if they are alleles.

cistron A genetic unit defined by the *cis-trans* test. For all practical purposes, synonymous with the word "gene."

clastogen An agent that causes DNA strand breaks.

clones Individuals formed by an asexual process so that they are genetically identical to the original individual.

coding strand The strand of a gene that has the same polarity and coding properties as the mRNA produced by the gene.

codominant Two alleles whose phenotypic effects are both expressed in the heterozygote.

codon A three-base sequence in mRNA that causes the insertion of a specific amino acid into protein or causes termination of translation.

cointegrate An intermediate in transposition of a transposon such as Tn3 from one replicon to another. The transposon replicates, and the cointegrate contains the two replicons joined through the two transposon copies.

colinearity A relationship between a gene and its protein product in which the distance between any two mutations is directly proportional to the distance between the corresponding altered amino acids.

colony hybridization A procedure for selecting a bacterial clone containing a gene of interest. DNAs from a large number of clones are simultaneously hybridized to a labeled probe for the gene of interest.

common garden experiment Growing organisms of different types in the same environment to determine whether the differences are genetic or environmental.

complementation test *See cis-trans* test.

composite transposon A bacterial transposon composed of two types of parts: two arms containing IS or IS-like elements, and a central region comprised of the genes for transposition and one or more antibiotic resistance genes.

concatemers DNAs of multiple genome length.

conditional lethal A mutation that is lethal under certain circumstances, but not under others (e.g., a temperature-sensitive mutation).

conditional probability The probability of an event occurring, contingent upon given circumstances.

conjugation (of bacteria) The movement of genetic material from an F^+ to an F^- bacterium.

consanguineous mating A mating between related individuals.

conservation genetics An effort to maintain genetic variation in a species by managing natural or seminatural populations or by collecting and storing diverse genetic types of a species.

conservative replication DNA (or RNA) replication in which both parental strands remain together, producing a progeny duplex, both of whose strands are new.

conservative transposition Transposition in which both strands of the transposon DNA are conserved as they leave their original location and move to a new site.

constitutive mutant An organism containing a constitutive mutation.

constitutive mutation A mutation that causes a gene to be expressed at all times, regardless of normal controls.

copia The prototype of a group of *copia*-like transposable elements found in *Drosophila*, which transpose via a retrovirus-like mechanism.

core particle The part of a nucleosome that remains after stringent DNase treatment. Contains only about 150 bp of DNA and no histone H1.

cos The cohesive ends of the linear λ phage DNA.

cosmid   A vector designed for cloning large DNA fragments. A cosmid contains the cos sites of λ phage, so it can be packaged into λ heads, and a plasmid origin of replication, so it can replicate as a plasmid.

cotransduction   Transduction of two or more genetic markers in the same virus.

crossing over   Physical exchange between homologous chromosomes that takes place in meiosis I, or during recombination in general.

crossover   The joining of corresponding DNA strands on homologous chromosomes by strand breakage and rejoining.

CRP   *See* CAP.

cyclic AMP (cAMP)   An adenine nucleotide with a cyclic phosphodiester linkage between the 3′ and 5′ carbons. Implicated in a variety of control mechanisms in prokaryotes and eukaryotes.

cytidine   A nucleoside containing the base cytosine.

cytosine (C)   The pyrimidine base that pairs with guanosine in DNA.

D

defective virus   A virus that is unable to replicate without a helper virus.

degenerate code   A genetic code, such as the one employed by all life on Earth, in which more than one codon can stand for a single amino acid.

degrees of freedom   An integer used to determine whether a chi-square value is statistically significant.

deleterious   An allele that causes some reduction in viability.

deletion   A mutation involving a loss of one or more base pairs; the case in which a chromosomal segment or gene is missing.

deletion mapping   Genetic mapping that employs deletion mutants. For example, one of the mating partners may contain a well-defined deletion, and the other an unknown point mutation. If recombination occurs, we know the point mutation lies outside the deleted area.

denaturation (DNA)   Separation of the two strands of DNA.

denaturation (protein)   Disruption of the three-dimensional structure of a protein without breaking any covalent bonds.

deoxyribose   The sugar in DNA.

determinant   A substance in an embryonic cell that contributes to determination of that cell.

determination   The irreversible commitment of a cell to differentiate along a given pathway.

dideoxyribonucleotide   A nucleotide, deoxy at both 2′ and 3′ positions, used to stop DNA chain elongation in DNA sequencing.

differentiation   Expression of the specialized characteristics of a given cell type.

dihybrid cross   A cross between two individuals that are heterozygous at two genes; for example, $AaBb \times AaBb$.

diploid   The chromosomal number of the zygote and other cells (except the gametes). Symbolized as 2N.

directional cloning   Insertion of foreign DNA into two different restriction sites of a vector, such that the orientation of the insert can be predetermined.

dispersive replication   A hypothetical mechanism in which the DNA becomes fragmented so that new and old DNA coexist in the same strand after replication.

D-loop   A loop formed when a free DNA end "invades" a double helix, base pairing with one of the strands and forcing the other to "loop out."

DNA (deoxyribonucleic acid)   A polymer composed of deoxyribonucleotides linked together by phosphodiester bonds. The material of which most genes are made.

DNA fingerprints   The use of highly variable genes (regions of DNA) to identify particular individuals.

DNA glycosylase   An enzyme that breaks the glycosidic bond between a damaged base and its sugar.

DNA gyrase   A topoisomerase that pumps negative superhelical turns into DNA. Relaxes the positive superhelical strain created by unwinding circular DNA during replication.

DNA ligase   An enzyme that joins two double-stranded DNAs end-to-end.

DNA melting   *See* denaturation (DNA).

DNA photolyase   The enzyme that carries out light repair by breaking thymine dimers.

DNA polymerase I   One of three different DNA-synthesizing enzymes in *E. coli;* used primarily in DNA repair.

DNA polymerase III holoenzyme   The enzyme within the replisome, which actually makes DNA during replication.

DNase   Deoxyribonuclease, an enzyme that degrades DNA.

dominant   An allele or trait that expresses its phenotype when heterozygous with a recessive allele; for example, *A* is dominant over *a* because the phenotypes of *AA* and *Aa* are the same. *Incomplete dominance* occurs when all genotypes, including the heterozygote, have different phenotypes.

double helix   The shape that two complementary DNA strands assume in a chromosome.

Drosophila melanogaster   A species of fruit fly used widely by geneticists.

Ds   A defective transposable element found in corn, which relies on an *Ac* element for transposition.

duplication   The situation in which a chromosomal segment or gene is represented twice.

E

electrophile   A molecule that seeks centers of negative charge in other molecules and attacks them there.

electrophoresis   A procedure in which voltage is applied to charged molecules, inducing them to migrate. This technique can be used to separate DNA fragments. It can also be used to separate and identify proteins or enzymes resulting from the presence of different alleles.

elongation factor   A protein that is necessary for either the aminoacyl-tRNA binding or the translocation step in the elongation phase of translation.

embryogenesis   Embryonic development.

endonuclease   An enzyme that makes cuts within a polynucleotide strand.

endoplasmic reticulum (ER)   Literally, "cellular network"; a network of membranes in the cell upon which proteins destined for export from the cell are synthesized.

enhancer   A DNA element that strongly stimulates transcription of a gene or genes. Enhancers are usually found upstream from the genes they influence, but they can also function if inverted or moved hundreds or even thousands of base pairs away.

enucleate egg   An egg without a nucleus.

env   The gene in a retrovirus that encodes the viral envelope (membrane) protein.

enzyme   A molecule—usually a protein, but sometimes an RNA—that catalyzes, or accelerates and directs, a biochemical reaction.

epistasis   The situation in which the alleles at one gene cover up or alter the expression of alleles at another gene.

equilibrium   A state at which there is no change in the allelic frequencies of the population.

error-prone repair A mechanism used by *E. coli* cells to replicate DNA that contains thymine dimers. DNA is replicated across from a dimer even though no base pairing is possible. This usually leads to mistakes.

Escherichia coli (E. coli) An intestinal bacterium; the favorite subject for bacterial genetics.

euchromatin Chromatin that is extended, accessible to RNA polymerase, and at least potentially active. These regions stain either lightly or normally and are thought to contain most of the genes.

eukaryote An organism whose cells have nuclei.

euploid *n.* A polyploid organism whose chromosome number is an integral multiple of the chromosome number of the organism from which it derived. *adj.* Having the normal number of chromosomes for the organism in question.

excision repair A DNA repair mechanism that removes damaged nucleotides, then replaces them with normal ones.

exon A region of a gene that is ultimately represented in that gene's mature transcript. The word refers to both the DNA and its RNA product.

exon shuffling A hypothetical process in which exons from different genes can combine to form new genes over evolutionary time.

exonuclease An enzyme that degrades a polynucleotide from the end inward.

expression site A locus on a chromosome where a gene can be moved to be expressed efficiently. For example, the expression site in trypanosomes is at the end (telomere) of a chromosome.

expression vector A cloning vector that allows expression of the cloned gene.

expressivity The level of phenotypic expression for a particular genotype.

F

F₁ The progeny of a cross between two parental types that differ at one or more genes; the first filial generation.

F₂ The progeny of a cross between two F₁ individuals or the progeny of a self-fertilized F₁; the second filial generation.

F′ An F plasmid that has picked up a piece of host DNA.

fate map A diagram that shows which parts of an embryo are destined to differentiate into each segment of the adult.

F-duction The transfer of bacterial genes from one cell to another on an F′ plasmid.

F pilus The thin projection, carried on the surface of an F⁺ bacterium, which allows bacterial cells to join during conjugation.

F plasmid The fertility plasmid that allows a donor bacterium to conjugate with a recipient.

fertility plasmid *See* F plasmid.

fine structure mapping Extensive genetic mapping within a gene, ideally down to the nucleotide level.

fingerprint (protein) The specific pattern of peptide spots formed when a protein is cut into pieces (peptides) with an enzyme (e.g., trypsin), and then the peptides are separated by chromatography.

5′-end The end of a polynucleotide with a free (or phosphorylated or capped) 5′-hydroxyl group.

fluctuation test A test to determine whether a given microbial trait is caused by a mutation. If it is, individual cultures should vary widely in the number of cells showing the trait, since mutation is a rare and random event and may occur early in the growth of some cultures and later in others. The test can also be used to measure mutation rates.

fMet *See* N-formyl methionine.

forced cloning *See* directional cloning.

forked-line approach A technique used to determine progeny types for multiple genes, based on independent assortment between the genes.

founder effect The chance changes in allelic frequency resulting from the initiation of a population by a small number of individuals.

frameshift mutation An insertion or deletion of one or two bases in the coding region of a gene, which changes the reading frame of the corresponding mRNA.

fushi tarazu (ftz) A homeobox-containing gene in *Drosophila* that helps define segmentation in the early embryo.

fusion protein A protein resulting from the expression of a recombinant DNA containing two open reading frames (ORFs) fused together. One or both of the ORFs can be incomplete.

G

gag The gene in a retrovirus that encodes the viral coat proteins.

gamete A haploid sex cell.

GC box A hexamer having the sequence GGGCGG on the coding strand, which occurs in a number of mammalian structural gene promoters. The binding site for the transcription factor Sp1.

gene The basic unit of heredity. Contains the information for making one RNA and, in most cases, one polypeptide.

gene cluster A group of related genes grouped together on a eukaryotic chromosome.

gene conversion The conversion of one gene's sequence to that of another.

gene expression The process by which gene products are made.

gene flow Movement of individuals from one population to another that results in the introduction of migrant alleles through mating and subsequent reproduction.

gene mutation A mutation confined to a single gene.

gene pool The total variety and number of alleles within a population or species.

generalized recombination Recombination that requires extensive sequence similarity between the recombining DNAs.

generalized transduction Use of P1 or other phage heads to carry bacterial genes from one cell to another.

genetic code The set of sixty-four codons and the amino acids (or terminations) they stand for.

genetic counseling Counseling based on the risks of producing genetically defective infants, usually given where some familial history of a genetic disease exists.

genetic drift The chance changes in allelic frequency that result from the sampling of gametes from generation to generation.

genetic marker A mutant gene or other peculiarity in a genome that can be used to "mark" a spot in a genome for mapping purposes.

genetic recombination *See* recombination.

genetic variance The variation of a phenotype in a population that results from genetic causes, symbolized by V_G. The extent of the additive genetic variance, symbolized by V_A, is important in determining the rate and amount of response to directional selection.

genome One complete set of genetic information from a genetic system; e.g., the single, circular chromosome of a bacterium is its genome.

genomic library A set of clones containing DNA fragments derived directly from a genome, rather than from mRNA.

genotype   The allelic constitution of a given individual. The genotypes at locus A may be *AA, Aa,* or *aa,* assuming only two possible alleles at this locus.

genotype-environment interaction   The effect on a phenotype value that results from a specific genotype and a specific environment and that is not predictable from either separately. Symbolized by GE_{ij}.

germinal mutation   *See* germ-line mutation.

germ-line mutation   A mutation that affects the sex cells, enabling it to be passed on to progeny.

Golgi apparatus   A membranous organelle that packages newly synthesized proteins for export from the cell.

G protein   A protein that binds GTP, which activates the protein to perform a function. When the G protein cleaves the GTP to GDP, it becomes inactivated until a new molecule of GTP binds.

guanine (G)   The purine base that pairs with cytosine in DNA.

guanosine   A nucleoside containing the base guanine.

H

hairpin   A structure resembling a hairpin (bobby pin), formed by intramolecular base pairing in an inverted repeat of a single-stranded DNA or RNA.

half chiasma   *See* chi structure.

half-life   The time it takes for half of a population of molecules to disappear.

haploid   The chromosomal number in the gamete. Symbolized as N.

haplotype   A set of linked genes or DNA sequences that tend to be inherited together.

Hardy-Weinberg principle   The principle stating that after one generation of random mating, single-locus genotypic frequencies can be represented as a function of the allelic frequencies.

helicase   An enzyme that unwinds the DNA double helix.

helix-turn-helix   A structural motif in certain DNA-binding proteins, especially those from prokaryotes, that fits into the DNA wide groove and gives the protein its binding capacity and specificity.

helper virus (or phage)   A virus that supplies functions lacking in a defective virus, allowing the latter to replicate.

hemoglobin   The red, oxygen-carrying protein in the red blood cells.

heritability   The proportion of phenotypic variance that is the additive genetic component. Symbolized as h^2. Realized heritability is an estimate of h^2 from a selection experiment.

hermaphrodite   An animal that has both male and female reproductive organs.

heterochromatin   Chromatin that is condensed, inactive, and usually darkly staining.

heteroduplex   A double-stranded polynucleotide whose two strands are not completely complementary.

heterogeneous nuclear RNA (hnRNA)   A class of large, heterogeneous-sized RNAs found in the nucleus, including unspliced mRNA precursors.

heterokaryon   A cell containing two or more nuclei of different origins.

heterokaryotype   Individuals that are heterozygous for two chromosomal types.

heterosis   Higher yield (or other phenotypic value) for F_1 individuals compared to the mean of the parents.

heterozygosity   The proportion of heterozygotes in a population.

heterozygote   A diploid genotype in which the two alleles for a given gene are different; for example, $A_1 A_2$.

heterozygote advantage   The situation in which the heterozygote has a higher fitness than either homozygote, leading to a stable equilibrium.

Hfr   A strain of *E. coli* that experiences a high frequency of recombination because of integration of the F plasmid into the bacterial chromosome, which mobilizes the chromosome.

histones   A class of five small, basic proteins intimately associated with DNA in most eukaryotic chromosomes.

hnRNA   *See* heterogeneous nuclear RNA.

homeobox   A sequence of about 180 base pairs found in homeotic genes and other development-controlling genes in *Drosophila* and other eukaryotes. Thought to encode a DNA-binding region of a protein.

homeotic gene   A gene in which a mutation causes the transformation of one body part into another. Sometimes called a **homoeotic** gene.

homologous chromosomes   Chromosomes that are identical in size, shape, and, except for allelic differences, in genetic composition.

homologous recombination   *See* generalized recombination.

homozygote   A diploid genotype in which both alleles for a given gene are identical; for example, $A_1 A_1$ or *aa*.

host range   The set of hosts that can be productively infected by a virus (or phage).

hot spots   Highly mutable sites within a genome—sites that accumulate more than the average number of mutations.

housekeeping genes   Genes that code for proteins used by all kinds of cells.

human immunodeficiency virus (HIV)   The agent that is thought to cause acquired immune deficiency syndrome (AIDS).

hybrid cell   A cell created by fusing two dissimilar cells. First, a heterokaryon forms; then, after nuclear fusion, a hybrid cell.

hybrid dysgenesis   A phenomenon observed in *Drosophila* in which the hybrid offspring of two certain parental strains suffer so much chromosomal damage that they are sterile, or dysgenic.

hybridization (of polynucleotides)   Forming a double-stranded structure from two polynucleotide strands (either DNA or RNA) from different sources.

hybrid polynucleotide   The product of polynucleotide hybridization.

hybrid vigor   *See* heterosis.

hyperchromic shift   The increase in a DNA solution's absorbance of 260 nm light upon denaturation.

hyphae   Strandlike chains of fungal cells.

I

illegitimate recombination   Recombination that requires little if any sequence similarity between the recombining DNAs and is therefore not site-specific.

imaginal disk   A group of insect larval cells that give rise to a body part, such as a wing, leg, or antenna.

immunity region   The control region of a λ or λ-like phage, containing the gene for the repressor as well as the operators recognized by this repressor.

immunoglobulin (antibody)   A protein that binds very specifically to an invading substance and alerts the body's immune defenses to destroy the invader.

inbreeding   Nonrandom mating in which the mating individuals are more closely related than individuals drawn by chance from the population.

inbreeding coefficient   The probability that the two homologous alleles in an individual are identical by descent. Symbolized as f.

inbreeding depression   The reduction of population fitness due to inbreeding.

in cis   A condition in which two genes are located on the same chromosome.

incision Nicking a DNA strand with an endonuclease.

incomplete dominance *See* dominant.

independent assortment A principle discovered by Mendel, which states that genes on different chromosomes behave independently.

independent culling levels Artificial selection in which animals must exceed a given value for each of several traits or be culled.

inducer A substance that releases negative control of an operon.

inducer (embryonic) A substance that one embryonic cell makes to influence the development of another cell.

initiation factor (IF) A protein that helps catalyze the initiation of translation.

inosine (I) A nucleoside containing the base hypoxanthine, which base-pairs with cytosine.

insertion sequence (IS) A simple type of transposon found in bacteria, containing only inverted terminal repeats and the genes needed for transposition.

interference The difference between the observed number of double recombinants and the number expected when recombination in one area does not influence that in another area.

interphase The stage of the cell cycle during which DNA is synthesized but the chromosomes are not visible.

intervening sequence (IVS) *See* intron.

intracistronic complementation Complementation of two mutations in the same gene. Can occur by cooperation among different defective monomers to form an active oligomeric protein.

in trans A condition in which two genes are located on separate chromosomes.

intron A region that interrupts the transcribed part of a gene. An intron is transcribed but is removed by splicing during maturation of the transcript. The word refers to the intervening sequence in both the DNA and its RNA product.

inversion An alteration in the sequence of genes in a chromosome. A pericentric inversion includes the centromere, while a paracentric inversion does not.

inverted repeat A symmetrical sequence of DNA, reading the same forward on one strand and backward on the opposite strand. For example:

GGATCC
CCTAGG

isoschizomers Two or more restriction endonucleases that recognize and cut the same restriction site.

J

joining region (J) The segment of an immunoglobulin gene encoding the last thirteen amino acids of the variable region. One of several joining regions is joined by a chromosomal rearrangement to the rest of the variable region, introducing extra variability into the gene.

K

karyotype A pictorial or photographic representation of all the chromosomes in a given individual.

Klenow fragment A fragment of DNA polymerase I, created by cleaving with a protease, that lacks the $5'{\rightarrow}3'$ exonuclease activity of the parent enzyme.

L

lac **operon** The operon encoding enzymes that permit a cell to metabolize the milk sugar lactose.

lariat The name given the lasso-shaped intermediate in certain splicing reactions.

leader A sequence of untranslated bases at the 5'-end of an mRNA.

leading strand The strand that is made continuously in semidiscontinuous DNA replication.

lethal An allele that causes mortality. A recessive lethal causes mortality only when homozygous.

light repair Direct repair of a thymine dimer by the enzyme DNA photolyase.

linkage The physical association of genes on the same chromosome.

linkage map A way to visualize the physical relationship of the genes on a chromosome in a linear arrangement.

locus (loci, pl.) The position of a gene on a chromosome, used synonymously with the term "gene" in many instances.

long terminal repeats (LTRs) Regions of several hundred base pairs of DNA found at both ends of the provirus of a retrovirus.

luxury genes Genes that code for specialized cell products.

Lyonized chromosome An X chromosome in a female mammal that is composed entirely of heterochromatin and is genetically inactive.

lysis Rupturing the membrane of a cell, as by a virulent phage.

lysogen A bacterium harboring a prophage.

M

map unit (centimorgan) The distance separating two genetic loci that recombine with a frequency of 1%.

maternal effect A phenotype governed by the genotype, not the phenotype, of the maternal parent, and unaffected by genes from the paternal parent.

maternal effect genes Genes active in an oocyte, which define a phenotype that is unaffected by genes from the male gamete.

maternal inheritance A nonMendelian pattern of inheritance that always depends only on the maternal parent. The paternal parent does not influence the F_1 or any succeeding generation.

maternal message An mRNA made in great quantity during oogenesis, then stored, complexed with protein in the egg cytoplasm.

mean The arithmetic average.

mean fitness The sum of the product of the relative fitness and the frequency of the genotypes in the population. Symbolized as w.

meiosis Cell division that produces gametes (or spores) having half the number of chromosomes of the parental cell.

merodiploid A bacterium that is only partially diploid, meaning diploid with respect to only some of its genes.

merozygote *See* merodiploid.

mesoderm The middle layer of embryonic cells, which develops into connective tissue, muscle, and blood, among other tissues.

message *See* mRNA.

messenger RNA *See* mRNA.

metacentric Chromosome with the centromere in the middle.

metal finger *See* zinc finger.

metaphase Intermediate stage of nuclear division in which the chromosomes move to the equatorial plate.

microconidium The small, male gamete of the mold *Neurospora*.

minimal medium A growth medium for microbes that contains only the simple substances (salts and an energy source such as glucose) needed for growth of prototrophs.

mismatch repair The correction of a mismatched base incorporated by accident—in spite of the proofreading system—into a newly synthesized DNA.

missense mutation A change in a codon that results in an amino acid change in the corresponding protein.

mitosis   Cell division that produces two daughter cells having nuclei identical to the parental cell.

modifier gene   A gene that modifies the phenotype of another gene.

molecular clock   A hypothesis suggesting that a regular turnover or replacement of molecular variants (amino acid or nucleotide substitution) takes place over time.

monohybrid cross   A cross between two individuals that are heterozygous at one gene; for example, $Aa \times Aa$.

monomorphic   The situation in which all individuals in a population are the same genetic type or have the same allele.

monosomy   Diploids having one missing chromosome ($2n - 1$).

morphological mutation   A mutation that affects the morphology, or appearance, of an organism.

mosaic embryo   An embryo in which the early blastomeres are different from one another.

mRNA (messenger RNA)   A transcript that bears the information for making one or more proteins.

multiple cloning site (MCS)   A region in certain cloning vectors, such as the pUC plasmids and M13 phage DNAs, that contains several restriction sites in tandem. Any of these can be used for inserting foreign DNA.

multiplicity of infection (m.o.i.)   The ratio of infectious virus (or phage) particles to infectable cells in an experiment.

mutagen   A mutation-causing agent.

mutant   An organism (or genetic system) that has suffered at least one mutation.

mutation   The original source of genetic variation, caused by a change in a DNA base or a chromosome, for example. Spontaneous mutations are those that appear without explanation, while induced mutations are those attributed to a particular mutagenic agent.

mutation rate   The rate of mutational change from one allelic form to another. Symbolized by u or v.

mutation-selection balance   The equilibrium that results when mutation introduces detrimental alleles into a population and selection eliminates them.

mutator strain   A strain of bacteria that makes more than the usual number of mistakes during DNA replication, yielding a higher than normal mutation rate.

mutually exclusive events   Events such that if one occurs, the other cannot; for example, heads and tails on a coin.

myc   A cellular oncogene, encoding a nuclear protein, that is rearranged in human Burkitt's lymphoma.

N

natural selection   *See* selection.

negative control   A control system in which gene expression is turned off unless a controlling element (e.g., a repressor) is removed.

Neurospora crassa   A common bread mold, developed by Beadle and Tatum as a subject for genetic investigation.

neutrality theory   The theory used to explain the existence of molecular variation when there is no differential selection (alleles are neutral with respect to each other) and when new variants are introduced into the population by mutation and eliminated by genetic drift.

N-formyl methionine (fMet)   The initiating amino acid in prokaryotic translation.

nick   A single-stranded break in DNA.

nondisjunction   Abnormal separation of chromosomes in meiosis (or mitosis).

nonsense mutation   A mutation that creates a premature stop codon within a gene's coding region. Includes amber mutations (UAG), ochre mutations (UAA), and opal mutations (UGA).

norm of reaction   The phenotypic pattern for a given genotype produced under different environmental conditions.

Northern blotting   Transferring RNA fragments to a support medium. (*See* Southern blotting.)

nucleic acid   A chainlike molecule (DNA or RNA) composed of nucleotide links.

nucleocapsid   A structure containing a viral genome (DNA or RNA) with a coat of protein.

nucleoid   A nonmembrane-bound area of a bacterial cell in which the chromosome resides.

nucleolus   A cell organelle found in the nucleus that disappears during part of cell division. Contains the rRNA genes.

nucleoside   A base bound to a sugar—either ribose or deoxyribose.

nucleosome   A repeating structural element in eukaryotic chromosomes, composed of a core of eight histone molecules with about 200 base pairs of DNA wrapped around the outside.

nucleotide   The subunit, or chainlink, in DNA and RNA, composed of a sugar, a base, and at least one phosphate group.

nutritional mutation   A mutation that makes an organism dependent on a substance, such as an amino acid or vitamin, that it previously could make for itself.

O

ochre codon   UAA, coding for termination.

ochre mutation   *See* nonsense mutation.

ochre suppressor   A tRNA bearing an anticodon that can recognize the ochre codon (UAA) and thereby suppress ochre mutations.

Okazaki fragment   Small DNA fragments, 1,000–2,000 bases long, created by discontinuous synthesis of the lagging strand.

oligomeric protein   A protein that contains more than one polypeptide subunit.

oligonucleotide   A short piece of RNA or DNA.

oncogene   A gene that participates in causing tumors.

oncogenesis   The formation of tumors.

oncogenic   Tumor-causing.

oocyte 5S rRNA genes   The 5S rRNA genes (haploid number about 19,500 in *Xenopus laevis*) that are expressed only in oocytes.

oogenesis   Development of an egg.

opal codon   UGA, coding for termination.

opal mutation   *See* nonsense mutation.

opal suppressor   A tRNA bearing an anticodon that can recognize the opal codon (UGA) and thereby suppress opal mutations.

open reading frame (ORF)   A reading frame that is uninterrupted by translation stop signals.

operator   A DNA element found in prokaryotes that binds tightly to a specific repressor and thereby regulates the expression of adjoining genes.

operon   A group of contiguous, coordinately controlled genes.

ORF   *See* open reading frame.

origin of replication   The unique spot in a replicon where replication begins.

outbreeding   Nonrandom mating in which the mating individuals are less closely related than individuals drawn by chance from the population.

P

pair-rule genes   Genes that, when mutated, give rise to insects missing segment-sized body parts at two-segment intervals.

palindrome   *See* inverted repeat.

paracentric inversion   *See* inversion.

parthenogenesis   Reproduction in which organisms produce offspring identical to themselves without fertilization.

paternity exclusion   The process of using genetic markers to exclude given men as fathers of a particular child.

pBR322 One of the original plasmid vectors for gene cloning.

pedigree A family tree illustrating the inheritance of particular genotypes or phenotypes.

penetrance The proportion of individuals with a given genotype that express it at the phenotypic level.

peptide bond The bond linking amino acids in a protein.

peptidyl transferase An enzyme that is an integral part of the large ribosomal subunit and catalyzes the formation of peptide bonds during protein synthesis.

pericentric inversion *See* inversion.

phage A bacterial virus.

phenocopy Occurs when environmental factors induce a particular abnormal phenotype that resembles a genetically determined phenotype.

phenotype The morphological, biochemical, behavioral, or other properties of an organism. Often only a particular trait of interest, such as weight, is considered.

phosphodiester bond The sugar-phosphate bond that links the nucleotides in a nucleic acid.

photoreactivating enzyme *See* DNA photolyase.

photoreactivation *See* light repair.

phylogenetic tree A diagram that organizes the relationship between species or populations, putatively indicating ancestral relationships.

physical map A genetic map based on physical characteristics of the DNA, such as restriction sites, rather than on locations of genes.

plaque A hole that a virus (or phage) makes on a layer of host cells by infecting and killing the cells.

plaque assay An assay for virus (or phage) concentration in which the number of plaques produced by a given dilution of virus is determined.

plaque-forming unit (pfu) A virus (or phage) capable of forming a plaque in a plaque assay.

plaque hybridization A procedure for selecting a phage clone that contains a gene of interest. DNAs from a large number of phage plaques are simultaneously hybridized to a labeled probe for the gene of interest.

plasmid A circular DNA that replicates independently of the cell's chromosome.

pleiotropic mutation A mutation that affects the expression of several other genes.

pleiotropy The situation in which a single gene affects two or more characteristics.

point mutation An alteration of one, or a very small number, of contiguous bases.

pol The gene in a retrovirus that encodes the viral reverse transcriptase and RNase H.

poly(A) Polyadenylic acid. The string of about two hundred A's added to the end of a typical eukaryotic mRNA.

poly(A) polymerase The enzyme that adds poly(A) to an mRNA or to its precursor.

polycistronic message An mRNA bearing information from more than one gene.

polygenic trait A phenotypic trait determined by a number of genes.

polymorphism The existence of two or more genetically determined forms (alleles) in a population in substantial frequency. In practice, a polymorphic gene is one at which the frequency of the most common allele is less than 0.99.

polynucleotide A polymer composed of nucleotide subunits; DNA or RNA.

polypeptide A single protein chain.

polyploids Organisms with three or more sets of chromosomes. Autopolyploids receive all their chromosomes from the same species, while allopolyploids obtain their chromosomes from two (or more) species.

polytene chromosomes Large chromosomes in diptera that contain many parallel replicates tightly held together.

population A group of individuals that exist together in time and space, and that can interbreed.

positive control A control system in which gene expression depends on the presence of a positive effector such as CAP (and cAMP).

positive strand The strand of a simple viral genome with the same sense as the viral mRNAs.

postreplication repair *See* recombination repair.

posttranscriptional control Control of gene expression that occurs during the posttranscriptional phase when transcripts are processed by splicing, clipping, and modification.

posttranslational modification The set of changes that occur in a protein after it is synthesized.

pp60^{src} The product of the *src* gene.

preinitiation complex The complex of transcription factors and RNA polymerase assembled at a promoter prior to initiation of transcription.

Pribnow box (–10 box) An *E. coli* promoter element centered about ten bases upstream from the start of transcription.

primary structure The sequence of amino acids in a polypeptide, or of nucleotides in a DNA or RNA.

primase The enzyme within the primosome that actually makes the primer.

primer A small piece of RNA that provides the free end needed for DNA replication to begin.

primosome A complex of about twenty polypeptides, which makes primers for *E. coli* DNA replication.

probability The proportion of the time that a particular event occurs.

processed pseudogene A pseudogene that has apparently arisen by retrotransposonlike activity: transcription of a normal gene, processing of the transcript, reverse transcription, and reinsertion into the genome.

processing (of RNA) The group of cuts that occur in RNA precursors during maturation, including splicing, 5'- or 3'-end clipping, or cutting rRNAs out of a large precursor.

product rule A rule stating that the probability of both of two independent events occurring is the product of their individual probabilities.

prokaryotes Microorganisms that lack nuclei. Synonymous with bacteria, including cyanobacteria (blue-green algae).

promoter The part of a gene to which RNA polymerase binds prior to initiation of transcription—usually found just upstream from the coding region of a gene.

proofreading The process of checking each nucleotide for complementarity with its base-pairing partner as it is inserted during DNA replication.

prophage A phage genome integrated into the host's chromosome.

prophase Early stage of nuclear division in which chromosomes coil, condense, and become visible.

protein A polymer, or polypeptide, composed of amino acid subunits. Sometimes the term "protein" denotes a functional collection of more than one polypeptide (e.g., the hemoglobin protein consists of four polypeptide chains).

proteolytic processing Cleavage of a protein into pieces.

proto-oncogene A cellular counterpart of a viral oncogene. Proto-oncogenes do not ordinarily participate in oncogenesis unless they are overexpressed or mutated.

protoperithecium   The large, female gamete of the mold *Neurospora*.

prototroph   An organism that can grow on minimal medium containing only salts and an energy source.

provirus   A double-stranded DNA copy of a retroviral RNA that inserts into the host genome.

pseudogene   A nonallelic copy of a normal gene that is mutated so that it cannot function.

pulse-chase   The process of giving a short period, or "pulse," of radioactive precursor so that a substance such as RNA becomes radioactive, then adding an excess of unlabeled precursor to "chase" the radioactivity out of the substance.

pure-breeding line   Identical individuals that always produce progeny like themselves.

purine   The parent base of guanine and adenine.

puromycin   An antibiotic that resembles an aminoacyl-tRNA and kills bacteria by forming a peptide bond with a growing polypeptide and then releasing the incomplete polypeptide from the ribosome.

pyrimidine   The parent base of cytosine, thymine, and uracil.

Q

quantitative trait   A trait that generally has a continuous distribution in a population and is usually affected by many genes and many environmental factors.

quaternary structure   The way two or more polypeptides interact in a complex protein.

quenching   Quickly chilling heat-denatured DNA to keep it denatured.

R

random mating population   A group of individuals in which the probability of members mating with individuals of particular types is equal to their frequency in the population.

ras   A family of oncogenes, found in several human tumors, that encode a protein called p21, which resembles a G protein.

rate of allelic substitution   The rate of replacement of alleles in a population. If amino acid or nucleotide sequences are known, the rate of amino acid or nucleotide substitution can be calculated.

reading frame   One of three possible ways the triplet codons in an mRNA can be translated. For example, the message CAGUGCUCGAC has three possible reading frames, depending on where translation begins: (1) CAG UGC UCG; (2) AGU GCU CGA; (3) GUG CUC GAC. A natural mRNA generally has only one correct reading frame.

recessive   An allele or trait that does not express its phenotype when heterozygous with a dominant allele; for example, *a* is recessive to *A* because the phenotype for *Aa* is like *AA* and not like *aa*.

reciprocal matings   Crosses that are identical genetically except that the male and female parents are switched.

recombinant DNA   The product of recombination between two (or more) fragments of DNA. Can occur naturally in a cell, or be fashioned by geneticists in vitro.

recombination   Reassortment of genes or alleles in new combinations. Technically, recombination can occur by either independent assortment or by crossing over between chromosomes. However, the term usually refers to recombination by crossing over.

recombination repair   A mechanism that *E. coli* cells use to replicate DNA containing thymine dimers. First, the two strands are replicated, leaving a gap across from the dimer. Next, recombination between the progeny duplexes places the gap across from normal DNA. Finally, the gap is filled in using the normal DNA as template.

recurrence risk   The probability that a given genetic disease will recur in a family, given that one individual (or more) in the family has already exhibited the disease.

regulative embryo   An embryo in which the early blastomeres are indistinguishable from one another.

regulatory gene   A gene that encodes a protein that regulates other genes.

relative fitness   The relative ability of different genotypes to pass on their alleles to future generations.

release factor   A protein that causes termination of translation at stop codons.

renaturation of DNA   *See* annealing of DNA.

repetitive DNA   DNA sequences that are repeated many times in a haploid genome.

rep helicase   The product of the *E. coli rep* gene; thought to be involved in unwinding the *E. coli* DNA during replication.

replicating fork   The point where the two parental DNA strands separate to allow replication.

replicative transposition   Transposition in which the transposon DNA replicates, so that one copy remains in the original location as another copy moves to the new site.

replicon   All the DNA replicated from one origin of replication.

replisome   The large complex of polypeptides, including the primosome, which replicates DNA in *E. coli*.

resolution   The second step in transposition through a cointegrate intermediate; it involves separation of the cointegrate into its two component replicons, each with its own copy of the transposon.

resolvase   The enzyme that catalyzes resolution of a cointegrate.

res sites   Sites on the two copies of a transposon in a cointegrate, between which a crossover occurs to accomplish resolution.

restriction endonuclease   An enzyme that recognizes specific base sequences in DNA and cuts at or near those sites.

restriction fragment   A piece of DNA cut from a larger DNA by a restriction endonuclease.

restriction fragment length polymorphism (RFLP)   A variation from one individual to the next in the number of cut sites for a given restriction endonuclease in a given locus. This results in a variation in the lengths of restriction fragments generated by that enzyme cutting within that locus. These variable lengths, which are really alleles, can be treated like any other genetic marker and used in mapping studies.

restriction site   A sequence of bases recognized and cut by a restriction endonuclease.

retrotransposon   Transposable elements such as *copia* and Ty that transpose via a retroviruslike mechanism.

retrovirus   An RNA virus whose replication depends on formation of a provirus by reverse transcription.

reverse transcriptase   RNA-dependent DNA polymerase; the enzyme that catalyzes reverse transcription, commonly found in retroviruses.

reverse transcription   Synthesis of a DNA using an RNA template.

reversion   A mutation that cancels the effects of an earlier mutation in the same gene.

RF (replicative form) The circular double-stranded form of the genome of a single-stranded DNA phage such as φX174. The DNA assumes this form in preparation for rolling circle replication.

RFLP *See* restriction fragment length polymorphism.

rho (ρ) A protein that is needed for transcription termination at certain terminators in *E. coli*.

ribonuclease H (RNase H) An enzyme that degrades the RNA part of an RNA-DNA hybrid.

ribose The sugar in RNA.

ribosome An RNA-protein particle that translates mRNAs to produce proteins.

ribozyme An RNA with enzymatic (catalytic) activity.

RNA (ribonucleic acid) A polymer composed of ribonucleotides linked together by phosphodiester bonds.

RNA-dependent DNA polymerase *See* reverse transcriptase.

RNA polymerase The enzyme that directs transcription, or synthesis of RNA.

RNA polymerase I The eukaryotic enzyme that makes the large ribosomal RNA precursors.

RNA polymerase II The eukaryotic enzyme that makes mRNA precursors (hnRNA).

RNA polymerase III The eukaryotic enzyme that makes tRNA, 5S rRNA, and a few other small RNAs.

RNA polymerase core The collection of subunits of a prokaryotic RNA polymerase having basic RNA chain elongation capacity but no specificity of initiation; all the RNA polymerase subunits except the σ factor.

rolling circle replication A mechanism of replication in which one strand of a double-stranded circular DNA remains intact and serves as the template for elongation of the other strand at a nick.

rRNA (ribosomal RNA) The RNA molecules contained in ribosomes.

S

screen A genetic sorting procedure that distinguishes desired organisms from unwanted ones, but does not automatically remove the latter.

secondary structure The local folding of a polypeptide or RNA. In the case of RNA, the secondary structure is defined by intramolecular base pairing.

sedimentation coefficient A measure of the rate at which a molecule or particle travels toward the bottom of a centrifuge tube under the influence of a centrifugal force.

segmentation genes Genes that establish the segmentation pattern of an animal, especially an insect.

segregation A principle discovered by Mendel, which states that heterozygotes produce equal numbers of gametes having the two different alleles.

selection The differential production of progeny genotypes or alleles because of differences in survival, mating, or fertility among the parental genotypes or their progeny. *Natural selection* occurs when these differences are due to intrinsic differences in the genotypes. *Artificial selection* occurs when these differences result because certain types are favored by humans in plant or animal breeding.

selection coefficient *See* selective disadvantage.

selection limit The point beyond which artificial selection is unable to change the mean phenotypic value.

selective disadvantage The reduction in relative fitness of genotype A_2A_2 as compared with that of A_1A_1; also known as the selection coefficient, symbolized by s.

self-fertilization Fertilization of eggs with sperm or pollen from the same individual. Sometimes called **selfing.**

semiconservative replication DNA replication in which the two strands of the parental duplex separate completely and pair with new progeny strands. One parental strand is therefore conserved in each progeny duplex.

semidiscontinuous replication A mechanism of DNA replication in which one strand is made continuously and the other is made discontinuously.

sequencing Determining the amino acid sequence of a protein, or the base sequence of a DNA or RNA.

sex-duction *See* F-duction.

sex-influenced Traits that are expressed differently in the two sexes.

sex-limited Traits that are only expressed in one sex.

sex-linked Alleles that are on the sex chromosomes.

Shine-Dalgarno (SD) sequence A G-rich sequence (consensus = AGGAGGU) that is complementary to a sequence at the 3′-end of *E. coli* 16S rRNA. Base pairing between these two sequences helps the ribosome bind an mRNA.

shuttle vector A cloning vector that can replicate in two or more different hosts, allowing the recombinant DNA to shuttle back and forth between hosts.

sickle-cell disease A genetic disease in which abnormal β-globin is produced. Because of a single amino acid change, this blood protein tends to aggregate under low oxygen conditions, distorting red blood cells into a sickle shape.

sigma (σ) The prokaryotic RNA polymerase subunit that confers specificity of transcription—that is, the ability to recognize specific promoters.

signal peptide A stretch of about twenty amino acids, usually at the amino terminus of a polypeptide, that helps to anchor the nascent polypeptide and its ribosome in the endoplasmic reticulum. Polypeptides with a signal peptide are destined for packaging in the Golgi apparatus and are usually exported from the cell.

silent mutation A mutation that causes no detectable change in an organism, even in a haploid organism or in a homozygote.

single-copy DNA DNA sequences that are present once, or only a few times, in a haploid genome.

site-directed mutagenesis A method for introducing specific, predetermined alterations into a cloned gene.

site-specific recombination Recombination that always occurs by crossovers in the same place and depends on limited sequence similarity between the recombining DNAs.

small nuclear RNAs (snRNAs) A class of nuclear RNAs a few hundred nucleotides in length. These RNAs, together with tightly associated proteins, make up snRNPs (**small nuclear ribonucleoproteins**) that participate in splicing, polyadenylation, and 3′-end maturation of transcripts.

somatic 5S rRNA genes The 5S rRNA genes (haploid number about 400 in *Xenopus laevis*) that are expressed in both somatic cells and oocytes.

somatic cell genetics Genetic manipulations involving fusing somatic cells instead of mating organisms.

somatic cell hybridization Fusion of cells from two different species; used to determine the chromosomal location of genes

somatic cell Nonsex cell.

somatic mutation A mutation that affects only somatic cells, so it cannot be passed on to progeny.

SOS response The activation of a group of genes, including *recA*, that helps *E. coli* cells respond to environmental insults such as chemical mutagens or radiation.

Southern blotting Transferring DNA fragments separated by gel electrophoresis to a suitable support medium such as nitrocellulose, in preparation for hybridization to a labeled probe.

spacer DNA DNA sequences found between, and sometimes within, repeated genes such as rRNA genes.

specialized transduction Use of λ or a similar phage to carry bacterial genes from one cell to another.

species A reproductively isolated group that generally has certain characteristics, such as morphology, behavior or molecular variants, which distinguish it from other species.

spliceosome The large RNA-protein body upon which splicing of nuclear mRNA precursors occurs.

splicing The process of linking together two exons while removing the intron that lies between them.

spore (1) A specialized haploid cell formed sexually by plants or fungi, or asexually by fungi. The latter can either serve as a gamete or germinate to produce a new haploid cell. (2) A specialized cell formed asexually by certain bacteria in response to adverse conditions. Such a spore is relatively inert and resistant to environmental stress.

sporulation Formation of spores.

src The oncogene of Rous sarcoma virus (RSV), which encodes a tyrosine protein kinase.

SSB Single-strand-binding protein, used during DNA replication. Binds to single-stranded DNA and keeps it from base pairing with a complementary strand.

standard deviation The square root of the variance.

statistically significant A chi-square value (or other statistic) large enough to establish that the results are not consistent with the hypothesis used.

sticky ends Single-stranded ends of double-stranded DNAs that are complementary and can therefore base-pair and stick together.

stop codon One of three codons (UAG, UAA, and UGA) that code for termination of translation.

streptomycin An antibiotic that kills bacteria by causing their ribosomes to misread mRNAs.

stringency (of hybridization) The combination of factors (temperature, salt and organic solvent concentration) that influence the ability of two polynucleotide strands to hybridize. At high stringency, only perfectly complementary strands will hybridize. At reduced stringency, some mismatches can be tolerated.

structural genes Genes that code for proteins.

sum rule A rule stating that the combined probability of two (or more) mutually exclusive events occurring is the sum of their individual probabilities.

supercoil *See* superhelix.

superhelix A form of circular double-stranded DNA in which the double helix coils around itself like a twisted rubber band.

suppressor mutation A mutation that reverses the effects of a mutation in the same or another gene.

SV40 Simian virus 40; a DNA tumor virus with a small circular genome, capable of causing tumors in certain rodents.

synapsis The process in meiosis I in which homologous chromosomes pair side by side and gene by gene with each other.

T

TAFs *See* TBP-associated factors.

TATA box An element with the consensus sequence TATAAAA, found about 25 base pairs upstream from the start of transcription in most eukaryotic promoters recognized by RNA polymerase II.

TATA box-binding protein (TBP) The polypeptide that serves as an organizing center in forming all three classes of eukaryotic preinitiation complexes. Binds to the TATA box of a promoter, if present. Part of the class II general factor TFIID.

tautomeric shift The reversible change of one isomer of a DNA base to another by shifting the locations of hydrogen atoms and double bonds.

TBP *See* TATA box-binding protein.

TBP-associated factors (TAFs) Polypeptides that associate with TBP in general transcription factors of each class (SL1, TFIID, and TFIIIB) and give these factors their promoter-specificity.

T-DNA The tumor-inducing part of the Ti plasmid.

telocentric Chromosome with the centromere very near one end.

telophase The last stage of nuclear division in which the nuclear membrane forms and encloses the chromosomes in the daughter cells.

temperate phage A phage that can enter a lysogenic phase in which a prophage is formed.

temperature-sensitive mutation A mutation that causes a defective product to be made at high temperature (the restrictive temperature) but yields a functional product at low temperature (the permissive temperature).

template A polynucleotide (RNA or DNA) that serves as the guide for making a complementary polynucleotide. For example, a DNA strand serves as the template for ordinary transcription.

terminal transferase An enzyme that adds deoxyribonucleotides, one at a time, to the 3'-end of a DNA.

terminator *See* transcription terminator.

tertiary structure The overall three-dimensional shape of a polypeptide or RNA.

testcross A cross between an F_1 individual (or an individual of unknown genotype) and the recessive parent of the F_1 (the tester).

tetrad The four products of meiosis synapsed during meiosis. In certain organisms, such as *Neurospora crassa*, these products are linearly arranged in four pairs of ascospores.

tetrad analysis The use of the four meiotic products from a single meiosis (a tetrad) to determine the behavior of genes in meiosis.

TFIIIA A transcription factor that helps activate the vertebrate 5S rRNA genes.

thalassemia A genetic disease marked by failure to produce a functional mRNA for one of the two major adult hemoglobin proteins, α-globin or β-globin.

3'-end The end of a polynucleotide with a free (or phosphorylated) 3'-hydroxyl group.

three-point cross Eukaryotic: A testcross in which one parent is heterozygous for three genes. Prokaryotic: A cross in which the two parental genotypes differ at three genes.

thymidine A nucleoside containing the base thymine.

thymine (T) The pyrimidine base that pairs with adenine in DNA.

thymine dimer Two adjacent thymines in one DNA strand linked covalently, interrupting their base pairing with adenines in the opposite strand. This is the main genetic damage caused by ultraviolet light.

Ti plasmid The tumor-inducing plasmid from *Agrobacterium tumefaciens.*

topoisomerase An enzyme that changes a DNA's superhelical form, or topology.

totipotent cell A cell that retains full genetic capability, including the ability to give rise to a complete organism under appropriate circumstances.

trailing strand The strand that is made discontinuously in semidiscontinuous DNA replication.

trans-acting A term that describes a genetic element, such as a repressor gene or transcription factor gene, that can be on a separate chromosome and still influence another gene. These *trans*-acting genes function by producing a diffusible substance that can act at a distance.

transcript An RNA copy of a gene.

transcription The process by which an RNA copy of a gene is made.

transcription factor A protein that stimulates (or sometimes inhibits) transcription by a eukaryotic RNA polymerase.

transcription factor, general A factor needed for transcription of all eukaryotic genes within a class; e.g., TFIID.

transcription factor, gene-specific A factor needed for transcription of only a set of genes within a class; e.g., GAL4.

transcription terminator A specific DNA sequence that signals transcription to terminate.

transducing virus (or phage) A virus that can carry host genetic information from one cell to another.

transductants Cells that have received genetic information from a transducing virus.

transduction The use of a phage (or virus) to carry host genes from one cell to another cell of different genotype.

transfer RNA *See* tRNA.

transformation (genetic) An alteration in a cell's genetic makeup caused by introducing exogenous DNA.

transformation (malignant) The process by which a normal cell begins behaving like a tumor cell.

transformed cell An immortal cell that displays many of the characteristics of a cancer cell, but that may or may not be tumorigenic.

transgenic organism An organism into which a new gene or set of genes has been transferred.

transition A mutation in which a pyrimidine replaces a pyrimidine, or a purine replaces a purine.

translation The process by which ribosomes use the information in mRNAs to synthesize proteins.

translocation The translation elongation step following the peptidyl transferase reaction that involves moving an mRNA one codon's length through the ribosome and bringing a new codon into the ribosome's A site.

translocation (chromosomal) Transfer of a chromosome segment from one chromosome to another nonhomologous chromosome.

transposable element A DNA element that can move from one genomic location to another.

transposase The name for the collection of proteins, encoded by a transposon, that catalyze transposition.

transposition Transfer of a segment of DNA from one location to another, with or without retention of a copy of the segment in the original location.

transposon *See* transposable element.

transversion A mutation in which a pyrimidine replaces a purine, or vice versa.

trisomy Diploids having one extra chromosome $(2N + 1)$.

tRNA (transfer RNA) Relatively small RNA molecules that bind amino acids at one end and "read" mRNA codons at the other, thus serving as the adapters that translate the mRNA code into a sequence of amino acids.

tumorigenicity The ability to cause tumors in whole animals.

Ty A yeast transposon that transposes via a retroviruslike mechanism.

U

ultimate carcinogen The most carcinogenic form to which a given substance can be metabolized.

undermethylated region A region of a gene or its flank that is relatively poor in, or devoid of, methyl groups.

unique DNA *See* single-copy DNA.

uracil (U) The pyrimidine base that replaces thymine in RNA.

uridine A nucleoside containing the base uracil.

V

variance A measure of the dispersion of the mean value in a population.

vector A piece of DNA (a plasmid or a phage DNA) that serves as a carrier in gene cloning experiments.

viroid An infectious agent (of plants) consisting simply of a small circular RNA.

virulent phage A phage that lyses its host.

W

wobble The ability of the third base of a codon to shift slightly to form a non-Watson-Crick base pair with the first base of an anticodon, thus allowing a tRNA to translate more than one codon.

wobble hypothesis Francis Crick's hypothesis that invoked wobble to explain how one anticodon could decode more than one codon.

wobble position The third base of a codon, where wobble base pairing is permitted.

Z

Z-DNA A left-handed helical form of double-stranded DNA whose backbone has a zigzag appearance. This form is stabilized by stretches of alternating purines and pyrimidines.

zinc finger A finger-shaped protein structural motif found in DNA-binding proteins and characterized by two cysteines and two histidines (or four cysteines) bound to a zinc ion. By inserting into the wide groove of the DNA, these fingers help the proteins bind DNA.

zygote A fertilized egg.

INDEX

A

AAUAAA, 226–27
ABO blood group, 47, 51–52, 71, 113, 414
Abortion, spontaneous, 94, 98
Ac element, 301–2
Acentric fragment, 121–22
Acetic acid, 133
Acetylcholinesterase gene, 410
Achondroplasia, 40–41, 46, 48, 279–80
Acid group, 238
Acridine dye, 280
Acrocentric chromosome, 77, 82, 91, 95
Adelberg, E. H., 320
Adenine, 15, 136, 138, 140
    alkylation of, 281
    deamination of, 278
    methylation of, 288
    tautomers of, 276–77
Adenosine, 138
Adenosine deaminase gene, 389, 410
Adenovirus
    introns in, 219–20
    promoter of, 216
Adenylate kinase gene, 410
Adjacent segregation, 93
Admixture, 422
Adrenoleukodystrophy, 115
Aflatoxin, 285
African–Americans
    Caucasian admixture in, 422
    sickle-cell allele in, 434
African sleeping sickness, 360
Agarose, 147, 380–81
Agouti color, 58
Agrobacterium tumefaciens, 391–93
AIDS, 303
AIDS vaccine, 388
Alanine, 238
Albinism, 116, 271, 418, 433, 442
    in humans, 39, 46
    in mice, 28–29
    ocular, 115
Alcaptonuria, 14, 241
Alcohol dehydrogenase genes, 410, 436–37
Alkaline phosphatase gene, 261
Alkylation, of nitrogenous bases, 281–82, 286
Alkyl group, 281
Allele, 11, 27
    deleterious, 48
    identical by descent, 416–18
    identical in kind, 416–17
    lethal. *See* Lethals
    multiple, 51–53, 414
    neutral, 435–37
Allelic frequency, 410–21, 424–26, 432–33, 442
    variance in, 420–21
Allelic substitution, rate of, 437
Allolactose, 186–88
Allopolyploid, 89, 95–96
Allosteric protein, 186
Allotetraploid, 95
α_1 Antitrypsin gene, 390
α-Galactosidase A deficiency. *See* Fabry's disease
α-Globin gene, 361, 381, 440
α-Helix, 197–98, 239–40
Alternate segregation, 93
*Alu*I, 364–65
Alzheimer's disease, 398
Amber mutation, 261, 272
Amber suppressor, 283
Ames, Bruce, 284
Ames test, 284–85
Amino acid, 239
    binding to tRNA, 250–51
    genetic code, 254
    structure of, 238
Aminoacyl-AMP, 251
Aminoacyl-tRNA, 251, Panel 7
Aminoacyl-tRNA synthetase, 247, 251

Amino group, 133, 238
Amish population, 435
Amniocentesis, 99
AMP, in polyadenylation, 226
Ampicillin resistance, 299, 366–68, 375
Anaphase
    meiosis I, 83–84, 87, Panel 4
    meiosis II, 83–84, 87
    mitotic, 80–82, Panels 2 & 3
Anderson, W. French, 17
Anemia
    Fanconi's, 290
    sickle-cell. *See* Sickle-cell disease
Aneuploid, 89, 94, 96–100, 270, 305
    in humans, 98–100
Angelman syndrome, 351
Animal breeding, 444–48
    endangered species in captivity, 443–44
Animal pole, 342
Antennapedia complex, 350, 352, 357
Anthocyanin, 56–57
Antibiotic, in livestock feed, 300
Antibiotic resistance, 296–97, 300, 309, 450
Antibiotic resistance plasmid, 299
Antibody, 52, 239–40. *See also* Immunoglobulin
    against A and B antigens, 52
Antibody probe, 376–78
Anticodon, 247–48, 250–51, 254–55
Antigen, of ABO blood group system, 52
Antirepressor, 195
Antiterminator, 195–96
Ape, chromosomes of, 100–101
3′ AP endonuclease, 286–87
5′ AP endonuclease, 286–87
A peptide, 231
AP site, 281, 286–87
Apurinic site. *See* AP site
Apyrimidinic site. *See* AP site
ara operon, 190
Arber, Werner, 364
Argentinian kidnap victims, 70
Arginine, 238
Arm, chromosome, 77
Arnberg, Annika, 223
Artificial selection, 66, 444–46
Ascospore, 119–21
Ascus, 119, 242
Asexual reproduction, 76
A site, ribosomal, 258–60, Panel 7
Asparagine, 238
Aspartate, 238
Aspergillus nidulans, chromosome number in, 77
Assortative mating
    negative, 415
    positive, 415
Ataxia telangiectasia, 290
ATP
    in transcription, 182–83
    in tRNA charging, 251
Atrazine resistance, 450–51
Attached X, 116–19
att site, 324
A-type particle, 304
Autohexaploid, 102
Autoimmune disease, 53
Autopolyploid, 89, 95–96
Autoradiography, 147, 164–65, 176, 371, 377–81, 385
Autosome, 12
Autotetraploid, 95
Avena fatua, inbreeding coefficient for, 417
Avery, Oswald, 14, 17, 134–35
Avirulent mutant, 132–35
Avise, Jon, 439–40
5–Azacytosine, 349–50

B

Bacillus subtilis
    phage SPO1, 191–92
    replication of, 161–62
    RNA polymerase of, 192
    sporulation in, 193–94
Backcross, 28, 37
Back-mutation. *See* Reversion

Backward mutation rate, 418
Bacteria. *See also Escherichia coli*
    conjugation in, 312–19, 333–34
    genetic notation for, 312
    transduction in, 323–26, 333
    transposable elements of, 296–300
Bacterial lawn, 328–29
Bacteriophage. *See* Phage
Baculovirus vector, 378
Ballou, Jon, 443
Baltimore, David, 303
*Bam*HI, 364–66, 369, 375
Banana, ploidy in, 96
Barbacid, Mariano, 308
Barley, genetic variation in, 410–11
Barnacle, sexual reproduction in, 76
Barnett, Leslie, 252
Barr, Murray, 99, 349
Barr body, 99, 349
Base. *See* Nitrogenous base
Base pairing, 15, 141–45, 149, 154, 182, 247–50, 258,
    Panels 5 & 6
    tautomerization and, 276–77
    wobble, 254–55, 283
Bateson, William, 29, 56, 106, 350
B chromosome. *See* Supernumerary chromosome
B-DNA, 143–44
Beadle, George, 14–15, 17, 241–42
Beaudry, Amber, 224
Behavioral trait, norm of reaction for, 59–60
Bell-shaped distribution. *See* Normal distribution
Benzer, Seymour, 261, 323, 331
Benzo[a]pyrene, 306–7
Berg, Paul, 17
β cells, 338–39
β-Galactosidase, 185–91, 367, 369, 371, 376–77
β-Galactosidic bond, 186
β-Globin gene, 210, 219, 221, 273–75, 345–46, 361,
    389, 407–8, 414, 426–27, 434
β-Lactamase, 298
β-Lactoglobulin gene, 390
β-Pleated sheet, 239
bGH. *See* Bovine growth hormone
*Bgl*II, 365
bicoid gene, 354–55, 357
Bidirectional replication, 156, 160–64
Binomial coefficient, 33–34
Binomial formula, 33
Binomial probability, 33–35, 40
Biochemical mutation. *See* Nutritional mutation
Biston betularia. See Peppered moth
Bisulfite, 280
Bithorax complex, 350, 352
Bivalent, 83–84, 96, Panel 4
Black bread mold. *See Aspergillus nidulans*
Blackburn, Elizabeth, 170
Bladder cancer, 307–8
Blakeslee, Alfred, 97
Blastoderm, 353, 355
Blastomere, 339
Blastula, 340–41, 348
Blattner, Fred, 369
Blending inheritance, 22, 25
Blood transfusion, 52
Blood type, 47, 51–52, 69, 71
Bloom's syndrome, 290
Bombyx mori. See Silk moth
Bone marrow transplantation, 389
Bottleneck, 420, 440
Bourgeois, Suzanne, 190
Boveri, Theodor, 17
Bovine growth hormone (bGH), 390
–10 Box, 184–85, 194, 210
–35 Box, 184–85, 194
Box A, 213
Box B, 213
Box C, 213
Boyer, Herbert, 17, 364–65
B peptide, 231
Branch migration, 172–73
BrdU. *See* 5-Bromodeoxyuridine
Brenner, Sydney, 17, 243, 261
Bridges, Calvin, 17, 87, 102
Briggs, Robert, 340
Brinster, Ralph, 390

5-Bromodeoxyuridine (BrdU), 280–81
Brown, Donald, 213, 216, 346–48
Burgess, Richard, 183, 194
Burkitt's lymphoma, 360

C

cI gene, 195–97
cII gene, 195, 197
cIII gene, 196–97
CAAT box. *See* CCAAT box
Cairns, John, 160–61, 165, 318
Calico cat, 349
Callan, H. G., 163
cAMP. *See* Cyclic AMP
Cancer, 290
    malignant transformation, 305–6
    mutations and, 270, 305–6
    radiation-induced, 306–7
    viruses and, 303
Cancer cells
    aneuploidy in, 305
    immortality of, 305
CAP. *See* Catabolite activator protein
Cap binding protein (CBP), 258
Capecchi, Mario, 262
Capping, of mRNA, 227–29, 258
Capping enzyme, 227
Carbonyl group, 133
Carboxyl group, 132–33, 238
Carcinogen, 281
    as mutagens, 306–7
    ultimate, 306
Carp, chromosome number in, 77
Carrier, 32, 40, 68, 412–13, 434
    identification of, 398
Casein mRNA, 229
Castle, William, 28–29, 411
Cat
    calico, 349
    chromosome number in, 77
    coat color in, 34–36, 50–51, 58, 272, 349
    hair length in, 34–36
    karyotype of, 78
    Manx, 48, 51
    Siamese, 50–51, 272
Catabolite activator protein (CAP), 190–91
Catabolite repression, 190–91
Cattanach, Bruce, 351
Cattle
    calf weight in, 448
    karyotype of, 78
Caucasian admixture, in African-Americans, 422
CBP. *See* Cap binding protein
CCAAT box, 210, 212, 216–17
CCAAT box transcription factor (CTF), 217–18
cDNA. *See* Complementary DNA
Cech, Thomas, 223–24
Cell culture, 219
Cell cycle, 79–80, Panel 2
Cell differentiation, 338–45
Cell division, 76–77
Cell fusion, 115
Cell morphology, 338–39
Cell plate, 82
Cell structure, Panel 1
Centric fission, 95
Centric fusion, 95, 100
Centriole, 81, Panels 1, 3, & 4
Centromere, 77, 113, 123, 208, Panels 3 & 4
Cepaea nemoralis. See Snail
C gene, 300–302
Chain-counting technique, 418
Chambon, Pierre, 210, 220, 343
Chaperone, 241
c-Ha-*ras* gene, 308–9
Chargaff, Erwin, 134, 140
Chargaff's rules, 140–41
Charon phage, 369–71
Chase, Martha, 134–37
Cheetah
    genetic variation in, 410–11
    inbreeding in, 443
Chemical analysis, of DNA, 135
Chemical mutagenesis, 280–81, 379

Chiasma, 83–84, 107, 115, Panel 4
    half. *See* Chi structure
Chicken
    β-globin gene of, 345–46
    chromosome number in, 77
    feather color in, 101
    feather development in, 342
    performance in, 445–46
    sex determination in, 101
Chimney sweep, 306
Chimpanzee
    chromosome number in, 77
    chromosomes of, 100–101
    evolution of, 439
Chironomus, polysomes of, 246
Chi-square test, 37–38, 41–42, 103, 413–14
Chi-square value, 37–38
Chi structure, 171–74
Chlamydomonas reinhardi, spontaneous mutation
        rate in, 280
Chloramphenicol, 258–59
Chloramphenicol resistance, 299
Chlorophyll, 343–44
Chloroplast
    DNA of, 450–51
    genes of, 351
Cholesterol, serum, 389
Chromatid. *See* Sister chromatids
Chromatin, 132, Panel 1
    folding of, 204–7
    gene expression and, 208
    histones in. *See* Histone
    nonhistone proteins in, 207
    structure of, 204–9
Chromosomal protein, 132
Chromosome. *See also* Meiosis; Mitosis
    acrocentric, 77, 82, 91, 95
    arms of, 77
    B. *See* Chromosome, supernumerary
    banding of, 78, 100–101, 125
    daughter, 82
    in different species, 100
    genes and, 12, 86–89
    homeologous, 95
    homologous, 13, 76, 82–84, 107, 171, Panel 3
    human, 78–79, 204, 206
    metacentric, 77, 82, 91, 95, 113
    morphology of, 77–79
    polytene, 78–79, 91
    pulsed-field gel electrophoresis of, 148
    ring, 93
    satellites, 77–78
    secondary constriction of, 78
    sex. *See* Sex chromosome
    sex determination and, 100–102
    staining of, 82
    structural changes in, 89–94, 406–7
    submetacentric. *See* Chromosome, acrocentric
    supernumerary, 77
    telocentric, 77, 82
Chromosome number, 77, 82
    changes in, 89, 94–100
Chromosome theory of inheritance, 11–12
Chronic granulomatous disease, 115
Chua, Nam-Hai, 344
Cigarette tar, 306
Ciliate, genetic code in, 256
cis-acting element, 323, 344
cis-dominant mutation, 189
cis-trans complementation test, 321–24, 331–33
Cistron, 186, 323, 333
*Cla*I, 365, 375–76
Clark, A. J., 390
Clarke, Cedric, 425
Clarkia, ring chromosomes in, 93
Classical genetics. *See* Mendelian genetics
Clastogen, 282
Clausen, J., 59
Cleft palate, 68
Clegg, Michael, 424
Clone, 59, 364
Cloned gene, 16. *See also* Gene cloning
    in agriculture, 390–93
    determining base sequence of, 385–87
    expression of, 374–78
    identification with specific probe, 372

    manipulation of, 378–87
    as probe, 380–85
    protein engineering with, 378–80
    rubisco gene, 344
Cloverleaf structure, of tRNA, 247–48
Codominance, 47, 53
Codon, 16, 248, 251. *See also* Genetic code
    initiation, 257–58, 387
    nonsense. *See* Codon, stop
    stop, 254, 256, 261, 272, 275
Codon-anticodon recognition, 248
Cohen, Stanley, 17, 364–65
Cohn, Melvin, 187
Cointegrate, 298, 300
Colchicine, 95–96
Colicin E1, 164
Collins, Francis, 17, 399
Colon cancer, 308
Colony hybridization, 371
Color blindness, 53–54, 114–15
Common garden experiment, 64
Complementarity, in DNA, 15
Complementary base pairing. *See* Base pairing
Complementary DNA (cDNA), 372
Complementary DNA (cDNA) cloning, 373–74
Complementary DNA (cDNA) library, 376
Complementary gene action, 57–58
Complementation, 333
    intracistronic, 323
Complementation test, *cis-trans,* 321–24, 331–33
Complete dominance, 46–47
Composite transposon, 299
Concatemer, 168
Conditional lethal, 50
Conditional mutation, 272–73
Conditional probability, 32, 40
Conjugation, in bacteria, 312–19
    mapping by, 317–18, 333–34
Conklin, E. G., 339
Conneally, Michael, 17, 394–99
Consanguineous mating, 38–39, 417–18, 427, 434, 442
Consensus sequence, 184–85, 194, 210, 221, 227
Conservation genetics, 441–44
Conservative replication, 154–55, 158
Conservative substitution, 204–5
Conservative transposition, 297
Constitutive heterochromatin, 78
Constitutive mutant, 188–90
copia, 302, 304
copia-like element, 302
Core element, 212
Core particle, of nucleosome, 204
Corn
    chromosome number in, 77
    DNA of, 146
    ear length in, 67
    genetic variation in, 441–42
    height of, 67
    hybrid, 448
    kernel color in, 30–31, 270, 300–302
    kernel shape in, 30–31
    oil and protein content of, 66
    recombination in, 13
    spontaneous mutation rate in, 280
    transposable elements in, 270, 300–302
    yield in, 67
Correlation, 65
Correns, Carl, 17, 23, 29, 46–47
Cosmid, 370–71
cos site, 370
Cotransduction, 326
Coumermycin, 166
Countable trait, 63
Coupling, 106, 108
Coupling double heterozygote, 108
Coupling gametes, 107
C peptide, 231
Creighton, Harriet, 17
Creutzfeld-Jacob disease, 388
Crick, Francis, 15, 17, 134, 140–41, 243–44, 247, 252
Cri du chat syndrome, 90
Criminal, identification of, 383–84
cro gene, 195–97
Crop plant
    genetically engineered, 390–93
    genetic variation in, 441–42

land races of, 441
  in monoculture, 448
  wild relatives of, 441–42
Crossover, 12–13, 82–83, 107, 171–74, Panel 4
  double, 111, 113–14
  four-stranded, 116–19
  quadruple, 113–14
  in transduction, 324–25
  triple, 114
  two-stranded, 116–17
  unequal, 121–24
Crown gall, 391–93
Cryptic mutant, 188
CTF. See CCAAT box transcription factor
CTP, in transcription, 182–83
Cuénot, Lucien, 28–29, 47–48
Culling level, 445
Cutting (plant), 76, 340
C-value paradox, 263
Cyclic AMP (cAMP), 190–91
Cysteine, 238
Cystic fibrosis, 17, 39, 42, 393, 432
  identification of carriers, 398
  mapping gene for, 399
Cytidine, 138
Cytochrome c, evolution of, 438
Cytosine, 15, 136, 138, 140
  deamination of, 277–79
  methylation of, 278–79
  tautomers of, 276–77

D

dADP, 139
Dalgarno, L., 258
dAMP, 138–39
Darbishire, 29
Darnell, James, 226
Darwin, Charles, 17, 422
dATP, 138–39, 154–55
Datura stramonium. See Jimson weed
Daughter chromosome, 82
dCMP, 138
dCTP, 154–55
DDT resistance, 63, 421, 423, 448–49
Deaf-mutism, 57
Deamination, of nitrogenous bases, 277–79
Deficiency. See Deletion
Degrees of freedom, 37
Degu, inbreeding in, 443
Delbrück, Max, 315, 327–28
Deleterious allele, 48
Deletion, 16, 89–91, 121–22, 252, 275–78, 284, 300–301
  interstitial, 90–91
  terminal, 90–91
Deletion heterozygote, 90–91
Deletion loop, 90
Deletion mapping, 115
Delta repeat, 303
Denaturation, of proteins, 272
Densitometer, 385
Deoxynucleoside, 138
Deoxyribonucleic acid. See DNA
Deoxyribose, 136, 138
Deprotonation reaction, 132–33
DES. See Diethylstilbestrol
Determinant, 340–42
Determination, 340
  in Drosophila melanogaster, 350–57
  nature of, 345–48
Detrimental allele. See Deleterious allele
Deutan color blindness, 115
Development
  mosaic, 339–40
  regulative, 339–41
Developmental genetics, 337–62
de Vries, Hugo, 17, 23
dGMP, 138
dGTP, 154–55
d'Herelle, F., 327
Diabetes
  genetic factors in, 68
  juvenile, 53
  threshold for, 63
Diakinesis, 83, Panel 4
Dicentric fragment, 121–22

Dickerson, Gordon, 445
Dicot, 392
Dideoxy method, of DNA sequencing, 385–86
Dideoxy ribonucleotide, 385–86
Diethylstilbestrol (DES), 229
Differentiation, 338–45
Diffusible gene product, 322
Dihybrid cross, 29, 32, 35, 38, 106
Dihydrodiol, 306
Dihydrouracil, 248
Diol epoxide, 306–7
Dipeptidyl-tRNA, 258–59, Panel 7
Diploid, 11, 76–77
Diplotene, 83, Panel 4
Directional cloning, 369
Discontinuous trait, 63
Displaced duplication, 89–90
Dizygotic twins, 65
D-loop, 172–74
DNA
  amplification of, 383–84
  annealing of. See DNA, renaturation of
  anticoding strand of, 262–63
  antiparallel strands in, 141–42, 154
  autoradiography of, 147, 164–65, 176
  bacterial exchange of, 312–14
  base composition of, 134, 140, 146
  base sequence of, 148–49, 291
  B-DNA, 143–44
  in chromatin, 204–9
  circular, 146, 164–66, 195, 314, 318
  coding strand of, 262–63
  complementary, 372–76
  content per haploid cell, 263
  denaturation of. See DNA, melting of
  direct repeat, 296–97, 305
  discovery of, 14, 132
  DNA-protein interactions, 197–98
  duplex. See Double helix
  electron microscopy of, 146
  electrophoresis of, 135, 146–49, 380–82
  elementary chemical analysis of, 135
  3′-end of, 139, 141–42, Panel 5
  5′-end of, 139, 141–42, Panel 5
  fingerprinting of, 69, 381–84
  GC content of, 144–45
  heteroduplex, 172–73
  hot spots in, 278, 331–32
  hybridization of. See Hybridization
  inverted repeat, 296, 301–2
  linker, 348
  long terminal repeat, 303–4
  major groove of, 142, 198, 217
  map distance along, 123–25
  melting of, 144–45
  methylation of, 278–79, 288, 349–52, 399
  minisatellite, 381–83
  minor groove of, 142
  mitochondrial, 146, 439–40
  molecular weight of, 146, 148
  mutations in. See Mutation
  negative strand, 372
  nicked, 157, 166, 171–74, 183–84, 286–87
  noncoding regions of, 263
  nontemplate strand of, 262–63
  palindromic, 365
  pericentromeric, 208
  physical chemistry of, 143–48
  positive strand, 372
  proof that it is genetic material, 14, 132–35
  recombinant, 365–66
  renaturation of, 145
  repair of. See DNA repair
  repeated, 208, 381
  replication of. See Replication
  selfish, 296
  sequencing of, 385–87, 400
  single-stranded, 371–72
  size of, genetic capacity and, 263
  sizes and shapes of, 146–48
  with sticky ends, 195, 365–67, 374
  structure of, 15, 17, 136–42
  supercoiled, 146, 165–66, 206
  T-DNA, 391–93
  telomeric, 169–70
  template strand of, 182–85, 262–63

of T-even phage, 327–28
transcription of. See Transcription
ultracentrifugation of, 135, 158
ultraviolet absorption spectrophotometry of, 135
undermethylated, 349
underwound, 165–66
variable number tandem repeat, 383
X-ray picture of, 140
Z-DNA, 143–44
DNA binding protein, 217
dnaB protein, 167
dnaC protein, 167
DNA glycosylase, 286–87
DNA gyrase, 165–66
DNA ligase, 155–57, 159–60, 171, 288–90, 365–67
DNA photolyase, 285–86
DNA polymerase, 154, 156, 289, 373, Panel 5
DNA polymerase I, 156–57, 287–88
DNA polymerase III, 290
DNA polymerase III holoenzyme, 167, 277
  proofreading activity of, 170
DNA primer, 384
DNA repair
  consequences of defects in, 290
  error-prone, 289–90, 307
  excision, 286–87, 290
  inaccurate, 307
  mismatch, 288
  photoreactivation, 285–86
  post-replication. See DNA repair, recombination
  recombination, 289
  single- and double-strand breaks, 282
DNase, 160, 345
DNase hypersensitivity site, 345–46
Dobzhansky, Theodosius, 406
Dog
  chromosome number in, 77
  hip dysplasia in, 63
  karyotype of, 78
Domain, of protein, 239
Dominance
  complete, 46–47
  incomplete, 46–47
Dominant trait, 11, 22, 25–28, 35–36, 46–47, 189
  in humans, 40–41
  with incomplete penetrance, 49
  pedigree for, 40–41
  X-linked, 55–56
Donkey
  chromosome number in, 77
  karyotype of, 78
Dorcas gazelle, inbreeding in, 443
Double crossover, 111, 113–14
Double fertilization, 86
Double helix, 15, 140–42, 250, Panel 5
  handedness of, 143–44
Down mutation, 184
Down syndrome, 98–100, 115
  translocation, 93–94
Drosophila melanogaster
  alcohol dehydrogenase genes in, 436–37
  attached X in, 116–19
  Bar eye mutant in, 89, 116–19, 121–23
  body size in, 67
  bristle number in, 67
  chromosome number in, 77
  chromosomes of, 100, 105
  determination in, 350–57
  DNA of, 146
  egg production in, 67
  embryogenesis in, 353–54
  eye color in, 12–13, 49, 87–88, 103, 106–8, 271, 351
  genetic map of, 109–13
  Glued mutant in, 424
  homeobox genes in, 352–54, 357
  homeotic genes in, 350, 352, 356–57
  hybrid dysgenesis in, 305
  maternal effect genes in, 354–55, 357
  modifier genes in, 51
  Morgan's experiments with, 11–13, 106–8
  morphological mutants in, 271
  recombination in, 13
  replication in, 162–63
  segmentation genes in, 354, 356–57
  sexual phenotypes in, 102
  spontaneous mutation rate in, 280

transposable elements in, 302–3
Ultrabar eye in, 121–23
wing morphology in, 12–13, 35–36, 49, 106–8, 271
X chromosome of, 53, 78–80, 103, 109–11, 351
X-linked traits in, 86–87
Drosophila pseudoobscura
genetic variation in, 410–11
inversions in, 92, 406–7
polytene chromosomes of, 91
Drosophila simulans, chromosomes of, 100
Ds element, 301–2
dTMP, 138
dTTP, 154–55
Duchenne muscular dystrophy, 115, 219, 393, 399
Duffy blood group, 422
Dulbecco, Renato, 285
Duplicate gene action, 57–58
Duplication, 89–90, 121–24, 361
displaced, 89–90
reverse, 89–90
tandem, 89–90
Dwarfism
achondroplasia, 40–41, 46, 48, 279–80
hypopituitary, 388
six-fingered, 435
Dystrophin, 399
Dystrophin gene, 219

E

East, Edward, 448
*Eco*RI, 164, 365–66, 369, 375, 377, 381
Ectoderm, 342
Edward syndrome, 98
EF. *See* Elongation factor
Egg cell, 9, 76, 85–86
enucleate, 340–41
gradient of determinants in, 342
Electron microscopy, of DNA, 146
Electrophile, 281, 306
Electrophoresis
of DNA, 135, 146–49, 380–82
of histones, 205
of proteins, 407–8
of ribosomal proteins, 245
Elephant
evolution of, 439
genetic variation in, 411
Elephant shrew, inbreeding in, 443
Ellis, E., 328
Ellis-van Crevald syndrome, 432, 435
Elongation factor (EF)
EF-G, 259
EF-Tu, 258–59, 308, Panel 7
Embryo
mosaic, 342
regulative, 340
Embryogenesis, 340
in *Drosophila melanogaster*, 353–54
Empirical recurrence risk, 68
EMS. *See* Ethylmethane sulfonate
Encephalomyopathy, 351
Endangered species, 419–21
breeding in captivity, 443–44
Endoplasmic reticulum, 230, Panel 1
Endospore, 193–94
engrailed gene, 352, 356
Enhancer, 210, 212–14, 343–44, 359
Enucleate egg, 340–41
Environment-genotype interactions, 59–60, 62, 64
Enzyme, 14, 241
one gene-one enzyme hypothesis, 14–15, 17
suicide, 286
Epigenesis, 338, 340
Epistasis, 56–59
EPSP synthase gene, 391–93
Equational division. *See* Meiosis II
Equatorial plane. *See* Metaphase plane
Equilibrium heterozygosity, 436
Error-prone repair, 289–90, 307
Escherichia coli
alkaline phosphatase gene of, 261
DNA of, 146, 263
genetic map of, 317–18, 320
Hershey-Chase experiment with, 134–37

lac operon of. *See lac* operon
mapping with, 320–23
mutator strains of, 275–77
nutritional mutations in, 271–72
phage λ-infected, 194–97
phage T7-infected, 192–93
promoters of, 184–85
replication in, 160–62, 164–71
as research organism, 312
RNA polymerase of, 183
spontaneous mutation rate in, 280
transcription and translation in, 246–47, 256–57
Esterase B1, 451
Ester bond, 250–51
Estrogen, 343
Ethanol, 133
Ethical issues
in gene cloning, 395
in genetic screening, 398
Ethylmethane sulfonate (EMS), 281–82
Euchromatin, 78, 208
Eukaryote, 13
DNA of, 146
expression of cloned genes, 377–78
gene expression in, 203–33
mRNA of, 258
polysomes in, 246
posttranscriptional control in, 228–29
posttranslational control in, 228, 230–31
promoters of, 209–13
replication in, 164
ribosomes of, 246
RNA polymerase of, 209, 214
transcriptional control in, 208, 213–18, 228–29, 343–45
transcription in, 209–18
translational control in, 228–30
transposable elements in, 300–303
Euploid, 94
Evolution
genomic imprinting and, 352
of histones, 204–5
molecular, 435–40
origin of life, 224, 255–56
of proteins, 437–38
Excision repair, 286–87, 290
Exon, 219–21, 231–32
Exon trapping, 397
Exonuclease
$3' \rightarrow 5'$, 170, 287, 373
$5' \rightarrow 3'$, 286–88
Expected numbers, 37–38
Expression site, 360
Expression vector, 374–78
eukaryotic genes, 377–78
inducible, 375
λ-based, 376–78
pβ-gal13C, 375–77
pIN-I-A1, 375
pKC30, 375, 377
producing fusion proteins, 375–77
ptrpL1, 375–76
with strong promoters, 375
Expressivity, 49–51
Extinction, 441
exuperantia gene, 355

F

Fabry's disease, 115
Factor IX, 390
Falconer, Douglas, 447
Fanconi's anemia, 290
Fate map, of *Drosophila* embryo, 354, 356–57
F-duction, 320–21, 324
Fedoroff, Nina, 301
Felsenfeld, Gary, 345–46, 349
Fertilization, 76–77, 86
double, 86
F_1 generation. *See* First filial generation
F_2 generation. *See* Second filial generation
Fibrinopeptide, evolution of, 438
Fine structure mapping, 387
in T-even phage, 331–32

Fingerprinting
of DNA, 69, 381–84
of protein, 273–74
Fingerprint-ridge count, 65, 67
Fink, Gerald, 304
First filial (F_1) generation, 11, 22, 27
Fitness
mean, 423
relative, 423, 425–27
F-*lac* plasmid, 320–21
Flax, rust resistance in, 451
Flour beetle
homeotic genes in, 124
pesticide resistance in, 449
Fluctuation test, 315
fMet. *See* N-Formyl methionine
Food poisoning, 299
Forced cloning. *See* Directional cloning
Forensic science, 383–84
Forked-line approach, 35–36
Formaldehyde, 133
Forward mutation, 283
Forward mutation rate, 418
Founder effect, 435
Four-o'clock, flower color in, 46–47
Four-stranded crossing over, 116–19
F pilus, 313, 316
F plasmid, 314–16, 320
F′ plasmid, 320–21
Frameshift mutation, 251–52, 275–78, 280, 283–85
Frankel-Conrat, H., 251
Franklin, Rosalind, 15, 17, 140–41
Free radical, 282
Frelinger, Jeffrey, 427
Frog
DNA of, 146, 263
nuclear transplantation in, 364
Fruit fly. See *Drosophila melanogaster*
F⁻ strain, 313
F⁺ strain, 313
Functional group, 132–33
Furrow, 82
fushi tarazu gene, 352, 356–57
Fusion protein, 374–77

G

Galactose-1-phosphate uridyl transferase gene, 410
Galactose transacetylase, 185–91
Galactoside permease, 185–91
gal operon, 190
Gamete, 9, 76
coupling, 107
repulsion, 107
Gametogenesis, 28, 30, 76–77, 352
Gamma radiation, as mutagen, 282
GAP. *See* GTPase activator protein
Garden pea
anatomy of, 24
chromosome number in, 77
flower characteristics in, 25–27, 37, 48
life cycle of, 24
linkage in, 110
Mendel's experiments with, 11, 22–31
pod characteristics in, 26–27
seed characteristics in, 11, 22, 26–30, 35, 38, 41–42, 48
stem characteristics in, 26–27
suitability for experimentation, 22
Garen, Alan, 261
Garrod, Archibald, 14, 17, 241
Gastrula, 348
GATC sequence, 288
G band, 78
GC box, 210, 212, 216, 218
GC content, of DNA, 144–45
Gehring, Walter, 356
Gel electrophoresis, 146–47. *See also* Electrophoresis
Gene, 10, 22, 27. *See also* DNA; specific genes
chromosomes and, 12, 86–89
cloned. *See* Cloned gene
DNA size and genetic capacity, 263
functions of, 15–16, 154
homeotic, 124
housekeeping, 338

interrupted, 219–22
jumping. *See* Transposable element
linked, 108
luxury, 338
master, 350, 353–54
monomorphic, 409–10
one gene-one enzyme hypothesis, 14–15, 17
one gene-one polypeptide hypothesis, 242
order on chromosome, 110–13
overlapping, 387–88
pair-rule, 356
polymorphic, 409–11
proteins and, 14–15, 241–42
rearrangements of
immunoglobulin genes, 358–60
oncogenes, 360
trypanosome coat genes, 360
regulatory, 186
structural, 186
structure of, 14
symbols for, 35–36
syntenic, 110
Gene action
complementary, 57–58
duplicate, 57–58
Gene cloning, 16, 364. *See also* Cloned gene
applications of, 387–93
cDNA cloning, 373–74
directional, 369
ethical issues in, 395
forced. *See* Gene cloning, directional
restriction endonucleases in, 364–66
safety of, 395
vectors for, 366–72
Gene cluster, 361
Gene expression. *See also* Transcriptional control; Translational control
chromatin structure and, 208
control of, 342–45
developmental activation of genes, 345–46
in eukaryotes, 203–33
measuring with DNA probes, 381–85
Gene family, 380–81
sequential activation of members of, 361
Gene flow, 421–22, 434, 444–45
Gene mutation. *See* Mutation
Gene pool, 406
Gene product, diffusible, 322
Generalized recombination. *See* Homologous recombination
Generalized transduction, 323–25
General transcription factor, 214–15
Gene-specific transcription factor, 214–17
Gene therapy, 17, 389
Genetic code, 16, 248, 264–65, 400
in ciliates, 256
deciphering of, 16, 253–54
degeneracy of, 248, 254, 275
lack of gaps in, 251–52
in mitochondria, 256
in *Mycoplasma*, 256
nonoverlapping of codons, 251
in plants, 256
triplet nature of, 252–53
universality of, 255–56
Genetic counseling, 68, 94, 398, 434
Genetic disease, 17
frequency of, 432–35
future incidence of, 433
mapping of, 393–99
mutation-selection balance for, 432–34
treatment of, 389
Genetic distance, 438
Genetic drift, 410, 419–21, 435–37, 440, 443–44
Genetic linkage. *See* Linkage
Genetic map, 109–10. *See also* Mapping
circular, 317–18, 320
of *Drosophila melanogaster*, 109–13
of *Escherichia coli*, 317–18, 320
of humans, 115, 125
of phage λ, 195
of phage Mu, 300
of phage φX174, 388
Genetic material, nature of, 132–36
Genetic recombination. *See* Recombination

Genetics
branches of, 10
classical. *See* Mendelian genetics
conservation, 441–44
developmental, 337–62
Mendelian. *See* Mendelian genetics
molecular, 10, 13–16
population, 10, 406, 432
reasons to study, 16–17
transmission. *See* Mendelian genetics
Genetic screening, 398
Genetic system, subcellular, 136, 146
Genetic variance, 65
estimation of, 65–67
Genetic variation. *See* Variation
Genome, 95
Genomic imprinting, 350–52
Genomic library, 369–71
Genotype, 12, 27
differential survival of, 422
relationship to phenotype, 59–60
Genotype-environment interactions, 59–60, 64
Genotypic frequency, 409–18, 423
Genotypic value, 64
Genus, 10
Germinal mutation. *See* Germ-line mutation
Germ-line mutation, 270
Germ plasm collection, 441
Gey, George, 219
Gilbert, Walter, 17, 219, 385–87
Globin genes, 89, 113, 123, 361
Globin mRNA, 228–29
Globular protein, 239–40
Glucocorticoid receptor, 343
Glucose
in catabolite repression, 190–91
repression of *lac* operon, 186
Glucose-6-phosphate dehydrogenase deficiency, 115
Glued mutant, 424
Glutamate, 238
Glutamate-oxaloacetate transaminase gene, 410
Glutamate-pyruvate transaminase gene, 410
Glutamine, 238
Glycine, 238
Glycosidic bond, 286
Glyphosate resistance, 390–93
Goat, karyotype of, 78
Golden lion tamarin, inbreeding in, 443
Golgi apparatus, 230–31, Panel 1
Gonadal dysgenesis, 115
Goodness of fit, 36–38
Gorilla
chromosomes of, 100–101
evolution of, 439
Gout, 115
gp28, 192
gp33, 192
gp34, 192
G_1 phase, 80, Panel 2
G_2 phase, 80, Panel 2
G protein, 308–9
Greater galago, inbreeding in, 443
Green sea turtle, mitochondrial DNA of, 439–40
Griffith, Frederick, 14, 132–35
GTP
capping of mRNA, 227
in transcription, 182–83
in translation, 258–59, Panel 7
GTPase, 308–9
GTPase activator protein (GAP), 308
Guanine, 15, 136, 138, 140
alkylation of, 281–82, 286
interaction with benzo[a]pyrene diol epoxide, 307
tautomers of, 276–77
Guanosine, 138
Gurdon, John, 340, 364
Gusella, James, 17, 394–99
Guthrie, Woody, 272, 393
Gyurasits, Elizabeth, 161–62

H

*Hae*III, 365, 381–82
Haemophilus influenzae, DNA of, 146
Haig, David, 352

Hair analysis, 383
Hairy ear rims, 55–56
Haldane, J. B. S., 114
Half chiasma. *See* Chi structure
Half-life, 229
Haplodiploid, 77
Haploid, 11, 76
Haplotype, 69–70
Ha-*ras* oncogene, 307–8
Hardy, Godfrey, 17, 410
Hardy-Weinberg principle, 17, 410–15
for more than two alleles, 414
for X-linked genes, 414–15
Harris, Harry, 410
Hartl, Daniel, 384
Hayes, William, 313–14
HD gene, 394–99
Heat-shock genes, 194
HeLa cells, 219, 305
Helicase, 164
Helix-turn-helix motif, 197–98
Helper phage, 323
Hemoglobin, 241, 338, 361, 438
sickle-cell. *See* Sickle-cell disease
Hemophilia, 54–55, 114–15, 270, 305, 390, 414–15
Herbicide resistance, 390–93
Heritability, 65–67, 71, 444
realized, 65
Herpes simplex virus, thymidine kinase gene of, 210, 212, 216–18
Hershey, A. D., 134–37, 329
Hessian fly resistance, 451
Heterochromatin, 78, 208–9, 349
constitutive, 78
facultative, 78
Heteroduplex DNA, 172–73
Heterogametic sex, 100
Heterogeneous nuclear RNA (hnRNA), 220, 226
Heterokaryotype, inversion, 92, 121–22
Heterokaryotypic individual, 89, 91
Heterosis, 448
Heterozygosity
average, 409–11
equilibrium, 436
Heterozygote, 11–12, 27–29, 32, 47, 272, 411–12. *See also* Carrier
coupling double, 108
deletion, 90–91
inversion, 91–92
repulsion double, 108
translocation, 91–94
Heterozygous advantage, 426–28, 434, 450
Hfr strain, 314–20
hGH. *See* Human growth hormone
Himalayan rabbit, 50
*Hin*dII, 364–65
*Hin*dIII, 365–66, 372, 375, 394–95
Histidine, 238
Histone, 204, 347–48
electrophoresis of, 205
evolution of, 204–5, 438
H1, 204–6, 347–49
H2A, 204–5
H2B, 204–5
H3, 204–5
H4, 204–5
Histone genes, 440
Histone mRNA, 229
HIV. *See* Human immunodeficiency virus
HLA. *See* Human leucocyte antigen
HML locus, 214
HMR locus, 214
hnRNA. *See* Heterogeneous nuclear RNA
Hogness, David, 356
Holliday, Robin, 171
Holmberg, Scott A., 299
Homeobox genes, 352–54, 357
expression during embryogenesis, 353–54
in higher organisms, 357–58
Homeologous chromosomes, 95
Homeosis, 350
Homeotic genes, 124, 350, 352, 356–57
Homogametic sex, 100, 114
Homologous chromosomes, 13, 76, 82–84, 107, 171, Panel 3
Homologous recombination, 171–74

Homozygote, 11–12, 27–29
    natural selection against, 423–26
Homunculus, 338
Hood, Leroy, 358
Hormone, 241
    gene activation by, 343
Hormone-response element (HRE), 343
Horse
    chromosome number in, 77
    karyotype of, 78
Horseshoe crab, genetic variation in, 410–11
Host-range mutant, of T-even phage, 329
Hotchkiss, Rollin, 134, 349–50
Hot spot, 278, 331–32
Housefly, pesticide resistance in, 449
Housekeeping gene, 338
*Hpa*I, 377
*Hpa*II, 364–65, 399
HPRT deficiency. *See* Lesch-Nyhan syndrome
HRE. *See* Hormone-response element
HTF island, 399
Huberman, J., 162
Human being. *See also* Genetic disease
    admixtures in, 422
    aneuploidy in, 98–100
    body size in, 62
    chromosome number in, 77
    chromosomes of, 78–79, 204, 206
    DNA of, 146
    dominant traits in, 40–41
    evolution of, 439–40
    genetic map of, 115, 125
    genetic variation in, 410–11
    genomic imprinting in, 351
    height of, 62, 65
    linkage in, 113–15
    mapping genes in, 115–16, 393–99
    Mendelian inheritance in, 38–41
    mitochondrial DNA of, 440
    positive assortative mating in, 415
    recessive traits in, 39–40
    spontaneous mutation rate in, 280
    transposable elements in, 305
    weight of, 62
    X chromosome in, 113–15
Human Genome Project, 17
Human growth hormone (hGH), 388, 390
Human immunodeficiency virus (HIV), 136, 388
Human leukocyte antigen (HLA), 52–53, 69–70,
    123–24, 440
Huntingtin, 397–99
Huntington disease, 17, 40–41, 46, 49–50, 272, 393–94
    identification of carriers, 398
    mapping gene with RFLPs, 394–99
    pedigree of family with, 395–97
Hybrid, 22
Hybrid dysgenesis, 305
Hybridization
    colony, 371
    of different polynucleotide chains, 145–46
    DNA-RNA, 145, 219–20
    plaque, 371
    somatic cell, 115–16
    stringency of, 372
Hybrid population, 422
Hybrid vigor, 448
Hydrocarbon, 132, 306
Hydrogen bond
    in DNA, 141–42
    in proteins, 239
Hydroxyl group, 133
Hypercholesterolemia, familial, 389
Hypopituitary dwarfism, 388
Hypoxanthine, 278
Hypoxanthine-guanine phosphoribosyl transferase
    deficiency. *See* Lesch-Nyhan syndrome

I

iab-2 gene, 357
Ichthyosis, 115
ICR. *See* Internal control region
IF. *See* Initiation factor
IGF. *See* Insulin-like growth factor

Illegitimate recombination, 171, 298
Imaginal disk, 353
Immunoglobulin, 239–40, 358
    constant region of, 358–59
    D region of, 359
    heavy chain of, 358–59
    joining region of, 358–59
    light chain of, 358–59
    variable region of, 358–59
Immunoglobulin genes, 440
    in Burkitt's lymphoma, 360
Imprinting, genomic, 350–52
Inborn errors of metabolism, 57, 68
Inbred line, 447
Inbreeding, 415–18, 434, 444
Inbreeding coefficient, 416–18, 427
Inbreeding depression, 442–44
Incision, 286, 290
Incomplete dominance, 46–47
Incomplete penetrance, 49–50
Independent assortment, 12, 29–31, 86–87, 106–7, 171
Independent culling level, 445
Index case, 40
Induced mutation, 418
Inducer
    embryonic, 342
    lac operon, 186–88
Induction, of expression vector, 375
Influenza virus, 136
Informational molecule, 132
Ingram, Vernon, 273
Inheritance
    blending, 22, 25
    chromosome theory of, 11–12
    particulate, 22
Initiation codon, 257–58, 387
Initiation complex, 257, Panel 7
Initiation factor (IF)
    IF-1, 256
    IF-2, 256–57
    IF-3, 256–57
Inosine, 254–55
Inouye, Masayori, 375
Insect, pesticide resistance in, 448–51
Insecticide resistance. *See* Pesticide resistance
Insertion, 16, 252, 275, 277–78, 284
Insertion sequence (IS), 296
    IS1, 296
Insulin, 241, 338–39, 387–88
Insulin-like growth factor (IGF), 352
Insurance coverage, 398
Integrated pest management, 451
Intelligence, race and, 67–68
Intercalation, 280
Interference, 113
Intergenic region, mutations in, 283
Intergenic suppression, 283
Interkinesis, 84
Intermediate, 14
Intermediate element, 213
Internal control region (ICR), 213, 216
Interphase, 80, Panels 2, 3, & 4
Interrupted gene, 219–22
Intersex, 102
Interstitial deletion, 90–91
Interstitial translocation, 91–92
Intervening sequence. *See* Intron
Intracellular signaling, 309
Intracistronic complementation, 323
Intragenic suppression, 283
Intron, 17, 219–23, 231–32, 263, 397
    adenovirus, 219–20
    catalytic, 223–24
    group I, 223
    group II, 223
    mutations in, 283
Inversion, 89, 91, 121–22, 300–301, 406–7
    paracentric, 91–92, 121
    pericentric, 91–92, 121
Inversion heterokaryotype, 92, 121–22
Inversion heterozygote, 91–92
Inversion loop, 91–92
IQ score, 59, 65, 67–68, 395
IS. *See* Insertion sequence
Island population, 421, 434–35

Islet of Langerhans, 338–39
Isoleucine, 238
Isoschizomer, 364

J

Jacob, François, 17, 187–90, 243–44, 315, 317
Jacobs, W., 132
Japanese serow, inbreeding in, 443
Jeffrey, Alec, 381, 383
Jimson weed, trisomy in, 97
Jones, Peter, 349
Joyce, Gerald, 224
Jumping gene. *See* Transposable element
Juvenile hormone, 450

K

Kanamycin resistance, 299
Karpechenko, G., 96
Karyotype, 77–78
Kazazian, Haig, 305
Kel Kummer Taureg tribe, 434
Kelner, Albert, 285
Khorana, Har Gobind, 16–17, 252
Kidwell, Margaret, 305
Kilobase pair, 146
Kimura, Motoo, 435
Kinetochore, 77
King, Thomas, 340
Klenow fragment, 373, 385–86
Klinefelter syndrome, 75, 98–99, 102
Klug, Aaron, 206
Kornberg, Arthur, 156, 166–67
*Kpn*I, 365
Kunkel, L. M., 399

L

lacA gene, 185–91
lacI gene, 186–87
lacY gene, 185–91
lacZ gene, 185–91
lac operator, 186–89
lac operon, 185–91, 198–99
    F-*lac*, 320–21
    mutations in, 188–89
    negative control of, 186–90
    positive control of, 190–91
Lac⁺ phenotype, 318, 320
lac promoter, 190, 375–76
lac repressor, 186–89, 375
Lacks, Henrietta, 219
Lactose, 186
Land race, 441
Lariat model, of RNA splicing, 222, 224
LDL receptor. *See* Low-density lipoprotein receptor
Leader region, 219
Leder, Philip, 352, 358
Lederberg, Joshua, 312–13
Leopard frog, chromosome number in, 77
Leptotene, 83, Panel 4
Lesch-Nyhan syndrome, 115, 389
Lethals, 47–48, 272, 424, 433
    conditional, 50, 272
Leucine, 238
Leukocyte hexokinase gene, 410
Levan, Albert, 78
Levene, P., 132
Lewontin, Richard, 384, 410
LexA protein, 289–90
Life, origin of, 224, 255–56
Life cycle
    of garden pea, 24
    of *Neurospora crassa*, 119, 242
Light-induced genes, in plants, 343–44
Light repair. *See* Photoreactivation repair
Lily, DNA of, 146, 263
Linkage, 31, 106
    in humans, 113–15
    physical explanation of, 106–8
Linkage disequilibrium, 434
Linkage map. *See* Genetic map

Linker DNA, 348
Linn, Stewart, 364
Locus, 27
Long terminal repeat (LTR), 303–4
Lovett, Michael, 164
Low-density lipoprotein (LDL) receptor, 389
lpp promoter, 375
LTR. *See* Long terminal repeat
Luciferase gene, 392–93
Lung cancer, 308
Luria, Salvador, 315
Luxury gene, 338
Lymphoma, Burkitt's, 360
Lyon, Mary, 349
Lyonization. *See* X chromosome, inactivation of
Lysenko, Trofim, 449
Lysine, 238
Lysogen, 195
Lysogenic infection, 194–98, 323
Lytic infection, 194–96, 323

M

McCarty, Maclyn, 17, 134–35
McClintock, Barbara, 13, 17, 300–301
MacLeod, Colin, 17, 134–35
Maize. *See* Corn
Major histocompatibility complex (MHC) genes, 123
Malaria, 275, 407–9, 426–28, 434
Mal de Meleda, 432, 434–35
Malignant transformation, 305–9
Mammoth, 438–39
Mannopine synthase, 392
Manx cat, 48, 51
Map distance, 109
    physical distance and, 123–25
Mapping, 12–13, 126
    by conjugation, 317–18, 333–34
    deletion, 115
    with *Escherichia coli,* 320–23
    fine structure, 331–32, 387
    in humans, 115–16, 393–99
    in *Neurospora crassa,* 119–21
    in T-even phage, 329–32
    transduction, 325–26
Map unit, 109, 317, 331
Marker, 312
Master gene, 350, 353–54
Maternal age, Down syndrome and, 100
Maternal effect gene, of *Drosophila melanogaster,* 354–55, 357
Maternal inheritance, 439–40
Maternal message, 229–30, 342
Mating
    consanguineous, 38–39, 417–18, 427, 434, 442
    negative assortative, 415
    nonrandom, 415–18
    positive assortative, 415
    random, 412–13, 417
    reciprocal, 22
Mating system, of yeast, 214
MAT locus, 214
Matthaei, Johann Heinrich, 252
Maxam, Alan, 385–87
*Mbo*I, 365
MCS. *See* Multiple cloning site
Mean, 62–63
Mean fitness, 423
Megabase, 148
Megasporogenesis, 86
Meiosis, 13, 28, 76–77, 82–84, Panel 4
    abnormal, 95–97, 99
Meiosis I, 82–84, 86, 97, 99, Panel 4
Meiosis II, 82–84, 97, 99, Panel 4
Melanin, 271, 425, 433
Melting temperature, of DNA, 144–45
Mendel, Gregor, 10–11, 17, 22–31
Mendelian genetics, 10
    extensions and applications of, 45–71
    fundamentals of, 21–41
    in humans, 38–41
Mendel's first law. *See* Segregation, principle of
Mendel's second law. *See* Independent assortment
Mercury resistance, 299
Meristic trait. *See* Countable trait

Merodiploid, 188–89, 313, 318, 320–24
Merozygote. *See* Merodiploid
Meselson, Matthew, 17, 158, 172–74, 243
Mesoderm, 342
Messenger RNA (mRNA), 16–17, 182. *See also* Translation
    base sequence of, 291
    bicoid, 354–55
    capping of, 227–29, 258
    casein, 229
    degradation of, 227, 229
    discovery of, 242–44
    3'-end of, 226–27, 246
    5'-end of, 227–28, 246, 258
    of eukaryotes, 258
    fushi tarazu, 356
    globin, 228–29
    half-life of, 229
    histone, 229
    immunoglobulin, 359
    maternal, 229–30, 342
    measurement with DNA probes, 381–85
    ovalbumin, 229
    polyadenylation of, 226, 229
    polycistronic, 186
    in polysomes, 246–47
    production of cDNA, 373
    ribosome binding sites, 258
    RNA splicing, 220–22
    synthesis of, 209
    synthetic, 252–53
    in translation, 256–62, Panel 7
Messing, Joachim, 371
Metabolism, inborn errors of, 57, 68
Metacentric chromosome, 77, 82, 91, 95, 113
Metafemale, 98, 102
Metamale, 102
Metaphase
    meiosis I, 83–84, Panel 4
    meiosis II, 83–84
    mitotic, 80–82, Panels 2 & 3
Metaphase plane, 81, 83–84
Methanol, 133
Methionine, 238
Methylamine, 133
Methylation, of DNA, 278–79, 288, 349–52, 399
5-Methylcytosine, 278–79, 349
Methyl group, 132–33
Methylguanine methyl transferase, 286–87
Methyl phosphate, 133
Methyl transferase, 227
MHC genes. *See* Major histocompatibility complex genes
Microchromosome, 77
Micrococcal nuclease, 204–5, 345
Microinjection, 340–41
Microsporogenesis, 86
Microtubules, 82, Panel 1
Miescher, Friedrich, 14, 17, 132
Migration, 421
Miller, Oscar, 222
Minichromosome, 204–6, 345
Minimal medium, 271, 312
Minisatellite DNA, 381–83
Minute (map unit), 317
Mismatch repair, 288
Mispairing, 121–23
Missense mutation, 251, 273–75, 285, 292
Mitochondria, Panel 1
    DNA of, 146, 439–40
    genes of, 351
    genetic code in, 256
    ribosomes of, 259
Mitochondrial malic enzyme gene, 410
Mitosis, 76–77, 79–82, Panels 2 & 3
MN blood group, 47, 69
Modifier gene, 51
m.o.i.. *See* Multiplicity of infection
Molecular clock, 437–38
Molecular evolution, 435–40
Molecular genetics, 10, 13–16
Monocot, 392
Monoculture, 448
Monod, Jacques, 187–90, 243–44
Monohybrid cross, 22, 25
Monomorphic gene, 409–10

Monosomy, 89, 97
Monozygotic twins, 65
Morgan, Thomas Hunt, 11–13, 17, 86–87, 106–8, 121–23
Morphological mutation, 271
Mosaic development, 339–40
Mosaic embryo, 342
Mosquito, DDT resistance in, 448–49, 451
Mouse
    body weight in, 67, 444–45
    chromosome number in, 77
    coat color in, 28–29, 47–50, 58–59, 272
    DNA of, 146
    genomic imprinting in, 351–52
    inbred lines of, 447
    litter size in, 67
    spontaneous mutation rate in, 280
    supermouse, 390
    tail length of, 67
    transgenic, 351–52
mRNA. *See* Messenger RNA
Muller, H. J., 17, 113
Mullis, Kary, 383
Multigene family, 123, 440
Multiple alleles, 51–53, 414
Multiple cloning site (MCS), 367–69, 371
Multiple drug resistance, 299–300
Multiple genes, 56–59, 71
Multiplicity of infection (m.o.i.), 329
Muscular dystrophy, 414–15
    Duchenne, 115, 219, 393, 399
Mutagen, 14, 275, 280–81
    carcinogens as, 306–7
    detection of, 283–85
    lysogen induction by, 196–97
Mutagenesis
    chemical, 280–81, 379
    site-directed, 210, 378–80
    transposon, 298–301
Mutant, 10, 12
    constitutive, 188–90
    cryptic, 188
    fixation in population, 436
    loss from population, 436
    neutral, 436
Mutation, 16, 154, 170
    amber, 261, 272
    biochemical. *See* Mutation, nutritional
    cancer and, 270, 305–6
    chemically-induced, 280–81
    chromosomal. *See* Chromosome, abnormalities of
    in cis, 189, 323
    cis-dominant, 189
    conditional, 272–73
    converting proto-oncogene to oncogene, 307–9
    DNA damage and repair, 269–93
    dominant, 189
    down, 184
    enhancement by transposons, 305
    forward, 283
    frameshift, 251–52, 275, 277–78, 280, 283–85
    germinal. *See* Mutation, germ-line
    germ-line, 270
    induced, 418
    in introns, 283
    lethal, 272, 424, 433
    missense, 251, 273–75, 285, 292
    in mitochondrial DNA, 439–40
    morphological, 271
    mutation-selection balance, 432–34
    mutator, 276–77
    nutritional, 271–72
    ochre, 261
    opal, 261
    operator constitutive, 189
    pleiotropic, 323
    point, 273–75
    radiation-induced, 281–82
    recessive, 188–89
    silent, 283
    somatic, 270, 360
    spontaneous, 275–80, 418
    temperature-sensitive, 272
    in trans, 189, 323
    transition, 275
    transposable elements, 295–310

transversion, 275
up, 184–85
variation and, 418–19
visible. *See* Mutation, morphological
Mutation rate, 279–80, 292, 433–34, 436–37
backward, 418
forward, 418
Mutator, 275–77
mut gene, 277
Mutually exclusive events, 31
myc oncogene, 360
Mycoplasma, genetic code in, 256
Myeloma cells, 358
Myoglobin, 239–41

N

Nail-patella syndrome, 113
Nalidixic acid, 166
Natal homing, 439–40
Natural selection, 48, 406, 422–27
against homozygote, 423–26
heterozygous advantage, 426–27
mutation-selection balance, 432–34
purifying, 423
Nature-nurture controversy, 67
Negative assortative mating, 415
Negative control, of *lac* operon, 186–90
Neurospora crassa
Beadle and Tatum experiments with, 14, 241–42
chromosome number in, 77
DNA of, 146
life cycle of, 119
nutritional mutations in, 272
spontaneous mutation rate in, 280
sporulation in, 242
tetrad analysis in, 119–21
Neutrality theory, 435–47
N-Formyl methionine (fMet), 257
N gene, 195–97
Nicholas II (Czar of Russia), 55
Nick translation, 157
Nilsson-Ehle, Herman, 61
Nirenberg, Marshall, 16–17, 252–53
Nitrocellulose filter, 371, 380–81
Nitrogenous base, 15, 136, 138
alkylation of, 281–82, 286
analogues of, 280–81
deamination of, 277–79
modified, 249, 278
tautomerization of, 276–77, 280–81
Nitrous acid, 280
Noll, Markus, 354
Nomura, Masayasu, 244
Nondisjunction, 87–88, 97–100
Nonhistone protein, 207
Nonparental chromosomal type, 107
Nonrandom mating, 415–18
Nonreciprocal recombination, 172
Nonrecombinant, 107, 109
Nonsense codon. *See* Stop codon
Nontranscribed spacer (NTS), 222, 225
Normal distribution, 62
Norm of reaction, 59–60
Northern blot, 385
Northern elephant seal, genetic variation in, 410–11, 419–20
*Not*I, 364–65
Novobiocin, 166
n′ protein, 167
NTS. *See* Nontranscribed spacer
Nuclear membrane, 81–84, Panels 3 & 4
Nucleic acid, 132. *See also* DNA; RNA
catalytic, 223–24
Nuclein, 14, 132
Nucleolus, 81–83, 222, 225, 246, Panels 1 & 3
Nucleoside, 138
Nucleosome, 345–49
core particle of, 204
folding of string of, 206–7
function of, 204–5
release from chromatin, 204–5
structure of, 204–6, 208
Nucleosome fiber, 206
Nucleotide, 14, 138

dideoxy, 385–86
forming DNA from, 154–55
Nucleus, transplantation of
in frogs, 364
in *Xenopus laevis,* 340–41
Nullisomy, 97
Nurse cells, 355
Nüsslein-Volhard, Christiane, 354–55
Nutritional mutation, 271–72

O

Observed numbers, 37–38
Ochoa, Severo, 261
Ochre mutation, 261
Ocular albinism, 115
O gene, 195
Okazaki, Reiji, 154, 159–60
Okazaki, Tuneko, 160
Okazaki fragment, 159–60, 168–69
Oligomeric protein, 323
Oligonucleotide, 160
Oligonucleotide primer, 379, 385–86
Oncogene, 307–9
rearrangement of, 360
Oncogenesis, 307
Oncogenic virus, 307
One-step growth experiment, 328–29
Oocyte
primary, 85
secondary, 85
Oocyte 5S rRNA genes, 347–48
Oogenesis, 84–86, 229–30
Oogonium, 85
Ootid, 85
Opal mutation, 261
Open reading frame, 264, 387
Operator
lac, 186–89
phage λ, 195, 198
Operator constitutive mutation, 189
Operon, 185–91. *See also specific operons*
Opine, 392
Orangutan
chromosomes of, 100–101
evolution of, 439
Organic molecule, 132–33
Organ transplantation, 53
Origin of life, 224, 255–56
Origin of replication, 161–64, 370
Outbreeding, 415
Outcrossing, 417
Ovalbumin gene, 343
Ovalbumin mRNA, 229
Ovary, 85
Overlapping genes, 387–88
Ovum. *See* Egg cell
Ozone layer, 282

P

p21 protein. *See* Ras protein
Pachytene, 83, Panel 4
Pair-rule genes, 356
Palindrome, 365
Palmiter, Richard, 390
Pancreatic amylase gene, 410
Pantothenic acid, synthesis of, 242
Paper chromatography, of peptides, 273–74
Paracentric inversion, 91–92, 121
Paramecium, genetic code in, 256
Pardee, Arthur, 188
Parentage testing, by DNA fingerprinting, 383–84
Parental chromosomal type, 107
Parthenogenesis, 76
Particulate inheritance, 22
Pascal's triangle, 33
Patau syndrome, 98
Paternity exclusion, 69–70
Pathway, 14, 185, 241–42
Pattern baldness, 55
Pauling, Linus, 139
pbsA protein, 450–51
PCR. *See* Polymerase chain reaction

Pedigree analysis
calculation of inbreeding coefficient, 417–18
with consanguineous mating, 417–18, 427, 434
of dominant trait, 40–41
of Huntington disease family, 395–97
of recessive disease caused by two genes, 57
of recessive trait, 39–40
symbols used in, 38–39
of X-linked dominant trait, 55
of X-linked recessive trait, 54–55
P element, 305
Penetrance, 49–51
incomplete, 49–50
Peppered moth, industrial melanism in, 52–53, 425
Pepsinogen gene, 410
Peptidase A gene, 410
Peptidase C gene, 410
Peptidase D gene, 410
Peptide bond, 239
formation of, 239, 258–60, Panel 7
Peptidyl transferase, 258–60, Panel 7
Pere David's deer, genetic variation in, 420
Performance, 445–46
Pericentric inversion, 91–92, 121
Pericentromeric DNA, 208
Permissive temperature, 272
Pero, J., 192
Pesticide resistance, 63, 421, 448–51
pfu. *See* Plaque-forming unit
P gene, 195
Phage
helper, 323
inadvertent sharing of, 330
infection by
burst size from, 328
latent period of, 328
rise period of, 328
temperate, 194, 323–24
T-even. *See* T-even phage
transducing, 323–26
as vectors, 369–71
virulent, 194
Phage fd, 327
Phage λ, 323–24
as cloning vector, 369–71, 374
DNA of, 146
genetic map of, 195
infection of *Escherichia coli,* 194–97
λ*dgal,* 323–24
λgt11, 374, 376–78
λ*pbio,* 323–24
lysogenic infection with, 194–97, 323
lytic infection with, 194–96, 323
promoters of, 375, 377
replication of, 168
repressor, 195–98, 375, 377
structure of, 327
Phage M13
as cloning vector, 371–72, 378–79, 385
replication of, 160
Phage MS2, 327
Phage Mu, 298–300
map of, 300
Phage P1, in transduction, 323–25
Phage φX174, 387
DNA of, 146–47
genetic map of, 388
structure of, 327
Phage PM2, DNA of, 146
Phage R17, 258
coat gene in, 261
Phage SPO1, 191–92
Phage T2
discovery of mRNA, 243–44
DNA of, 146, 263
h and *r* genes of, 329–30
Hershey-Chase experiment with, 134–37
plaque assay for, 329
spontaneous mutation rate in, 280
Phage T4
DNA of, 146, 263
rapid lysis mutants of, 271
replication of, 159–60
rII locus of, 331–33
structure of, 327–28
Phage T5, 327

Phage T7, 327
    RNA polymerase of, 192–93
Phenocopy, 51
Phenotype, 10, 12, 22
    relationship to genotype, 59–60
Phenotypic value, 64
Phenotypic variance, 64–65
Phenylalanine, 238, 241, 391
Phenylketonuria (PKU), 48–49, 59, 432–33
Phosphate group, 132–33, 138
Phosphodiester bond, 139, 154–55, 182–83
Phospho-ester bond, 138
Phosphoglucomutase-1 gene, 410
Phosphoglucomutase-3 gene, 410
6-Phosphogluconate dehydrogenase gene, 410
Phosphoglycerate kinase deficiency, 115
Phosphoribosylpyrophosphate synthetase deficiency, 115
Phosphoric acid, 136
Photoreactivating enzyme. *See* DNA photolyase
Photoreactivation repair, 285–86
Photoreactivation, 343–44, 450–51
Phylogenetic tree, 438–39
Pig
    back fat thickness in, 67, 445–46
    growth rate of, 445–46
    karyotype of, 78
    litter size in, 67
    superpig, 390
    weight of, 67
Pigeon, transferrin in, 427
Pigweed, atrazine resistance in, 450–51
Pine, chromosome number in, 77
Pink bread mold. See *Neurospora crassa*
Pisum sativum. See Garden pea
PKU. *See* Phenylketonuria
Placental alkaline phosphatase gene, 410
Plant
    aneuploidy in, 97
    genetically engineered, 390–93
    genetic code in, 256
    light-induced genes in, 343–44
    polyploidy in, 95–96
    regeneration in, 340
    transgenic, 344, 390
Plant breeding, 96, 390–93, 444–48
Plantibody, 390
Plaque, phage, 323, 328
Plaque assay, 328–29
Plaque-forming unit (pfu), 328
Plaque hybridization, 371
Plasmid. *See also* Conjugation
    antibiotic resistance genes on, 299, 450
    F factor. *See* F plasmid
    replication of, 164
    as vectors, 366–69
Plasmid pBR322, 366–67
Plasmid pIN, 375
Plasmid pKC30, 375, 377
Plasmid pSC101, 365–66
Plasmid pUC, 367–69, 374
Plasmid RSF1010, 365–66
Pleiotropic mutation, 323
Pleiotropy, 48–49
P-M system, 305
Point mutation, 273–75
Polar body, 85
Pollen, 76, 86
Polyadenylation. *See* Poly(A) tail
Poly(A) polymerase, 226
Poly(A) tail, of mRNA, 226–27, 229
Polycistronic message, 186
Polygenic trait, 61
Polyhedrin, 378
Polymerase chain reaction (PCR), 383–84
Polymorphic gene, 409–11
Polynucleotide probe, 372
Polypeptide, 14, 239
    one gene-one polypeptide hypothesis, 242
Polyphenylalanine, 252–53
Polyploid, 94–96
Polyribosome. *See* Polysome
Polyserine, 253
Poly(serine-leucine), 252–53
Polysome, 246–47, Panel 1
Polytene chromosome, 78–79, 91

Poly(tyr-leu-ser-ile), 253
Poly(U), 253
Poly(UAUC), 253
Poly(UC), 252–53
Poly(UUC), 252–53
Population, 409
    hybrid, 422
Population genetics, 10, 406, 432
Positive assortative mating, 415
Positive control, of *lac* operon, 190–91
Post-replication repair. *See* Recombination repair
Posttranscriptional control, 343
    in eukaryotes, 228–29
Posttranslational control, in eukaryotes, 228, 230–31
Potato, ploidy in, 95
Potentilla glandulosa, phenotypes at different altitudes, 59–60
Pott, Sir Percival, 306
Poultry. *See also* Chicken
    egg weight in, 67
    viability of, 67
    weight of, 67
Prader-Willi syndrome, 351
Preinitiation complex
    class I, 215
    class II, 215
    class III, 216
Preproinsulin, 230–31
Pribnow, David, 184
Pribnow box. *See* -10 Box
Primary transcript, 220
Primase, 166–68
Primer
    DNA, 384
    oligonucleotide, 385–86
    RNA. *See* RNA primer
Primosome, 166–67
Primrose, flower color in, 50
Probability, 31–38
    binomial, 33–35, 40
    conditional, 32, 40
Proband, 40
Probe, 371–72
    antibody, 376–78
    cloned gene as, 380–85
    counting genes with, 380–81
    fingerprinting DNA with, 381–82
    measuring gene expression with, 381–85
    polynucleotide, 372
Product rule, 31–33
Proinsulin, 230–31
Prokaryote, 13
    DNA of, 146
    polysomes in, 246–47
    replication in, 164
    RNA polymerase in, 209
    transcriptional control in, 181–200, 343
    translation in, 247
Prolactin, 229
Proline, 238
Promoter, 183, 190, 214, 323, Panel 6
    adenovirus, 216
    base sequence of, 184–85
    Escherichia coli, 184–85
    in eukaryotes, 209–13
        recognized by polymerase I, 212
        recognized by polymerase II, 210–12, 215
        recognized by polymerase III, 213
    expression vector, 375
    ftz gene, 357
    lac, 190, 375–76
    lpp, 375
    mannopine synthase, 392
    mutations in, 184–85
    phage λ, 195–97, 375, 377
    RNA polymerase binding to, 184
    rRNA, 212, 215
    5S rRNA, 213
    strength of, 184–85
    SV40, 210, 212, 216, 218
    tRNA, 213
    trp, 375–76
Promoter complex
    closed, 184, Panel 6
    open, 184, Panel 6

Proofreading, in replication, 277
Prophage, 195
Prophase
    meiosis I, 83, Panel 4
    meiosis II, 84
    mitotic, 80–82, Panels 2 & 3
Propidium iodide, 82
Protan color blindness, 115
Protein, 14
    amino terminus of, 230
    denaturation of, 272
    DNA-protein interactions, 197–98
    domain of, 239
    electrophoresis of, 407–8
    evolution of, 437–38
    exported, 230–31
    fingerprint of, 273–74
    functions of, 241
    fusion, 374–77
    genes and, 14–15, 241–42
    genetically engineered, 378–80, 387–88
    globular, 239–40
    oligomeric, 323
    posttranslational modification of, 230–31
    primary structure of, 239, 291–92, 400
    quaternary structure of, 239–40
    secondary structure of, 239
    sequencing of, 273
    synthesis of. *See* Translation
    tertiary structure of, 239–40
    from transgenic organisms, 390
Protenor, sex determination in, 100
Proto-oncogene, 307–9
    translocations of, 360
Prototroph, 271, 312
Protozoa, genetic code in, 256
Provirus, 303–4
Pseudodominance, 90
Pseudogene, 361
Pseudouridine, 248–49
Pseudo-wild-type revertant, 252
P site, ribosomal, 258–60, Panel 7
*Pst*I, 364–67
Pulsed-field gel electrophoresis, 148
Pulse labeling, 159
Punnett, R. C., 30, 56, 106
Punnett square, 30
Purifying selection, 423
Purine, 136, 138
Puromycin, 260
*Pvu*I, 364–65
Pygmy hippopotamus, inbreeding in, 443
Pyloric stenosis, 68
Pyrimidine, 136, 138
Pyrimidine dimer, 289–90

Q

Q band, 78
Q gene, 195–96
Quadrivalent, 96
Quadruple crossover, 113–14
Quantitative trait, 60–64, 71
    mean and variance of, 62–63
    model of, 64–65
Quenching, 145

R

rII locus, 331–33
Rabbit
    coat color in, 50
    Himalayan, 50
Race
    admixture estimates, 422
    intelligence and, 67–68
Radding, 172–74
Radial loop model, of chromatin folding, 206–8
Radiation
    cancer induction by, 306–7
    lysogen induction by, 196–97
    mutations induced by, 281–82
Ralls, Kathy, 443
Random mating, 412–13, 417

Rapid-lysis mutant, of T-even phage, 329
Rapist, identification of, 383
ras genes, 307–9
Ras protein, 308–9
Rat
  inbred lines of, 447
  maze-learning ability in, 59–60, 64
  pesticide resistance in, 449
  warfarin resistance in, 426, 450–51
R band, 78
Reading frame, 252, 275, 277, 284, 387–88
Realized heritability, 65
recA gene, 196–97, 289–90
RecA protease, 197, 289–90
Receptor. *See specific receptors*
Recessive trait, 11, 22, 25–28, 35, 188–89
  in humans, 39–40
  mutation-selection balance for, 432–34
  pedigree for, 39–40
  X-linked, 54–56
Reciprocal mating, 22
Reciprocal recombination, 171–72
Reciprocal translocation, 91–92, 95
Recognition helix, 197
Recombinant, 13, 107, 109
  double, 111
Recombinant DNA, 365–66
Recombination, 12–13, 83, 107
  between T-even phages, 329–33
  bimolecular, 171
  in conjugation, 313, 318, 320
  in corn, 13
  in *Drosophila melanogaster,* 13
  generalized. *See* Recombination, homologous
  homologous, 171–74
  illegitimate, 171, 298
  intramolecular, 171–72
  inversions and, 121–22
  mechanism of, 171–74
  nonreciprocal, 172
  physical evidence for, 13
  rate of, 109, 114, 119–21
  reciprocal, 171–72
  site-specific, 171
  in transduction, 324–25
  transposon-mediated, 300–301
Recombination frequency, 113
Recombination repair, 289
Recon, 331
Recurrence risk, 68
  empirical, 68
Red blood cells, 338–39
  antigens of, 47
  sickled. *See* Sickle-cell disease
Reductional division. *See* Meiosis I
Regeneration, in plants, 340
Regulative development, 339–41
Regulative embryo, 340
Regulatory gene, 186
Regulatory mutant, 189
Relative fitness, 423, 425–27
Release factor (RF)
  eRF, 262
  RF-1, 262
  RF-2, 262
  RF-3, 262
Renaturation, of DNA, 145
Replica plating, 367
Replication, 15–17, 80, 143, Panels 2 & 5
  in *Bacillus subtilis,* 161–62
  bidirectional, 154, 156, 160–64
  of colE1, 164
  conservative, 154–55, 158
  direction of, 154, Panel 5
  in *Drosophila melanogaster,* 162–63
  elongation stage of, 167–68
  enzymology of, 164–71
  errors in, 275–77
  in *Escherichia coli,* 160–62, 164–71
  in eukaryotes, 164
  fidelity of, 170–71
  initiation of, 166–67
  lagging strand in, 154–55, 157, 167–69
  leading strand in, 154–55, 157, 167–69
  mechanism of, 154–64

  origin of, 161–64, 370
  of phage λ, 168
  of phage M13, 160
  of phage T4, 159–60
  in prokaryotes, 164
  proofreading in, 277
  rate of, 175–76
  RNA primer for, 154–57, 160, 166–67, 169–71
  rolling circle, 168–69, 316, 372, 379
  semiconservative, 143, 154–59, 174–75
  semidiscontinuous, 154–55, 159–60, 171
  strand separation in, 154, 157, 164–66
  termination of, 168–70
  theta mode of, 161
  transcription compared to, 185
  in *Triturus,* 163
  unidirectional, 156, 164
Replication fork, 161–63
Replicative form (RF), 371–72, 379
Replicative transposition, 297–98
Replicon, 163–64
Replisome, 167–68
Repression, 186
Repressor
  lac, 186–89, 375
  λ, 195–97, 375, 377
  LexA, 289–90
Repulsion, 106, 108
Repulsion double heterozygote, 108
Repulsion gametes, 107
Resolution, of cointegrate, 298
Resolvase, 298
Response, 65
res site, 298
Restriction endonuclease, 364–66
Restriction fragment length polymorphism (RFLP), 125, 394–95
  mapping Huntington disease gene, 394–99
Restriction site, 365–66
Restrictive temperature, 272
Retinoblastoma, 49
Retinoic acid receptor, 343
Retinoschisis, 115
Retroposon. *See* Retrotransposon
Retrotransposon, 304
Retrovirus, 303–5
Retrovirus-like particle, 304
Reverse duplication, 89–90
Reverse transcriptase, 303–4, 373
Reverse transcription, 303
Reversion, 270, 283–84
  second-site, 283
  true, 283
Revertant, 261
RF. *See* Release factor; Replicative form
RFLP. *See* Restriction fragment length polymorphism
Rheumatoid arthritis, 53
Ribonuclease H (RNase H), 373
Ribonucleic acid. *See* RNA
Ribonucleotide triphosphate, 182
Ribose, 136, 138
Ribosomal proteins, 244–45
Ribosomal RNA (rRNA)
  human, 225–26
  5S, 209, 213, 216, 244
  5.8S, 223–26, 246
  16S, 244, 258
  18S, 223–26, 246
  23S, 244
  28S, 223–26, 246
  synthesis of, 209
    self-splicing precursors, 223–24
    trimming precursors of, 222–26
Ribosomal RNA (rRNA) genes, 212, 222–26, 246, 440
  introns in, 219
  5S
    oocyte, 347–48
    somatic, 347–48
    of *Xenopus laevis,* 346–48
Ribosome, 16, 243–47, Panels 1 & 7
  A site on, 258–60, Panel 7
  binding sites for mRNA, 258
  of eukaryotes, 246
  mitochondrial, 259
  in polysomes, 246–47

  P site on, 258–60, Panel 7
  reconstitution of, 244–45
  70S, 243–45
  self-assembly of, 245
  30S particle, 243–45, 256–57
  50S particle, 243–45, 256–57
  subunits of, 243–45
  in translation, 256–62
Ribosome binding site, 375–76
Ribozyme, 223–24
Ribulose bisphosphate carboxylase (rubisco), 344
Rich, Alexander, 143
Ricin gene, 372
Rifampicin, 160
Ring chromosome, 93
Riordan, Jack, 17, 399
R-looping technique, 219–20
RNA, 14, 132
  as genetic material, 136
  heterogeneous nuclear. *See* Heterogeneous nuclear RNA
  hybridization of. *See* Hybridization
  messenger. *See* Messenger RNA
  processing of, 222–27
  ribosomal. *See* Ribosomal RNA
  small nuclear. *See* Small nuclear RNA
  spacer, 222–23
  splicing of, 220–22, 231–32
  stable, 224
  structure of, 136–39
  synthesis of. *See* Transcription
  transfer. *See* Transfer RNA
  U1, 221
RNA polymerase, 160, 182–83, 207, Panel 6. *See also* Sigma factor
  of *Bacillus subtilis,* 192
  binding to promoter, 184
  core polymerase, 183–84
  of *Escherichia coli,* 183
  in eukaryotes, 209, 214
  modification of, 191–92
  mutations in, 272
  in phage T7, 192–93
  in prokaryotes, 209
RNA polymerase I, 209, 212, 215, 225
RNA polymerase II, 209–12, 215, 226, 231, 303
RNA polymerase II genes, 348–49
RNA polymerase III, 209, 213, 216, 347
RNA primer, Panel 5
  removal of, 169–70
  in replication, 154–57, 160, 166–67, 169–71
RNA virus, 303–5
Roberts, Richard, 17, 444
Roeder, Robert, 209, 216, 346–48
Rolling circle replication, 168–69, 316, 372, 379
Roux, Wilhelm, 339
rRNA. *See* Ribosomal RNA
Rubisco. *See* Ribulose bisphosphate carboxylase
Rust resistance, 451
Rutter, William, 209

S

Saccharomyces cerevisiae. See Yeast
S-Adenosyl methionine (SAM), 227
Safety, of gene cloning, 395
*Sal*I, 365–66
Salmonella newport, 299–300
Salmonella typhimurium
  in Ames test, 284–85
  in cattle, 299–300
  DNA of, 146
Salmonellosis, 299–300
SAM. *See* S-adenosyl methionine
Sanger, Frederick, 17, 273, 385–87
Satellite, chromosomal, 77–78
Scaffold, chromosomal, 207
Screening, by replica plating, 367
Scrotal cancer, 306
SD sequence. *See* Shine-Dalgarno sequence
Sea urchin
  fertilization in, 86
  germ cells of, 9
  maternal messages in eggs, 230

Second filial (F_2) generation, 11, 25, 27, 106
Second-site reversion, 283
Sedimentation coefficient, 243
Segmentation genes, of *Drosophila melanogaster*, 354, 356–57
Segregation, principle of, 22, 28–29, 41–42, 86
Selection, 325
    artificial, 66, 444–46
    natural. *See* Natural selection
Selection coefficient. *See* Selective disadvantage
Selection differential, 65, 444
Selection limit, 444–45
Selective disadvantage, 423–26, 432–33
Self-assembly, of ribosomes, 245
Self-fertilization, 22, 24, 28, 76, 95, 415–17, 447
Selfing. *See* Self-fertilization
Selfish DNA, 296
Self-splicing, 223–24
Semen, DNA fingerprinting, 383
Semiconservative replication, 143, 154–59, 174–75
Semidiscontinuous replication, 154–55, 159–60, 171
Sendai virus, 116
Serine, 238
Severe combined immunodeficiency, 17, 389
Sex chromosome, 12
    of domestic mammals, 78
    nondisjunction of, 98–99
Sex determination, 95, 100–102
Sex-duction. *See* F-duction
Sex-influenced trait, 55–56
Sex-limited trait, 55
Sex-linked gene, 12, 53–56, 351
Sex pilus, 313
Sex-reversed individual, 102
Sexual reproduction, 76–77
Shapiro, James, 296
Shapiro, Lawrence, 349
Sharp, Phillip, 17, 219–20
Sheep
    horned versus polled, 55–56
    karyotype of, 78
    transgenic, 390
Shepherd's purse, pod characteristics in, 58
Sheppard, Phillip, 425
Shine, J., 258
Shine-Dalgarno sequence (SD sequence), 258, 375–76
Shull, George, 448
Shuttle vector, 378
Siamese cat, 50–51, 272
Sickle-cell disease, 16, 47, 68, 273–75, 389, 406–9, 423, 426–28, 432, 434
    identification of carriers, 398
Sigma factor, 183–84, Panel 6
    heat-shock specific, 194
    phage-encoded, 192
    phage SPO1-specific, 191–92
    sporulation-specific, 193–94
Signaling, intracellular, 309
Signal peptidase, 230
Signal peptide, 230–31
Silencer, 213–14
Silent mutation, 283
Silk moth, recombination in, 114
Single-strand-binding protein (SSB), 164–65, 167
Sister chromatids, 81–82, 84, Panel 4
Site-directed mutagenesis, 210, 378–80
Site-specific recombination, 171
Six-fingered dwarfism, 435
Skin cancer, 290, 307
7SL RNA, 213
*Sma*I, 364–65
Small nuclear ribonucleoprotein (snRNP), 221–22
Small nuclear RNA (snRNA), 221
Smith, Hamilton, 364
Snail, shell banding pattern in, 406–7, 423
snRNA. *See* Small nuclear RNA
snRNP. *See* Small nuclear ribonucleoprotein
⟨...⟩ial maturity score, 65
⟨...⟩id, 206, 208
⟨...⟩ 5S rRNA genes, 347–48
⟨...⟩cell hybridization, 115–16
⟨...⟩utation, 270, 360
⟨...⟩se, 196–97, 289–90
⟨...⟩n clawed frog. See *Xenopus laevis*
⟨...⟩81

Southern blotting, 380–82
Spacer RNA, 222–23
Specialized cells, 338–39
Specialized transduction, 323–24
Species, 10
Species name, 10
Sperm, 9, 76, 85
Spermatid, 85
Spermatocyte
    primary, 84–85
    secondary, 84–85
Spermatogenesis, 84–86
Spermatogonium, 84–85
Spermatozoa. *See* Sperm
S phase, 80, Panel 2
Spina bifida, 68
Spindle apparatus, 81–82, Panels 3 & 4
Splice junction, 221
    cryptic sites, 221
Spliceosome, 221–22
Splicing, RNA, 220–22
Splicing signal, 220–21
Spontaneous mutation, 275–80, 418
Sporogenesis, 76–77
Sporulation
    in *Bacillus subtilis*, 193–94
    in *Neurospora crassa*, 242
    transcription during, 193–94
SSB. *See* Single-strand-binding protein
Stable equilibrium frequency, 421, 426
Stable RNA, 224
Stahl, Franklin, 17, 158
Stalker, David, 391–92
Standard deviation, 63
Standard type, 12
Stasis, 410
Statistical significance, 37–38, 413
staufen gene, 355
Steitz, Joan, 221, 258
Steroid hormone, 343
Steroid receptor, 343
Steroid sulfatase deficiency. *See* Ichthyosis
Steward, F.C., 340
Steyla
    determinants in, 342
    mosaic development in, 339–40
Stop codon, 254, 256, 261, 272, 275
Strand exchange. *See* Crossover
Strand invasion, 171
Strawberry, ploidy in, 95
Streptococcus pneumoniae
    colony morphology in, 271
    melting curve of DNA of, 144
    transformation in, 132–35
Streptomycin resistance, 299, 366
Stringency, of hybridization, 372
Structural gene, 186
Struhl, 355
Sturtevant, Alfred, 13, 17, 109–10, 121–23
Submetacentric chromosome. *See* Acrocentric chromosome
Suicide enzyme, 286
Sulfonamide resistance, 299
Sum rule, 31–33
Sunlight exposure, 282, 290, 307
Supermouse, 390
Supernumerary chromosome, 77
Superpig, 390
Suppressor, 261–62, 272
    amber, 261, 265, 283
    intergenic, 283
    intragenic, 283
Surf clam, maternal messages in eggs, 230
Sutton, Walter, 17
SV40 virus
    DNA of, 146, 263
    minichromosome of, 204–6, 345
    promoter of, 210, 212, 216, 218
    T antigen of, 214
swallow gene, 355
Sweet pea
    Bateson and Punnett experiments with, 106
    flower color in, 56–57
Symbols
    bacterial genetic notation, 312

gene, 35–36
    pedigree analysis, 38–39
Synapsis, 83
Synaptonemal complex, 83
Syntenic gene, 110

T

TAF. *See* TBP-associated factor
Tandem duplication, 89–90
T antigen, 214
TATA box, 210–13, 215, 344
TATA box-binding protein (TBP), 215–16
Tatum, E. L., 14–15, 17, 241–42, 312–13
Tautomerization, 280–81
Taylor, Austin, 300
Tay-Sachs disease, 68, 432
TBP. *See* TATA box-binding protein
TBP-associated factor (TAF), 215–16
T-DNA, 391–93
Telocentric chromosome, 77, 82
Telomerase, 169–70
Telomere, 78
Telomeric DNA, 169–70
Telophase
    meiosis I, 83, Panel 4
    meiosis II, 84
    mitotic, 80–82, Panels 2 & 3
Temin, Howard, 303
Temperate phage, 194, 323–24
Temperature effect, on phenotype, 50–51
Temperature-sensitive mutation, 272
Terminal deletion, 90–91
Terminal transferase, 373–74
Termination factor, Panel 7
Terminator, 185, Panel 6
    ρ-dependent, 195–96
Testcross, 28–29
Testes, 84
Testicular feminization, 102, 115
Tetracycline, 300
Tetracycline resistance, 299, 366–67
Tetrad, 119
Tetrad analysis, 119–21, 126
Tetrahymena
    rRNA precursor of, 223–24
    telomerase of, 170
Tetramer, 323
Tetraploid, 95–96
Tetrasomy, 97
T-even phage
    DNA of, 263
    genetics of, 326–33
    growth of, 327–28
    host-range mutants of, 329
    mapping in, 329–32
    rapid lysis mutants of, 329
    recombination between, 329–33
    structure of, 327–28
Thalassemia, 221, 440
Thalidomide, 51
Theta mode of replication, 161
Three-factor cross, 326, 334
Three-point cross, 110–13
Threonine, 238
Thymidine, 138, 280–81
Thymidine kinase gene
    of herpes simplex virus, 210, 212, 216–18
    of mouse, 116
Thymine, 15, 136, 138, 140, 278
    tautomers of, 276–77
Thymine dimer, 281–82, 286–87
Thyroid hormone receptor, 343
Tijo, Joe-Hin, 78
Ti plasmid, 391–93
Tjian, Robert, 212, 215–16
tnpA gene, 298
tnpR gene, 298
Tobacco
    chromosome number in, 77
    transgenic, 390–93
Tomato, genetic variation in, 441–42
Tomizawa, J., 318
Tonegawa, Susumu, 358
Topoisomerase, 165

Totipotent cells, 340, 342, 361
tra gene, 102
trans-acting element, 323
Transcribed spacer, 222
Transcript, 243
    primary, 220
Transcription, 239, 243, Panel 6
    asymmetry of, 182, 185
    direction of, 185
    DNA template for, 182–85
    elongation phase of, 183, 185
    in eukaryotes, 209–18
    initiation of, 182–85
    mechanism of, 182–85
    of phage λ, 195
    replication compared to, 185
    reverse, 303
    termination of, 183, 185
    in vitro and in vivo systems, 210–11
Transcriptional control
    in eukaryotes, 208, 213–18, 228–29, 343–45
    in prokaryotes, 181–200, 343
    in sporulation, 193–94
    temporal, 191–97
Transcription bubble, 185
Transcription factor, 213–17, 343, 346, 348–49, 352
    activation of *ftz* gene, 356
    bicoid protein, 355
    general, 214–15
    gene-specific, 214–17
    SL1, 215
    Sp1, 216
    TFIIA, 215
    TFIIB, 215
    TFIID, 215
    TFIIE, 215
    TFIIF, 215
    TFIIH, 215
    TFIIJ, 215
    TFIIIA, 216–17, 347–48
    TFIIIB, 216, 347
    TFIIIC, 216, 347
Transduction, in bacteria, 323–26, 333
    generalized, 323–25
    mapping by, 325–26
    specialized, 323–24
Transferrin, 427
Transfer RNA (tRNA), 247–51
    acceptor stem of, 248, 250
    amino acid binding to, 250–51
    anticodon loop of, 248, 250
    charged, 250–51
    cloverleaf structure of, 247–48
    dihydrouracil loop (D-loop) of, 248, 250
    for initiation, 257–58, Panel 7
    isoaccepting species of, 254
    secondary structure of, 248–49
    suppressor, 261–62, 272, 283
    synthesis of, 209, 213
    three-dimensional structure of, 249–50
    TψC loop of, 248, 250
    in translation, 256–62
    variable loop of, 249
Transfer RNA (tRNA) genes, 216, 219, 261, 440
Transformation
    in bacteria, 14, 132–35
    malignant, 305–9
Transforming substance, 134–35
Transgenic animal, 351–52, 390
Transgenic plant, 344, 390
Transition, 275
Translation, 182, 239, 243, Panel 7
    direction of, 258
    elongation step in, 258–60, Panel 7
    in *Escherichia coli*, 256–57
    initiation of, 256–58, Panel 7
    in prokaryotes, 247
    termination of, 261–62, Panel 7
    translocation step in, 259
Translational control, 343
    in eukaryotes, 228–30
Translocation, 89, 91–94, 121, 307
    interstitial, 91–92
    of proto-oncogenes, 360
    reciprocal, 91–92, 95

Translocation heterozygote, 91–94
Transmission genetics. *See* Mendelian genetics
Transplantation, organ, 53
Transposable element, 296
    in bacteria, 296–300
    in corn, 270, 300–302
    in *Drosophila melanogaster*, 302–3
    in eukaryotes, 300–303
    in humans, 305
    in yeast, 302–3
Transposase, 296, 298, 301–2
Transposition, 309
    conservative, 297
    impacts on public health, 299–300
    mechanism of, 297–98
    replicative, 297–98
    tracking with antibiotic resistance markers, 296–97
Transposon, 296
    composite, 299
    copia, 302
    mutability of chromosomes and, 305
    as mutagenic agent, 298–301
    Tn3, 296–99, 309
    Tn4, 299
    Tn9, 299
    Tn10, 299
    Tn55, 299
    Ty, 302–5, 309
Transversion, 275, 308
Travers, Andrew, 183, 194
Tree frog, genetic variation in, 411
Tribolium. See Flour beetle
Trinucleotide, 139
Triphosphate linkage, 227
Triple crossover, 114
Triploid, 95–96
Trisomy, 89, 97, 115
Trisomy 13. *See* Patau syndrome
Trisomy 18. *See* Edward syndrome
Trisomy 21. *See* Down syndrome
Triturus
    development in, 342
    replication in, 163
Trivalent, 96
tRNA. *See* Transfer RNA
trp promoter, 375–76
True-breeding line, 22, 27
True reversion, 283
Trypanosome coat genes, rearrangement of, 360
Tryptophan, 238, 391
Tsai, A., 162
Tschermak, Erich von, 17, 23, 29
Tsugita, A., 251
Tsui, Lap-Chee, 17, 399
Tumor virus, 214
Turner syndrome, 98–99, 102
Twins
    correlations of traits between, 65
    dizygotic, 65
    monozygotic, 65
Twort, F. W., 326–27
Two-stranded crossing over, 116–17
tyb gene, 304
Tyrosinase gene, 271
Tyrosine, 238, 391
Ty transposon, 302–5, 309

U

UBF. *See* Upstream binding factor
Ubx gene, 352, 356–57
UCE. *See* Upstream control element
Ultimate carcinogen, 306
Ultracentrifugation, of DNA, 135, 158
Ultraviolet radiation
    absorption spectrophotometry of DNA, 135
    as mutagen, 281–82, 290
umuDC operon, 289–90
Undermethylation, of DNA, 349
Unequal crossing over, 121–24
Unidirectional replication, 156, 164
Univalent, 96
Universal code, 255–56
Up mutation, 184–85
Upstream binding factor (UBF), 215

Upstream control element (UCE), 212, 215
Upstream promoter elements, 210–12
Uracil, 136, 138
Uracil-DNA-glycosidase, 278–79
Uridine, 138
UTP, in transcription, 182–83
uvrABC endonuclease, 287
uvr genes, 287, 290

V

Vaccine, genetically engineered, 388
Vaccinia virus, 388
Valine, 238
Variable number tandem repeat (VNTR), 383
Variance, 62–63
    in allelic frequency, 420–21
Variation, 10
    conservation genetics, 441–44
    measurement of, 409–10
    mutation and, 418–19
    selection and, 423
    types of, 406–9
Variegated colony, 318
Vector, 366–72
    baculovirus, 378
    expression. *See* Expression vector
    phage, 369–71
    plasmid, 366–69
    shuttle, 378
Vegetal pole, 342
Vegetative state, 193
Vernalization, 449
Victoria (Queen of England), 54–55, 270
Viroid, 136
Virulence, 132–35
Virulent phage, 194
Virus, 136
    cancer and, 303
    oncogenic, 307
Visible mutation. *See* Morphological mutation
Vitamin D_3 receptor, 343
VNTR. *See* Variable number tandem repeat

W

Wake, R. B., 161–62
Wang, D., 249
Warfarin resistance, 426, 450–51
Watson, James, 15, 17, 134, 140–41
W chromosome, 101
Weed, pesticide resistance in, 449–51
Weigert, Martin, 261
Weinberg, Robert, 307
Weinberg, Wilhelm, 17, 410
Weintraub, Harold, 349
Western blot, 385
Wexler, Nancy, 17, 394–99
Wheat
    Hessian fly resistance in, 451
    kernel color in, 61, 64
    ploidy in, 95–96
White blood cells, cell cycle in, 80
Whittaker, J. R., 342
Wigler, Michael, 307
Wild-type, 12, 35–36, 46
Wilkins, Maurice, 15, 17, 140–41
Williamson, Robert, 399
Wobble hypothesis, 254–55
Wobble position, 283
Wollman, Elie, 315, 317
Worton, Ronald, 399
Wu, T. T., 326

X

X chromosome, 12, 53–56, 100–102
    attached X, 116–19
    of domestic mammals, 78
    of *Drosophila melanogaster*, 53, 78–80, 103, 109–11, 351
    of humans, 113–15
    inactivation of, 99, 349–50
Xenopus laevis

nuclear transplantation in, 340–41
protein *Xfin*, 217
5S rRNA genes of, 213, 346–48
Xeroderma pigmentosum, 290, 307
Xfin protein, 217
Xg blood group system, 115
X-linked trait, 53–55
    dominant, 55–56
    in *Drosophila melanogaster*, 86–87
    Hardy-Weinberg principle for, 414–15
    recessive, 54–56
*Xma*I, 364–65
X ray
    as mutagen, 282
    picture of DNA, 140

XX male, 102
XY female, 102
XYY male, 98–99

Y

Y chromosome, 53–56, 78, 100–102
Yeast
    chromosome number in, 77
    chromosomes of, 148
    DNA of, 146
    mating system of, 214
    petite, 271
    transposable elements in, 302–3

Z

Z chromosome, 101
Z-DNA, 143–44
Zea mays. See Corn
Zinc finger, 217
Zoo animals, 443–44
Zubay, Geoffrey, 190
Zygote, 76, 340
Zygotene, 83, Panel 4